Lecture Notes in Computer Science   10726

Commenced Publication in 1973
Founding and Former Series Editors:
Gerhard Goos, Juris Hartmanis, and Jan van Leeuwen

More information about this series at http://www.springer.com/series/7410

Xiaofeng Chen · Dongdai Lin
Moti Yung (Eds.)

Information Security and Cryptology

13th International Conference, Inscrypt 2017
Xi'an, China, November 3–5, 2017
Revised Selected Papers

 Springer

Editors
Xiaofeng Chen
Xidian University
Xi'an
China

Moti Yung
Columbia University
New York, NY
USA

Dongdai Lin
SKLOIS, Institute of Information
  Engineering
Chinese Academy of Sciences
Beijing
China

ISSN 0302-9743   ISSN 1611-3349 (electronic)
Lecture Notes in Computer Science
ISBN 978-3-319-75159-7   ISBN 978-3-319-75160-3 (eBook)
https://doi.org/10.1007/978-3-319-75160-3

Library of Congress Control Number: 2018931889

LNCS Sublibrary: SL4 – Security and Cryptology

Printed on acid-free paper

This Springer imprint is published by the registered company Springer International Publishing AG
part of Springer Nature
The registered company address is: Gewerbestrasse 11, 6330 Cham, Switzerland

Preface

The 13th International Conference on Information Security and Cryptology (Inscrypt 2017) was held during November 3–5, 2017, in Xi'an, China. This volume contains the papers presented at Inscrypt 2017. The Program Committee also invited five distinguished researchers to deliver keynote talks. The keynote speakers were Elisa Bertino from Purdue University, USA; Yang Xiang from Swinburne University of Technology, Australia; Mirosław Kutyłowski from Wroclaw University of Technology, Poland; Yunlei Zhao from Fudan University, China; and Kui Ren from University at Buffalo, USA. Inscrypt is a well-recognized annual international forum for security researchers and cryptographers to exchange ideas and present their work, and is held every year in China.

The conference received 80 submissions. Each submission was reviewed by at least three Program Committee members. The committee accepted 27 papers to be included in the conference program. The proceedings contain revised versions of the accepted papers. While revisions are expected to take the reviewers' comments into account, this was not enforced and the authors bear full responsibility for the content of their papers.

Inscrypt 2017 was held in cooperation with the International Association for Cryptologic Research (IACR), and was co-organized by the School of Cyber Engineering, Xidian University, the State Key Laboratory of Information Security (SKLOIS) of the Institute of Information Engineering of Chinese Academy of Science, and the Chinese Association for Cryptologic Research (CACR). Furthermore, Inscrypt 2017 was sponsored by the State Key Laboratory of Integrated Services Networks (ISN) and National 111 Center of Mobile Internet Security (111 project No. B16037), Xidian University. The conference would not have been a success without the support of these organizations, and we sincerely thank them for their continued assistance and support.

We would also like to thank the authors who submitted their papers to Inscrypt 2017, and the conference attendees for their interest and support. We thank the Organizing Committee for their time and effort dedicated to arranging the conference. This allowed us to focus on selecting papers and dealing with the scientific program. We thank the Program Committee members and the external reviewers for their hard work in reviewing the submissions; the conference would not have been possible without their expert reviews. Finally, we thank the EasyChair system and its operators, for making the entire process of managing the conference convenient.

November 2017

Xiaofeng Chen
Dongdai Lin
Moti Yung

Inscrypt 2017

13th International Conference on Information Security and Cryptology

Xi'an, China
November 3–5, 2017

Sponsored and organized by

State Key Laboratory of Integrated Services Networks (ISN)
National 111 Project for Mobile Internet Security (Xidian University)
State Key Laboratory of Information Security
(Chinese Academy of Sciences)
Chinese Association for Cryptologic Research

in cooperation with

International Association for Cryptologic Research

Honorary Chairs

Xinbo Gao	Xidian University, China
Dongdai Lin	Chinese Academy of Sciences, China

Steering Committee

Feng Bao	Huawei International, Singapore
Kefei Chen	Hangzhou Normal University, China
Dawu Gu	Shanghai Jiao Tong University, China
Xinyi Huang	Fujian Normal University, China
Hui Li	Xidian University, China
Dongdai Lin	Chinese Academy of Sciences, China
Peng Liu	Pennsylvania State University, USA
Wenfeng Qi	National Digital Switching System Engineering and Technological Research Center, China
Meiqin Wang	Shandong University, China
Xiaofeng Wang	Indiana University at Bloomington, USA
Xiaoyun Wang	Tsinghua University, China
Jian Weng	Jinan University, China

Moti Yung	Snapchat Inc. and Columbia University, USA
Fangguo Zhang	Sun Yat-Sen University, China
Huanguo Zhang	Wuhan University, China

Technical Program Committee

Erman Ayday	Bilkent University, USA
Ioana Boureanu	University of Surrey, UK
Donghoon Chang	NIST, USA
Kai Chen	Chinese Academy of Sciences, China
Kefei Chen	Hangzhou Normal University, China
Xiaofeng Chen	Xidian University, China
Cunsheng Ding	Hong Kong University of Science and Technology, Hong Kong, SAR China
Jintai Ding	University of Cincinnati, USA
Karim Eldefrawy	SRI International, USA
Chun-I Fan	National Sun Yat-sen University, Taiwan, China
Debin Gao	Singapore Management University, Singapore
Dawu Gu	Shanghai Jiao Tong University, China
Huaqun Guo	Institute for Infocomm Research, Singapore
Jian Guo	Nanyang Technological University, Singapore
Weili Han	Fudan University, China
Lucjan Hanzlik	Wrocław University of Technology, Poland
Lei Hu	Institute of Information Engineering of CAS, China
Xinyi Huang	Fujian Normal University, China
Miroslaw Kutylowski	Wroclaw University of Technology, Poland
Kwangsu Lee	Sejong University, South Korea
Tieyan Li	Huawei International, Singapore
Yingjiu Li	Singapore Management University, Singapore
Dongdai Lin	Chinese Academy of Sciences, China
Zhe Liu	University of Waterloo, Canada
Florian Mendel	TU Graz, Austria
Mridul Nandi	Indian Statistical Institute, India
Josef Pieprzyk	Queensland University of Technology, Australia
Kouichi Sakurai	Kyushu University, Japan
Willy Susilo	University of Wollongong, Australia
Qiang Tang	Cornell University, USA
Qian Wang	Wuhan University, China
Wenling Wu	Chinese Academy of Science, China
Shouhuai Xu	University of Texas at San Antonio, USA
Yu Yu	Shanghai Jiao Tong University, China
Moti Yung	Columbia University, USA
Fangguo Zhang	Sun Yat-sen University, China
Xianfeng Zhao	Chinese Academy of Sciences, China
Yongjun Zhao	The Chinese University of Hong Kong, Hong Kong, SAR China
Cliff Zou	University of Central Florida, USA

Additional Reviewers

Agrawal, Megha
Alkadri, Nabil
Anada, Hiroaki
Bao, Zhenzhen
Blaskiewicz, Przemyslaw
Chen, Huashan
Chen, Yi
Chow, Sherman S. M.
Cui, Tingting
Dai, Ting
Ding, Ning
Dobraunig, Christoph
Fang, Chengfang
Feng, Yaokai
Gao, Xinwei
Garcia Lebron, Richard
Garg, Surabhi
Guo, Jiale
Guo, Qian
Huang, Yan
Jia, Haoyang
Jiang, Linzhi
Kim, Hyoseung
Kim, Jonghyun
Kumar Chauhan, Amit
Kumar, Abhishek
Larangeira, Mario
Lee, Youngkyung
Li, Huige
Li, Lingchen
Li, Sisi
Li, Xiangxue
Li, Zengpeng
Lin, Hsiao Ying
Liu, Guozhen
Liu, Jianghua
Liu, Ximing

Liu, Zhen
Long, Yu
Lu, Yuan
Mishra, Sweta
Nakano, Yuto
Nogami, Yasuyuki
Pan, Yanbin
Ruj, Sushmita
Rv, Sara
Seo, Minhye
Shahandashti, Siamak
Song, Ling
Sui, Han
Sun, Siwei
Tian, Yangguang
Unterluggauer, Thomas
Wang, Daibin
Wang, Fuqun
Wang, Haoyang
Wang, Huige
Wang, Lei
Wang, Liangliang
Wang, Weijia
Wang, Yuntao
Xie, Shaohao
Xu, Jiayun
Xu, Lingling
Yang, Anjia
Yang, Shao-Jun
Yu, Yong
Yuen, Tsz Hon
Zha, Mingming
Zhang, Huang
Zheng, Yafei
Zhong, Chen
Zong, Peiyuan

Keynote Speeches

AI-Driven Cyber Security

Yang Xiang

Swinburne University of Technology, Hawthorn VIC 3122, Australia
yxiang@swin.edu.au

Today we have evidenced massive cyber-attacks, such as WannaCry ransomware, having hit millions of people in more than 150 countries with billions of dollars lose. Cyber security has become one of the top priorities globally in the research and development agenda [1].

Recent years, Artificial Intelligence (AI) [2] has been widely used in numerous fields and industries, including finance, healthcare, education, and transportation, support-ed by a diversity of datasets from a huge number of sources in different domains [3, 4]. These datasets consist of multiple modalities, each of which has a different representation, distribution, scale, and density [5–8].

In addition, with the increase of AI based software like digital services and prod-ucts, software vulnerability detection [8] that certify the security of using the AI-based Software has become an important research area in both academia and industries [9]. The number of vulnerabilities has been reported to be positively correlated to the volume of the software copies. For example, in 2010, there were about only 4500 vulnerabilities registered in the well-known CVE (Common Vulnerabilities and Expo-sures) database [10], however, this number increased to 17265 in 2017. In another example, more than 43000 software vulnerabilities have been reported via NVD (National Vulnerability Database) since 1997 [11]. These vulnerabilities affected more than 17000 software services and caused about 266 billion dollars losses a year [12]. The trend seems to be increasing with the increase of AI-based software services.

People have recognized that AI technologies are some of the most effective defenses against cyber intrusions [13]. Cyber security companies are increasingly looking to AI to improve defense systems and create the next generation of cyber protection. In this respect, machine learning-based software vulnerable detection techniques are becoming an important research area with the increasingly rich of vulnerability related data. A few important questions have been asked, such as:

– How AI models learn and understand what is normal and what is abnormal on a system?
– How AI that uses machine learning and other technologies can differentiate benign or harmful binary or source codes?
– How can hackers bypass AI-driven security solutions?

Although AI has been talked as one of the game-changing technologies for cyber security, many doubts still persist. New methods and tools, consequently, must follow up in order to adapt to this emerging security paradigm. In this talk, we will discuss the concept of AI-Driven Cyber Security and how data analytics can be used to ad-dress the security and privacy problems in cyberspace. We will outline how deep learning can learn high-level representations based on the source code we collected and labelled. Deep learning is in part due to an ability to learn feature representations and complex non-linear structure in datasets. Deep learning has achieved particular successes in data domains such as vision, speech and natural language, which each exhibit hierarchies of patterns at fine to coarse scales. Software vulnerability detection is ready for similar success owing to its complex, hierarchical, non-linear detection tasks. For example, in one of our research that using deep learning for software vulnerability detection on cross-project scenario. We first collected datasets from three open-source projects: Libtiff, LibPNG and FFmpeg [8]. Then, the raw features are extracted from the Abstract Syntax Trees (ASTs) of functions. Afterwards, a stacked LSTM network is designed and a proxy for learning ASTs representations of functions is introduced. Finally, classification models are built based on the feature representation learned from the LSTM network.

References

1. Lee, K.-C., Hsieh, C.-H., Wei, L.-J., Mao, C.-H., Dai, J.-H., Kuang, Y.-T.: Sec-Buzzer: cyber security emerging topic mining with open threat intelligence retrieval and time-line event annotation. Soft. Comput. **21**(11), 2883–2896 (2017)
2. Nilsson, N.J.: Principles of Artificial Intelligence. Morgan Kaufmann (2014)
3. Zhang, J., Xiang, Y., Wang, Y., Zhou, W., Xiang, Y., Guan, Y.: Network traffic classification using correlation information. IEEE Trans. Parallel Distrib. Syst. **24**(1), 104–117 (2013)
4. Chen, C., Wang, Y., Zhang J., Xiang, Y., Zhou, W., Min, G.: Statistical features-based real-time detection of drifted twitter spam. IEEE Trans. Inf. Forensics Secur. **12**(4), 914–925 (2017)
5. Wen, S., Zhou, W., Xiang, Y., Zhou, W.: CAFS: a novel lightweight cache-based scheme for large-scale intrusion alert fusion. Concurrency Comput. Pract. Experience **24**(10), 1137–1153 (2012)
6. Liu, S., Zhang, J., Xiang, Y., Zhou, W.: Fuzzy-based information decomposition for incomplete and imbalanced data learning. IEEE Trans. Fuzzy Syst. **25**(6), 1476–1490 (2017)
7. Ghaffarian, S.M., Shahriari, H.R.: Software vulnerability analysis and discovery using machine-learning and data-mining techniques: a survey. ACM Comput. Surv. (CSUR) **50**(4), 56 (2017)
8. Lin, G., Zhang, J., Luo, W., Pan, L., Xiang, Y.: Poster: vulnerability discovery with function representation learning from unlabeled projects. In: Proceedings of the ACM SIGSAC Conference on Computer and Communications Security (CCS), pp. 2539–2541. ACM (2017)
9. Zhang, Z.-K., Cho, M.C.Y., Wang, C.-W., Hsu, C.-W., Chen, C.-K., Shieh, S.: IoT security: ongoing challenges and research opportunities. In: IEEE 7th International Conference on Service-Oriented Computing and Applications (SOCA), pp. 230–234. IEEE (2014)
10. Perl, H., et al.: Vccfinder: finding potential vulnerabilities in open-source projects to assist code audits. In: Proceedings of the 22nd ACM SIGSAC Conference on Computer and Communications Security, pp. 426–437. ACM (2015)
11. Zhang, S., Ou, X., Caragea, D.: Predicting cyber risks through national vulnerability database. Inf. Secur. J. Global Perspect. **24**(4–6), 194–206 (2015)

12. Alves, H., Fonseca, B., Antunes, N.: Experimenting machine learning techniques to predict vulnerabilities. In: Seventh Latin-American Symposium on Dependable Computing (LADC), pp. 151–156. IEEE (2016)
13. Yampolskiy, R.V., Spellchecker, M.: Artificial intelligence safety and cybersecurity: a timeline of AI failures (2016). arXiv preprint, arXiv:1610.07997

Generic and Efficient Lattice-Based Key Exchange from Key Consensus with Noise

Yunlei Zhao

School of Computer Science, Fudan University, Shanghai, China

Lattice-based cryptography is promising in the post-quantum era. For cryptographic usage, compared with the classic hard lattice problems such as SVP and CVP, the learning with errors (LWE) problem and its variants are proven to be much more versatile. Based upon them, a large number of impressive works are developed in recent years, with key exchange (KE) as the focus of this work.

For KE and public-key encryption (PKE) schemes from LWE and its variants, a key ingredient is the key reconciliation mechanisms. However, they were only previously used and analyzed in a *non-black-box* way. This means, for new KE or PKE schemes developed in the future, we need to analyze from scratch. Also, for the various parameters involved in key reconciliation, the bounds on what could or couldn't be achieved are unclear.

In this work, we abstract and study this key ingredient. Specifically, we formalize the building tool, referred to as key consensus (KC) and its asymmetric variant AKC. KC and AKC allow two communicating parties to reach consensus from close values obtained by some secure information exchange (such as exchanging LWE samples). KC and AKC are fundamental to lattice based cryptography, in the sense that a list of cryptographic primitives based on LWE and its variants can be constructed from them *in a modular and black-box way*. As a conceptual contribution, this much simplifies the design and analysis of these cryptosystems in the future.

Abstracting KC and AKC also allows us to study and prove the inherent upper-bounds among the parameters. In particular, we discover the upper-bounds on parameters for any KC and AKC. This allows us to understand what can or cannot be achieved with any KC and AKC, and guides our actual protocol design. These upper-bounds also guide parameter choosing for various trade-offs, and are insightful in performance comparison.

Guided by, and motivated for reaching, these proved upper-bounds, we then design and analyze both general and highly practical KC and AKC schemes, which are referred to as OKCN and AKCN respectively for presentation simplicity. Both OKCN and AKCN almost meet the proved upper-bounds in general, and can be instantiated to tightly match these upper-bounds. Moreover, they are the first multi-bit reconciliation

Extended abstract of the work joint with Zhengzhong Jin, which originally appeared at arXiv: https://arxiv.org/abs/1611.06150. This research was supported in part by NSFC (Grant Nos. 61472084 and U1536205), National Key R&D Program of China (No.2017YFB0802000), Shanghai innovation action project No. 16DZ1100200, and Shanghai science and technology development funds No.16JC1400801.

mechanisms, to the best of our knowledge. We note that OKCN and AKCN have already been influential, and are used in some concurrent subsequent works. For example, some versions of AKCN were used in the schemes of Lizard (Cryptology ePrint Archive, 2016/1126) and Kyber (Cryptology ePrint Archive, 2017/634).

Based on KC and AKC, we present generic constructions of key exchange from LWE and its variants: LWR, RLWE and MLWE, with delicate analysis of error probabilities. Then, for the instantiations of these generic constructions with our OKCN and AKCN schemes, we elaborate on evaluating and choosing the concrete parameters in order to achieve a well-balanced performance among security, computational efficiency, bandwidth efficiency, error rate, and operation simplicity. At a high level, OKCN-based KE corresponds to Diffie-Hellman in the lattice world, while AKCN-based KE is not. Specifically, with AKCN, the responder can predetermine and set the shared-key at its wish. But AKCN-based KE can be directly used for CPA-secure PKE. We suggest that OKCN-based KE is more versatile, and is more appropriate for incorporating into the existing standards like IKE and TLS.

We propose the first construction of key exchange *merely* based on the LWR problem with concrete analysis and evaluation, to the best of our knowledge. In particular, we provide a delicate approach to calculating its error rate. Specifically, for the LWR-based KE protocol, the main difficulty here is the error probability analysis: the rounding operation in LWR brings new noises, yet these noises are *deterministic*, because they are completely determined by the public matrix and the secret vector. In the formula calculating the error probability, the deterministic noises will multiply the secret vector. However, they are correlated. This correlation prevents us from calculating the error probability efficiently. This is a new difficulty we encounter in LWR-based KE. Our contribution is to provide an analysis breaking the correlation, and design an algorithm to calculate the error probability numerically. When applied to LWE-based cryptosystems, OKCN can directly result in more practical or well-balanced schemes of key exchange. The comparisons between OKCN-based KE and Frodo, proposed by Bos et al. at ACMCCS2016, are briefly summarized in Table 1.

Table 1. Brief comparison between OKCN-LWE/LWR and Frodo. $|\mathbf{K}|$ refers to the size in bits of the shared key; "bw.(kB)" refers to bandwidth in kilo bytes; "err." refers to the error rate, and "pq-sec" refers to the best known quantum attack against the underlying lattice problem.

	$|\mathbf{K}|$	bw.(kB)	err.	pq-sec
OKCN-LWR	256	16.19	2^{-30}	130
OKCN-LWE	256	18.58	2^{-39}	134
Frodo	256	22.57	$2^{-38.9}$	130

When applying OKCN/AKCN to MLWE-based KE, they result in the (up-to-date) most efficient lattice-based key exchange protocols for 256-bit shared-key. MLWE is a variant between LWE and RLWE. On the one hand, MLWE-based protocols are more efficient than LWE-based; And on the other hand, they are more secure than RLWE-based, as the MLWE problem has fewer algebraic structures than RLWE.

Table 2. Brief comparison between OKCN/AKCN-MLWE and Kyber.

	$|\mathbf{K}|$	bw.(B)	err.	pq-sec
OKCN-MLWE-KE	256	1856	$2^{-50.1}$	183
OKCN-MLWE-PKE	256	2048	$2^{-166.4}$	171
AKCN-MLWE-PKE (Kyber)	256	2272	$2^{-142.7}$	171

The comparisons between OKCN/AKCN-MLWE and CPA-secure Kyber are briefly summarized in Table 2.

When applied to RLWE-based cryptosystems, AKCN can lead to the most efficient KE protocols with shared-key of size of at least 512 bits, which may be prudent for ensuring 256-bit post-quantum security in reality. For RLWE-based KE, we develop new approaches to lowering the error probability. Firstly, we make a key observation on RLWE-based key exchange, by proving that the errors in different positions in the shared-key are almost independent. This can play a fundamental basis for the approach to lowering error rate of RLWE-based KE with error-correction codes. Then, based upon this observation, we present a super simple and fast code, referred to as *single-error correction* (SEC) code, to correct at least one bit error. By equipping OKCN/AKCN with the SEC code, we achieve the simplest (up to now) RLWE-based KE for much longer shared-key size with error rate that can be viewed as negligible in practice. To further improve the bandwidth, error rate and post-quantum security simultaneously, we develop new lattice code in E_8. Note that sphere packing is optimal with the lattice E_8. The comparisons with NewHope, proposed by Alkim et al at USENIX Security 2016, are briefly summarized in Table 3.

Table 3. Brief comparison between OKCN/AKCN-RLWE and NewHope.

	$|\mathbf{K}|$	bw.(B)	err.	pq-sec
OKCN-RLWE-SEC	765	3392	2^{-61}	258
NewHope	256	3872	2^{-61}	255
AKCN-RLWE-SEC	765	3520	2^{-61}	258
AKCN-RLWE-E8	512	3360	$2^{-63.3}$	262
NewHope-Simple	256	4000	2^{-61}	255

Contents

Encryption

Cryptanalysis and Attack

Applications

Keynote Speeches

Security and Privacy in the IoT

Elisa Bertino[(✉)]

Purdue University, West Lafayette, IN, USA
`bertino@purdue.edu`

Abstract. Deploying existing data security solutions to the Internet of Things (IoT) is not straightforward because of device heterogeneity, highly dynamic and possibly unprotected environments, and large scale. In this paper, we first outline IoT security and privacy risks and critical related requirements in different application domains. We then discuss aspects of a roadmap for IoT security and privacy with focus on access control, software and firmware, and intrusion detection systems. We conclude the paper by outlining a few challenges.

1 Introduction

Internet of Things (IoT) refers to the network of physical objects or "things" embedded with electronics, software, sensors, and connectivity to enable objects to exchange data with servers, centralized systems, and/or other connected devices based on a variety of communication infrastructures [2]. IoT makes it possible to sense and control physical objects creating opportunities for more direct integration between the physical world and computer-based systems. IoT is ushering automation in a large number of application domains, ranging from manufacturing and energy management (e.g. SmartGrid), to healthcare management and urban life (e.g. SmartCity). However, because of its fine-grained, continuous and pervasive data acquisition, and control and actuation capabilities, and the lack of adequate security, IoT expands the cyber attack surface. An analysis by HP on several common devices has shown an average of 25 vulnerabilities per device [4]. For example, 80% of devices failed to require passwords of sufficient complexity and length, 70% did not encrypt local and remote traffic communications, and 60% contained vulnerable user interfaces and/or vulnerable firmware. Multiple attacks have been reported in the past against different embedded devices [5,6] and we can expect many more in the IoT domain. Therefore comprehensive and articulated security solutions are required that address the specific security and privacy requirements of IoT systems. In this paper, we first discuss security risks for IoT and specific security and privacy requirements for different IoT application domains. We then discuss a preliminary research roadmap in which we focus on specific security building blocks and discuss initial approaches and research directions. We conclude the paper by outlining a few additional challenges.

© Springer International Publishing AG, part of Springer Nature 2018
X. Chen et al. (Eds.): Inscrypt 2017, LNCS 10726, pp. 3–10, 2018.
https://doi.org/10.1007/978-3-319-75160-3_1

2   Security Risks for IoT

IoT systems are at high security risks for several reasons [3]. They do not have well defined perimeters, are highly dynamic, and continuously change because of mobility. In addition IoT systems are highly heterogeneous with respect to communication medium and protocols, platforms, and devices. IoT devices may be autonomous entities that control other IoT devices. IoT systems may also include "objects" not designed to be connected to the Internet. IoT systems, or portions of them, may be physically unprotected and/or controlled by different parties. Finally user interactions with IoT devices are not scalable and therefore, unlike current applications running on mobile phones in which users are asked permissions before installing applications, such interactions may not be possible to the scale and number of these IoT devices. Attacks, against which there are established defense techniques in the context of conventional information systems and mobile environments, are thus much more difficult to protect against in the IoT. The OWASP Internet of Things Project [7] has identified the most common IoT vulnerabilities and has shown that many vulnerabilities arise because of the lack of adoption of well-known security techniques, such as encryption, authentication, access control and role-based access control. A reason for the lack of adoption may certainly be security unawareness by companies involved in the IoT space and by end-users. However another reason is that existing security techniques, tools, and products may not be easily deployed to IoT devices and systems, for reasons such as the variety of hardware platforms and limited computing resources on many types of IoT devices.

3   Application Domains

As IoT covers many different application domains, types of devices, and systems, we can expect that security solutions have different requirements for different types of domains. We outline below some significant classes of domains and identify related relevant security requirements. We would like to emphasize that in all these areas it is critical to adopt good security practices, like the ones required to address the vulnerabilities mentioned in the previous section. However each area may have additional specific requirements that may require the design of novel security techniques or extensions of existing techniques.

– *Consumer IoT*. It refers to use of IoT devices for end-user applications, such as devices deployed in smart houses. Architectures in this type of applications are simpler and use low-end devices. In terms of security, data confidentiality is critical in order to assure privacy. Authentication of devices and of users to devices is also critical to ensure physical security, for example, for making sure that an attacker cannot unlock the door of a house when smart locks are used [8]. In addition, in many consumer IoT applications several users may be involved in requesting services from the same IoT device; for example, several users may have the authorization to open a door by activating a smart lock [8].

However, some of those users may be revoked this authorization, for example house guests. Therefore, protocols for selective authorization and revocation are a critical requirement.

– *Medical IoT.* It refers to the use of personal medical devices (such as pacemakers) and wellness devices. In addition this category includes devices which are not implanted but are used to inject drugs (such as insulin pumps, and pumps used in hospital for intravenous drug administration) and patient monitoring devices. Safety, that is, to make sure that the devices do not harm the patients, is critical in addition to privacy. Security is critical also in that one must assure that a device continues to work even under attack [9]. Therefore, making sure that the software deployed on these devices correctly works and is resilient under attack is critical. A major challenge is that, because of safety requirements, many such devices are "closed" and it is not possible to install additional software, such as software for data encryption.

– *Industrial IoT (IIoT).* It refers to the convergence of different technologies, namely intelligent machines, industrial analytics, and sensor-driven computing. The main goal is not only to enhance operational efficiency but also to introduce novel information services [10]. Intelligent machines refer to machines, such as industrial robots, that have not only mechanical functions but are also able to communicate with other equipment and learn how to lower their own operating costs. Such machines have embedded systems and rich communication capabilities. As discussed in [10], critical requirements include: (i) the need for continuous operations, which requires protection against attacks aiming at disrupting the operations; (ii) and assuring human health and safety. In addition IIoT systems pose specific challenges, such as the long lifecycle of systems, legacy equipment, need for regulatory compliance, that complicate the design of suitable security solutions. A specific subarea within IIoT is represented by control systems that will be increasingly connected to sensors, devices, other systems and Internet via digital communication capabilities (we refer to them as IoT-enhanced control systems). Unlike IIoT applications mainly focusing on improving operations and on information services, control systems often have real-time requirements and must ensure the continuity of the controlled processes, which again makes the design of security solutions more complex (see [11] for a detailed discussion about the security landscape for industrial control systems). It is important to mention that the security of IoT-enhanced control systems is (will be) critical for the protection of many critical infrastructures (energy, transportation, waste management, etc.).

4   Roadmap

The development of a proper roadmap for comprehensive solutions for IoT requires: (a) discussing the security building blocks and identifying the challenges to be addressed in order to extend these building blocks for deployment within IoT systems; (b) identifying novel functions that would be specific to IoT systems. In what follows we focus on a number of such building blocks. In our discussion we do not cover the cloud component of the IoT ecosystem, mainly because cloud security has been already widely investigated.

4.1   Access Control

Access control is basic security technique that is critical when sensitive data needs to be shared among different parties [1]. Fundamental access control models include: mandatory access control models (based on access classes of subjects and protected objects); discretionary access control models based on permissions granted to subjects for executing specific actions on protected objects; role-based access control (RBAC) models in which permissions are granted to functional entities called roles and not directly to users - notable extensions of RBAC include spatial RBAC and temporal RBAC; and attribute-based access control (ABAC) in which authorizations are expressed in terms of identity attributes of subjects and protected objects (e.g. only the researchers that have completed the HIPAA training course can access sensitive data about cancer patients) - XACML is a widely known standard for ABAC. Different approaches also exist for enforcing access control policies, including cryptographic mechanisms and access control lists.

In the context of IoT, access control is required from four different orthogonal purposes:

- Controlling which end-user (or application on behalf of which end-user) can access which IoT devices for which purpose (and possibly for how long and/or in which context). One example would be to allow guests in a house to operate a smart lock only for a given weekend.
- Controlling which IoT device can invoke functions on other IoT devices or get data from other IoT devices. One example would be not to allow a device regulating plant watering in a back yard communicating with a smart lock device as the former would typically be an unprotected device and thus more vulnerable to attacks.
- Controlling which information an IoT device may collect from a given physical environment and under which circumstances. An example would be preventing an IoT device with a sound recorder to record sounds when one is at the premises of a customer. An access control mechanism supporting this type of access control policies has been developed for mobile smart phones [12]. Under such approach, the applications running on the mobile phone are dynamically revoked the permission to use the sound recorder based on the context. However its application to IoT devices may be challenging as IoT devices may not be able to acquire contextual information (such as location).
- Controlling which applications running on a gateway or on the cloud or cloud users can access data from a given IoT device. We mention this purpose for completeness even though cloud security it outside the scope of the discussion. We notice that enforcing access control policies for IoT originated data may require the IoT system to provide relevant meta-data needed for determining

the applicable access control policies (such the location where the data was collected by the IoT device).

The definition, implementation and deployment of access control policies for IoT systems entail addressing a lot of issues ranging from the access control model (e.g. whether it should be ABC and context-based), the language used for specifying the authorizations, the architecture for managing and enforcing control. Given the various purposes for access control, it is easy to see that different access control enforcers may have to be used for different purposes. For example, end-user permissions may be checked at the gateway, whereas controlling which information an IoT device may acquire from the physical environment (through the use of its own peripherals) must be executed in the IoT device itself. An initial limited approach for access control for devices communicating via Bluetooth Low Energy has been recently proposed [13]. Even though this approach is very limited, it is interesting in that it shows the use of device identity attributes for access control in IoT systems.

The specification and administration of access control policies will however be the major challenge. Users already have problems in managing permissions on mobile phones; managing permissions in IoT systems will be undoubtedly much more complex. In order to address such management complexity and design proper administration tools, understanding which access control purposes are relevant for specific types of IoT systems is critical.

4.2 Software and Firmware Security

Software is always the critical element of security as many attacks exploit errors in software. Because of software criticality for security, a huge number of approaches have been proposed for "software security", such as for ensuring memory safety and the integrity of the application execution control flow. Approaches range from compiler techniques that detect errors (such as buffer overflow) in the source code to approaches that instrument or randomize binaries to make sure that even if errors exist in the code these errors cannot be exploited by attackers. The latter are typically adopted when the application source code is not available or it cannot be modified. Other approaches do not require modifying neither the source code nor the binaries and rely on monitoring the execution of the application to verify that the execution control flow is not modified by the attacker. However, adapting such techniques to IoT systems is unfeasible without extensive redesign [14] as they have heavy requirements in storage, memory, presence of a memory management unit (MMU), and hardware memory-protection. In addition the performance overhead that existing solutions impose is not acceptable for energy-constrained devices. An approach to spatial memory safety in embedded devices has been developed that greatly reduces overhead [14]. However, research is needed to extend other available techniques for use in IoT devices.

Recent research has also focused on firmware that is susceptible not only to a wide range of software errors and attacks, such as attacks exploiting firmware updates [16], and memory errors, but also to logical flaws. Two categories of such logical flaws that widely occur in IoT devices are insufficient authentication, including authentication bypass, and insufficient authorization. Notice that these are logical flaws and not software errors; therefore they cannot be detected by techniques used to detect common software errors or by techniques used to ensure the integrity of the execution control flow. Automatically detecting such types of logical flaws requires addressing several challenges including the ability of reverse engineering the firmware [17].

Understanding which specific techniques or combination techniques should be used in IoT systems for securing software is quite challenging due to the wide diversity of IoT applications and scenarios. Also the selection of one or more techniques may depend on the processes in place for managing software. While we can reasonably expect that an organization using IoT devices for some critical application may be willing to perform firmware analysis, we cannot expect that such an analysis be carried out in the context of consumer IoT. In the end the selection must also be based on a risk assessment analysis. One important advantage to keep in mind, however, is that several IoT devices have very specialized functions and limited and constrained input, and therefore their behavior is predictable. So one could design techniques able to predict actions executed by the IoT devices based on input parameters and use these prediction techniques within monitoring tools to detect anomalies with respect to the predicted behavior. One such approach has been recently developed for the more complex case of database applications [18].

4.3   Intrusion Detection Systems

Intrusion detection systems (IDS) are a widely used class of defense techniques. Many IDS have been proposed for conventional systems and networks. By contrast there is almost no work concerning IDS for IoT systems and there are not even simple tools, like logging tools. The only IDS specifically designed for IoT systems is SVELTE [19] that focuses on protecting against network layer and routing attacks. SVELTE is both centralized (at the hub of an IoT system/group of devices) and distributed (at each device). It is composed of a module, called 6Mapper, that reconstructs the topology of the IoT devices with respect to the hub, and an intrusion detection module that analyzes data and detects incidents. Even though SVELTE is an interesting IDS, it does not support comprehensive anomaly detection capabilities. It also lacks functionality for real-time fine-grained diagnoses of attacks and anomalies. Such functionality is critical in environments that rapidly change in order to be able to patch and strengthen the system of interest to prevent further attacks. A more advanced IDS addressing such requirement is the Kalis system [15] that exploits knowledge about the IoT system network in order to support accurate diagnosis of attacks and anomalies. However IDS suitable for IoT systems must be designed able to: support flexible configurations on IoT systems on a wide variety of settings (from settings with rich devices to settings with very limited devices) and collaborative intrusion and anomaly detection by the deployment of multiple "agents";

exploit data analytics (both locally at the IDS system itself and remotely on some cloud); exploit security information services (such as those supported by security information aggregation services such as VirusTotal); and provide services for IoT device discovery and identification. Finally a critical issue is where to deploy IDS and other tools like anomaly detectors within an IoT system. Gateways appear a logical choice especially when IoT devices are limited and it is not possible to install software on them. Another interesting alternative is to deploy such an IDS, perhaps also extended with device profiling capabilities, on its own hardware (possibly with hardware tamper-proof capabilities) resulting in a "security-specialized IoT device".

5   Concluding Remarks

IoT systems pose formidable challenges with respect to security techniques and security management because of the diversity of devices, communication media, communication protocols, software systems, and very large scale. So even testing IoT systems may be very difficult. It is important to mention that today, most IoT systems are closed systems and tailored to very specific applications, and thus perhaps reasoning on the security of these systems is feasible. However, when moving towards interoperability, understanding security in interoperating IoT systems may be difficult if at all possible. As concluding remark, we would like to mention that perhaps the main reason of the difficulties in deploying comprehensive security solutions is that IoT ecosystems are characterized by many different parties, each involved in performing security relevant functions (e.g. assigning an identifier to an IoT device, patching the software on the IoT device, etc.). Keeping track of actions and information, such as device cryptographic keys, in massive distributed systems with multiple security/administration domains is challenging. Perhaps technologies like BlockChain can be investigated as a basis for novel implementations of security building blocks [20].

References

1. Bertino, E., Ghinita, G., Kamra, A.: Access control for databases: concepts and systems. Found. Trends Databases **3**(1–2), 1–148 (2011)
2. Bertino, E.: Data security and privacy in the IoT. In: Proceedings of the 19th International Conference on Extending Database Technology, EDBT 2016, Bordeaux, France, March 15–16, 2016, Bordeaux, France, 15–16 March 2016
3. Bertino, E., Islam, N.: Botnets and Internet of Things security. IEEE Comput. **50**(2), 76–79 (2017)
4. Rawlinson, K.: HH Study Reveals 70 Percent of Internet of Things Devices Vulnerable to Attack. http://www8.hp.com/us/en/hp-news/
5. Bansal, S.K.: Linux Worm targets Internet-enabled Home Appliances to Mine Cryptocurrencies, March 2014. http://thehackernews.com/2014/03/linux-worm-targets-internet-enabled.html
6. Wright, A.: Hacking cars. Commun. ACM **54**(11), 18–19 (2011)
7. https://www.owasp.org/index.php/OWASP_Internet_of_Things_Project

8. Ho, G., Leung, D., Mishra, P., Hosseini, A., Song, D., Wagner, D.: Smart locks: lessons for securing commodity Internet of Things devices. In: Proceedings of the 11th ACM on Asia Conference on Computer and Communications Security, AsiaCCS 2016, Xi'an, China, May 30–June 3 2016

9. Sametinger, J., Rozenblit, J.W., Lysecky, R.L., Ott, P.: Security challenges for medical devices. Commun. ACM **58**(4), 74–82 (2015)

10. Accenture. Driving the Unconventional Growth through the Industrial Internet of Things (2015). https://www.accenture.com/us-en/_acnmedia/Accenture/ next-gen/reassembling-industry/pdf/Accenture-Driving-Unconventional-Growth-through-IIoT.pdf

11. McLaughin, S., et al.: The cybersecurity landscape in industrial control systems. Proc. IEEE **104**(5), 1039–1057 (2016)

12. Shebaro, B., Oluwatimi, O., Bertino, E.: Context-based access control systems for mobile devices. IEEE Trans. Dependable Secure Comput. **12**(2), 150–163 (2015)

13. Levy, A., Long, J., Riliskis, L., Levis, P., Winstein, K.: Beetle: flexible communication for bluetooth low energy. In: Proceedings of the 14th Annual International Conference on Mobile Systems, Applications, and Services, MobiSys 2016, Singapore, 26–30 June 2016

14. Midi, D., Payer, M., Bertino, E.: Memory safety for embedded devices with nesCheck. In: Proceedings of the 2017 ACM on Asia Conference on Computer and Communications Security, AsiaCCS 2017, Abu Dhabi, United Arab Emirates, 2–6 April 2017

15. Midi, D., Rullo, A., Mudgerikar, A., Bertino, E.: Kalis - a system for knowledge-driven adaptable intrusion detection for the Internet of Things. In: 37th IEEE International Conference on Distributed Computing Systems, ICDCS 2017, Atlanta, GA, USA, 5–8 June 2017

16. Cui, A., Costello, M., Stolfo, S.: When firmware modifications attack: a case study of embedded exploitation. In: 20th Annual Network and Distributed System Security Symposium, NDSS 2013, San Diego, California, USA, 24–27 February 2013

17. Shoshitaishvili, Y., Wang, R., Hauser, C., Kruegel, C., Vigna, G.: Firmalice - automatic detection of authentication bypass vulnerabilities in binary firmware. In: 22nd Annual Network and Distributed System Security Symposium, NDSS 2015, San Diego, California, USA, 8–11 February 2015

18. Bossi, L., Bertino, E., Hussain, S.R.: A system for profiling and monitoring database access patterns by application programs for anomaly detection. IEEE Trans. Software Eng. **43**(5), 415–431 (2017)

19. Raza, S., Wallgren, L., Voigt, T.: SVELTE: real-time intrusion detection in the Internet of Things. Ad Hoc Netw. **11**, 2661–2674 (2013)

20. Won, J.H., Singla, A., Bertino, E.: Blockchain-based Public Key Infrastructure for Internet-of-Things (2017, Submitted for Publication)

On Crossroads of Privacy Protection

Mirosław Kutyłowski[(✉)]

Faculty of Fundamental Problems of Technology,
Wrocław University of Science and Technology, Wrocław, Poland
miroslaw.kutylowski@pwr.edu.pl

Abstract. The privacy protection has recently become a hot topic because of the increase in cyber-crime (using personal data for mounting attacks) as well as legal obligations for parties controlling personal data (eg. GDPR regulation of European Union). This creates a big market for pragmatic technical solutions.

In this paper we discuss a few general issues related to these problems, focused on current challenges and the necessity of paradigm shifting in the construction of IT systems, which should be secure-by-design in a demonstrable way.

The necessity of privacy protection has been gaining acceptance in the recent years, as it is no longer considered merely as means for hiding personal activities, but as one of security measures along the rule of data minimization. A growing wave of cyber crime provides firm arguments justifying this view, as personal data are particularly valuable for criminal activities in the cyber space.

1 Legal Trends

There are substantial changes to be noted in the approach to personal data protection in Europe. Nowadays, privacy protection is not only recognized as necessity due to criminality problems, but it is considered a fundamental human right – Article 16(1) of the Treaty on the Functioning of the European Union (TFEU).

Following the Treaty, there has been a Directive (95/46/EC) describing a framework of privacy protection in the Member States. Then the Member States have been obliged to implement this framework in their national legal systems. This approach turned out to be problematic due to a few reasons:

- national parliaments can implement the rules in a way compatible with the Directive but incompatible with each other,
- there are cross-border problems such as supervision of cross-border activities and different interpretations of Personal Data Commissioners in different countries.

Supported by Polish National Science Centre grant OPUS, no. 2014/15/B/ST6/ 02837.

It was these reasons, among others, that gave rise to the new GDPR Regulation [1] which now must be directly accommodated by the legal systems in each of the Member States. The new regulation introduces many deep changes in the approach to personal data. From the technological perspective, there are two fundamental changes:

security-by-design: a data controller is not only responsible for the misuse of personal data (including their leakage): it must create a system for processing personal data, which is secure in its very nature. Moreover, the security of the system must be demonstrable. This is a deep change, since the previous ruling in that regard allowed processing the data in insecure way, and the only means of discouragement against this were administrative fines.
right-to-be-forgotten: the new right of a citizen is to withdraw its consent to process personal data.

Already the first requirement is a big technical challenge. For the second one, the problem is even deeper, as nowadays most of the requests are processed manually, generating huge costs and potentially overlooking many data that need to be erased.

The GDPR Regulation has been reflected in another important regulation of the European Union - the eIDAS regulation [2] concerning mostly identification, authentication and digital signatures. Pseudonymization has been explicitly pointed to in this regulation, making room for introducing these techniques.

Unfortunately, despite the fact that the rules of GDPR regulation start to apply in May 2018, the level of readiness of the IT systems in the end of 2017 is still very low.

2   Mechanisms

One of the pragmatic approaches to dealing with the requirements of the GDPR regulation is to move outside of its scope. This is possible since the regulation applies only to data concerning *an identified or identifiable natural person.* If the identity of a person is not deducible, then automatically a sufficient protection level according to GDPR has been achieved. For this reason, the anonymous identification and authentication protocols gain importance.

2.1   Anonymous Credentials

The idea of anonymous credentials is that in many cases we do not have to identify ourselves but only prove that we have certain rights (e.g. to get access to something) due to our certain attributes. A simple example would be a vending machine selling cigarettes, where we do not have to provide fill identification data (the birth date among them), but only a proof that we have reached the legal age.

An anonymous credentials system is run by an Issuer that:

- verifies attributes of a person applying to join the system (the actual verification process is outside the scope of the core anonymous credentials system),
- issues a cryptographic witness for this person,
- runs revocation process, e.g. by maintaining a cryptographic accumulator disabling credential presentation of each revoked person.

After getting a witness, the user may present anonymous credentials to a verifier in the following way:

- choosing a set of attributes S to be used (there is no need to present all attributes – thereby implementing the minimal information rule), and a formula ϕ over the attributes form S,
- using the witness obtained from the Issuer to create an anonymous credential for the value of the formula $\phi(S)$.

The attributes S have to be confirmed in the witness obtained by the user - otherwise the construction is deemed to fail.

It depends on a concrete scheme which kind of formulas can be chosen by the user. As always, more flexibility results in more advanced constructions.

The main problem with the existing technical realizations [3,4] is their complexity (meaning the cost of the final application) as well as fairly complicated design, not easily understandable even for specialists.

The anonymous credential systems are focused on attributes and not on pseudonyms (even if some of the attributes might be the chosen pseudonyms). In some sense, these systems are static: the set of attributes remains the same until the user gets a new witness. Therefore, it can be used as a source for pseudonyms, but only for a fixed number of cases.

2.2   Pseudonymous Identity

The concept of pseudonyms is parallel to the concept of anonymous credentials. The user gets a possibility to create pseudonyms so that:

- potentially there is an unlimited number of domains,
- for each *domain* the user can create his pseudonym,
- it is infeasible for the user to create two pseudonyms in the same domain.

A domain is here a virtual concept, but its realizations might be for instance different virtual services.

The above properties are based on two assumptions:

- a user should be given the possibility of creating a pseudonym in any domain (in particular, a domain that emerges after the cryptographic token for issuing his pseudonyms has been created),
- the scheme should prevent Sybil attacks – appearing in the same domain under different pseudonymous. Otherwise, pseudonymous identity would become a perfect tool supporting misbehavior – after getting bad reputation one could reset the reputation and continue to cheat new partners.

A direct consequence of the above rules is that a user or group of users should not have the possibility to create a new virtual user that would be able to produce valid pseudonyms (*seclusiveness* property).

The main application area of pseudonymous identity scheme is the management of online services: within a service the user should have exactly one identity in order to maintain their account. On the other hand, there should be no possibility of linking data from different services as it may create personal data protection threats. This *unlinkability* property may be roughly expressed as follows: given two systems:

– one where the pseudonyms are created according to a given scheme,
– one where the pseudonyms are created independently at random,

an external observer cannot distinguish between these two systems based on the observed pseudonyms.

Despite the unlinkability requirements there are certain situations when linking might be necessary. For instance, if an account is used for criminal activities it should be possible to deanonymize a user by linking his pseudonym with a real identity. This option exists for Restricted Identification scheme implemented on German personal identity cards [5]. In cases such as pseudonymization used to protect databases with health data records, it might be necessary to translate a pseudonym from domain A to a pseudonym of the same user in domain B. The question is who should be able to perform such a translation. The scheme proposed in [6] is based on an authority that performs translations blindly (that is without knowing which pseudonym is translated and what is the result). For practical purposes there might be different use cases here.

2.3   Pseudonymous Authentication and Pseudonymous Signatures

Pseudonyms alone might be of limited use. A more advanced option would be to have authentication means coupled with the anonymous identity. A standard choice here would be a kind of digital signature – a *domain signature*. Apart from the standard assumptions, the following properties should be fulfilled:

binding to pseudonyms: each signature should be attributed to a single pseudonymous identity,
seclusiveness: it should be based on a system of pseudonyms satisfying the seclusiveness property, that is, it should be infeasible to create a signature corresponding to a pseudonymous identity not related to a legitimate user.
single private key: the user should use the same private key for all domains. On the other hand, there should be no corresponding public key for more than one domain, as it would immediately break the unlinkability property.

For domain signatures there are a few issues that are considerably harder than for standard signatures. Apart from the seclusiveness there is a problem of creating a revocation framework that does not violate unlinkability properties. Indeed, revocation might be necessity in case when the signatory is a victim of an attack. In this case deanonymization would be an additional harm for him.

A few schemes of this kind have been proposed so far. Pseudonymous Signature aimed for personal identity documents and developed by German information security authority [5] is simple and easy to implement. However, there are critical drawbacks: the Issuer knows the signing keys of each user, breaking into just two ID cards enables him to learn the system keys and create new pseudonymous identities. Also, it is possible to prepare the signing keys so that a third party can link the pseudonyms, and there is no revocation scheme. Other designs [7,8] use pairings in order to escape seclusiveness problems. This is a major practical complication due to missing support for pairings on smart cards.

3   Privacy-by-Design and E2E Systems

Implementing the privacy-by-design concept is a challenge due to the fact that contemporary IT systems are deeply rooted in early systems of the 20th century, when their scale was small (in most cases local), the amount of private information processed was very limited and much better control over users' behavior was possible. Consequently, there are many basic decisions that are not compatible with the concept of the privacy-by-design, but which are irreversible for economic reasons. Therefore, a privacy-aware system must be built so that it is secure regardless of these of its components which do not comply with the privacy rules.

Due to the obligation of demonstrability of privacy protection, the concept must not be based on unjustified trust assumptions. By unjustified assumption we should mean here in particular declarations of a manufacturer as it does not fulfil demonstrability condition.

The problem of systems secure-by-design in potentially hostile environment has been already considered for e-voting schemes (see e.g. [9]). The central concept of this approach is public verifiability of security properties during the system operation. The lack of trust assumption is expressed as *End-to-End* security.

3.1   Black-Box Devices

One of the crucial issues are black-bock devices – so frequently used in cryptographic protocols despite the fact that there is an abundance of kleptographic techniques that can be used to leak the information from them in an invisible way. Moreover, it has been shown that these methods can be used in areas where privacy issues play a fundamental role [10].

Basically any nondeterminism in operation of black-box devices may be a source of privacy leakage. Unfortunately, a device can never be purely deterministic as it can use such side channels as response time or fault messages to transmit data.

The only approach that seems to be applicable to black-box devices is the following:

1. a black-box device works in slave mode only,
2. it performs only deterministic functions with verifiable output, or ...
3. randomness used is verifiable by external components.

3.2   Subversion and Watchdog Concepts

Recently, the possibility of an adversary taking control over a part of a system (in particular, replacing the original software with malicious one) has been incorporated into the security model ([11]. Unlike in the traditional approach, security protection is not based on the search for malware installed in the software by system administrators (which may hardly be considered as a secure-by-design approach), but based on *watchdog* components. Such a watchdog interacts with the inspected system (or just analyses its output) in order to detect any irregularities. The watchdog activities may be extended to such sensitive operations as generation of private keys (by definition involving a lot of randomness) [12]). It has been shown that in some cases a watchdog can be created on top of already existing and widely used schemes implemented in black box devices [13].

3.3   (Un)trusted Parties and Under-Specification

In case of privacy protection there should be no assumptions about trusted parties. The problem is that even a trustworthy institution may be under attack and for instance a system administrator of such an institution may mount a privacy leakage attack. So, no party should be considered a secure black box and external control measures in the sense of E2E systems should be implemented.

One of the crucial issues is under-specification of the systems used when there are two extreme situations possible:

– the system is implemented as intended, without creating backdoors,
– the system fully complies with the specification, however there are backdoors based on certain details.

An example of this situation is U-Prove revocation system, where it has been shown that both options are possible [14]. However, the same problem concerns the specification of the German personal identity card [5]. Therefore the proofs should concern not only the specified system, but also any system that *looks the same* from the point of view of an external verifier.

4   Challenges

4.1   Unification

Privacy protection, in order to be successful, must be a cross-border initiative. The example of ICAO and biometric passports shows that such a world-wide deployment is possible despite the lack of formal agreement between all countries. The necessity for solving its internal problems may potentially set the European Union up for a big role in this enterprise.

A decision should made which mechanism to choose for implementation and what architecture of privacy protection model is best suited for the most urgent needs. The basic choice is between anonymous credentials, identification with pseudonyms and/or pseudonymous signatures.

4.2   Legal Traps

Diversity of means of identification and authentication in European countries leads to a solution adopted in eIDAS: once a citizen initiates an identification and authentication process outside its home country, their request is redirected to his country of origin and executed in the regular way. Finally, the result of execution is provided to the server in the country, where the procedure has been initiated.

This architecture attempts to integrate incompatible protocols. However, it generates serious privacy issues. In the regular case, an identification attempt is executed with the party that has to learn the (anonymous) identity of the citizen as it has to process his data. In the case of the framework introduced by eIDAS, a central server of the country of origin learns the full track of identification attempts abroad. This is a direct violation of the concept of data minimization in privacy protection, and even a violation of the German constitution (due to limitations of gathering data about own citizens by central authorities). However, it seems the only solution available at the moment. For the future, it seems necessary to rebuild the protocols so that the central server of the country of origin may perform the essential steps of authentication, however in a fully blind way. The main challenge is that involvement of the server in the foreign country should be minimal.

4.3   Lifecycle Issues

One of the critical issues concerning applications utilising secret keys is implementing them in secure hardware. Inevitably, there are effects of hardware aging as well as cases of destroying or loosing secure hardware. For this reason, there should be ways of either replacing the old keys by new ones, or uploading the old keys to new devices. This is much harder than, say, in case of regular electronic signatures, since

- the identities resulting from old keys have to be linked with identities resulting from the new ones without providing other linking information (such as via using the same digital evidence for different pseudonyms),
- if the old keys can be uploaded to the new devices, there must be a way to prevent doing this more than once, thus creating cloned devices.

The second case is more tricky than the case of digital signatures, where no problem arises as long as the clone does not create digital signatures. In case of privacy protection the attack might be purely passive – the clone might be used only for breaking anonymity during offline attacks. Such activities are not observable in public.

This problem has been treated for anonymous credentials [15], but no solution has been proposed so far for domain pseudonyms and signatures.

4.4   Limitations of Cryptographic Protection

Long time protection. Each cryptographic mechanism secure at the present moment may weaken due to progress in cryptanalytic tools. In particular, the mechanisms based on hardness of discrete logarithm problem may break down as soon as decisional Diffie-Hellman Problem can be solved.

In case of attacks on encrypted data the scope of the attack is in some sense limited: it concerns only the past data. In case of anonymization this is not the case, as pseudonymous identity by definition should persist over a long time. In case when we expect weakening of the cryptographic mechanism we cannot just replace the old pseudonymous identity by a new one for two important reasons:

- proliferation of the pseudonym might be hard to control,
- replacement might leave many traces which in turn enable linking the old compromised identity with a new one.

Limited protection scope. Cryptographic mechanisms presented above enable hiding the real identity. However, the protection mechanism is limited to removing the explicit identity information and replacing them by strings that are not distinguishable from random ones. However, this does not protect at all against attacks that are based on identity recovery from the available data. This creates a lot of challenges as it requires careful data analysis and processing them accordingly. Unfortunately, in many cases (like medical research) such data modifications should not be allowed.

4.5   Transparency

The privacy protection techniques should be, to some extent, understandable to the general public. The number of specialists that would be able to analyze them in a more or less complete way should be fairly large. If this were not the case, then we would essentially retreated back to the black box devices and/or to solutions based on blindly trusting a small group of people.

Unfortunately, many of the designs proposed so far are so complicated and involved that this condition cannot be satisfied. A good example of this is the design of U-Prove [3], otherwise a powerful anonymous credentials tool.

Importantly, a complicated design also significantly increases the chance of implementation faults, overlooking intentional or non-intentional backdoors.

4.6   Formal Security Proofs

As privacy protection must be *by-design* [1], there should be an effective technical framework enabling building compliant systems.

Currently, formal proof methods are quite limited for the following reasons:

- It is not always clear what privacy protection means, there is abundance of models constructed mainly case by case, almost separately for each proposed protocol. In many cases they are quite complicated and confusing to the end user (see e.g. [7]).
- Scaling problems – the approaches that may be applied to isolated components are not suited to big systems with a large number of interacting parties and components.
- Reusing the old proofs after a system update. In most cases, the whole formal analysis should be redone and a risk remains that a detailed investigation is replaced by a declaration that the "proof works in exactly the same way".
- *Universal composability framework* might be a solution of the above problem, however its major disadvantage is that the protocols get more complicated. This might be unacceptable for practical reasons.
- Finally, the proofs have to take into account that there are components of the system that are out of control, misfunctioning or simply malicious. So far, most of the proofs have been presented without such concerns. Interestingly, there are cases where these issues have been taken into account – a good example is PACE password authentication protocol [16].

References

1. The European Parliament and the Council of the European Union: Regulation (EU) 2016/679 of the European Parliament and of the Council of 27 April 2016 on the protection of natural persons with regard to the processing of personal data and on the free movement of such data, and repealing Directive 95/46/ec (General Data Protection Regulation). Off. J. Eur. Union **119**
2. The European Parliament and the Council of the European Union: Regulation (EU) No 910/2014 of the European Parliament and of the Council of 23 July 2014 on electronic identification and trust services for electronic transactions in the internal market and repealing Directive 1999/93/ec (2014). http://eur-lex.europa. eu/legal-content/EN/TXT/?uri=uriserv:OJ.L_.2014.257.01.0073.01.ENG
3. Microsoft: U-Prove. Webpage of the project. Accessed 2017
4. IBM: Idemix. Webpage of the project. Accessed 2017
5. BSI: Technical guideline tr-03110 v2.21 - advanced security mechanisms for machine readable travel documents and eidas token (2016). https://www.bsi.bund. de/EN/Publications/TechnicalGuidelines/TR03110/BSITR03110.html
6. Camenisch, J., Lehmann, A.: (Un)linkable pseudonyms for governmental databases, pp. 1467–1479. [17]
7. Bringer, J., Chabanne, H., Lescuyer, R., Patey, A.: Efficient and strongly secure dynamic domain-specific pseudonymous signatures for ID documents. In: Christin, N., Safavi-Naini, R. (eds.) FC 2014. LNCS, vol. 8437, pp. 255–272. Springer, Heidelberg (2014)

8. Kluczniak, K.: Anonymous authentication using electronic identity documents. Ph.D. dissertation, Institute of Computer Science, Polish Academy of Sciences (2016)
9. Popoveniuc, S., Kelsey, J., Regenscheid, A., Vora, P.L.: Performance requirements for end-to-end verifiable elections. In: Jones, D.W., Quisquater, J., Rescorla, E. (eds.) 2010 Electronic Voting Technology Workshop/Workshop on Trustworthy Elections, EVT/WOTE 2010, Washington, D.C., USA, 9–10 August 2010. USENIX Association (2010)
10. Gogolewski, M., Klonowski, M., Kubiak, P., Kutyłowski, M., Lauks, A., Zagórski, F.: Kleptographic attacks on e-voting schemes. In: Müller, G. (ed.) ETRICS 2006. LNCS, vol. 3995, pp. 494–508. Springer, Heidelberg (2006). https://doi.org/10.1007/11766155_35
11. Ateniese, G., Magri, B., Venturi, D.: Subversion-resilient signature schemes, pp. 364–375. [17]
12. Tang, Q., Yung, M.: Cliptography: post-snowden cryptography. In: Thuraisingham, B.M., Evans, D., Malkin, T., Xu, D. (eds.) Proceedings of the 2017 ACM SIGSAC Conference on Computer and Communications Security, CCS 2017, Dallas, TX, USA, 30 October–03 November 2017, pp. 2615–2616. ACM (2017)
13. Hanzlik, L., Kluczniak, K., Kutyłowski, M.: Controlled randomness – a defense against backdoors in cryptographic devices. In: Phan, R.C.-W., Yung, M. (eds.) Mycrypt 2016. LNCS, vol. 10311, pp. 215–232. Springer, Cham (2017). https://doi.org/10.1007/978-3-319-61273-7_11
14. Hanzlik, L., Kubiak, P., Kutylowski, M.: Tracing attacks on U-prove with revocation mechanism: tracing attacks for U-prove. In: Bao, F., Miller, S., Zhou, J., Ahn, G. (eds.) Proceedings of the 10th ACM Symposium on Information, Computer and Communications Security, ASIA CCS 2015, Singapore, 14–17 April 2015, pp. 603–608. ACM (2015)
15. Baldimtsi, F., Camenisch, J., Hanzlik, L., Krenn, S., Lehmann, A., Neven, G.: Recovering lost device-bound credentials. In: Malkin, T., Kolesnikov, V., Lewko, A.B., Polychronakis, M. (eds.) ACNS 2015. LNCS, vol. 9092, pp. 307–327. Springer, Cham (2015). https://doi.org/10.1007/978-3-319-28166-7_15
16. ISO/IEC JTC1 SC17 WG3/TF5 for the International Civil Aviation Organization: Supplemental access control for machine readable travel documents. Technical report (2014) version 1.1, April 2014
17. Ray, I., Li, N., Kruegel, C. (eds.): Proceedings of the 22nd ACM SIGSAC Conference on Computer and Communications Security, Denver, CO, USA, 12–6 October 2015. ACM (2015)

The Dual Role of Smartphones in IoT Security

Kui Ren[(✉)]

Institute of Cyber Security Research, Zhejiang University, Hangzhou, China
kuiren@zju.edu.cn

Abstract. The world is entering the era of Internet of Things (IoT), where the interconnected physical devices of various forms, often embedded with electronics, software, sensors, actuators, etc., jointly perform sophisticated sensing and computing tasks and provide unprecedented services. Centering around this new paradigm is the ubiquitous smartphone. Equipped with abundant sensing, computing and networking capabilities, the smartphone is widely recognised as one of the key enablers towards IoT and the driving force that brings a great many innovative services under the way.

Despite the promising aspects, along with the rise of IoT is the increasing concerns on cybersecurity. The smartphone in this new context, however, plays a very intriguing dual role, due to the fact that it is deeply interleaved into almost every aspect of our daily living. On the one hand, it could be used as a low-cost attacking device, trying to penetrate into the scenarios that have never been considered before. On the other hand, it is also the first line of defense in the security forefront. In both cases, we need to carefully study and comprehensively understand the capability of smartphones, as well as their security implications. In this talk, we will use two examples to illustrate this observation and hopefully promote further researches along this line.

1 Introduction

Incorporating IoT into our lives revolutionizes the way people interact with the physical world, and brings enormous benefits to areas such as health-care, transportation and manufacturing. However, the rise of IoT also raises increasing concerns about the threat of cyberattacks. According to a Gemalto survey [1], 65% of consumers are concerned about a hacker controlling their IoT devices while 60% of consumers are concerned about data leakage. Unfortunately, due to the fact that IoT devices are diversified in communication protocols, internetworking architectures, and application scenarios, managing and securing massive heterogeneous IoT devices have always been a challenging task.

Among various IoT devices, the smartphone, with its sufficient computational power, various device connection capabilities, and convenient user interface, is generally regarded as the brain that manages and controls other interconnected physical devices [2]. In the context of IoT security, smartphones play vital roles as well. On one hand, it could be used as a low-cost attacking device. Compared

© Springer International Publishing AG, part of Springer Nature 2018
X. Chen et al. (Eds.): Inscrypt 2017, LNCS 10726, pp. 21–24, 2018.
https://doi.org/10.1007/978-3-319-75160-3_3

with professional equipments, smartphones are more commonly used in our daily life. An adversary using a smartphone can inconspicuously launch his attack because of the portability and pervasiveness of smartphones. On the other hand, it could also be used as a security hub to provide the first line of defense for IoT devices. Using the smartphone as the security hub not only empowers the users to enforce security policy across devices, but also eases the way users manage their personal information.

In this talk, we use two examples to illustrate the potential of smartphones in attacking and securing IoT devices. We first introduce a smartphone-based side-channel attack [3] that steals the designs from a 3-D printer. It uses an off-the-shelf smartphone to collect the acoustic noise and electromagnetic radiation the printer emits and then reconstruct the object being printed with high accuracy. Next, we present a voice impersonation defense system [4] that uses the smartphone to defeat machine-based impersonation attacks. Leveraging the magnetometer in modern smartphones, the system can accurately detect the magnetic field produced by conventional loudspeakers and thus defeat machine-based voice impersonation attacks.

2   Side-Channel Attacks Against 3D Printers

The 3D printer is a manufacturing platform that could efficiently transfer a digital 3D blueprint into a 3D physical object. After decades of development, the 3D printer is increasingly being used to fabricate highly intellectual property (IP) sensitive products, which makes the digital design being printed on the printer high value targets for cybercrime.

While 3-D printers are often equipped with sophisticated defenses for hacking and other online threats [5,6], most of them are vulnerable against side channel attacks. During the manufacturing process, multiple electromechanical components in the 3D printer generate diverse side-channel signals according to the movement of the nozzle. By collecting those signals, an adversary can reconstruct the movement of the nozzle and build up the blueprint being printed on the 3D printer accordingly.

Being equipped with rich set of on-board sensors, modern smartphones are able to capture ambient signals from audio to magnetic field to light conditions, which makes them good candidates to conduct side-channel attacks. Using off-the-shelf smartphones, we developed a system that can capture the blueprint being printed on a 3D printer utilizing magnetic and acoustic signals [3]. During the attacking process, the system first runs a recording application that collects the magnetic and acoustic data simultaneously. It applies the Savitzky-Golay filter [7] on the collected data to remove noise and integrates the obtained data into parameter sequences in time series. The system then uses the support vector machines (SVM) as the classifiers to infer the printer operation parameters (Time stamp/Distance/Device info) for each frame. Finally, the system derives the blueprint from the printer status set using a G-code reconstruction algorithm. Experimental results show that the system can successfully reconstruct the blueprints and their G-code with more than 90% accuracy.

3   Defending Against Voice Impersonation Attacks

IoT devices are heterogeneous in size, scale, and form. The management and security of the diversified IoT devices calls for universal security interface that supports friendly interaction, convenient update, and continuous protection. Voice control is one of the top candidates. Compared with traditional device interface such as touch screen and button, voice control requires only audio I/O components that are quite small and inexpensive [8]. It provides a convenient and non-intrusive way for communication as well as command control.

However, voice control is vulnerable against impersonation attacks, where an adversary uses a prerecorded [9], manipulated [10,11] or synthesized [12] audio signal to spoof or impersonate the identity of a legitimate user [13]. For instance, an adversary who has collected a certain amount of audio clips from a legitimate user's online social networking website could concatenate speech samples and conduct replay attacks. Although several detection mechanisms [9,14,15] have been proposed to defeat voice impersonation attacks, most of them suffer from high false acceptance rate.

Leveraging the fact that the loudspeaker used in the machine-based voice impersonation attack generates considerable magnetic field, we propose to use smartphones to detect voice impersonation attacks through checking the magnet nearby during the process of voice authentication [4]. Our system consists of four verification components: (1) Sound source distance verification, which uses acoustic and motion data to calculate the distance between the smartphone and the sound source. (2) Sound field verification, which verifies the characteristics of the sound field to detect if the sound is formed and articulated by a sound source of a size close to a human mouth. (3) Loudspeaker verification, which crosschecks the magnetometer data and motion trajectory to verify if the sound is produced by a human speaker or a loudspeaker. (4) Speaker identity verification, which verifies the speaker's identity through analyzing the spectral and prosodic features of the acoustic data. Integrating above components, our detection mechanism achieves 100% accuracy at 0% equal error rates in detecting the machine-based voice impersonation attacks.

4   Conclusion

With the above systems in mind and with the IoT security issues summarized herein, we conclude that the role of modern smartphones needs to be carefully studied to understand their capability as well as security implications. In the context of IoT security, the capability of smartphone is a promising research area of paramount importance and challenges. We believe considerable research efforts will be conducted along this line in the coming years.

References

1. Gemalto: The state of IoT security. https://www.gemalto.com/m2m/documents/iot-security-report/
2. Hausenblas, M.: Smart phones and the internet of things. https://mapr.com/blog/smart-phones-and-internet-things/
3. Song, C., Lin, F., Ba, Z., Ren, K., Zhou, C., Xu, W.: My smartphone knows what you print: exploring smartphone-based side-channel attacks against 3D printers. In: Proceedings of the ACM SIGSAC Conference on Computer and Communications Security (CCS), pp. 895–907. ACM (2016)
4. Chen, S., Ren, K., Piao, S., Wang, C., Wang, Q., Weng, J., Su, L., Mohaisen, A.: You can hear but you cannot steal: defending against voice impersonation attacks on smartphones. In: Proceedings of the 37th IEEE International Conference on Distributed Computing Systems (ICDCS), pp. 183–195 (2017)
5. Anderson, P., Sherman, C.A.: A discussion of new business models for 3D printing. Int. J. Technol. Mark. **2**, 280–294 (2007)
6. Hou, J.U., Kim, D.G., Choi, S., Lee, H.K.: 3D print-scan resilient watermarking using a histogram-based circular shift coding structure. In: Proceedings of the 3rd ACM Workshop on Information Hiding and Multimedia Security, pp. 115–121 (2015)
7. Savitzky, A., Golay, M.J.: Smoothing and differentiation of data by simplified least squares procedures. Anal. Chem. **36**, 1627–1639 (1964)
8. Lee, K.B., Grice, R.A.: The design and development of user interfaces for voice application in mobile devices. In: Proceedings of IEEE International Professional Communication Conference, pp. 308–320 (2006)
9. Villalba, J., Lleida, E.: Detecting replay attacks from far-field recordings on speaker verification systems. In: Vielhauer, C., Dittmann, J., Drygajlo, A., Juul, N.C., Fairhurst, M.C. (eds.) BioID 2011. LNCS, vol. 6583, pp. 274–285. Springer, Heidelberg (2011). https://doi.org/10.1007/978-3-642-19530-3_25
10. Stylianou, Y.: Voice transformation: a survey. In: Proceedings of IEEE International Conference on Acoustics, Speech and Signal Processing (ICASSP), pp. 3585–3588 (2009)
11. Wu, Z., Li, H.: Voice conversion and spoofing attack on speaker verification systems. In: Proceedings of IEEE Asia-Pacific Signal and Information Processing Association Annual Summit and Conference (APSIPA), pp. 1–9 (2013)
12. Alegre, F., Vipperla, R., Evans, N., Fauve, B.: On the vulnerability of automatic speaker recognition to spoofing attacks with artificial signals. In: Proceedings of the 20th European Signal Processing Conference (EUSIPCO), pp. 36–40 (2012)
13. Evans, N., Yamagishi, J., Kinnunen, T.: Spoofing and countermeasures for speaker verification: a need for standard corpora, protocols and metrics. IEEE Signal Processing Society Speech and Language Technical Committee Newsletter, May 2013
14. Villalba, J., Lleida, E.: Preventing replay attacks on speaker verification systems. In: Proceedings of IEEE International Carnahan Conference on Security Technology (ICCST), pp. 1–8 (2011)
15. Wang, Z.F., Wei, G., He, Q.H.: Channel pattern noise based playback attack detection algorithm for speaker recognition. In: Proceedings of IEEE International Conference on Machine Learning and Cybernetics (ICMLC), vol. 4, pp. 1708–1713 (2011)

Cryptographic Protocols and Algorithms

Implementing Indistinguishability Obfuscation Using GGH15

Zheng Zhang[1,2], Fangguo Zhang[1,2(✉)], and Huang Zhang[1,2]

[1] School of Data and Computer Science, Sun Yat-sen University,
Guangzhou 510006, China
isszhfg@mail.sysu.edu.cn

[2] Guangdong Key Laboratory of Information Security, Guangzhou 510006, China

Abstract. Obfuscation is an extraordinarily powerful object that has been shown to enable a whole set of new cryptographic possibilities. Because of the impossibility of the general-purpose virtual black-box (VBB) obfuscation, Barak *et al.* suggested to implement a weak variant which is called the indistinguishability obfuscation (iO). The iO is the substrate of various cryptographic primitives such as the universal function encryption, the self-bilinear map and so on. However, current obfuscation is too cumbersome to implement in practice.

In this paper, we implement an obfuscation for NC1 circuits by using the GGH15 multilinear map. Several techniques are proposed to improve the efficiency and adaptability of the implementation. We reduce the matrix dimension and the depth of encoding graph to increase the speed of confusion. Splitting the matrix into block matrix and encoding each block instead of using the entire matrix will reduce the size of matrix effectively. The plaintext matrix will be one block of the matrix. Besides, we put matrices into groups and encode one group on path $u \rightsquigarrow v$. Then the depth of the graph depends on the number of groups rather than the number of matrices. Those methods have led to a significant reduction in the rate of obfuscation.

Keywords: Multilinear map · iO · Implementation · GGH15

1 Introduction

Obfuscation is an useful technique to hide the specifics of a program while preserving its functionality. The earliest obfuscation schemes are often based on replacement and padding [25,29]. Because those obfuscation schemes are short of rigorous security proof, they provide the limited guarantees of security for programmers. Fortunately, the first mathematical definition, called virtual black-box (VBB) obfuscation, was given by Barak *et al.* [4] in 2001. However, they also proved that the general-purpose VBB obfuscator is impossible to be constructed. As a compromise, they suggested to construct a less intuitive, but realizable, computationally indistinguishable definition, notion of indistinguishability obfuscation (iO), which requires only that on input any two programs with

© Springer International Publishing AG, part of Springer Nature 2018
X. Chen et al. (Eds.): Inscrypt 2017, LNCS 10726, pp. 27–43, 2018.
https://doi.org/10.1007/978-3-319-75160-3_4

the same size and functionality. From then on, to design a general-purpose iO becomes a long standing open problem. Several works for obfuscating specific functions were made during that time [8,9]. The first iO for all polynomial-size circuits, was proposed by Garg *et al.* [15]. The core of their construction is to design an obfuscator for any NC1 circuits. Once it is done, we will obtain an obfuscator for polynomial-size circuits by using a fully homomorphic encryption [17] and NC1 obfuscator as the subroutines. However, the success is still in the distance.

Current general-purpose obfuscation schemes are based on another important primitive in cryptography, called the multilinear map (a.k.a. the graded encoding scheme) [13,14,18]. For now, only three constructions of multilinear maps were well known – GGH13 [14], CLT13 [13], and GGH15 [18]. However, current constructions of graded encoding scheme (GES) have got several severe cryptanalysis on their security. Since the zerozing attack [7,10–12,22] and the annihilation attack [27], amounts of obfuscation schemes were not secure when they used GGH13 and CLT13 multilinear maps.

GGH15 is the multilinear map proposed by Gentry *et al.* at TCC 2015, which seems to be the most secure one of all current multilinear maps [18]. There is a cryptanalysis of GGH15 [30], which breaks the multipartite key-agreement protocol in polynomial time. GGH15 is still security because the security of its multipartite key-agreement protocol is based on the combination of ciphertext but not on the encode. The construction of GGH15 involves many differences with GGH13 and CLT13, because of its graph-induced setting. In GGH15, a multilinear map is tightly associated with a directed acyclic graph $G = (V, E)$ and each edge implicitly fixes a preimage "group". An encoding of a secret message on edge $u \rightsquigarrow v$ is a short matrix $D \in \mathbb{Z}_q^{m \times m}$ satisfying $D \cdot A_u = A_v \cdot S + E$ for some small error $E \in (\chi)^{m \times n}$. The addition and multiplication between encodings imply the corresponding operations between underlying messages. Besides the construction of multilinear map, the authors also show how to design an iO scheme for matrix branching programs (MBP) using their multilinear map.

Our research began with the following observations. First, the security of GGH15 is measured by the dimension of matrices assigned to the vertexes of the graph and hence the obfuscation scheme in [18] is limited to obfuscate the MBP whose width is large enough. Second, the matrices of the multilinear map and that of the MBP should have the same dimension so that obfuscating one program implies initiating a fresh multilinear map. Third, as computational models, the MBP is weaker than the boolean circuit, so that the obfuscator in [18] is not sufficient enough to satisfy our purpose. All the efforts made in our work is to improve the performance of the foregoing aspects.

Contributions: In this paper, we implement a boolean-function obfuscator with the graph-induced multilinear map. We show both the theoretical designs and the practical simulation results. Besides that, several improvements are give to reduce the complexity of obfuscation.

Let L be the depth of the encoding graph, n be the dimension of (message/plaintext) matrices. In our scheme, the vertexes on a path will be divided into several sets in order, and hence message will be encoded set by set rather than vertex by vertex. Furthermore, we improve the procedures of GGH15 multilinear map to handle the matrix, whose dimension is not equivalent to n. Namely, we split the matrix into block matrix and encode each block instead of the entire matrix. The plaintext matrix will be one block of the matrix. Benefiting from above improvement, the amount of calculation in our implementation has been reduced. We can reach that the parameters of multilinear map are fixed.

Organization: The rest of our paper is organized as follows. The preliminaries used in our paper are introduced in Sect. 2. The specificity of the boolean-function obfuscation will be described in Sect. 3. Some improvements are given in Sect. 4. The last section is a conclusion.

2   Preliminaries

In this section, we briefly recall several notions and techniques used in our paper. Let $[n]$ denote the set $\{1, 2, \ldots, n\}$. We denote probabilistic polynomial time Turing machines by PPT. The term negligible is used for denoting functions that are (asymptotically) smaller than any inverse polynomial. We use boldface to denote vectors, for example, $\mathbf{u}, \mathbf{v}, \mathbf{c}$ etc., and use uppercase to denote matrixes.

2.1   (Matrix) Branching Program

Any boolean circuit can be represented by a branching program. A branching program (BP) is a directed acyclic graph whose vertices are partitioned into disjoint layers, defined as follows:

Definition 1 (branching program). *A branching program of width ω and length m is a directed acyclic graph with s vertices, such that the following constraints hold:*

1. *Each vertex is either a source vertex, an internal vertex or a final vertex.*
2. *Each non-final vertex is labeled with an element $l \in \{x_1, \ldots, x_n\}$ and has out-degree two, with one of the outgoing edges labeled 0 and the other 1.*
3. *There are two final vertices; both have out-degree zero and one is labeled '0' and the other 1.*
4. *There is a unique source vertex with in-degree zero.*
5. *The vertices are partitioned into m layers L_j, where $|L_j| \le \omega$. All vertices in a layer have the same label x_i.*
6. *Edges starting in layer L_j end in some layer $L_{j'}$ with $j' > j$.*
7. *Both final vertices are in the same layer L_m.*

Once the branching program gets an input, several edges, corresponding the labels and input bits, will construct a road from source vertex to one of final vertices, where the label of the final vertex is the output of the boolean circuit.

Further more, the branching program can be converted to matrix branching program (MBP), where we can operate by matrix calculation. The MBP is defined as follows:

Definition 2 (matrix branching program). *A matrix branching program of width ω and length $\omega - 2$ for n-bit inputs is given by a tuple*

$$MBP(f) = (inp, \mathbf{s}, (B_{j,0}, B_{j,1})_{j \in [\omega-2]}, \mathbf{t})$$

where $B_{j,b} \in \{0,1\}^{\omega \times \omega}$, for $j \in [\omega - 2]$ and $b \in \{0,1\}$, $inp : [\omega - 2] \to [n]$ is a function mapping layer indices j to input-bit indices i, $\mathbf{s} := (1, 0, \ldots, 0)$ and $\mathbf{t} := (0, \ldots, 0, 1)^T$.

On the matrix branching program, we can easily get the output of $f(\mathbf{x})$ with input $\mathbf{x}$. The corresponding matrixes, chosen from the input bit x_i and $inp(x_i)$ function, can be calculated by $MBP_f(\mathbf{x}) = \mathbf{s} \cdot (\prod_{j=1}^{m} B_{j,x_{inp(j)}}) \cdot \mathbf{t} \in \{0,1\}$. And we can see that $MBP_f(\mathbf{x}) = f(\mathbf{x})$. Note that the output of a MBP is the 0/1 entry in the first row and last column in the iterated matrix product, and the vectors $\mathbf{s}$ and $\mathbf{t}$ simply select this entry as the output.

2.2   Indistinguishability Obfuscation

The indistinguishability obfuscator ($i\mathcal{O}$) is an efficient randomized algorithm that makes circuits C_0 and C_1 computationally indistinguishable if they have the same functionality. We recall the notion of indistinguishability obfuscation for a class of circuit defined by [14].

Definition 3 (indistinguishability obfuscator). *A uniform PPT machine $i\mathcal{O}$ is called an indistinguishability obfuscator for a circuit class $\{C_\lambda\}$ if the following conditions are satisfied:*

1. *For all security parameters $\lambda \in \mathbb{N}$, for all inputs x, we have that*

$$Pr[C'(x) = C(x) : C' \leftarrow i\mathcal{O}(\lambda, C)] = 1.$$

2. *For any PPT distinguisher D, there exists a negligible function α such that the following holds: For all security parameters $\lambda \in \mathbb{N}$, for all pairs of circuits $C_0, C_1 \in C_\lambda$, we have that if $C_0(x) = C_1(x)$ for all inputs x, then*

$$|Pr[D(i\mathcal{O}(\lambda, C_0)) = 1] - Pr[D(i\mathcal{O}(\lambda, C_1)) = 1]| \leq \alpha(\lambda).$$

2.3   GGH15 Multilinear Maps

Boneh and Silverberg defined the notion of multilinear maps [6], which is a generalization of bilinear maps.

Definition 4 (multilinear map). *For $k+1$ cyclic groups $G_1, \ldots G_k, G_T$ of the same order p, the map $e : G_1 \times G_2 \times \cdots \times G_k \to G_T$ is called a multilinear map if the following properties are satisfied*

1. $\forall a \in \mathbb{Z}_p, e(g_1, g_2, \ldots, g_i^a, \ldots, g_k) = e(g_1, g_2, \ldots, g_k)^a$.
2. If g_i is the generator of the cyclic group G_i for $i = 1, \ldots, k$, then $e(g_1, \ldots, g_k)$ is the generator of G_T.

In 2015, Gentry et al. [18] proposed the graph-induced multilinear map from lattices (GGH15). Such a multilinear map is tightly associated with an underlying directed acyclic graph (DAG) $G = (V, E)$. Based on the graph, there is a random matrix $A_v \in \mathbb{Z}_q^{n \times m}$ with each node $v \in V$. When we need to encode a matrix S on the path $u \rightsquigarrow v$, we choose an error $E \in \mathbb{Z}^{n \times m}$ and calculate $A_u \cdot D = S \cdot A_v + E \mod q$. The encoded matrix D is the encoded matrix.

In more detail, the scheme consists of following efficient procedures.

$\mathcal{G}_{es} = (PrmGen, InstGen, Sample, Enc, Add, Neg, Mult, ZeroTest, Extract)$:

- $PrmGen(1^\lambda, G, \mathcal{C})$: On input the security parameter λ, an underlying DAG $G = (V, E)$ and class $\mathcal{C}$ of supported circuits, it computes some global parameters like LWE parameters, error distribution and so on.
- $InstGen(gp)$: Based on the global parameters, we use $TrapSamp(1^n, 1^m, q)$ to compute public parameters and private parameters.
- $Sample(pp)$: We sample a matrix S as the plaintext.
- $Enc(sp, p, S)$: we set $V = S \cdot A_v + E \in \mathbb{Z}_q^{n \times m}$, and use the trapdoor τ_u [1,20,26] to compute the encoding D s.t. $A_u \cdot D = V \pmod{q}$.
- $Add, Neg, Mult$: It will show $D + D', -D, D \cdot D'$ are also the encodings of $S + S', -S, S \cdot S'$.
- $ZeroTest(pp, D)$: The zero-test procedure outputs 1 if and only if $\|A_u \cdot D\| < q/2^{t+1}$.
- $Extract(pp, D)$: This procedure output $\omega := RandExt_\beta(msb_t(A_u \cdot D + \Delta))$

In GGH15, it uses $TrapSamp()$ to get pairs (A_i, τ_i), where A_i is a random node matrix and τ_i is a trapdoor of A_i. In our applications, we consider one special kind of lattices called primitive lattices [26] for effectiveness. A lattice $\Lambda^\perp(G)$, which is built by primitive matrix G, will be called primitive lattice. And we say that a matrix $G \in \mathbb{Z}_q^{n \times \omega}$ is primitive if $G \cdot \mathbb{Z}^\omega = \mathbb{Z}_q^n$.

Formally, a primitive matrix $G \in \mathbb{Z}_q^{n \times \omega}$ has the following characters:

1. The lattice $\Lambda^\perp(G)$ has a known basis $S \in \mathbb{Z}^{\omega \times \omega}$ with $\|\widetilde{S}\| \leq \sqrt{5}$ and $\|S\| \leq max\{\sqrt{5}, \sqrt{k}\}$.
2. Both G and S are sparse and highly structured.
3. Inverting $g_G(\mathbf{s}, \mathbf{e}) = \mathbf{s}^T G + \mathbf{e}^T \mod q$ can be performed in quasilinear time $O(n \cdot \log^c n)$.
4. Preimage sampling for $f_G(\mathbf{x}) = G\mathbf{x} \mod q$ can be performed in quasilinear time $O(n \cdot \log^c n)$.

The primitive lattice $\Lambda^\perp(G)$ helps one to get (A_i, τ_i) effectively, where τ_i is a G-trapdoor for A_i. First of all, we need to know how to build a primitive matrix G. Let $q \geq 2$ be an integer modulus and $k \geq 1$ be an integer dimension.

The construction will begin with a primitive vector $\mathbf{g} \in \mathbb{Z}_q^k$ and it sets $\mathbf{g} = (1, 2, 4, \ldots, 2^{k-1})^T \in \mathbb{Z}_q^k$. Then there is a low-dimensional lattice $\Lambda^\perp(\mathbf{g}^T)$, which have a basis $S_k \in \mathbb{Z}^{k \times k}$ where $\mathbf{g}^T \times S_k = \mathbf{0} \in \mathbb{Z}_q^{1 \times k}$ and $|det(S_k)| = q$. And we get a primitive dimension $G = I_n \otimes \mathbf{g}^T \in \mathbb{Z}_q^{n \times nk}$. After that, The pair (A_i, τ_i) can be got by Algorithm 1 about TrapSamp function.

Algorithm 1. TrapSamp Function

Input : $H = I$, G, n, m, and distribution $\mathcal{D}$ over $\mathbb{Z}^{\overline{m} \times \omega}$
Output : (A, τ_i)
1. Sample $\overline{A} \in \mathbb{Z}^{n \times \overline{m}}$ uniformly
2. Choose a matrix $\tau_i \in \mathbb{Z}^{\overline{m} \times \omega}$ from the distribution $\mathcal{D}$
3. Output $A = [\overline{A} | HG - \overline{A}\tau_i] \in \mathbb{Z}_q^{n \times m}$ and trapdoor $\tau_i \in \mathbb{Z}^{\overline{m} \times \omega}$

---

3 Implementation

We will show the details of implementing indistinguishability obfuscation for boolean circuits with polynomial size. Our implementation is based on the general construction using multilinear map by [15,18]. From the implementation, we find some problems and these problems will be solved in Sect. 4.

3.1 The Framework of Obfuscation

Depending on the techniques in [15,18], the framework of obfuscating a boolean circuit is as follows.

– **C(f)** $\Rightarrow$ **BP(f)**: Convert the original boolean circuit $C(f)$ into a branching program $BP(f)$.
– **BP(f)** $\Rightarrow$ **MBP(f)**: Encode $BP(f)$ as the matrix branching program MBP. Generate a dummy program, which has the same format with MBP and consists of some simple matrices like E. The MBP and the dummy program constitute $MBP(f)$.
– **MBP(f)** $\Rightarrow$ $\widetilde{\textbf{MBP}}$**(f)**: Randomize $MBP(f)$ to get $\widetilde{MBF}(f)$ through filling and mapping.
– $\widetilde{\textbf{MBP}}$**(f)** $\Rightarrow$ **En(f)**: Encode the randomized program $\widetilde{MBF}(f)$ with multilinear maps to get $En(f)$.
– **En(f)** $\Rightarrow$ **O(f)**: Randomize $En(f)$ to get the obfuscated program $O(f)$.

3.2 Implementation Process

From $C(f)$ to $BP(f)$. In order to get $BP(f)$, we can use Barrington's Theorem [5] or the approach of Sauerhoff *et al.* [28]. Comparing two methods, we can find that, for the same boolean function, the length of BP generated from Barrington's theorem is longer while the size of each matrix in MBP with

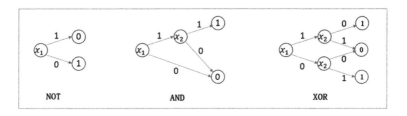

Fig. 1. Branching program of each gate

the approach of Sauerhoff *et al.* is bigger. However, the approach of Sauerhoff *et al.* is more efficient both asymptotically [2] and in practice, and thus we choose this approach.

Sauerhoff *et al.* [28] put forward a way to get a branching program of boolean function f. Given a boolean function f, it reorganizes the expression of f using AND gate, NOT gate and XOR gate. According the Fig. 1, each gate can be converted to the corresponding subgraph and the branching program consists of these subgraphs. We suppose that the input of $BP(f)$ is n-bit and there are ω nodes in the BP.

From $BP(f)$ to $MBP(f)$. The matrix branching program (MBP) can be easily got from branching program. If the $BP(f)$, what is got from step 1, is a directed acyclic $G = (V, E)$ and $|V| = \omega$, the size of $B_{i,b}$ is $\omega \times \omega$. For any $i \in [\omega - 2]$, $b \in \{0, 1\}$, let the elements on the diagonal of matrix $B_{i,b}$ be 1 and the element $b_{k,j}$ of $B_{i,b}$ be 1 if the edge labeled b of node i is connected to i and j. The function $inp(i) = j$ means that the node i is labeled the j-th bit input. Meanwhile, $\mathbf{s} := (1, 0, \ldots, 0)$ and $\mathbf{t} := (0, \ldots, 0, 1)^T$. Then we can get the corresponding $MBP(f) = \{inp, \mathbf{s} \in \mathbb{Z}_q^\omega, \mathbf{t} \in \mathbb{Z}_q^\omega, (B_{i,b})_{B \in \mathbb{Z}_q^{\omega \times \omega}, i \in [\omega-2], b \in \{0,1\}}\}$. What's more, the output of f can be calculated by Algorithm 2.

Algorithm 2. calculate $f(x)$

Input : $x \in \{0, 1\}^n$, $MBP(f)$
Output : $m \in \{0, 1\}$
1. for i from 0 to $\omega - 2$
2. $k = inp(i)$ to find the corresponding input bit
3.   choose B_{i,x_k} and put it in a queue
4. $m = \mathbf{s} \cdot \prod_0^{\omega-2} B_{i,x_k} \cdot \mathbf{t}$ where the B_{i,x_k} is fond from the queue
5. return m

Before we randomize the $MBP(f)$, we should build a dummy program. The *dummy program* (DP) has such definition:

$$DP = \{inp, \mathbf{s}', (D_{j,b})_{D \in \mathbb{Z}_q^{\omega \times \omega}, j \in [\omega - 2], b \in \{0,1\}}, \mathbf{t}'\},$$

where $\mathbf{s}'$ and $\mathbf{t}'$ in dummy program are exactly the same as in $MBP(f)$. The elements on the diagonal of $D_{1,0}$ and $D_{1,1}$ are 1. The element $d_{0,\omega-1}$ in $D_{1,0}$ and the element $d'_{0,\omega-1}$ in $D_{1,1}$ are 1. For the rest of $(D_{j,0}, D_{j,1})_{j\in[1,\omega-2]}$, we set each one as $E \in \mathbb{Z}_q^{\omega\times\omega}$. We can easily know the $m = \mathbf{s} \cdot \prod_{i=1}^{\omega-2} D_{i,x} \cdot \mathbf{t}$ has the same result with $MBP(f)$ when $f(x) = 1$.

$$B'_{i,b} = \begin{bmatrix} B_{i,b} & 0 & 0 \\ 0 & \$ & 0 \\ 0 & 0 & \$ \end{bmatrix}_{(\omega+2)\times(\omega+2)} \qquad D'_{i,b} = \begin{bmatrix} D_{i,b} & 0 & 0 \\ 0 & \$ & 0 \\ 0 & 0 & \$ \end{bmatrix}_{(\omega+2)\times(\omega+2)}$$

$$\bar{s} = \bar{s}' = [\mathbf{s} \mid 0\ 0] \qquad \bar{T} = \bar{T}' = \begin{bmatrix} \mathbf{t} & 0 & \cdots & 0 & \$ & \$ \\ \$ & \$ & \cdots & \$ & \$ & \$ \\ \$ & \$ & \cdots & \$ & \$ & \$ \end{bmatrix}_{(\omega+2)\times(\omega+2)}$$

Fig. 2. Filling $B_{i,b}$, $D_{i,b}$, $\mathbf{s}$ and $\mathbf{t}$

From $MBP(f)$ to $\widetilde{MBP}(f)$. The randomization is divided into two parts: filling and mapping.

On one side, the dimensions of the matrices $B_{i,b}$ and $D_{i,b}$ are expanded from $\omega \times \omega$ to $(\omega + 2) \times (\omega + 2)$ [16]. The $\mathbf{s}$ and $\mathbf{t}$ are expanded and illustrated in Fig. 2. The purpose of filling is to put some random elements in $MBP(f)$ and $DP(f)$.

On the other side, we use mapping to disrupt the structure and the Kilian-randomization [23] can hit it. For Kilian-randomization, we will get $R_0, R_1, \ldots, R_{n-2}$ at random and R_i is $(\omega + 2) \times (\omega + 2)$ reversible matrix. Then, we will do $\widetilde{B}_{i,b} = R_{i-1} B'_{i,b} R_i^{-1}$, $\tilde{s} = \bar{s} \cdot R_0^{-1}$ and $\tilde{T} = R_{n-2} \cdot \bar{T}$. What's more, we do the same for $D'_{i,b}$, $\bar{s}'$ and $\bar{T}'$ to get $\widetilde{DP}$.

After these two steps, we will get the randomized matrix branching program $(\widetilde{MBP}(f))$ and randomized dummy program $(\widetilde{DP})$ where

$$\widetilde{MBP}(f) = \{inp, \tilde{s} \in \mathbb{Z}_q^{\omega+2}, \tilde{T} \in \mathbb{Z}_q^{(\omega+2)\times(\omega+2)}, (\tilde{B}_{i,b})_{\tilde{B}\in\mathbb{Z}_q^{(\omega+2)\times(\omega+2)}, i\in[\omega-2], b\in\{0,1\}}\}$$

$$\widetilde{DP} = \{inp, \tilde{s}' \in \mathbb{Z}_q^{\omega+2}, \tilde{T}' \in \mathbb{Z}_q^{(\omega+2)\times(\omega+2)}, (\tilde{D}_{i,b})_{\tilde{D}\in\mathbb{Z}_q^{(\omega+2)\times(\omega+2)}, i\in[\omega-2], b\in\{0,1\}}\}.$$

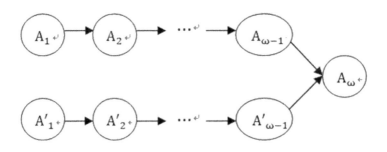

Fig. 3. $G = (V, E)$

From $\widetilde{MBP}(f)$ to $En(f)$. Next, the $\widetilde{MBP}(f)$ and $\widetilde{DP}$ are encoded with GGH15 multilinear maps. First of all, we should build a graph $G = (V, E)$ as Fig. 3, which contains two chains, one for $\widetilde{MBP}(f)$ and the other for $\widetilde{DP}$. The length of each chain is equal to ω. Then we can encode these programs $\widetilde{MBP}(f)$ and $\widetilde{DP}$ by Algorithm 3.

Algorithm 3. Encoding

Input : security parameter λ, $\widetilde{MBP}(f)$, $\widetilde{DP}$ and $G = (V, E)$
Output : $En(f)$, $En(DP)$
1. $gp \leftarrow \text{PrmGen}(1^\lambda, G, \mathcal{C})$
2. for v from 1 to ω
3. $\quad (A_v, \tau_v) \leftarrow TrapSamp(1^{\omega+2}, 1^m, q)$
4. for i from 1 to $\omega - 2$
5. $\quad A_i \cdot \widetilde{C}_{i,b} = \widetilde{B}_{i,b} \cdot A_{i+1} + E_{i,b}$
6. $\quad A'_i \cdot \widetilde{C}'_{i,b} = \widetilde{D}_{i,b} \cdot A'_{i+1} + E'_{i,b}$
7. $A_{\omega-1} \cdot C_T = \widetilde{T} \cdot A_\omega + E_T$ and $A'_{\omega-1} \cdot C'_T = \widetilde{T}' \cdot A_\omega + E'_T$
8. $\mathbf{a} = \tilde{\mathbf{s}} \cdot A_1 + \mathbf{e_t}$ and $\mathbf{a}' = \tilde{\mathbf{s}}' \cdot A'_1 + \mathbf{e'_t}$
9. return $En(f) = \{inp, C_T, \mathbf{a}, (\widetilde{C}_{i,b})\}$ and $En(DP) = \{inp, C'_T, \mathbf{a}', (\widetilde{C}'_{i,b})\}$

In step 3, the trapdoor τ_v of a random matrix $A_v \in \mathbb{Z}_q^{(\omega+2)\times m}$, which can be got by $TrapSamp$ function, is seen as a secret parameter. And we call a matrix $\tau_v \in \mathbb{Z}^{(m-\omega-2)\times(\omega+2)}$ is a G-trapdoor for A, if $A[\frac{\tau_v}{I}] = HG$, where H is a invertible matrix (we set $H = I$ in our implementing) and G is a primitive matrix. In step 5, 6 and 7, we want to get C from the equation $A_u \cdot C = M \cdot A_v + E$ generally. It is easy to get $T = M \cdot A_v + E$, while the problem is how to calculate $A_u \cdot C = T$ and each vector in C drawn from a distribution within $O(\epsilon)$ statistical distance of $D_{A_\mathbf{u}^\perp(A_u), r \cdot \sqrt{\Sigma}}$. Based on [26], it will be solved by Algorithm 4.

Algorithm 4. Find C for $A_u \cdot C = T$

Input : (A_u, τ_u) where $A_u = [\overline{A}|HG - \overline{A}\tau_u] \in \mathbb{Z}_q^{(\omega+2)\times m}$, T, G and σ_x
Output : A matrix C where c_i is drawn from a distribution within $O(\epsilon)$
$\qquad\qquad$ statistical distance of $D_{A_\mathbf{u}^\perp(A_u), r \cdot \sqrt{\Sigma}}$

---

1. for i from 0 to m-1
2. $\quad$ choose a perturbation $\mathbf{p}_i \leftarrow D_{\mathbb{Z}^m, r\sqrt{\Sigma_{\mathbf{p}_i}}}$, where $\Sigma_{\mathbf{p}_i} = \sigma_x^2 I - \sigma_x^2[\frac{\tau_u}{I}][\tau_u|I]$.
3. $\quad$ let $\mathbf{p}_i = [\frac{\mathbf{p}_{i1}}{\mathbf{p}_{i2}}]$ for $\mathbf{p}_{i1} \in \mathbb{Z}^{\overline{m}}$, $\mathbf{p}_{i2} \in \mathbb{Z}^\omega$.
4. $\quad$ compute $\overline{\mathbf{w}}_i = \overline{A}(\mathbf{p}_{i1} - \tau_u \cdot \mathbf{p}_{i2})$ and $\mathbf{w}_i = G \cdot \mathbf{p}_{i2}$.
5. $\quad$ Let $\mathbf{v}_i \leftarrow \mathbf{t}_i - \overline{\mathbf{w}}_i - \mathbf{w}_i$
6. $\quad$ choose $\mathbf{z}_i \leftarrow D_{A_{\mathbf{v}_i}^\perp(G), r \cdot \sqrt{\Sigma}}$ and $\mathbf{c}_i \leftarrow \mathbf{p}_i + [\frac{\tau_u}{I}]\mathbf{z}_i$
7. Return $C = \{\mathbf{c}_i\}$

From $En(f)$ to $O(f)$. For the encoded matrix, we will use Kilian-style randomization to mix them again and multiplicative bundling to preserve input consistency. We should get random reversible matrix P_0, P_1, ..., $P_{\omega-2}$ and P'_0, P'_1, ..., $P'_{\omega-2}$, where $P_i, P'_i \in \mathbb{Z}_q^{m \times m}$. Then we sample scalars $\{\beta_{i,b}, \beta'_{i,b} : i \in [\omega-2], b \in \{0,1\}\}$ modulo q. And they have $\prod_{I_j} \beta_{i,b} = \prod_{I_j} \beta'_{i,b} = 1$ where $I_j = \{i \in [\omega-2] : inp(i) = j\}$.

Finally, we get $O(f) =$

$$\{inp, \widehat{C}_T = P_{\omega-2}^{-1} C_T, \widehat{\mathbf{a}} = \mathbf{a} P_0, (\widehat{C}_{i,b} = P_{i-1}^{-1} \widetilde{C}_{i,b} P_i \cdot \beta_{i,b})_{i \in [\omega-2]}\}$$

and $O(DP) =$

$$\{inp, \widehat{C}'_T = P_{\omega-2}'^{-1} C'_T, \widehat{\mathbf{a}}' = \mathbf{a} P'_0, (\widehat{C}'_{i,b} = P_{i-1}'^{-1} \widetilde{C}'_{i,b} P'_i \cdot \beta'_{i,b})_{i \in [\omega-2]}\}.$$

3.3   Correctness

For the boolean function $f(x)$, we have $MBP(f) = (inp, \mathbf{s}, (B_{j,0}, B_{j,1})_{j \in [\omega-2]}, \mathbf{t})$. And it have $F(x) = \mathbf{s} \cdot (\prod_{i=0}^{\omega-2} B_{i,x_{inp(i)}}) \cdot \mathbf{t}$. For the dummy program, we have $DP = (inp, \mathbf{s}', (D_{j,0}, D_{j,1})_{j \in [\omega-2]}, \mathbf{t}')$. And it have $1 = \mathbf{s}' \cdot (\prod_{i=0}^{\omega-2} D_{i,x_{inp(i)}}) \cdot \mathbf{t}'$. It means that $f(x) = 1$ for any input $x \in \{0,1\}^n$, while the calculation of $MBP(f)$ is the same as it of DP.

Then, if there is a boolean function $f(x)$, and an indistinguishability obfuscation $\{O(f), O(DP)\}$ of $f(x)$. For any input $x \in \{0,1\}^n$, we have $O(f) = \widehat{\mathbf{a}} \cdot \prod_{i=1}^{\omega-2} \widehat{C}_{i,inp(x)} \cdot \widehat{C}_T$ and $O(DP) = \widehat{\mathbf{a}}' \cdot \prod_{i=1}^{\omega-2} \widehat{C}'_{i,inp(x)} \cdot \widehat{C}'_T$.

we have that

$$O(f) = \widehat{\mathbf{a}} \cdot \prod_{i=1}^{\omega-2} \widehat{C}_{i,inp(x)} \cdot \widehat{C}_T$$

$$= \mathbf{a} \cdot P_0 (\prod_{i=1}^{\omega-2} P_{i-1}^{-1} \cdot \widetilde{C}_{i,inp(x)} \cdot P_i \cdot \beta_{i,inp(x)}) \cdot P_{\omega-2}^{-1} \cdot C_T$$

$$= \mathbf{a} \cdot (\prod_{i=1}^{\omega-2} \widetilde{C}_{i,inp(x)}) \cdot C_T = \mathbf{s} \cdot A_1 \cdot (\prod_{i=1}^{\omega-2} \widetilde{C}_{i,inp(x)}) \cdot C_T$$

$$\approx \mathbf{s} \cdot (\prod_{i=1}^{\omega-2} \widetilde{B}_{i,inp(x)}) \cdot A_{\omega-1} \cdot C_T \approx (\mathbf{s} \cdot \prod_{i=1}^{\omega-2} \widetilde{B}_{i,inp(x)} \widetilde{T}) \cdot A_{\omega}$$

$$= [f(\mathbf{x}), \varepsilon, \dots, \varepsilon] \cdot A_{\omega}$$

and

$$O(DP) = \widehat{\mathbf{a}}' \cdot \prod_{i=1}^{\omega-2} \widehat{C}'_{i,inp(x)} \cdot \widehat{C}'_T$$

$$= \mathbf{a}' \cdot P'_0 (\prod_{i=1}^{\omega-2} P_{i-1}'^{-1} \cdot \widetilde{C}'_{i,inp(x)} \cdot P'_i \cdot \beta'_{i,inp(x)}) \cdot P_{\omega-2}'^{-1} \cdot C'_T$$

$$= \mathbf{a}' \cdot (\prod_{i=1}^{\omega-2} \widetilde{C}'_{i,inp(x)}) \cdot C'_T = \mathbf{s}' \cdot A'_1 \cdot (\prod_{i=1}^{\omega-2} \widetilde{C}'_{i,inp(x)}) \cdot C'_T$$

$$\approx \mathbf{s}' \cdot \left(\prod_{i=1}^{w-2} \tilde{D}_{i,inp(x)}\right) \cdot A'_{w-1} \cdot C'_T \approx \left(\mathbf{s}' \cdot \prod_{i=1}^{w-2} \tilde{D}_{i,inp(x)}\widetilde{T'}\right) \cdot A_w$$

$$= [1, \varepsilon, \ldots, \varepsilon] \cdot A_w$$

So we can get the output of $f(x)$ by comparing $msb_t(O(f))$ and $msb_t(O(DP))$.

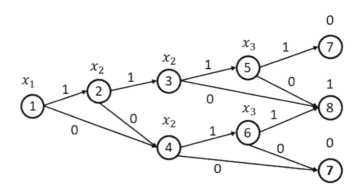

Fig. 4. BP for $f(\mathbf{x}) = (x_1 \wedge x_2) \oplus (x_2 \wedge x_3)$

3.4 Analysis

Some problems will be fond by a small example $f(\mathbf{x}) = (x_1 \wedge x_2) \oplus (x_2 \wedge x_3)$. Using the approach of Sauerhoff, the branching program is shown in Fig. 4. Then $MBP(f) = \{inp, \mathbf{s} \in \mathbb{Z}_q^8, \mathbf{t} \in \mathbb{Z}_q^8, (B_{i,b})_{B \in \mathbb{Z}_q^{8 \times 8}, i \in [6], b \in \{0,1\}}\}$ where $\mathbf{s} = [\,1\,0\,0\,0\,0\,0\,0\,0\,]$ and $\mathbf{t} = [\,0\,0\,0\,0\,0\,0\,0\,1\,]^T$. At the same time, A DP will be constructed where $DP = \{inp, \mathbf{s}' \in \mathbb{Z}_q^8, \mathbf{t}' \in \mathbb{Z}_q^8, (D_{i,b})_{D \in \mathbb{Z}_q^{8 \times 8}, i \in [6], b \in \{0,1\}}\}$. After the filling and Kilian-randomization, the size will be expanded to 10.

In [21], Shai *et al.* show the encoding parameters as Table 1. Besides, there are some fixed parameters. $P_1, P_2, \ldots, P_n$ are prime from 71 to 181 and $e = 3$ for $q = \prod_{i<k} P_i^e$. Each entry of E is chosen from a Gaussian distribution with parameter 2^7. The parameter $r = 4$ [3,26], which r is used for sampling the trapdoor R over $\mathbb{Z}$.

If $n = 10$ and $L = 8$, we can set $\log_2 q = 90$, $m = 3932$ and $\sigma_x = 371245$. Based on these parameters, we will have $En(f) = \{inp, \mathbf{a} \in \mathbb{Z}_q^{3932}, C_T \in \mathbb{Z}_q^{3932 \times 3932}, (\tilde{C}_{i,b})_{\tilde{C} \in \mathbb{Z}_q^{3932 \times 3932}, i \in [6], b \in \{0,1\}}\}$. It is easy to find the size of these encoded matrixes $\tilde{C}$ are too large to compute effectively.

What's more, these parameters depend on different circuits. If we want obfuscate different circuits, It need to build different graphs and pick matrix A_i with different size for each node. Its characteristic is not conducive to the standard design and use for obfuscator.

So, how to fix the parameters for GGH15 and reduce the size of encoded matrixes?

Table 1. Encoding parameter

Parameter	Explain	Setting
λ	Security parameter	$\lambda = 80$
n	$A \in \mathbb{Z}_q^{n \times m}$	Based on plaintext size
$\overline{m}$	$\overline{A} \in \mathbb{Z}_q^{n \times \overline{m}}$	$\overline{m} \geq \sqrt{\lambda + 2} \cdot \sqrt{n \log_2 q}$
m	$A \in \mathbb{Z}_q^{n \times m}$	$m \geq (\log_2 q - 7)(\lambda + 110)/7.2$ [19]
q	Module parameter	$q = \prod_{i<k} P_i^e$
L	Encoded graph depth	$\log_2 q \geq 7 + \log_2 \sigma_x (L-1) + \frac{L}{2} \log_2 m + (L-1) + 10$
σ_x	Sample $Ax = u$ with Gaussian parameter σ_x	$\sigma_x > 2900 \cdot (\sqrt{\overline{m}} + \sqrt{\omega} + 6)$ and $\omega = nek$
σ_z	Sample $Gz = v$ with Gaussian parameter σ_z	$\sigma_x = r \cdot max_i(p_i)$

4   Improvements

In this part, we will show the difference between our implementstiom and Shai Halevi's implementation [21] with GGH15. At the same time, the solution of two problems, mentioned in last section, will be explained.

4.1   Fix the Parameter L

In original obfuscation, we encode each $\tilde{B}_{j,b}$ on $A_j \rightsquigarrow A_{j+1}$. If the length of branching program is $\omega - 2$, the depth of $G = (V, E)$ will be $\omega - 1$ with another path to encode matrix T. Now we want to fix L in a small size, which is the depth of the graph G.

One solution can be that we encode $\prod \tilde{B}_{i,b}$ on $A_j \rightsquigarrow A_{j+1}$ where encode the product of these adjacent matrices on it instead of one matrices. Obviously, it can fix the depth of G on a smaller integer. From the Table 1, we can get smaller q based on a small depth L. And then, a smaller m will be get.

At the same time, we need to encode more matrices. For example, suppose there is a set $\{(\tilde{B}_{j,b})_{i\in[6],b\in\{0,1\}}\}$, we put them into two or three groups for comparison. For two groups, we have $g_1 = \{\tilde{B}_{1,0}, \tilde{B}_{1,1}, \tilde{B}_{2,0}, \tilde{B}_{2,1}, \tilde{B}_{3,0}, \tilde{B}_{3,1}\}$ and $g_2 = \{\tilde{B}_{4,0}, \tilde{B}_{4,1}, \tilde{B}_{5,0}, \tilde{B}_{5,1}, \tilde{B}_{6,0}, \tilde{B}_{6,1}\}$. For three groups, we have $g_1' = \{\tilde{B}_{1,0}, \tilde{B}_{1,1}, \tilde{B}_{2,0}, \tilde{B}_{2,1}\}$, $g_2' = \{\tilde{B}_{3,0}, \tilde{B}_{3,1}, \tilde{B}_{4,0}, \tilde{B}_{4,1}\}$ and $g_3' = \{\tilde{B}_{5,0}, \tilde{B}_{5,1}, \tilde{B}_{6,0}, \tilde{B}_{6,1}\}$. If we do not group them, we need to encode 12 matrices. If we group them into 2 or 3, we need to encode 16 or 12 matrices, while each encoding process has smaller amount of calculation. So, it is important to find a appropriate group number through optimization algorithm.

Formally, we want to reduce L to a smaller integer so that reduce the number of multiplications for $\widetilde{MBP}(f) = \{inp, \tilde{s}, \tilde{T}, (\tilde{B}_{j,0}, \tilde{B}_{j,1})_{j\in[\omega-2]}\}$. Because matrix multiplication has the associative law, $\{\tilde{B}_{j,b}\}$ can be divided into g groups on average. If $\omega - 2$ can not be divided exactly by g, we will put the minimum number of E with the same size to $\{\widetilde{B}_{j,b}\}_{j\in[\psi]}$ and the number ψ of new set can be divided by g where $k = \psi/g$. Then, the $group_i = \{\widetilde{B}_{(i-1)\psi/g+1,b}, \widetilde{B}_{(i-1)\psi/g+2,b}, \ldots, \widetilde{B}_{i\psi/g,b}\}$. We define

$$\widetilde{M}_{j,d} = \prod_{i=1}^{\psi/g} \widetilde{B}_{(j-1)\psi/g+i,b_i} \ and \ d = \sum_{k=1}^{\psi/g} b_i \cdot 2^{\psi/g-k}$$

When we want to know the output of $f(x)$, we should choose $\widetilde{M}_{j,d}$ according to the input x and then put them for multiplication. So we need a new $inp' : [g] \rightarrow [2^{\psi/g}]$ defined $inp'(x) = \sum_{i=1}^{\psi/g} inp((x-1)\psi/g+i) \cdot 2^{\psi/g-i}$ for chosen matrices. After all above steps, we will get a new structure of $\widetilde{MBP}(f)$ where $\widetilde{MBP}(f) = \{inp', \widetilde{s}, \widetilde{T}, (\widetilde{M}_{j,d})_{j\in[g],d\in[2^{\psi/g}]}\}$. And we can get the same new structure of $\widetilde{DP}(f)$ where $\widetilde{DP}(f) = \{inp', \widetilde{s}', \widetilde{T}', (\widetilde{N}_{j,d})_{j\in[g],d\in[2^{\psi/g}]}\}$. Now, we can encode $\widetilde{M}_{j,d}$ on the path from A_j to A_{j+1} and $\widetilde{N}_{j,d}$ on the path from A'_j to A'_{j+1}, where the depth of the graph is $g+1$ and $g+1 < L$.

4.2   Reduce the Size of Encoded Matrixes

In order to further reduce the amount of calculation, we put forward a method of encoding partitioned matrix. The $BP(f)$ which we get from the approach of Sauerhoff et $al.$, has ω nodes in graph. And the size of each matrix in $MBP(f)$ is ω, too. From the Table 1, we find the slight decrease in n has little effect on the parameters. If we have a lager n, we will reduce n to at least one order of magnitude and the parameter will decrease obviously. So the splitting of the matrix is an effective improvement method for matrix with large size.

Before splitting the matrix, Filling matrix dimension to the multiple of n, which can simply use the filling method in Sect. 3.4. Suppose we get each matrix is in $\mathbb{Z}^{N\times N}$, which $N = kn$ and $k \in \mathbb{Z}$. And then, the new matrix will be split as

$$\widetilde{M}_{i,d} = \begin{bmatrix} F_{11}^i & F_{12}^i & \cdots & F_{1k}^i \\ F_{21}^i & F_{22}^i & \cdots & F_{2k}^i \\ \vdots & \vdots & \ddots & \vdots \\ F_{k1}^i & F_{k2}^i & \cdots & F_{kk}^i \end{bmatrix}$$

If we want to get $\begin{bmatrix} A_i & & \\ & \ddots & \\ & & A_i \end{bmatrix} \cdot \widetilde{C}_{i,d}$ as before, we can get

$$\begin{bmatrix} A_i \cdot D_{11}^i & \cdots & A_i \cdot D_{1k}^i \\ A_i \cdot D_{21}^i & \cdots & A_i \cdot D_{2k}^i \\ \vdots & \ddots & \vdots \\ A_i \cdot D_{k1}^i & \cdots & A_i \cdot D_{kk}^i \end{bmatrix} = \begin{bmatrix} F_{11}^i \cdot A_{i+1} + E_{11}^i & \cdots & F_{1k}^i \cdot A_{i+1} + E_{1k}^i \\ F_{21}^i \cdot A_{i+1} + E_{21}^i & \cdots & F_{2k}^i \cdot A_{i+1} + E_{2k}^i \\ \vdots & \ddots & \vdots \\ F_{k1}^i \cdot A_{i+1} + E_{k1}^i & \cdots & F_{kk}^i \cdot A_{i+1} + E_{kk}^i \end{bmatrix}$$

which is equivalent to $\widetilde{M}_{i,d} \cdot \begin{bmatrix} A_{i+1} & & \\ & \ddots & \\ & & A_{i+1} \end{bmatrix} + \begin{bmatrix} E^i_{11} & \cdots & E^i_{1k} \\ \vdots & \ddots & \vdots \\ E^i_{k1} & \cdots & E^i_{kk} \end{bmatrix}$. That is to say, we have finished the original encoding process.

4.3   Comparison

In this section, we want to show our impelmentation by comparing the time with improvements and without improvements. All of our implementing was done on an $Intel(R)\ Core(TM)\ i5 - 3330s\ CPU$ @2.70 GHZ. This machine has 6 GB RAM and 64 CPUs. The compiler was $Microsoft\ Visual\ Studio\ Community$ 201514.0.25431.01. What we want to record will be divided into three parts - Initialization, obfuscation and calculation.

The first part is initialization. It include parameter generation, graph building and node-matrix building. Given fixed n and L, we choose appropriate $\log_2 q$, $\overline{m}$ and m to satisfy the previous requirements. The result will be shown as Table 2. In addition, we can calculate them off line and then put them in memory as system parameters, once the n and L can be fixed. Then the time of this part can be neglected.

Table 2. Initialization

n	L	$\log_2 q$	k	$\overline{m}$	m	σ_x	Running time
10	4	90	4	2070	2190	281302	0.244
10	8	156	7	3272	3932	371245	1.43
20	9	177	8	4007	4487	384505	6.474
20	18	343	16	7906	8866	533133	57.012

The second part is obfuscation. From a boolean function $f : \{0,1\}^n \to \{0,1\}$ to the obfuscated program $\widetilde{\mathcal{O}}(f)$, the whole process are included in this part. Specially, we fix the parameter n = 10. Then the simulating result are in Table 3. The third part is Calculation. For an obfuscated program $\widetilde{\mathcal{O}}(f)$, we want to know the speed of the calculation. If the program get an input $\mathbf{x} \in \{0,1\}^n$, we will choose corresponding matrices and multiply them to get the input. Compared the original construction with the improve construction, the result will be shown that the smaller integers for L and n are more effective at Table 3.

Table 3. Obfuscating $f_1(\mathbf{x}) = (x_1 \bigwedge x_2) \bigoplus (x_2 \bigwedge x_3)$

$f(x)$	ω	L	$\log_2 q$	k	$\overline{m}$	m	σ_x	Obfuscation	Calculation
$f_1(x)$	8	6	112	5	2620	2770	314326	8437	953
$f_1(x)$	8	3	68	3	5320	5620	195487	1673	174

5   Conclusion

In this work, we proposed an obfuscator for the NC1 circuit by using the GGH15 multilinear map. Several improvements are involved to reduce the size of parameters so that our construction is more efficient than the one given in [18]. By encoding message set by set and splitting matrices into block matrices, the dimension of matrices in our GGH15 multilinear map is independent of the particular MBP. This lead to a more flexible obfuscating process, since an instantiated multilinear map can be reused for several times. Moreover, the MBP whose width does not reach the security parameter n (the dimension of matrix) can still be obfuscated by our obfuscator. The performance of the simulation gives evidences to ensure that the theoretic analysis for our obfuscation is correct.

However, even our advanced obfuscation is not efficient enough to implement in practice. How to improve the efficiency of current obfuscation scheme is still a worthwhile work to do.

Acknowledgements. This work is supported by the National Natural Science Foundation of China (No. 61672550, No. 61379154) and the Fundamental Research Funds for the Central Universities (No. 17lgjc45).

References

1. Ajtai, M.: Generating hard instances of the short basis problem. In: Wiedermann, J., van Emde Boas, P., Nielsen, M. (eds.) ICALP 1999. LNCS, vol. 1644, pp. 1–9. Springer, Heidelberg (1999). https://doi.org/10.1007/3-540-48523-6_1
2. Ananth, P., Gupta, D., Ishai, Y., Sahai, A.: Optimizing obfuscation: avoiding Barrington's theorem. In: 2014 ACM SIGSAC CCS, pp. 646–658. ACM (2014)
3. Arora, S., Ge, R.: New algorithms for learning in presence of errors. In: Aceto, L., Henzinger, M., Sgall, J. (eds.) ICALP 2011. LNCS, vol. 6755, pp. 403–415. Springer, Heidelberg (2011). https://doi.org/10.1007/978-3-642-22006-7_34
4. Barak, B., Goldreich, O., Impagliazzo, R., Rudich, S., Sahai, A., Vadhan, S., Yang, K.: On the (Im)possibility of obfuscating programs. In: Kilian, J. (ed.) CRYPTO 2001. LNCS, vol. 2139, pp. 1–18. Springer, Heidelberg (2001). https://doi.org/10.1007/3-540-44647-8_1
5. Barrington, D.A.: Bounded-width polynomial-size branching programs recognize exactly those languages in NC^1. In: ACM STOC 1986, pp. 1–5. ACM (1986)
6. Boneh, D., Silverberg, A.: Applications of multilinear forms to cryptography. Contemp. Mathe. **324**(1), 71–90 (2003)
7. Brakerski, Z., Gentry, C., Halevi, S., Lepoint, T., Sahai, A., Tibouchi, M.: Cryptanalysis of the quadratic zero-testing of GGH. Cryptology ePrint Archive, Report 2015/845 (2015)
8. Canetti, R.: Towards realizing random oracles: hash functions that hide all partial information. In: Kaliski, B.S. (ed.) CRYPTO 1997. LNCS, vol. 1294, pp. 455–469. Springer, Heidelberg (1997). https://doi.org/10.1007/BFb0052255
9. Canetti, R., Rothblum, G.N., Varia, M.: Obfuscation of hyperplane membership. In: Micciancio, D. (ed.) TCC 2010. LNCS, vol. 5978, pp. 72–89. Springer, Heidelberg (2010). https://doi.org/10.1007/978-3-642-11799-2_5

10. Cheon, J.H., Fouque, P.-A., Lee, C., Minaud, B., Ryu, H.: Cryptanalysis of the new CLT multilinear map over the integers. In: Fischlin, M., Coron, J.-S. (eds.) EUROCRYPT 2016. LNCS, vol. 9665, pp. 509–536. Springer, Heidelberg (2016). https://doi.org/10.1007/978-3-662-49890-3_20
11. Cheon, J.H., Han, K., Lee, C., Ryu, H., Stehlé, D.: Cryptanalysis of the multilinear map over the integers. In: Oswald, E., Fischlin, M. (eds.) EUROCRYPT 2015. LNCS, vol. 9056, pp. 3–12. Springer, Heidelberg (2015). https://doi.org/10.1007/978-3-662-46800-5_1
12. Coron, J.-S., Gentry, C., Halevi, S., Lepoint, T., Maji, H.K., Miles, E., Raykova, M., Sahai, A., Tibouchi, M.: Zeroizing without low-level zeroes: new MMAP attacks and their limitations. In: Gennaro, R., Robshaw, M. (eds.) CRYPTO 2015. LNCS, vol. 9215, pp. 247–266. Springer, Heidelberg (2015). https://doi.org/10.1007/978-3-662-47989-6_12
13. Coron, J.-S., Lepoint, T., Tibouchi, M.: Practical multilinear maps over the integers. In: Canetti, R., Garay, J.A. (eds.) CRYPTO 2013. LNCS, vol. 8042, pp. 476–493. Springer, Heidelberg (2013). https://doi.org/10.1007/978-3-642-40041-4_26
14. Garg, S., Gentry, C., Halevi, S.: Candidate multilinear maps from ideal lattices. In: Johansson, T., Nguyen, P.Q. (eds.) EUROCRYPT 2013. LNCS, vol. 7881, pp. 1–17. Springer, Heidelberg (2013). https://doi.org/10.1007/978-3-642-38348-9_1
15. Garg, S., Gentry, C., Halevi, S., Raykova, M., Sahai, A., Waters, B.: Candidate indistinguishability obfuscation and functional encryption for all circuits. In: FOCS 2013, pp. 40–49, October 2013
16. Garg, S., Gentry, C., Halevi, S., Sahai, A., Waters, B.: Attribute-based encryption for circuits from multilinear maps. In: Canetti, R., Garay, J.A. (eds.) CRYPTO 2013. LNCS, vol. 8043, pp. 479–499. Springer, Heidelberg (2013). https://doi.org/10.1007/978-3-642-40084-1_27
17. Gentry, C.: Fully homomorphic encryption using ideal lattices. In: ACM STOC 2009, pp. 169–178. ACM (2009)
18. Gentry, C., Gorbunov, S., Halevi, S.: Graph-induced multilinear maps from lattices. In: Dodis, Y., Nielsen, J.B. (eds.) TCC 2015. LNCS, vol. 9015, pp. 498–527. Springer, Heidelberg (2015). https://doi.org/10.1007/978-3-662-46497-7_20
19. Gentry, C., Halevi, S., Smart, N.P.: Homomorphic evaluation of the AES circuit. In: Safavi-Naini, R., Canetti, R. (eds.) CRYPTO 2012. LNCS, vol. 7417, pp. 850–867. Springer, Heidelberg (2012). https://doi.org/10.1007/978-3-642-32009-5_49
20. Gentry, C., Peikert, C., Vaikuntanathan, V.: Trapdoors for hard lattices and new cryptographic constructions. In: ACM STOC 2008, pp. 197–206. ACM (2008)
21. Halevi, S., Halevi, T., Shoup, V., Stephens-Davidowitz, N.: Implementing BP-obfuscation using graph-induced encoding. Cryptology ePrint Archive, Report 2017/104 (2017)
22. Hu, Y., Jia, H.: Cryptanalysis of GGH map. In: Fischlin, M., Coron, J.-S. (eds.) EUROCRYPT 2016. LNCS, vol. 9665, pp. 537–565. Springer, Heidelberg (2016). https://doi.org/10.1007/978-3-662-49890-3_21
23. Kilian, J.: Founding crytpography on oblivious transfer. In: ACM STOC 1988, pp. 20–31. ACM (1988)
24. Klein, P.: Finding the closest lattice vector when it's unusually close. In: ACM-SIAM SODA 2000, pp. 937–941. SIAM (2000)
25. Linn, C., Debray, S.: Obfuscation of executable code to improve resistance to static disassembly. In: ACM CCS 2003, pp. 290–299. ACM (2003)

26. Micciancio, D., Peikert, C.: Trapdoors for lattices: simpler, tighter, faster, smaller. In: Pointcheval, D., Johansson, T. (eds.) EUROCRYPT 2012. LNCS, vol. 7237, pp. 700–718. Springer, Heidelberg (2012). https://doi.org/10.1007/978-3-642-29011-4_41

27. Miles, E., Sahai, A., Zhandry, M.: Annihilation attacks for multilinear maps: cryptanalysis of indistinguishability obfuscation over GGH13. In: Robshaw, M., Katz, J. (eds.) CRYPTO 2016. LNCS, vol. 9815, pp. 629–658. Springer, Heidelberg (2016). https://doi.org/10.1007/978-3-662-53008-5_22

28. Sauerhoff, M., Wegener, I., Werchner, R.: Relating branching program size and formula size over the full binary basis. In: Meinel, C., Tison, S. (eds.) STACS 1999. LNCS, vol. 1563, pp. 57–67. Springer, Heidelberg (1999). https://doi.org/10.1007/3-540-49116-3_5

29. Wroblewski, G.: General method of program code obfuscation. Ph.D. thesis, Institute of Engineering Cybernetics, Wroclaw University of Technology (2002)

30. Coron, J.-S., Lee, M.S., Lepoint, T., Tibouchi, M.: Cryptanalysis of GGH15 multilinear maps. In: Robshaw, M., Katz, J. (eds.) CRYPTO 2016. LNCS, vol. 9815, pp. 607–628. Springer, Heidelberg (2016). https://doi.org/10.1007/978-3-662-53008-5_21

From Attack on Feige-Shamir to Construction of Oblivious Transfer

Jingyue Yu[1,2,3]([✉]) [ID], Yi Deng[1,2,3], and Yu Chen[1,2,3] [ID]

[1] State Key Laboratory of Information Security, Institute of Information
Engineering, Chinese Academy of Sciences, Beijing 100093, China
{yujingyue,deng,chenyu}@iie.ac.cn
[2] School of Cyber Security, University of Chinese Academy of Sciences,
Beijing 100093, China
[3] State Key Laboratory of Cryptology, P.O. Box 5159, Beijing 100878, China

Abstract. Following the work of [Deng, Eurocrypt 2017], under the assumption of the existence of injective one way function, we prove that at least one of the following statements is true:

- (Infinitely-often) Oblivious transfer exists.
- For every inverse polynomial ϵ, the 4-round Feige-Shamir protocol is ϵ-distributional concurrent zero knowledge for any hard distribution over sparse OR-relation.

Both these statements have been shown to be unprovable [Gertner et al. FOCS 2000; Canetti et al. STOC 2001] via black-box reductions.

We show how to transform the magic adversary who breaks the ϵ-distributional concurrent zero knowledge of the classic Feige-Shamir protocols into oblivious transfer under the existence of injective one way function. As a key ingredient, we introduce the concept of distributional witness encryption to achieve the encryption scheme in which "public keys" can be sampled separately of "private keys", and show that if there exists a magic adversary breaking the ϵ-distributional concurrent zero knowledge of Feige-Shamir protocols over a hard distribution, it can be transformed to an (infinitely-often) distributional witness encryption based on injective one way function.

Keywords: Concurrent zero knowledge · Feige-Shamir protocol
Oblivious transfer · (Distributional) witness encryption
Black-box separations

1 Introduction

In cryptography, when demonstrating that constructions of a primitive or protocol are secure, we usually come up with a reduction algorithm that can turn any adversary breaking the target scheme into a successful adversary against the underlying assumptions (such as one way functions, factoring). Today, the most commonly used proof method is the black-box reduction in which treats the adversary as an oracle and does not use its any internal workings.

© Springer International Publishing AG, part of Springer Nature 2018
X. Chen et al. (Eds.): Inscrypt 2017, LNCS 10726, pp. 44–64, 2018.
https://doi.org/10.1007/978-3-319-75160-3_5

Impagliazzo and Rudich [20] were the first to offer a formal methodology for studying the limitations of black-box reduction, by showing that there are no black-box reductions of key agreement to one way functions. The formal methodology has subsequently been used to derive a plenty of black-box separations. For example, Simon [26] gave a black-box separation between one way functions and collision-resistant hash functions. Gertner et al. [17] showed the black-box separation between oblivious transfer and public key encryption/key agreement. They gave a further result that there are no black-box reductions of oblivious transfer (even for honest parties) to injective trapdoor functions.

Another important research line is to lower the round-complexity of protocols. A notable example is the round-complexity of concurrent zero knowledge. Dwork et al. [13] proposed the concept of concurrent zero knowledge, and showed that the classic constant-round zero knowledge protocols (including Feige-Shamir protocols [14] and Goldreich-Kahan protocols [19]) do not preserve zero-knowledge under concurrent execution with a traditional black-box simulator. Subsequent work [5,25] showed that concurrent zero-knowledge with black-box simulation requires a logarithmic number of rounds for NP languages. In [3], Barak put forward a new powerful reduction technique—the non-black-box reduction which makes use of the code of the adversary. The reduction technique breaks several lower bounds for black-box zero-knowledge. Based on some strong assumptions (such as indistinguishability obfuscation), Chung et al. [7] gave the first public-coin constant-round zero knowledge protocols with a non-black-box simulator. Up to now, it is still not known whether there exists constant-round concurrent zero knowledge under standard assumptions.

Recently, Deng [10] presented an unexpected connection between the round complexity of concurrent zero knowledge and the complexity of public key encryption. Specifically, if there exists a magic concurrent adversary who breaks 4-round Feige-Shamir protocol, they showed how to transform it into public key encryption under the existence of injective one way function. That means, the known impossibility results about constant-round concurrent zero knowledge and public key encryption cannot coexist unconditionally, and there must be a new reduction technique that can help to bypass at least one of black-box lower bounds.

In this paper, we revisit the black-box lower bound on the assumption of oblivious transfer, and on the distributional concurrent zero knowledge of Feige-Shamir protocol. We show these black-box impossibility results cannot coexist unconditionally. That is, if there exists a magic adversary breaking the distributional concurrent zero knowledge of Feige-Shamir protocol, we can transform it into constructions of oblivious transfer.

1.1   Our Results

We mainly follow the strategy in [10], transforming a magic adversary V^* which breaks the distributional concurrent zero knowledge of Feige-Shamir protocol into construction of a special type of strong one way functions. As observed in [10], this special type of one way functions can be extracted with the corresponding witnesses to the statement (proved in Feige-Shamir protocol).

Let X_n denote the uniform distribution over $\{0,1\}^n$, and L_{OR} be the OR language of $L \subset \{0,1\}^*$ and its associated relation $R_{L_{OR}} = \{(x_1, x_2, w) : w \in R_L(x_1) \cup R_L(x_2)\}$. Consider $\{D_n = (D_n^x, D_n^w)\}_{n \in \mathbb{N}}$ as an arbitrary efficiently samplable distribution ensemble over R_L. We also define $\{(D_n^x \times D_n^x, D_n^w)\}_{n \in \mathbb{N}}$ over $R_{L_{OR}}$ in the following way: sample $(x_1, w_1) \leftarrow D_n$, $(x_2, w_2) \leftarrow D_n$, $b \xleftarrow{\text{R}} \{1,2\}$; output (x_1, x_2, w). $\{(X_n \times D_n^x, D_n^w)\}_{n \in \mathbb{N}}$ is defined in the following way: sample $x_1 \leftarrow X_n$, $(x_2, w) \leftarrow D_n$; output a *random order* of (x_1, x_2) and w.[1]

We observe that $\{(D_n^x \times D_n^x, D_n^w)\}_{n \in \mathbb{N}}$ over $R_{L_{OR}}$ is computationally distinguishable from $\{(X_n \times D_n^x, D_n^w)\}_{n \in \mathbb{N}}$, if $\{D_n\}_{n \in \mathbb{N}}$ is a (decisional) hard distribution ensemble over R_L. Moreover, we instantiate the hard distribution over sparse relation with the language induced by PRG.

Furthermore, we prove that if there exists a PPT adversary that can break ϵ-distributional concurrent zero knowledge of Feige-Shamir protocol over $\{(D_n^x \times D_n^x, D_n^w)\}_{n \in \mathbb{N}}$ over $R_{L_{OR}}$, then it also can break ϵ-distributional concurrent zero knowledge of Feige-Shamir protocol over $\{(X_n \times D_n^x, D_n^w)\}_{n \in \mathbb{N}}$, assuming $\{D_n\}_{n \in \mathbb{N}}$ is a hard distribution ensemble over R_L, by Lemma 1.

Distributional Witness Encryption. Based on the construction of the special type of one way functions, we introduce the concept of distributional witness encryption schemes and provide a construction. The construction of distributional witness encryption is an important ingredient of OT. It provides a way of sampling "public keys" separately of "private keys". In the previous work, the best possible constructions [11] of witness encryption scheme (proposed in [16]) are implemented by smooth projective hash functions, which can be efficiently constructed under specific assumptions with various algebraic structures [9]. And other constructions can be seen in [1,15,16].

Oblivious Transfer. Subsequently, we present semantically secure oblivious transfer protocols against semi-honest adversaries and malicious receivers respectively, based on the distributional witness encryption, without additional assumptions. These protocols break the black-box lower bounds of oblivious transfer [17]. Our semi-honest OT protocols are 2-round, as well as those based on trapdoor permutations [21].

We follow the framework of [23] (where oblivious transfers are based on the DDH assumption) to present our malicious receiver OT protocol. The sender offers m_b to the receiver, by encrypting (m_0, m_1) with the received "public-key" pair (pk_0, pk_1). In [23], it's easy to show that the ciphertext ct_{1-b} is uniformly distributed, though the adversary maliciously set some DDH tuples into pk_{1-b} (i.e. The adversary may know partial secret key of pk_{1-b}, but it is still not able to decrypt ct_{1-b}. Because ct_{1-b} is uniformly distributed, only if there exists a non-DDH tuple in pk_{1-b}).

In our case, the underlying encryption scheme is the distributional witness encryption based on injective one way functions. The "public key" contains q

---

[1] In this setting, $\{(X_n \times D_n^x, D_n^w)\}_{n \in \mathbb{N}} = \{((X_n \times D_n^x) \cup (D_n^x \times X_n), D_n^w)\}_{n \in \mathbb{N}}$.

instances, while the active instances used to generate the ciphertext are only q_2 (which is a polynomial factor of q). This leads to a possibility that the malicious receiver may know both m_0 and m_1. To complete the proof, we show that if the adversary is able to decrypt within $q/2$ witnesses, then the adversary can be used to break the hard distribution over sparse relation. If the "public key" pk_σ contains more than $q/2$ Yes-instances, the simulator can extract the bit σ and construct a view of the malicious receiver in the ideal model which is indistinguishable from its view in the real model. This proof technique may find applications in other places.

1.2   Related Work

This work mainly follows the basic strategy in [10]. They showed that if the 4-round Feige-Shamir protocol is not ϵ-distributional concurrent zero knowledge, then the public-key encryption can be based on (injective) one way functions. Key agreement also can be based on injective one way function, as a consequence of the fact public key encryption implies key agreement.

In this paper, we exhibit more consequences—oblivious transfer protocols can be based on injective one way functions. In [17], they studied the relationship between public key encryption and oblivious transfer. There are black-box separations between public key encryption and oblivious transfer. They also showed that (honest) oblivious transfer can be implied by the strong version of public key encryption, in which public keys are sampled "separately of private keys". Towards it, we construct distributional witness encryption and then (malicious receiver) oblivious transfer protocol, under the existence of injective one way function.

Previous Work. There are several constructions of oblivious transfer, following the strong version of public key encryption [17] in which the public keys sampled separately of private key are (computationally) indistinguishable from the public keys generated by the key-generation algorithm. An immediate example is the construction of OT according to the ElGamal encryption scheme whose public keys can be directly sampled at random from the cyclic group. In [24], they presented an OT protocol in the CRS model, via dual model encryption under the specific assumptions, such as the DDH, quadratic residuosity, Learning With Errors (LWE). In the dual model encryption, there are two indistinguishable types of public keys—the real public keys and the messy (or lossy) keys. The messy key producer still gains no information (statistically) about the encrypted message from the ciphertext encrypted under the messy key. Our OT protocol is constructed according to the distributional witness encryption scheme in the standard model under the general assumption—injective one way function.

Concurrent Work. In a concurrent and independent work, Badrinarayanan et al. [2] presented a construction of weak OT from witness encryption and a non-interactive commitment scheme. However, their focus is on achieving indistinguishability based security against the sender, and super-polynomial simulation against a malicious receiver, whereas our focus is on achieving the (polynomial) simulation security for both the sender and receiver. Our paper also provides an OT protocol which satisfies the (polynomial) simulation security against a malicious receiver.

1.3   Organization

In Sect. 2, we present the definitions of ϵ-distributional concurrent zero knowledge, oblivious transfer and other related notations and protocols. We give an overview of Deng's approach to public-key encryption based on injective one way function in Sect. 3. Our main ingredient of OT—distributional witness encryption is presented in Sect. 5. In Sect. 6, we give the constructions of OT based on injective one way functions, if there exists a magic adversary breaking ϵ-distributional concurrent zero knowledge of Feige-Shamir protocols.

2   Preliminaries

2.1   Basic Notations

Throughout the paper, n denotes the security parameter. A function $\mathsf{negl}(n)$ is said to be negligible if for any polynomial $\mathsf{poly}(n)$ there exists an N such that for all $n \geq N$, $\mathsf{negl}(n) \leq \frac{1}{\mathsf{poly}(n)}$. We will abbreviate probabilistic polynomial-time with PPT.

For a positive integer κ, $[\kappa]$ denotes $\{1, 2, \ldots, \kappa\}$. Vectors are written using bold lower-case letters, e.g. $\boldsymbol{x} := \{x_i\}_{i \in [\kappa]}$. The i-th component of $\boldsymbol{x}$ is denoted by x_i. For a set S, we write $x \xleftarrow{\mathrm{R}} S$ to denote that x is chosen uniformly at random from S. For a distribution D over a finite set $S \subseteq \{0, 1\}^*$, we denote by $x \leftarrow D$ the process that the sample $x \in S$ is drawn according to the distribution D. And $\boldsymbol{x} \leftarrow D^{\otimes \kappa}$ denotes the process of sampling κ times x_i from D independently. For any function f, $f(\boldsymbol{x}) = \boldsymbol{y}$ is denoted $(f(x_1), \ldots, f(x_\kappa)) = (y_1, \ldots, y_\kappa)$. For any algorithm M, $\mathsf{M}(\boldsymbol{x}) = \boldsymbol{y}$ works as follows: Run $\mathsf{M}(x_i) = y_i$ independently, for $i \in [\kappa]$.

2.2   The Relations

Consider X as the set $\{0, 1\}^*$, $L \subset X$ as the NP language with its associated relation R_L. L_{OR} denotes the OR language of $L \subset X$ with its associated OR-relation $R_{L_{OR}} = \{(x_1, x_2, w) : w \in R_L(x_1) \cup R_L(x_2)\}$.

Let X_n denote the uniform distribution over $\{0, 1\}^n$, and L_n denote all n-length instances in L language, with its associated relation $R_L^n = \{(x, w) \in R_L : |x| = n\}$. Also define L_{OR}^n as the OR composition of L_n, with its associated relation $R_{L_{OR}}^n = \{(x_1, x_2, w) : |x_1| = |x_2| = n, \ w \in R_L^n(x_1) \cup R_L^n(x_2)\}$.

We call R is a *sparse relation*, if the size of the support of R is a negligible fraction of the size of the support of X_n.

2.3   Interactive Arguments

An interactive argument system $\langle P, V \rangle$ for an NP language L with its associated relation R_L consists of a pair of interactive PPT Turing machines P and V. The prover P wants to convince the verifier V of some statement $x \in L$. We denote by $\mathsf{Trans}_V \langle P(w), V(z) \rangle(x)$ the transcript of an execution of $\langle P, V \rangle$ on common input x, P's private input w and V's auxiliary input z. Naturally, $\mathsf{Trans}_V \langle P(W_n), V(Z_n) \rangle(X_n)$ denotes the distribution over V's all possible view, for a joint distribution (X_n, W_n, Z_n). For convenience, $\mathsf{Trans}_V \langle P(W_n), V(Z_n) \rangle(X_n)$ is being written as $\mathsf{Trans}_V \langle P, V \rangle(X_n, W_n, Z_n)$.

Definition 1 (Witness indistinguishability). *Let L be an NP language defined by R_L. An interactive argument $\langle P, V \rangle$ is said to be witness indistinguishable for relation R_L if for every PPT V^*, every auxiliary input $z \in \{0,1\}^*$ and every sequence $\{(x, w, w')\}_{x \in L}$, where $(x, w), (x, w') \in R_L$, the following two distribution ensembles are computationally indistinguishable:*

$$\{\mathsf{Trans}_{V^*} \langle P(w), V^*(z) \rangle(x)\}_{x \in L, z \in \{0,1\}^*} \overset{c}{\approx} \{\mathsf{Trans}_{V^*} \langle P(w'), V^*(z) \rangle(x)\}_{x \in L, z \in \{0,1\}^*}$$

A zero knowledge argument system is an interactive argument in which for any PPT adversarial verifier, there exists a PPT simulator (without the corresponding witnesses) that can reconstruct the view of the verifier. We consider a weak variant of ZK, namely, distributional zero knowledge [8,12,18], where the instances are chosen from some efficiently samplable distribution over the language. Furthermore, we allow that the simulator's running time depends on the distinguishable ability ϵ of the distinguisher D. This kind of ϵ-distributional zero knowledge can also be considered as distributional weak zero knowledge [8,12,22], in which the simulator is allowed to depend on the distinguisher D (besides on the distribution of instances).

We follow the definition of ϵ-distributional concurrent zero knowledge in [10]. Let $L_n = \{x : x \in L \cap \{0,1\}^n\}$, and its corresponding relation $R_L^n = \{(x, w) \in R_L : |x| = n\}$. The joint distribution (D_n^x, D_n^w, Z_n) is an arbitrary efficiently samplable distribution over $R_L^n \times \{0,1\}^*$.

Definition 2 (ϵ-Distributional Concurrent Zero Knowledge). *An interactive argument $\langle P, V \rangle$ for language L is said to be ϵ-Distributional Concurrent Zero Knowledge if for every PPT concurrent adversary V^*, and every distribution ensemble $\{(D_n^x, D_n^w, Z_n)\}_{n \in \mathbb{N}}$ over $R_L \times \{0,1\}^*$, there exists a PPT Sim such that for all PPT distinguisher D and sufficient large n, satisfying:*

$$\Pr[\mathsf{D}(\mathsf{Trans}_{V^*} \langle P, V^* \rangle(D_n^x, D_n^w, Z_n)] - \Pr[\mathsf{D}(\mathsf{Sim}(V^*, D_n^x, Z_n)) = 1] < \epsilon$$

The Feige-Shamir ZK Argument for L. The Feige-Shamir ZK protocol proceeds in the following two stages, on a common input $x \in L$.

1. In Stage 1, the verifier V chooses $\alpha_1, \alpha_2 \overset{\mathrm{R}}{\leftarrow} \{0,1\}^n$ randomly and independently. Then V computes their images $\beta_1 = f(\alpha_1), \beta_2 = f(\alpha_2)$ under the

injective one way function $f : \{0,1\}^n \to \{0,1\}^{l(n)}$. V attempts to convince P of that he knows one of α_1, α_2, by executing the 3-round n-parallel-repetition of the Blum's protocol [4].

2. In Stage 2, the prover P proves to V the fact that either x is in the language L, or he knows one of α_1, α_2.

2.4 Oblivious Transfer

Oblivious transfer is a protocol between two parties—a sender S with a pair of inputs (m_0, m_1) and a receiver R with a choice bit $b \in \{0,1\}$. At the end of this protocol, the receiver R obtains m_b while the sender S learns nothing about b. Formally, let $\Pi = (S, R)$ denote the protocol that computes the oblivious transfer functionality, $f_{\mathsf{OT}}((m_0, m_1), b) = (\bot, m_b)$.

Recall the definitions of OT. We start with the definition of oblivious transfer against semi-honest adversaries.

Definition 3 (Semi-honest oblivious transfer). *An OT protocol $\Pi = (S, R)$, where the sender S on inputs $(m_0, m_1) \in \{0,1\}$, outputs nothing, and the receiver R on input $b \in \{0,1\}$ outputs m, is said to be a semi-honest oblivious transfer, satisfying the follows:*

– *(Correctness) If both parties are honest, we have*

$$\Pr[m = m_b] \geq 1 - \mathsf{negl}(n).$$

– *(Security for the sender) For an honest sender and an honest (but curious) receiver, there exists a PPT simulator Sim such that*

$$\{View_R^\Pi(S(m_0, m_1), R(b))\} \overset{c}{\approx} \{Sim(b, m_b)\}$$

– *(Security for the receiver) For an honest receiver and an honest (but curious) sender, there exists a PPT simulator Sim such that*

$$\{View_S^\Pi(S(m_0, m_1), R(b))\} \overset{c}{\approx} \{Sim(m_0, m_1)\}$$

Now, we consider oblivious transfer against malicious adversaries. Parts of the definition of (malicious) oblivious transfer are verbatim from [23].

Oblivious Transfer in the Ideal Model. An ideal oblivious transfer execution denoted by $\mathrm{Ideal}_{Sim}^{f_{OT}}(m_0, m_1, b)$, proceeds as follows:

Inputs: The sender S takes as inputs $m_0, m_1 \in \{0,1\}$, and the receiver R takes as input $b \in \{0,1\}$.

Send inputs to trusted party: An honest party always sends its inputs unchanged to the trusted party. A malicious party may either abort, in which case it sends $\bot$ to the trusted party, or sends some other inputs to the trusted party.

Trusted party computes output: If the trusted party receives $\bot$ from one of the parties, then it sends $\bot$ to both parties and halts. Otherwise, upon receiving some (m'_0, m'_1) from S and a bit b' from R, the trusted party sends $m'_{b'}$ to R and halts.

Outputs: An honest party always outputs the message it has obtained from the trusted party. A malicious party may output an arbitrary function of its initial input and the message obtained from the trusted party.

Execution in the Real Model. Consider the real model in which a real two-party protocol without the trusted third party. In this case, a malicious party may follow any strategy implementable by non-uniform PPT machine $\mathcal{A}$. In contrast, an honest party always follows the instructions of Π. We denote by $\mathrm{Real}_{\mathcal{A}}^{\Pi}(m_0, m_1, b)$ the above execution in the real model.

Definition 4 ((Malicious) Oblivious Transfer). *Let f_{OT} denote the oblivious transfer functionality and let Π denote the two-party protocol that computes f_{OT}. A protocol $\Pi = (S, R)$ is said to be a secure oblivious transfer protocol against the malicious adversaries if for every non-uniform PPT adversary $\mathcal{A}$ for the real model, there exists a non-uniform expected polynomial-time simulator Sim for the ideal model, such that for every $m_0, m_1 \in \{0, 1\}$ and every $b \in \{0, 1\}$, such that*

$$\{\mathrm{Ideal}_{Sim}^{f_{OT}}(m_0, m_1, b)\} \overset{c}{\approx} \{\mathrm{Real}_{\mathcal{A}}^{\Pi}(m_0, m_1, b)\}.$$

3   Overview of Deng's Approach to PKE Under OWF

In [10], they introduced a dissection procedure for concurrent adversaries to connect the complexity of public-key encryption with the round-complexity of concurrent zero knowledge. Specifically, if there is a magic adversary who breaks the ϵ-concurrent zero knowledge of the Feige-Shamir protocol, then it can be transformed to construct a public-key encryption and key agreement based on injective one way functions. On the other side, if it could be proved that there are no reductions of public key encryption to the existence of injective one way functions, then all possible concurrent adversaries for the Feige-Shamir protocol share a common computational structure.

Formally, let $L_{OR} = L \vee L$ be the OR language, and with its corresponding relation $R_{L_{OR}} = \{(x_1, x_2, w) : w \in R_L(x_1) \cup R_L(x_2)\}$. Given an arbitrary hard distribution ensemble $\{D_n (= D_n^x, D_n^w)\}_{n \in \mathbb{N}}$ over R_L and an arbitrary efficiently samplable distribution Z_n over $\{0, 1\}^*$, we define the joint distribution ensemble $\{(D_n^x \times D_n^x, D_n^w, Z_n)\}_{n \in \mathbb{N}}$ over $R_{L_{OR}} \times \{0, 1\}^*$: Sample $(x_1, w_1) \leftarrow D_n$, $(x_2, w_2) \leftarrow D_n$, $z \leftarrow Z_n$, $b \leftarrow \{1, 2\}$ and output (x_1, x_2) and w_b.

They proved that at least one of the following statements is true, assuming that the injective one way function $f : \{0, 1\}^n \rightarrow \{0, 1\}^{l(n)}$ exists, (the main theorem in [10]):

- (Infinitely-often) Public-key encryption and key agreement can be constructed from the injective one way function f.

– For every inverse polynomial ϵ, the Feige-Shamir protocol based on f is ϵ-distributional concurrent zero knowledge for $\{(D_n^x \times D_n^x, D_n^w, Z_n)\}_{n \in \mathbb{N}}$ (defined as above).

As the proof strategy shown in [10], they first transformed a magic adversary V^* which breaks ϵ-distributional concurrent zero knowledge into constructing a special type of weak one way functions, which can be efficiently inverted with the witness (of the statement used in Feige-Shamir protocol). Then they used Yao's hardness amplification to construct the special type of strong one way functions. The specific proceeds as follows:

- **Transformation.** For infinitely many n, the PPT algorithm M, on input $(x_1, x_2, w, z) \leftarrow (D_n^x \times D_n^n, D_w^n, Z_n)$, executes the concurrent version of Feige-Shamir protocol $\langle P(w), V^*(z) \rangle (x_1 \vee x_2)$, and extracts one-pre-images of f output by V^* at its key step. (At the key step of V^*, V^* sends a pair of images which can be extracted by P (with w, by rewinding), while either of them cannot be extracted by any PPT simulator (without w)). M outputs the extracted pre-image α and its corresponding image β, together with some auxiliary information st.
- **Weak one way functions.** $M(x_1 \vee x_2, w, z)$ successfully generates a preimage/image pair (α, β) of the special weak one way function f and st, with noticeable probability over $(x_1, w_1) \leftarrow D_n, (x_2, w_2) \leftarrow D_n, w \xleftarrow{\text{R}} \{w_1, w_2\}$, $z \xleftarrow{\text{R}} \{0,1\}^*$. For any non-uniform PPT adversary $\mathcal{A}$, the probability that it can efficiently invert β is bounded away from 1, if not knowing the witness of $x_1 \vee x_2$; while with the help of the corresponding witness, there exists a PPT algorithm Find which can efficiently invert β, with overwhelming probability. Formally, there exists a PPT M, for *infinitely many n*

$$\Pr[M(x_1 \vee x_2, w, z) \to (\alpha, \beta, st) : f(\alpha) = \beta \wedge (2), (3) \text{ holds}] = \frac{1}{p(n)} \quad (1)$$

For every n.u. PPT $\mathcal{A}$, for any polynomial poly, for infinitely many n,

$$\Pr[(f(\alpha), st, w) \leftarrow G(1^n) : \mathcal{A}(1^n, f(\alpha), st) \in f^{-1}(f(\alpha))] \leq 1 - \frac{1}{\text{poly}(n)} \quad (2)$$

There exists a PPT algorithm Find (which works exactly like M), such that,

$$\Pr[(f(\alpha), st, w) \leftarrow G(1^n) : \text{Find}(1^n, f(\alpha), st, w) \in f^{-1}(f(\alpha))] = 1 - \text{negl}(n) \quad (3)$$

Where G is a PPT algorithm, defined as follows: Draw (x_1, x_2, w, z) from $(D_n^x \times D_n^x, D_n^w, Z_n)$; Feed M with x_1, x_2, w, z and obtain (α, β, st); Finally, output (β, st, w).
- **One way functions.** $M(\boldsymbol{x_1} \vee \boldsymbol{x_2}, \boldsymbol{w}, \boldsymbol{z})$ successfully generates a preimage/image pair $(\boldsymbol{\alpha}, \boldsymbol{\beta})$ of the special (strong) one way function f and the state $\boldsymbol{st}$, with overwhelming probability, where $M(\boldsymbol{x_1} \vee \boldsymbol{x_2}, \boldsymbol{w}, \boldsymbol{z}) \to (\boldsymbol{\alpha}, \boldsymbol{\beta}, \boldsymbol{st})$ works as follows: Run $M(x_{1,j} \vee x_{2,j}, w_j, z_j)$, for $j \in [q]$ independently,

and set the first q_2 pair $\{(\alpha_i, \beta_i, st_i)\}_{i \in [q_2]}$ as $(\boldsymbol{\alpha}, \boldsymbol{\beta}, \boldsymbol{st})$; Output $(\boldsymbol{\alpha}, \boldsymbol{\beta}, \boldsymbol{st})$, where $f(\boldsymbol{\alpha}) = \boldsymbol{\beta}$.

Formally, there exists a PPT M, for *infinitely many* n

$$\Pr[\mathsf{M}(\boldsymbol{x}_1 \vee \boldsymbol{x}_2, \boldsymbol{w}, \boldsymbol{z}) \to (\boldsymbol{\alpha}, \boldsymbol{\beta}, \boldsymbol{st}) : f(\boldsymbol{\alpha}) = \boldsymbol{\beta} \wedge (5), (6) \text{ holds}] = 1 - \mathsf{negl}(n) \tag{4}$$

For every PPT $\mathcal{A}$, for infinitely many n,

$$\Pr[(f(\boldsymbol{\alpha}), \boldsymbol{st}, \boldsymbol{w}) \leftarrow \mathsf{G}(1^n) : \mathcal{A}(1^n, f(\boldsymbol{\alpha}), \boldsymbol{st}) \in f^{-1}(f(\boldsymbol{\alpha}))] \leq \mathsf{negl}(n) \tag{5}$$

There exists a PPT algorithm Find (which works exactly like M), such that,

$$\Pr[(f(\boldsymbol{\alpha}), \boldsymbol{st}, \boldsymbol{w}) \leftarrow \mathsf{G}(1^n) : \mathsf{Find}(1^n, f(\boldsymbol{\alpha}), \boldsymbol{st}, \boldsymbol{w}) \in f^{-1}(f(\boldsymbol{\alpha}))] = 1 - \mathsf{negl}(n) \tag{6}$$

Where G is a PPT algorithm, defined as before.

Based on this type of special one way functions, they constructed a CPA-secure (infinitely-often) public key encryption scheme based on injective one way functions [10], for infinitely many security parameter n.

Key Generation PKE.Gen(1^n): Draw $(\boldsymbol{x}_1, \boldsymbol{w}_1) \leftarrow D_n^{\otimes q}$, and output $pk = \boldsymbol{x}_1, sk = \boldsymbol{w}_1$.

Encryption PKE.Enc($pk = \boldsymbol{x}_1, m \in \{0, 1\}$):

1. $(\boldsymbol{x}_2, \boldsymbol{w}_2) \leftarrow D_n^{\otimes q}$, $\boldsymbol{z} \leftarrow Z_n^{\otimes q}$.
2. $(\boldsymbol{\beta}, \boldsymbol{\alpha}, \boldsymbol{st}) \leftarrow \mathsf{M}(\boldsymbol{x}_1 \vee \boldsymbol{x}_2, \boldsymbol{w}_2, \boldsymbol{z})$. Note that the number q_2 of components in $\boldsymbol{\beta}/\boldsymbol{\alpha}/\boldsymbol{st}$ is a factor of q, and M takes a *random order* of $(x_{1,k}, x_{2,k})$ as inputs, for $k \in [q]$.
3. $\boldsymbol{r} \xleftarrow{\text{R}} \{0, 1\}^{|\boldsymbol{\alpha}|}$, $h \leftarrow \langle \boldsymbol{\alpha}, \boldsymbol{r} \rangle \in \{0, 1\}$.
4. Output $ct = (\boldsymbol{x}_1, \boldsymbol{x}_2, \boldsymbol{z}, \boldsymbol{\beta}, \boldsymbol{st}, \boldsymbol{r}, h \oplus m)$.

Decryption PKE.Dec($sk = \boldsymbol{w}_1, ct$):

1. Parse ct into $(\boldsymbol{x}_1, \boldsymbol{x}_2, \boldsymbol{z}) || (\boldsymbol{\beta}, \boldsymbol{st}) || \boldsymbol{r} || c$.
2. $\boldsymbol{\alpha} \leftarrow \mathsf{Find}(\boldsymbol{x}_1 \vee \boldsymbol{x}_2, \boldsymbol{w}_1, \boldsymbol{z}, \boldsymbol{\beta}, \boldsymbol{st})$.
3. $h \leftarrow \langle \boldsymbol{\alpha}, \boldsymbol{r} \rangle$.
4. Output $m = h \oplus c$.

4 (Decisional) Hard Distributions over OR-Relation

Let X denote the set $\{0, 1\}^*$, and L_{OR} denote the OR language of $L \subset X$ with its associated relation $R_{L_{OR}} = \{(x_1, x_2, w) : w \in R_L(x_1) \cup R_L(x_2)\}$. Consider X_n as the uniform distribution over $\{0, 1\}^n$ and L_n as all n-length instances in L language, with its associated relation $R_L^n = \{(x, w) \in R_L : |x| = n\}$. Let $D_n = (D_n^x, D_n^w)$ be an efficiently samplable over R_L^n. We say $D_n = (D_n^x, D_n^w)$ is a (decisional) *hard distribution* over R_L^n, if for any non-uniform PPT machine D, for $n \in \mathbb{N}$,

$$|\Pr[\mathsf{D}(D_n^x)] - \Pr[\mathsf{D}(X_n)]| \leq \mathsf{negl}(n).$$

Remark 1. If the distribution D_n^x (defined as above) is sparse in X_n, then for any non-uniform PPT machine D, the above hard distribution $D_n = (D_n^x, D_n^w)$ also holds, for any non-uniform PPT machine D,

$$| \Pr[\mathsf{D}(D_n^x)] - \Pr[\mathsf{D}(X_n/L_n)]| \leq \mathsf{negl}(n).$$

Now we define the distribution over OR-relation $R_{L_{OR}}$. Let $\{(D_n^x \times D_n^x, D_n^w)\}_{n\in\mathbb{N}}$ over $R_{L_{OR}}$ be sampled in the following way: draw $(x_1, w_1) \leftarrow D_n$, $(x_2, w_2) \leftarrow D_n$, $b \xleftarrow{\mathsf{R}} \{1, 2\}$; output (x_1, x_2, w). And also, we define $\{(X_n \times D_n^x, D_n^w)\}_{n\in\mathbb{N}}$ in the following way: sample $x_1 \leftarrow X_n$, $(x_2, w) \leftarrow D_n$; output a *random order* of (x_1, x_2) and w. Note that in this setting, $\{(X_n \times D_n^x, D_n^w)\}_{n\in\mathbb{N}}$ is identical to $\{((X_n \times D_n^x) \cup (D_n^x \times X_n), D_n^w)\}_{n\in\mathbb{N}}$.

If $D_n = (D_n^x, D_n^w)$ is a (decisional) *hard distribution* over R_L^n, then for any non-uniform PPT machine D, we have

$$| \Pr[\mathsf{D}(D_n^x \times D_n^x, D_n^w)] - \Pr[\mathsf{D}(X_n \times D_n^x, D_n^w)]| \leq \mathsf{negl}(n).$$

Recall that, [10] introduced a dissection procedure for the magic adversary V^* who breaks ϵ-distributional concurrent zero knowledge of Feige-Shamir protocol over the joint distributions $\{(D_n^x \times D_n^x, D_n^w, Z_n)\}_{n\in\mathbb{N}}$.[2]

We now show that the dissection procedure in [10] is also suitable for the adversary who breaks ϵ-distributional concurrent zero knowledge of Feige-Shamir protocol over distributions $\{(X_n \times D_n^x, D_n^w, Z_n)\}_{n\in N}$.

Lemma 1. *Assume that $\{D_n = (D_n^x, D_n^w)\}_{n\in\mathbb{N}}$ is a hard distribution ensemble over R_L. For an arbitrary inverse polynomial ϵ, if there exists V^* breaking ϵ-distributional concurrent zero knowledge of Feige-Shamir protocol over distributions $\{(D_n^x \times D_n^x, D_n^w, Z_n)\}_{n\in N}$, then it also can break $(\epsilon - \mathsf{negl}(n))$-distributional concurrent zero knowledge of Feige-Shamir protocol over distributions $\{(X_n \times D_n^x, D_n^w, Z_n)\}_{n\in N}$.*

Proof. V^* breaks ϵ-distributional concurrent zero knowledge of Feige-Shamir protocol over distribution $\{(D_n^x \times D_n^x, D_n^w, Z_n)\}_{n\in N}$, concretely, $\forall\epsilon, \exists V^*, \exists\{(D_n^x \times D_n^x, D_n^w, Z_n)\}_{n\in N}$ over $R_{L_{OR}} \times \{0, 1\}^*$(defined as above), $\forall\mathsf{Sim}\ \exists\mathsf{D}$, and infinitely many n, such that

$$\Pr[\mathsf{D}(\mathsf{Trans}_{V^*}\langle P, V^*\rangle(D_n^x \times D_n^x, D_n^w, Z_n)] - \Pr[\mathsf{D}(\mathsf{Sim}(V^*, D_n^x \times D_n^x, Z_n)) = 1] > \epsilon(n)$$

Since $\{D_n = (D_n^x, D_n^w)\}_{n\in\mathbb{N}}$ is a hard distribution ensemble over R_L, we have $\{(D_n^x \times D_n^x, D_n^w)\}_{n\in\mathbb{N}}$ is a hard distribution ensemble over $R_{L_{OR}}$, i.e. for any PPT distinguisher D'(considered as $P, V^*, \mathsf{Sim}, \mathsf{D}$), with auxiliary input $z \leftarrow Z_n$,

$$| \Pr[\mathsf{D}'(X_n \times D_n^x, D_n^w, Z_n) = 1] - \Pr[\mathsf{D}'(D_n^x \times D_n^x, D_n^w, Z_n) = 1]| \leq \mathsf{negl}(n)$$

By the hybrid arguments, we can see that V^* can break $(\epsilon - \mathsf{negl}(n))$-distributional concurrent zero knowledge of Feige-Shamir protocol over distribution $\{(D_n^x \times D_n^x, D_n^w, Z_n)\}_{n\in N}$.   □

---

[2] The joint distribution $(D_n^x \times D_n^x, D_n^w, Z_n)$ (resp. $(X_n \times D_n^x, D_n^w, Z_n)$) over $R_{L_{OR}} \times \{0, 1\}^*$ is sampled in the natural way: sample $(x_1, x_2, w) \leftarrow (D_n^x \times D_n^x, D_n^w)$ (resp. $(X_n \times D_n^x, D_n^w)$) and $z \leftarrow Z_n$; output (x_1, x_2, w, z).

Now we present a couple of examples of (decisional) hard distribution over OR-relation.

Example 1: Hard distributions over OR-PRG relation.

X_n denotes the uniform distribution over $\{0,1\}^n$. Let $R^n_{\mathsf{PRG}} := \{(x,w) : x = \mathsf{PRG}(w) \wedge |x| = n\}$ where $\mathsf{PRG} : \{0,1\}^{n/2} \to \{0,1\}^n$.

For any efficiently samplable distribution D_n over R^n_{PRG}, we have $\{D_n\}_{n \in \mathbb{N}}$ is a hard distribution ensemble. This is, for all PPT distinguisher D, such that $\forall n \in N$

$$| \Pr[x \leftarrow X_n, \mathsf{D}(x) = 1] - \Pr[x \leftarrow D^x_n, \mathsf{D}(x) = 1]| \leq \mathsf{negl}(n)$$

Note that D_n is also a *sparse* relation over X_n, since R^n_{PRG} is sparse.

Then we have the hard distribution $(D^x_n \times D^x_n, D^w_n)$ over OR-PRG (sparse) relation over $R_{\mathsf{PRG}_{OR}}$, sampled as before.

Example 2: Hard distributions over OR-DDH relation.

DDH assumption: Let Gen be a randomized algorithm that on security parameter 1^n outputs (G, g, q), where G is a cyclic group of order q with generator g. Then for a randomly chosen triplet (a, b, c), for every PPT algorithm D and sufficient large n, there exists a negligible function $\mathsf{negl}(n)$ such that

$$| \Pr[\mathsf{D}(1^n, (G, g, q), g^a, g^b, g^{ab}) = 1] - \Pr[\mathsf{D}(1^n, (G, g, q), g^a, g^b, g^c) = 1]| \leq \mathsf{negl}(n)$$

We define a witness relation

$$R_{\mathsf{DDH}} = \{(x = ((G, g, q), g_1, g_2, g_3), w = (a_1, a_2, a_3)) : g_i = g^{a_i}, i = 1, 2, 3 \wedge a_3 = a_1 a_2\}$$

For any efficiently samplable distribution D_n over R^n_{DDH}, we have $\{D_n\}_{n \in \mathbb{N}}$ is a hard distribution ensemble. D_n can be sampled in the following way: generate a group $(G, g, q) \leftarrow \mathsf{Gen}(1^n)$, and choose $a, b \leftarrow_R \mathbb{Z}_q$, set $x = (g^a, g^b, g^{ab})$ and $w = (a, b, ab)$; Output (x, w).

Then, using the above D_n distribution, we have a hard distribution ensemble $\{D^x_n \times D^x_n, D^w_n\}_{n \in \mathbb{N}}$ over the OR-DDH relations $R_{\mathsf{DDH}_{OR}}$, sampled as before.

Note that besides the above two examples, there are many examples of hard distributions, for instance, the distribution distinguish problem in [6].

5 Main Tool: Distributional Witness Encryption

Witness encryption (WE) is proposed by Garg et al. [16]. In witness encryption scheme, one can use some instance x (instead of an encryption key) in the language L to produce a ciphertext ct for a message $m \in \{0,1\}$. And the users with the witness $w \in R_L(x)$ can decrypt the ciphertext ct. For security, if x is not in the language, then there are no PPT adversaries can distinguish the encryptions of 0 or 1.

There have been several constructions of WE for NP languages over the past few years. Garg et al. [16] gave us the first candidate construction of WE, based on the NP-complete EXACT COVER problem and approximate multilinear maps (MLMs). Garg et al. [15] showed that indistinguishability obfuscation implies witness encryption. Abusalah et al. [1] introduced the notation of offline witness encryption, and gave a corresponding construction which still requires obfuscation. Derler et al. [11] showed that witness encryption schemes for L can be constructed from smooth projective hash functions for L. Current known efficient instantiations of the smooth projective hash functions are all based on highly structured algebraic assumptions, such as DDH [9], bilinear maps [11].

In the following, we define a variant of WE, namely, distributional witness encryption, in which the instances are drawn from a prefixed hard distribution. For generating the distribution, we require a setup program at the beginning of the distributional witness encryption schemes. In [1,11,27], they also defined WE with a setup algorithm.

Definition 5 (Distributional Witness Encryption (DWE)). *A distributional witness encryption scheme for an* NP *language $L \subset X$ consists of the following three algorithms:*

- Setup(1^n): *on input the security parameter n, the key generation algorithm outputs public parameters pp including $X_n, D_n(\subset R_L^n)$.*
- Enc(pp, x, m): *The encryption algorithm takes as inputs pp, an instance $x \in$ Support(X_n) and a message $m \in \{0,1\}$, and outputs the ciphertext ct.*
- Dec(ct, w): *The encryption algorithm takes as inputs a ciphertext ct and a witness w, and outputs a message m or $\perp$.*

- **Correctness.** *For any security parameter n, for any message $m \in \{0,1\}$, and for any $(x, w) \leftarrow D_n$, we have that*

$$\Pr\left[\mathsf{Dec}(\mathsf{Enc}(pp, x, m), w) = m\right] = 1 - \mathsf{negl}(n)$$

- **Adaptive soundness.** *For all PPT adversary $\mathcal{A}$, there is a negligible function $\mathsf{negl}(\cdot)$ such that for any $x \notin L$,*

$$\Pr\left[\begin{array}{l} pp \leftarrow \mathsf{Setup}(1^n), \\ (m_0, m_1, st) \leftarrow \mathcal{A}(pp, x), b \xleftarrow{R} \{0,1\}, : \\ ct \leftarrow \mathsf{Enc}(pp, x, m_b), b' \leftarrow \mathcal{A}(ct, st) \end{array} \middle| \begin{array}{l} b = b' \wedge \\ m_0, m_1 \in \{0,1\} \end{array} \right] = \frac{1}{2} + \mathsf{negl}(n)$$

5.1   Construction of DWE from Injective One Way Functions

Let L be an NP language, and with sparse relation R_L (L can be considered as the PRG language (defined in Sect. 4)). Consider $\{D_n\}_{n \in \mathbb{N}}$ as a hard distribution ensemble over R_L. The notations of $X_n, (D_n^x \times D_n^x, D_n^w, Z_n), (X_n \times D_n^x, D_n^w, Z_n)$ are defined as in Sect. 4.

Here, we give a construction of DWE for L (with sparse relation R_L), based on injective one way function. Let the above PKE = (PKE.Gen, PKE.Enc, PKE.Dec) (based on injective one way function) be the ingredients of the following construction.

- **The setup algorithm** Setup(1^n):
  Choose a proper polynomial pair (q_1, q_2) and $q = q_1 \times q_2$;
  Choose an efficient sampleable hard distribution on R_L;
  Set $\boldsymbol{D}_n = D_n^{\otimes q}$ and $\boldsymbol{X}_n = X_n^{\otimes q}$; where D_n, X_n are defined as before.
  Output $pp = (n, q_1, q_2, D_n, X_n)$
- **The encryption algorithm** Enc($pp, \boldsymbol{x}_1, m$):
  Run $ct \leftarrow$ PKE.Enc($\boldsymbol{x}_1, m$);
  Output ct.
- **The decryption algorithm** Dec($ct, \boldsymbol{w}_1$):
  Run $m \leftarrow$ PKE.Dec($\boldsymbol{w}_1, ct$);
  Output m.

Theorem 1. *Assume injective one way function f exists, at least one of the following statements is true:*

- *(Infinitely-often) Distributional witness encryption can be constructed from injective one way function f.*
- *For every inverse polynomial ϵ, the Feige-Shamir protocol based on f is ϵ-distributional concurrent zero knowledge for $\{(D_n^x \times D_n^x, D_n^w, Z_n)\}_{n \in \mathbb{N}}$ (defined as Sect. 4 on* PRG *language).*

Now we prove Theorem 1, by showing that the above instantiation is a secure distributional witness encryption.

Proof (of sketch). The *correctness* of this scheme follows the correctness of the underlying public key encryption scheme.

To prove *adaptive soundness*, we have to demonstrate the functionality of the encryption algorithm Enc under the No-instances $\boldsymbol{x}_N \notin L^{\otimes q}$. By Lemma 1, we have that if V^* breaks ϵ-distributional concurrent zero knowledge of Feige-Shamir protocol over distribution $\{(D_n^x \times D_n^x, D_n^w, Z_n)\}_{n \in N}$, then it also can break $(\epsilon - \mathsf{negl}(n))$-distributional concurrent zero knowledge of Feige-Shamir protocol over distribution $\{(X_n \times D_n^x, D_n^w, Z_n)\}_{n \in N}$. So, even for $\boldsymbol{x}_N \xleftarrow{\text{R}} (X/L)^{\otimes q}$ (sparse relation), the encryption algorithm Enc works well with probability $1 - \mathsf{negl}(n)$.

Next we show that $\{\mathsf{Enc}(\boldsymbol{x}_N, m)\} \stackrel{c}{\approx} \{\mathsf{Enc}(\boldsymbol{x}_Y, m)\}$, for $m = 0, 1$, where $\boldsymbol{x}_N \xleftarrow{\text{R}} (X/L)^{\otimes q}$, $\boldsymbol{x}_Y \leftarrow D_n^{x \otimes q}$. If there exists a PPT adversary $\mathcal{A}$ can distinguish the above two distributions, then we can construct a simulator $R^{\mathcal{A}}$ can distinguish D_n^x from X_n, by running the encryption algorithm Enc($\cdot, m$).

By the CPA security of Yes-instances in $D_n^{x \otimes q}$, we have $\{\mathsf{Enc}(\boldsymbol{x}_Y, 0)\} \stackrel{c}{\approx} \{\mathsf{Enc}(\boldsymbol{x}_Y, 1)\}$. Thus we have $\{\mathsf{Enc}(\boldsymbol{x}_N, 0)\} \stackrel{c}{\approx} \{\mathsf{Enc}(\boldsymbol{x}_N, 1)\}$, as desired.   □

6   Oblivious Transfer

In the following, we show more consequences of breaking the ϵ-distributional concurrent zero knowledge of 4-round Feige-Shamir protocol. As observed in [17], there are no black box reductions of (even honest) oblivious transfer to public-key encryption. As a consequence, oblivious transfer cannot be reduced to injective trapdoor functions via black-box reductions. They also showed that OT can be implied by the strong version of public key encryption, in which public keys are sampled "separately of private keys".

Here we show there exist oblivious transfer protocols based on injective one way functions, if Feige-Shamir is not ϵ-distributional concurrent zero knowledge.

Theorem 2. *Assume injective one way function f exists, at least one of the following statements is true:*

- *(Infinitely-often) Oblivious transfer can be constructed from injective one way function f.*
- *For every inverse polynomial ϵ, the Feige-Shamir protocol based on f is ϵ-distributional concurrent zero knowledge for $\{(D_n^x \times D_n^x, D_n^w, Z_n)\}_{n \in \mathbb{N}}$ (defined as Sect. 4 on PRG language).*

To prove this theorem, we construct OT protocols against semi-honest adversaries and malicious receiver, which are implemented by the witness encryption schemes in Sect. 5.

6.1   Construction of Semi-honest Oblivious Transfer

Now we focus on the case—the adversaries are honest but curious.

Protocol 1: (2-message) Semi-honest OT protocol

1. The receiver R samples $(\boldsymbol{x}_b, \boldsymbol{w}_b) \leftarrow D_n^{\otimes q}$ and $\boldsymbol{x}_{1-b} \leftarrow X_n^{\otimes q}$ and sends $(\boldsymbol{x}_0, \boldsymbol{x}_1)$ to the sender S.
2. The sender S using $\boldsymbol{x}_0, \boldsymbol{x}_1$ as "public keys" encrypts m_0, m_1 respectively and independently, and sends $ct_0 = \mathsf{Enc}(\boldsymbol{x}_0, m_0)$ and $ct_1 = \mathsf{Enc}(\boldsymbol{x}_1, m_1)$ to the receiver R.
3. The receiver R using its "secret key" $\boldsymbol{w}_b$ decrypts ct_b to get m_b.

Proposition 1. *Protocol 1 is a semi-honest OT protocol, if the underlying encryption scheme is a secure distributional witness encryption and $\{(D_n^x \times D_n^x, D_n^w)\}_{n \in \mathbb{N}}$ is a hard distribution ensemble over sparse OR-relation.*

Proof. (Proof of Proposition 1)

Correctness. If both parties are honest, the correctness of the above OT protocol follows by the correctness of the underlying encryption schemes and the functionality of Enc under No-instances, with probability $1 - \mathsf{negl}(n)$.

Security for the Sender. For an honest sender and an honest (but curious) receiver, there exists a PPT simulator Sim such that

$$\{View_R^\Pi(S(m_0, m_1), R(b))\} \stackrel{c}{\approx} \{Sim(b, m_b)\}$$

The simulator Sim upon receiving 1^n, the receiver R's input b and its output m_b, works as follows:

1. Sample $(\boldsymbol{x}_b, \boldsymbol{w}_b) \leftarrow D_n^{\otimes q}$ and $\boldsymbol{x}_{1-b} \leftarrow X_n^{\otimes q}$;
2. Compute $ct_b = \mathsf{Enc}(\boldsymbol{x}_b, m_b)$, and $ct_{1-b} = \mathsf{Enc}(\boldsymbol{x}_{1-b}, \underline{0})$;
3. Output: $\boldsymbol{x}_b, \boldsymbol{w}_b, \boldsymbol{x}_{1-b}, ct_b$, and also ct_{1-b}.

The view of the receiver $View_R^\Pi(S(m_0, m_1), R(b))$ can be written as $(\boldsymbol{x}_b, \boldsymbol{w}_b, \boldsymbol{x}_{1-b}, ct_b, ct_{1-b})$, where $(\boldsymbol{x}_b, \boldsymbol{w}_b) \leftarrow D_n^{\otimes q}$ and $\boldsymbol{x}_{1-b} \leftarrow X_n^{\otimes q}$; $ct_b = \mathsf{Enc}(\boldsymbol{x}_b, m_b)$, and $ct_{1-b} = \mathsf{Enc}(\boldsymbol{x}_{1-b}, \underline{m_{1-b}})$.

For distributions $\{Sim(b, m_b)\}$ and $\{View_R^\Pi(S(m_0, m_1), R(b))\}$, the only difference is due to the ciphertext ct_{1-b}. The two distributions are computationally indistinguishable, following the adaptive soundness of distributional witness encryption.

Security for the Receiver. For an honest receiver and an honest (but curious) sender, there exists a PPT simulator such that

$$\{View_S^\Pi(S(m_0, m_1), R(b))\} \stackrel{c}{\approx} \{Sim(m_0, m_1)\}$$

Given the sender's inputs (m_0, m_1) and its empty output, the simulator Sim works as follows:

1. Sample $\boldsymbol{x}_0 \leftarrow X_n^{\otimes q}$ and $\boldsymbol{x}_1 \leftarrow X_n^{\otimes q}$;
2. Compute $ct_0 = \mathsf{Enc}(\boldsymbol{x}_0, m_0)$, and $ct_1 = \mathsf{Enc}(\boldsymbol{x}_1, m_1)$;
3. Output: $\boldsymbol{x}_0, \boldsymbol{x}_1, ct_0, ct_1$.

The view of the sender $View_S^\Pi(S(m_0, m_1), R(b))$:

1. The sender S receives $\boldsymbol{x}_0, \boldsymbol{x}_1$ which are sampled as the following way: $(\underline{\boldsymbol{x}_b}, \boldsymbol{w}_b) \leftarrow D_n^{\otimes q}$ and $\boldsymbol{x}_{1-b} \leftarrow X_n^{\otimes q}$;
2. Compute $ct_0 = \mathsf{Enc}(\boldsymbol{x}_0, m_0)$, and $ct_1 = \mathsf{Enc}(\boldsymbol{x}_1, m_1)$;
3. Output: $\boldsymbol{x}_0, \boldsymbol{x}_1, ct_0, ct_1$.

For distributions $\{Sim(m_0, m_1)\}$ and $\{View_S^\Pi(S(m_0, m_1), R(b))\}$, the only difference is due to $\boldsymbol{x}_b$. In the real mode, $\boldsymbol{x}_b$ is sampled from $D_n^{\otimes q}$, while in ideal mode it is sampled from $X_n^{\otimes q}$. Following the hard distribution D_n over sparse relation R_L, we have $\{D_n^{\otimes q}\}_{n \in \mathbb{N}} \stackrel{c}{\approx} \{X_n^{\otimes q}\}_{n \in \mathbb{N}}$. $\square$

6.2 Construction of OT Against Malicious Receiver

Now, we consider the receiver is malicious in the oblivious transfer, but the sender is always honest. Following the framework of [23], we have the below oblivious transfer protocol against malicious receiver, under the existence of injective one way function.

Protocol 2: OT against malicious receiver based on PRG languages

1. The receiver R samples $\{x_{b_k,k}, w_{b_k,k}\}_{k=1}^\ell \leftarrow D_n^{\otimes \ell}$, and sends $\{com(x_{b_k,k}; R_k)\}_{k=1}^\ell$, where $b_k \stackrel{R}{\leftarrow} \{0,1\}$, for each $k = 1, \ldots, \ell$. Note that com is statistically binding.

2. S sends $r \xleftarrow{\text{R}} \{0,1\}^{n \otimes \ell}$ to the receiver.
3. R computes $\{x_{1-b_k,k}\}_{k=1}^{\ell} = \{x_{b_k,k} \oplus r_k\}_{k=1}^{\ell}$ and sends x_0, x_1 to S.
4. S first checks whether $x_0 \oplus x_1 = r$. If not, S halts and outputs $\bot$. Otherwise, S chooses $e \in \{0,1\}^{\ell}$ at random and sends e to R.
5. R sends the opening of $com(x_{b_k,k}; R_k)$ and its corresponding witness $w_{b_k,k}$, for $e_k = 1$. For $e_k = 0$, R sends the reorderings denoted by $\{y_{0,k}, y_{1,k}\}_{k \in K}$, where $y_{b,k} := x_{b_k,k}, y_{1-b,k} := x_{1-b_k,k}$, for $k \in K = \{k : e_k = 0\}$. Until now, R uses its choice bit $b \in \{0,1\}$.
6. Then S independently encrypts m_0, m_1 using $\{y_{0,k}, y_{1,k}\}_{k \in K}$ as public keys respectively, and sends $ct_0 = \text{Enc}(\{y_{0,k}\}_{k \in K}, m_0)$ and $ct_1 = \text{Enc}(\{y_{1,k}\}_{k \in K}, m_1)$ to the receiver R.
7. The receiver R using its "secret key" decrypts ct_b to get m_b.

Remark 2. Set $\ell = 3q$. The probability of $(|K|/3q) * (3q) > q$ is a constant high (Normal distribution), where $K = \{k | e_k = 0\}$, since e is chosen at random. For consistency, when $|K| > q$, the encrypter takes the first q instances in $\{y_{0,k}, y_{1,k}\}_{k \in K}$ as "public keys" y_0, y_1.

Proposition 2. *The above Protocol is an OT protocol against malicious receiver, if the underlying encryption scheme is a secure distributional witness encryption, $\{(D_n^x \times D_n^x, D_n^w)\}$ is a hard distribution ensemble over sparse OR-relation, and the commitment scheme com is statistically binding.*

Proof. The receiver R is corrupted. Let $\mathcal{A}$ be any non-uniform PPT adversary controlling R. Now we construct an expected polynomial-time simulator Sim to extract the bit used by $\mathcal{A}$ by rewinding it.

The simulator Sim works as follows:

1. Sim receives commitments $\{com_k\}_{k=1}^{\ell}$ from $\mathcal{A}$.
2. Sim sends $\{r_k\}_{k=1}^{\ell} \xleftarrow{\text{R}} \{0,1\}^{n \otimes \ell}$ and receives back the instances $\{x_{0,k}, x_{1,k}\}_{k=1}^{\ell}$.
3. Sim sends $e \xleftarrow{\text{R}} \{0,1\}^{\ell}$ to $\mathcal{A}$ and receives back the decommitments c_k with the corresponding witness w_k, when $e_k = 1$. For $e_k = 0$, $\mathcal{A}$ sends the reorderings. Set (y_0, y_1) to be the reorderings $\{y_{0,k}, y_{1,k}\}_{k \in K}$.

   If the values sent by $\mathcal{A}$ are not valid (i.e. there exists a k such that $x_{0,k} \oplus x_{1,k} \neq r_k$, or the decommitments to c_k is not the right decommitments, or w_k is not a valid witness), Sim sends $\bot$ to the trusted party and outputs whatever $\mathcal{A}$ outputs.
4. For $i = 1, 2 \ldots, l_s$, Sim rewinds $\mathcal{A}$ to step 3.
   (a) It sends $\tilde{e}^i \xleftarrow{\text{R}} \{0,1\}^{\ell}$ to $\mathcal{A}$, then receives back the decommitments c_k and its corresponding witness w_k for $\tilde{e}_k^i = 1$ and the reorderings for $\tilde{e}_k^i = 0$. If there exists any invalid value (invalid values means there exists a k such that $x_{0,k} \oplus x_{1,k} \neq r_k$ or the decommitments to c_k is not the right decommitments, or w_k is not a valid witness), Sim repeats this step using fresh randomness.

    (b) If $e = \tilde{e}^i$, then Sim outputs "fail" and halts. Otherwise, Sim searches for the first value t_i such that $e_{t_i} \neq \tilde{e}^i_{t_i}$ and extracts the bit σ_i. If $t < t_{i-1}$, set $t := t_i$; otherwise, $t := t_{i-1}$.

5. If $e_t = 0$, Sim rewinds $\mathcal{A}$ to step 3, and resends the same messages e as step 3. Otherwise, Sim rewinds $\mathcal{A}$ to the beginning of this protocol.

6. Set σ to be the majority of $\{\sigma_i\}_{i \in [l_s]}$. Sim sends σ to the trusted party and receives back m_σ. Sim encrypts $m_\sigma, 0$ using $\{y_{\sigma,k}\}_{k \in K}$ and $\{y_{1-\sigma,k}\}_{k \in K}$ respectively. Finally, Sim outputs whatever $\mathcal{A}$ outputs and halts.

Now, we prove that the output distribution of $\mathcal{A}$ in a real execution with an honest sender S is computationally indistinguishable from the output distribution of Sim in an ideal execution.

Due to the underlying encryption scheme, a decrypter only needs to be equipped with at least as q_2 *active* witnesses (active witness means, its corresponding instance successfully generates α_i). The adversary controlling R may maliciously generate a "public-key" pair (y_0, y_1). Both y_0 and y_1 could contain part of L instances and part of X/L instances. Without loss of generality, we assume that at least one of y_0, y_1 contains at least $(1/2 + \mu(n))q$ instances in L, where μ is an arbitrary inverse polynomial. (Otherwise, the adversary obtains neither of m_0, m_1 with overwhelming probability, as in case 2.) Let y_b denote the "public key" whose a *better half* instances are in L language.

First we show the extracted value σ equals b with overwhelming probability. Set $l_s = \frac{16}{\mu^4}$. Define the random variable $\eta_i = 1$ iff $\sigma_i = b$, for $i \in [l_s]$. By the Chernoff bound, $\Pr[|\sum_{i=1}^{l_s} \eta_i - (\frac{1}{2} + \mu)q| > \frac{\mu}{2}(\frac{1}{2} + \mu)q] < 2 \exp^{\frac{1}{16\mu^2} \cdot (\frac{1}{2} + \mu) \cdot l_s}$, since η_i is independent, and $\mathrm{E}[\eta_i] = \Pr[\eta_i = 1] \geq 1/2 + \mu(n)$. Note that if $\sum_{i=1}^{l_s} \eta_i \geq (1/2 + \mu/2)q$, then $\sigma = b$. Therefore, we have $\Pr[\sigma \neq b] = \mathsf{negl}(n)$.

We observe that the view of $\mathcal{A}$ consists of r, e (uniformly), and the last message sent by the sender. The only difference between the ideal and real distributions is due to the ciphertexts in the last step. They are generated by using y_0, y_1 to encrypt m_0, m_1 in the real model, while to encrypt $m_\sigma, 0$ in the ideal world. Consider the following two cases:

- Case 1: The adversary is done honestly. The probability $y_{1-b} \notin L^{\otimes q}$ is overwhelming, since R_L is a sparse relation. It's oblivious the two distributions are computationally indistinguishable (based on the adaptive soundness of DWE).
- Case 2: The adversary sets $y_{1-b,k} = x_{b_k,k}$, for some k such that $e_k = 0$, as its strategy. In this case, the adversary will know partial "secret key" of y_{1-b}.

In Case 2, we assume the adversary $\mathcal{A}$ with its strategy can distinguish the ciphertext ct^*_{1-b} encrypted to m_{1-b} from 0, by contradiction. This means, $\mathcal{A}$ can correctly decrypt ct^*_{1-b} only using less than $q/2$ instances in L with noticeable probability. Most likely, $\mathcal{A}$ precisely knows all the witnesses whose associated instances are used to successfully generate a valid (α, β) pair. Otherwise, it will break the special type of strong one way functions induced by the magic adversary V^* which breaks the ϵ-distributional concurrent zero knowledge of Feige-Shamir protocols over $\{(D_n^x \times D_n^x, D_n^w, Z_n)\}_{n \in \mathbb{N}}$.

In the following, we show that it's impossible that $\mathcal{A}$ provides some $\boldsymbol{y}$ in which only half instances are in L, and all $\{(\alpha_i, \beta_i)\}_{i \in [q_2]}$ pairs are generated by the Yes-instances (instances in L) of $\boldsymbol{y}$, with noticeable probability. By the fact that the view of $\mathcal{A}$ up to the last message (only including $\boldsymbol{r}, \boldsymbol{e}$) can be generated uniformly and the rest of inputs $(\boldsymbol{y}_2, \boldsymbol{w}_2, \boldsymbol{z})$ used in the algorithm M are generated after $\boldsymbol{y}$, we have that the probability M successfully generates (β, α) such that $\beta = f(\alpha)$ using the Yes-instance (provided by $\mathcal{A}$) is negligibly close to the probability using the No-instances (instances not in L). Otherwise, the PPT machine M can easily distinguish the Yes-instances provided by $\mathcal{A}$ from the No-instances (provided by $\mathcal{A}$). That is, M breaks the hard distribution sampled by $\mathcal{A}$ (with uniformly auxiliary input) over R_L.

Furthermore, $\Pr[Sim \text{ outputs "fail"}] = \sum_{i=1}^{l_s} \Pr[e = \tilde{e}^i] = \mathsf{negl}(n)$. Since the sender is honest, e and $\tilde{e}^i$ are uniformly and independently generated.

Therefore, the output distribution of $\mathcal{A}$ in a real execution with an honest sender S is computationally indistinguishable from the output distribution of Sim in an ideal execution.

To complete the proof, we have to show that Sim's running time is in excepted polynomial time. First, we analyze the running time of that Sim (in step 4(a)) successfully obtains σ_i, for $i \in [l_s]$. Denote by δ the probability that $\mathcal{A}$ outputs valid values upon receiving a uniformly string $\boldsymbol{e} \xleftarrow{\text{R}} \{0,1\}^\ell$. With probability δ, there are rewinding attempts, and in such case there are an excepted $1/\delta$ iterations. We now focus on the number of times that Sim is expected to start from the beginning of this protocol. Since $\boldsymbol{e}$ are drawn uniformly, at any position t, the probability $e_t = 0$ is $1/2$. So, the expected number of times that Sim have to restart the protocol is $\mathcal{O}(1)$. To sum up, the expected running time of Sim is $\delta \cdot l_s \cdot (1/\delta) \cdot \mathcal{O}(1)$, which is a polynomial. $\qquad\square$

Acknowledgements. We thank Yanyan Liu, Shunli Ma, Hailong Wang for discussions and careful proofreading. We also thank the anonymous reviewers and editors for helpful comments.

The first and second authors were supported in part by the National Natural Science Foundation of China (Grant No. 61379141). The third author was supported in part by the National Key Research and Development Plan (Grant No. 2016YFB0800403), the National Natural Science Foundation of China (Grant No. 61772522) and Youth Innovation Promotion Association CAS. All authors were also supported by Key Research Program of Frontier Sciences, CAS (Grant No. QYZDB-SSW-SYS035), and the Open Project Program of the State Key Laboratory of Cryptology.

References

1. Abusalah, H., Fuchsbauer, G., Pietrzak, K.: Offline witness encryption. In: Manulis, M., Sadeghi, A.-R., Schneider, S. (eds.) ACNS 2016. LNCS, vol. 9696, pp. 285–303. Springer, Cham (2016). https://doi.org/10.1007/978-3-319-39555-5_16

2. Badrinarayanan, S., Garg, S., Ishai, Y., Sahai, A., Wadia, A.: Two-message witness indistinguishability and secure computation in the plain model from new assumptions. In: Advances in Cryptology - ASIACRYPT 2017 (2017, to appear)

3. Barak, B.: How to go beyond the black-box simulation barrier. In: Proceedings of the 42th Annual IEEE Symposium on Foundations of Computer Science - FOCS 2001, pp. 106–115. IEEE Computer Society (2001)

4. Blum, M.: How to prove a theorem so no one else can claim it. In: Proceedings of International Congress of Mathematicians - ICM 1986 (1986)

5. Canetti, R., Kilian, J., Petrank, E., Rosen, A.: Black-box concurrent zero-knowledge requires omega(log n) rounds. In: Proceedings of the 33rd Annual ACM Symposium Theory of Computing - STOC 2001, pp. 570–579. ACM Press (2001)

6. Chen, Y., Zhang, Z., Lin, D., Cao, Z.: Generalized (identity-based) hash proof system and its applications. Secur. Commun. Netw. 9(12), 1698–1716 (2016)

7. Chung, K.-M., Lin, H., Pass, R.: Constant-round concurrent zero-knowledge from indistinguishability obfuscation. In: Gennaro, R., Robshaw, M. (eds.) CRYPTO 2015. LNCS, vol. 9215, pp. 287–307. Springer, Heidelberg (2015). https://doi.org/10.1007/978-3-662-47989-6_14

8. Chung, K.-M., Lui, E., Pass, R.: From weak to strong zero-knowledge and applications. In: Dodis, Y., Nielsen, J.B. (eds.) TCC 2015. LNCS, vol. 9014, pp. 66–92. Springer, Heidelberg (2015). https://doi.org/10.1007/978-3-662-46494-6_4

9. Cramer, R., Shoup, V.: Universal hash proofs and a paradigm for adaptive chosen ciphertext secure public-key encryption. In: Knudsen, L.R. (ed.) EUROCRYPT 2002. LNCS, vol. 2332, pp. 45–64. Springer, Heidelberg (2002). https://doi.org/10.1007/3-540-46035-7_4

10. Deng, Y.: Magic adversaries versus individual reduction: science wins either way. In: Coron, J.-S., Nielsen, J.B. (eds.) EUROCRYPT 2017. LNCS, vol. 10211, pp. 351–377. Springer, Cham (2017). https://doi.org/10.1007/978-3-319-56614-6_12

11. Derler, D., Slamanig, D.: Practical witness encryption for algebraic languages and how to reply an unknown whistleblower. IACR Cryptology ePrint Arch. 2015, 1073 (2015)

12. Dwork, C., Naor, M., Reingold, O., Stockmeyer, L.J.: Magic functions. J. ACM 50(6), 852–921 (2003)

13. Dwork, C., Naor, M., Sahai, A.: Concurrent zero-knowledge. In: Proceedings of the 30rd Annual ACM Symposium Theory of Computing- STOC 1998, pp. 409–418. ACM Press (1998)

14. Feige, U., Shamir, A.: Zero knowledge proofs of knowledge in two rounds. In: Brassard, G. (ed.) CRYPTO 1989. LNCS, vol. 435, pp. 526–544. Springer, New York (1990). https://doi.org/10.1007/0-387-34805-0_46

15. Garg, S., Gentry, C., Halevi, S., Raykova, M., Sahai, A., Waters, B.: Candidate indistinguishability obfuscation and functional encryption for all circuits. SIAM J. Comput. 45(3), 882–929 (2016)

16. Garg, S., Gentry, C., Sahai, A., Waters, B.: Witness encryption and its applications. In: Proceedings of the Forty-Fifth Annual ACM Symposium on Theory of Computing, pp. 467–476. ACM (2013)

17. Gertner, Y., Kannan, S., Malkin, T., Reingold, O., Viswanathan, M.: The relationship between public key encryption and oblivious transfer. In: 2000 Proceedings of the 41st Annual Symposium on Foundations of Computer Science, pp. 325–335. IEEE (2000)
18. Goldreich, O.: A uniform-complexity treatment of encryption and zero-knowledge. J. Cryptology **6**(1), 21–53 (1993)
19. Goldreich, O., Kahan, A.: How to construct constant-round zero-knowledge proof systems for NP. J. Cryptology **9**(3), 167–190 (1996)
20. Impagliazzo, R., Rudich, S.: Limits on the provable consequences of one-way permutations. In: Proceedings of the 21th Annual ACM Symposium on the Theory of Computing - STOC 1989, pp. 44–61. ACM Press (1989)
21. Ishai, Y., Kushilevitz, E., Lindell, Y., Petrank, E.: Black-box constructions for secure computation. In: Proceedings of the Thirty-Eighth Annual ACM Symposium on Theory of Computing, pp. 99–108. ACM (2006)
22. Jain, A., Kalai, Y.T., Khurana, D., Rothblum, R.: Distinguisher-dependent simulation in two rounds and its applications. In: Katz, J., Shacham, H. (eds.) CRYPTO 2017. LNCS, vol. 10402, pp. 158–189. Springer, Cham (2017). https://doi.org/10.1007/978-3-319-63715-0_6
23. Lindell, A.Y.: Efficient fully-simulatable oblivious transfer. In: Malkin, T. (ed.) CT-RSA 2008. LNCS, vol. 4964, pp. 52–70. Springer, Heidelberg (2008). https://doi.org/10.1007/978-3-540-79263-5_4
24. Peikert, C., Vaikuntanathan, V., Waters, B.: A framework for efficient and composable oblivious transfer. In: Wagner, D. (ed.) CRYPTO 2008. LNCS, vol. 5157, pp. 554–571. Springer, Heidelberg (2008). https://doi.org/10.1007/978-3-540-85174-5_31
25. Prabhakaran, M., Rosen, A., Sahai, A.: Concurrent zero knowledge with logarithmic round-complexity. In: Proceedings of the 43th Annual IEEE Symposium on Foundations of Computer Science - FOCS 2002, pp. 366–375. IEEE Computer Society (2002)
26. Simon, D.R.: Finding collisions on a one-way street: can secure hash functions be based on general assumptions? In: Nyberg, K. (ed.) EUROCRYPT 1998. LNCS, vol. 1403, pp. 334–345. Springer, Heidelberg (1998). https://doi.org/10.1007/BFb0054137
27. Zhandry, M.: How to avoid obfuscation using witness PRFs. In: Kushilevitz, E., Malkin, T. (eds.) TCC 2016. LNCS, vol. 9563, pp. 421–448. Springer, Heidelberg (2016). https://doi.org/10.1007/978-3-662-49099-0_16

A New Lattice Sieving Algorithm Base on Angular Locality-Sensitive Hashing

Ping Wang[1(✉)] and Dongdong Shang[2]

[1] College of Information Engineering, Shenzhen University,
Shenzhen 518060, China
wangping@szu.edu.cn
[2] College of Computer Science and Software, Shenzhen University,
Shenzhen 518060, China
shangdongszu@gmail.com

Abstract. Currently, the space requirement of sieving algorithms to solve the shortest vector problem (SVP) grows as $2^{0.2075n+o(n)}$, where n is the lattice dimension. In high dimensions, the memory requirement makes them uncompetitive with enumeration algorithms. Shi Bai et al. presents a filtered triple sieving algorithm that breaks the bottleneck with memory $2^{0.1887n+o(n)}$ and time $2^{0.481n+o(n)}$.

Benefiting from the angular locality-sensitive hashing (LSH) method, our proposed algorithm runs in time $2^{0.4098n+o(n)}$ with the same space complexity $2^{0.1887n+o(n)}$ as the filtered triple sieving algorithm. Our experiment demonstrates that the proposed algorithm achieves the desired results. Furthermore, we use the proposed algorithm to solve the closest vector problem (CVP) with the lowest space complexity as far as we know in the literature.

Keywords: Filtered triple sieving
Angular locality-sensitive hashing · Shortest vector problem
Closest vector problem

1 Introduction

A lattice $\mathcal{L}$ is a discrete additive subgroup of $\mathbb{R}^n$. It have been widely used in cryptology and can be generated by a basis $\mathbf{B} = \{\mathbf{b}_1, \mathbf{b}_2, ..., \mathbf{b}_n\} \subset \mathbb{R}^n$ of linear independent vectors. We assume that the vectors $\mathbf{b}_1, \mathbf{b}_2, ..., \mathbf{b}_n$ form the columns of the $n \times n$ matrix $\mathbf{B}$. That is,

$$\mathcal{L}(\mathbf{B}) = \mathbb{Z}^n \cdot \mathbf{B} = \left\{ \sum_{i=1}^n x_i \cdot \mathbf{b}_i | x_1, ..., x_n \in \mathbb{Z} \right\}.$$

For $n \geq 2$, a lattice $\mathcal{L}$ has infinitely many bases, while a basis can only form a unique lattice. For a lattice basis $\mathbf{B}$, the fundamental parallelepiped $\mathcal{P}(\mathbf{B})$ of the lattice is defined as:

$$\mathcal{P}(\mathbf{B}) = \{\mathbf{B}\mathbf{x} | \mathbf{x} = (x_1, ..., x_n) \in \mathbb{R}^n, \ 0 \leq x_i < 1, i = 1, 2, ..., n\}.$$

© Springer International Publishing AG, part of Springer Nature 2018
X. Chen et al. (Eds.): Inscrypt 2017, LNCS 10726, pp. 65–80, 2018.
https://doi.org/10.1007/978-3-319-75160-3_6

The *determinant* $\det(\mathcal{L})$ of a lattice $\mathcal{L}$ is defined as the volume of the fundamental parallelepiped $\mathcal{P}(\mathbf{B})$ by selecting any basis $\mathbf{B}$. More precisely, for any basis $\mathbf{B}$ of a lattice $\mathcal{L}$, the determinant of $\mathcal{L}$ is computed as:

$$\det(\mathcal{L}) = \sqrt{\det\left(\mathbf{B}^T\mathbf{B}\right)} = \sqrt{\det\left(\langle\mathbf{b}_i,\mathbf{b}_j\rangle\right)_{1\leq i,j\leq n}}.$$

The determinant of a lattice is well-defined in the sense that the determinant does not depend on the choice of the basis. The ith *successive minimum* $\lambda_i(\mathcal{L})$ of a lattice $\mathcal{L}$ implies the smallest radium of a sphere within which there are i linearly independent lattice points, i.e.,

$$\lambda_i(\mathcal{L}) = \inf\left\{r \in \mathbb{R}^n \mid \dim\left\{\mathrm{span}\left(\mathcal{L}\cap\mathcal{B}_n\left(\mathbf{O},r\right)\right)\right\} = i\right\}.$$

Owing to the discreteness character of lattices, there exists two special lattice vectors in lattices. One is the nonzero vector with the smallest non-zero Euclidean norm in each lattice, designated as λ_1, and the other is the lattice vector closest to a given target vector $\boldsymbol{t}$ in $\mathbb{R}^m$. Finding the two special lattice vectors leads to the two famous computational problems regarding lattices:

- Shortest Vector Problem (SVP): Given a lattice $\mathcal{L}$, find the shortest nonzero vector λ_1 in the lattice.
- Closest Vector Problem (CVP): Given a lattice $\mathcal{L}$ and a target vector $\boldsymbol{t}$, find the lattice vector closest to the target vector $\boldsymbol{t}$.

The two famous computational problems in lattice have been played an prime importance role in the lattice cryptography for the past 20 years. Since the day it was born, lattice-based cryptography has been emerged as a credible alternative to classical public-key cryptography based on factorization and discrete logarithm and heir security is as hard as solving the worst-case hardness of the variants of SVP and CVP.

1.1   Related Work

There are two main types of algorithms for solving the shortest (short) lattice vectors problems according to the space size they used. One is the polynomial space algorithms, and the other is the exponential space algorithms. For polynomial-space algorithms, there are two different strategies are used. one is lattice reduction, including the famous LLL algorithm [1], HKZ reduction [2] and BKZ reduction [3]. The other important technique is enumeration technique which is an exact algorithm to find the shortest vectors. The first polynomial-space (enumeration) algorithm was provided by Kannan [2] in 1980s with time complexity of $2^{o(n\log n)}$. Another popular polynomial-space algorithm is Schnorr-Euchner enumeration, the theory of this enumeration algorithm is based on the BKZ reduction. In 2010, Gama et al. [4] presented an extreme pruning enumeration algorithms with exponential speedup. Chen and Nguyen used the extreme pruning technique to improve the BKZ algorithm named BKZ 2.0 [5]. Further improvements of BKZ are also achieved in result predictions [6] and progressive strategy [7] in recent years.

The first exponential space algorithm is the randomized sieve algorithm (AKS) that proposed in 2001 by Ajtai et al. [8] but the sieve method was widely believed impractical in fact. In 2008, Nguyen and Vidick [9] present a heuristic variant of AKS with $(4/3)^{n+o(n)}$ time and $(4/3)^{2n+o(n)}$ space requirement. Wang et al. [10] exhibited a version of the Nguyen-Vidick sieve which requires a running-time of $2^{0.3836n+o(n)}$ and space of $2^{0.2557n+o(n)}$. Micciancio and Voulgaris [11] proposed two famous algorithms: ListSieve and a heuristic derivation GaussSieve, which GaussSieve is thought to be the most practical sieving algorithm at present. In 2009, Pujol and Stehle [12] illustrated the time complexity bounds of ListSieve [11] was reduced to $2^{2.465n+o(n)}$ at the cost of $2^{1.233n+o(n)}$ space by using the birthday paradox. Micciancio and Voulgaris [11] illustrated that ListSieve and GaussSieve can obtain substantial speedups by taking advantage of the additional structure present in ideal lattices.

Another important exponential space algorithm is the deterministic algorithm with $O(2^{2n})$ time and $O(2^n)$ space given by Micciancio and Voulgaris [13]. The latest progress is the randomized algorithm with $O(2^n)$ time using the discrete Gaussian sampling method [14], the significance of the algorithm is obvious as it is the first randomized algorithm (without heuristic assumption) faster than the deterministic algorithm [6].

1.2   Contributions

In this paper we optimize the filtered triple sieving algorithm with a technique of angular locality-sensitive hashing (LSH) [15,16]. In the filtered triple sieving algorithm, searching the nearby vectors is brute-force which costs a lot of time. Our proposed algorithm can overcome the shortcoming effectively. When searching nearby vectors, we just need to consider those vectors that have at least one matching hash value in one of the hash tables. Such searching for nearby vectors can be done exponentially faster. Using this strategy, the time complexities of our proposed algorithm is bounded by $2^{0.4098n+o(n)}$.

According to the traditional angular locality-sensitive hashing (LSH) method, we need to store exponentially many hash tables in memory at the same time. We found that the hash tables can be processed sequentially, and we only need to store and use one hash table at a time. With this simple but crucial modification, our new algorithm's space complexities can be reduced to $2^{0.1887n+o(n)}$, and the asymptotic time complexity remains the same heuristic speed-up.

To date, the best space complexity of lattice sieving algorithms for solving the CVP is $2^{0.2075n+o(n)}$. In this paper we will show how our proposed algorithm can be modified to heuristically solve the CVP using $2^{0.1887n+o(n)}$ bits which is the best space complexity as far as we know.

1.3   Roadmap

In Sect. 2 we present some basic concepts and principles on the angular locality-sensitive hashing technique for solving nearest neighbor search problem (NNS).

In Sect. 3 we show how to apply these techniques to the filtered triple sieving algorithm and analysis its optimal time and space complexities. In Sect. 4 we shows experimental results of our implementation. In Sect. 5 we show how to using the filtered triple sieving algorithm's deformation solving CVP.

2   Locality-Sensitive Hashing

2.1   Locality-Sensitive Hash Families

The nearest neighbor problem is the following [16]: If L is a list of n-dimensional vectors, $L = \{w_1, w_2, ..., w_N\} \subset \mathbb{R}^n$, preprocess L in such a way that, when given a target vector $v \notin L$, one can efficiently find an element $w \in L$ which is close(st) to v.

If the list L has a certain structure, or if there is a significant gap between what is meant "nearby" and "far away", there are many methods can replace the brute-force list search of time complexity $O(N)$. One method for solving nearest neighbor problem in time sub-linear in N is base on the use of locality-sensitive hash functions h. Locality-sensitive hash functions h map n-dimensional vectors w to low-dimensional sketches $h(w)$. The most obvious characteristics of h is that nearby vectors are more likely to be mapped to the same output value (i.e. $h(v) = h(w)$) than distant pairs of vectors. To better understand this property, Let us firstly know about the definition of a locality-sensitive hash family $\mathcal{H}$. Here D is a similarity measure on $\mathbb{R}^n$, and U is commonly a finite subset of N.

Definition 1 ([16]). *A family $\mathcal{H} = \{h : \mathbb{R}^n \rightarrow U\}$ is called (r_1, r_2, p_1, p_2)-sensitive for similarity measure D if for any $v, w \in \mathbb{R}^n$:*

- *If $D(v, w) \leq r_1$ then $\mathbb{P}_{h \in \mathcal{H}}[h(v) = h(w)] \geq p_1$;*
- *If $D(v, w) \geq r_2$ then $\mathbb{P}_{h \in \mathcal{H}}[h(v) = h(w)] \leq p_2$.*

Note that if there exists an LSH family $\mathcal{H}$ is (r_1, r_2, p_1, p_2)-sensitive with $p_1 \gg p_2$, then (without computing $D(v, w)$) we can use $\mathcal{H}$ to distinguish between vectors which are at most r_1 away from v, and vectors which are at least r_2 away from v with non-negligible probability.

2.2   Amplification

Here we will show how such hash families can be constructed actually. Firstly constructing an (r_1, r_2, p_1, p_2)-sensitive hash family $\mathcal{H}$ with $p_1 \approx p_2$, Second uses several AND- and OR-compositions to turn it into an (r_1, r_2, p_1', p_2')-sensitive hash family $\mathcal{H}'$ with $p_1' > p_1$ and $p_2' < p_2$, thereby amplifying the gap between p_1 and p_2.

AND-composition. Given an (r_1, r_2, p_1, p_2)-sensitive hash family $\mathcal{H}$. we can construct an (r_1, r_2, p_1^k, p_2^k)-sensitive hash family $\mathcal{H}'$ by taking k different, pairwise independent functions $h_1, ..., h_k \in \mathcal{H}$ and a bijective function $\alpha : U^k \rightarrow U$ and defining $h \in \mathcal{H}'$ by the relation $h(v) = h(w)$ iff $h_i(v) = h_i(w)$ for all $i \in [k]$.

OR-composition. Given an $(r_1, r_2, p_1, p_2)-$sensitive hash family $\mathcal{H}$. we can construct an $(r_1, r_2, (1-p_1)^t, (1-p_2)^t)-$sensitive hash family $\mathcal{H}'$ by taking t different, pairwise independent functions $h_1, ..., h_t \in \mathcal{H}$ and defining $h \in \mathcal{H}'$ by the relation $h(\boldsymbol{v}) \neq h(\boldsymbol{w})$ iff $h_i(\boldsymbol{v}) \neq h_j(\boldsymbol{w})$ for all $i \in [t]$.

We can turn an $(r_1, r_2, p_1, p_2)-$sensitive hash family $\mathcal{H}$ into an $(r_1, r_2, 1 - (1 - p_1^k)^t, 1 - (1 - p_2^k)^t)-$sensitive hash family $\mathcal{H}'$ by combing a $k-$ wise AND-composition with a $t-$wise OR-composition:

$$(r_1, r_2, p_1, p_2) \xrightarrow{k-AND} (r_1, r_2, p_1^k, p_2^k) \xrightarrow{t-OR} (r_1, r_2, 1 - (1 - p_1^k)^t, 1 - (1 - p_2^k)^t).$$

It is obvious that we can find suitable k and t make $p_1^* = 1 - \left(1 - p_1^k\right)^t \approx 1$ and $p_1^* = 1 - \left(1 - p_2^k\right)^t \approx 0$ under the condition of $p_1 > p_2$.

2.3   Finding nearest neighbors

The following method that using these hash families to find nearest neighbors was first shown in [16]. First, choose $t \cdot k$ random hash functions $h_{i,j} \in \mathcal{H}$, and use the AND-composition to combine k of them at a time to build t different hash functions $h_1, ..., h_t$. Then, given the list L, we insert every $\boldsymbol{w} \in L$ into the bucket labeled $h_i(\boldsymbol{w})$ for each hash table T, in this way we can get t different hash tables $T_1, ..., T_t$. Finally, given the vector $\boldsymbol{v}$, we gather all the candidate vectors that collide with $\boldsymbol{v}$ in at least one of these hash tables in a list of candidates by computing its t images $h_i(\boldsymbol{v})$. Search this set of candidates, we can find a nearest neighbor for the given vector $\boldsymbol{v}$.

The quality of the underlying hash family $\mathcal{H}$ and the parameters k and t play an important role in the quality of this finding nearest neighbors algorithm. The following lemma shows how to balance k and t so that the overall time complexity is minimized.

Lemma 1 ([16]). *Suppose there exists an $(r_1, r_2, p_1, p_2)-$sensitive family $\mathcal{H}$. For a list $\mathcal{L}$ of size N, let*

$$\rho = \frac{\log(1/p_1)}{\log(1/p_2)}, \qquad k = \frac{\log(N)}{\log(1/p_2)}, \qquad t = O(N^\rho).$$

Then with high probability we can either (a) find an element $\boldsymbol{w}^* \in L$ that satisfies $D(\boldsymbol{v}, \boldsymbol{w}^*) \leq r_2$, or (b) conclude that with high probability, no elements $\boldsymbol{w} \in L$ with $D(\boldsymbol{v}, \boldsymbol{w}) > r_1$ exist, with the following costs:

1. *Time for preprocessing the list: $\tilde{O}(kN^{1+\rho})$.*
2. *Space complexity of the preprocessed data: $\tilde{O}(N^{1+\rho})$.*
3. *Time for answering a query $\boldsymbol{v}$: $\tilde{O}(N^\rho)$.*
   (a) Hash evaluations of the query vector $\boldsymbol{v}$: $\tilde{O}(N^\rho)$.
   (b) List vectors to compare to the query vector $\boldsymbol{v}$: $\tilde{O}(N^\rho)$.

We can tune k and t to obtain different time and space. In a way, this algorithm can be seen as a generalization of the naive brute-force search solution for finding nearest neighbors. When $k = 0$ and $t = 1$, this algorithm corresponds to checking the whole list for nearby vectors.

2.4   Angular Hashing

There are many hash families for the similarity measure D such as Terasawa's cross-polytope LSH family [17], Charikar's cosine hash family [15] and Becker's LSH family [18]. In this section what we are interested in is Charikar's cosine hash family defined on $\mathbb{R}^n$ as:

$$D(v, w) = \theta(v, w) = \arccos(\frac{v^T w}{\|w\| \cdot \|v\|}).$$

This similarity measure is under the condition that two vectors have similar Euclidean norms. Obviously, if their common angle is small, two vectors are nearby; if their angle is large, two vectors are far apart. Paper [15] introduces this hash family $\mathcal{H}$:

$$\mathcal{H} = \{h_a : a \in \mathbb{R}^n, \|a\| = 1\}, \quad h_a(v) \overset{def}{=} \begin{cases} 1 & if \ a^T v \geq 0; \\ 0 & if \ a^T v < 0. \end{cases}$$

For the two vectors $v, w \in \mathbb{R}^n$ that form a two-dimensional plane that passing through the origin. A hash vector does not lie on this plane (for $n > 2$) with probability 1, that is say that this plane intersects the hyperplane defined by a in some line l. Maps v and w to different hash values iff ℓ separates v and w in the plane. This hash function h_a maps v and w to different hash values iff ℓ separates v and w in the plane. So the probability that $h(v) = h(w)$ is:

$$\mathbb{P}_{h_a \in \mathcal{H}} [h_a(v) = h_a(w)] = 1 - \frac{\theta(v, w)}{\pi}.$$

The family $\mathcal{H}$ is $(\theta_1, \theta_2, 1 - \frac{\theta_1}{\pi}, 1 - \frac{\theta_2}{\pi})$–sensitive if the two angle θ_1 and θ_2 satisfy $\theta_1 < \theta_2$.

3   From the Filtered Triple Sieving to the FT-HashSieve

In this section we first introduce the filtered triple sieving algorithm. Then we show how we can speed up the filtered triple sieving of Shi Bai and Thijs Laarhoven [19] using angular LSH. Lastly, we show how to chose parameters k and t to optimal its time and space complexities.

3.1   The Filtered Triple Sieving

For better understanding of the filtered triple sieving algorithm, we should know about the TripleSieve algorithm. In short, the TripleSieve is to reduce triple of vectors in L to get a new list L'. The vectors in L have norms at most constant R; The vectors in L' have norms at most constant $\gamma \cdot R$ ($\gamma < 1$). This factor γ ensures that in each iteration of the sieve the norms of the list vectors become shorter and shorter.

In the analysis of the TripleSieve [19], we saw that a lot of the sphere was covered by vectors with angle θ significantly smaller than arccos $(1/3)$, but these vectors do not appear often, so we can take advantage of this feature to reduce time complexity. Vectors contribute to $1 - O(1)$ of the found collisions among triples of vectors when their angle $\theta \approx \arccos(1/3)$; vectors rarely lead to reductions even though this case appears very often when their angle larger than arccos $(1/3)$.

Algorithm 1. Filtered Triple Sieve

Input: An input list L_m of $(27/16)^{n/4}$ lattice vectors of norm at most $R =$ $\max_{v \in L_m} \|v\|$

Output: The output list $L_{(m+1)}$ has size $(27/16)^{n/4}$ and only contains lattice vectors of norm at most $\gamma \cdot R$

1: Initialize an empty list L_{m+1}
2: **for** each $v, w \in L$ **do**
3:     **if** $|\langle v, w \rangle| \geq \frac{1}{3}$ **then**
4:         **for** each $x \in L$ **do**
5:             **if** $\|v \pm w \pm x\| \leq \gamma \cdot R$ **then**
6:                 $L_{m+1} \leftarrow L_{m+1} \cup \{v \pm w \pm x\}$
7:             **end if**
8:         **end for**
9:     **end if**
10: **end for**
11: Output List L_{m+1} .

All in all, only those with pairwise angle less than $\theta_1 = \arccos(1/3)$ survive. After a trigonometric exercise, a fraction $p = (2\sqrt{2}/3)^n$ of all pairs of vectors survive the first round, so the probability of pairs that survive is proportional to $(\sin \theta)^n$. The time cost of the algorithm is $N^2 (1 + p \cdot N)$, as N is much larger than $1/p$, leads to the following result (time and space complexity).

Proposition 1. *Under the aforementioned heuristic assumptions, the Filtered Triple Sieve solves SVP in time* $(27/16)^{3n/4} \cdot (2\sqrt{2}/3)^n = 2^{0.4812n}$ *and space* $(27/16)^{n/4} = 2^{0.1887n}$.

3.2   The FT-HashSieve Algorithm

In this section we apply the technique of angular LSH to the sieve step of the filtered triple sieving algorithm. The core idea of Algorithm 2 is that replacing the third brute-force list search in the original algorithm with the technique of angular locality-sensitive hashing (LSH). Note that the setup costs of locality-sensitive hashing (building the hash tables) are only paid once, rather than once for each search.

Algorithm 2. FT-HashSieve

Input: An input list L_m of $(27/16)^{n/4}$ lattice vectors of norm at most $R =$ $\max_{v \in L_m} \|v\|$

Output: The output list $L_{(m+1)}$ has size $(27/16)^{n/4}$ and only contains lattice vectors of norm at most $\gamma \cdot R$

1: Initialize an empty list L_{m+1}
2: Initialize k empty hash tables T_i and sample $t \cdot k$ random hash functions $h_{i,j} \in H$
3: **for** each $y \in L$ **do**
4:     Add y to all k hash tables T_i, to the buckets $T_i[h_i(w)]$
5: **end for**
6: **for** each $v, w \in L$ **do**
7:     **if** $|\langle v, w \rangle| \geq \frac{1}{3}$ **then**
8:         Obtain the set of candidates $C = \bigcup_{i=1}^{t} T_i[h_i(v \pm w)]$
9:         **for** each $x \in C$ **do**
10:             **if** $\|v \pm w \pm x\| \leq \gamma \cdot R$ **then**
11:                 $L_{m+1} \leftarrow L_{m+1} \cup \{v \pm w \pm x\}$
12:             **end if**
13:         **end for**
14:     **end if**
15: **end for**
16: Output List L_{m+1} .

First, we sample $t \cdot k$ random hash functions $h_{i,j} \in \mathcal{H}$, use t random hash functions to initialize empty hash tables T_i. Then we add every $y \in L$ to all k hash tables T_i. This hash functions ensure that nearby vectors are more likely to be putted in the same bucket than distant pairs of vectors, the reason why we have stated in the second part. So when searching the third vectors x, we just need to consider those vectors that have at least one matching hash value $h_i(v \pm w)$ in one of the hash tables instead of brute-force search all of vectors in the list. Obviously, the time complexity can be reduced in this way. The increased space complexity of sieving with LSH comes purely from using many "rerandomized" hash tables, and not from an increased list size. In the Sect. 3.4 we will show how to reduce its space complexity.

3.3 Heuristically Solving SVP in Time and Space $2^{0.4098n+O(n)}$

The core idea of the FT-HashSieve algorithm is that replacing the third brute-force list search in the original algorithm with the technique of angular locality-sensitive hashing. Following, we will show the way how to choose the parameters k and t to optimize FT-HashSieve's asymptotic time complexity.

Suppose $N = 2^{c_n \cdot n}$, $p_2^* = N^{-\alpha 1}$, $p = 2^{c_f \cdot n}$ and $t = 2^{c_t \cdot n}$. To return the actual consequences for the complexity of the scheme, recall that the overall time and

---

[1] The average probability that a distant (non-reducing) vector w collides with v in at least one of the t hash tables [20].

space complexities are heuristically given by:

- Time (hashing): $O(N \cdot t) = 2^{(c_n + c_t)n + o(n)}$.
- Time (searching): $O(N^2 \cdot p \cdot N \cdot p_2^*) = 2^{(c_n + (2-\alpha)c_n + c_f)n + o(n)}$.
- Time (overall): $2^{(c_n + \max\{c_t, (2-\alpha)c_n\} + c_f)n + o(n)}$.
- Space: $O(N \cdot t) = 2^{(c_n + c_t)n + o(n)}$.

The overall time complexity and the space complexity can be written respectively as $2^{c_{time}n + o(n)}$ and $2^{c_{space}n + o(n)}$, that is to say:

$$c_{time} = c_n + \max\{c_t, (2-\alpha)c_n + c_f\}, \qquad c_{space} = c_n + c_t.$$

Recall that from bounds on the kissing constant in high dimensions, we expect that $N = 2^{0.1887n}$. From the sect. 3.1, we know that $p = (2\sqrt{2}/3)^n$. $p_2^* \leqslant O(N^{-0.3782})$ and $k = \log_{3/2}(t) - \log_{3/2}(\ln 1/\varepsilon)$ can been known from [20]. In order to balance the asymptotic time complexities for hashing and searching, so that the time and space complexities are the same and the time complexity is minimized, we solve $(2-\alpha)c_n + c_f = c_t$ numerically for c_t to obtain the following corollary.

Corollary 1. *Taking $c_t = 0.2211406$ leads to:*

$$\alpha \approx 0.378163, \qquad c_{time} = 0.4098406, \qquad c_{space} = 0.4098406.$$

That is to say, using $t \approx 2^{0.2211406n}$ hash tables and a hash length of $k \approx 0.3505$, the heuristic time complexities of the algorithm is $2^{0.4098406n + o(n)}$; the heuristic space complexities of the algorithm is $2^{0.4098406n + o(n)}$.

Note that $c_t = 0$ leads to the original filtered triple sieving, while $c_t = 0.2211406$ minimizes the heuristic time complexity at the cost of more space. One can also obtain a continuous time-memory trade-off between the filtered triple sieving and the FT-HashSieve by considering values $c_t \in (0, 0.2211406)$.

3.4 Heuristically Solving SVP in Time $2^{0.4098406n + o(n)}$ and Space $2^{0.1887n + o(n)}$

Being able to handle the increased memory complexity is crucial to making this method practical in higher dimensions. For the filtered triple sieving [19], we take the method of processing the hash tables sequentially to eliminate storing exponentially many hash tables in memory at the same time. We process the hash tables one by one in this algorithm; we first construct the first hash table, and add all vectors in L_m to this hash table, then look for short difference vectors to add to L_{m+1}. Finally deleting this hash table from memory and building a new hash table. Repeat the process $t = 2^{0.2211406n + o(n)}$ times until we have found the exact same set of short vectors for the next iteration. With this strategy, we do not need to store all hash tables in memory at the same time, and the memory need of the modified algorithm is almost the same as the filtered triple sieving.

Algorithm 3. FT-HashSieve (space-efficient)

Input: An input list L_m of $(27/16)^{n/4}$ lattice vectors of norm at most $R = \max_{v \in L_m} \|v\|$

Output: The output list $L_{(m+1)}$ has size $(27/16)^{n/4}$ and only contains lattice vectors of norm at most $\gamma \cdot R$

1: Initialize an empty list L_{m+1}
2: **for** each $i \in \{1, \ldots, t\}$ **do**
3:     Initialize k empty hash tables T_i and sample k random hash functions $h_{i,j} \in H$
4:     **for** each $y \in L$ **do**
5:         Add y to all k hash tables T_i, to the buckets $T_i[h_i(w)]$
6:     **end for**
7:     **for** each $v, w \in L$ do **do**
8:         **if** $|\langle v, w \rangle| \geq \frac{1}{3}$ **then**
9:             Obtain the set of candidates $C = \bigcup\limits_{i=1}^{t} T_i[h_i(v \pm w)]$
10:             **for** each $x \in C$ **do**
11:                 **if** $\|v \pm w \pm x\| \leq \gamma \cdot R$ **then**
12:                     $L_{m+1} \leftarrow L_{m+1} \cup \{v \pm w \pm x\}$
13:                 **end if**
14:             **end for**
15:         **end if**
16:     **end for**
17: **end for**
18: Output List L_{m+1} .

The core idea of the Algorithm 3 is that we process hash tables sequentially rather than in parallel, to prevent getting an increased space complexity. In order to decrease the memory requirement of the hash tables, Panigraphy suggested that one could also check several hash buckets in each hash table for nearby vectors, and use fewer hash tables overall to get a similar quality for the list of candidates, using significantly less memory. In this way, we can avoid using many hash tables and checking only one hash bucket in each table for candidate nearby vectors. For more details, please see e.g. [21]. Using this technology in this algorithm is leaved for our future work.

4 Experiment Results

For implementing the FT-HashSieve algorithm, we note that we can use following two simple tweaks to further improve the algorithm's performance. These include:

(a) The hash of $-v$ can be computed for free from $h_i(v)$ as $h_i(-v) = -h_i(v)$.
(b) Instead of comparing $\pm v$ to all candidate vectors w, we only compare $+v$ to the vectors in the bucket $h_i(v)$ and $-v$ to the vectors in the bucket $-h_i(v)$. This further reduces the number of comparisons by a factor 2 compared to the filtered triple sieving, where both comparisons are done for each potential reduction.

In this paper, our experiments were conducted with random lattices generated by a efficient method proposed by Goldstein-Mayer [22]. The TU Darmstadt Lattice Challenge project provides Goldstein-Meyer style lattices. For more details, please refer to [23]. To verify the claimed speedups, Firstly, our experiments implement the TripleSieve. Secondly, we implement the filtered triple sieving. At last, for the FT-HashSieve we chose k and t by rounding the theoretical estimates of Theorem 1 to the nearest integers, i.e. $k = \lfloor 0.5233n + o(n) \rceil$ and $t = \lfloor 2^{0.306n + o(n)} \rceil$.

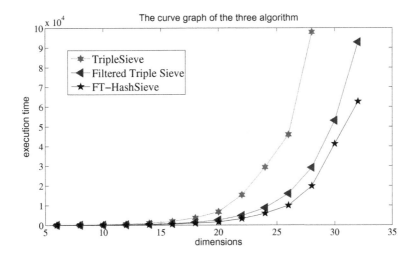

Fig. 1. The execution time of the three algorithms.

Figure 1 shows the execution time of the three algorithm. Usually, the bigger the dimensions are, the more time are. From the graph, we know the black line is the best effect. Next, listing detailed values by a table.

Table 1. The execution time of the three algorithms.

Curve\dimensions	18	20	22	24	26	28
Tuple-Lattice	3677.063	6776.219	15300.124	29353.377	45924.026	97931.823
Tuple-Lattice (Filed)	1393.837	2545.148	4823.057	8767.854	15973.555	29183.686
(TLS-)HashSieve	994.926	1547.202	3192.386	5973.372	9977.236	19819.679

Table 1 lists execution time of the three algorithms. We can see the difference between each curve. Because program has recording function, so the results of

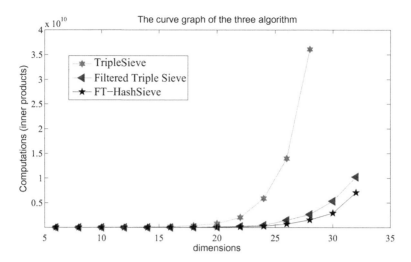

Fig. 2. The Computations (inner products) numbers of the three algorithms.

running could be different every time. We can know that the last one is the best. This directly verifies the effectiveness of our algorithm.

Figure 2 shows the number of computations (inner products) under different dimensions. Usually, the bigger the dimensions are, the more the number of computations (inner products) are. As a whole, the graph will present a rising trend. Of course, it still have detailed difference listed in Table 2. The number of black line is always the minimal.

Table 2. The Computations (inner products) numbers of the three algorithms.

Curve\dimensions	20	22	24	26	28
Tuple-Lattice	803954894	2047673117	5885012538	13953364729	36097354553
Tuple-Lattice (Filed)	96372676	232740013	462082022	1474394968	2661592452
(TLS-)HashSieve	51088412	133585980	246065376	716050246	1855261135

Table 2 lists detailed the numbers of computations (inner products) of the three algorithm. The number of computations for the second case is often less than the first one due to the filter technology. The third one is usually less than the first two. This verifies the validity of our algorithm from the side.

Our main idea is that combining Charikar's angular locality-sensitive hashing (LSH) method with the filtered triple sieving algorithm. We use the technique of angular locality-sensitive hashing (LSH) instead of brute-force list search, it is not surprising that the new algorithm is better than the previous algorithms. From the above experiment we can know that the new algorithm work well than the previous algorithm in practical.

5   FT-HashSieve for the Closest Vector Problem

In this section we will show how the FT-HashSieve algorithm can be modified to solve CVP firstly. In order to obtain the best overall space complexity, we adapt the entire sieving algorithm to the closest Vector Problem instance. When solving several CVP instances, the costs scale is linearly with the number of instances.

Using one list. The main idea behind this method is to translate the SVP algorithm by the target vector t; we generate a list of lattice vectors close to t instead of generating a long list of lattice vectors reasonably close to 0, and then combine lattice vectors to find lattice vectors that more and more closer to t. Finally, finding a list that hopefully contains a closest vector to t.

It is obvious that this does not work, as the fundamental property of lattices does not hold for the lattice coset $t+\mathcal{L}$: if $w_1, w_2, w_3 \in t+\mathcal{L}$, not all combinations $w_1 \pm w_2 \pm w_3 \notin t + \mathcal{L}$. In other words, three lattice vectors close to t can be combined to form lattice vectors close to t, $-t$ or $3t$. So if we start with a list of vectors close to t, and combine vectors in this list as in the TripleSieve [19] sieve, then after one iteration we will end up with a list $\mathcal{L}'$ of lattice vectors close to t, $-t$ or $3t$.

Using two list. The strategy that translating the whole problem by t work for the FT-HashSieve is as follows:

For the given two lists L_0 (lattice vectors close to 0) and L_t (lattice vectors close to t), we construct a sieve which maps two input lists L_0, L_t to two output lists L_0', L_t' of lattice vectors slightly closer to 0 and t. The radius of the two initial lists L_0, L_t is R; the two finally lists L_0' and L_t' with lattice vectors at distance at most approximately $\lambda_1 (\mathcal{L})$ from 0 and t. This algorithm's arguments are almost identical to algorithm that solving SVP. Algorithm 4 describe this program in detail.

Theorem 1. *The adaptive filtered triple sieving with angular LSH (using in step 3 and 8) can solve CVP in time T and space S, with*

$$S = 2^{0.1887n+o(n)}, \qquad T = 2^{0.4098n+o(n)}.$$

Using the idea of FT-HashSieve, the exponential time complexity bounds of Algorithm 4 can be reduced to $2^{0.4098n+o(n)}$ at the cost of $2^{0.1887n+o(n)}$ space. $2^{0.1887n+o(n)}$ is the best exponential space complexity of lattice sieving algorithms for solving CVP as far as we know in the literature.

Algorithm 4. The adaptive filtered triple sieving for finding closest vectors

---

Input: An input list $L_0, L_t \subset \mathcal{L}$ containing $(27/16)^{n/4}$ lattice vectors at distance $\leq R$ from $\mathbf{0}, \mathbf{t}$

Output: An output list $(27/16)^{n/4}$ lattice vectors at distance $\leq \gamma R$ from $\mathbf{0}, \mathbf{t}$

1: Initialize an empty list L_0', L_t'
2: **for** each $(\mathbf{w}_1, \mathbf{w}_2, \mathbf{w}_3) \in L_0 \times L_0 \times L_0$ **do**
3:     **if** $|\langle \mathbf{w}_1, \mathbf{w}_2 \rangle| \geq \frac{1}{3}$ **then**
4:         **if** $\|\mathbf{w}_1 \pm \mathbf{w}_2 \pm \mathbf{w}_3\| \leq \gamma \cdot R$ **then**
5:             Add $\mathbf{w}_1 \pm \mathbf{w}_2 \pm \mathbf{w}_3$ to the list L_0'
6:         **end if**
7:     **end if**
8: **end for**
9: **for** each $(\mathbf{w}_1, \mathbf{w}_2, \mathbf{w}_3) \in L_0 \times L_0 \times L_t$ **do**
10:     **if** $|\langle \mathbf{w}_1, \mathbf{w}_2 \rangle| \geq \frac{1}{3}$ **then**
11:         **if** $\|(\mathbf{w}_1 \pm \mathbf{w}_2 \pm \mathbf{w}_3) - \mathbf{t}\| \leq \gamma \cdot R$ **then**
12:             Add $\mathbf{w}_1 \pm \mathbf{w}_2 \pm \mathbf{w}_3$ to the list L_t'
13:         **end if**
14:     **end if**
15: **end for**
16: Output List L_0', L_t'.

---

6   Future Work

Michael Schneider [24] have shown that it is indeed possible to make use of the special structure of ideal lattices when searching for shortest vectors. The gained speedup does not affect the asymptotic runtime of the SVP algorithms, but it allows for some improvements in practice. This provides us with a new research direction of the filtered triple sieving. One of the properties of ideal lattices is that for each lattice vector $\mathbf{v}$, rotations of this vector are also contained in the lattice. We can use this property in the filtered triple sieving.

The filtered triple sieving is a generalization of ListSieve [11]. The different of IdealListSieve [24] and ListSieve is that if the rotations of every vector in the list L are regard as the elements of the list L or not. So we guess that filtered triple sieving can take the similar approach to reducing the time and space complexity. Note that the speedups predicted may not hold for the filtered triple sieving in practice, this is leaved for our future work.

References

1. Lenstra, H.W., Lenstra, A.K., Lovfiasz, L.: Factoring polynomials with rational coefficients. Mathematische Annalen **261**, 515–534 (1982)
2. Kannan, R.: Improved algorithms for integer programming and related lattice problems. In: ACM Symposium on Theory of Computing, 25–27 April 1983, Boston, Massachusetts, USA, pp. 193–206 (1983)

3. Schnorr, C.P., Euchner, M.: Lattice basis reduction: improved practical algorithms and solving subset sum problems. In: Budach, L. (ed.) FCT 1991. LNCS, vol. 529, pp. 68–85. Springer, Heidelberg (1991). https://doi.org/10.1007/3-540-54458-5_51

4. Gama, N., Nguyen, P.Q., Regev, O.: Lattice enumeration using extreme pruning. In: Gilbert, H. (ed.) EUROCRYPT 2010. LNCS, vol. 6110, pp. 257–278. Springer, Heidelberg (2010). https://doi.org/10.1007/978-3-642-13190-5_13

5. Chen, Y., Nguyen, P.Q.: BKZ 2.0: better lattice security estimates. In: Lee, D.H., Wang, X. (eds.) ASIACRYPT 2011. LNCS, vol. 7073, pp. 1–20. Springer, Heidelberg (2011). https://doi.org/10.1007/978-3-642-25385-0_1

6. Micciancio, D., Walter, M.: Practical, predictable lattice basis reduction. In: Fischlin, M., Coron, J.-S. (eds.) EUROCRYPT 2016. LNCS, vol. 9665, pp. 820–849. Springer, Heidelberg (2016). https://doi.org/10.1007/978-3-662-49890-3_31

7. Aono, Y., Wang, Y., Hayashi, T., Takagi, T.: Improved progressive BKZ algorithms and their precise cost estimation by sharp simulator. In: Fischlin, M., Coron, J.-S. (eds.) EUROCRYPT 2016. LNCS, vol. 9665, pp. 789–819. Springer, Heidelberg (2016). https://doi.org/10.1007/978-3-662-49890-3_30

8. Ajtai, M., Kumar, R., Sivakumar, D.: A sieve algorithm for the shortest lattice vector problem. In: ACM Symposium on Theory of Computing, pp. 601–610 (2002)

9. Nguyen, P.Q., Vidick, T.: Sieve algorithms for the shortest vector problem are practical. J. Math. Cryptology **2**(2), 181–207 (2008)

10. Wang, X., Liu, M., Tian, C., Bi, J.: Improved Nguyen-Vidick heuristic sieve algorithm for shortest vector problem. In: ACM Symposium on Information, Computer and Communications Security, ASIACCS 2011, Hong Kong, China, March 2011, pp. 1–9 (2011)

11. Micciancio, D., Voulgaris, P.: Faster exponential time algorithms for the shortest vector problem. In: ACM-SIAM Symposium on Discrete Algorithms, pp. 1468–1480 (2010)

12. Pujol, X., Stehl, D.: Solving the shortest lattice vector problem in time 2 2.465n. IACR Cryptology ePrint Archive, vol. 2009 (2006)

13. Micciancio, D., Voulgaris, P.: A deterministic single exponential time algorithm for most lattice problems based on Voronoi cell computations. In: ACM Symposium on Theory of Computing, pp. 351–358 (2010)

14. Aggarwal, D., Dadush, D., Regev, O., Stephens-Davidowitz, N.: Solving the shortest vector problem in 2 n time using discrete Gaussian sampling: extended abstract. In: Forty-Seventh ACM Symposium on Theory of Computing, pp. 733–742 (2015)

15. Charikar, M.S.: Similarity estimation techniques from rounding algorithms. In: Thiry-Fourth ACM Symposium on Theory of Computing, pp. 380–388 (2002)

16. Indyk, P., Motwani, R.: Approximate nearest neighbors: towards removing the curse of dimensionality. In: Theory of Computing, no. 11, pp. 604–613 (2000)

17. Becker, A., Laarhoven, T.: Efficient (ideal) lattice sieving using cross-polytope LSH. In: Pointcheval, D., Nitaj, A., Rachidi, T. (eds.) AFRICACRYPT 2016. LNCS, vol. 9646, pp. 3–23. Springer, Cham (2016). https://doi.org/10.1007/978-3-319-31517-1_1

18. Becker, A., Ducas, L., Gama, N., Laarhoven, T.: New directions in nearest neighbor searching with applications to lattice sieving. In: Twenty-Seventh ACM-SIAM Symposium on Discrete Algorithms, pp. 10–24 (2016)

19. Shi, B.: Tuple lattice sieving. LMS J. Comput. Math. **19**(A), 146–162 (2016)

20. Laarhoven, T.: Sieving for shortest vectors in lattices using angular locality-sensitive hashing. In: Gennaro, R., Robshaw, M. (eds.) CRYPTO 2015. LNCS, vol. 9215, pp. 3–22. Springer, Heidelberg (2015). https://doi.org/10.1007/978-3-662-47989-6_1

21. Panigrahy, R.: Entropy based nearest neighbor search in high dimensions. In: SODA 2006: Proceedings of the Seventeenth Annual ACM-SIAM Symposium on Discrete Algorithms, pp. 1186–1195 (2005)
22. Goldstein, D., Mayer, A.: On the equidistribution of hecke points. Forum Mathematicum **15**(2), 165–189 (2003)
23. Goldstein, D.M.A.: SVP challenge (2010). http://www.latticechallenge.org
24. Schneider, M.: Sieving for shortest vectors in ideal lattices. In: Youssef, A., Nitaj, A., Hassanien, A.E. (eds.) AFRICACRYPT 2013. LNCS, vol. 7918, pp. 375–391. Springer, Heidelberg (2013). https://doi.org/10.1007/978-3-642-38553-7_22

A Simpler Bitcoin Voting Protocol

Haibo Tian$^{(\boxtimes)}$, Liqing Fu, and Jiejie He

Guangdong Key Laboratory of Information Security,
School of Data and Computer Science,
Sun Yat-Sen University, Guangzhou 510275,
Guangdong, People's Republic of China
tianhb@mail.sysu.edu.cn

Abstract. Recently, Zhao and Chan proposed a Bitcoin voting protocol where n voters could fund Bitcoins to one of two candidates determined by majority voting. Their protocol preserves the privacy of individual ballot while they cast ballots by Bitcoin transactions. However, their protocol supports only two candidates and relies on a threshold signature scheme. We extend their method to produce a ballot by a voter selecting at least k_{min}, at most k_{max} from L candidates. We also redesign a vote casting protocol without threshold signatures to reduce transaction numbers and protocol complexities. We also introduce new polices to make the Bitcoin Voting protocol more fair.

Keywords: Bitcoin · Voting · Privacy

1 Introduction

A trusted third party (TTP) is a powerful component in a security protocol. It is usually the magic part of a protocol. For example, a certificate authority (CA) is a kind of TTP that could prove the binding between an identity and a public key. A tally authority (TA) in a voting system is another kind of TTP to count votes faithfully. However, a TTP may incur attacks for its magic function. An adversary may attack a CA to obtain faked certificates or attack a TA to modify a voting result. These attacks may be technically or socially, and are usually difficult to defend against.

The Bitcoin technique provides a decentralized TTP based on a large dynamic point-to-point network. Roughly, the TTP records transactions and executes scripts in the transactions faithfully. Technically, nodes in the network run the proof-of-work consensus mechanism to produce a new block and the node who created the new block is expected to record recent Bitcoin transactions and execute the related scripts. The expectation of node behavior is based on the hypothesis of rational economic man since there are transaction fees. The Bitcoin system works well as a trustable ledger of Bitcoin transactions since 2009 [17]. Ethereum, as a successor of Bitcoin system, provides a Turing complete script language, and can support more complex applications [22].

© Springer International Publishing AG, part of Springer Nature 2018
X. Chen et al. (Eds.): Inscrypt 2017, LNCS 10726, pp. 81–98, 2018.
https://doi.org/10.1007/978-3-319-75160-3_7

The vivid decentralized TTP certainly attracts protocol designers to build their protocols based on the Bitcoin-like systems. We found about 100 decentralized applications available on Ethereum including voting, contract signing, auction and so on [7]. However, no privacy was enabled. Zhao and Chan [23] proposed a Bitcoin voting protocol where the voters cast ballots by Bitcoin transactions. Their protocol do claim the privacy of a ballot, which induces us to study their constructions. They rely on secret sharing, commitment, and zkSNARK [21] to produce a ballot that could be opened, summed and verified. They use separately the "Claim-or-Refund" [4] and "Joint Transaction" [2] to cast their ballots by Bitcoin transactions. To tackle with the transaction malleability problem [6], they use an $(n\text{-}n)$ threshold signature to make sure that no voter could claim their deposit back prematurely.

Zhao and Chan's protocol [23] mainly has three shortcomings. One is that it only supports two candidates. The other is that their vote casting protocol is a bit complex. A more general approval voting protocol should allow a voter to select candidates in a range. Suppose there are L candidates and the range is $[k_{min}, k_{max}]$. Then Zhao and Chan's protocol could be viewed as a special case with $k_{min} = k_{max} = 1$ and $L = 2$. The complexity of their protocol mainly lies in the usage of $n\text{-}n$ threshold signature scheme. To produce such a signature, all participants should use their secret shares to compute part of the final signature and all of them should exchange messages. To securely produce such a signature, one has to add verification part for their messages to prove their good behaviours, which introduces new complexity. The last is about the policy to distribute the prize. If some voters stops prematurely, the prize could not be used anymore. That is, some Bitcoin coins are lost. Honest voters also lose their funds under this case. A more natural way is to return the funds back to voters.

This paper supports voters to select at least k_{min}, at most k_{max} from L candidates by exploiting Zhao and Chan's vote commitment protocol as a subroutine, adding a zkSNARK proof, and giving new transactions. The new design removes threshold signatures and uses the recent Bitcoin improvement proposal 65 [20] to produce a more fair and simple vote casting protocol.

1.1   Related Works

Bitcoin System. The Bitcoin system provides a natural way to implement the "fairness by compensation" idea. Barder et al. [4] provides a fair Bitcoin mixer by their "Claim-or-Refund" pattern. Bentov and Kumaresan [5] formalize the pattern in a universal composable model. Andrychowicz et al. [3] provide a fair gambling protocol with a new "Timed-Commitment" pattern. They [2] further give a "Simultaneous-Timed-Commitment" pattern to provide mutual fairness. Zhao and Chan [23] use both patterns to design two vote casting protocols in their Bitcoin voting protocol. They call the "Simultaneous-Timed-Commitment" pattern as "Joint Transaction". Kiayias et al. [16] use the "Claim-or-Refund" pattern to construct a general fair multiparty computation protocol.

A transaction malleability problem [6] appears in protocols based on Bitcoin system. Basically, a Bitcoin transaction's identity $TxID$ is a hash value of the whole transaction, especially including the input script part where a signature is included. But the signature in the transaction does not cover the input script. One could insert an "NOP" opration in the input script without changing the validity of the transaction. But the $TxID$ is changed even that only an "NOP" is inserted. One could also launch this attack based on the signature itself exploiting the presentation form of a elliptic curve point used in the Bitcoin system. Decker and Wattenhofer [6] analyzed the problem and the failure of MtGox. Andrychowicz et al. [3] showed that the "Claim-or-Refund" pattern is under the attack of transaction malleability. So they proposed the "Timed-Commitment" pattern. This pattern is supported by the LOCKTIME field of the Bitcoin system and is immune to malleability attacks [3]. The Bitcoin improvement proposal (BIP) 65 [20] proposed an opcode $CHECKLOCKTIMEVERIFY$ for the Bitcoin scripting system that allows a transaction output to be made unspendable until some point in the future. It could be used to further simplify the "Timed-Commitment" pattern.

The Bitcoin itself is developing. Some observations affects the protocols based on it. Karame et al. [10] showed the double spending problem of fast payments in Bitcoin. Heilman et al. [13] showed the eclipse attack where an adversary could control the network connections of a victim node to make double spending possible. Gervais et al. [9] showed that an adversary could delay Bitcoin transactions to a victim node arbitrarily to launch double spending attack with a small cost. If an adversary could launch the double spending attack, the protocols based on the Bitcoin system are generally problematic. The lucky thing is that there are some proposals to mitigate the eclipse attack [8,14], and the authors [9] also show countermeasures to defend against their observations.

Voting System. E-voting has a long history. A typical e-voting protocol replies on a trustable tally to protect the voter's privacy. For a cautious voter, a verifiable tally is more appreciated since the voter could verify whether their ballot is counted. To reduce the risk of single trustable tally, a group of tallying authorities are used to fulfil the same task where threshold cryptography is usually employed. Helios [1] is such a system where more than one hundred thousand votes have been cast.

Kiayias and Yung [15] first introduced a self-tallying voting protocol for boardroom voting. A self tallying protocol has an open procedure that allows anybody to perform the tally computation once all ballots are cast. There is no tallying authority at all. Groth [11], Hao [12] proposed different self-tallying protocols. A ideal self-tallying protocol should protect the privacy of a voter, and should ensure the secrecy of partial voting results before all voters casted their ballots, and should ensure the open verification of the correctness of the protocol, and should ensure that a corrupted voter could not prevent the election result from being announced. It is not easy to achieve both of these properties. In fact, the final voter usually could know the voting result and could stop their casting by simply leaving the voting procedure in a self-tallying protocol.

Recent Works. Hao [18] proposed a smart contract for boardroom voting with maximum voter privacy. It combines their previous self-tallying protocols and the "Claim and Refund" paradigm [4]. If the final voter stops casting, the voter will lose their deposit. Obviously, if their smart contract is modified to add funds for candidates, it turns into an *ether* (Ethereum coins) voting protocol. However, our proposal here is different to their work:

- The ballot in [18] is an element in a multiplicative cyclic group with a zero knowledge proof. Our ballot here is in fact some hash values similar to Zhao and Chan's scheme [23]. The different choices lead to different implementation complexity. Hao et al. [18] includes an external library to perform cryptographic computations, which lead to their voting contract becoming too large to store on the Blockchain. They finally separated their program into two smart contracts. And the cryptographic computations on the Blockchain are expensive. As a contrast, our proposal here uses only hash functions which is supported by most Blockchain instances including Bitcoin and Ethereum. Certainly, Hao et al. [18] has a simple procedure to produce a ballot. Comparatively, the method we used needs zkSNARK proofs and several round communications. So our proposal has a "fat" client with a "thin" Blockchain code. The protocol in [18] is a mirror image.

Very recently, Tarasov and Tewari [19] proposed an Internet voting protocol using Zcash. Roughly, they require each voter to be authenticated to obtain a Zcash vote token. The voters then give their Zcash vote token to candidates at their will privately. And finally the candidates show their votes by sending their total Zcash vote tokens to a tally address publicly. This is certainly a different and interesting approach. However, it highly relies on the honesty of candidates and voters, and it is not easy to track abnormal behaviours.

1.2   Contributions

We present a simple Bitcoin Voting protocol. It allows n voters to select at least k_{min}, at most k_{max} winners from L candidates, and fund the winners with Bitcoins. Technically, there are two improvements:

- We extend the ballot generation method in [23] to allow multiple candidates greater than 2. When $k_{min} = k_{max} = 1$, $L = 2$, our extension method has the same effect as that in [23].
- We design a simpler vote casting protocol with fewer transactions. It uses "Simul taneous-Timed-Commitment" [2] and an opcode in BIP 65 [20] to prevent transaction malleability attacks and obtain fairness for voters and candidates.

Our Bitcoin voting protocol introduces new policies for multiple winners and dishonest voters. Roughly, if there are multiple winners, they equally share the fund of all voters. If some voters stop their casting procedure before the final result could be revealed, after an expected time period, their deposits will be shared by all candidates and their funds will be returned. Comparing with the unspendable prize in [23], our policy is more fair.

2   Preliminaries

This section includes a brief overview of Bitcoin system and basic assumptions about it, the Bitcoin transaction symbols, and the BIP 65 proposal with its application in the "Timed-Commitment" design pattern.

2.1   Overview of Bitcoin

Bitcoin System. The Bitcoin system contains a chain of blocks. A block contains a header and some Bitcoin transactions. Each transaction has an identity which is the hash value of the entire transaction. All identities of the transactions in a block form the leaves of a Merkle tree. The root of the Merkle tree is in a field in the block header. So the transactions are bound together with a block header. Another field in the block header is the hash value of the previous block header. This field makes the blocks a chain. New blocks are generated under the proof of work consensus mechanism. Minters compute hash values of header candidates to generate and broadcast a new block. When most minters verify and accept the new block, it is added to the chain, and the transactions in the new block are confirmed once.

Basic Assumptions of Bitcoin System. There are some common assumptions about the Bitcoin system [3, 16].

1. Assume that the parties are connected by an insecure channel.
2. Assume that each party can access the current contents of the Bitcoin blockchain.
3. Assume that each party can post transactions on the Bitcoin blockchain within a maximum delay max_D.
4. The confirmed transactions on the blockchain are tamper resistant.

The assumptions 2 and 3 state that each party could connect to the Bitcoin blockchain. These assumptions are not true if the observations of Heilman et al. [13] and Gervais et al. [9] could not be fixed. Their attacks could lead to double spending, which breaks the baseline of any electronic cash system. We believe the Bitcoin system will take advices in [8,9,13,14] to defend against such attacks. So as long as the Bitcoin system is usable, the assumptions 2 and 3 hold.

Bitcoin Transactions. A Bitcoin transaction has some inputs and outputs. We follow the work in [3] to express a transaction in a box. In Fig. 1, we give two transactions T^A and T^B. The producer of the transactions is expressed as A and B. They are actually two public keys. If an input of T^B contains the $TxID$ of T^A and an output index of T^A, the input of T^B should be connected to the output of T^A. The connection is represented as a line with an arrow in Fig. 1. T^B is usually called as the redeeming transaction of T^A.

The input of T^B contains the input script (is) that matches the output script (os) of T^A. Roughly, the input script of T^B contains a signature Sig_B

and the public key B. The signature message of T^B contains the output script of T^A and all of the contents of T^B except the input script. This signature message is denoted by $[T^B]$. Sometimes, one need to express the content of a transaction excluding its input script, which is called as a simplified transaction [23]. It is not equivalent to the signature message of a transaction that includes some information of its connected transaction. The output script of T^A contains a hash value of the public key B and a signature verification command. To connect the T^B to T^A, minters match the hash value in the output script of T^A to the hash value of the public key in the input script of T^B, and verify the signature in the input script of T^B. The connection requirement of T^A is denoted by $os(body, \delta_B)$. That is, the output script requires a signature message and a signature to form verification conditions. Here the verification condition is mainly to verify a signature, which is denoted by $ver_B(body, \delta_B)$. The signature message $body$ here needs to be $[T^B]$. An output of a transaction also includes a value representing Bitcoins. The symbol C and δ_C in Fig. 1 denote the public key of any user C and their signature.

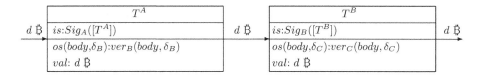

Fig. 1. Two connected transactions.

The Bitcoin system include many other aspects such as chain forking, mining, networking and so on. These contents are not closely related to the voting protocol and is omitted here.

2.2   BIP 65 Proposal

The $LOCKTIME$ field in a transaction prevents the transaction from being mined until a certain block height or time has been reached. It can be used to prove that it is possible to redeem a transaction output in the future. For example, if the T^B transaction in Fig. 1 has set the $LOCKTIME$ field, T^B transaction could redeem the output of T^A in the future. This feature is used both in the "Claim-or-Refund" [4] and the "Timed-Commitment" [3] patterns.

However, the $LOCKTIME$ field could not prove that it is impossible to spend a transaction output until sometime in the future. That is, an output in a transaction could not express a time constraint for any redeeming transactions. BIP 65 noticed this problem, and proposed an opcode $CHECKLOCKTIMEVERIFY$ for the Bitcoin scripting system that allows a transaction output to be made unspendable until some point in the future. This opcode compares its argument in a transaction against the $LOCKTIME$ field of another redeeming transaction. If the argument is greater than the

$LOCKTIME$ field, the opcode returns false and the verification fails. It indirectly verify that the desired block height or time has been reached.

Since the argument to $CHECKLOCKTIMEVERIFY$ opcode determines the possible time to redeem an output, we use a symbol $CH(t)$ to denote the opcode, where t is the argument and CH is the opcode. We rewrite the "Timed-Commitment" pattern using the $CH(\cdot)$ function in Fig. 2.

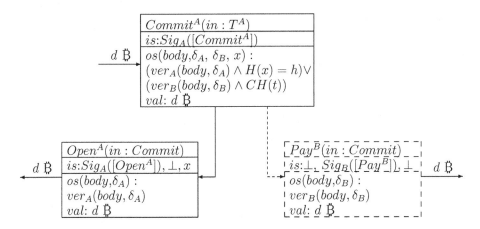

Fig. 2. Timed-Commitment scheme using the $CH(\cdot)$ function

In Fig. 2, A and B are public keys of two parties. The Bitcoin blockchain contains an unredeemed transaction T^A, which can be redeemed with key A, having value d ฿.

- In the commitment phase, the committer with key A computes $h = H(x)$. He sends to the Bitcoin blockchain the transaction $Commit^A$. This means that he reveals h as it is a part of the transaction. It also uses the $CH(t)$ function to specify a future point t after which the function returns true.
- In the open phase, the committer sends to the Bitcoin blockchain the transaction $Open^A$, which reveals the secret x. If within time t, the transaction $Open^A$ does not appear on the blockchain, then the party with key B sends the transaction Pay^B to the blockchain and earns d ฿.

Apparently, there is no need to create a Pay^B transaction by the committer. This reduces the communication cost and simplifies the overall logic of the timed commitment pattern.

3   Bitcoin Voting Protocol Review

Zhao and Chan [23] proposed the Bitcoin voting scenario. There are n voters, each of which wishes to fund exactly one of two candidates with 1 ฿. The winning candidate is selected by majority voting (assuming n is odd). The voting protocol should have the following properties:

– Privacy and Verifiability. Only the number of votes received by each candidate is known, while individual votes are kept private. However, each voter can still prove that he follows the protocol. For instance, no voter can vote for the same candidate twice.
– Irrevocability. Once the final outcome of the voting is revealed, the winner is guaranteed to receive the total fund n ฿. No voter can withdraw his funding even if the candidate he voted for does not win.

The interesting property is about ballot privacy. Zhao and Chan [23] proposed a solution where a ballot is in fact a hash value. Their solution includes two protocols. One is the vote commitment protocol. The voters in the protocol provably share some random numbers that sum to zero. Each voter provably constructs a ballot using their random number and commits to the ballot. The other is the vote casting protocol, where each voter opens their commitment fairly. Since the ballot looks like a random number, the Bitcoin nodes could only sum the ballots together to count the number of vote for a candidate. They use both the "Claim-or-Refund" and "Simultaneous-Timed-Commitment" patterns to satisfy the irrevocability property. In the short review, we only show their vote casting protocol using the "Simultaneous-Timed-Commitment" pattern since it is more efficient in communication than that using the "Claim-or-Refund" pattern.

3.1   Vote Commitment Protocol Review

There are n voters $\{P_1, \ldots, P_n\}$. Each voter has a secret vote $O_i \in \{0, 1\}$. Each voter has proving and verification keys of zkSNARKs. For each $i \in \{1, \ldots, n\}$, we use $VC_i(O_i)$ to denote the procedure of P_i making a committed ballot with the vote O_i. The $VC_i(O_i)$ procedure is as follows:

1. **Random Number Generation.** Generate n secret random numbers $r_{ij} \in \mathbb{Z}_N$, for $j \in \{1, \ldots, n\}$, such that they sum to 0, where N is the least power of 2 that is greater than the number n. For $j \in \{1, \ldots, n\}$, compute the commitment of the random number $(c_{ij}, k_{ij}) \leftarrow Commit(r_{ij})$ where k_{ij} is the opening key to the commitment c_{ij}. Generate zero-knowledge proofs that prove $\sum_j r_{ij} = 0$ for $j \in \{1, \ldots, n\}$ using the zkSNARK. Broadcast the commitments and zero-knowledge proofs to all voters.
2. **Random Number Proof Verification.** Receive commitments and verify the zero-knowledge proofs of all other parties.
3. **Open Key Distribution.** For all $j \in \{1, \ldots, n\} \setminus \{i\}$, send to P_j the opening key k_{ij}.
4. **Provable Ballot Generation.** For all $j \in \{1, \ldots, n\} \setminus \{i\}$, wait for the opening key k_{ji} from P_j, and check that $r_{ji} = Open(c_{ji}, k_{ji}) \neq \bot$. Compute $R_i \leftarrow \sum_j r_{ji}$ and $\hat{O}_i \leftarrow R_i + O_i$, and commit $(C_i, K_i) \leftarrow Commit(R_i)$ and

$(\hat{C}_i, \hat{K}_i) \leftarrow Commit(\hat{O}_i)$, where K_i, $\hat{K}_i$ are the opening keys. Generate the zero-knowledge proofs for the following constrains using the zkSNARK:
(a) $R_i = \sum_j r_{ji}$.
(b) The committed value in $\hat{C}_i$ minus that in C_i is either 0 or 1.

5. **Ballot Proof Distribution.** Broadcast the commitment C_i, $\hat{C}_i$ and the proof of the zkSNARK.
6. **Ballot Proof Verification.** Receive and verify all proofs from other parties.

Note that the open key distribution step needs a secure channel between each voters. For a voter P_i, the $VC_i(O_i)$ procedure needs $n - 1$ secure unicast and 2 times public broadcast. It also needs to run the zkSNARK 3 times to produce proofs and $3(n - 1)$ times to verify proofs.

3.2   Vote Casting Protocol Review

With all the committed ballots, a voter could produce or check the constraints in a joint transaction. After the joint transaction is confirmed on the Bitcoin blockchain, a voter opens its committed ballot to get its deposit back or stops to pay its deposit as penalty. Zhao and Chan [23] consider the transaction malleability problem, and take a threshold signature scheme to stop a voter claiming its deposit back prematurely. Note that a simplified transaction is a transaction excluding its input script [23]. We next describe their protocol on the viewpoint of P_i.

1. **Threshold Signature Key Generation.** P_i runs an $(n\text{-}n)$ threshold signature key generation algorithm with other voters to jointly generate a group address $\hat{pk}$ and to get its private key share $\hat{sk}_i$.
2. **Lock Transaction.** P_i creates a transaction $LOCK_i$. Its input is $(1 + d)$ ฿ owned by P_i, and its output is the address of the group public key $\hat{pk}$. The value d for deposit is suggested to be $2n$. The extra 1 ฿ is to fund the winner.
3. **Back Transaction.** P_i creates a simplified transaction $BACK_i$ that transfers the money from $LOCK_i$ back to a Bitcoin address pk_i of P_i. P_i broadcast the simplified $BACK_i$ to all other voters. On receiving $BACK_j$ for $j \in \{1, \ldots, n\} \setminus \{i\}$, P_i checks that the hash value referred to by its input is not $hash(LOCK_i)$. That is, the $BACK_j$ does't redeem the deposit of P_i. For each $j \in \{1, \ldots, n\}$, P_i runs the threshold signature algorithm to sign $BACK_j$ using his secret key share $\hat{sk}_i$.
4. **Joint Transaction.** P_i produces a simplified transaction $JOIN$ as follows:
   - The transaction has n inputs, each of which refers to a $LOCK_j$ transaction for $1 \leq j \leq n$ that contributes $(1 + d)$ ฿. Note that P_i knows all $LOCK_j$ identities from the Back transaction generation procedure.
   - The transaction has $n + 1$ outputs:
     • $out\text{-}deposit_j, j \in \{1, \ldots, n\}$: Each has value d ฿, and requires either (1) the opening key $\hat{K}_i$ and a signature verifiable with P_i's Bitcoin key pk_i, or (2) a valid signature verifiable with the threshold signature public key $\hat{pk}$.

- *out-prize* : The output has value n ฿, and requires all opening keys $\hat{K}_i$s and a signature from the winning candidate. The output script includes expressions to identify a winner. Readers could check that there is no way to spend the prize if some opening keys are unavailable.

P_i then runs the threshold signature algorithm with other voters to produce n signatures redeeming the n $LOCK_j$ transaction for $1 \leq j \leq n$.

5. **Pay Transaction.** P_i produces n time locked transaction PAY_j that may be used to redeem the *out-deposit$_i$* output for $1 \leq j \leq n$. The output script of PAY_j shares P_j's deposit between candidates. That is, P_j may be punished in the future point. P_i runs the threshold signature algorithm with other voters to sign each PAY_j transaction.

6. **Lock-Joint-Back Procedure.** The voter P_i submits $LOCK_i$ to the bitcoin blockchain. After all $LOCK_j$ for $1 \leq j \leq n$ have appeared on the Bitcoin blockchain, the $JOIN$ transaction is submitted to the blockchain. If no any $JOIN$ transaction appears on the blockchain after the time locked in the $BACK$ transaction has passed, P_i terminates the whole protocol by submitting $BACK_i$ to get back the $(1 + d)$ ฿ in $LOCK_i$.

7. **Claim Transaction.** P_i creates $CLAIM_i$ transaction that provides the opening key $\hat{K}_i$ to reveal the committed ballot $\hat{O}_i$, and to get its deposit back.

8. **Winner Transaction.** The winning candidate produces a $Winner$ transaction to redeem n ฿ from the *out-prize* output.

9. **Out Revealing Procedure.** After a $JOIN$ transaction appears on the Bitcoin blockchain, each voter submits their $CLAIM$ transaction. If all voters have submitted their $CLAIM$ transactions, a winning candidate is determined who can submits the $Winner$ transaction. Otherwise, the PAY transactions are used to share some voters' deposits.

Except the threshold signature key generation algorithm, the vote casting protocol includes six transactions and two procedures. A voter P_i needs to create the $BACK$, $Joint$ and PAY transactions with other voters to create $2n + 1$ signatures. The protocol needs $n(2n + 1)$ threshold signatures to be produced, which means large communication and computation costs among voters. The Bitcoin rounds of this protocol are 4 rounds. The first round is for the $LOCK$ transactions. The second round is for the $JOIN$ transactions. The third round is for the $BACK$ or $CLAIM$ transactions. The last round is for the $Winner$ transactions or the PAY transactions.

4   General Bitcoin Voting Protocol

Zhao and Chan's Bitcoin protocol is obviously a typical "YES/NO" election. It is a special case of approval voting. In approval voting, a voter is allowed to select at least k_{min}, at most k_{max} from L candidates. A valid single vote satisfies that

$$\{v_1, \ldots, v_L : v_i \in \{0, 1\}, \sum_{i=1}^{L} v_i \in [k_{min, k_{max}}]\}. \tag{1}$$

We use the $VC_i(O_i)$ procedure as a subroutine, design a flexible vote commitment protocol and redesign the vote casting protocol.

The "Timed-Commitment" pattern is enough to defend against the transaction malleability attack since there is no need to sign any transaction that is not on the Bitcoin blockchain. With the BIP 65 proposal, it is more efficient to express this logic. When we use joint transaction with the BIP 65 proposal, the joint transaction is indeed a timed-constrain enabled contract that is signed by all parties. Anyone could be the initiator to create the body of a joint transaction, signs it, and sends it to its next player. The next player could verify the body of the transaction. If it agrees all the constrains in the transaction, the next player signs and sends a joint transaction with two signatures to its next one. When the final player receives the transaction and signs it, the transaction could be submitted to the Bitcoin blockchain. If the joint transaction is confirmed, it redeemed all players' unspent Bitcoins. If the joint transaction does not appear on the Bitcoin blockchain, no one will lose its coins. If one of the unspent output referred by an input in the joint transaction is redeemed, the joint transaction as a whole is invalid, and will not appear on the blockchain. We here use this approach to design a simpler bitcoin casting protocol.

4.1 Flexible Vote Commitment Protocol

There are L candidates listed in an order denoted by $\{1, \ldots, L\}$. There are n voters $\{P_1, \ldots, P_n\}$. A voter P_i for a candidate c has a secret vote $v_i^c \in \{0, 1\}$. Each voter has proving and verification keys for zkSNARK. For each $i \in \{1, \ldots, n\}$, the procedure for P_i is as follows.

1. Call $VC_i(v_c^i)$ for $1 \leq c \leq L$ to produce L committed ballots $\hat{C}_i^c$ and C_i^c. Note that $(C_i^c, K_i^c) \leftarrow Commit(R_i^c)$, $(\hat{C}_i^c, \hat{K}_i^c) \leftarrow Commit(\hat{O}_i^c)$, $\hat{O}_i^c = R_i^c + v_i^c$ and R_i^c is collaboratively produced by all voters.
2. Generate and broadcast publicly the zero-knowledge proofs for the following using the zkSNARK:
   (a) The sum of committed values in $\hat{C}_i^c$ for $c \in \{1, \ldots, L\}$ minus that in C_i^c is in $[k_{min}, k_{max}]$.
3. Receive and verify all proofs from other parties.

That is, the protocol uses Zhao and Chan's vote commitment protocol to vote "YES/NO" for a candidate. There are L candidates, and there are L calls. In each call, the $V_C(v_c^i)$ procedure has proven $v_c^i \in \{0, 1\}$. The additional proof shows that the voter has formed a valid single vote satisfying Eq. (1). If $L = 1, k_{min} = 0, k_{max} = 1$, it is a "YES/NO" approval for a candidate. If $L = 2, k_{min} = k_{max} = 1$, it has the same functionality as Zhao and Chan's protocol, while the general method is inefficient as expected. It supports the cases of $L > 2$ with linearly increased complexity.

4.2 Simpler Vote Casting Protocol

With all the committed vote, a voter could produce or check the constraints in a joint transaction. From the viewpoint of P_i, the vote casting protocol is executed as follows.

1. **Transaction Exchange.** P_i broadcast a transaction T^{P_i} having value $d+1$ ฿. P_i receives all voters' transactions and check that each transaction has an unspent output with value at least $d+1$ ฿.
2. **Joint Commitment Transaction.** P_i generates a simplified transaction $JOINT\_COM$ as follows:
   - It has n inputs, each of which refers to a transaction with $(1+d)$ ฿ owned by a voter.
   - It has $n+1$ outputs.
     - $out\text{-}deposit_i$, $i \in \{1,\ldots,n\}$: Each has value d ฿, and requires either (1) the opening keys $\{\hat{K}_i^c : c \in \{1,\ldots,L\}\}$ and a signature δ_{P_i} verifiable with P_i's public key pk_i or (2) a $CH(t)$ function returns $true$ and L valid signatures $\{\delta_{C_1},\ldots,\delta_{C_L}\}$ from the candidates with Bitcoin public keys $\{C_1,\ldots,C_L\}$.
     - $out\text{-}prize$: It has value n ฿, and requires all opening keys $\{\hat{K}_i^c : c \in \{1,\ldots,L\}, i \in \{1,\ldots,n\}\}$ and signatures from winning candidates or all voters' signatures. If possible, the winning candidates are calculated as a winner set $WinSet$. A candidate C_c is put into the set if the sum of ballots about it $\sum_i Open(\hat{C}_i^c, \hat{K}_i^c)$ is equivalent to the max value of all candidates' records. That is,

       $$WinSet = \{c|\sum_i Open(\hat{C}_i^c, \hat{K}_i^c) = max(\sum_i Open(\hat{C}_i^v, \hat{K}_i^v)|v = \{1,\ldots,L\})\}.$$

       For all winning candidates in the winner set, their signatures should be taken as input and they share all the funds. However, if there are no enough opening keys to select winners, the funds are returned back to each voter when the $CH(t + max_D)$ function returns $true$.
3. **Joint Commitment Procedure.** P_i receives a partial $JOINT\_COM$ transaction signed by $P_1,\ldots,P_{i-1}$ from P_{i-1}. If $i = 1$, it receives nothing and creates the simplified $JOINT\_COM$ transaction. P_i then checks the previous signatures, $TxIDs$, and constraints in the $JOINT\_COM$ transaction. If everything is fine, P_i signs $JOINT\_COM$ transaction, and sends it to P_{i+1}. If $i = n$, it submits $JOINT\_COM$ transaction to to the Bitcoin blockchain.
4. **Claim Transaction.** P_i generates a $CLAIM^{P_i}$ transaction to get its deposit back, which provides the opening keys $\{\hat{K}_i^c : c \in \{1,\ldots,L\}\}$ to reveal his committed vote.
5. **Winners Transaction.** When the voting result is available, the $WinSet$ is determined. The winners in the $WinSet$ jointly create a WIN transaction to share the $out\text{-}prize$ output. The input script of the WIN transaction includes all opening keys of all ballots and the signatures of all expected winners. The output of the WIN transaction includes the address and Bitcoin values for each winner, which requires each winner to check.

6. **Joint Claim Transaction.** L candidates jointly create a $JOINT\_CLAIM$ transaction to share the penalty of a voter. The transaction has one input referred to an output of the $JOINT\_COM$ transaction. It has L outputs. Each output has $\lfloor d/L \rfloor$ ฿ and requires a valid signature of a candidate. Note that when we compute the $\lfloor \cdot \rfloor$ function, it is natural to use the smallest Bitcoin unit "satoshi".

7. **Failing Return Transaction.** After $JOINT\_COM$ transaction is confirmed, some voters may refuse to open his ballots. While these voters will lose their deposits, they really have a chance to get their funds back by this transaction. The transaction has one input referred to the $n + 1^{th}$ output of the $JOINT\_COM$ transaction. It has n outputs. Each output has 1 ฿ and requires all signatures of voters.

8. **Claim-Win and Joint Claim-Failing Return Procedure.** After the $JOINT\_COM$ transaction is confirmed, P_i submits its $CLAIM^{P_i}$ transaction. If all voters submit their claim transactions, the candidates calculate the winners locally. The winners create the WIN transaction and submits it to the Bitcoin blockchain. If there are some voters stops without submitting their claim transactions, their deposits will be shared among all candidates by their $JOINT\_CLAIM$ transactions. After candidates get their winning prize or collect deposits of dishonest voters in a max_D time period, the funds may be returned to all voters by their $FAIL\_RETURN$ transaction.

The time line of the protocol is as follows:

– Form the viewpoint of a voter P_i, we suppose that at time t_i^0, P_i receives a partial $JOINT\_COM$ transaction. It should check the parameter t in the $CH(\cdot)$ function to make sure that there is enough time for other parties to jointly create the transaction and to claim their deposit back. Let Δ be a voter's time to deal with the partial transaction. P_i should check that $t - t_i^0 \geq (n + 1 - i)\Delta + 2max_D$. At time $t_i^1 \geq t_i^0 + (n + 1 - i)\Delta + max_D$, if P_i does not see the $JOINT\_COM$ transaction on the Bitcoin blockchain, it redeems its transaction T^{P_i} to firmly stop the protocol. Otherwise, P_i submits the $CLAIM^{P_i}$ transaction to get back its deposit. At time $t_i^2 \geq t + max_D$, if some voters does not claim their deposits, P_i will receive a partial $FAIL\_RETURN$ transaction. P_i should check that their address is in the output script, signs it, and gives it to another voter.

– From the viewpoint of a candidate C_w, after the $JOINT\_COM$ transaction is fully confirmed on the Bitcoin blockchain, the candidate will notice the time parameter t in the transaction. At time $t_w^0 > t$, the candidate checks whether all voters have revealed their votes. If there are some voters that do not reveal their votes, all candidates have a chance to jointly collect these voters' deposits one by one by $JOINT\_CLAIM$ transaction. Otherwise, the candidates checks that whether it is a winner. If the candidate is not a winner, nothing is left to do. If it is in the winner set, the candidate has a chance to jointly create the WIN transaction to share the prize.

Except the transaction exchange procedure, the simpler vote casting protocol includes five transactions and two procedures. No threshold signatures are required. We show the main steps of the protocol in Fig. 3. This protocol needs at most 4 Bitcoin rounds. The first round is for the $JOINT\_COM$ transaction. The second round is for the $CLAIM$ transaction. The third round is for the WIN or the $JOINT\_CLAIM$ transaction. A possible final round is for the $FAIL\_RETURN$ transaction.

4.3 Protocol Analysis

We analyze the expected properties of a Bitcoin voting protocol and consider some potential problems.

Privacy. In the vote casting protocol, the revealed vote of P_i is $O_i^c \leftarrow R_i^c + v_i^c$ for $c \in \{1, \dots, L\}$, where v_c^i is the real vote for the candidate C_c. Since R_i^c is a

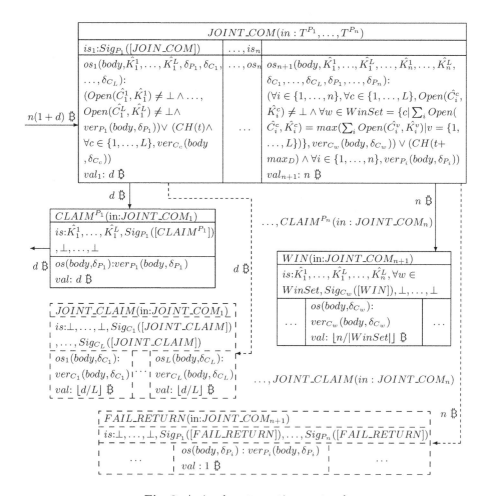

Fig. 3. A simple vote casting protocol

distributed generated random value for this vote and the random value is used only once, no one could guess the v_i^c with a non-negligible advantage. Thus, individual votes are kept private.

Verifiability. In the vote casting protocol, each voter has only one chance to cast its vote on the Bitcoin blockchain. If the vote is constructed correctly, a voter does not follow the protocol will only pay their deposit. In the vote commitment protocol, there are $n(2L + 1)$ proofs of zkSNARK to ensure that each voter behaves correctly. Basically, the proofs show that random numbers are generated correctly, random numbers are summed correctly, the ballot for a candidate are formed correctly, and the vote for all candidates are formed correctly.

Irrevocability. When the final ballot of the voting is revealed, the winner set $WinSet$ is determined. The only unspent output in the $JOINT\_COM$ transaction is the $n + 1^{th}$ output, which could be spent by all winners in the $WinSet$ or all voters. Note that when the $JOINT\_COM$ transaction is confirmed, all voters' deposits and funds have been redeemed. As long as the Bitcoin blockchain could stop double spending, no voter can withdraw his funding even if the candidates he voted for do not win expect that all voters agree to get their funds back. Even when all voters want their funds back, they have to wait a max_D time period for winners to jointly claim their prize at first.

Secure Channel Requirement. As we have pointed out, the vote commitment protocol needs secure unicast channels between any two voters. The secure channels may be established roughly as follows using Bitcoin transactions. Suppose the step "Transaction Exchange" in the vote cast protocol is executed before the vote commitment protocol. And we require that the exchanged transaction of P_i should be a transaction signed by P_i. It is natural for P_i to create a transaction with the required $d + 1$ ₿ as a new output. Next, P_i could produce a dynamic shared secrete key with P_j by the Diffie-Hellman protocol and the current block header hash value. That is, P_i and P_j use the public keys in their exchanged transactions to establish a fixed secret value, and update the secrete by the current block header hash value. As the block chain grows, the secret is continuously updated. This secret could be used as a symmetric key to establish a secure channel between two voters.

Uncooperative Voters. Our proposal needs voters to cooperatively produce $JOINT\_COM$ transaction. If some of them do not, other voters could quit without financial damage. If some voters refuse to open their ballots, the voters' deposit will be shared by candidates. Now all voters could produce the $FAIL\_RETURN$ transaction to get their funds back. However, there is no method to get their funds back if some of the voters is uncooperative to produce the transaction. To solve the problem, a trivial method is to split the $n + 1^{th}$

output of the $JOINT\_COM$ transaction into n outputs, each output has 1 ฿. Although this method increases the transaction numbers of winners, it is more fair for honest voters.

Comparison. We give a Table 1 to summarize the main differences of our protocol and that of Zhao and Chan [23]. Our protocol is general since it supports $L \geq 1$ candidates. And our protocol is simpler since we remove the threshold signatures and reduce transaction numbers. The transaction numbers on the Blockchain may vary according to different events. At least, in [23], there should be Lock, Joint, Claim, Winner transactions. The number of transactions on the Blockchain is $2n + 2$. Comparatively, our protocol needs Joint Commitment, Claim, Winners transaction. The number of transactions on the Blockchain is $n + 2$.

Table 1. Comparisons between our protocol and the protocol in [23]

	Transaction numbers	Threshold signatures	Candidate numbers
[23]	$2n + 2$	$n(2n + 1)$	2
Ours	$n + 2$	0	$L \geq 1$

5 Conclusion

In this paper, we present a simple Bitcoin voting protocol. It relies on zkSNARK proofs off the blockchain to produce verified ballots for L candidates. The simultaneous timed commitment pattern is then used for voters to confirm their ballots and to make sure that the ballots should be open. Comparing with the recent Ethereum voting smart contract [18], our protocol shows a method to design a lightweight blockchain code to fulfil the Bitcoin voting task. And obviously, if the fund is set to be zero ฿, our protocol serves as a pure decentralized voting protocol. Comparing with Zhao and Chan's work [23], our protocol is more fair, simple and efficient.

Acknowledgment. This work is supported by the National Key R&D Program of China (2017YFB0802503), Natural Science Foundation of China (61672550, 61379154), Natural Science Foundation of Guangdong Province, China (2015A030313133), and Fundamental Research Funds for the Central Universities (No. 17lgjc45).

References

1. Adida, B.: Helios: web-based open-audit voting. In: Proceedings of the 17th Conference on Security Symposium, pp. 335–348 (2008)
2. Andrychowicz, M., Dziembowski, S., Malinowski, D., Mazurek, Ł.: Fair two-party computations via bitcoin deposits. In: Böhme, R., Brenner, M., Moore, T., Smith, M. (eds.) FC 2014. LNCS, vol. 8438, pp. 105–121. Springer, Heidelberg (2014). https://doi.org/10.1007/978-3-662-44774-1_8
3. Andrychowicz, M., Dziembowski, S., Malinowski, D., Mazurek, L.: Secure multiparty computations on bitcoin. In: 2014 IEEE Symposium on Security and Privacy, pp. 443–458, May 2014
4. Barber, S., Boyen, X., Shi, E., Uzun, E.: Bitter to better — how to make bitcoin a better currency. In: Keromytis, A.D. (ed.) FC 2012. LNCS, vol. 7397, pp. 399–414. Springer, Heidelberg (2012). https://doi.org/10.1007/978-3-642-32946-3_29
5. Bentov, I., Kumaresan, R.: How to use bitcoin to design fair protocols. In: Garay, J.A., Gennaro, R. (eds.) CRYPTO 2014. LNCS, vol. 8617, pp. 421–439. Springer, Heidelberg (2014). https://doi.org/10.1007/978-3-662-44381-1_24
6. Decker, C., Wattenhofer, R.: Bitcoin transaction malleability and MtGox. In: Kutyłowski, M., Vaidya, J. (eds.) ESORICS 2014. LNCS, vol. 8713, pp. 313–326. Springer, Cham (2014). https://doi.org/10.1007/978-3-319-11212-1_18
7. EtherCasts: a curated collection of decentralized apps (2017). http://dapps.ethercasts.com. Accessed 4 Apr 2017
8. Germanus, D., Ismail, H., Suri, N.:. Pass: an address space slicing framework for p2p eclipse attack mitigation. In: 2015 IEEE 34th Symposium on Reliable Distributed Systems (SRDS), pp. 74–83, September 2015
9. Gervais, A., Ritzdorf, H., Karame, G.O., Capkun, S.: Tampering with the delivery of blocks and transactions in bitcoin. In: Proceedings of the 2015 ACM SIGSAC Conference on Computer and Communications Security (CCS 2015), pp. 692–705. ACM (2015)
10. Ghassan, O.K., Elli, A., Srdjan, C.: Double-spending fast payments in bitcoin. In: Proceedings of the 2012 ACM Conference on Computer and communications security, pp. 906–917. ACM (2012)
11. Groth, J.: Efficient maximal privacy in boardroom voting and anonymous broadcast. In: Juels, A. (ed.) FC 2004. LNCS, vol. 3110, pp. 90–104. Springer, Heidelberg (2004). https://doi.org/10.1007/978-3-540-27809-2_10
12. Hao, F., Ryan, P.Y.A., Zielinski, P.: Anonymous voting by two-round public discussion. IET Inf. Secur. 4(2), 62–67 (2010)
13. Heilman, E., Kendler, A., Zohar, A., Goldberg, S.: Eclipse attacks on bitcoin's peer-to-peer network. In: Proceedings of the 24th USENIX Conference on Security Symposium (SEC 2015), pp. 129–144. USENIX Association Berkeley, CA, USA (2015)
14. Ismail, H., Germanus, D., Suri, N.: Detecting and mitigating p2p eclipse attacks. In: 2015 IEEE 21st International Conference on Parallel and Distributed Systems (ICPADS), pp. 224–231, December 2015
15. Kiayias, A., Yung, M.: Self-tallying elections and perfect ballot secrecy. In: Naccache, D., Paillier, P. (eds.) PKC 2002. LNCS, vol. 2274, pp. 141–158. Springer, Heidelberg (2002). https://doi.org/10.1007/3-540-45664-3_10
16. Kiayias, A., Zhou, H.-S., Zikas, V.: Fair and robust multi-party computation using a global transaction ledger. In: Fischlin, M., Coron, J.-S. (eds.) EUROCRYPT 2016. LNCS, vol. 9666, pp. 705–734. Springer, Heidelberg (2016). https://doi.org/10.1007/978-3-662-49896-5_25

17. Nakamoto, S.: Bitcoin: a peer-to-peer electronic cash system (2008). https://bitcoin.org/bitcoin.pdf. Accessed 4 Apr 2017
18. McCorry, S.S.P., Hao, F.: A smart contract for boardroom voting with maximum voter privacy. In: Financial Cryptography and Data Security 2017, pp. 1–18 (2017)
19. Tarasov, H.T.P.: Internet voting using zcash (2017). Accessed 23 June 2017
20. Peter, T.: Op_checklocktimeverify (2014). https://github.com/bitcoin/bips/blob/master/bip-0065.mediawiki. Accessed 4 Apr 2017
21. Sasson, E.B., Chiesa, A., Garman, C., Green, M., Miers, I., Tromer, E., Virza, M.: Zerocash: decentralized anonymous payments from bitcoin. In: 2014 IEEE Symposium on Security and Privacy, pp. 459–474, May 2014
22. Wood, D.G.: Ethereum: a secure decentralised g generalised transaction ledger homestead (2014). http://gavwood.com/paper.pdf. Accessed 4 Apr 2017
23. Zhao, Z., Chan, T.-H.H.: How to vote privately using bitcoin. In: Qing, S., Okamoto, E., Kim, K., Liu, D. (eds.) ICICS 2015. LNCS, vol. 9543, pp. 82–96. Springer, Cham (2016). https://doi.org/10.1007/978-3-319-29814-6_8

Post-Quantum Secure Remote Password Protocol from RLWE Problem

Xinwei Gao[1], Jintai Ding[2(✉)], Jiqiang Liu[1(✉)], and Lin Li[1]

[1] Beijing Key Laboratory of Security and Privacy in Intelligent Transportation,
Beijing Jiaotong University, Beijing 100044, People's Republic of China
{xinweigao,jqliu,lilin}@bjtu.edu.cn
[2] Department of Mathematical Sciences, University of Cincinnati,
Cincinnati 45219, USA
jintai.ding@gmail.com

Abstract. Secure Remote Password (SRP) protocol is an augmented Password-based Authenticated Key Exchange (PAKE) protocol based on discrete logarithm problem (DLP) with various attractive security features. Compared with basic PAKE protocols, SRP does not require server to store user's password and user does not send password to server to authenticate. These features are desirable for secure client-server applications. SRP has gained extensive real-world deployment, including Apple iCloud, 1Password etc. However, with the advent of quantum computer and Shor's algorithm, classic DLP-based public key cryptography algorithms are no longer secure, including SRP. Motivated by importance of SRP and threat from quantum attacks, we propose a RLWE-based SRP protocol (RLWE-SRP) which inherit advantages from SRP and elegant design from RLWE key exchange. We also present parameter choice and efficient portable C++ implementation of RLWE-SRP. Implementation of our 209-bit secure RLWE-SRP is more than 3x faster than 112-bit secure original SRP protocol, 5.5x faster than 80-bit secure J-PAKE and 14x faster than two 184-bit secure RLWE-based PAKE protocols with more desired properties.

Keywords: Post-quantum · RLWE · SRP · PAKE · Protocol
Implementation

1 Introduction

1.1 Key Exchange

Key exchange (KE) is an important and fundamental cryptographic primitive. It allows two or multiple parties to agree on same session key, which is later utilized in encryption and other cryptographic primitives. With the ground-breaking Diffie-Hellman key exchange proposed in 1976 [14], public key cryptography came into reality and it has been widely deployed in real world applications. Since public key computations are rather expensive compared with symmetric-based ones, symmetric encryption is adopted to encrypt actual communication

© Springer International Publishing AG, part of Springer Nature 2018
X. Chen et al. (Eds.): Inscrypt 2017, LNCS 10726, pp. 99–116, 2018.
https://doi.org/10.1007/978-3-319-75160-3_8

data instead of public key encryption. The shared key generated during key exchange is extremely important, especially in constructing real-world security protocols and applications. Important applications of key exchange include Transport Layer Security (TLS), Secure Shell (SSH), Internet Key Exchange (IKE), Internet Protocol Security (IPsec), Virtual Private Network (VPN) etc.

However, Diffie-Hellman and other unauthenticate key exchange protocols are vulnerable to Man-In-The-Middle (MITM) attack, where an adversary in the middle between communicating parties can intercept and tamper messages and pretend himself as legit counterpart. An important line of key exchange protocols that can defeat such attack is authenticated key exchange (AKE). In AKE, authentication mechanisms can ensure one or both sides of key exchange are securely authenticated. HMQV [22] is an example of AKE. There are various approaches to achieve authentication, including public key infrastructure (PKI)-based (using signatures and verified public key with valid certificates), password (and its variants)-based AKE protocol (PAKE) etc. PAKE is an important approach to realize AKE. Examples of PAKE protocols are PAK & PPK [10], J-PAKE [20], EKE [7], SPAKE2 [2] etc. Some additional works include [17,19,24] etc.

In most network security protocols, PKI-based authentication (certificate and signature) is more popular mostly because it is "secure enough". In most cases, server side can be securely authenticated using certificate but client side is not since generally client does not have a valid certificate. This highlights one advantage of PAKE protocols - simpler mutual authentication. In PAKE, mutual authentication can be securely achieved using pre-shared value or tokens for both parties (in most cases, such pre-shared value is password or hash of password). This also saves some rather expensive computations related to public key operations (e.g., compute/verify signature). A shortcoming for basic PAKE protocols is that these constructions directly using password or hash of password as pre-shared value. We can foresee that once server is compromised and actual user password (or its hash value) is leaked, adversary can easily impersonate as the client to authenticate himself. This stresses an crucial problem and challenge for basic PAKE protocols.

A solution to this issue is augmented PAKE, since it only requires the server to store a pre-shared verifier (or a token) which is generated usingsword and other elements, instead of simply storing actual password or hash of password. In execution of augmented PAKE protocol, client needs to enter correct password in order to compute intermediate values correctly and authenticate himself. Server uses the stored verifier to authenticate user. Meanwhile, actual password is not sent to server. The trick is that these intermediate values can be only computed with correct password. Adversary cannot compute such intermediate values thanks to delicate design and hard problems like discrete logarithm problem etc. The advantage of augmented PAKE protocols is that even if attacker owns the verifier, he cannot impersonate as client by sending the verifier he captured since he does not know actual password. Examples of augmented PAKE protocols are SRP [31], PAK-Z [10] etc.

1.2 Post-Quantum World

Diffie-Hellman key exchange protocol and various current public key cryptosystems are constructed based on discrete logarithm problem (DLP), integer factorization problem (IFP), elliptic curve variant of DLP (ECDLP) etc. Currently, there are no public known algorithms on classic computers that can solve these hard problems with large parameters. However, Shor introduced a quantum algorithm in 1994 [29], which suggested that DLP, IFP and ECDLP can be solved efficiently on a sufficient large quantum computer. This is horrible for current public key cryptography since most current widely deployed public key algorithms are constructed based on these hard problems. Once sufficient large quantum computer is built, most current public key algorithms can be broken. We also know that the development of quantum computers within past decades is incredibly fast. In 2015, National Security Agency (NSA) announced their plan of switching to post-quantum cryptography in near future. At PQCrypto 2016, National Institute of Standards and Technology (NIST) announced call for post-quantum cryptography standards. All these facts stress the importance of post-quantum cryptography and severity of deploying post-quantum cryptographic constructions in real world applications.

There are several approaches to build post-quantum cryptography primitives, including lattice-based, multivariate-based, hash-based, code-based, supersingular elliptic curve-based etc. Lattice-based ones are very popular due to strong security and high efficiency. Among all lattice-based constructions, Learning With Errors (LWE) and Ring-LWE based ones are more practical and outstanding due to much better efficiency, robust security and versatility. There are several works have demonstrated real-world efficiency of post-quantum cryptographic primitives, including experimenting "NewHope" RLWE key exchange [5] in Chrome canary build [11], deploying BLISS lattice-based signature in strongSwan VPN [1] etc.

1.3 Related Works

Secure Remote Password (SRP) Protocol and Important Real-World Applications. Thomas Wu proposed the Secure Remote Password (SRP) protocol in 1998 [31]. SRP protocol is an augmented PAKE protocol designed based on DLP. Compared with basic PAKE protocols, advantages of SRP are: (1) Server only stores a securely pre-shared verifier. Neither user password nor hash of password are stored for both client and server; (2) SRP can stop adversary from impersonating as client even if server is compromised; (3) No one (adversary, malicious server etc.) can recover user's password from verifier; (4) SRP does not require user sending actual password or its variants to servers to authenticate himself. These are major advantages of SRP compared with other PAKE protocols. SRP is also a key exchange protocol which provides mutual authentication and forward secrecy. SRP is standardized in RFC 2945, ISO 11770-4 and IEEE P1363.

SRP has been widely deployed in industry and critical real-world applications. Here we list a few of them:

1. Apple iCloud. SRP is adopted in iCloud to authenticate user when attempting to access iCloud account using iCloud security code (iCSC), where iCSC is set and only known by user (i.e., so-called "password" in SRP). According to Apple security white paper [6], SRP protocol keeps Apple servers away from acquiring information about user's actual iCSC. Moreover, for HomeKit accessories, which are developed for smart home automation are also using 3072-bit SRP to authenticate between iOS device and HomeKit accessory. SRP stops Apple from knowing the setup code printed on HomeKit accessories.
2. 1Password. 1Password is the leading password manager. It adopts SRP to handle user's master password and login attempts. In fact, 1Password security team claims that user's security is not affected in recent "Cloudbleed" bug thanks to the adoption of SRP [18].
3. ProtonMail. Highly-secure email service ProtonMail adopts SRP for secure login since 2017, protecting the security of user's account even if server is compromised.
4. Blizzard. The leading gaming company also uses SRP to authenticate user login. In 2012, their servers were compromised in a security breach, but SRP protects safety of user's password [25].
5. TLS. There are SRP ciphersuites for TLS protocol (e.g., [27]). OpenSSL has supported TLS-SRP ciphersuites.

To the best of our knowledge, there are no practical attacks against SRP while formal security proof is not presented in original SRP paper [31]. They claim that SRP is secure against several attacks with security analysis. Moreover, security of SRP protocol is heavily relied on hardness of discrete logarithm problem, which can be broken by quantum computers in near future. Importance of SRP and threats from quantum computers directly motivate this work.

Post-Quantum Key Exchange Protocols from RLWE. Jintai Ding et al. proposed the first LWE and RLWE based key exchange protocols which are analogues of Diffie-Hellman key exchange in 2012 (denoted as DING12) [16]. DING12 proposed a "robust extractor" (i.e., error reconciliation mechanism), which allows two parties to agree on same key over approximately equal values with overwhelming probability. This work is known as the foundation of error reconciliation-based LWE & RLWE key exchange protocols. There are various variants (e.g., [5,8,9,26]) that share similar protocol structure and error reconciliation techniques as DING12. It is proven secure under passive probabilistic polynomial-time (PPT) attack and enjoys very high efficiency.

Jiang Zhang et al. proposed the first RLWE-based AKE protocol which is a RLWE analogue of HMQV in 2015 [32], where core idea of DING12 and HMQV are well inherited and applied. This protocol is proven secure under Bellare-Rogaway model with enhancements to capture weak perfect forward secrecy. They also present parameter choices and proof-of-concept implementation.

Jintai Ding et al. proposed the first RLWE-based PAKE protocol in 2017 (denoted as PAKE17) [15]. This work also follows idea in DING12 and gives RLWE analogues of classic PAK and PPK protocols. To the best of our knowledge, this is the only work up to now that gives RLWE-based PAKE constructions and proof-of-concept implementation. RLWE-PAK and RLWE-PPK are proven secure under extended security model of PAK and PPK (random oracle model, ROM). Proof-of-concept implementation shows that PAKE17 is efficient. Another lattice-based PAKE construction is given in [21]. It is constructed based on common reference ring (CRS), therefore it is more complex and less efficient.

1.4   Our Contributions

Directly motivated by threats from quantum computers and widely deployed SRP protocol, we first propose a RLWE-based secure remote password protocol which enjoys elegant design and high efficiency (denoted as RLWE-SRP). Our RLWE-SRP protocol can be regarded as RLWE variant of original SRP protocol. Same as SRP, in our RLWE-SRP, user only need to remember his password and server only stores a verifier. The verifier is not actual password and no information about password is revealed. During key exchange, no actual password is transmitted. Even if server is compromised and verifier is captured, adversary cannot impersonate as user to authenticate with server, nor recover actual password. RLWE-SRP enjoys several desired features, including: mutual authentication, high efficiency and resistant to quantum attacks. Leakage of session key of previous sessions will not help adversary to identify user's password and verifier. We also present security analysis following same approach as original SRP paper.

Second, we present practical parameter choice and efficient portable C++ implementation of RLWE-SRP. With current state-of-the-art cryptanalysis tool for RLWE-based cryptosystems, our parameter choice offers 209-bit security. Benchmark shows that our construction and implementation are truly practical. Compared with implementation of original SRP protocol (112-bit security), RLWE-SRP is 3x faster with higher security level. Compared with other PAKE protocols including 80-bit secure J-PAKE [20], 184-bit secure RLWE-PAKE and RLWE-PPK from PAKE17 [15], our implementation is 5.53x, 13.96x and 13.96x faster respectively.

2   Preliminaries

2.1   Ring Learning with Errors

Oded Regev presented LWE problem in 2005 [28] and was extended by Lyubashevsky et al. to Ring-LWE in 2010 [23]. RLWE problem is a direct variant of LWE problem in ring setting. LWE and RLWE have become most popular primitives to build post-quantum cryptography. Hardness of LWE and RLWE problem is directly reduced to NP-hard lattice problems, including Shortest Vector

Problem (SVP) etc. in regular lattice and ideal lattice respectively. Compared with LWE, one significant advantage of RLWE is much reduced key size and communication overhead due to the fact that large matrix is replaced by degree n polynomial in R_q in RLWE. This reduces at least quadratic computation and communication overhead. Currently, there are no algorithms can solve LWE, RLWE and underlying lattice problems with properly chosen parameters on both classical and quantum computers.

Let $n \in Z$ be power of 2, $q \in Z$, ring $R_q = Z_q[x]/(x^n + 1)$ where $Z_q = Z/qZ$, $D_{Z^n,\sigma}$ be discrete Gaussian distribution on Z^n with standard deviation σ. For uniform randomly generated public parameter $a \in R_q$, small and fixed term $s \leftarrow D_{Z^n,\sigma} \in R_q$, $e \leftarrow D_{Z^n,\sigma} \in R_q$, let $A_{s,D_{Z^n,\sigma}}$ be distribution over pairs $(a, b = as + e) \in R_q \times R_q$. RLWE assumption implies that for fixed s, distribution $A_{s,D_{Z^n,\sigma}}$ is indistinguishable from uniform distribution on R_q^2 given polynomial many samples. Search version of RLWE problem is to recover s given RLWE samples and decision version is to distinguish RLWE samples from uniform random ones. There are security reductions from hard problems in ideal lattice to RLWE. If one can solve RLWE problem, then he can solve underlying hard lattice problems as well.

RLWE has been applied to construct various cryptography primitives, including public key encryption, digital signatures, key exchange, attribute-based encryption, homomorphic encryption etc.

2.2 Revisit DING12 RLWE Key Exchange Protocol

In 2012, Jintai Ding et al. proposed LWE and RLWE key exchange protocols [16]. This is the first work that gives simple and provable secure LWE and RLWE key exchange. DING12 takes advantage of commutative property and constructs key exchange protocols that are analogues of Diffie-Hellman key exchange. One technical challenge for building key exchange protocol over LWE & RLWE is how to reconcile errors, since public key and key exchange materials k_i, k_j are perturbed by small error terms, therefore values for both parties are approximately equal, not rigorously equal $(g^a)^b \bmod q = (g^b)^a \bmod q$ in Diffie-Hellman key exchange. To solve this problem, DING12 gives a new error reconciliation mechanism called "robust extractor" (i.e., error reconciliation mechanism). In order to reconcile errors, it is required that the difference between key exchange materials of two parties is approximately equal. One side needs to send a carefully computed "signal" value to the other side. Signal value implies which "region" does coefficient of polynomial belongs to and this will leads to correct error reconciliation and key computation. This work sets the foundation of LWE & RLWE key exchange. Various RLWE-based unauthenticated key exchange constructions including [5,8,9,26] follow this idea. [15,32] and this work also use this notion to construct AKE and PAKE protocols respectively.

Here we recall several core functions and properties

Signal Function. Let $q > 2$ be a prime. Hint functions $\sigma_0(x)$, $\sigma_1(x)$ from Z_q to $\{0, 1\}$ are defined as:

$$\sigma_0(x) = \begin{cases} 0, x \in [-\lfloor\frac{q}{4}\rfloor, \lfloor\frac{q}{4}\rfloor] \\ 1, otherwise \end{cases}, \quad \sigma_1(x) = \begin{cases} 0, x \in [-\lfloor\frac{q}{4}\rfloor + 1, \lfloor\frac{q}{4}\rfloor + 1] \\ 1, otherwise \end{cases}$$

Signal function Cha() is defined as: For any $y \in Z_q$, Cha$(y) = \sigma_b(y)$, where $b \xleftarrow{\$} \{0,1\}$. If Cha$(y) = 1$, we denote y is in outer region, otherwise y is in inner region. Signal value from an execution of key exchange is indistinguishable from uniform random bits.

Robust extractor. Mod$_2$() is a deterministic function with error tolerance δ. Mod$_2$ is defined as: Mod$_2(x, w) = (x + w \cdot \frac{q-1}{2} \mod q) \mod 2$.

Input of Mod$_2$() are: (1) $x \in Z_q$ and (2) Signal w. Output of Mod$_2$ is key bit k. Denote error tolerance as δ. For any $x, y \in Z_q$, if $\|x - y\|_\infty \le \delta$, Mod$_2(x, w) =$ Mod$_2(y, w)$, where $w =$ Cha(y). Error tolerance $\delta = \frac{q}{4} - 2$ for reconciliation mechanism in DING12, which is the key to ensure correctness of key exchange over LWE and RLWE with overwhelming probability.

Randomness. For any odd $q > 2$, if x is uniformly random in Z_q, then Mod$_2(x, w)$ is uniformly random conditioned on signal w.

DING12 RLWE key exchange is illustrated in Fig. 1:

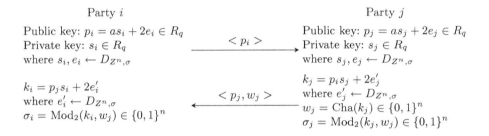

Fig. 1. DING12 RLWE key exchange protocol

3   Post-Quantum Secure Remote Password Protocol

Our RLWE-SRP protocol is a direct RLWE variant of original SRP [31]. Our design inherits advantages of SRP and strengthen its hardness further by constructing based on RLWE problem.

RLWE-SRP has 3 phases: Verifier setup, Authenticated key exchange and Verification.

– In "Verifier setup" phase, client generates verifier v and sends to server. v is used for authentication and it is the replacement of "pre-shared password" for basic PAKE protocols. Client only need to remember the correct password.

- In "Authenticated key exchange" phase, client and server share session key and authenticate each other. Client does not send password to server to authenticate himself.
- In "Verification" phase, both parties compute hash value of some elements to verify key exchange and authentication are indeed successful. After three phases, a mutual authenticated secure channel is established.

We note that "Verifier setup" is only executed once when user registers to the system, "Authenticated key exchange" and "Verification" are executed for each login attempt.

It is known that SRP is an augmented PAKE protocol, where augmentation refers to server stores a verifier, instead of storing password or hash value of password like basic PAKE protocols. For basic PAKE constructions, if server is compromised, adversary can directly impersonate as user to authenticate with server using captured password or hash of password. SRP solves this problem using "verifier". Our RLWE-SRP inherits this novel property and enhances the security of SRP by constructing based on RLWE problem. Only user with correct password can compute a particular value and use this value in key exchange to authenticate himself.

3.1   Protocol Construction

In this section, we present our post-quantum secure remote password protocol RLWE-SRP. Let $D_{Z^n,\sigma}$ be a discrete Gaussian distribution with standard deviation σ. H is standard secure hash function (e.g., SHA3-256), XOF is Extendable-Output Functions (e.g., SHAKE-128) and "$\|$" denotes concatenation. Definition of Cha() and $\mathrm{Mod}_2()$ are exactly the same as we recalled in Sect. 2.2. We assume that client and server execute the protocol honestly and have no deliberate malicious behaviours (e.g., client deliberately reveals password, server reveals verifier, leakage of variables stored in memory etc.).

Phase 0: Verifier Setup

Verifier	$v = a \cdot s_v + 2e_v$
Password	pwd
Salt	$salt$
Username	I

$a \leftarrow R_q$ is public parameter in RLWE which is shared by both parties. To compute v, client computes $s_v \leftarrow D_{Z^n,\sigma}$ and $e_v \leftarrow D_{Z^n,\sigma}$. Note that this is different from other random sampling operations, since it is required that s_v and e_v are sampled using specific $seed1 = $ SHA3-256($salt\|$SHA3-256($I\|pwd$)) and $seed2 = $ SHA3-256($seed1$) as seed respectively for pseudorandom generator (PRNG). Purpose for this design is to stop attackers from recovering password from leaked verifier. We will elaborate this later.

Client need to send username I, salt $salt$ and verifier v to server to complete verifier setup. Setting up verifier has to go through secure channels (e.g., strong

TLS or other security measures). *salt* and v are stored and indexed by I in server's database. After v is generated, client should delete all variables from memory to prevent potential leakage. If v is stolen, client needs to setup a new verifier and proceed this phase from scratch.

We note that similar setup phase also exists in other PAKE protocols. For other PAKE protocols, pre-shared password or hash of password is stored on server in this phase, where our RLWE-SRP stores verifier. This phase is only executed once for each user during registration process. For each key exchange and authentication process, only phase 1 and 2 are executed.

Phase 1: Authenticated Key Exchange

Client initiation. Randomly samples s_1 and e_1 from $D_{Z^n,\sigma}$ and computes ephemeral public key $p_i = a \cdot s_1 + 2e_1$. Send username I and ephemeral public key p_i to server.

Server Response. Search for client's salt *salt* and verifier v according to username I in database. Randomly samples s_1', e_1' and e_1''' from $D_{Z^n,\sigma}$ and computes ephemeral public key $p_j = a \cdot s_1' + 2e_1' + v$ and $u = $ SHAKE-128(SHA3-256($p_i\|p_j$)). Compute key exchange material $k_j = v \cdot s_1' + 2e_1''' + p_i \cdot s_1' + u \cdot v$, signal $w_j = $ Cha(k_j). Send *salt*, p_j and w_j to client.

Client finish. Compute $v = a \cdot s_v + 2e_v$ where s_v and e_v are sampled from $D_{Z^n,\sigma}$ using seed $seed1 = $ SHA3-256($salt\|$SHA3-256($I\|pwd$)) and $seed2 = $ SHA3-256($seed1$) for PRNG respectively. Randomly samples e_1'' from distribution $D_{Z^n,\sigma}$ and compute $u = $ SHAKE-128(SHA3-256($p_i\|p_j$)). Compute key exchange material $k_i = (p_j - v) \cdot s_v + 2e_1'' + (p_j - v) \cdot s_1 + u \cdot v$. Compute final shared session key $sk_i = $ SHA3-256(Mod$_2(k_i, w_j)$). Delete all variables except p_i, p_j and sk_i from memory.

Server finish. Compute final shared session key $sk_j = $ SHA3-256(Mod$_2$ (k_j, w_j)). Delete all variables except p_i, p_j and sk_j from memory.

RLWE-SRP protocol is illustrated in Fig. 2:

Phase 2: Verification

Verification steps of RLWE-SRP are identical to original SRP protocol and given as follows:

$C \rightarrow S$: Client computes $M_1 = $ SHA3-256($p_i\|p_j\|sk_i$) and sends to server. Server computes $M_1' = $ SHA3-256($p_i\|p_j\|sk_j$). If $M_1 = M_1'$, key exchange is successful and client is authenticated.

$S \rightarrow C$: Server computes $M_2' = $ SHA3-256($p_i\|M_1'\|sk_j$) and sends to client. Client computes $M_2 = $ SHA3-256($p_i\|M_1\|sk_i$). If $M_2 = M_2'$, key exchange is successful and mutual authentication is achieved.

Note that reader may consider that verification step gains additional round-trip at first glance. However in practice, this might not be the case. For verification step from $C \rightarrow S$, client can send encrypted data alongside M_1. Server first verifies identity with above approach, then decrypt the data using shared

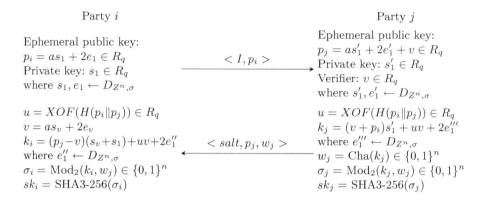

Party i Party j

Ephemeral public key:

Party j — Ephemeral public key:
$p_j = as_1' + 2e_1' + v \in R_q$
Private key: $s_1' \in R_q$
Verifier: $v \in R_q$
where $s_1', e_1' \leftarrow D_{Z^n,\sigma}$

Party i — Ephemeral public key:
$p_i = as_1 + 2e_1 \in R_q$
Private key: $s_1 \in R_q$
where $s_1, e_1 \leftarrow D_{Z^n,\sigma}$

$< I, p_i >$

$u = XOF(H(p_i\|p_j)) \in R_q$
$v = as_v + 2e_v$
$k_i = (p_j - v)(s_v + s_1) + uv + 2e_1''$
where $e_1'' \leftarrow D_{Z^n,\sigma}$
$\sigma_i = \mathrm{Mod}_2(k_i, w_j) \in \{0,1\}^n$
$sk_i = \mathrm{SHA3\text{-}256}(\sigma_i)$

$< salt, p_j, w_j >$

$u = XOF(H(p_i\|p_j)) \in R_q$
$k_j = (v + p_i)s_1' + uv + 2e_1'''$
where $e_1''' \leftarrow D_{Z^n,\sigma}$
$w_j = \mathrm{Cha}(k_j) \in \{0,1\}^n$
$\sigma_j = \mathrm{Mod}_2(k_j, w_j) \in \{0,1\}^n$
$sk_j = \mathrm{SHA3\text{-}256}(\sigma_j)$

Fig. 2. Post-quantum secure remote password protocol

key generated from authenticated key exchange phase. If server fails to verify the identity of client, then he cannot decrypt the data. Same approach applies to server, i.e., for verification step from $S \rightarrow C$, server can send encrypted data alongside M_2. Client first verifies identity of server, then decrypt the data.

3.2 Correctness

Correctness of RLWE-SRP is guaranteed with properly chosen parameter. We know that $sk_x = \mathrm{Mod}_2(k_x, w_j)$ ($x = i$ or j) and core notion of error reconciliation in DING12 is the difference between key exchange materials k_i and k_j is very small. If k_i and k_j are sufficiently close (i.e., $\|k_i - k_j\|_\infty \leq$ error tolerance δ), correctness of RLWE-SRP can be guaranteed with overwhelming probability. As we recalled in Sect. 2.2 and [16], if $\|k_i - k_j\|_\infty \leq \frac{q}{4} - 2$, then both sides can derive same output from $\mathrm{Mod}_2()$, i.e., same session key.

For k_i and k_j:

$$k_i = (p_j - v)s_v + (p_j - v)s_1 + uv + 2e_1''$$
$$= \boxed{as_v s_1'} + 2e_1' s_v + \boxed{as_1 s_1'} + 2s_1 e_1' + \boxed{uv} + 2e_1''. \tag{1}$$

$$k_j = vs_1' + p_i s_1' + uv + 2e_1'''$$
$$= \boxed{as_v s_1'} + 2e_v s_1' + \boxed{as_1 s_1'} + 2e_1 s_1' + \boxed{uv} + 2e_1'''. \tag{2}$$

Note that in (1) and (2), terms in boxes have significantly larger l_∞-norm than other small error terms. If $\|k_i - k_j\|_\infty = \|(2e_1' s_v + 2s_1 e_1' + 2e_1'') - (2e_v s_1' + 2e_1 s_1' + 2e_1''')\|_\infty \leq \|8se + 4e\|_\infty \leq \frac{q}{4} - 2$, where s and e are sampled from same distribution $D_{Z^n,\sigma}$. Correctness of key exchange is guaranteed with overwhelming probability if above inequality holds.

Lemma 1 ([30], **lemma 2.5**). *For $\sigma > 0$, $r \geq 1/\sqrt{2\pi}$, $\Pr[\|x\|_2 > r\sigma\sqrt{n}; x \leftarrow D_{Z^n,\sigma}] < (\sqrt{2\pi e r^2} \cdot e^{-\pi r^2})^n$.*

Our technique for choosing parameters is exact same as [15, 16, 32]. Moreover, $\|k_i - k_j\|_\infty$ and error tolerance are very similar with above three constructions, therefore we omit details here. Parameter choice which guarantees overwhelming success probability (much higher than $1 - 2^{-1024}$) is presented in Sect. 4.1. We note that much more compact parameter q can be chosen easily using Lemma 1.

3.3 Security of RLWE-SRP

We are able to prove the security of our RLWE-SRP protocol under Universally Composable security model (UC-secure) [12]. UC-secure is one of the strongest simulation-based security model. It overcomes shortcomings in game-based models and takes more general simulation-based techniques. It allows secure arbitrary compositions with other primitives once they are proven secure. However, due to page limitation, we do not present formal security proofs in conference version of this paper. Instead, we take similar approach as original SRP paper to show that our RLWE-SRP remain secure with same attacks as original SRP paper suggested. In original SRP paper [31], they do not present formal security analysis for SRP protocol. They show that SRP remain secure under various practical attacks. We also consider such attacks for our RLWE-SRP and categorize them into four major aspects. Moreover, we consider threat from quantum adversary, which attempts to break RLWE-SRP by attacking RLWE problem.

Reduction to RLWE. Fortunately, our RLWE-SRP is built based on RLWE problem, whose hardness is directly reduced to hard lattice problems. Public key, secret key and verifier are constructed as RLWE problem requires. Since no public known algorithms can solve lattice/LWE/RLWE problems on classic and quantum computers with properly chosen parameters, our construction remain secure against such adversaries. In comparison with original SRP, which relies on hardness Diffie-Hellman problem, can be solved easily by quantum adversaries. For RLWE-based constructions (including RLWE-SRP), adversary cannot recover secret key or error term from ephemeral public key. Properly chosen parameters guarantee hardness of RLWE instance. If adversary can break our protocol by recovering secret key, then he can also break RLWE problem and underlying lattice problems.

Hardness of password recovery. Practically, it is possible that server is compromised, therefore adversary may get hands on the verifier. Adversary may attempt to recover user's password from verifier. However, this will not work for both SRP and RLWE-SRP. This is also an advantage for SRP and RLWE-SRP. From construction of RLWE-SRP, we can see that: (1) Only the client who knows correct password can be only computed with correct s_v and e_v, i.e., correct *seed*1 and *seed*2. *salt* is 256-bit long and uniform randomly generated by client. Client does not need to store s_v and e_v since they can be computed with *salt*, I and correct *pwd*, where *pwd* is only known to himself. I and *salt* are publicly transmitted; (2) Adversary who obtain $seed1$ or $seed2$ cannot compute pwd reversely thanks to strong SHA3-256 function, randomness from 256-bit $salt$ and strong PRNG; If one can compute pwd with public-known elements, then he can at least break hash function and PRNG. (3) v is indistinguishable from uniform random since $v = a \cdot s_v + 2e_v$ is a standard RLWE sample.

If adversary wishes to recover password from v, he first need to solve RLWE problem to recover s_v, then break PRNG in order to find correct seeds, and find preimage of hash value. If one can distinguish v from uniform random, then he can solve decision-RLWE problem, which implies that adversary can solve hard lattice problems. To summarize, If one can recover pwd from verifier v, then he can at least solve RLWE problem, break SHA3-256 hash function and PRNG simultaneously.

No password leakage from key exchange. In original SRP paper, they consider password leakage from final shared key and key exchange executions. For leakage of password, we discuss this issue in above subsection, showing that if the adversary can recover password, they he can break various cryptographic primitives. For leakage from shared session key, since we use error reconciliation mechanism in [16] and they proved that signal value and public key leak no information about final shared key, therefore our construction is strong. Output of $\text{Mod}_2()$ and signal w are indistinguishable from uniform random, which is also proven in [16]. Ephemeral public key is also indistinguishable from uniform random, whose security is guaranteed by hardness of RLWE problem and properly chosen parameters.

We note that public keys in RLWE-SRP are ephemeral, namely all parties need to generate fresh keys for key exchange and authentication, therefore our RLWE-SRP is also forward secure like ephemeral Diffie-Hellman key exchange. Adversary cannot decrypt past sessions if current session is compromised.

RLWE-SRP is also resistant to Denning-Sacco attack [13], where attacker tries to impersonate user with captured session key or recover password with brute-force search. Analysis in original SRP implies that such attack can be detected and defeated in verification phase. Since RLWE-SRP uses exact same verification phase as original SRP without any modification, our RLWE-SRP also enjoys this nice property.

No impersonation even if verifier is stolen. Since the pre-shared verifier v can only be computed with correct password, adversary cannot impersonate as user to authenticate himself to the server if server is compromised and verifier is leaked. In k_i computation, adversary has to generate correct s_v (not verifier $v = a \cdot s_v + e_v$) in order to impersonate as legit user. However, s_v and e_v can only be computed with correct password, therefore adversary cannot impersonate as user even if he owns v. However, the adversary can impersonate as server for user to login. This works for almost all security protocols and applications once such information is leaked. In RLWE-SRP, since user does not send actual password to server and password cannot be recovered from verifier or during key exchange, adversary cannot acquire user's actual password even if he impersonates as legit server. This nice property also holds for original SRP. Adversary can recover s_v

and e_v only by solving RLWE problem given lost verifier $v = a \cdot s_v + 2e_v$, which is known be very hard.

Rest of security analysis in original SRP paper are mathematical constraints on choosing parameters regarding to discrete logarithm computations since SRP is built based on DLP. Our RLWE-SRP is built based on RLWE problem, therefore we can safely ignore these analysis. Concrete parameter choice and security level estimation are presented in Sect. 4.

Compared with the RLWE-based PAKE protocol in [15], advantages of our RLWE-SRP come from the usage of verifier. Using actual password or hash of password directly is not strong enough as we discussed above. In [15], if password or hash of password is stolen, adversary can immediately impersonate as legit user and authenticate to user, while this attack does not work for RLWE-SRP. Another advantage is that if the hash function is broken, i.e., one can find preimage of hash value, password can be recovered successfully in [15]. For RLWE-SRP, we first use $seed1 = \text{SHA3-256}(salt\|\text{SHA3-256}(I\|pwd))$ to generate a seed using hash function, then we generate $s_v \in R_q \leftarrow D_{Z^n,\sigma}$ with $seed1$ for PRNG. If adversary can recover password from RLWE-SRP, then he can break hash function, PRNG and RLWE problem.

4   Instantiation, Implementation and Performance

In this section, we introduce parameter choice, security level estimation, implementation details and benchmark of RLWE-SRP. In order to demonstrate efficiency of our protocol and implementation, we also compare performance with original SRP protocol [31] and J-PAKE [20] protocols which are vulnerable to quantum attacks, and two RLWE-based PAKE protocols in [15].

4.1   Parameter Choice and Security Level Estimation

We first choose practical parameter for our protocol. We choose $\sigma = 3.192$, $n = 1024$ and $q = 1073479681$ (approximately 30 bits). Our choice of q can guarantee the correctness of our protocol with extremely low failure probability (much lower than 2^{-1024}). Our 30-bit q is 2-bit smaller than modulus in [9,15], where [9] claims roughly $2^{-131072}$ failure probability with very similar $k_i - k_j$ norm and error tolerance for reconciliation mechanism. Moreover, our choice of q can instantiate NTT for quick polynomial multiplication efficiently since it holds $q \equiv 1 \mod 2n$. This technique is also adopted in various implementations of RLWE-based protocols (e.g., [5]) as well. $\sigma = 3.192$ and $n = 1024$ follow parameter choices of various RLWE-based key exchange protocols (e.g., [9,15,32] etc.). We set statistical distance between sampled distribution and discrete Gaussian distribution to be 2^{-128} to preserve high statistical quality and security.

Security level of our parameter choice is estimated using LWE estimator in [4]. The state-of-the-art LWE estimator gives a thorough security estimation for both LWE and RLWE-based cryptosystems. It evaluates security level of cryptosystems by computing attack complexity of exhaustive search, BKW, lattice reduction, decoding, reducing BDD to unique-SVP and MITM attacks. Due to the fact that currently there are no known attacks that take advantage of the ring structure in RLWE, therefore LWE estimator can be used to estimate security level of RLWE-based constructions. LWE estimator is regarded as most accurate, robust and easy-to-use tool of security level estimation for LWE and RLWE-based constructions. Given any parameters, LWE estimator outputs computation and space complexity of various attacks. [9] and various works also use this tool to estimate security of their parameter choice.

Instruction for estimation is given as follows:

– load("https://bitbucket.org/malb/lwe-estimator/raw/HEAD/estimator. py")
– n, alpha, q = 1024, alphaf(8, 1073479681), 1073479681
– set_verbose(1)
– _ = estimate_lwe(n, alpha, q, skip=["arora-gb"])

With above estimation approach, our parameter choice offers 209-bit security. We note that since LWE estimator is constantly updated, estimation result from latest version of LWE estimator may different from what we present here.

We note that above security claim is actually pessimistic, since public and private keys in our RLWE-SRP are ephemeral-only, while the attacks listed in LWE estimator requires large amount of RLWE samples to work. Given only one sample from our protocol execution, it is much more difficult to break the protocol by attempting to solve RLWE problem and recover s.

4.2   Implementation, Performance and Comparison

We present portable C++ implementation of RLWE-SRP. Our implementation is done using our modified version of NFLlib library [3]. NFLlib is an efficient NTT-based library for RLWE-based cryptography. We modify latest version of NFLlib to adapt to our design for s_v and e_v, i.e., passing certain seed to random number generator when generating random numbers in s_v and e_v. Our implementation only runs on single core and does not utilize parallel computing techniques. Efficiency of our non-constant time implementation benefits from efficient protocol design and SSE-optimized discrete Gaussian sampling and NTT computations. It is portable and can run on more outdated devices.

For hash functions, we choose Keccak sponge function family (standardized and known as SHA3-*). Hashing computations are done using SHA3-256. We also take advantage of an extendable output function (XOF) in Keccak family - SHAKE-128. It can be used as hash function with desired output length. It is used in generating u in our implementation, i.e., extending SHA3-256 output of $(p_i\|p_j)$ to 4096 bytes with SHAKE-128 ($u = $ SHAKE-128(SHA3-256($p_i\|p_j$), 4096)). Each coefficient is stored in "uint32_t" type variable, therefore we use SHAKE-128 to extend SHA3-256($p_i\|p_j$) to 4096 bytes to instantiate polynomial coefficients of u.

We test implementation of this work, two PAKE protocols that vulnerable to quantum attacks (original SRP [31] and J-PAKE [20]), the only two quantum-resistant RLWE-based PAKE protocols till now in [15] on same server equipped with 3.4 GHz Intel Xeon E5-2687W v2 processor and 64 GB memory, running 64-bit Ubuntu 14.04. All implementations are compiled by gcc or g++ compiler version 4.8.4 with same optimization flags "-O3 -march=native -m64". We report average runtime over 10,000 executions of all protocols in Table 1:

Table 1. Performance comparison of multiple PAKE protocols

	Security level	Security assumption	Client (ms)	Server (ms)
RLWE-SRP	209-bit	RLWE	0.286	0.257
Original SRP [31]	112-bit	DLP	0.805 (2.81x)	0.804 (3.13x)
J-PAKE [20]	80-bit	DLP	1.499 (5.24x)	1.495 (5.82x)
RLWE-PAK [15]	184-bit	RLWE	3.472 (12.14x)	4.053 (15.77x)
RLWE-PPK [15]	184-bit	RLWE	3.488 (12.20x)	4.041 (15.72x)

Number in parentheses is number of times of corresponding runtime for each protocol compared with this work as baseline. Compared with original SRP protocol, our implementation achieves 3x speedup. Compared with J-PAKE, RLWE-PAK and RLWE-PPK, our implementation is 5.53x, 13.96x and 13.96x faster respectively. Benchmark proves the efficiency of our protocol and implementation. We remark that for our RLWE-SRP protocol, client side computation costs slightly more time than server side since client needs to compute verifier v (i.e., sampling s_v and e_v then NTT) while server does not.

Communication cost of all PAKE protocols compared in Table 1 is given in Table 2. Here we assume that for all PAKE protocols, size of username is 64 bytes, salt is 256 bits, output length of hash function is 256 bits. For original SRP and J-PAKE, 80-bit security implies choosing 1024-bit prime modulus, 112-bit security implies choosing 2048-bit prime modulus.

Table 2. Communication cost comparison of multiple PAKE protocols

	Security level	Client→Server (Byte)	Server→Client (Byte)
RLWE-SRP	209-bit	3936	4032
Original SRP [31]	112-bit	352	320
J-PAKE [20]	80-bit	640	640
RLWE-PAK [15]	184-bit	4192	4256
RLWE-PPK [15]	184-bit	4192	4224

We can see that due to mathematical structure of RLWE and parameter choice, RLWE-based protocols have much larger communication cost than discrete logarithm-based ones. In fact, all RLWE-based constructions have much larger communication cost compared with current public key algorithms (e.g., Diffie-Hellman, ECDH, RSA etc.) due to public key in RLWE-based constructions is degree n polynomial in R_q modulus q. Size of public key is estimated to be $n \cdot \lceil \log_2 q \rceil / 8$ bytes. We can also choose smaller q by increasing key exchange failure probability to more practical value (e.g., 2^{-60}) and get roughly 18-bit modulus q. Moreover, choosing slightly smaller σ can reduce size of modulus q by 1 bit. With much smaller q and slightly smaller σ, security level of parameter choice can be increased. Compared with two RLWE-based PAKE protocols in [15], RLWE-SRP has smaller communication cost due to 2-bit smaller modulus q.

5   Conclusions

In this paper, we present the first practical and efficient RLWE-based post-quantum secure remote protocol. Our RLWE-SRP enjoys various nice security properties, including: (1) Resistant to quantum attacks; (2) Mutual authenticated key exchange; (3) No actual password is sent to server; (4) Attacker cannot impersonate client to authenticate even if verifier is stolen; (5) Attacker cannot recover user password with stolen verifier etc. Our 209-bit secure parameter choice and efficient implementation further highlight the practicality of this work. Compared with PAKE protocols that are vulnerable to quantum attacks and two RLWE-PAKE protocols, our RLWE-SRP implementation is much faster. We believe our RLWE-SRP is one step forward for secure remote password protocol towards post-quantum world.

Acknowledgement. We would like to thank anonymous reviewers for valuable feedbacks. This work is supported by China Scholarship Council, National Natural Science Foundation of China (Grant No. 61672092) and Fundamental Research Funds for the Central Universities (Grant No. 2017YJS038). Jintai Ding is partially supported by NSF grant DMS-1565748 and US Air Force grant FA2386-17-1-4067.

References

1. Bliss - strongSwan. https://wiki.strongswan.org/projects/strongswan/wiki/BLISS
2. Abdalla, M., Pointcheval, D.: Simple password-based encrypted key exchange protocols. In: Menezes, A. (ed.) CT-RSA 2005. LNCS, vol. 3376, pp. 191–208. Springer, Heidelberg (2005). https://doi.org/10.1007/978-3-540-30574-3_14
3. Aguilar-Melchor, C., Barrier, J., Guelton, S., Guinet, A., Killijian, M.-O., Lepoint, T.: NFLLIB: NTT-based fast lattice library. In: Sako, K. (ed.) CT-RSA 2016. LNCS, vol. 9610, pp. 341–356. Springer, Cham (2016). https://doi.org/10.1007/978-3-319-29485-8_20
4. Albrecht, M.R., Player, R., Scott, S.: On the concrete hardness of learning with errors. J. Math. Cryptology **9**(3), 169–203 (2015)

5. Alkim, E., Ducas, L., Pöppelmann, T., Schwabe, P.: Post-quantum key exchange-a new hope. IACR Cryptology ePrint Archive 2015, 1092 (2015)
6. Apple: iOS Security. https://www.apple.com/business/docs/iOS_Security_Guide. pdf
7. Bellovin, S.M., Merritt, M.: Encrypted key exchange: password-based protocols secure against dictionary attacks. In: Proceedings of 1992 IEEE Computer Society Symposium on Research in Security and Privacy, pp. 72–84. IEEE (1992)
8. Bos, J., Costello, C., Ducas, L., Mironov, I., Naehrig, M., Nikolaenko, V., Raghunathan, A., Stebila, D.: Frodo: take off the ring! practical, quantum-secure key exchange from lwe. In: Proceedings of the 2016 ACM SIGSAC Conference on Computer and Communications Security, pp. 1006–1018. ACM (2016)
9. Bos, J.W., Costello, C., Naehrig, M., Stebila, D.: Post-quantum key exchange for the tls protocol from the ring learning with errors problem. In: 2015 IEEE Symposium on Security and Privacy (SP), pp. 553–570. IEEE (2015)
10. Boyko, V., MacKenzie, P., Patel, S.: Provably secure password-authenticated key exchange using Diffie-Hellman. In: Preneel, B. (ed.) EUROCRYPT 2000. LNCS, vol. 1807, pp. 156–171. Springer, Heidelberg (2000). https://doi.org/10.1007/3-540-45539-6_12
11. Braithwaite, M.: Experimenting with Post-Quantum Cryptography. https://security.googleblog.com/2016/07/experimenting-with-post-quantum.html
12. Canetti, R.: Universally composable security: a new paradigm for cryptographic protocols. In: Proceedings of 42nd IEEE Symposium on Foundations of Computer Science 2001, pp. 136–145. IEEE (2001)
13. Denning, D.E., Sacco, G.M.: Timestamps in key distribution protocols. Commun. ACM **24**(8), 533–536 (1981)
14. Diffie, W., Hellman, M.: New directions in cryptography. IEEE Trans. Inf. Theory **22**(6), 644–654 (1976)
15. Ding, J., Alsayigh, S., Lancrenon, J., Saraswa, R.V., Snook, M.: Provably secure password authenticated key exchange based on RLWE for the post-quantum world. In: Handschuh, H. (ed.) CT-RSA 2017. LNCS, vol. 10159, pp. 183–204. Springer, Cham (2017). https://doi.org/10.1007/978-3-319-52153-4_11
16. Ding, J., Xie, X., Lin, X.: A simple provably secure key exchange scheme based on the learning with errors problem. IACR Cryptology EPrint Archive 2012, 688 (2012)
17. Dousti, M.S., Jalili, R.: Forsakes: a forward-secure authenticated key exchange protocol based on symmetric key-evolving schemes. Adv. Math. Commun. **9**(4), 471–514 (2015). http://aimsciences.org/journals/displayArticlesnew.jsp?paperID=11939
18. Goldberg, J.: Three layers of encryption keeps you safe when ssl/tls fails. https://blog.agilebits.com/2017/02/23/three-layers-of-encryption-keeps-you-safe-when-ssltls-fails/
19. Gonzláez, S., Huguet, L., Martínez, C., Villafañe, H.: Discrete logarithm like problems and linear recurring sequences. Adv. Math. Commun. **7**(2), 187–195 (2013). http://aimsciences.org/journals/displayArticlesnew.jsp?paperID=8550
20. Hao, F., Ryan, P.Y.A.: Password authenticated key exchange by juggling. In: Christianson, B., Malcolm, J.A., Matyas, V., Roe, M. (eds.) Security Protocols 2008. LNCS, vol. 6615, pp. 159–171. Springer, Heidelberg (2011). https://doi.org/10.1007/978-3-642-22137-8_23

21. Katz, J., Vaikuntanathan, V.: Smooth projective hashing and password-based authenticated key exchange from lattices. In: Matsui, M. (ed.) ASIACRYPT 2009. LNCS, vol. 5912, pp. 636–652. Springer, Heidelberg (2009). https://doi.org/10.1007/978-3-642-10366-7_37

22. Krawczyk, H.: HMQV: a high-performance secure Diffie-Hellman protocol. In: Shoup, V. (ed.) CRYPTO 2005. LNCS, vol. 3621, pp. 546–566. Springer, Heidelberg (2005). https://doi.org/10.1007/11535218_33

23. Lyubashevsky, V., Peikert, C., Regev, O.: On ideal lattices and learning with errors over rings. In: Gilbert, H. (ed.) EUROCRYPT 2010. LNCS, vol. 6110, pp. 1–23. Springer, Heidelberg (2010). https://doi.org/10.1007/978-3-642-13190-5_1

24. Micheli, G.: Cryptanalysis of a noncommutative key exchange protocol. Adv. Math. Commun. **9**(2), 247–253 (2015). http://aimsciences.org/journals/displayArticlesnew.jsp?paperID=11174

25. Morhaime, M.: Important security update. http://us.blizzard.com/en-us/securityupdate.html

26. Peikert, C.: Lattice cryptography for the internet. In: Mosca, M. (ed.) PQCrypto 2014. LNCS, vol. 8772, pp. 197–219. Springer, Cham (2014). https://doi.org/10.1007/978-3-319-11659-4_12

27. Perrin, T., Wu, T., Mavrogiannopoulos, N., Taylor, D.: Using the secure remote password (SRP) protocol for TLS authentication. https://tools.ietf.org/html/rfc5054

28. Regev, O.: On lattices, learning with errors, random linear codes, and cryptography. J. ACM (JACM) **56**(6), 34 (2009)

29. Shor, P.W.: Polynomial-time algorithms for prime factorization and discrete logarithms on a quantum computer. SIAM Rev. **41**(2), 303–332 (1999)

30. Stephens-Davidowitz, N.: Discrete gaussian sampling reduces to CVP and SVP. In: Proceedings of the Twenty-Seventh Annual ACM-SIAM Symposium on Discrete Algorithms, pp. 1748–1764. Society for Industrial and Applied Mathematics (2016)

31. Wu, T.D., et al.: The secure remote password protocol. In: NDSS, vol. 98, pp. 97–111 (1998)

32. Zhang, J., Zhang, Z., Ding, J., Snook, M., Dagdelen, Ö.: Authenticated key exchange from ideal lattices. In: Oswald, E., Fischlin, M. (eds.) EUROCRYPT 2015. LNCS, vol. 9057, pp. 719–751. Springer, Heidelberg (2015). https://doi.org/10.1007/978-3-662-46803-6_24

Hashing into Twisted Jacobi Intersection Curves

Xiaoyang He[1,2,3], Wei Yu[1,2,3($\boxtimes$)], and Kunpeng Wang[1,2,3]

[1] State Key Laboratory of Information Security, Institute of Information
Engineering, Chinese Academy of Sciences, Beijing, China
{hexiaoyang,wangkunpeng}@iie.ac.cn, yuwei_1_yw@163.com
[2] Data Assurance and Communication Security Research Center,
Chinese Academy of Sciences, Beijing, China
[3] University of Chinese Academy of Sciences, Beijing, China

Abstract. By generalizing Jacobi intersection curves introduced by
D. V. Chudnovsky and G. V. Chudnovsky, Feng et al. proposed twisted
Jacobi intersection curves, which contain more elliptic curves. Twisted
Jacobi intersection curves own efficient arithmetics with regard to their
group law and are resistant to timing attacks. In this paper, we proposed
two hash functions indifferentiable from a random oracle, mapping binary
messages to rational points on twisted Jacobi intersection curves. Both
functions are based on deterministic encodings from $\mathbb{F}_q$ to twisted Jacobi
intersection curves. There are two ways to construct such encodings: (1)
utilizing the algorithm of computing cube roots on $\mathbb{F}_q$ when $3 \mid q + 1$;
(2) using Shallue-Woestijne-Ulas algorithm when $4 \mid q + 1$. In both cases,
our encoding methods are more efficient than existed ones. Moreover, we
estimate the density of images of both encodings by Chebotarev theorem.

Keywords: Elliptic curves · Twisted Jacobi intersection curves
Character sum · Hash function · Random oracle

1 Introduction

Hashing into elliptic curves is a key step in a myriad of cryptographic protocols and schemes. The password authenticated key exchange protocols [11] and
simple password exponential key exchange [10] protocols are examples of the
utilization of such hashing algorithms. Moreover, identity-based schemes like
Lindell's universally-composable scheme [7], encryption schemes [1,2] and signature schemes [3,4] also contain hashing algorithms as vital components.

Canonical hash functions map bit-string $\{0,1\}^*$ to finite field $\mathbb{F}_q$. To construct
a hash function into elliptic curves, we need an encoding from $\mathbb{F}_q$ to rational
points on an elliptic curves. Boneh and Franklin [8] firstly proposed an example
of such encodings, which is probabilistic and fails to return a rational point on
an elliptic curve at the probability of $1/2^k$, where k is a predetermined bound.

X. He—This work is supported in part by National Natural Science Foundation of
China under Grant No. 61502487, 61672030.

X. Chen et al. (Eds.): Inscrypt 2017, LNCS 10726, pp. 117–138, 2018.
https://doi.org/10.1007/978-3-319-75160-3_9

Thus the total running steps depend on the input $u \in \mathbb{F}_q$ and are not deterministic. Therefore, their algorithm would be vulnerable facing timing attacks [9]. So finding algorithms running in constant number of operations other than Boneh and Franklin's is of importance.

For short Weierstrass-form elliptic curves, there exist several routes to achieve deterministic encodings. When $q \equiv 3 \pmod 4$, Shallue and Woestijne provided an algorithm [12] based on Skalba's equality [13], using a modification of Tonelli-Shanks algorithm to calculate square roots efficiently as $x^{1/2} = x^{(q+1)/4}$. Fouque and Tibouchi [14] simplified this encoding by applying brief version of Ulas' function [16]. Moreover, they generalized Shallue and Woestijne's method so as to hash into some special hyperelliptic curves. When $q \equiv 2 \pmod 3$, Icart [17] gave an algorithm based on calculating cube roots efficiently as $x^{1/3} = x^{(2q-1)/3}$ in Crypto 2009. Fouque et al. construct an encoding function on a special case of hyperelliptic curves called odd hyperelliptic curves [15]. Farashahi et al. [18], Alasha [22] and Yu et al. [19–21] proposed further works on deuterogenic curves.

Jacobi intersection curve is the intersection of two quadratic surfaces with a rational point on it. First introduced by D. V. Chudnovsky and G. V. Chudnovsky [24], Jacobi intersection curves have been researched for its competitive efficiency in point addition and scalar multiplication, such as faster doubling and triplings. The previous work can be found in [25–28]. Twisted Jacobi intersection curves, as the generalization of Jacobi intersection curves, are first introduced by Feng et al. [23] in 2010. Feng et al. show that every elliptic curve over the prime field with 3 points of order 2 is isomorphic to a twisted Jacobi intersection curve. In addition, they proposed new addition formulas more effective than previous ones. Wu and Feng studied the complete set of addition laws for twisted Jacobi intersection curves [29]. Cao and Wang constructed skew-Frobenius mapping on twisted Jacobi intersection curves to speed up the scalar multiplications [30].

We construct two hash function efficiently mapping binary messages into twisted Jacobi intersection curves, which are both indifferentiable from a random oracle. Each of the hash functions is based on deterministic encodings directly from $\mathbb{F}_q$ to twisted Jacobi intersection curves: brief Shallue-Woestijne-Ulas (SWU) encoding and cube root encoding. Based on Skalba's equality [13], brief SWU encoding costs three field squarings and five multiplications less than birational equivalence from short Weierstrass curve to twisted Jacobi intersections curve composed with Ulas' original encoding [16]. It costs three squarings less than birational equivalence from short Weierstrass curve to twisted Jacobi intersection curve composed with simplified Ulas map [33]. To prove our encoding's B-well-distributed property, we estimate the character sum of an arbitrary non-trivial character defined over twisted Jacobi intersections curve through brief SWU encoding. We also estimate the size of image of brief SWU encoding. Based on calculating cube root of elements in $\mathbb{F}_q$, cube root encoding saves one field inversion compared with Alasha's encoding at the price of one field squaring and three field multiplications. We estimate the relevant character sum and the size of image of cube root encoding in similar way.

We do experiments over $192-$bit prime field $\mathbb{F}_{P192}$ and 384-bit prime field $\mathbb{F}_{P384}$ recommended by NIST in the elliptic curve standard [32]. On both fields, there exist efficient algorithms to compute the square root and cube root for each element. On $\mathbb{F}_{P192}$, our cube root encoding f_2 saves 11.70% running time compared with Alasha's encoding function f_A, on $\mathbb{F}_{P384}$, f_2 saves 7.13% compared with f_A. Our brief SWU encoding f_1 also runs faster than f_U, birational equivalence composed with Ulas' encoding function and f_E, birational equivalence composed with Fouque and Tibouchi's brief encoding. Experiments show that f_1 saves 9.89% compared with f_U and 8.72% with f_E on $\mathbb{F}_{P192}$, while it saves 6.17% compared with f_U and 6.23% with f_E on $\mathbb{F}_{P384}$.

Organization of the paper:

- In Sect. 2, we recall some basic facts of twisted Jacobi intersection curves.
- Section 3 contains the definition of brief SWU encoding, and its B-well-distributed property is proved by estimating the character sum of this encoding. We also calculate the density of image of the encoding.
- In Sect. 4, we provided the cube root encoding, prove its B-well-distributed property and calculate the density of image of the encoding by similar methods.
- In Sect. 5, we construct two hash functions indifferentiable from random oracle.
- Time complexity of given algorithms is analysed in Sect. 6, where we furthermore presented the experimental results.
- Section 7 concludes the paper.

2   Twisted Jacobi Intersection Curves

Suppose $\mathbb{F}_q$ is a finite field whose characteristic is greater than 2. In this chapter we firstly recall the definition proposed by Feng et al. in [23]:

Definition 1. *Twisted Jacobi intersection curves can be written as:*

$$au^2 + v^2 = bu^2 + w^2 = 1,$$

where $a, b \in \mathbb{F}_q$ with $ab(a - b) \neq 0$.

A Jacobi intersection curve is a twisted Jacobi intersection curve with $a = 1$. The twisted Jacobi intersection curve $E_{a,b}$: $au^2 + v^2 = bu^2 + w^2 = 1$ is a quadratic twist of the Jacobi intersection curve $E_{1,b/a}$: $u'^2 + v'^2 = (1/a)u'^2 + w'^2 = 1$. The map $(u, v, w) \mapsto (u'/\sqrt{a}, v', w')$ is an isomorphism from $E_{a,b}$ to $E_{1,b/a}$ over $\mathbb{F}_q(\sqrt{a})$.

Remark: Consider Jacobi model $J_{(r,s)}$ of the elliptic curve over $\mathbb{F}_q$, constructed by quartic intersection in $\mathbb{P}^3$ [28]:

$$rX_0^2 + X_1^2 = X_2^2,$$
$$sX_0^2 + X_2^2 = X_3^2,$$
$$tX_0^2 + X_3^2 = X_1^2,$$

where $r + s + t = 0$, with identity point $O = (0 : 1 : 1 : 1)$ and $2-$tostion points

$$T_1 = (0 : -1 : 1 : 1), T_2 = (0 : 1 : -1 : 1), T_3 = (0 : 1 : 1 : -1).$$

Then a twisted Jacobi intersection curve $au^2 + v^2 = bu^2 + w^2 = 1$ is an affine model for a curve in Jacobi model family for $(r, s, t) = (a, -b, b - a)$, with the embedding

$$e : (u, v, w) \mapsto (u : v : 1 : w),$$

and the neutral point of twisted Jacobi intersection curve is $e^{-1}(O) = (0, 1, 1)$ while all $2-$torsion points are $(0, -1, -1), (0, -1, 1)$ and $(0, 1, -1)$.

Feng et al. have shown that each elliptic curve with 3 rational points of order 2 is $\mathbb{F}_q$-isomorphic to a twisted Jacobi intersection curve [23]. In particular, $au^2 + v^2 = bu^2 + w^2 = 1$ and curve $y^2 = x(x - a)(x - b)$ are isomorphic.

3   Brief SWU Encoding

In the case that $q = p^n \equiv 3 \pmod 4$, Ulas constructed an encoding function from $\mathbb{F}_q$ to curve $y^2 = x^n + ax^2 + bx$ [16]. We give our deterministic encoding function f_1, called brief SWU encoding, by generalizing his algorithm, mapping $t \in \mathbb{F}_q$ to $(u, v, w) \in E_{a,b}(\mathbb{F}_q)$.

3.1   Algorithm

Input: t and parameter $a, b \in \mathbb{F}_q$, $ab(a - b) \neq 0$.
Output: a point $(u, v, w) \in E_{a,b}(\mathbb{F}_q)$.

1. If $t = 0$ then return $(0, 1, 1)$ directly.
2. $X(t) = -\dfrac{ab}{a + b}(t^2 - 1)$.
3. Calculate $g(X(t))$ where $g(s) = s^3 - (a + b)s^2 + abs$.
4. $Y(t) = \dfrac{ab}{a + b}(1 - \dfrac{1}{t^2})$.
5. Calculate $g(Y(t))$.
6. If $g(X(t))$ is a quadratic residue, then $(x, y) = \left(X(t), -\sqrt{g(X(t))} \right)$,
   otherwise $(x, y) = \left(Y(t), \sqrt{g(Y(t))} \right)$.
7. $(u, v, w) = (0, 1, 1) + \dfrac{1}{x^2 - ab}(-2y, 2a(b - x), 2b(a - x))$.

According to [14], there exists a function $U(t) = t^3 g(Y(t))$, such that the equality

$$U(t)^2 = -g(X(t))g(Y(t)) \tag{1}$$

holds. Thus either $g(X(t))$ or $g(Y(t))$ is a quadratic residue. Select the one whose square roots are in $\mathbb{F}_q$. Since $q \equiv 3 \pmod 4$, the standard square root can be calculated efficiently by $\sqrt{a} = a^{(q+1)/4}$. Then the mapping $t \mapsto (x, y)$ satisfying

$y^2 = g(t)$ can be constructed. In the final step of the algorithm, we transfer (x, y) to $(u, v, w) \in E_{a,b}(\mathbb{F}_q)$, using a birational equivalence. It is easy to check that this birational equivalence is one-to-one and onto when it is extended to a map between projective curves. $(x, y) = \infty$ corresponds to $(u, v, w) = (0, 1, 1)$ while $(x, y) = (0, 0), (a, 0)$ and $(b, 0)$ corresponds to $(0, -1, -1), (0, -1, 1)$ and $(0, 1, -1)$ respectively. Denote the map $t \mapsto (x, y)$ by ρ, and denote the map $(x, y) \mapsto (u, v, w)$ by ψ, we call the composition $f_1 = \psi \circ \rho$ brief SWU encoding. Therefore given $(x, y) \in Im(\rho)$, either $y = \sqrt{g(x)}$ hence x is the image of $Y(t)$ and has at most 2 preimages, or $y = -\sqrt{g(x)}$ hence x is the image of $X(t)$ and has still at most 2 preimages. Moreover, it is easy to check that ψ is one-to-one. Therefore for points $(u, v, w) = (0, 1, -1)$ and $(0, -1, 1)$, f_1 has at most 4 preimages while for other points on $E_{a,b}$, f_1 has at most 2 preimages.

3.2 Theoretical Analysis of Time Cost

Let S denote field squaring, M denote field multiplication, I field inversion, E_S the square root, E_C the cube root, D the determination of the square residue. Suppose $a, b \in \mathbb{F}_q$. In this paper we make the assumption that $S = M$, $I = 10M$ and $E_S = E_C = E$.

The cost of f_1 can be calculated as follows:

1. Calculating t^2 costs S, multiplying $t^2 - 1$ by $-\dfrac{ab}{a+b}$ costs M, and it is enough to calculate $X(t)$.
2. To compute $Y(t)$, we need to calculate the inversion of t^2 for $I + M$.
3. When x is known, computing $g(x) = x(x-a)(x-b)$ takes $2M$. To make sure that the algorithm be run in constant time, both $g(X(t))$ and $g(Y(t))$ must be calculated and it requires $4M$.
4. In general case, exact one of $g(X(t))$ and $g(Y(t))$ is a quadratic residue. We only need to check once and it takes D, then compute the square root E_S of the quadratic residue. Then values of x and y are known.
5. Finally, we calculate the inverse of $x^2 - ab$, which requires $S + I$. Then multiplying the inverse by $-2y$, $2b(a-x)$ and $2a(b-x)$ costs $3M$, then this step requires $I + S + 3M$.

Therefore, f_1 requires $E_S + 2I + 9M + 2S + D = E + 31M + D$ in all.

3.3 B-well-distributed Property of Brief SWU Encoding

In this section, we will show that the brief SWU encoding has a B-well-distributed property for a constant B, which leads to the construction of the hash function indifferentiable from random oracle in Sect. 5.2.

Definition 2 (Character Sum). *Suppose f is an encoding from $\mathbb{F}_q$ into a smooth projective elliptic curve E, and $J(\mathbb{F}_q)$ denotes the Jacobian group of E. Assume that E has an $\mathbb{F}_q$ − rational point O, by sending $P \in E(\mathbb{F}_q)$ to the $\deg 0$*

divisor $(P) - (O)$, we can regard f as an encoding to $J(\mathbb{F}_q)$. Let χ be an arbitrary character of $J(\mathbb{F}_q)$. We define the character sum

$$S_f(\chi) = \sum_{s \in \mathbb{F}_q} \chi(f(s)).$$

We say that f is B-well-distributed if for any nontrivial character χ of $J(\mathbb{F}_q)$, the inequality $|S_f(\chi)| \leqslant B\sqrt{q}$ holds [34].

Lemma 1 (Corollary 2, Sect. 3, [34]). *If f is a B-well-distributed encoding into a curve E, then the statistical distance between the distribution defined by $f^{\otimes s}$ on $J(\mathbb{F}_q)$ and the uniform distribution is bounded as:*

$$\sum_{D \in J(\mathbb{F}_q)} |\frac{N_s(D)}{q^s} - \frac{1}{\#J(\mathbb{F}_q)}| \leqslant \frac{B^s}{q^{s/2}} \sqrt{\#J(\mathbb{F}_q)},$$

where

$$f^{\otimes s}(u_1, \ldots, u_s) = f(u_1) + \ldots + f(u_s),$$

$$N_s(D) = \#\{(u_1, \ldots, u_s) \in (\mathbb{F}_q)^s | D = f(u_1) + \ldots + f(u_s)\},$$

i.e., $N_s(D)$ is the size of preimage of D under $f^{\otimes s}$. In particular, when s is greater than the genus of E, the distribution defined by $f^{\otimes s}$ on $J(\mathbb{F}_q)$ is statistically indistinguishable from the uniform distribution. Especially, in the elliptic curves' case, $g_E = 1$, let $s = g_E + 1 = 2$, the hash function construction

$$m \mapsto f^{\otimes 2}(h_1(m), h_2(m))$$

is indifferentiable from random oracle if h_1, h_2 are seen as independent random oracles into $\mathbb{F}_q$ (See [34]).

Hence, it is of great importance to estimate the character sum of an encoding into an elliptic curve, and we will study the case of twisted Jacobi intersection curves.

Definition 3 (Artin Character). *Let E be a smooth projective elliptic curve, $J(\mathbb{F}_q)$ be Jacobian group of E. Let χ be a character of $J(\mathbb{F}_q)$. Its extension is a multiplicative map $\overline{\chi} : Div_{\mathbb{F}_q}(E) \to \mathbb{C}$,*

$$\overline{\chi}(n(P)) = \begin{cases} \chi(P)^n, & P \in S, \\ 0, & P \notin S. \end{cases}$$

Here P is a point on $E(\mathbb{F}_q)$, S is a finite subset of $E(\mathbb{F}_q)$, usually denotes the ramification locus of a morphism $Y \to X$. Then we call $\overline{\chi}$ an Artin character of X.

Theorem 1. *Let $h : \tilde{X} \to X$ be a nonconstant morphism of projective curves, and χ is an Artin character of X. Suppose that $h^*\chi$ is unramified and nontrivial, φ is a nonconstant rational function on $\tilde{X}$. Then*

$$\left| \sum_{P \in \tilde{X}(\mathbb{F}_q)} \chi(h(P)) \right| \leqslant (2\tilde{g} - 2)\sqrt{q},$$

$$\left| \sum_{P \in \tilde{X}(\mathbb{F}_q)} \chi(h(P)) \left(\frac{\varphi(P)}{q} \right) \right| \leqslant (2\tilde{g} - 2 + 2\deg\varphi)\sqrt{q},$$

where $\left(\dfrac{\cdot}{q} \right)$ denotes Legendre symbol, and $\tilde{g}$ is the genus of $\tilde{X}$.

Proof. See Theorem 3, [34].

Theorem 2. *Let f_1 be the brief SWU encoding from $\mathbb{F}_q$ to twisted Jacobi intersection curve $E_{a,b}$, $q \equiv 3 \pmod 4$. For any nontrivial character χ of $E_{a,b}(\mathbb{F}_q)$, the character sum $S_{f_1}(\chi)$ satisfies:*

$$|S_{f_1}(\chi)| \leqslant 16\sqrt{q} + 45.$$

Proof. Let $S = \{0\} \bigcup \{\text{roots of } g(X(t)) = 0\} \bigcup \{\text{roots of } g(Y(t)) = 0\}$ where $X(\cdot)$ and $Y(\cdot)$ are defined as in Sect. 3.1. For any $t \in \mathbb{F}_q \backslash S$, $X(t)$ and $Y(t)$ are both well defined and nonzero. Let $C_X = \{(t, x, y) \in \mathbb{F}_q^3 | x = X(t), y = -\sqrt{g(X(t))}\}, C_Y = \{(t, x, y) \in \mathbb{F}_q^3 | x = Y(t), y = \sqrt{g(Y(t))}\}$ be the smooth projective curves. It is trivial to see there exist one-to-one map $P_X : t \mapsto (t, x \circ \rho_X(t), y \circ \rho_X(t))$ from $\mathbb{P}^1(\mathbb{F}_q)$ to C_X and $P_Y : t \mapsto (t, x \circ \rho_Y(t), y \circ \rho_Y(t))$ from $\mathbb{P}^1(\mathbb{F}_q)$ to C_Y. Let h_X and h_Y be the projective maps on C_X and C_Y satisfying $\rho_X(t) = h_X \circ P_X(t)$ and $\rho_Y(t) = h_Y \circ P_Y(t)$. Let $g_X = P_X^{-1}$, $g_Y = P_Y^{-1}$, $S_X = g_X^{-1}(S \bigcup \{\infty\}) = P_X(S) \bigcup P_X(\infty)$, $S_Y = g_Y^{-1}(S \bigcup \{\infty\}) = P_Y(S) \bigcup P_Y(\infty)$.

To estimate $S_{f_1}(\chi)$,

$$S_{f_1}(\chi) = \left| \sum_{t \in \mathbb{F}_q \backslash S} (f_1^*\chi)(t) + \sum_{t \in S} (f_1^*\chi)(t) \right|$$

$$\leqslant \left| \sum_{t \in \mathbb{F}_q \backslash S} (f_1^*\chi)(u) \right| + \#S,$$

we deduce as follows,

$$\left| \sum_{t \in \mathbb{F}_q \backslash S} (f_1^*\chi)(u) \right| = \left| \sum_{\substack{P \in C_Y \backslash S_Y \\ \left(\frac{y(P)}{q} \right) = 1}} (h_Y^*\psi^*\chi)(P) + \sum_{\substack{P \in C_X \backslash S_X \\ \left(\frac{y(P)}{q} \right) = -1}} (h_X^*\psi^*\chi)(P) \right|$$

$$\leqslant \#S_Y + \#S_X + \left| \sum_{\substack{P \in C_Y \\ \left(\frac{y(P)}{q} \right) = +1}} (h_Y^*\psi^*\chi)(P) \right| + \left| \sum_{\substack{P \in C_X \\ \left(\frac{y(P)}{q} \right) = -1}} (h_X^*\psi^*\chi)(P) \right|,$$

and

$$2 \left| \sum_{\substack{P \in C_Y \\ \left(\frac{y(P)}{q}\right)=+1}} (h_Y^* \psi^* \chi)(P) \right|$$

$$= \left| \sum_{P \in C_Y} (h_Y^* \psi^* \chi)(P) + \sum_{P \in C_Y} (h_Y^* \psi^* \chi)(P) \cdot \left(\frac{y(P)}{q}\right) \right.$$

$$\left. - \sum_{\left(\frac{y(P)}{q}\right)=0} (h_Y^* \psi^* \chi)(P) \right|$$

$$\leqslant \left| \sum_{P \in C_Y} (h_Y^* \psi^* \chi)(P) \right| + \left| \sum_{P \in C_Y(\mathbb{F}_q)} (h_Y^* \psi^* \chi)(P) \cdot \left(\frac{y(P)}{q}\right) \right|$$

$$+ \#\{\text{roots of } g(Y(t)) = 0\}.$$

From the covering $\psi \circ h_Y : C_Y \to E$, $Y(t) = s \circ \psi^{-1}(u, v, w)$, which implies

$$T(t) = (aw + bw + a + vb)t^2 - (v - 1)b = 0.$$

$$\Leftrightarrow t^2 = \frac{(v-1)b}{(a+b)w + a + vb}.$$

Indeed, $\psi \circ h_Y$ is ramified if and only if $T(t)$ has multiple roots, which occurs when $t = 0$ or at infinity. Hence by Riemann-Hurwitz formula,

$$2g_{C_Y} - 2 = 0 + 1 + 1 = 2.$$

Hence curve C_Y is of genus 2. Similarly, C_X is also of genus 2.

Observe that

$$\deg y = [\mathbb{F}_q(t, x, y) : \mathbb{F}_q(y)] = [\mathbb{F}_q(t, x, y) : \mathbb{F}_q(x, y)][\mathbb{F}_q(x, y) : \mathbb{F}_q(y)] = 2 \cdot 3 = 6.$$

Further more, by Theorem 1, $\left| \sum_{P \in C_Y} (h_Y^* \psi^* \chi)(P) \right| \leqslant (2g_{C_Y} - 2)\sqrt{q} = 2\sqrt{q}$,

$\left| \sum_{P \in C_Y} (h_Y^* \psi^* \chi)(P) \cdot \left(\frac{y(P)}{q}\right) \right| \leqslant (2g_{C_Y} - 2 + 2 \det y)\sqrt{q} = 14\sqrt{q}$, and

$g(Y(t)) = 0$ is a sextic polynomial, we can derive

$$\left| \sum_{\substack{P \in C_Y \\ \left(\frac{y(P)}{q}\right)=+1}} (h_Y^* \psi^* \chi)(P) \right| \leqslant 8\sqrt{q} + 3.$$

And

$$\left| \sum_{\substack{P \in C_X \\ \left(\frac{y(P)}{q}\right)=-1}} (h_X^* \psi^* \chi)(P) \right|$$

has the same bound.

Hence $|S_{f_1}(x)| \leqslant 16\sqrt{q} + 6 + \#S_Y + \#S_X + \#S$. Note that $g(X(t)) = 0$ and $g(Y(t)) = 0$ have common roots, we can deduce that $\#S \leqslant 1 + 6 = 7$. Thus $\#S_X \leqslant 2(\#S + 1) \leqslant 16$. By the same reason, $\#S_Y \leqslant 16$. Then $|S_{f_1}(x)| \leqslant 16\sqrt{q} + 45$. Thus f_1 is well-distributed encoding using the Theorem 3 in [34]. ∎

3.4 Calculating the Density of the Image

Ideally, a hash function has better evaluation, if its image covers more points. Hence it is an interesting problem to know how many points there are on the image. When dealing with short Weierstrass curves, Icart conjectured that the density of image $\dfrac{\#Im(f)}{\#E(\mathbb{F}_q)}$, is near $\dfrac{5}{8}$, see [17]. Fouque and Tibouchi proved this conjecture [14] using Chebotarev density theorem. Now we apply this theorem onto twisted Jacobi intersection curves, and estimate the sizes of images of deterministic encodings.

Theorem 3 (Chebotarev, [36]). *Let K be an extension of $\mathbb{F}_q(x)$ of degree $n < \infty$ and L a Galois extension of K of degree $m < \infty$. Assume $\mathbb{F}_q$ is algebraically closed in L, and fix some subset φ of $Gal(L/K)$ stable under conjugation. Let $s = \#\varphi$ and $N(\varphi)$ the number of places v of K of degree 1, unramified in L, such that the Artin symbol $\left(\dfrac{L/K}{v}\right)$ (defined up to conjugation) is in φ. Then*

$$|N(\varphi) - \frac{s}{m}q| \leqslant \frac{2s}{m}((m + g_L) \cdot q^{1/2} + m(2g_K + 1) \cdot q^{1/4} + g_L + nm)$$

where g_K and g_L are genera of the function fields K and L.

Theorem 4. *Let $E_{a,b}$ be the Jacobi Intersection curve over $\mathbb{F}_q$ defined by equation $au^2 + v^2 = bu^2 + w^2 = 1, ab(a - b) \neq 0$, f_1 is the corresponding brief SWU encoding function. Then*

$$|\#Im(f_1) - \frac{1}{2}q| \leqslant 4q^{1/2} + 6q^{1/4} + 27.$$

Proof. K is the function field of $E_{a,b}$ which is the quadratic extension of $\mathbb{F}_q(t)$, hence $n = 2$, and by the property of elliptic curve, $g_K = 1$.

By Algorithm 3.1, step 2 and 4, it is trivial to find $Gal(L/K) = S_2$, hence $m = \#S_2 = 2$. φ is the subset of $Gal(L/K)$ consisting a fixed point, which is just $(1)(2)$, then $s = 1$.

Let W be the preimage of the map ψ, $W(\mathbb{F}_q)$ be the corresponding rational points on W. By the property that ψ is one-to-one rational map, $\#Im(f_1) = \#Im(\psi^{-1} \circ f_1) = I_X + I_Y + I_0$, where $I_X = \#\{(x, y) \in W(\mathbb{F}_q)|\exists t \in \mathbb{F}_q, x = X(t), y = -\sqrt{g(X(t))} \neq 0\}$, $I_Y = \#\{(x, y) \in W(\mathbb{F}_q)|\exists t \in \mathbb{F}_q, x = Y(t), y = \sqrt{g(Y(t))} \neq 0\}$, $I_0 = \#\{(s, 0) \in W(\mathbb{F}_q)|g(X(t)) = 0 \text{ or } g(Y(t)) = 0\}$. It is trivial to see that $I_0 \leqslant 3$.

Let N_X denote the number of rational points on the curve W with an x-coordinate of the form $X(t)$ and N_Y denote the number of rational points on the curve W with an x-coordinate of the form $Y(t)$, we have

$$2I_X \leqslant N_X \leqslant 2I_X + I_0 \leqslant 2I_X + 3,$$
$$2I_Y \leqslant N_Y \leqslant 2I_Y + 3.$$

Hence $I_X + I_Y \leqslant \dfrac{1}{2}(N_X + N_Y) \leqslant I_X + I_Y + 3.$

Since the place v of K of degree 1 correspond to the projective unramified points on $E_{a,b}(\mathbb{F}_q)$, hence $|N_X - N(\varphi)| \leqslant 12 + 3 = 15$, where 3 represents the number of infinite points, 12 represents the number of ramified points. Then we have

$$|N_X - \frac{1}{2}q| \leqslant |N_X - N(\varphi)| + |N(\varphi) - \frac{1}{2}q|$$
$$\leqslant 15 + (4q^{1/2} + 6q^{1/4} + 6) = 4q^{1/2} + 6q^{1/4} + 21.$$

Analogously, $|N_Y - \frac{1}{2}q| \leqslant 4q^{1/2} + 6q^{1/4} + 21.$

Therefore, we have

$$|\#Im(f_1) - \frac{1}{2}q| \leqslant |\#Im(f_1) - \frac{N_X + N_Y}{2}| + |\frac{N_X + N_Y}{2} - \frac{1}{2}q|$$
$$\leqslant I_0 + |I_X - \frac{N_X}{2}| + |I_Y - \frac{N_Y}{2}| + (4q^{1/2} + 6q^{1/4} + 21)$$
$$\leqslant 3 + \frac{3}{2} + \frac{3}{2} + (4q^{1/2} + 6q^{1/4} + 21)$$
$$= 4q^{1/2} + 6q^{1/4} + 27.$$

∎

4   Cube Root Encoding

4.1   Algorithm

In the case that $q = p^n \equiv 2 \pmod 3$ is a power of odd prime number, we give our deterministic construction $f_2 : t \mapsto (u, v, w)$ in the following way:

Input: a, b and $t \in \mathbb{F}_q$, $ab(a - b) \neq 0$.
Output: A point $(u, v, w) \in E_{a,b}(\mathbb{F}_q)$.

1. If $t = 0$, return $(u, v, w) = (0, 1, 1)$ directly.
2. $\alpha = \dfrac{a + b + t^2}{3}.$
3. $\beta = \dfrac{ab - 3\alpha^2}{2}.$
4. $\gamma = \alpha t + \sqrt[3]{t\beta^2 - (\alpha t)^3}.$
5. $(u, v, w) = (0, 1, 1) - \dfrac{2t}{\gamma^2 - abt^2}(t\gamma + \beta, a(\gamma - bt), b(\gamma - at)).$

In step 4, we can calculate the cube root by $\sqrt[3]{x} = x^{(2q-1)/3}$ efficiently, since $q \equiv 2 \pmod 3$.

4.2   Theoretical Analysis of Time Cost

Let M, S, I and E_C represent the same as in Sect. 3.2. The cost of encoding function f_2 can be estimated as follows:

1. Computing t^2 costs an S. Then α can be calculated.
2. To compute β, we need an S.
3. We use an M to compute αt, then use $S + M$ to calculate $t\beta^2$ and another $S + M$ to calculate $(\alpha t)^3$, an E_C to compute the value of γ.
4. In the end, to calculate the value of $\dfrac{2t}{\gamma^2 - abt^2}$, we need $S + M + I$. An M to calculate $t\gamma$, then $3M$ to calculate u,v and w respectively.

Therefore, f_2 requires $E_C + I + 5S + 8M = E + 23M$.

4.3   Properties of Cube Root Encodings

Lemma 2. *Suppose $P(u, v, w)$ is a point on twisted Jacobi intersection curve $E_{a,b}$, then equation $f_2(t) = P$ has solutions satisfying $H(t; u, v, w) = 0$, where*

$$H(t; u, v, w) = (-bv + b + aw - a)t^4 + (-2\,a^2 - 2\,b^2v + 2\,b^2 + 2\,a^2w + 4\,bva \\ - 4\,awb)t^2 + (6\,b^2ua - 6\,ba^2u)t + (b^2wa - a^2wb + b^2va + b^3 - b^3v \\ - a^2vb - a^3 + a^3w - 2\,b^2a + 2\,ba^2).$$

Proof. By the algorithm in Sect. 4.1, we have

$$\begin{cases} \dfrac{u}{v-1} = \dfrac{t\gamma + \beta}{a(\gamma - bt)} \\ \dfrac{u}{w-1} = \dfrac{t\gamma + \beta}{b(\gamma - at)} \end{cases}$$

$$\Rightarrow \gamma = -\frac{(v-1)\,t^4 + (2\,va - 2\,a - 2\,b + 2\,bv)\,t^2 - 6\,uabt + (a^2 + b^2 - ab)v + ab - a^2 - b^2}{6(-tv + t + ua)}$$

$$= -\frac{(w-1)t^4 + (2aw - 2a - 2b + 2wb)t^2 - 6uabt + (a^2 + b^2 - ab)w + ab - a^2 - b^2}{6(-tw + t + ub)}$$

$$\Rightarrow (-bv + b + aw - a)\,t^4 + \left(-2\,a^2 - 2\,b^2v + 2\,b^2 + 2\,a^2w + 4\,bva - 4\,awb\right)t^2 \\ + \left(6\,b^2ua - 6\,ba^2u\right)t + (b^2wa - a^2wb + b^2va + b^3 - b^3v - a^2vb \\ - a^3 + a^3w - 2\,b^2a + 2\,ba^2) = 0.$$

4.4   The Genus of Curve C

Denote F by the algebraic closure of $\mathbb{F}_q$. We consider the graph of f_2:

$$
\begin{aligned}
C &= \{(t, u, v, w) \in E_{a,b} \times \mathbb{P}^1(F)|\quad f_2(u) = (u, v, w)\} \\
&= \{(t, u, v, w) \in E_{a,b} \times \mathbb{P}^1(F)|\quad H(t; u, v, w) = 0\},
\end{aligned}
$$

which is the subscheme of $E_{a,b} \times \mathbb{P}^1(F)$.

Now we calculate the genus of C. The projection $g : C \to E_{a,b}$ is a morphism of degree 4, hence the fiber at each point of $E_{a,b}$ contains 4 points. The branch points of $E_{a,b}$ are points $(u, v, w) \in E_{a,b}$ where $H(t; u, v, w)$ has multiple roots, which means the discriminant $D = disc(H)$ vanishes at (u, v, w). By substituting $u^2 = \dfrac{1 - v^2}{a}$ and $w^2 = 1 - \dfrac{b}{a}(1 - v^2)$ into D, it can be represented as

$$
D = -144ab^2(P(v)w + Q(v)) = 0 \Rightarrow w = -\frac{Q(v)}{P(v)},
$$

where $P(v)$ is a polynomial of degree 5, $Q(v)$ is a polynomial of degree 6. Substituting $w = -\dfrac{Q(v)}{P(v)}$ into $\dfrac{b}{a}(1 - v^2) = 1 - w^2$, we find that v satisfies $R(v) = 0$, where $R(v)$ is a polynomial of degree 12. For each v, the value of w is uniquely determined while there are 2 values of corresponding u, one is opposite to the other. Hence there are at most 24 branch points on $E_{a,b}$. If $H(t; u, v, w)$ has triple roots at (u, v, w), we have:

$$
\begin{cases}
E(u, v, w) = 0 \\
H(t; u, v, w) = 0 \\
\dfrac{d}{dt}H(t; u, v, w) = 0 \\
\dfrac{d^2}{dt^2}H(t; u, v, w) = 0.
\end{cases}
\tag{2}
$$

In general cases, (2) has no solution, thus all 24 branch points have ramification index 2. By Riemann-Hurwitz formula,

$$
2g_C - 2 \leqslant 4 \cdot (2 \cdot 1 - 2) + 24 \cdot (2 - 1),
$$

we get $g_C \leqslant 13$.

Hence we have

Theorem 5. *The genus of curve C is at most* 13.

Next, we will utilize this theorem to estimate the upper bound of the character sum for an arbitrary nontrivial character of $E_{a,b}(\mathbb{F}_q)$.

4.5   Estimating Character Sums on the Curve

Theorem 6. *Let f_2 be the cube root encoding from $\mathbb{F}_q$ to twisted Jacobi intersection curve E, $q \equiv 3 \pmod 4$. For any nontrivial character χ of $E_{a,b}(\mathbb{F}_q)$, the character sum $S_{f_2}(\chi)$ satisfies $|S_{f_2}(\chi)| \leqslant 24\sqrt{q} + 3$.*

Proof. Let $K = \mathbb{F}_q(x, y)$ be the function field of E. Recall that a point $(u, v, w) \in E_{a,b}$ is the image of t if and only if

$$H(t; u, v, w) = 0.$$

Then a smooth projective curve $C = \{(u, v, w, t) | (u, v, w) \in E_{a,b}, H(t; u, v, w) = 0\}$ is introduced, whose function field is the extension $L = K[t]/(H)$. By field inclusions $\mathbb{F}_q(u) \subset L$ and $K \subset L$ we can construct birational maps $g : C \to \mathbb{P}^1(\mathbb{F}_q)$ and $h : C \to E$. Then g is a bijection and $f_2(u) = h \circ g^{-1}(u)$.

Since the genus of curve C is at most 13, by Theorem 1, we have

$$|S_{f_2}(\chi) + \sum_{P \in C(\mathbb{F}_q), t(P) = \infty} \chi \circ h(P)| = | \sum_{P \in C(\mathbb{F}_q)} \chi \circ h(P)| \leqslant (2 \cdot 13 - 2)\sqrt{q} = 24\sqrt{q}.$$

For any point on $E_{a,b}$ except $(u, v, w) = (0, 1, 1)$, the coefficient of t^4 in $H(t; u, v, w)$ is not 0, hence $H(t; u, v, w)$ has 4 finite solutions; $t = 0$ corresponds to $(u, v, w) = (0, 1, 1)$, therefore there exist at most 3 infinite solutions of t when $(u, v, w) = (0, 1, 1)$. Thus $|\sum_{P \in C(\mathbb{F}_q), t(P) = \infty} \chi \circ h(P)| \leqslant 3$. Hence $|S_{f_2}(\chi)| \leqslant 24\sqrt{q} + 3$. ∎

4.6   Galois Group of Field Extension

Let $K = F(u, v, w)$ be the function field of twisted Jacobi intersection curve $E_{a,b}$, L be the function field of C. To estimate the character sum of any character of Jacobian group of $E_{a,b}$, or to estimate the size of image of f_2, we need know the structure of $Gal(L/K)$. By [36], when L/K is a quartic extension, then $Gal(L/K) = S_4$ if and only if

1. $H(t)$ is irreducible over K.
2. Let $R(t)$ be the resolvent cubic of $H(t)$, then $R(t)$ is irreducible over K.
3. The discriminant of $R(t)$ is not a square in K.

Thus we have to prove 3 following lemmas:

Lemma 3. *The polynomial $H(t)$ is irreducible over K.*

Proof. Substitute $u = -\dfrac{2y}{x^2 - ab}$, $v = \dfrac{x^2 - 2ax + b}{x^2 - ab}$ and $w = \dfrac{x^2 - 2bx + ab}{x^2 - ab}$ into $H(t; u, v, w)$, we only need to show

$$\tilde{H}(t; x, y) = t^4 + (2b + 2a - 6x)t^2 + 6yt + a^2 - ab + b^2$$

is irreducible over $F(x, y) = F(u, v, w) = K$. Let σ be the non trivial Galois automorphism in $Gal(F(x, y)/F(x))$, which maps y to $-y$, it remains to show $\tilde{H}_0(t; x, y) = \tilde{H}(t; x, y)\tilde{H}(t; x, y)^\sigma$ is irreducible over $F(x)$. Let $s = t^2$, Note that $\tilde{H}_0(t)$ can be represented as polynomial of sv:

$$
\begin{aligned}
J_0(s) = {}& s^4 + (4\,b + 4\,a - 12\,x)s^3 + (6\,a^2 + 6\,ab + 6\,b^2 - 24\,bx - 24\,ax + 36\,x^2)s^2 \\
& + (4\,a^3 - 12\,a^2 x - 24\,abx + 4\,b^3 - 12\,b^2 x - 36\,x^3 + 36\,bx^2 + 36\,ax^2)s \\
& + a^4 - 2\,a^3 b + 3\,a^2 b^2 - 2\,b^3 a + b^4.
\end{aligned}
$$

$$(3)$$

From (3), by Theorem 1.2.3 in [36], if $J_0(s)$ is reducible over $F(x)$, then either it can be decomposed as

$$
\begin{aligned}
J_0(s) &= (s + A)(s^3 + Bs^2 + Cs + D) \\
&= s^4 + (A + B)s^3 + (AB + C)s^2 + (AC + D)s + AD,
\end{aligned}
$$

or it can be decomposed as

$$
\begin{aligned}
J_0(s) &= (s^2 + As + B)(s^2 + Cs + D) \\
&= s^4 + (A + C)s^3 + (B + AC + D)s^2 + (BC + AD)s + BD,
\end{aligned}
$$

where $A, B, C, D \in F[x]$.

In the first case, note that $AD = a^4 - 2\,a^3 b + 3\,a^2 b^2 - 2\,b^3 a + b^4$, we can deduce that A and D are both constant. Since $A + B = 4\,b + 4\,a - 12\,x$, B is of degree 1. Since the coefficient of s^2 is 2, degree of C is 2, which can lead to the inference that the coefficient of v is also degree-2, a contradiction to the fact it is 3.

In the second case, B and D are constants. Hence summation of the degree of A and the degree of C equals to 2, which shows that the coefficient of s is at most degree-2, also a contradiction.

Then we have shown that $J_0(s)$ is irreducible over $F(x)$. Let z be a root of $H_0(t)$. Then

$$
[F(x, z) : F(x)] = [F(x, z) : F(x, z^2)] \cdot [F(x, z^2) : F(x)] = 4[F(x, z) : F(x, z^2)].
$$

Since $\tau \in Gal(F(x, z)/F(x, z^2))$ which maps z to $-z$ is not an identity, hence $Gal(F(x, z)/F(x, z^2)) \neq \{\iota\}$, thus $[F(x, z) : F(x, z^2)] \geqslant 2$. Hence $[F(x, z) : F(x)] \geqslant 8$, which shows that $H_0(t)$ is irreducible over $F(x)$. ∎

Lemma 4. *The resolvent polynomial* $R(t; u, v, w)$ *is irreducible over* K.

Proof. The resolvent cubic of H is

$$
\begin{aligned}
R(t; u, v, w) = {} & (b - a + wa - vb)^2 t^3 - (b - a + wa - vb)(-2a^2 - 4abw + 4abv + 2b^2 \\
& - 2b^2 v + 2a^2 w)t^2 - 4(b - a + wa - vb)(-a^2 bw - 2b^2 a - a^3 + ab^2 v - a^2 vb \\
& + b^2 wa - b^3 v + a^3 w + 2a^2 b + b^3)t + (8b^2 a^3 - 16b^4 a - 16a^5 w - 16b^5 v \\
& + 8b^5 v^2 + 8a^5 w^2 - 16a^4 b + 8a^2 b^3 + 8b^5 + 8a^5 + 8a^4 vbw + 8b^4 wav \\
& + 24a^2 b^3 v^2 - 24a^4 bw^2 - 16a^3 v^2 b^2 + 32a^2 b^3 w + 40b^4 av - 48b^2 a^3 w \\
& - 8b^4 wa - 8a^4 bv + 40a^4 bw + 24a^3 b^2 w^2 - 16b^3 w^2 a^2 - 36a^2 b^4 u^2 \\
& + 72a^3 b^3 u^2 - 36a^4 u^2 b^2 + 32a^3 b^2 v - 24ab^4 v^2 - 48b^3 a^2 v)
\end{aligned}
$$

(4)

Similar to previous lemma, we only need to show $\tilde{R}(t; x, y)$, the transformation of $R(t; u, v, w)$ such that it is defined on $\psi^{-1}(E)$, is irreducible over K. Represent u, v, w with variable x, y, we have

$$
\begin{aligned}
\tilde{R}(t; x) = {} & t^3 + (-2a - 2b + 6x)t^2 + (-4a^2 + 4ab - 4b^2)t + 8a^3 - 24b^2 x \\
& - 24a^2 x + 36ax^2 - 12bax + 8b^3 - 36x^3 + 36bx^2
\end{aligned}
$$

(5)

If $\tilde{R}(t; x)$ is reducible, it must have a degree 1 factor $t + A$, where $A \in F[x, y]$. If $A \notin F[x]$, then $(t + A)^\sigma$ is a factor of $\tilde{R}(t; x)^\sigma = \tilde{R}(t; x)$. Hence $\dfrac{\tilde{R}(t; x)}{(t + A)(t + A)^\sigma} \in F[x]$. Without loss of generality, we suppose $A \in F[x]$. Hence $\tilde{R}(t; x) = (t + A)(t^2 + Bu + C)$, $A, B, C \in F[x]$. In this case, $\tilde{R}(t; x)$ has a solution in $F[x]$ whose degree is 1, since when the value of t is a polynomial with degree $\neq 1$, $\tilde{R}(t; x)$ will be equal to a polynomial whose degree greater than 0. Suppose $A = Px + Q$, $P, Q \in F$, then

$$
\begin{cases}
B & = 6x - 2a - 2b - (Px + Q) \\
C & = 4ab - 4a^2 - 4b^2 - AB \\
AC & = 8a^3 - 24b^2 x \\
& \quad -24a^2 x + 36ax^2 - 12bax + 8b^3 - 36x^3 + 36bx^2.
\end{cases}
$$

Then P and Q satisfies

$$
\begin{aligned}
& (P^3 - 6P^2 + 36)x^3 + (-36b - 12PQ + 3P^2 Q + 2bP^2 - 36a + 2aP^2)x^2 \\
& + (4bPQ - 6Q^2 + 24a^2 + 12ab + 24b^2 + 4aPQ + 4abP - 4a^2 P - 4b^2 P \\
& + 3PQ^2)x - 8a^3 - 8b^3 + 2aQ^2 - 4a^2 Q - 4b^2 Q + Q^3 + 4abQ + 2bQ^2 = 0.
\end{aligned}
$$

(6)

where x is the variable. When $char(F) \geqslant 3$, it can be checked that solutions of P and Q do not exist. $\blacksquare$

Lemma 5. *Let $D(u, v, w)$ be the discriminant of $R(t; u, v, w)$, then $D(u, v, w)$ is not a square in K.*

Proof. Similar to previous proof, we only need to show that

$$\tilde{D}(x, y) = D(u(x, y), v(x, y), w(x, y))$$

is not a square in K. After simplification,

$$\tilde{D}(x,y) = 36864(\frac{ab(a-b)}{ab-x^2})^8(-27\,x^6 + (54\,b + 54\,a)\,x^5 - (27a^2 + 108ab + 27b^2)\,x^4$$
$$+ (16\,a^3 + 30\,a^2b + 30\,b^2a + 16\,b^3)\,x^3 + (-24\,a^3b + 69\,b^2a^2 - 24\,b^3a)\,x^2$$
$$+ (-24\,b^2a^3 - 24\,a^2b^3)\,x + 16\,a^4b^2 - 16\,b^3a^3 + 16\,b^4a^2)$$

$$(7)$$

In fact, we only need to show that $\tilde{G}(x, y) = -\dfrac{1}{36864}(\dfrac{ab-x^2}{ab(a-b)})^8\tilde{D}(x, y)$ is not a square over K.

Suppose $\tilde{G}$ is a square in $F(x, y)$, then $F(x, y) \supseteq F(x, \sqrt{\tilde{G}}) \supseteq F(x)$. Note that $[F(x, y) : F(x)] = 2$, either $F(x, \sqrt{\tilde{G}}) = F(x, y)$ or $F(x, \sqrt{\tilde{G}}) = F(x)$.

In the first case, $\tilde{G}$ is $x(x-a)(x-b) = y^2$ times a square in $F(x)$. But divide $\tilde{G}$ by $x(x-a)(x-b)$, the remainder does not vanish in general case.

In the second case, $\tilde{G}$ is a square over $F(x)$. Suppose

$$\tilde{G}(x) = \left(\sqrt{27}x^3 + Bx^2 + Cx \pm 4ab\sqrt{a^2 + b^2 - ab}\right)^2,$$

expand the right hand side of this equation and compare its coefficients of x^i, $i = 1$ to 5 with the left hand side, and it is checked there are no $B, C \in F$ s.t the equality holds. ∎

Summarize these lemmas, we directly deduce:

Theorem 7. *Let $K = \mathbb{F}_q(u, v, w)$ be the function field of $E_{a,b}$. The polynomial $H(t; u, v, w)$ is irreducible over K and its Galois group is S_4.*

In Sect. 5.2, we will use this theorem to construct a hash function indifferentiable from random oracle.

4.7   Calculating the Density

Similar to Sect. 3.4, we apply Chebotarev density theorem to estimate the size of image of f_2.

Theorem 8. *Let $E_{a,b}$ be the twisted Jacobi intersection curve over $\mathbb{F}_q$ defined by equation $au^2 + v^2 = bu^2 + w^2 = 1, ab(a-b) \neq 0$, f_2 is the corresponding hash function defined in Sect. 4.1. Then we have*

$$|\#Im(f_2) - \frac{5}{8}q| \leq \frac{5}{4}(31q^{1/2} + 72q^{1/4} + 65).$$

Proof. K is the function field of $E_{a,b}$ which is the quadratic extension of $\mathbb{F}_q(u)$, hence $n = 2$, and by the property of elliptic curve, $g_K = 1$. In Theorem 7 we see $Gal(L/K) = S_4$, then $m = \#S_4 = 24$. φ is the subset of $Gal(L/K)$ consisting at least 1 fixed point, which are conjugates of $(1)(2)(3)(4), (12)(3)(4)$ and $(123)(4)$, then $s = 1 + 6 + 8 = 15$. Since the place v of K of degree 1 correspond to the projective unramified points on $E_{a,b}(\mathbb{F}_q)$, hence $|\#Im(f_2) - N(\varphi)| \leqslant 12$, where 12 represents the number of ramified points. Then we have

$$|\#Im(f_2) - \frac{5}{8}q| \leqslant |\#Im(f_2) - N(\varphi)| + |N(\varphi) - \frac{5}{8}q|$$

$$\leqslant 12 + \frac{5}{4}(31q^{1/2} + 72q^{1/4} + 55)$$

$$\leqslant \frac{5}{4}(31q^{1/2} + 72q^{1/4} + 65).$$

∎

5 Construction of Hash Function Indifferentiable from Random Oracle

Suppose h is a canonical hash function from messages in $\{0,1\}^*$ to finite field $\mathbb{F}_q$. For $i = 1, 2$, we can show that each $f_i \circ h$ is one-way and collision-resistance according to the fact that through f_i, every point on $E_{a,b}$ has finite preimage [17]. Therefore $f_i \circ h, i = 1, 2$ are hash functions $\{0,1\}^* \to E_{a,b}(\mathbb{F}_q)$. However, since f_i are not surjective, $f_i \circ h$ are easy to be distinguished from a random oracle even when h is modeled as a random oracle to $\mathbb{F}_q$ [37]. Thus, we propose 2 new constructions of hash functions which are indifferentiable from a random oracle.

5.1 First Construction

Suppose $f : \mathbb{S} \to \mathbb{G}$ is a weak encoding [33] to a cyclic group $\mathbb{G}$, where $\mathbb{S}$ denotes prime field $\mathbb{F}_q$, $\mathbb{G}$ denotes $E(\mathbb{F}_q)$ which is of order N with generator G, $+$ denotes elliptic curve addition. According to the proof of random oracle, we can construct a hash function $H : \{0,1\}^* \to \mathbb{G}$:

$$H(m) = f(h_1(m)) + h_2(m)G,$$

where $h_1 : \{0,1\}^* \to \mathbb{F}_q$ and $h_2 : \{0,1\}^* \to \mathbb{Z}/N\mathbb{Z}$ are both canonical hash functions. $H(m)$ is indifferentiable from a random oracle in the random oracle model for h_1 and h_2.

We only need to show for $i = 1, 2$, f_i are both weak encodings to prove that $H_i(m) = f_i(h_1(m)) + h_2(m)G, i = 1, 2$ are indifferentiable from a random oracle in the random oracle model for h_1 and h_2. By the definition of weak encoding [33], f_1 is a $\frac{2N}{q}$-weak encoding and f_2 is a $\frac{4N}{q}$-weak encoding, both $\frac{2N}{q}$ and $\frac{4N}{q}$ are polynomial functions of the security parameter.

5.2   Second Construction

Another construction is as follows:

$$H_{i'} = f_i(h_1(m)) + f_i(h_2(m)), i = 1, 2.$$

We have proved that f_1 and f_2 are both well distributed encodings in Sects. 3.3 and 4.5. According to corollary 2 of [34], $H_{1'}$ and $H_{2'}$ are both indifferentiable from a random oracle, where h_1 and h_2 are regarded as independent random oracles with values in $\mathbb{F}_q$.

6   Time Comparison

When $q \equiv 3 \pmod 4$, the critical step of an encoding function is calculating square root for given element of $\mathbb{F}_q$. For convenience to make comparisons, we first introduce a birational map between twisted Jacobi intersection curve $E_{a,b}$ and Weierstrass curve

$$E_W : y^2 = x(x - a)(x - b),$$

via maps

$$\vartheta : E_{a,b} \rightarrow E_W :$$

$$(u, v, w) \mapsto (x, y) = \left(-\frac{a(w + 1)}{v - 1}, \frac{au}{v - 1}(x - b) \right),$$

$$\varsigma : E_W \rightarrow E_{a,b} :$$

$$(x, y) \mapsto (u, v, w) = \left(-\frac{2y}{x^2 - ab}, \frac{x^2 - 2ax + ab}{x^2 - ab}, \frac{x^2 - 2bx + ab}{x^2 - ab} \right).$$

(8)

Therefore, we compare our encoding f_1 with 2 encodings: birational equivalence ς in (8) composed with Ulas' encoding function [16], denoted by f_U; ς composed with simplified Ulas map given by Eric Brier et al. [33], denoted by f_E.

When $q \equiv 2 \pmod 3$, the essential of an encoding function is calculating the cube root for elements of $\mathbb{F}_q$. We compare our encoding f_2 with Alasha's work [22] denoted by f_A.

We have shown that f_1 costs $E + D + 31M$, f_2 costs $E + 23M$. For comparison, f_U costs $(E_S + I + 4S + 11M + D) + (I + 3M + S) = E + D + 39M$ by Theorem 2.3(2), [16] and the map ς in (8), while f_E costs $(E_S + I + 4S + 6M + D) + (I + 3M + S) = E + D + 34M$ by [14]. Alasha's encoding f_A costs $E_C + 9M + 4S + 2I = E + 33M$.

We do experiments on prime fields $\mathbb{F}_{P192}$ and $\mathbb{F}_{P384}$ (see Table 2). General Multiprecision PYthon project (GMPY2) [38], which supports the GNU Multiple Precision Arithmetic Library (GMP) [39] is used for big number arithmetic. The experiments are operated on an Intel(R) Core(TM) $i5-4570$, $3.20\,\mathrm{GHz}$ processor. We ran f_1, f_U, f_E, f_2 and f_A $1,000,000$ times each, where t is randomly chosen on $\mathbb{F}_{P192}$ and $\mathbb{F}_{P384}$ (Table 1).

Table 1. Theoretic time cost of different deterministic encodings

Encoding	Cost	Converted cost
f_1	$E_S + 2I + D + S + 10M$	$E + D + 31M$
f_U	$E_S + 2I + D + 5S + 14M$	$E + D + 39M$
f_E	$E_S + 2I + D + 5S + 9M$	$E + D + 34M$
f_2	$E_C + I + 5S + 8M$	$E + 23M$
f_A	$E_C + 2I + 4S + 9M$	$E + 33M$

Table 2. NIST primes

Prime	Value	Residue (mod 3)	Residue (mod 4)
$P192$	$2^{192} - 2^{64} - 1$	2	3
$P384$	$2^{384} - 2^{128} - 2^{96} + 2^{32} - 1$	2	3

Table 3. Time cost (ms) of different square root methods on NIST

Prime	$P192$	$P384$
f_1	0.052	0.233
f_E	0.057	0.247
f_U	0.059	0.249

Table 4. Time cost (ms) comparison between f_2 and f_A

Prime	$P192$	$P384$
f_2	0.050	0.230
f_A	0.059	0.249

From the average running times listed in Table 3, f_1 is the fastest among encodings which need calculate square roots. On $\mathbb{F}_{P192}$, it saves 9.89% running time compared with f_U, 8.72% running time compared with f_E. On $\mathbb{F}_{P384}$, f_1 saves 6.17% running time compared with f_U and 6.23% running time compared with f_E. f_2 is also the fastest among encodings which need to calculate cube roots. On $\mathbb{F}_{P192}$, it saves 13.85% of running time compared with f_A. On $\mathbb{F}_{P384}$, the relevant percentages is 7.63% (see Table 4).

7   Conclusion

We provide two constructions of deterministic encoding into twisted Jacobi intersection curves over finite fields, namely, brief SWU encoding and cube root encoding. We do theoretical analysis and practical implementations to show that when $q \equiv 3 \pmod 4$, SWU encoding is the most efficient among existed methods mapping $\mathbb{F}_q$ into twisted Jacobi intersections curve $E_{a,b}$, while cube root encoding is the most efficient one when $q \equiv 2 \pmod 3$. When $q \equiv 11 \pmod{12}$, in which case f_1 and f_2 can be both applied, it can be checked f_2 is faster than f_1. For any nontrivial character χ of $E(\mathbb{F}_q)$, we estimate the upper bound of the character sums of both encodings. As a corollary, hash functions indifferentiable from random oracle are constructed. We also estimate image sizes of our encodings by applying Chebotarev density theorem.

References

1. Baek, J., Zheng, Y.: Identity-based threshold decryption. In: Bao, F., Deng, R., Zhou, J. (eds.) PKC 2004. LNCS, vol. 2947, pp. 262–276. Springer, Heidelberg (2004). https://doi.org/10.1007/978-3-540-24632-9_19
2. Horwitz, J., Lynn, B.: Toward hierarchical identity-based encryption. In: Knudsen, L.R. (ed.) EUROCRYPT 2002. LNCS, vol. 2332, pp. 466–481. Springer, Heidelberg (2002). https://doi.org/10.1007/3-540-46035-7_31
3. Boneh, D., Gentry, C., Lynn, B., Shacham, H.: Aggregate and verifiably encrypted signatures from bilinear maps. In: Biham, E. (ed.) EUROCRYPT 2003. LNCS, vol. 2656, pp. 416–432. Springer, Heidelberg (2003). https://doi.org/10.1007/3-540-39200-9_26
4. Zhang, F., Kim, K.: ID-based blind signature and ring signature from pairings. In: Zheng, Y. (ed.) ASIACRYPT 2002. LNCS, vol. 2501, pp. 533–547. Springer, Heidelberg (2002). https://doi.org/10.1007/3-540-36178-2_33
5. Boyen, X.: Multipurpose identity-based signcryption. In: Boneh, D. (ed.) CRYPTO 2003. LNCS, vol. 2729, pp. 383–399. Springer, Heidelberg (2003). https://doi.org/10.1007/978-3-540-45146-4_23
6. Libert, B., Quisquater, J.-J.: Efficient signcryption with key privacy from gap Diffie-Hellman groups. In: Bao, F., Deng, R., Zhou, J. (eds.) PKC 2004. LNCS, vol. 2947, pp. 187–200. Springer, Heidelberg (2004). https://doi.org/10.1007/978-3-540-24632-9_14
7. Lindell, Y.: Highly-efficient universally-composable commitments based on the DDH assumption. In: Paterson, K.G. (ed.) EUROCRYPT 2011. LNCS, vol. 6632, pp. 446–466. Springer, Heidelberg (2011). https://doi.org/10.1007/978-3-642-20465-4_25
8. Boneh, D., Franklin, M.: Identity-based encryption from the weil pairing. In: Kilian, J. (ed.) CRYPTO 2001. LNCS, vol. 2139, pp. 213–229. Springer, Heidelberg (2001). https://doi.org/10.1007/3-540-44647-8_13
9. Boyd, C., Montague, P., Nguyen, K.: Elliptic curve based password authenticated key exchange protocols. In: Varadharajan, V., Mu, Y. (eds.) ACISP 2001. LNCS, vol. 2119, pp. 487–501. Springer, Heidelberg (2001). https://doi.org/10.1007/3-540-47719-5_38
10. Jablon, D.P.: Strong password-only authenticated key exchange. SIGCOMM Comput. Commun. Rev. **26**(5), 5–26 (1996)
11. Boyko, V., MacKenzie, P., Patel, S.: Provably secure password-authenticated key exchange using Diffie-Hellman. In: Preneel, B. (ed.) EUROCRYPT 2000. LNCS, vol. 1807, pp. 156–171. Springer, Heidelberg (2000). https://doi.org/10.1007/3-540-45539-6_12

12. Shallue, A., van de Woestijne, C.E.: Construction of rational points on elliptic curves over finite fields. In: Hess, F., Pauli, S., Pohst, M. (eds.) ANTS 2006. LNCS, vol. 4076, pp. 510–524. Springer, Heidelberg (2006). https://doi.org/10.1007/11792086_36

13. Skalba, M.: Points on elliptic curves over finite fields. Acta Arith. **117**, 293–301 (2005)

14. Fouque, P.-A., Tibouchi, M.: Estimating the size of the image of deterministic hash functions to elliptic curves. In: Abdalla, M., Barreto, P.S.L.M. (eds.) LATIN-CRYPT 2010. LNCS, vol. 6212, pp. 81–91. Springer, Heidelberg (2010). https://doi.org/10.1007/978-3-642-14712-8_5

15. Fouque, P.-A., Tibouchi, M.: Deterministic encoding and hashing to odd hyperelliptic curves. In: Joye, M., Miyaji, A., Otsuka, A. (eds.) Pairing 2010. LNCS, vol. 6487, pp. 265–277. Springer, Heidelberg (2010). https://doi.org/10.1007/978-3-642-17455-1_17

16. Ulas, M.: Rational points on certain hyperelliptic curves over finite fields. Bull. Polish Acad. Sci. Math. **55**, 97–104 (2007)

17. Icart, T.: How to hash into elliptic curves. In: Halevi, S. (ed.) CRYPTO 2009. LNCS, vol. 5677, pp. 303–316. Springer, Heidelberg (2009). https://doi.org/10.1007/978-3-642-03356-8_18

18. Farashahi, R.R.: Hashing into Hessian curves. In: Nitaj, A., Pointcheval, D. (eds.) AFRICACRYPT 2011. LNCS, vol. 6737, pp. 278–289. Springer, Heidelberg (2011). https://doi.org/10.1007/978-3-642-21969-6_17

19. Yu, W., Wang, K., Li, B., Tian, S.: About hash into montgomery form elliptic curves. In: Deng, R.H., Feng, T. (eds.) ISPEC 2013. LNCS, vol. 7863, pp. 147–159. Springer, Heidelberg (2013). https://doi.org/10.1007/978-3-642-38033-4_11

20. Yu, W., Wang, K., Li, B., He, X., Tian, S.: Hashing into Jacobi quartic curves. In: Lopez, J., Mitchell, C.J. (eds.) ISC 2015. LNCS, vol. 9290, pp. 355–375. Springer, Cham (2015). https://doi.org/10.1007/978-3-319-23318-5_20

21. Yu, W., Wang, K., Li, B., He, X., Tian, S.: Deterministic encoding into twisted edwards curves. In: Liu, J.K., Steinfeld, R. (eds.) ACISP 2016. LNCS, vol. 9723, pp. 285–297. Springer, Cham (2016). https://doi.org/10.1007/978-3-319-40367-0_18

22. Alasha, T.: Constant-time encoding points on elliptic curve of diffierent forms over finite fields (2012). http://iml.univ-mrs.fr/editions/preprint2012/files/tammam_alasha-IML_paper_2012.pdf

23. Feng, R., Nie, M., Wu, H.: Twisted Jacobi intersections curves. In: Kratochvíl, J., Li, A., Fiala, J., Kolman, P. (eds.) TAMC 2010. LNCS, vol. 6108, pp. 199–210. Springer, Heidelberg (2010). https://doi.org/10.1007/978-3-642-13562-0_19

24. Chudnovsky, D.V., Chudnovsky, G.V.: Sequences of numbers generated by addition in formal groups and new primality and factorization tests. Adv. Appl. Math. **7**, 385–434 (1986)

25. Bernstein, D.J., Lange, T.: Explicit-formulae database. http://www.hyperelliptic.org/EFD

26. Hisil, H., Carter, G., Dawson, E.: New formulae for efficient elliptic curve arithmetic. In: Srinathan, K., Rangan, C.P., Yung, M. (eds.) INDOCRYPT 2007. LNCS, vol. 4859, pp. 138–151. Springer, Heidelberg (2007). https://doi.org/10.1007/978-3-540-77026-8_11

27. Hisil, H., Koon-Ho Wong, K., Carter, G., Dawson, E.: Faster group operations on elliptic curves. In: Brankovic, L., Susilo, W. (eds.) Proceedings of Seventh Australasian Information Security Conference (AISC 2009), Wellington, New Zealand. CRPIT, vol. 98. pp. 7–19. ACS (2009)

28. Liardet, P.-Y., Smart, N.P.: Preventing SPA/DPA in ECC systems using the Jacobi form. In: Koç, Ç.K., Naccache, D., Paar, C. (eds.) CHES 2001. LNCS, vol. 2162, pp. 391–401. Springer, Heidelberg (2001). https://doi.org/10.1007/3-540-44709-1_32

29. Wu, H., Feng, R.: A complete set of addition laws for twisted Jacobi intersection curves. Wuhan Univ. J. Nat. Sci. **16**(5), 435–438 (2011)

30. Hongyu, C.A.O., Kunpeng, W.A.N.G.: Skew-frobenius mapping on twisted Jacobi intersection curve. Comput. Eng. **41**(1), 270–274 (2015)

31. Elmegaard-Fessel, L.: Efficient Scalar Multiplication and Security against Power Analysis in Cryptosystems based on the NIST Elliptic Curves Over Prime Fields. Eprint, 2006/313. http://eprint.iacr.org/2006/313

32. Standards for Efficient Cryptography, Elliptic Curve Cryptography Ver. 0.5 (1999). http://www.secg.org/drafts.htm

33. Brier, E., Coron, J.-S., Icart, T., Madore, D., Randriam, H., Tibouchi, M.: Efficient indifferentiable hashing into ordinary elliptic curves. In: Rabin, T. (ed.) CRYPTO 2010. LNCS, vol. 6223, pp. 237–254. Springer, Heidelberg (2010). https://doi.org/10.1007/978-3-642-14623-7_13

34. Farashahi, R.R., Fouque, P.-A., Shparlinski, I.E., Tibouchi, M., Voloch, J.F.: Indifferentiable deterministic hashing to elliptic and hyperelliptic curves. Math. Comp. **82**, 491–512 (2013)

35. Farashahi, R.R., Shparlinski, I.E., Voloch, J.F.: On hashing into elliptic curves. J. Math. Cryptology **3**(4), 353–360 (2009)

36. Roman, S.: Field Theory. Graduate Texts in Mathematics, vol. 158, 2nd edn. Springer, New York (2011)

37. Tibouchi, M.: Impossibility of surjective Icart-Like encodings. In: Chow, S.S.M., Liu, J.K., Hui, L.C.K., Yiu, S.M. (eds.) ProvSec 2014. LNCS, vol. 8782, pp. 29–39. Springer, Cham (2014). https://doi.org/10.1007/978-3-319-12475-9_3

38. GMPY2, General Multiprecision Python (Version 2.2.0.1). https://gmpy2.readthedocs.org

39. GMP: GNU Multiple Precision Arithmetic Library. https://gmplib.org/

Digital Signatures

Identity-Based Key-Insulated Aggregate Signatures, Revisited

Nobuaki Kitajima[1,2(✉)], Naoto Yanai[3,4], and Takashi Nishide[2]

[1] Tata Consultancy Services Japan, Ltd., Tokyo, Japan
nobuaki.kitajima.t@gmail.com
[2] University of Tsukuba, Tsukuba, Japan
[3] Osaka University, Osaka, Japan
[4] National Institute of Advanced Industrial Science and Technology (AIST),
Tsukuba, Japan

Abstract. Identity-based key-insulated cryptography is a cryptography which allows a user to update an exposed secret key by generating a temporal secret key as long as the user can keep any string as its own public key. In this work, we consider the following question; namely, can we construct aggregate signatures whereby individual signatures can be aggregated into a single signature in an identity-based key-insulated setting? We call such a scheme *identity-based key-insulated aggregate signatures (IBKIAS)*, and note that constructing an IBKIAS scheme is non-trivial since one can aggregate neither each signer's randomness nor components depending on the temporal secret keys. To overcome this problem, we utilize the synchronized technique proposed by Gentry and Ramzan (PKC'06) for both aas state information and a partial secret key generated by a secure device. We then show that the proposed scheme is still provably secure under an adaptive security model of identity-based aggregate signatures.

Keywords: Key-insulated signatures · Identity-based signatures
Aggregate signatures

1 Introduction

1.1 Background and Motivation

Digital signatures ensure the validity of digital documents. One of the main approaches to ensure the security of digital signatures is to reduce them to computationally difficult problems, such as an integer factoring problem and a discrete logarithm problem. However, it is typically easier for an adversary, who tries to forge a signature, to obtain a secret key from a naive user than from breaking computationally difficult problems. Nowadays, such threats seem to have increased with respect to deployments of digital signatures on insecure devices, e.g. mobile devices or Internet-of-Things (IoT) devices, since these can

© Springer International Publishing AG, part of Springer Nature 2018
X. Chen et al. (Eds.): Inscrypt 2017, LNCS 10726, pp. 141–156, 2018.
https://doi.org/10.1007/978-3-319-75160-3_10

be stolen or lost easily. Another standpoint is downgrading of underlying hardness assumptions. Mobile or IoT devices often need to use a shorter security parameter because a longer parameter cannot be used due to a small size of memories. However, such a shorter parameter may give an adversary an advantage in the sense that their underlying hard problems become easier. The problem described above is often called *secret key-exposure* (or simply *key-exposure*). In this paper, we focus on *key-insulated public key cryptosystems* proposed by Dodis et al. [2] to reduce the damage caused by the secret key-exposure.

In the model in [2], a user begins by registering a single public key pk, and stores a "master" secret key sk^* on a physically secure device. The device is resistant to compromise. Cryptographic operations, such as decryption in public key encryptions or signing in digital signatures, are done on a user's insecure devices such that the key-exposure is expected. The lifetime of a protocol is divided into distinct periods $1, \ldots, N$ (for simplicity, we assume that these time periods have an equal length; e.g. one day). At the beginning of each period, a user interacts with the secure device to obtain a temporary secret key sk_k for period k, but a public key pk does not change at each period. Instead, ciphertexts or signatures are labeled with the time period k. The user's insecure device is assumed to be vulnerable to repeated key-exposures. Namely, up to $\mathcal{T} < N$ periods, keys can be compromised. When a secret key sk_k is exposed, an adversary will be able to decrypt ciphertexts or sign messages for period k. However, the goal in [2] is to *minimize* the effect of such compromises; an adversary will be unable to decrypt ciphertexts or forge signatures for any remaining $N - \mathcal{T}$ periods. A scheme satisfying this notion is called $(\mathcal{T}, N)$-*key-insulated*.

Our main motivation for this work is to consider *identity-based aggregate signatures (IBAS)* [4] in the key-insulated setting described above. In aggregate signatures, proposed by Boneh et al. [1], multiple signatures can be aggregated into a single short signature even if these signatures are generated by different signers on different messages. This is useful since, for mobile or IoT devices whose memory size and bandwidth are small, the size of storages and communication overhead can become small. Moreover, owing to identity-based setting, IBAS allow us to use an arbitrary string as its public keys (e.g. email address. In a real world scenario, this is where many users have their own mobile devices). In fact, the identity-based setting is known as a suitable setting for mobile or IoT communication [12]. However, key-exposure in insecure devices becomes an unavoidable threat in the environments as described above. If a secret key of a signer in aggregate signatures is compromised, then the whole of the aggregate signatures become untrustworthy. Moreover, in an identity-based environment, a signer corresponding to the exposed secret key cannot use his/her string as the public key, and removing these keys from its system is necessary. Thus, reducing the damage caused by key-exposure in IBAS is important.

1.2 Contribution

We propose *identity-based key-insulated aggregate signatures (IBKIAS)* which are IBAS with the key-insulated property. Although Reddy and Gopal [9] first considered the concept of IBKIAS, unfortunately, their scheme is *not* an aggregate signature scheme. In their scheme, one cannot aggregate each signer's randomness used in an individual signing. Namely, the size of aggregate signatures is linear with respect to the number of signers. *Compactness*, where the size of signatures is fixed with respect to the number of signers, is the most important feature, and our scheme can achieve the compactness. Thus, our scheme is the first IBKIAS scheme in the sense of the compactness of aggregate signatures.

We note that constructing an IBKIAS scheme from an identity-based key-insulated signature scheme [11,14] is non-trivial since both components depend on each signer's randomness, and components depending on a temporal secret key become linear with respect to the number of signers, similarly to the scheme by Reddy and Gopal [9]. To overcome this problem, we utilize a synchronized technique proposed by Gentry and Ramzan [4] where each signer shares a common random string in advance. Moreover, by sharing not only the common random string but also a single physically secure device between signers, our scheme completely achieves the compactness. That is, we utilize the synchronized technique for both signatures themselves and the physically secure device. We show that IBKIAS in such a setting is still provably secure in the adaptive security model of IBAS suggested by [4]. We also revisit a security model of IBKIAS. Our model is able to discuss an adaptive-ID security where a target identity is not designated in advance, while the existing model [9] can only deal with a selective-ID model. We compare the scheme by Reddy and Gopal [9] with our proposed scheme IBKIAS$_{\text{Ours}}$, and present the results in Table 1.

Table 1. Comparison of different schemes. Size of aggregate signature (Agg Size) counts group elements, and we denote by I the number of signers and by $|\mathbb{G}|$ a bit length of an element in a group $\mathbb{G}$. We also assume that the size of state information w (described later) is equal to a security parameter λ.

	Security model	Agg size	Compactness
Reddy and Gopal [9]	Selective-ID	$(I+1)|\mathbb{G}|$	No
IBKIAS$_{\text{Ours}}$	Adaptive-ID	$4|\mathbb{G}| + \lambda$	Yes

Our future work proposes an IBKIAS scheme without the synchronized technique and/or random oracles.

1.3 Related Work

Due to page limitations, we briefly review some security notions of key-insulated signatures in Sect. 2. Some identity-based key-insulated cryptosystems also have been proposed such as hierarchical encryption schemes [5,10] or signature schemes [11,14]. We consider a similar and efficient instantiations of IBKIAS that may be constructed by applying our technique.

Zhao et al. [13] considered key-insulated aggregate signatures, but their scheme is not an aggregate signature scheme because the compactness does not hold. Instead of their construction, our construction essentially implies a secure key-insulated aggregate signature although we omit the details. The works in [6–8] are recent and significant result in deriving (identity-based) aggregate signatures from any standard signatures. We consider that the security of our scheme can be proven by introducing these methodologies.

2   Preliminaries

2.1   Notation and Security Assumption

If X is a finite set, we use $x \xleftarrow{\$} X$ to denote that x is chosen uniformly at random from X. If $\mathcal{A}$ is an algorithm, we use $x \leftarrow \mathcal{A}(y)$ to denote that x is output by $\mathcal{A}$ whose input is y. Let $\mathbb{Z}_q$ denote $\{0, 1, \ldots, q-1\}$ and $\mathbb{Z}_q^*$ denote $\mathbb{Z}_q \backslash \{0\}$. We say that a function $f(\cdot) : \mathbb{N} \to [0, 1]$ is negligible if, for all positive polynomials $p(\cdot)$ and all sufficiently large $\lambda \in \mathbb{N}$, we have $f(\cdot) < 1/p(\lambda)$. We say that a randomized algorithm $\mathcal{A}$ runs in probabilistic polynomial time (PPT) if there exists a polynomial p such that execution time of $\mathcal{A}$ with input length λ is less than $p(\lambda)$.

Let $\mathbb{G}$ and $\mathbb{G}_T$ be groups with the same prime order q. We then define bilinear maps and the computational Diffie-Hellman (CDH) assumption in $\mathbb{G}$ as follows. Here, we write $\mathbb{G}$ additively and $\mathbb{G}_T$ multiplicatively.

Definition 1 *(Bilinear maps). A bilinear map $e : \mathbb{G} \times \mathbb{G} \to \mathbb{G}_T$ is a map such that the following conditions hold: (1) Bilinearity: for all $P, Q \in \mathbb{G}$ and $a, b \in \mathbb{Z}_q$, $e(aP, bQ) = e(P, Q)^{ab}$. (2) Non-degeneracy: there exist $P \in \mathbb{G}$ such that $e(P, P) \neq 1$. (3) Computability: for all $P, Q \in \mathbb{G}$, there exists an efficient algorithm to compute $e(P, Q)$.*

Definition 2 *(ϵ-Computational Diffie-Hellman assumption in $\mathbb{G}$). Let us consider a probabilistic polynomial time (PPT) algorithm $\mathcal{A}$ which takes as input $(e, q, \mathbb{G}, \mathbb{G}_T, P, aP, bP)$ and outputs $abP \in \mathbb{G}$. The advantage of $\mathcal{A}$ is defined as a probability $\epsilon := \Pr[P \xleftarrow{\$} \mathbb{G}, (a, b) \xleftarrow{\$} \mathbb{Z}_q : abP \leftarrow \mathcal{A}(e, q, \mathbb{G}, \mathbb{G}_T, P, aP, bP)]$. We say that ϵ-computational Diffie-Hellman (CDH) assumption in $\mathbb{G}$ holds against $\mathcal{A}$ if $\mathcal{A}$ cannot output $abP \in \mathbb{G}$ for a given $(e, q, \mathbb{G}, \mathbb{G}_T, P, aP, bP)$ with a success probability greater than ϵ.*

2.2   Security Notions of Key-Insulated Signatures

Here, we briefly review three types of key-exposure and countermeasures against them: (1) ordinary key-exposure to model (repeated) compromise of insecure devices; (2) master key-exposure to model compromise of physically secure devices; (3) key-update exposure to model (repeated) compromise of insecure devices during a key-updating step. One can minimize the damage caused by ordinary key-exposure if the scheme satisfies $(\mathcal{T}, N)$-key-insulation described in Sect. 1. Moreover, if $\mathcal{T} = N - 1$ holds, then we say that the scheme satisfies *perfectly key-insulation* [3]. Against the remaining attacks (2) and (3), as advanced security, Dodis et al. [2,3] gave the notions of *strong key-insulation* and *secure key-updates*, respectively. They formalized the security notions of $(\mathcal{T}, N)$-key-insulation, strong key-insulation and secure key-updates for standard key-insulated signatures.

Strong key-insulation. If physically secure devices are completely trusted, they may generate (pk, sk^*). Then they keep sk^* and publish pk on behalf of users where sk^* and pk denote the user's master secret key and public key, respectively. When a user requests a secret key for a time period k, the device computes sk_k and sends it. On the other hand, a solution when physically secure devices are *not* trusted is to generate (pk, sk), publish pk and then derive keys sk^* and sk_0. The user then sends sk^* to the device and stores sk_0 on its insecure device. When the user wants to update its key to that of a period j, the physically secure devices compute and send a "partial" key $sk'_{j,k}$ to the user, who may then compute the "actual" key sk_j by using sk_k and $sk'_{j,k}$. A scheme meeting this level of security is termed *strong*.

Secure key-updates. Consider attacks in which an adversary breaks into the user's storage while a key update is taking place (i.e., exposure occurs between two periods k and j). Dodis et al. [2,3] call this a *key-update exposure at period j*. In this case, the adversary receives sk_k, $sk'_{j,k}$ and (can compute) sk_j. They say a scheme has *secure key-updates* if a key-update exposure at period j is equivalent to key-exposures at periods k and j and no more.

3   Identity-Based Key-Insulated Aggregate Signatures

We define a syntax and the security of IBKIAS. Our security model captures not only security notions described in Sect. 2, but also some desirable properties for key-insulated signatures described later. Moreover, one can see that our model is better than that of Reddy and Gopal [9] in terms of adaptiveness. Namely, in our model, an adversary does not need to designate a target identity in advance. In addition, our model also allows an adversary to issue either key-extraction queries or temporary secret key queries (described later) for an output of a forgery, although an adversary in the model by Reddy and Gopal [9] cannot issue both key-extraction queries and temporary secret key queries for the designated identity.

3.1   Syntax

We consider the following two cases: (1) there is a single physically secure device and it is shared by signers; (2) each signer has his/her own physically secure device and it is not shared. The following syntax captures both these conditions.

Definition 3 *(Identity-based key-insulated aggregate signatures). An IBKIAS scheme is a tuple of polynomial-time algorithms* (Setup, Ext, Upd*, Upd, Sign, Agg, Vrfy) *such that:*

$\mathsf{Setup}(1^\lambda, N) \to (PP, MSK)$: *On input of a security parameter λ and (possibly) a total number of time periods N, output a public parameter PP and a master key MSK.*

$\mathsf{Ext}(PP, MSK, ID_i) \to (sk_i^*, sk_i^0)$: *On input of a public parameter PP, a master key MSK and an identity ID_i, output a device key sk_i^* and an initial secret key sk_i^0 for ID_i.*

$\mathsf{Upd}^*(PP, j, k, ID_i, sk_i^*) \to sk_i'^{j,k}$: *On input of a public parameter PP, indices j, k for time periods $(1 \le k, j \le N)$, an identity ID_i and a device key sk_i^*, output a partial secret key $sk_i'^{j,k}$.*

$\mathsf{Upd}(PP, j, k, ID_i, sk_i^k, sk_i'^{j,k}) \to sk_i^j$: *On input of a public parameter PP, indices j, k, a secret key sk_i^k and a partial secret key $sk_i'^{j,k}$, output a secret key sk_i^j for the time period j.*

$\mathsf{Sign}(PP, j, sk_i^j, m_i, w) \to (j, s_i)$: *On input of a public parameter PP, a time period j, a secret key sk_i^j, a message $m_i \in M$, and (possibly) some state information w, output a partial signature s_i for the time period j. Here, M is a message space.*

$\mathsf{Agg}(PP, j, \{s_i\}_{i=1}^I) \to (j, S)$: *On input of a public parameter PP, a time period j and partial signatures $\{s_i\}_{i=1}^I$, output an aggregate signature S for the time period j. If any error occurs, output an error symbol $\perp$.*

$\mathsf{Vrfy}(PP, j, \{(ID_i, m_i)\}_{i=1}^I, S) \to \{0, 1\}$: *On input of a public parameter PP, a time period j, identities $\{ID_i\}_{i=1}^I$, messages $\{m_i\}_{i=1}^I$ and an aggregate signature S, output 0 or 1.*

Here, we assume that Setup and Ext are executed by a Private Key Generator (PKG) similarly to conventional identity-based cryptosystems. For an IBKIAS scheme to satisfy the following condition as correctness, we require the following:

Definition 4 *(Correctness of IBKIAS scheme). We say that an aggregate signature is valid if* $\mathsf{Vrfy}(PP, j, \{(ID_i, m_i)\}_{i=1}^I, S) \to 1$. *Every honestly generated aggregate signature is valid, i.e., for any $\lambda \in \mathbb{N}$, time period $j \in \{1, \ldots, N\}$, identity $ID_i \in \{0,1\}^*$ and message $m_i \in M$, we have* $\Pr[\mathsf{Vrfy}(PP, j, \{(ID_i, m_i)\}_{i=1}^I, S) = 1] = 1$, *where* $(PP, MSK) \leftarrow \mathsf{Setup}(1^\lambda, N)$, $(sk_i^*, sk_i^0) \leftarrow \mathsf{Ext}(PP, MSK, ID_i)$, $(j, s_i) \leftarrow \mathsf{Sign}(PP, j, sk_i^j, m_i, w)$ *and* $(j, S) \leftarrow \mathsf{Agg}(PP, j, \{s_i\}_{i=1}^I)$.

Next, we define *random-access key-updates*.

Definition 5 *(Random-access key-updates). The above definition corresponds to schemes supporting random-access key-updates in which one can update sk_i^k to sk_i^j in one step for any $k, j \in \{1, \ldots, N\}$. A weaker definition only allows $j = k + 1$.*

Moreover, we define *unbounded number of time periods*. The scheme without this property suffers from a shortcoming such that when all the time periods are used up, the scheme cannot work unless it is re-initialized.

Definition 6 *(Unbounded number of time periods). If the total number of time periods N is not fixed in the Setup, then we say that it supports an unbounded number of time periods.*

Finally, we require an IBKIAS scheme to satisfy the following condition:

Definition 7 *(Compactness). We say that an IBKIAS scheme satisfies compactness if the size of an aggregate signature S is the same as that of a partial signature s_i generated in* Sign.

The compactness is for efficiency and is a main motivation of an aggregate signature scheme.

3.2 Security Model

We give a security model of IBKIAS. Note that the following game captures two cases; that is, (1) there is a single physically secure device shared by signers; and (2) each signer owns a physically secure device and the device is not shared. An IBKIAS scheme is called *perfectly key-insulated* if, for any polynomial-time adversary $\mathcal{F}$, an advantage in the following game is negligible:

Setup: The challenger $\mathcal{C}$ executes Setup($1^\lambda, N$) to generate PP and MSK. $\mathcal{C}$ gives PP to $\mathcal{F}$ and keeps MSK.

Queries: $\mathcal{F}$ adaptively issues a series of queries as follows:

- Key-extraction query (ID_i): $\mathcal{C}$ executes Ext(PP, MSK, ID_i) to generate (sk_i^*, sk_i^0) and gives (sk_i^*, sk_i^0) to $\mathcal{F}$. In the case of a single device, $\mathcal{C}$ gives only sk_i^0 to $\mathcal{F}$.
- Temporary secret key query (j, ID_i): $\mathcal{C}$ executes Upd($PP, j, k, ID_i, sk_i^k, sk_i'^{j,k}$) to generate sk_i^j and gives sk_i^j to $\mathcal{F}$.
- Signing query (j, ID_i, m_i): $\mathcal{C}$ executes Sign(PP, j, sk_i^j, m_i, w) to generate (j, s_i) and gives (j, s_i) to $\mathcal{F}$.

Output: $\mathcal{F}$ outputs a tuple $(j^*, \{(ID_i^*, m_i^*)\}_{i=1}^I, S^*)$. We say that $\mathcal{F}$ wins this game if the following conditions hold: Vrfy($PP, j^*, \{(ID_i^*, m_i^*)\}_{i=1}^I, S^*) = 1$; there exists at least one ID_i^* such that either key-extraction for ID_i^* or temporary secret key for (j^*, ID_i^*) have never been queried; and there exists at least one ID_i^* such that a signature on (j^*, ID_i^*, m_i^*) has never been queried.

   In the above game, a device-key query is not explicitly provided for $\mathcal{F}$. However, $\mathcal{F}$ can obtain the device key adaptively by issuing key-extraction queries in the case of (2), i.e., each signer has his/her own physically secure device and it is not shared (but we require at least one ID_i^* whose key-extraction has never been queried). Moreover, we do not give the definition of $(\mathcal{T}, N)$-key-insulation for IBKIAS in this paper since we directly address the perfectly key-insulated security, which does not bound the number of temporary secret key queries for $\mathcal{F}$. Namely, $\mathcal{F}$ is allowed to issue any temporary secret key query (but we require at least one ID_i^* whose temporary secret key has never been queried). The following definition corresponds to the *perfectly key-insulation* in IBKIAS.

Definition 8 *(Perfectly key-insulation). We say that an IBKIAS scheme is perfectly key-insulated if, for any polynomial-time adversary $\mathcal{F}$, the advantage ϵ' is negligible.*

Next, we address attacks that compromise a physically secure device. This includes attacks by the device itself which are untrusted. We model this attack by giving a device key to an adversary (i.e., we can also deal with this kind of attack by allowing the adversary to issue device-key queries for any identity). Note that the adversary is also allowed to obtain the temporary secret keys for any identity in any time period since these temporary secret keys can be derived from key-extraction queries implicitly in the case of (2), i.e., each signer owns a physically secure device which is not shared. On the other hand, in the case of (1), i.e., there is a single physically secure device shared by signers, the adversary can obtain temporary secret keys by issuing key-extraction queries and device-key queries (but we require that there exist at least one identity whose key-extraction has never been queried). Therefore, we do not provide temporary secret key queries for an adversary in the following game. An IBKIAS scheme is called *strong key-insulated* if, for any polynomial time adversary $\mathcal{F}$, an advantage in the following game is negligible:

Setup: The challenger $\mathcal{C}$ executes $\mathsf{Setup}(1^\lambda, N)$ to generate PP and MSK. $\mathcal{C}$ gives PP to $\mathcal{F}$ and keeps MSK.

Queries: $\mathcal{F}$ adaptively issues a series of queries as follows:

- Key-extraction query (ID_i): $\mathcal{C}$ executes $\mathsf{Ext}(PP, MSK, ID_i)$ to generate (sk_i^*, sk_i^0) and gives (sk_i^*, sk_i^0) to $\mathcal{F}$. In the case of a single device, $\mathcal{C}$ gives only sk_i^0 to $\mathcal{F}$.
- Device-key query (ID_i): $\mathcal{C}$ executes $\mathsf{Ext}(PP, MSK, ID_i)$ to generate sk_i^* and gives sk_i^* to $\mathcal{F}$.
- Signing query (j, ID_i, m_i): $\mathcal{C}$ executes $\mathsf{Sign}(PP, j, sk_i^j, m_i, w)$ to generate (j, s_i) and gives (j, s_i) to $\mathcal{F}$.

Output: $\mathcal{F}$ outputs a tuple $(j^*, \{(ID_i^*, m_i^*)\}_{i=1}^I, S^*)$. We say that $\mathcal{F}$ wins this game if the following conditions hold: $\mathsf{Vrfy}(PP, j^*, \{(ID_i^*, m_i^*)\}_{i=1}^I, S^*) = 1$; there exists at least one ID_i^* whose key-extraction has never been queried; and there exists at least one ID_i^* such that a signature on (j^*, ID_i^*, m_i^*) has never been queried.

One might think that device-key queries are redundant in the above game since key-extraction queries imply the device-key queries. However, it is reasonable to provide device-key queries for $\mathcal{F}$ in this game since $\mathcal{F}$ is disallowed to issue key-extraction queries such that there is *no* ID_i^* whose key-extraction has never been queried, whereas $\mathcal{F}$ can issue device-key queries for all ID_i^*. The following definition corresponds to the *strong key-insulation* in IBKIAS.

Definition 9 *(Strong key-insulation). We say that an IBKIAS scheme is strong key-insulated if, for any polynomial-time adversary $\mathcal{F}$, the advantage ϵ'' is negligible.*

Finally, we address an adversary who compromises a user's storage while a key is being updated from sk_i^k to sk_i^j. We call this attack a key-update exposure at (k, j) in IBKIAS. When this occurs, the adversary can obtain sk_i^k, $sk_i'^{j,k}$

and sk_i^j (actually, sk_i^j can be computed by using sk_i^k and $sk_i^{\prime j,k}$). We say that an IBKIAS scheme has secure key-updates if a key-update exposure at (k, j) is of no more help to the adversary than temporary secret key exposures at both time periods k and j. The formal definition is as follows.

Definition 10 *(Secure key-updates). We say that an IBKIAS scheme has secure key-updates if the view of any adversary $\mathcal{F}$ making a key-update exposure at time periods (k, j) can be perfectly simulated by an adversary $\mathcal{F}'$ making temporary secret key queries at time periods k and j.*

4 Proposed Scheme

We propose a concrete construction of IBKIAS. In this scheme, we assume that there is a single physically secure device shared by signers. As described in Sect. 1, extending an identity-based key-insulated signature scheme [11,14] to an IBKIAS scheme is *non-trivial*. An aggregate signature requires *compactness* such that the size of aggregate signatures must be the same as that of individual signatures *regardless of the number of signers*. Therefore, we have to aggregate each signer's randomness used in individual signing. To do this, we focus on the technique proposed by Gentry and Ramzan [4]. In their technique, a signing algorithm first creates a "dummy message" w that is mapped to an element P_w in $\mathbb{G}$, whose scalar coefficient provides a *place* to aggregate randomness for any individual signature. Moreover, we utilize the similar technique in a physically secure device to output a common key T between different signers. That is, by using a single physically secure device and aggregating only signatures with the same T, our scheme *completely* achieves the compactness. The proposed construction IBKIAS$_{\mathsf{Ours}}$ consists of the following algorithms:

Setup($1^\lambda, N$): It chooses two cyclic groups $\mathbb{G}$ and $\mathbb{G}_T$ with a prime order q. Let P be an arbitrary generator of $\mathbb{G}$ and e be bilinear maps such that $e : \mathbb{G} \times \mathbb{G} \to \mathbb{G}_T$. It chooses four cryptographic hash functions such that $H_1, H_2, H_3 : \{0, 1\}^* \to \mathbb{G}$ and $H_4 : \{0, 1\}^* \to \mathbb{Z}_q$. It chooses $s \xleftarrow{\$} \mathbb{Z}_q$ and computes $Q := sP$. It outputs a public parameter $PP := (\mathbb{G}, \mathbb{G}_T, e, P, Q, H_1, H_2, H_3, H_4)$ and a master key $MSK := s$.

Ext(PP, MSK, ID_i): It chooses $(sk_0^*, sk_1^*) \xleftarrow{\$} \mathbb{Z}_q^2$ and outputs a device secret key $sk_i^* := (sk_0^*, sk_1^*)$. It computes $T_0 := sk_0^* \cdot P$ and $T_1 := sk_1^* \cdot P$ and sets $T := (T_0, T_1)$. It computes $D_{i,0,0} := sH_1(ID_i, 0) + sk_0^* \cdot H_2(ID_i, T_0, 0)$ and $D_{i,0,1} := sH_1(ID_i, 1) + sk_1^* \cdot H_2(ID_i, T_1, 0)$ and sets $D_{i,0} := (D_{i,0,0}, D_{i,0,1})$. It outputs an initial secret key $sk_i^0 := (D_{i,0}, T)$.

Upd*(PP, j, k, ID_i, sk_i^*): For time periods j and k, it computes $T_0 := sk_0^* \cdot P$ and $T_1 := sk_1^* \cdot P$ and sets $T := (T_0, T_1)$. It computes $sk_{i,0}^{\prime j,k} := sk_0^*(H_2(ID_i, T_0, j) - H_2(ID_i, T_0, k))$ and $sk_{i,1}^{\prime j,k} := sk_1^*(H_2(ID_i, T_1, j) - H_2(ID_i, T_1, k))$ and outputs a partial secret key $sk_i^{\prime j,k} := (sk_{i,0}^{\prime j,k}, sk_{i,1}^{\prime j,k})$.

Upd($PP, j, k, ID_i, sk_i^k, sk_i^{\prime j,k}$): Parse $sk_i^k = (D_{i,k}, T)$. It computes $D_{i,j,0} := D_{i,k,0} + sk_{i,0}^{\prime j,k}$ and $D_{i,j,1} := D_{i,k,1} + sk_{i,1}^{\prime j,k}$ and sets $D_{i,j} := (D_{i,j,0}, D_{i,j,1})$. It outputs $sk_i^j := (D_{i,j}, T)$.

$\mathsf{Sign}(PP, j, sk_i^j, m_i, w)$: It first chooses a string $w \in \{0,1\}^*$ that has never been used before. It computes $P_w := H_3(w)$. It chooses $r_i \xleftarrow{\$} \mathbb{Z}_q$ and computes $R_i := r_i P$. It computes $c_i := H_4(j, ID_i, m_i, w)$ and $\sigma_i := r_i P_w + D_{i,j,0} + c_i D_{i,j,1}$. It sets $s_i := (\sigma_i, R_i, T, w)$ and outputs a partial signature (j, s_i).

$\mathsf{Agg}(PP, j, \{s_i\}_{i=1}^I)$: It parses s_i for any $i \in [1, I]$ as (σ_i, R_i, T, w), and computes $G := \sum_{i=1}^I \sigma_i$ and $R := \sum_{i=1}^I R_i$. If there exist any T^* and w^* which are distinct from the other T's and w's, respectively, then output $\perp$. Otherwise, it keeps one pair of T and w. It sets $S := (G, R, T, w)$ and outputs (j, S).

$\mathsf{Vrfy}(PP, j, \{(ID_i, m_i)\}_{i=1}^I, S)$: It outputs 1 if all of ID_i are distinct and

$$e(G, P) = e(R, P_w)e(Q, \sum_{i=1}^I (H_1(ID_i, 0) + c_i H_1(ID_i, 1)))$$

$$\cdot e(T_0, \sum_{i=1}^I H_2(ID_i, T_0, j))e(T_1, \sum_{i=1}^I c_i H_2(ID_i, T_1, j))$$

holds where $P_w = H_3(w)$ and $c_i = H_4(j, ID_i, m_i, w)$. Else, it outputs 0.

Remark 1. We assume that the algorithm Upd^* is executed by a physically secure device.

Remark 2. Our proposed scheme $\mathsf{IBKIAS}_{\mathsf{Ours}}$ supports an unbounded number of time periods since the total number of time periods is not fixed in the algorithm Setup. Algorithm Upd further shows that our scheme supports random-access key-updates.

Remark 3. Our proposed scheme $\mathsf{IBKIAS}_{\mathsf{Ours}}$ achieves the compactness as follows: first, G and R are closed under the modular operations and can be combined into single values, respectively; second, T and w are output if all of the values given to the aggregate algorithm are the same. The former statement obviously holds. Moreover, the latter statement is guaranteed as long as the synchronized setting is utilized.

5   Security

We analyze the security of our proposed scheme $\mathsf{IBKIAS}_{\mathsf{Ours}}$.

Theorem 1. $\mathsf{IBKIAS}_{\mathsf{Ours}}$ *is perfectly key-insulated in the random oracle model under the ϵ-CDH assumption in $\mathbb{G}$. Concretely, given an adversary $\mathcal{F}$ that has advantage ϵ' against the key-insulated security of $\mathsf{IBKIAS}_{\mathsf{Ours}}$ by asking at most q_{H_a} ($a = 1, 2, 3, 4$) hash queries, q_E key-extraction queries, $q_{E'}$ temporary secret key queries and q_S signing queries, there exists an algorithm $\mathcal{B}$ that breaks the ϵ-CDH assumption in $\mathbb{G}$ with the probability $\epsilon \geq \left(1 - \frac{1}{q}\right)\left(\frac{4}{e(q_E + q_{E'} + q_S + q_{H_4} + 4)}\right)^4 \epsilon'$, where e denotes the base of the natural logarithm.*

Proof. We show how to construct an algorithm $\mathcal{B}$ against the ϵ-CDH assumption in $\mathbb{G}$. $\mathcal{B}$ is given an instance (P, Q, P', e) and will interact with an adversary $\mathcal{F}$ as follows in an attempt to compute sP' where $Q = sP$ holds.

Setup: $\mathcal{B}$ sets the public parameter $PP := (\mathbb{G}, \mathbb{G}_T, e, P, Q, H_1, H_2, H_3, H_4)$ and transmits PP to $\mathcal{F}$. Here, the H_a's ($a = 1, 2, 3, 4$) are random oracles controlled by $\mathcal{B}$.

Hash Queries: $\mathcal{F}$ can issue H_1-query, H_2-query, H_3-query or H_4-query at any time. $\mathcal{B}$ then gives identical responses to the given queries in the following manner, maintaining lists related to its previous responses for consistency. $\mathcal{B}$ also maintains H_4-list2, which addresses certain special cases of the H_4 simulation. For $\mathcal{F}$'s H_1-query on (ID_i, α) for $\alpha \in \{0, 1\}$:

1. If ID_i was in a previous H_1-query, then the previously defined value is returned.
2. Else, $\mathcal{B}$ generates a random H_1-$coin_i \in \{0, 1\}$ so that $\Pr[H_1$-$coin_i = 0] = \delta$. If H_1-$coin_i = 0$, $\mathcal{B}$ generates $(\beta_{i0}, \beta_{i1}) \xleftarrow{\$} \mathbb{Z}_q^2$ and sets $\beta_{i0}' = \beta_{i1}' = 0$; else, $\mathcal{B}$ generates $(\beta_{i0}, \beta_{i0}', \beta_{i1}, \beta_{i1}') \xleftarrow{\$} \mathbb{Z}_q^4$. $\mathcal{B}$ records $(ID_i, H_1$-$coin_i, \beta_{i0}, \beta_{i0}', \beta_{i1}, \beta_{i1}')$ in its H_1-list.
3. $\mathcal{B}$ responds $\mathcal{A}$ with $H_1(ID_i, \alpha) := P_{i\alpha} = \beta_{i\alpha}P + \beta_{i\alpha}'P'$.

For $\mathcal{F}$'s H_2-query on (ID_i, T_γ, k) for $\gamma \in \{0, 1\}$:

1. If (ID_i, T_γ, k) was in a previous H_2-query, then the previously defined value is returned.
2. Else, $\mathcal{B}$ generates a random H_2-$coin_{i\gamma} \in \{0, 1\}$ so that $\Pr[H_2$-$coin_{i\gamma} = 0] = \delta$. $\mathcal{B}$ generates $(\eta_{i0k}, \eta_{i1k}) \xleftarrow{\$} \mathbb{Z}_q^2$.
3. $\mathcal{B}$ runs H_1-query on (ID_i, α) to recover β_{i0}' and β_{i1}' from its H_1-list. If H_1-$coin_i = 0$ or H_2-$coin_{i\gamma} = 1$, $\mathcal{B}$ records $(ID_i, T_\gamma, k, H_2$-$coin_{i\gamma}, \eta_{i0k}, \eta_{i1k}, \cdot, \cdot, \cdot)$ and responds with $H_2(ID_i, T_\gamma, k) = P_{i\gamma k} = \eta_{i\gamma k}P$. If H_1-$coin_i = 1$ and H_2-$coin_{i\gamma} = 0$, $\mathcal{B}$ generates $r' \xleftarrow{\$} \mathbb{Z}_q$ and records $(ID_i, T_\gamma, k, H_2$-$coin_{i\gamma}, \cdot, \cdot, \beta_{i0}', \beta_{i1}', r')$. Then, $\mathcal{B}$ responds with $H_2(ID_i, T_\gamma, k) := P_{i\gamma k} = \eta_{i\gamma k}P + (\beta_{i\gamma}'/r')P'$.

Remark 4. Note that r' is independent of H_2-query and it is fixed value. We also note that, even if we introduce such fixed r', the signature is probabilistic owing to $\eta_{i\gamma k}$.

For $\mathcal{F}$'s H_3-query on w_z:

1. If w_z was in a previous H_3-query, then the previously defined value is returned.
2. Else, $\mathcal{B}$ generates a random H_3-$coin_z \in \{0, 1\}$ so that $\Pr[H_3$-$coin_z = 0] = \delta$. $\mathcal{B}$ then generates $g_z \xleftarrow{\$} \mathbb{Z}_q^*$ and records $(w_z, H_3$-$coin_z, g_z)$ in its H_3-list.
3. If H_3-$coin_z = 0$, $\mathcal{B}$ responds with $H_3(w_z) := P_{w_z} = g_zP'$. Otherwise, $\mathcal{B}$ responds with $H_3(w_z) := P_{w_z} = g_zP$.

For $\mathcal{F}$'s H_4-query on (k, ID_i, m_y, w_z):

1. If (k, ID_i, m_y, w_z) was in a previous H_4-query, then the previously defined value is returned.
2. Else, $\mathcal{B}$ runs an H_1-query on (ID_i, α) to recover β'_{i0} and β'_{i1} from its H_1-list. $\mathcal{B}$ generates a random H_4-$coin_{iyz} \in \{0,1\}$ so that $\Pr[H_4\text{-}coin_{iyz} = 0] = \delta$.
   (a) If H_1-$coin_i = 1$, H_2-$coin_{i\gamma} = 1$, H_3-$coin_z = 1$, and H_4-$coin_{iyz} = 0$, $\mathcal{B}$ checks whether H_4-list2 contains a tuple $(k, ID_{i'}, m_{y'}, w_{z'}) \neq (k, ID_i, m_y, w_z)$ with $ID_{i'} = ID_i$. If so, aborts. If not, $\mathcal{B}$ adds (k, ID_i, m_y, w_z) to H_4-list2 and sets $d_{iyz} = -\beta'_{i0}/\beta'_{i1} \mod q$.
   (b) If H_1-$coin_i = 0$, H_2-$coin_{i\gamma} = 0$, H_3-$coin_z = 0$, or H_4-$coin_{iyz} = 1$, $\mathcal{B}$ generates $d_{iyz} \xleftarrow{\$} \mathbb{Z}_q^*$.
   (c) $\mathcal{B}$ records $(k, ID_i, m_y, w_z, H_4\text{-}coin_{iyz}, d_{iyz})$ in its H_4-list.
3. $\mathcal{B}$ responds with $H_4(k, ID_i, m_y, w_z) := d_{iyz}$.

Key-extraction Queries: When $\mathcal{F}$ requests the secret key corresponding to ID_i, $\mathcal{B}$ acts as follows:

1. $\mathcal{B}$ recovers H_1-$coin_i$. If H_1-$coin_i = 0$, $\mathcal{B}$ responds with $(sP_{i0}, sP_{i1}) = (\beta_{i0}Q, \beta_{i1}Q)$. If H_1-$coin_i = 1$, $\mathcal{B}$ aborts the process.
2. $\mathcal{B}$ recovers H_2-$coin_{i\gamma}$ and generates $(sk_0^*, sk_1^*) \xleftarrow{\$} \mathbb{Z}_q^2$.
   (a) If H_2-$coin_{i\gamma} = 0$, $\mathcal{B}$ computes $T_\gamma := sk_\gamma^* \cdot P - r'Q$ and sets $D_{i,0,\gamma} := \beta_{i\gamma}Q + sk_\gamma^* \cdot \eta_{i\gamma 0}P - r'\eta_{i\gamma 0}Q = \beta_{i\gamma}Q + (sk_\gamma^* - r's)\eta_{i\gamma 0}P$.
   (b) If H_2-$coin_{i\gamma} = 1$, $\mathcal{B}$ computes $T_\gamma := sk_\gamma^* \cdot P$ and sets $D_{i,0,\gamma} := \beta_{i\gamma}Q + sk_\gamma^* \cdot \eta_{i\gamma 0}P$.

$\mathcal{B}$ sets $D_{i,0} := (D_{i,0,0}, D_{i,0,1})$, $T := (T_0, T_1)$, and responds with $sk_i^0 := (D_{i,0}, T)$.

Temporary secret key Queries: When $\mathcal{F}$ requests a temporary secret key query corresponding to ID_i for time period j, $\mathcal{B}$ acts as follows:

1. $\mathcal{B}$ recovers H_1-$coin_i$. If H_1-$coin_i = 0$, $\mathcal{B}$ responds with $(sP_{i0}, sP_{i1}) = (\beta_{i0}Q, \beta_{i1}Q)$. If H_1-$coin_i = 1$, $\mathcal{B}$ aborts the process.
2. $\mathcal{B}$ recovers H_2-$coin_{i\gamma}$.
   (a) If H_2-$coin_{i\gamma} = 0$, $\mathcal{B}$ computes $D_{i,j,\gamma} := \beta_{i\gamma}Q + sk_\gamma^* \cdot \eta_{i\gamma j}P - r'\eta_{i\gamma j}Q = \beta_{i\gamma}Q + (sk_\gamma^* - r's)\eta_{i\gamma j}P$.
   (b) If H_2-$coin_{i\gamma} = 1$, $\mathcal{B}$ computes $D_{i,j,\gamma} := \beta_{i\gamma}Q + sk_\gamma^* \cdot \eta_{i\gamma j}P$.

$\mathcal{B}$ sets $D_{i,j} := (D_{i,j,0}, D_{i,j,1})$, $T := (T_0, T_1)$, and responds with $sk_i^j := (D_{i,j}, T)$.

Signing Queries: When $\mathcal{F}$ requests a signature on (j, ID_i, m_y, w_z), $\mathcal{B}$ first confirms that $\mathcal{F}$ has not previously requested a signature by ID_i on w_z (otherwise, it is an improper query). Then, $\mathcal{B}$ proceeds as follows:

1. If H_1-$coin_i = H_2$-$coin_{i\gamma} = H_3$-$coin_z = H_4$-$coin_{iyz} = 1$, $\mathcal{B}$ aborts the process.
2. If H_1-$coin_i = 0$, $\mathcal{B}$ generates $r \xleftarrow{\$} \mathbb{Z}_q$ and outputs the signature $s_i := (\sigma_i, R_i, T, w_z)$, where $\sigma_i = \beta_{i0}Q + sk_0^* \cdot P_{i0j} + d_{iyz}(\beta_{i1}Q + sk_1^* \cdot P_{i1j}) + rP_{w_z} = D_{i,j,0} + d_{iyz}D_{i,j,1} + rP_{w_z}$ and $R_i = rP$.

3. If $H_1\text{-}coin_i = 1$ and $H_2\text{-}coin_{i\gamma} = 0$, $\mathcal{B}$ recovers $r' \in \mathbb{Z}_q$ from H_2-list, and generates $r \xleftarrow{\$} \mathbb{Z}_q$. $\mathcal{B}$ then outputs the signature $s_i := (\sigma_i, R_i, T, w_z)$, where $R_i = rP$ and

$$\sigma_i = \beta_{i0}Q + \frac{\beta'_{i0}sk_0^*P'}{r'} + d_{iyz}(\beta_{i1}Q + \frac{\beta'_{i1}sk_1^*P'}{r'}) + rP_{w_z} + \eta_{i0k}T_0 + d_{iyz}\eta_{i1k}T_1$$

$$= \beta_{i0}Q + \beta'_{i0}sP' + (sk_0^* - sr')\frac{\beta'_{i0}}{r'}P' + \eta_{i0k}T_0$$

$$+ d_{iyz}(\beta_{i1}Q + \beta'_{i1}sP' + (sk_1^* - sr')\frac{\beta'_{i1}}{r'}P' + \eta_{i1k}T_1) + rP_{w_z}$$

$$= D_{i,j,0} + d_{iyz}D_{i,j,1} + rP_{w_z}.$$

4. If $H_1\text{-}coin_i = H_2\text{-}coin_{i\gamma} = 1$ and $H_3\text{-}coin_z = 0$, $\mathcal{B}$ generates $r \xleftarrow{\$} \mathbb{Z}_q$ and outputs the signature $s_i := (\sigma_i, R_i, T, w_z)$ where

$$\sigma_i = \beta_{i0}Q + sk_0^* \cdot \eta_{i0k}P + d_{iyz}(\beta_{i1}Q + sk_1^* \cdot \eta_{i1k}P) + rg_zP$$

$$= \beta_{i0}Q + \beta'_{i0}sP' + sk_0^* \cdot \eta_{i0k}P + d_{iyz}(\beta_{i1}Q + \beta'_{i1}sP' + sk_1^* \cdot \eta_{i1k}P)$$

$$+ (r - (\beta'_{i0} + d_{iyz}\beta'_{i1})sg_z^{-1})g_zP$$

$$= D_{i,j,0} + d_{iyz}D_{i,j,1} + (r - (\beta'_{i0} + d_{iyz}\beta'_{i1})sg_z^{-1})P_{w_z} \text{ and}$$

$$R_i = rP - (\beta'_{i0} + d_{iyz}\beta'_{i1})g_z^{-1}Q = (r - (\beta'_{i0} + d_{iyz}\beta'_{i1})sg_z^{-1})P.$$

5. If $H_1\text{-}coin_i = H_2\text{-}coin_{i\gamma} = H_3\text{-}coin_z = 1$ and $H_4\text{-}coin_{iyz} = 0$, $\mathcal{B}$ generates $r \xleftarrow{\$} \mathbb{Z}_q$ and outputs the signature $s_i := (\sigma_i, R_i, T, w_z)$, where $R_i = rP$ and

$$\sigma_i = \beta_{i0}Q + sk_0^* \cdot \eta_{i0k}P - \frac{\beta'_{i0}}{\beta'_{i1}}(\beta_{i1}Q + sk_1^* \cdot \eta_{i1k}P) + rg_zP$$

$$= \beta_{i0}Q + \beta'_{i0}sP' + sk_0^* \cdot \eta_{i0k}P - \frac{\beta'_{i0}}{\beta'_{i1}}(\beta_{i1}Q + \beta'_{i1}sP' + sk_1^* \cdot \eta_{i1k}P) + rg_zP$$

$$= D_{i,j,0} + d_{iyz}D_{i,j,1} + rP_{w_z}.$$

Output: Finally, with probability ϵ', $\mathcal{F}$ outputs $(j^*, \{ID_i^*\}_{i=1}^I, \{m_y^*\}_{y=1}^I, S^*)$ with $I \leq I'$, such that there exist $K, Y \in [1, I']$ such that either key-extraction for ID_K^* or temporary secret key for (j^*, ID_K^*) has never been queried or requested a signature for (ID_K^*, m_Y^*, w_z) for time period j^*. Here, S^* is defined as $S^* := (G_I, R_I, T, w_Z)$, and it satisfies the equation

$$e(G_I, P) = e(R_I, P_{w_z})e(Q, \sum_{i=1}^I (P_{i,0} + c_iP_{i,1}))e(T_0, \sum_{i=1}^I P_{i0j^*})e(T_1, \sum_{i=1}^I c_iP_{i1j^*}),$$

where $P_{i,\rho} = H_1(ID_i^*, \rho), P_{i\mu j^*} = H_2(ID_i^*, T_\mu, j^*), P_{w_z} = H_3(w_Z)$ and $c_i = H_4(j^*, ID_i^*, m_y^*, w_z)$ are required. If it is not the case that (K, Y, Z) can satisfy $H_1\text{-}coin_K = H_2\text{-}coin_{K\gamma} = H_3\text{-}coin_Z = H_4\text{-}coin_{KYZ} = 1$, then aborts. Otherwise, $\mathcal{B}$ can solve the instance of CDH with probability $1 - 1/q$ as follows.

$\mathcal{F}$'s forgery has the form (G_I, R_I, T, w_Z), where $R_I = rP$ and $G_I = rP_{w_Z} + \sum_{i=1}^{I}(D_{i,j^*,0} + c_i D_{i,j^*,1})$, where we let $c_i = H_4(j^*, ID_i^*, m_y^*, w_Z)$ be the hash of the tuple "signed" by the entity with identity ID_i^*. Since $H_3\text{-}coin_z = 1$, $\mathcal{B}$ knows the discrete logarithm g_Z of P_{w_Z} with respect to P. Therefore, it can compute:

$$
\begin{aligned}
G_I - g_Z R_I &= \sum_{i=1}^{I}(D_{i,j^*,0} + c_i D_{i,j^*,1}) \\
&= \sum_{i=1}^{I}(s(\beta_{i0}P + \beta'_{i0}P') + sk_0^* \cdot \eta_{i,0,j^*}P + c_i(s(\beta_{i1}P + \beta'_{i1}P') \\
&\quad + sk_1^* \cdot \eta_{i,1,j^*}P)) \\
&= s(\sum_{i=1}^{I}(\beta_{i0} + c_i\beta_{i1}))P + (\sum_{i=1}^{I}(sk_0^* \cdot \eta_{i,0,j^*} + c_i \cdot sk_1^* \cdot \eta_{i,1,j^*}))P \\
&\quad + s(\sum_{i=1}^{I}(\beta'_{i0} + c_i\beta'_{i1}))P'.
\end{aligned}
$$

If $H_1\text{-}coin_i = H_4\text{-}coin_{iyz} = 1$ for at least one of the signed tuples, then the probability that $\sum_{i=1}^{I}(\beta'_{i0} + c_i\beta'_{i1}) \neq 0$ is $1 - 1/q$; if $\sum_{i=1}^{I}(\beta'_{i0} + c_i\beta'_{i1}) \neq 0$, $\mathcal{B}$ can easily derive sP' from the expression above.

We want to bound $\mathcal{B}$'s success probability. $\mathcal{B}$ aborts in four situations: (1) $\mathcal{F}$ makes a key-extraction query on an ID_i for which $H_1\text{-}coin_i = 1$; (2) $\mathcal{F}$ makes a temporary secret key query on a tuple (j, ID_i) for which $H_1\text{-}coin_i = 1$; (3) $\mathcal{F}$ makes a signing query on a tuple (j, ID_i, m_y, w_z) for which $H_1\text{-}coin_i = H_2\text{-}coin_{i\gamma} = H_3\text{-}coin_z = H_4\text{-}coin_{iyz} = 1$; (4) $\mathcal{F}$ makes two H_4-queries on tuples $(j, ID_i, m_{y'}, w_{z'})$ and (j, ID_i, m_y, w_z) for which $H_1\text{-}coin_i = 1$, $H_2\text{-}coin_{i\gamma} = 1$, $H_3\text{-}coin_z = H_3\text{-}coin_{z'} = 1$, and $H_4\text{-}coin_{iyz} = H_4\text{-}coin_{iy'z'} = 0$. For the first situation, obviously we can see that the probability that $\mathcal{B}$ does not abort during a key-extraction query if $\mathcal{F}$ is permitted to make at most q_E queries is at least δ^{q_E} since $\mathcal{B}$ only aborts when $H_1\text{-}coin_i = 1$. Similarly, for the second situation, the probability that $\mathcal{B}$ does not abort during a temporary secret key query if $\mathcal{F}$ is permitted to make at most $q_{E'}$ queries is at least $\delta^{q_{E'}}$. In considering the third situation, the probability that $\mathcal{B}$ does not abort during a signing query if $\mathcal{F}$ is permitted to make at most q_S queries is at least δ^{q_S} since, if $H_1\text{-}coin_i = 0$, $\mathcal{B}$ never aborts. Similarly, for the fourth situation, the probability that $\mathcal{B}$ does not abort during an H_4-query if $\mathcal{F}$ is permitted to make at most q_{H_4} queries is at least $\delta^{q_{H_4}}$. Finally, we consider the probability that $\mathcal{B}$ does not abort in the output phase. This probability is $(1 - \delta)^4$ since all coins should be 1.

From the above, we have $\delta^{q_E + q_{E'} + q_S + q_{H_4}}(1 - \delta)^4$, which is maximized at $\delta_{opt} = \frac{q_E + q_{E'} + q_S + q_{H_4}}{q_E + q_{E'} + q_S + q_{H_4} + 4}$. Using δ_{opt}, $\mathcal{B}$'s success probability ϵ is

$$
\begin{aligned}
\epsilon &> \left(1 - \frac{1}{q}\right)\left(\frac{q_E + q_{E'} + q_S + q_{H_4}}{q_E + q_{E'} + q_S + q_{H_4} + 4}\right)^{q_E + q_{E'} + q_S + q_{H_4}}\left(\frac{4}{q_E + q_{E'} + q_S + q_{H_4} + 4}\right)^4 \epsilon' \\
&\geq \left(1 - \frac{1}{q}\right)\left(\frac{4}{e(q_E + q_{E'} + q_S + q_{H_4} + 4)}\right)^4 \epsilon'.
\end{aligned}
$$

This is because, denoting $q_E + q_{E'} + q_S + q_{H_4}$ by x, $(x/(x+4))^x = (1+4/x)^{-x} = ((1 + \frac{1}{x/4})^{x/4})^{-4}$ tends to e^{-4} for large x. This concludes the proof. □

Theorem 2. IBKIAS$_{\mathsf{Ours}}$ *is strong key-insulated in the random oracle model under the ϵ-CDH assumption in $\mathbb{G}$. Concretely, given an adversary $\mathcal{F}$ that has advantage ϵ'' against the strong key-insulated security of IBKIAS$_{\mathsf{Ours}}$ by asking at most q_{H_a} ($a = 1, 2, 3, 4$) hash queries, q_E key-extraction queries, q_D device-key queries and q_S signing queries, there exists an algorithm $\mathcal{B}$ that breaks the ϵ-CDH assumption in $\mathbb{G}$ with the probability $\epsilon \geq \left(1 - \frac{1}{q}\right) \left(\frac{4}{e(q_E + q_S + q_{H_4} + 4)}\right)^4 \epsilon''$, where e denotes the base of the natural logarithm.*

This proof is similar to that of Theorem 1, and hence we omit here due to the page limitation.

Theorem 3. IBKIAS$_{\mathsf{Ours}}$ *has secure key-updates.*

Proof. This theorem follows from the fact that for any time period indices k, j and any identity ID_i, the partial secret key $sk_i'^{j,k}$ can be derived from sk_i^k and sk_i^j. □

Acknowledgment. We would like to thank the anonymous reviewers for their helpful comments. This work was supported in part by JSPS KAKENHI Grant Numbers 16K16065 and 17K00178, and Secom Science and Technology Foundation.

References

1. Boneh, D., Gentry, C., Lynn, B., Shacham, H.: Aggregate and verifiably encrypted signatures from bilinear maps. In: Biham, E. (ed.) EUROCRYPT 2003. LNCS, vol. 2656, pp. 416–432. Springer, Heidelberg (2003). https://doi.org/10.1007/3-540-39200-9_26
2. Dodis, Y., Katz, J., Xu, S., Yung, M.: Key-insulated public key cryptosystems. In: Knudsen, L.R. (ed.) EUROCRYPT 2002. LNCS, vol. 2332, pp. 65–82. Springer, Heidelberg (2002). https://doi.org/10.1007/3-540-46035-7_5
3. Dodis, Y., Katz, J., Xu, S., Yung, M.: Strong key-insulated signature schemes. In: Desmedt, Y.G. (ed.) PKC 2003. LNCS, vol. 2567, pp. 130–144. Springer, Heidelberg (2003). https://doi.org/10.1007/3-540-36288-6_10
4. Gentry, C., Ramzan, Z.: Identity-based aggregate signatures. In: Yung, M., Dodis, Y., Kiayias, A., Malkin, T. (eds.) PKC 2006. LNCS, vol. 3958, pp. 257–273. Springer, Heidelberg (2006). https://doi.org/10.1007/11745853_17
5. Hanaoka, Y., Hanaoka, G., Shikata, J., Imai, H.: Identity-based hierarchical strongly key-insulated encryption and its application. In: Roy, B. (ed.) ASIACRYPT 2005. LNCS, vol. 3788, pp. 495–514. Springer, Heidelberg (2005). https://doi.org/10.1007/11593447_27
6. Hohenberger, S., Koppula, V., Waters, B.: Universal signature aggregators. In: Oswald, E., Fischlin, M. (eds.) EUROCRYPT 2015. LNCS, vol. 9057, pp. 3–34. Springer, Heidelberg (2015). https://doi.org/10.1007/978-3-662-46803-6_1
7. Hohenberger, S., Sahai, A., Waters, B.: Full domain hash from (leveled) multi-linear maps and identity-based aggregate signatures. In: Canetti, R., Garay, J.A. (eds.) CRYPTO 2013. LNCS, vol. 8042, pp. 494–512. Springer, Heidelberg (2013). https://doi.org/10.1007/978-3-642-40041-4_27

8. Liang, B., Li, H., Chang, J.: The generic transformation from standard signatures to identity-based aggregate signatures. In: Lopez, J., Mitchell, C.J. (eds.) ISC 2015. LNCS, vol. 9290, pp. 21–41. Springer, Cham (2015). https://doi.org/10.1007/978-3-319-23318-5_2

9. Reddy, P.V., Gopal, P.V.S.S.N.: Identity-based key-insulated aggregate signature scheme. J. King Saud University Comput. Inf. Sci. **2015**, 1–8 (2015). https://doi.org/10.1016/j.jksuci.2015.09.003

10. Watanabe, Y., Shikata, J.: Identity-based hierarchical key-insulated encryption without random oracles. In: Cheng, C.-M., Chung, K.-M., Persiano, G., Yang, B.-Y. (eds.) PKC 2016. LNCS, vol. 9614, pp. 255–279. Springer, Heidelberg (2016). https://doi.org/10.1007/978-3-662-49384-7_10

11. Weng, J., Liu, S., Chen, K., Li, X.: Identity-based key-insulated signature with secure key-updates. In: Lipmaa, H., Yung, M., Lin, D. (eds.) Inscrypt 2006. LNCS, vol. 4318, pp. 13–26. Springer, Heidelberg (2006). https://doi.org/10.1007/11937807_2

12. Zhao, S., Aggarwal, A., Frost, R., Bai, X.: A survey of applications of identity-based cryptography in mobile ad-hoc networks. IEEE Commun. Surv. Tutorials **14**(2), 380–400 (2012)

13. Zhao, H., Yu, J., Duan, S., Cheng, X., Hao, R.: Key-insulated aggregate signature. Front. Comput. Sci. **8**(5), 837–846 (2014). https://doi.org/10.1007/s11704-014-3244-1

14. Zhou, Y., Cao, Z., Chai, Z.: Identity based key insulated signature. In: Chen, K., Deng, R., Lai, X., Zhou, J. (eds.) ISPEC 2006. LNCS, vol. 3903, pp. 226–234. Springer, Heidelberg (2006). https://doi.org/10.1007/11689522_21

A New Constant-Size Accountable Ring Signature Scheme Without Random Oracles

Sudhakar Kumawat[1(✉)] (iD) and Souradyuti Paul[2]

[1] Indian Institute of Technology Gandhinagar, Gandhinagar, India
sudhakar.bm07@gmail.com
[2] Indian Institute of Technology Bhilai, Datrenga, India
souradyuti.paul@gmail.com

Abstract. Accountable ring signature (ARS), introduced by Xu and Yung (CARDIS 2004), combines many useful properties of ring and group signatures. In particular, the signer in an ARS scheme has the flexibility of choosing an ad hoc group of users, and signing on their behalf (like a ring signature). Furthermore, the signer can designate an opener who may later reveal his identity, if required (like a group signature). In 2015, Bootle et al. (ESORICS 2015) formalized the notion and gave an efficient construction for ARS with signature-size logarithmic in the size of the ring. Their scheme is proven to be secure in the random oracle model. Recently, Russell et al. (ESORICS 2016) gave a construction with constant signature-size that is secure in the standard model. Their scheme is based on q-type assumptions (q-SDH).

In this paper, we give a new construction for ARS having the following properties: signature is constant-sized, secure in the standard model, and based on indistinguishability obfuscation ($i\mathcal{O}$) and one-way functions. To the best of our knowledge, this is the first $i\mathcal{O}$-based ARS scheme. Independent of this, our work can be viewed as a new application of *puncturable programming* and *hidden sparse trigger* techniques introduced by Sahai and Waters (STOC 2014) to design $i\mathcal{O}$-based deniable encryption.

Keywords: Accountable ring signatures
Indistinguishability obfuscation · Punturable PRFs

1 Introduction

The notion of group signature, introduced by Chaum and Van Heyst [9], allows a user to sign anonymously on behalf of a group of users without revealing his identity. Here, the membership of the group is controlled exclusively by the group manager, that is, he can include as well as expel users in the group. Moreover, the group manager can revoke the anonymity of the signer using a secret tracing key. Unlike a group signature, a ring signature, introduced by Rivest et al. [18], allows a signer to choose an ad hoc group of users (called the ring) and to sign on their behalf. Also, there is no group manager in a ring signature that controls the group and traces the signer. The notion of accountable ring signature

© Springer International Publishing AG, part of Springer Nature 2018
X. Chen et al. (Eds.): Inscrypt 2017, LNCS 10726, pp. 157–179, 2018.
https://doi.org/10.1007/978-3-319-75160-3_11

(ARS) [20] was borne out of the above two notions. It provides the flexibility of choosing a group of users, and ensures accountability by allowing an opener – designated by the signer – to later reveal the signer's identity. An ARS scheme must be unforgeable, anonymous, traceable and traceably sound. Applications of accountable ring signature include anonymous e-cash schemes [20], anonymous forums and auction systems [4].

Related Work. In 2004, Xu and Yung [20] introduced the notion of accountable ring signature (ARS). Their construction relies on the existence of a trapdoor permutation and uses a threshold decryption scheme to reveal the signer's identity. Their main trick lies in the use of tamper-resistant smart cards to retain some footprint of the identity of the signer in the signature. The size of the signature in their scheme grows linearly with that of the ring. In 2015, Bootle et al. [4] formalized the notion of ARS, and gave an efficient construction secure in the random oracle model by combining Camenisch's group signature scheme [8] with a generalized version of Groth and Kohlweiss's one-out-of-many proofs of knowledge [15]. Their construction relies on the hardness of Decision Diffie-Hellman (DDH) problem. The signature-size of their scheme grows logarithmically with the ring-size. Recently, Lai et al. [17] gave the first constant-size ARS scheme in standard model by combining the ring signature scheme of [5], with the structure-preserving encryption scheme of [7]. Their construction relies on q-type assumptions, in particular the q-SDH assumption.

Table 1. Comparison between several *accountable ring signature* schemes. The symbol R denotes the ring.

Paper	Signature size	Security model	Unforgeability	Anonymity	Hardness assumption
[20]	$O(|R|)$	RO	Fully adaptive	Fully adaptive	Trapdoor permutations
[4]	$O(\log|R|)$	RO	Fully adaptive	Fully adaptive	DDH
[17]	$O(1)$	Standard	q-Adaptive	q-Adaptive	q-type
This paper	$O(1)$	Standard	Selective	Selective	$i\mathcal{O}$+OWF

Our Contribution. We propose a new constant-size accountable ring signature scheme. Our scheme relies on the security of indistinguishability obfuscation ($i\mathcal{O}$) and one-way functions. Our approach is inspired by the *puncturable programming* and *hidden sparse trigger* techniques introduced, and used by Sahai and Waters [19] to construct deniable encryption schemes. Two important features of our scheme are as follows: Unlike the schemes in [4,20], (1) our scheme is proven secure in the standard model, and (2) the signature-size of our scheme is constant, and does not increase with the ring-size. While the scheme in [17] achieves stronger security guarantees with respect to unforgeability and anonymity, it does not achieve full adaptiveness (i.e. q-adaptive). This is because, its security proof relies on q-type assumption (q-SDH) that restricts the number of queries made to the signing oracle to a fixed parameter q (determined in the setup phase). However, this does not work in the real world as an adversary can always get more than q valid signatures. Note that setting q to a very large number makes the assumption stronger and, also, drastically degrades the practical efficiency of the scheme. In contrast, our scheme is secure against arbitrary number (polynomial) of signing queries. One major disadvantage of our scheme is that it is selective secure. Thus, constructing a fully unforgeable and anonymous, and constant-size ARS scheme in the standard model still remains an interesting open problem. In Table 1, we compare our scheme with previous ones.

In this paper, we stress that, due to the low efficiency of the existing $i\mathcal{O}$ candidates, our focus is mainly on the existence of a constant signature-size ARS scheme (secure without random oracles) based on $i\mathcal{O}$, but not on their practicability. However, note that, with the recent publication of results on efficient implementation of $i\mathcal{O}$-based cryptographic primitives [2], we hope that our scheme may be close to being practical.

Technical Overview. First, a trusted authority runs the setup algorithm *once* that gives two obfuscated programs Sign and Verify. Two random keys K_1 and K_2 are hardwired into these programs, and are known only to the trusted authority. It is important for any ARS scheme that the footprint of the signer be retained in the signature so that the signer's identity could be revealed by a designated party, if required. To achieve this, we use the *hidden sparse trigger* technique introduced by Sahai and Waters [19], and briefly describe it in the Sign and Verify programs below.

– Sign: This program takes as input a message m, a signing-verification key pair (sk, vk), a ring R (a set of verification keys arranged in a specified order), a public key ek of an opener and some randomness r. Then, it checks if $vk = f(sk)$ and $vk \in R$. If so, it sets $\alpha = F_1(K_1, (m\|vk\|\mathsf{Hash}(R)\|ek\|\mathsf{PRG}(r)))$ and $\beta = F_2(K_2, \alpha) \oplus (m\|vk\|\mathsf{Hash}(R)\|ek\|\mathsf{PRG}(r))$; and outputs $\sigma = (\alpha, \beta)$ as a signature of (m, sk, R, ek, r). The security relies on the following assumptions: (1) $F_1(K_1, \cdot)$ is *injective pucturable PRF*; $F_2(K_2, \cdot)$ is pucturable PRF; $\mathsf{Hash}(\cdot)$ is collision-resistant; and $f(\cdot)$ is one-way. (These are discussed in detail in Sect. 5).

– Verify: This program takes as input a message m, signature $\sigma = (\alpha, \beta)$, a ring R and the encryption key ek of an opener. Then it checks if the claimed signature σ is a valid encoding of these inputs under the encoding scheme as described in the Sign program above. More concretely, it checks if $F_2(K_2, \alpha) \oplus \beta = (m'\|vk'\|h'\|ek'\|r')$, $m = m'$, $\mathsf{Hash}(R) = h'$, $ek = ek'$ and $f_1(\alpha) = f_1(F_1(K_1, (m'\|vk'\|h'\|ek'\|r')))$. If so, it outputs verification key vk' of the signer encrypted under the public key ek'. Otherwise it outputs $\bot$. Here $f_2(\cdot)$ is a one-way function.

In addition, we require two more functionalities, namely Open and Judge, to reveal and prove the signer's identity. Open works as follows: it invokes Verify to obtain the encrypted verification key. Next, it decrypts it. Finally, it produces a NIZK proof of correct decryption. Judge verifies if the proof returned by Open is correct.

Notation. We use $x \leftarrow S$ to denote that x is sampled uniformly from the set S; $y := \mathcal{A}(x)$ denotes that $\mathcal{A}$ is a deterministic algorithm whose output is assigned to y; when $\mathcal{A}$ is randomized, the process is denoted by $y \leftarrow \mathcal{A}(x)$ (or $y = A(x; r)$). "PPT" stands for probabilistic polynomial time.

Paper Organization. The rest of the paper is organized as follows. In Sect. 2, we recall the definition and security model of ARS scheme from [4]. In Sect. 3, we recall the definitions of various cryptographic primitives used in our construction. In Sect. 4, we present our $i\mathcal{O}$-based ARS scheme. In Sect. 4, we give a proof of security of our construction. Finally, we conclude in Sect. 6 giving some open problems in the area.

2   Accountable Ring Signature Scheme

2.1   Syntax

Assume that each user in the system is uniquely identified by an index $i \in [n], n \in \mathbb{N}$. In addition, imagine that the PKI maintains a public registry reg of the registered users.

Definition 1 (Accountable Ring Signature (ARS) [4]). *An ARS scheme* ARS = (ARS.UKGen, ARS.OKGen, ARS.Sign, ARS.Verify, ARS.Open, ARS.Judge) *over a* PPT *setup* ARS.Setup *is a 6-tuple of algorithms.*

- *ARS.Setup(1^λ): On input the security parameter 1^λ,* ARS.Setup *outputs a list of public parameters params consisting of a class $\{\mathcal{SK}, \mathcal{VK}, \mathcal{EK}, \mathcal{DK}\}$ – denoting key spaces for signing, verification, encryption and decryption – along with poly-time algorithms for sampling and deciding memberships.*
- ARS.UKGen(*params*): *On input params, the* PPT *algorithm* ARS.UKGen *outputs a signing key sk and corresponding verification key vk for a user.*
- ARS.OKGen(*params*): *On input params, the* PPT *algorithm* ARS.OKGen *outputs an encryption/decryption key pair (ek, dk) for an opener.*
- ARS.Sign(*params, m, sk, R, ek*): *On input params, message m, signer's secret key sk, a ring R – which is a set of verification keys – and an opener's public key ek, the* PPT *algorithm* ARS.Sign *outputs the ring signature σ.*
- ARS.Verify(*params, m, σ, R, ek*): *On input params, message m, signature σ, a ring R and a public key ek,* ARS.Verify *outputs 1/0.*
- ARS.Open(*params, m, σ, R, dk*): *On input params, message m, signature σ, a ring R and the opener's secret key dk,* ARS.Open *outputs the verification key vk of signer and a proof ϕ that the owner of vk generated σ.*
- ARS.Judge(*params, m, σ, R, ek, vk, ϕ*): *On input params, message m, signature σ, a ring R, the opener's public key ek, the verification key vk and the proof ϕ,* ARS.Judge *outputs 1 if the proof is correct and 0 otherwise.*

Correctness. An ARS scheme is correct if for any PPT adversary $\mathcal{A}$:

$$\Pr \begin{bmatrix} params \leftarrow \mathsf{ARS.Setup}(1^\lambda); \ (vk, sk) \leftarrow \mathsf{ARS.UKGen}(params); \\ (m, R, ek) \leftarrow \mathcal{A}(params, sk); \ \sigma \leftarrow \mathsf{ARS.Sign}(m, sk, R, ek) \\ : vk \in R \land \mathsf{ARS.Verify}(m, \sigma, R, ek) = 1 \end{bmatrix} \approx 1$$

2.2 Security Model

Unforgeability. Unforgeability requires that an adversary cannot falsely accuse an honest user of creating a ring signature even if some of the members of the ring are corrupt and that the adversary controls the opener. More concretely, consider the following game between a PPT adversary $\mathcal{A}$ and a challenger $\mathcal{C}$.

1. **Setup Phase:** $\mathcal{C}$ runs the algorithms ARS.Setup to generate public parameters $params$. Then, it chooses a set $S \subset [n]$ of signers, runs ARS.UKGen to generate key-pairs $\{(sk_i, vk_i)\}_{i \in S}$ and registers them with the PKI, and sends $params, \{vk_i\}_{i \in S}$ to $\mathcal{A}$.
2. **Query Phase:** $\mathcal{A}$ can make the following three types of queries to $\mathcal{C}$. $\mathcal{C}$ answers these queries via oracles $\mathcal{O}_{Reg}$, $\mathcal{O}_{Cor}$ and $\mathcal{O}_{Sig}$.
   - **Registration query:** $\mathcal{A}$ runs the algorithm ARS.UKGen to generate a signing-verification key pair $(sk_i, vk_i), i \notin [n] \setminus S$ and interacts with the oracle $\mathcal{O}_{Reg}$ to register vk_i with the PKI. Let $\mathcal{Q}_{Reg}$ be the set of verification keys registered by $\mathcal{A}$.
   - **Corruption query:** $\mathcal{A}$ queries a verification key $vk_i, i \in S$ to the oracle $\mathcal{O}_{Cor}$. The oracle returns the corresponding signing key sk_i. Let $\mathcal{Q}_{Cor}$ be the set of verification keys vk_i for which the corresponding signing keys has been revealed.
   - **Signing query:** $\mathcal{A}$ queries (m, vk_i, R, ek, r) to the oracle $\mathcal{O}_{Sign}$. The oracle returns a signature $\sigma_i = \mathsf{Sign}(m, vk_i, R, ek, r)$ if $vk_i \notin \mathcal{Q}_{Reg} \cup \mathcal{Q}_{Cor}$. Let $\mathcal{Q}_{Sign}$ be the set of queries and their responses $(m, vk, R, ek, r, \sigma)$.
3. **Forgery Phase:** $\mathcal{A}$ outputs a signature σ^* w.r.t some $(m^*, vk^*, R^*, ek^*, r^*)$.

$\mathcal{A}$ wins the above game if ARS.Verify outputs 1 on input $(m^*, \sigma^*, R^*, ek^*)$.

Definition 2. *We say that the ARS scheme is unforgeable, if for any PPT adversary $\mathcal{A}$, its advantage in the game above is negligible i.e.* $\mathbf{Adv}_{\mathcal{A}}^{Unforge} = \Pr[\mathcal{A} \text{ wins}] \leq \mathsf{negl}(\lambda).$

Remark 1. In our scheme we consider a selective variant of the above definition where adversary $\mathcal{A}$ is required to commit to a forgery input $(m^*, vk^*, R^*, ek^*, r^*)$ in the setup phase. Then $\mathcal{A}$ cannot make signing query $(m^*, vk^*, R^*, ek^*, r^*)$ to the oracle $\mathcal{O}_{Sign}$.

Anonymity. Anonymity requires that the signature keeps the identity of signer (who is a ring member) anonymous unless the opener explicitly want to open the signature and reveal the signer's identity. Our definition also captures unlinkability (i.e. anonymity breaks if an adversary can link signatures from same signer) and anonymity against full key exposure attacks (i.e. the signatures remain anonymous even if the secret signing keys were revealed). More concretely, consider the following game between a PPT adversary $\mathcal{A}$ and a challenger $\mathcal{C}$.

1. **Setup Phase:** $\mathcal{C}$ runs the algorithm ARS.Setup to generate public parameters $params$. Then, $\mathcal{C}$ runs the algorithm ARS.KGen to generate signing-verification key pairs $\{(sk_i, vk_i)\}_{i \in [n]}$ for the users and registers them with the PKI. In addition, it also generates a private-public key pair (dk^*, ek^*) of an opener. Then, it sends $params, ek^*, \{(sk_i, vk_i)\}_{i \in [n]}$ to $\mathcal{A}$.
2. **Challenge Phase:** $\mathcal{A}$ submits a message m^*, a ring $R^* \subseteq \{vk_i\}_{i \in [n]}$ and two secret signing keys $sk_{i_0}^*, sk_{i_1}^* \in \{sk_i\}_{i \in [n]}, i_0 \neq i_1$, such that $vk_{i_0}^*, vk_{i_1}^* \in R^*$. Next, $\mathcal{C}$ chooses $b \leftarrow \{0, 1\}$ and produces an accountable ring signature $\sigma_{i_b}^* = $ ARS.Sign$(m^*, sk_{i_b}^*, R^*, ek^*; r)$ and returns $\sigma_{i_b}^*$ to $\mathcal{A}$.
3. **Guess Phase:** $\mathcal{A}$ guesses b and outputs $b' \in \{0, 1\}$.

Definition 3. *We say that the ARS scheme is anonymous, if for any* PPT *adversary* $\mathcal{A}$, *its advantage in the game above is negligible i.e.* $\mathbf{Adv}_{\mathcal{A}}^{Anony} = |\Pr[b = b'] - \frac{1}{2}| \leq \mathsf{negl}(\lambda)$.

Remark 2. In our scheme we consider a selective variant of the above definition where adversary $\mathcal{A}$ is required to commit to a target input $(m^*, sk_{i_0}^*, sk_{i_1}^*, R^*, ek^*)$ in the setup phase.

Traceability. Traceability requires that the specified opener is always able to identify the signer in the ring and produce a valid proof that the signer actually produced that particular signature; i.e. for any PPT adversary $\mathcal{A}$:

$$\Pr \left[\begin{array}{c} params \leftarrow \mathsf{ARS.Setup}(1^\lambda); \ (m, \sigma, R, ek, dk) \leftarrow \mathcal{A}(params); \\ (vk, \phi) \leftarrow \mathsf{ARS.Open}(m, \sigma, R, dk) \\ : \mathsf{ARS.Verify}(m, \sigma, R, ek) = 1 \wedge \mathsf{ARS.Judge}(m, \sigma, R, ek, vk, \phi) = 0 \end{array} \right] \leq \mathsf{negl}(\lambda)$$

Tracing Soundness. Tracing soundness requires that a signature can be traced to only one user; even when all users as well as the opener are corrupt; i.e. for any PPT adversary $\mathcal{A}$:

$$\Pr \left[\begin{array}{c} params \leftarrow \mathsf{ARS.Setup}(1^\lambda); \ (m, \sigma, R, ek, vk_1, vk_2, \phi_1, \phi_2) \leftarrow \\ \mathcal{A}(params) : \forall i \in \{1, 2\}, \mathsf{ARS.Judge}(m, \sigma, R, ek, vk_i, \phi_i) = 1 \\ \wedge vk_1 \neq vk_2 \end{array} \right] \leq \mathsf{negl}(\lambda)$$

3   Preliminaries

3.1   Indistinguishability Obfuscation

Definition 4 (Indistinguishability Obfuscator ($i\mathcal{O}$) **[1]).** *A* PPT *algorithm* $i\mathcal{O}$ *is said to be an indistinguishability obfuscator for a collection of circuits* $\{\mathcal{C}_\lambda\}$ *(λ is the security parameter), if it satisfies the following two conditions:*

1. **Functionality:** $\forall \lambda \in \mathbb{N}$ and $\forall C \in \mathcal{C}_\lambda$, $\Pr[\forall x : i\mathcal{O}(1^\lambda, C)(x) = C(x)] = 1$
2. **Indistinguishability:** For all PPT distinguisher $\mathcal{D}$, there exists a negligible function negl, such that for all $\lambda \in \mathbb{N}$ and for all pairs of circuits C_1, $C_2 \in \mathcal{C}_\lambda$, if $\Pr[\forall x : C_1(x) = C_2(x)] = 1$, then we have $\boldsymbol{Adv}_\mathcal{D}^{Obf} \leq \mathsf{negl}(\lambda)$ i.e.

$$| \Pr[\mathcal{D}(i\mathcal{O}(1^\lambda, C_1)) = 1] - \Pr[\mathcal{D}(i\mathcal{O}(1^\lambda, C_2)) = 1]| \leq \mathsf{negl}(\lambda)$$

Here, we will consider polynomial-size circuits only. In 2013, Garg et al. [11] proposed the first candidate construction of an efficient *indistinguishability obfuscator* for any general purpose boolean circuit. Their construction is based on the multilinear map candidates of [10,11].

3.2   Puncturable Pseudorandom Functions

Definition 5 (Puncturable Pseudorandom Functions (PPRFs) [19]). Let $\ell(\cdot)$ and $m(\cdot)$ be two polynomially bounded length functions. Let $\mathcal{F} = \{F_K : \{0,1\}^{\ell(\lambda)} \rightarrow \{0,1\}^{m(\lambda)} | K \leftarrow \{0,1\}^\lambda, \lambda \in \mathbb{N}\}$. $\mathcal{F}$ is called a family of puncturable PRFs if it is associated with two turing machines $K_\mathcal{F}$ and $P_\mathcal{F}$, such that $P_\mathcal{F}$ takes as input a key $K \leftarrow \{0,1\}^\lambda$ and a point $u^* \in \{0,1\}^{\ell(\lambda)}$, and outputs a punctured key K_{u^*}, so that the following two conditions are satisfied:

1. **Functionality preserved under puncturing:** For every $u^* \in \{0,1\}^{\ell(\lambda)}$,

$$\Pr\left[K \leftarrow K_\mathcal{F}(1^\lambda); K_{u^*} = P_\mathcal{F}(K, u^*) : \forall u \neq u^*, F_K(u) = F_{K_{u^*}}(u)\right] = 1$$

2. **Indistinguishability at punctured points:** $\forall$ PPT distinguisher $\mathcal{D}$, below ensembles are computationally indistinguishable i.e. $\boldsymbol{Adv}_\mathcal{D}^{PPRF} \leq \mathsf{negl}(\lambda)$:
   - $\{u^*, K_{u^*}, F_K(u^*) : K \leftarrow K_\mathcal{F}(1^\lambda); K_{u^*} = P_\mathcal{F}(K, u^*)\}$
   - $\{u^*, K_{u^*}, x : K \leftarrow K_\mathcal{F}(1^\lambda); K_{u^*} = P_\mathcal{F}(K, u^*); x \leftarrow \{0,1\}^{\ell(\lambda)}\}$

Recently, [3,6,16] observed that puncturable PRFs can easily be constructed from GGM's PRFs [13] which are based on one-way functions. We will use the following lemma on statistical injective PPRF in our construction.

Lemma 1 ([19]). If one-way functions exist, then for all efficiently computable functions $\ell(\lambda)$, $m(\lambda)$, and $e(\lambda)$ such that $m(\lambda) \geq 2\ell(\lambda) + e(\lambda)$, there exists a statistically injective PPRF family with failure probability $2^{-e(\lambda)}$ that maps $\ell(\lambda)$ bits to $m(\lambda)$ bits.

3.3   IND-CPA Secure Public Key Encryption Scheme

Definition 6 (Public Key Encryption Scheme (PKE)). A *public key encryption scheme* $\mathsf{PKE} = (\mathsf{PKE.KGen}, \mathsf{PKE.Encrypt}, \mathsf{PKE.Decrypt})$ *over a PPT setup* $\mathsf{PKE.Setup}$ *is a 3-tuple of algorithms.*

- $\mathsf{PKE.Setup}(1^\lambda)$: On input the security parameter 1^λ, $\mathsf{PKE.Setup}$ outputs a list of public parameters params as follows: key spaces $(\mathcal{EK}, \mathcal{DK})$; plaintext and ciphertext spaces $\mathcal{P}$ and $\mathcal{C}$.

- PKE.KGen(*params*): *On input params, the* PPT *algorithm* PKE.KGen *outputs a pair of public and private keys* (ek, dk).
- PKE.Encrypt($ek, m \in \mathcal{M}$): *On input public key ek and message m, the* PPT *algorithm* PKE.Encrypt *outputs a ciphertext* $c \in \mathcal{C}$.
- PKE.Decrypt(dk, c): *On input secret key dk and ciphertext c,* PKE.Decrypt *outputs a message* $m \in \mathcal{M}$.

Correctness. For all key pairs $(ek, dk) \leftarrow$ PKE.KGen(*params*), and for all messages $m \in \mathcal{M}$: $\Pr[\text{PKE.Decrypt}(dk, \text{PKE.Encrypt}(ek, m)) = m] = 1$.

Security. For all PPT adversaries $\mathcal{A}$:

$$\Pr \begin{bmatrix} params \leftarrow \text{PKE.Setup}(1^\lambda); \ (ek, dk) \leftarrow \text{PKE.KGen}(params); \\ (m_0, m_1) \leftarrow \mathcal{A}(ek, params); \ b \leftarrow \{0,1\}; \\ c \leftarrow \text{PKE.Encrypt}(ek, m_b) : \mathcal{A}(c) = m_b \end{bmatrix} \leq \frac{1}{2} + \mathsf{negl}(\lambda)$$

3.4   One-Way Function

Definition 7 (One-way Function (OWF) [12]). *A function* $f : \mathcal{X} \to \mathcal{Y}$ *(over* PPT *setup* OWF.Setup*, which defines* $f, \mathcal{X}, \mathcal{Y}$ *is one-way if* f *is polynomial-time computable and is hard to invert, i.e. for all* PPT *adversaries* $\mathcal{A}$, $\boldsymbol{Adv}_{\mathcal{A}}^{OWF} \leq$ $\mathsf{negl}(\lambda)$ *i.e.* $\Pr\left[x \leftarrow \mathcal{X}; \ y := f(x) : \mathcal{A}(y) = x\right] \leq \mathsf{negl}(\lambda)$.

3.5   Pseudorandom Generator

Definition 8 (Pseudorandom Generator (PRG) [14]). *A function* PRG : $\{0,1\}^\lambda \to \{0,1\}^{p(\lambda)}$ *(over a setup* PRG.Setup*, which defines the function* PRG*, its domain and range) is* PRG *if it is polynomial-time computable, length expanding and is pseudorandom, i.e. for all* PPT *adversaries* $\mathcal{A}$, $\boldsymbol{Adv}_{\mathcal{A}}^{PRG} \leq \mathsf{negl}(\lambda)$ *i.e.*

$$\left| \Pr[\mathcal{A}(\text{PRG}(s)) = 1 : s \leftarrow \{0,1\}^\lambda] - \Pr[\mathcal{A}(r) = 1 : r \leftarrow \{0,1\}^{p(\lambda)}] \right| \leq \mathsf{negl}(\lambda)$$

3.6   Collision Resistant Hash Function

Definition 9 (Collision Resistant Hash Function (CRHF)). *A family* $\mathcal{H}$ *of collection of functions* Hash : $\{0,1\}^* \to \{0,1\}^{\ell_h}$ *(over* PPT *setup* CRHF.Setup*, which defines the function* Hash*, its domain and range) is called a family of collision resistant hash functions if* Hash *is polynomial-time computable, length compressing and it is hard to find collisions in* Hash*, i.e. for all* PPT *adversaries* $\mathcal{A}$, $\boldsymbol{Adv}_{\mathcal{A}}^{CRHF} \leq \mathsf{negl}(\lambda)$ *i.e.*

$$\Pr_{\text{Hash} \leftarrow \mathcal{H}} \begin{bmatrix} params \leftarrow \text{CRHF.Setup}(1^\lambda); \ (x_0, x_1) \leftarrow \mathcal{A}(params, \text{Hash}) \\ : x_0 \neq x_1 \wedge \text{Hash}(x_0) = \text{Hash}(x_1) \end{bmatrix} \leq \mathsf{negl}(\lambda)$$

4   A New Accountable Ring Signature Scheme

We now present the details of our construction. We assume that each user in the system is uniquely identified by an index $i \in [n], n \in \mathbb{N}$. Let $\mathcal{M} = \{0,1\}^{\ell_m}$ be our message space. Let $\mathcal{SK} = \{0,1\}^{\ell_s}$, $\mathcal{VK} = \{0,1\}^{\ell_v}$, $\mathcal{EK} = \{0,1\}^{\ell_e}$ and $\mathcal{DK} = \{0,1\}^{\ell_d}$ respectively be our signing, verification, encryption and decryption key spaces. Here $\ell_m, \ell_s, \ell_v, \ell_e$ and ℓ_d are polynomials in security parameter λ. We assume that the set of verification keys in the ring R are always arranged in *increasing order* of their indexes, in order to enable the hash function to compute on R. We will use following primitives in our construction:

- Pseudo-random generator $\mathsf{PRG} : \{0,1\}^\lambda \rightarrow \{0,1\}^{\ell_p}$ with $\ell_p \geq 2\lambda$.
- One-way function $f : \{0,1\}^{\ell_s} \rightarrow \{0,1\}^{\ell_v}$.
- IND-CPA secure PKE scheme as defined in Sect. 3.3.
- Collision resistant hash function $\mathsf{Hash} : \{0,1\}^{n \cdot \ell_v} \rightarrow \{0,1\}^{\ell_h}$. Here n is size of the ring and $n \cdot \ell_v > \ell_h$.
- A statistically injective PPRF $F_1(K_1, \cdot)$ that accepts inputs of length $\ell_m + \ell_s + \ell_h + \ell_e + \ell_p$, and outputs strings of length ℓ_1. Note that this type of PPRF exists, and is easy to construct from Lemma 1.
- A PPRF $F_2(K_2, \cdot)$ that accepts inputs of length ℓ_1, and outputs strings of length ℓ_2 such that $\ell_2 = \ell_m + \ell_s + \ell_h + \ell_e + \ell_p$.
- One-way function $f_1 : \{0,1\}^{\ell_1} \rightarrow \{0,1\}^{\ell_3}$.
- A PPRF $F_3(K_3, \cdot)$ that accepts inputs of length $\ell_e + \ell_c + \ell_v$, and outputs strings of length ℓ_ϕ. Here ℓ_c is the length of ciphertext as output by the PKE scheme and ℓ_ϕ is the length of NIZK proof.
- One-way function $f_2 : \{0,1\}^{\ell_\phi} \rightarrow \{0,1\}^{\ell_4}$.
- An indistinguishability obfuscator $i\mathcal{O}$ as defined in Sect. 3.1.

Construction of our ARS scheme from $i\mathcal{O}$

- $\mathsf{ARS.Setup}(1^\lambda)$: A trusted authority runs this algorithm for once. It takes security parameter 1^λ as input and generates key spaces as defined above. Next, it

---

$$\mathcal{P}_S$$

Hardwired: Keys K_1, K_2
Input: Message m, key pair (sk, vk) , ring R, key ek and randomness r
1 **if** $(f(sk) = vk \wedge vk \in R)$ **then**
2   Set $\alpha := F_1(K_1, (m\|vk\|\mathsf{Hash}(R)\|ek\|\mathsf{PRG}(r)))$
3   Set $\beta := F_2(K_2, \alpha) \oplus (m\|vk\|\mathsf{Hash}(R)\|ek\|\mathsf{PRG}(r))$
4   **return** $(\sigma := (\alpha, \beta))$
5 **else**
6   **return** $\bot$

Fig. 1. The circuit Sign

$$\mathcal{P}_V$$

Hardwired: Keys K_1, K_2
Input: Message m, signature $\sigma = (\alpha, \beta)$, ring R and opener's public key ek
1 Compute $F_2(K_2, \alpha) \oplus \beta = (m' \| vk' \| h' \| ek' \| r')$
2 **if** $\big(m = m' \wedge \mathsf{Hash}(R) = h' \wedge ek = ek' \wedge f_1(\alpha) =$
   $f_1(F_1(K_1, (m' \| vk' \| h' \| ek' \| r')))) $ **then**
3 | **return** $(c := \mathsf{PKE.Encrypt}(ek, vk'; r'))$
4 **else**
5 | **return** $\bot$

Fig. 2. The circuit Verify

$$\mathcal{P}_{NP}$$

Hardwired: Keys K_3
Input: Statement $x = (ek \| c \| vk)$ and witness dk
1 **if** $\Big(\big(ek = \mathsf{PKE.KGen}(params, dk) \in \mathcal{EK} \big) \wedge \big(vk =$
   $\mathsf{PKE.Decrypt}(dk, c) \big) \wedge \big(vk \in \mathcal{VK} \big) \Big) = 1$ **then**
2 | **return** $\phi := F_3(K_3, x)$
3 **else**
4 | **return** $\bot$

Fig. 3. The circuit NIZKprove

$$\mathcal{P}_{NV}$$

Hardwired: Keys K_3
Input: Statement $x = (ek \| c \| vk)$ and proof ϕ
1 **if** $f_2(\phi) = f_2(F_3(K_3, x))$ **then**
2 | **return** 1
3 **else**
4 | **return** $\bot$

Fig. 4. The circuit NIZKverify

chooses keys K_1, K_2, K_3 for PPRFs F_1, F_2, F_3 respectively, and creates obfuscated programs $\mathsf{Sign} = i\mathcal{O}(\mathcal{P}_S)$, $\mathsf{Verify} = i\mathcal{O}(\mathcal{P}_V)$, $\mathsf{NIZKprove} = i\mathcal{O}(\mathcal{P}_{NP})$ and $\mathsf{NIZKverify} = i\mathcal{O}(\mathcal{P}_{NV})$ (shown in Figs. 1, 2, 3 and 4). Finally, it outputs public parameters $params = (\mathcal{SK}, \mathcal{VK}, \mathcal{DK}, \mathcal{EK}, \mathsf{Sign}, \mathsf{Verify}, \mathsf{NIZKprove}, \mathsf{NIZKverify})$.

- ARS.UKGen($params$): This algorithm takes $params$ as input, chooses a secret signing key $sk \leftarrow \mathcal{SK}$ and computes verification key as $vk = f(sk) \in \mathcal{VK}$.
- ARS.OKGen($params$): This algorithm takes $params$ as input and outputs a key pair (ek, dk) for the PKE scheme. This algorithm is same as PKE.KGen.

- ARS.Sign($params, m, sk, vk, R, ek, r$): This algorithm takes $params$, message m, a key pair (sk, vk), a ring R, an opener's public key ek and randomness r as input, runs Sign on these inputs and returns ring signature σ.
- ARS.Verify($params, m, \sigma, R, ek$): This algorithm takes $params$, message m, signature σ, a ring R and an opener's public key ek as input, runs Verify on these inputs and returns a verification key of the signer encrypted under the public key ek of the opener if the signature is accepted, else it returns $\bot$.
- ARS.Open($params, m, \sigma, R, ek, dk$): This algorithm takes $params$, message m, signature σ, a ring R and an opener's PKE key pair (ek, dk) as input, runs Verify on these inputs to retrieve c. Next, it decrypts c using secret key dk to output verification key vk of the signer. In addition, it also outputs a proof ϕ of correct decryption using the program NIZKprove.
- ARS.Judge($params, m, \sigma, R, ek, vk, \phi$): This algorithm takes $params$, message m, signature σ, a ring R and an opener's public key ek, verification key vk, proof ϕ as input, runs Verify on these inputs to retrieve c and outputs 1 if NIZKverify$((ek\|c\|vk), \phi) = 1$ else 0.

5   Proof of Security

We will use the following lemma in our security proofs.

Lemma 2. *For any fixed input* (m, sk, R, ek, r) *to the* Sign *program, there can exist at most one valid signature string* $\sigma = (\alpha, \beta)$. *This happens with high probability if* K_1 *is chosen such that the* $F_1(K_1, \cdot)$ *is statistically injective and* Hash *is a collision resistant hash function.*

Proof. Since $F_1(K_1, \cdot)$ is injective over the choice of K_1 and Hash is collision resistant therefore, there can exist at most one string α (which implies unique (α, β)) when $F_1(K_1, \cdot)$ is applied to a unique string $(m, vk, \text{Hash}(R), \text{PRG}(r))$. Furthermore, if $\sigma = (\alpha, \beta)$ is choses at random then there can exists at most $2^{\ell_m + \ell_v + \ell_h + \ell_e + \ell_p}$ valid values of σ. (This lemma is adapted from a similar lemma of [19] to include an additional hash function.)

Theorem 1. *The* ARS *scheme of Sect. 4 is selectively unforgeable if the primitives of Sect. 3 satisfy their respective security properties.*

Proof. We prove the security by a sequence of hybrid experiments and show that if there exists a PPT adversary $\mathcal{A}$ that can break the selective unforgeability of our ARS scheme, then we can construct a challenger $\mathcal{B}$ who can break the security of one-way function.

Hybrid 0. This hybrid is same as the unforgeability game of Sect. 2.

1. **Setup Phase:**
   (a) $\mathcal{C}$ chooses a set $S \subset [n]$ and runs the algorithm ARS.UKGen to generate signing-verification key pairs $\{(sk_i, vk_i)\}_{i \in S}$ for the users and registers them with the PKI. Finally, it sends $\{vk_i\}_{i \in S}$ to the adversary $\mathcal{A}$.

$$\mathcal{P}_S^*$$

Hardwired: Keys $K_1, K_2(\{\alpha^*\})$ and values $m^*, vk^*, R^*, ek^*, r^*, \beta^*$

Input: Message m, key pair (sk, vk), ring R, key ek and randomness r

1 **if** $(m = m^* \land f(sk) = vk^* \land f(sk) \in R^* \land R = R^* \land ek = ek^* \land r = r^*)$

   then

2 | Set $\alpha^* := F_1(K_1, (m^*\|vk^*\|\mathsf{Hash}(R^*)\|ek^*\|\mathsf{PRG}(r^*)))$

3 | **return** $(\sigma^* := (\alpha^*, \beta^*))$

4 **else if** $(f(sk) = vk \land f(sk) \in R)$ **then**

5 | Set $\alpha := F_1(K_1, (m\|vk\|\mathsf{Hash}(R)\|ek\|\mathsf{PRG}(r)))$

6 | Set $\beta := F_2(K_2, \alpha) \oplus (m\|vk\|\mathsf{Hash}(R)\|ek\|\mathsf{PRG}(r))$

7 | **return** $(\sigma := (\alpha, \beta))$

8 **else**

9 | **return** $\bot$

Fig. 5. The circuit Sign

(b) $\mathcal{A}$ sends a message m^*, a ring $R^* \subseteq \{vk_i\}_{i \in S}$, a verification key $vk^* \in R^*$, an opener's public key ek^* and randomness r^* to challenger $\mathcal{C}$, claiming it can forge an accountable ring signature σ^* w.r.t $(m^*, vk^*, R^*, ek^*, r^*)$.

(c) The challenger $\mathcal{C}$ first chooses keys K_1, K_2 for PPRFs F_1, F_2 respectively, and produces two obfuscated programs $\mathsf{Sign} = i\mathcal{O}(\mathcal{P}_S)$ and $\mathsf{Verify} = i\mathcal{O}(\mathcal{P}_V)$ (Figs. 1 and 2). Then, it sends these programs to the adversary $\mathcal{A}$.

2. **Query Phase:** $\mathcal{A}$ can make the following three types of queries to $\mathcal{C}$. $\mathcal{C}$ answers these queries via oracles $\mathcal{O}_{Reg}$, $\mathcal{O}_{Cor}$ and $\mathcal{O}_{Sig}$.

   – **Registration query:** $\mathcal{A}$ runs the algorithm $\mathsf{ARS.UKGen}$ to generate a signing-verification key pair $(sk_i, vk_i), i \notin [n] \setminus S$ and interacts with the oracle $\mathcal{O}_{Reg}$ to register vk_i with the PKI. Let $\mathcal{Q}_{Reg}$ be the set of verification keys registered by $\mathcal{A}$.

   – **Corruption query:** $\mathcal{A}$ queries a verification key $vk_i, i \in S$ such that $vk_i \neq vk^*$ to the oracle $\mathcal{O}_{Cor}$. The oracle returns the corresponding signing key sk_i. Let $\mathcal{Q}_{Cor}$ be the set of verification keys vk_i for which the corresponding signing keys has been revealed.

   – **Signing query:** $\mathcal{A}$ queries (m, vk_i, R, ek, r) to the oracle $\mathcal{O}_{Sign}$. The oracle returns a signature $\sigma_i = \mathsf{Sign}(m, vk_i, R, ek, r)$ if $vk_i \neq vk^*$ and $vk_i \notin \mathcal{Q}_{Reg} \cup \mathcal{Q}_{Cor}$. Let $\mathcal{Q}_{Sign}$ be the set of queries and their responses $(m, vk, R, ek, r, \sigma)$.

3. **Forgery Phase:** $\mathcal{A}$ outputs a signature σ^* w.r.t $(m^*, vk^*, R^*, ek^*, r^*)$.

$\mathcal{A}$ wins the above game if $\mathsf{ARS.Verify}(m, \sigma, R, ek) = 1$. Let $\boldsymbol{Adv}_{\mathcal{A}}^{Hyb0}$ denote the advantage of $\mathcal{A}$ in this hybrid.

Hybrid 1. This hybrid is identical to *Hybrid 0* with the exception that $\mathcal{C}$ gives out the two obfuscated programs as $\mathsf{Sign} = i\mathcal{O}(\mathcal{P}_S^*)$ and $\mathsf{Verify} = i\mathcal{O}(\mathcal{P}_V^*)$ (Figs. 5 and 6). The details of the modifications are as follows: $\mathcal{C}$ punctures the

$$\mathcal{P}_V^*$$

Hardwired: Keys $K_1, K_2(\{\alpha^*\})$ and values $m^*, vk^*, R^*, ek^*, r^*$
Input: Message m, signature $\sigma = (\alpha, \beta)$, ring R and opener's public key ek
1 **if** $\big(m = m^* \wedge R = R^* \wedge ek = ek^* \wedge f_1(\alpha) =$
   $f_1(F_1(K_1, (m^*\|vk^*\|\mathsf{Hash}(R^*)\|ek^*\|\mathsf{PRG}(r^*)))))$ **then**
2 $\quad$ **return** $(c^* := \mathsf{PKE.Encrypt}(ek, vk^*; r^*))$
3 **else if** $\big(F_2(K_2, \alpha) \oplus \beta = (m'\|vk'\|h'\|ek'\|r') \wedge m = m' \wedge \mathsf{Hash}(R) =$
   $h' \wedge ek = ek' \wedge f_1(\alpha) = f_1(F_1(K_1, (m'\|vk'\|h'\|ek'\|r'))))$ **then**
4 $\quad$ **return** $(c := \mathsf{PKE.Encrypt}(ek, vk'; r'))$
5 **return** $\bot$

Fig. 6. The circuit Verify

key K_2 at the point $\alpha^* = F_1(K_1, (m^*\|vk^*\|\mathsf{Hash}(R^*)\|ek^*\|\mathsf{PRG}(r^*))$ and hardwires this punctured key $K_2(\{\alpha^*\})$ along with message m^*, verification key vk^*, ring R^*, public key ek^* and randomness r^* in programs $\mathcal{P}_S^*$ and $\mathcal{P}_V^*$. In addition, it hardwires value $\beta^* = F_2(K_2, \alpha^*) \oplus (m^*\|vk^*\|\mathsf{Hash}(R^*)\|ek^*\|\mathsf{PRG}(r^*))$ in program $\mathcal{P}_S^*$. Furthermore, a "if" condition is added at line 1 in program $\mathcal{P}_S^*$ to check $(m, vk, R, ek, r) = (m^*, vk^*, R^*, ek^*, r^*)$. If so, output the signature as $\sigma^* = (\alpha^*, \beta^*)$. Similarly, $\mathcal{P}_V^*$ is modified to output c^* if $(m, \sigma, R, ek) = (m^*, \sigma^*, R^*, ek^*)$. Let $\boldsymbol{Adv}_{\mathcal{A}}^{Hyb\,1}$ denote the advantage of $\mathcal{A}$ in this hybrid.

Hybrid 2. This hybrid is identical to *Hybrid 1* with the exception that β^* is chosen uniformly at random from the set $\{0,1\}^{\ell_2}$. Let $\boldsymbol{Adv}_{\mathcal{A}}^{Hyb\,2}$ denote the advantage of $\mathcal{A}$ in this hybrid.

Hybrid 3. This hybrid is identical to *Hybrid 2* with the exception that $\mathcal{C}$ gives out the two obfuscated programs as $\mathsf{Sign} = i\mathcal{O}(\mathcal{P}_S^{**})$ and $\mathsf{Verify} = i\mathcal{O}(\mathcal{P}_V^{**})$ (Figs. 7 and 8). The details of the modifications are as follows: $\mathcal{C}$ punctures the key K_1 at the point $(m^*\|vk^*\|\mathsf{Hash}(R)\|ek^*\|\mathsf{PRG}(r^*))$ and hardwires this punctured key $K_1(\{(m^*\|vk^*\|\mathsf{Hash}(R)\|ek^*\|\mathsf{PRG}(r^*))\})$ in programs $\mathcal{P}_S^{**}$ and $\mathcal{P}_V^{**}$. Finally, it hardwires values $\alpha^* = F_1(K_1, (m^*\|vk^*\|\mathsf{Hash}(R)\|ek^*\|\mathsf{PRG}(r^*)))$ and $u^* = f_1(\alpha^*)$ in programs $\mathcal{P}_S^{**}$ and $\mathcal{P}_V^{**}$ respectively. Let $\boldsymbol{Adv}_{\mathcal{A}}^{Hyb\,3}$ denote the advantage of $\mathcal{A}$ in this hybrid.

Hybrid 4. This hybrid is identical to *Hybrid 3* with the exception that α^* is chosen uniformly at random from the set $\{0,1\}^{\ell_1}$. Let $\boldsymbol{Adv}_{\mathcal{A}}^{Hyb\,4}$ denote the advantage of $\mathcal{A}$ in this hybrid.

The indistinguishability of successive hybrid games (discussed above) is shown in Lemmas 3, 4, 5, 6 and 7.

Lemma 3. *If $\mathcal{A}$ can distinguish Hybrid 0 from Hybrid 1 in the unforgeability game, i.e. $\boldsymbol{Adv}_{\mathcal{A}}^{Hyb\,0} - \boldsymbol{Adv}_{\mathcal{A}}^{Hyb\,1}$ is non-negligible then there exists an adversary $\mathcal{B}$ who can break the security of $i\mathcal{O}$.*

$$\mathcal{P}_S^{**}$$

Hardwired: Keys $K_1(\{m^*\|vk^*\|\mathsf{Hash}(R^*)\|ek^*\|\mathsf{PRG}(r^*)\})$, $K_2(\{\alpha^*\})$ and
values $m^*, vk^*, R^*, ek^*, r^*, \alpha^*, \beta^*$
Input: Message m, key pair (sk, vk), ring R, key ek and randomness r
1 **if** $(m = m^* \wedge f(sk) = vk^* \wedge f(sk) \in R^* \wedge R = R^* \wedge ek = ek^* \wedge r = r^*)$
then
2 $\quad$ **return** $(\sigma^* := (\alpha^*, \beta^*))$
3 **else if** $(f(sk) = vk \wedge f(sk) \in R)$ **then**
4 $\quad$ Set $\alpha := F_1(K_1, (m\|vk\|\mathsf{Hash}(R)\|ek\|\mathsf{PRG}(r)))$
5 $\quad$ Set $\beta := F_2(K_2, \alpha) \oplus (m\|vk\|\mathsf{Hash}(R)\|ek\|\mathsf{PRG}(r))$
6 $\quad$ **return** $(\sigma := (\alpha, \beta))$
7 **else**
8 $\quad$ **return** $\bot$

Fig. 7. The circuit Sign

$$\mathcal{P}_V^{**}$$

Hardwired: Keys $K_1, K_2(\{\alpha^*\})$ and values $m^*, vk^*, R^*, ek^*, r^*, u^*$
Input: Message m, signature $\sigma = (\alpha, \beta)$, ring R and opener's public key ek
1 **if** $(m = m^* \wedge R = R^* \wedge ek = ek^* \wedge f_1(\alpha) = u^*)$ **then**
2 $\quad$ **return** $(c^* := \mathsf{PKE.Encrypt}(ek, vk^*; r^*))$
3 **else if** $(F_2(K_2, \alpha) \oplus \beta = (m'\|vk'\|h'\|ek'\|r') \wedge m = m' \wedge \mathsf{Hash}(R) =$
$h' \wedge ek = ek' \wedge f_1(\alpha) = f_1(F_1(K_1, (m'\|vk'\|h'\|ek'\|r'))))$ **then**
4 $\quad$ **return** $(c := \mathsf{PKE.Encrypt}(ek, vk'; r'))$
5 **return** $\bot$

Fig. 8. The circuit Verify

Proof. We argue that if there is a non-negligible difference in advantages of $\mathcal{A}$ in hybrids *Hybrid 0* and *Hybrid 1* then we can construct an adversary $\mathcal{B}$ that breaks the security of $i\mathcal{O}$. Firstly, observe that the programs $\mathcal{P}_S$ and $\mathcal{P}_S^*$ as well as the programs $\mathcal{P}_V$ and $\mathcal{P}_V^*$, are functionally equivalent. Now, consider a scenario where the attacker $\mathcal{B}$ interacts with an $i\mathcal{O}$ challenger while acting as a challenger for $\mathcal{A}$. $\mathcal{B}$ submits the program pairs $P_0 = (\mathcal{P}_S, \mathcal{P}_V)$ and $P_1 = (\mathcal{P}_S^*, \mathcal{P}_V^*)$ to the $i\mathcal{O}$ challenger. The $i\mathcal{O}$ challenger chooses $b \leftarrow \{0,1\}$ at random, obfuscates the programs in P_b, and returns the output to $\mathcal{B}$. $\mathcal{B}$ plugs in these obfuscated programs in *Hybrid 0* and plays the rest of the game. Note that if the $i\mathcal{O}$ challenger chooses P_0 then we are in *Hybrid 0*. If it chooses P_1 then we are in *Hybrid 1*. $\mathcal{B}$ will output 1 if $\mathcal{A}$ successfully forges. Therefore, if $\mathcal{A}$ have different advantages in hybrids *Hybrid 0* and *Hybrid 1* then $\mathcal{B}$ can attack $i\mathcal{O}$ security with high probability i.e. $\mathbf{Adv}_{\mathcal{A}}^{Hyb\,0} - \mathbf{Adv}_{\mathcal{A}}^{Hyb\,1} = \mathbf{Adv}_{\mathcal{B}}^{i\mathcal{O}} \leq \mathsf{negl}(\lambda)$ (Sect. 3.1). $\quad\square$

Lemma 4. *If $\mathcal{A}$ can distinguish* Hybrid 1 *from* Hybrid 2 *in the* unforgeability game, *i.e.* $\boldsymbol{Adv}_{\mathcal{A}}^{Hyb\,1} - \boldsymbol{Adv}_{\mathcal{A}}^{Hyb\,2}$ *is non-negligible then there exists an adversary* $\mathcal{B}$ *who can break the security of* PPRF F_2.

Proof. We argue that if there is a non-negligible difference in advantages of $\mathcal{A}$ in hybrids *Hybrid 1* and *Hybrid 2* then we can construct an adversary $\mathcal{B}$ who can break the security of PPRF F_2. Consider a scenario where the attacker $\mathcal{B}$ interacts with a PPRF challenger while acting as a challenger for $\mathcal{A}$. First, $\mathcal{B}$ receives $(m^*, vk^*, R^*, ek^*, r^*)$ from $\mathcal{A}$. Then, it chooses $b \leftarrow \{0,1\}, K_1 \leftarrow \{0,1\}^\lambda$, computes $\alpha^* = F_1(K_1, (m^*\|vk^*\|\mathsf{Hash}(R^*)\|ek^*\|\mathsf{PRG}(r^*)))$, and submits α^* to the PPRF challenger. Next, the PPRF challenger choses $K_2 \leftarrow \{0,1\}^\lambda$, punctures K_2 at point α^*, and sends this punctured PRF key $K_2(\{\alpha^*\})$ along with a challenge $w^* \in \{0,1\}^{\ell_2}$ to $\mathcal{B}$. Then, $\mathcal{B}$ continues to run the game in *Hybrid 1* except that it plugs in β^* as $\beta^* = w^* \oplus (m^*\|vk^*\|\mathsf{Hash}(R^*)\|ek^*\|\mathsf{PRG}(r^*))$. Note that if the PPRF challenger outputs w^* as $w^* = F_2(K_2, \alpha^*)$ then we are in *Hybrid 1*. If it outputs w^* as $w^* \leftarrow \{0,1\}^{\ell_2}$ then we are in *Hybrid 2*. $\mathcal{B}$ will output 1 if $\mathcal{A}$ successfully forges. Therefore, if $\mathcal{A}$ have different advantages in hybrids *Hybrid 1* and *Hybrid 3* then $\mathcal{B}$ can attack the PPRF security with high probability i.e. $\boldsymbol{Adv}_{\mathcal{A}}^{Hyb\,1} - \boldsymbol{Adv}_{\mathcal{A}}^{Hyb\,2} = \boldsymbol{Adv}_{\mathcal{B}}^{PPRF} \leq \mathsf{negl}(\lambda)$ (Sect. 3.2). $\quad\square$

Lemma 5. *If $\mathcal{A}$ can distinguish* Hybrid 2 *from* Hybrid 3 *in the* unforgeability game, *i.e.* $\boldsymbol{Adv}_{\mathcal{A}}^{Hyb\,2} - \boldsymbol{Adv}_{\mathcal{A}}^{Hyb\,3}$ *is non-negligible then there exists an adversary* $\mathcal{B}$ *who can break the security of* $i\mathcal{O}$.

Proof. We argue that if there is a non-negligible difference in advantages of $\mathcal{A}$ in hybrids *Hybrid 2* and *Hybrid 3* then we can construct an adversary $\mathcal{B}$ that breaks the security of $i\mathcal{O}$. Firstly, observe that the programs $\mathcal{P}_S^*$ and $\mathcal{P}_S^{**}$ as well as the programs $\mathcal{P}_V^*$ and $\mathcal{P}_V^{**}$, are functionally equivalent. Now, consider a scenario where the attacker $\mathcal{B}$ interacts with an $i\mathcal{O}$ challenger while acting as a challenger for $\mathcal{A}$. $\mathcal{B}$ submits the program pairs $P_0 = (\mathcal{P}_S^*, \mathcal{P}_V^*)$ and $P_1 = (\mathcal{P}_S^{**}, \mathcal{P}_V^{**})$ to the $i\mathcal{O}$ challenger. The $i\mathcal{O}$ challenger chooses $b \leftarrow \{0,1\}$ at random, obfuscates the programs in P_b, and returns the output to $\mathcal{B}$. $\mathcal{B}$ plugs in these obfuscated programs in *Hybrid 2* and plays the rest of the game. Note that if the $i\mathcal{O}$ challenger chooses P_0 then we are in *Hybrid 2*. If it chooses P_1 then we are in *Hybrid 3*. $\mathcal{B}$ will output 1 if $\mathcal{A}$ successfully forges. Therefore, if $\mathcal{A}$ have different advantages in hybrids *Hybrid 2* and *Hybrid 3* then $\mathcal{B}$ can attack $i\mathcal{O}$ security with high probability i.e. $\boldsymbol{Adv}_{\mathcal{A}}^{Hyb\,2} - \boldsymbol{Adv}_{\mathcal{A}}^{Hyb\,3} = \boldsymbol{Adv}_{\mathcal{B}}^{i\mathcal{O}} \leq \mathsf{negl}(\lambda)$ (Sect. 3.1). $\quad\square$

Lemma 6. *If $\mathcal{A}$ can distinguish* Hybrid 3 *from* Hybrid 4 *in the* unforgeability game, *i.e.* $\boldsymbol{Adv}_{\mathcal{A}}^{Hyb\,3} - \boldsymbol{Adv}_{\mathcal{A}}^{Hyb\,4}$ *is non-negligible then there exists an adversary* $\mathcal{B}$ *who can break the security of* PPRF F_1.

Proof. We argue that if there is a non-negligible difference in advantages of $\mathcal{A}$ in hybrids *Hybrid 1* and *Hybrid 2* then we can construct an adversary $\mathcal{B}$ who can break the security of PPRF F_1. Consider a scenario where the attacker $\mathcal{B}$ interacts with a PPRF challenger while acting as a challenger for $\mathcal{A}$. First, $\mathcal{B}$ receives $(m^*, vk^*, R^*, ek^*, r^*)$ from $\mathcal{A}$ and submits it to the PPRF challenger.

Next, the PPRF challenger chooses a PPRF key $K_1 \leftarrow \{0,1\}^\lambda$ and punctures K_1 at point $(m^*\|vk^*\|\mathsf{Hash}(R^*)\|ek^*\|\mathsf{PRG}(r^*))$ and sends this punctured key $K_1(\{(m^*\|vk^*\|\mathsf{Hash}(R^*)\|ek^*\|\mathsf{PRG}(r^*))\})$ and a challenge z^* to $\mathcal{B}$. Then, $\mathcal{B}$ continues to run the game in *Hybrid 3* except it plugs in $\alpha^* = z^*$. Note that if the PPRF challenger outputs z^* as $z^* = F_1(K_1, (m^*\|vk^*\|\mathsf{Hash}(R^*)\|ek^*\|\mathsf{PRG}(r^*)))$ then we are in *Hybrid 3*. If it outputs z^* as $z^* \leftarrow \{0,1\}^{\ell_1}$ then we are in *Hybrid 2*. $\mathcal{B}$ will output 1 if $\mathcal{A}$ successfully forges. Therefore, if $\mathcal{A}$ have different advantages in hybrids *Hybrid 3* and *Hybrid 4* then $\mathcal{B}$ can attack the PPRF security with high probability i.e. $\boldsymbol{Adv}_{\mathcal{A}}^{Hyb\,3} - \boldsymbol{Adv}_{\mathcal{A}}^{Hyb\,4} = \boldsymbol{Adv}_{\mathcal{B}}^{PPRF} \leq \mathsf{negl}(\lambda)$ (Sect. 3.2). □

Lemma 7. *If $\mathcal{A}$'s advantage $\boldsymbol{Adv}_{\mathcal{A}}^{Hyb\,4}$ in* Hybrid 4 *is non-negligible, then there exists an adversary $\mathcal{B}$ who can break the security of one-way function f_1.*

Proof. We argue that if $\mathcal{A}$ successfully forges an ARS signature in *Hybrid 4* then we can construct an adversary $\mathcal{B}$ who can break the security of OWF f_1. Consider a scenario where the attacker $\mathcal{B}$ interacts with a OWF challenger while acting as a challenger for $\mathcal{A}$. First, $\mathcal{B}$ receives $(m^*, vk^*, R^*, ek^*, r^*)$ from $\mathcal{A}$ and submits it to the OWF challenger. Next, $\mathcal{B}$ receives a OWF challenge y^* and sets u^* as $u^* = y^*$ and continues to run the game in *Hybrid 4*. Now, if $\mathcal{A}$ successfully forges a signature on $(m^*\|vk^*\|\mathsf{Hash}(R^*)\|ek^*\|\mathsf{PRG}(r^*))$, then by definition it has computed a σ^* such that $f_1(\sigma^*) = u^*$. Therefore, if f_1 is secure, then $\mathcal{A}$ cannot forge with non-negligible advantage i.e. $\boldsymbol{Adv}_{\mathcal{A}}^{Hyb\,4} = \boldsymbol{Adv}_{\mathcal{B}}^{OWF} \leq \mathsf{negl}(\lambda)$ (Sect. 3.4). □

Hence, $\boldsymbol{Adv}_{\mathcal{A}}^{Unforge} = \boldsymbol{Adv}_{\mathcal{A}}^{Hyb\,0} - \boldsymbol{Adv}_{\mathcal{A}}^{Hyb\,4} \leq \mathsf{negl}(\lambda)$.

Theorem 2. *The ARS scheme of Sect. 4 is anonymous if the primitives of Sect. 3 satisfy their respective security properties.*

Proof. Our proof proceeds through a sequence of hybrids. The hybrids are chosen such that each successive hybrid game is indistinguishable from each other.

Hybrid 0. This hybrid is the *selective* variant of the *anonymity game* of Sect. 2.

1. **Setup Phase:** $\mathcal{C}$ runs the algorithm ARS.UKGen to generate signing- verification key pairs $\{(sk_i, vk_i)\}_{i\in[n]}$ for the users and registers them with the PKI. In addition, it also generates a private-public key pair (dk^*, ek^*) of an opener. Then, it sends $ek^*, \{(sk_i, vk_i)\}_{i\in[n]}$ to $\mathcal{A}$.
2. **Challenge Phase:** $\mathcal{A}$ submits a message m^*, a ring $R^* \subseteq \{vk_i\}_{i\in[n]}$ and two secret signing keys $sk_{i_0}^*, sk_{i_1}^* \in \{sk_i\}_{i\in[n]}, i_0 \neq i_1$, such that $vk_{i_0}^*, vk_{i_1}^* \in R^*$. Next, $\mathcal{C}$ chooses $b \leftarrow \{0,1\}$ and produces an ARS signature σ_{i_b} as follows:
   (a) Choose $r^* \leftarrow \{0,1\}^\lambda$ and let $t^* := \mathsf{PRG}(r^*)$ where $t^* \in \{0,1\}^{\ell_p}$.
   (b) Set $\alpha_{i_b} := F_1(K_1, (m^*\|vk_{i_b}\|\mathsf{Hash}(R^*)\|ek^*\|t^*))$
   (c) Set $\beta_{i_b} := F_2(K_2, \alpha_{i_b}) \oplus (m^*\|vk_{i_b}\|\mathsf{Hash}(R^*)\|ek^*\|t^*)$
   (d) Let $\sigma_{i_b} := (\alpha_{i_b}, \beta_{i_b})$

Finally, $\mathcal{C}$ chooses keys K_1, K_2 for PPRFs F_1, F_2 respectively, and produces two obfuscated programs $\mathsf{Sign} = i\mathcal{O}(\mathcal{P}_S)$ and $\mathsf{Verify} = i\mathcal{O}(\mathcal{P}_V)$ (Figs. 1 and 2). Then, it sends these programs with the signature $\sigma_{i_b} = (\alpha_{i_b}, \beta_{i_b})$ to $\mathcal{A}$.

3. **Guess Phase:** $\mathcal{A}$ guesses b and outputs $b' \in \{0,1\}$. Let $\boldsymbol{Adv}_{\mathcal{A}}^{Hyb\,0}$ denote the advantage of $\mathcal{A}$ in this hybrid.

Hybrid 1. This hybrid is identical to *Hybrid 0* with the exception that in the *challenge phase* t^* is chosen at random from $\{0,1\}^{\ell_p}$. Let $\boldsymbol{Adv}_{\mathcal{A}}^{Hyb\,1}$ denote the advantage of $\mathcal{A}$ in this hybrid.

Hybrid 2. This hybrid is identical to *Hybrid 1* with the exception that, in the *challenge phase*, $\mathcal{C}$ gives out the two obfuscated programs as $\mathsf{Sign} = i\mathcal{O}(\mathcal{P}_S^\dagger)$ and $\mathsf{Verify} = i\mathcal{O}(\mathcal{P}_V^\dagger)$ (Figs. 9 and 10). The details of the modifications are as follows: $\mathcal{C}$ punctures the key K_2 at the point $\alpha_{i_b}^* = F_1(K_1, (m^*\|vk_{i_b}^*\|R^*\|ek^*\|t^*))$, and hardwires this punctured key $K_2(\{\alpha_{i_b}^*\})$ in $\mathcal{P}_S^\dagger$ and $\mathcal{P}_V^\dagger$. In addition, it also hardwires message m^*, verification key $vk_{i_b}^*$, ring R^*, public key ek^* and randomness t^* in $\mathcal{P}_V^\dagger$. Furthermore, $\mathcal{P}_V^\dagger$ is modified to output c_b^* if $(m, \sigma, R, ek) = (m^*, \sigma_b^*, R^*, ek^*)$. Let $\boldsymbol{Adv}_{\mathcal{A}}^{Hyb\,1}$ denote the advantage of $\mathcal{A}$ in this hybrid.

Hybrid 3. This hybrid is identical to *Hybrid 2* with the exception that in the *challenge phase* $\beta_{i_b}^*$ is chosen uniformly at random from the set $\{0,1\}^{\ell_2}$. Let $\boldsymbol{Adv}_{\mathcal{A}}^{Hyb\,3}$ denote the advantage of $\mathcal{A}$ in this hybrid.

Hybrid 4. This hybrid is identical to *Hybrid 3* with the exception that in the *challenge phase*, $\mathcal{C}$ gives out the two obfuscated programs as $\mathsf{Sign} = i\mathcal{O}(\mathcal{P}_S^\dagger)$ and $\mathsf{Verify} = i\mathcal{O}(\mathcal{P}_V^\dagger)$ (Figs. 11 and 12). The details of the modifications are as follows: $\mathcal{C}$ punctures the key K_1 at point $(m^*\|vk_{i_b}^*\|\mathsf{Hash}(R)\|ek^*\|t^*)$, and hardwires

---

$$\mathcal{P}_S^\dagger$$

Hardwired: Keys $K_1, K_2(\{\alpha_{i_b}^*\})$
Input: Message m, key pair (sk, vk), ring R, key ek and randomness r
1   **if** $(f(sk) = vk \wedge f(sk) \in R)$ **then**
2   Set $\alpha := F_1(K_1, (m\|vk\|\mathsf{Hash}(R)\|ek\|\mathsf{PRG}(r)))$
3   Set $\beta := F_2(K_2, \alpha) \oplus (m\|vk\|\mathsf{Hash}(R)\|ek\|\mathsf{PRG}(r))$
4   **return** $(\sigma := (\alpha, \beta))$
5   **else**
6   **return** $\perp$

Fig. 9. The circuit Sign

$$\mathcal{P}_V^\dagger$$

Hardwired: Keys $K_1, K_2(\{\alpha_{i_b}^*\})$ and values $m^*, vk_{i_b}^*, R^*, ek^*, t^*$
Input: Message m, signature $\sigma = (\alpha, \beta)$, ring R and opener's public key ek

1 **if** $\Big(m = m^* \wedge R = R^* \wedge ek = ek^* \wedge f_1(\alpha) =$

    $f_1(F_1(K_1, (m^*\|vk_{i_b}^*\|\mathsf{Hash}(R^*)\|ek^*\|t^*)))\Big)$ **then**

2 $\quad$ **return** $(c_b^* := \mathsf{PKE.Encrypt}(ek, vk_{i_b}^*; t^*))$

3 **else if** $\Big(F_2(K_2, \alpha) \oplus \beta = (m'\|vk'\|h'\|ek'\|r') \wedge m = m' \wedge \mathsf{Hash}(R) =$

    $h' \wedge ek = ek' \wedge f_1(\alpha) = f_1(F_1(K_1, (m'\|vk'\|h'\|ek'\|r'))))$ **then**

4 $\quad$ **return** $(c := \mathsf{PKE.Encrypt}(ek, vk'; r'))$

5 **return** $\perp$

Fig. 10. The circuit Verify

$$\mathcal{P}_S^\ddagger$$

Hardwired: Keys $K_1(\{(m^*\|vk_b^*\|\mathsf{Hash}(R^*)\|ek^*\|t^*)\}), K_2(\{\alpha_{i_b}^*\})$
Input: Message m, key pair (sk, vk), ring R, key ek and randomness r

1 **if** $(f(sk) = vk \wedge f(sk) \in R)$ **then**

2 $\quad$ Set $\alpha := F_1(K_1, (m\|vk\|\mathsf{Hash}(R)\|ek\|\mathsf{PRG}(r)))$

3 $\quad$ Set $\beta := F_2(K_2, \alpha) \oplus (m\|vk\|\mathsf{Hash}(R)\|ek\|\mathsf{PRG}(r))$

4 $\quad$ **return** $(\sigma := (\alpha, \beta))$

5 **else**

6 $\quad$ **return** $\perp$

Fig. 11. The circuit Sign

$$\mathcal{P}_V^\ddagger$$

Hardwired: Keys $K_1(\{(m^*\|vk_b^*\|\mathsf{Hash}(R^*)\|ek^*\|t^*)\}), K_2(\{\alpha_{i_b}^*\})$ and
            values $m^*, R^*, ek^*, t^*, \alpha_{i_b}^*$
Input: Message m, signature $\sigma = (\alpha, \beta)$, ring R and opener's public key ek

1 **if** $\Big(m = m^* \wedge R = R^* \wedge ek = ek^* \wedge f_1(\alpha) = f_1(\alpha_{i_b}^*)\Big)$ **then**

2 $\quad$ **return** $(c_b^* := \mathsf{PKE.Encrypt}(ek, vk_{i_b}^*; t^*))$

3 **else if** $\Big(F_2(K_2, \alpha) \oplus \beta = (m'\|vk'\|h'\|ek'\|r') \wedge m = m' \wedge \mathsf{Hash}(R) =$

    $h' \wedge ek = ek' \wedge f_1(\alpha) = f_1(F_1(K_1, (m'\|vk'\|h'\|ek'\|r'))))$ **then**

4 $\quad$ **return** $(c := \mathsf{PKE.Encrypt}(ek, vk'; r'))$

5 **return** $\perp$

Fig. 12. The circuit Verify

this punctured key $K_1(\{(m^*\|vk_{i_b}^*\|\mathsf{Hash}(R)\|ek^*\|t^*)\})$ in programs $\mathcal{P}_S^\ddagger$ and $\mathcal{P}_V^\ddagger$. In addition, it also hardwires message m^*, ring R^*, public key ek^*, and value $\alpha_{i_b}^* = F_1(K_1, (m^*\|vk_{i_b}^*\|\mathsf{Hash}(R)\|ek^*\|t^*))$ in program $\mathcal{P}_V^\ddagger$. Let $\boldsymbol{Adv}_\mathcal{A}^{Hyb\,4}$ denote the advantage of $\mathcal{A}$ in this hybrid.

Hybrid 5. This hybrid is identical to *Hybrid 4* with the exception that in the *challenge phase*, $\alpha_{i_b}^*$ is chosen at random from $\{0,1\}^{\ell_1}$ instead of computing them using PPRF. Let $\boldsymbol{Adv}_\mathcal{A}^{Hyb\,5}$ denote the advantage of $\mathcal{A}$ in this hybrid.

The indistinguishability of successive hybrid games (discussed above) is shown in Lemmas 8, 9, 10, 11, 12 and 13.

Lemma 8. *If $\mathcal{A}$ can distinguish* Hybrid 0 *from* Hybrid 1 *in the anonymity game, i.e. $\boldsymbol{Adv}_\mathcal{A}^{Hyb\,0} - \boldsymbol{Adv}_\mathcal{A}^{Hyb\,1}$ is non-negligible then there exists an adversary $\mathcal{B}$ who can break the security of pseudo-random generator* PRG.

Proof. We argue that if there is a non-negligible difference in advantages of $\mathcal{A}$ in hybrids *Hybrid 0* and *Hybrid 1* then we can construct an adversary $\mathcal{B}$ that breaks the security of PRG. Now, consider a scenario where the attacker $\mathcal{B}$ interacts with PRG challenger while acting as a challenger for $\mathcal{A}$. $\mathcal{B}$ receives a challenge y^* from the PRG challenger, plugs in $t^* = y^*$ in *Hybrid 0* and plays the rest of the game. Note that if the PRG challenger computes $t^* = \mathsf{PRG}(r^*), r^* \leftarrow \{0,1\}^\lambda$ then we are in *Hybrid 0*. If it chooses $t^* \leftarrow \{0,1\}^{\ell_p}$ then we are in *Hybrid 1*. $\mathcal{B}$ will output 1 if $\mathcal{A}$ successfully guesses. Therefore, if $\mathcal{A}$ have different advantages in hybrids *Hybrid 0* and *Hybrid 1* then $\mathcal{B}$ can attack $i\mathcal{O}$ security with high probability i.e. $\boldsymbol{Adv}_\mathcal{A}^{Hyb\,0} - \boldsymbol{Adv}_\mathcal{A}^{Hyb\,1} = \boldsymbol{Adv}_\mathcal{B}^{PRG} \leq \mathsf{negl}(\lambda)$ (Sect. 3.5). □

Lemma 9. *If $\mathcal{A}$ can distinguish* Hybrid 1 *from* Hybrid 2 *in the anonymity game, i.e. $\boldsymbol{Adv}_\mathcal{A}^{Hyb\,1} - \boldsymbol{Adv}_\mathcal{A}^{Hyb\,2}$ is non-negligible then there exists an adversary $\mathcal{B}$ who can break the security of $i\mathcal{O}$.*

Proof. We argue that if there is a non-negligible difference in advantages of $\mathcal{A}$ in hybrids *Hybrid 1* and *Hybrid 2* then we can construct an adversary $\mathcal{B}$ that breaks the security of $i\mathcal{O}$. Firstly, observe that the programs $\mathcal{P}_S$ and $\mathcal{P}_S^\ddagger$ as well as the programs $\mathcal{P}_V$ and $\mathcal{P}_V^\ddagger$, are functionally equivalent. This is because when t^* is chosen at random, with probability $1 - \frac{1}{2^\lambda}$, t^* is not in the image of the PRG. Therefore neither program will evaluate. Thus, puncturing excising t^* out from the key will not make a difference in input/output behavior. Now, consider a scenario where the attacker $\mathcal{B}$ interacts with an $i\mathcal{O}$ challenger while acting as a challenger for $\mathcal{A}$. $\mathcal{B}$ submits the program pairs $P_0 = (\mathcal{P}_S, \mathcal{P}_V)$ and $P_1 = (\mathcal{P}_S^\ddagger, \mathcal{P}_V^\ddagger)$ to the $i\mathcal{O}$ challenger. The $i\mathcal{O}$ challenger chooses $b \leftarrow \{0,1\}$ at random, obfuscates the programs in P_b, and returns the output to $\mathcal{B}$. $\mathcal{B}$ plugs in these obfuscated programs in *Hybrid 0* and plays the rest of the game. Note that if the $i\mathcal{O}$ challenger chooses P_0 then we are in *Hybrid 1*. If it chooses P_1 then we are in *Hybrid 2*. $\mathcal{B}$ will output 1 if $\mathcal{A}$ successfully guesses. Hence, if $\mathcal{A}$ have different advantages in hybrids *Hybrid 1* and *Hybrid 2* then $\mathcal{B}$ can attack

iO security with high probability i.e. $\boldsymbol{Adv}_{\mathcal{A}}^{Hyb\,1} - \boldsymbol{Adv}_{\mathcal{A}}^{Hyb\,2} = \boldsymbol{Adv}_{\mathcal{B}}^{iO} \leq \mathsf{negl}(\lambda)$ (Sect. 3.1). □

Lemma 10. *If* $\mathcal{A}$ *can distinguish* Hybrid 2 *from* Hybrid 3 *in the anonymity game, i.e.* $\boldsymbol{Adv}_{\mathcal{A}}^{Hyb\,2} - \boldsymbol{Adv}_{\mathcal{A}}^{Hyb\,3}$ *is non-negligible then there exists an adversary* $\mathcal{B}$ *who can break the security of* PPRF F_2.

Proof. We argue that if there is a non-negligible difference in advantages of $\mathcal{A}$ in hybrids *Hybrid 2* and *Hybrid 3* then we can construct an adversary $\mathcal{B}$ who can break the security of PPRF F_2. Consider a scenario where the attacker $\mathcal{B}$ interacts with a PPRF challenger while acting as a challenger for $\mathcal{A}$. First, $\mathcal{B}$ receives $(m^*, R^*, sk_{i_0}^*, sk_{i_1}^*)$ from $\mathcal{A}$. Then, it chooses $b \leftarrow \{0,1\}, K_1 \leftarrow \{0,1\}^\lambda, t^* \leftarrow \{0,1\}^{\ell_p}$, computes $\alpha_{i_b}^* = F_1(K_1, (m^*\|vk_{i_b}^*\|R^*\|ek^*\|t^*))$, and submits $\alpha_{i_b}^*$ to the PPRF challenger. Next, the PPRF challenger choses $K_2 \leftarrow \{0,1\}^\lambda$, punctures K_2 at point $\alpha_{i_b}^*$, and sends this punctured PRF key $K_2(\{\alpha_{i_b}^*\})$ along with a challenge $w^* \in \{0,1\}^{\ell_2}$ to $\mathcal{B}$. Then, $\mathcal{B}$ continues to run the game in *Hybrid 2* except that it plugs in $\beta_{i_b}^*$ as $\beta_{i_b}^* = z_{i_b}^* \oplus (m^*\|vk_{i_b}^*\|R^*\|ek^*\|t^*)$. Note that if the PPRF challenger outputs $z_{i_b}^*$ as $z_{i_b}^* = F_2(K_2, \alpha_{i_b}^*)$ then we are in *Hybrid 2*. If it outputs z^* as $z_{i_b}^* \leftarrow \{0,1\}^{\ell_2}$ then we are in *Hybrid 3*. $\mathcal{B}$ will output 1 if $\mathcal{A}$ successfully guesses. Therefore, if $\mathcal{A}$ have different advantages in hybrids *Hybrid 2* and *Hybrid 3* then $\mathcal{B}$ can attack the PPRF security with high probability i.e. $\boldsymbol{Adv}_{\mathcal{A}}^{Hyb\,2} - \boldsymbol{Adv}_{\mathcal{A}}^{Hyb\,3} = \boldsymbol{Adv}_{\mathcal{B}}^{PPRF} \leq \mathsf{negl}(\lambda)$ (Sect. 3.2). □

Lemma 11. *If* $\mathcal{A}$ *can distinguish* Hybrid 3 *from* Hybrid 4 *in the anonymity game, i.e.* $\boldsymbol{Adv}_{\mathcal{A}}^{Hyb\,3} - \boldsymbol{Adv}_{\mathcal{A}}^{Hyb\,4}$ *is non-negligible then there exists an adversary* $\mathcal{B}$ *who can break the security of* iO.

Proof. We argue that if there is a non-negligible difference in advantages of $\mathcal{A}$ in hybrids *Hybrid 3* and *Hybrid 4* then we can construct an adversary $\mathcal{B}$ that breaks the security of iO. Firstly, observe that the programs $\mathcal{P}_S^\dagger$ and $\mathcal{P}_S^\ddagger$ as well as the programs $\mathcal{P}_V^\dagger$ and $\mathcal{P}_V^\ddagger$, are functionally equivalent. Now, consider a scenario where the attacker $\mathcal{B}$ interacts with an iO challenger while acting as a challenger for $\mathcal{A}$. $\mathcal{B}$ submits the program pairs $P_0 = (\mathcal{P}_S^\dagger, \mathcal{P}_V^\dagger)$ and $P_1 = (\mathcal{P}_S^\ddagger, \mathcal{P}_V^\ddagger)$ to the iO challenger. The iO challenger chooses $b \leftarrow \{0,1\}$ at random, obfuscates the programs in P_b, and returns the output to $\mathcal{B}$. $\mathcal{B}$ plugs in these obfuscated programs in *Hybrid 3* and plays the rest of the game. Note that if the iO challenger chooses P_0 then we are in *Hybrid 3*. If it chooses P_1 then we are in *Hybrid 4*. $\mathcal{B}$ will output 1 if $\mathcal{A}$ successfully guesses. Therefore, if $\mathcal{A}$ have different advantages in hybrids *Hybrid 3* and *Hybrid 4* then $\mathcal{B}$ can attack iO security with high probability i.e. $\boldsymbol{Adv}_{\mathcal{A}}^{Hyb\,3} - \boldsymbol{Adv}_{\mathcal{A}}^{Hyb\,4} = \boldsymbol{Adv}_{\mathcal{B}}^{iO} \leq \mathsf{negl}(\lambda)$ (Sect. 3.1). □

Lemma 12. *If* $\mathcal{A}$ *can distinguish* Hybrid 4 *from* Hybrid 5 *in the anonymity game, i.e.* $\boldsymbol{Adv}_{\mathcal{A}}^{Hyb\,4} - \boldsymbol{Adv}_{\mathcal{A}}^{Hyb\,5}$ *is non-negligible then there exists an adversary* $\mathcal{B}$ *who can break the security of* PPRF F_1.

Proof. We argue that if there is a non-negligible difference in advantages of $\mathcal{A}$ in hybrids *Hybrid 4* and *Hybrid 5* then we can construct an adversary $\mathcal{B}$ who can break the security of PPRF F_1. Consider a scenario where the attacker $\mathcal{B}$ interacts with a PPRF challenger while acting as a challenger for $\mathcal{A}$. First, $\mathcal{B}$ receives $(m^*, R^*, sk_{i_0}^*, sk_{i_1}^*)$ from $\mathcal{A}$. Then, it chooses $b \leftarrow \{0,1\}, t \leftarrow \{0,1\}^{\ell_p}$ and submits $(m^*, R^*, vk_{i_b}^*, t^*)$ to the PPRF challenger. The PPRF challenger chooses a PPRF key K_1 at random and punctures K_1 at point $(m^* \| vk_{i_b}^* \| R^* \| ek^* \| t^*)$ and sends this punctured key $K_1(\{(m^* \| vk_{i_b}^* \| R^* \| ek^* \| t^*)\})$ and a challenge z^* to $\mathcal{B}$. Then, $\mathcal{B}$ continues to run the game in *Hybrid 4* except it plugs in $\alpha_{i_b}^* = z^*$. Note that if the PPRF challenger outputs z^* as $z^* = F_1(K_1, (m^* \| vk_{i_b}^* \| R^* \| ek^* \| t^*))$ then we are in *Hybrid 4*. If it outputs z^* as $z^* \leftarrow \{0,1\}^{\ell_1}$ then we are in *Hybrid 5*. $\mathcal{B}$ will output 1 if $\mathcal{A}$ successfully guesses. Therefore, if $\mathcal{A}$ has different advantages in hybrids *Hybrid 4* and *Hybrid 5* then $\mathcal{B}$ can attack the PPRF security with high probability i.e. $\boldsymbol{Adv}_{\mathcal{A}}^{Hyb\,4} - \boldsymbol{Adv}_{\mathcal{A}}^{Hyb\,5} = \boldsymbol{Adv}_{\mathcal{B}}^{PPRF} \leq \mathsf{negl}(\lambda)$ (Sect. 3.2). $\square$

Lemma 13. *$\mathcal{A}$'s advantage in* Hybrid 5 *is zero.*

Proof. We observe that the variables $\alpha_{i_b}^*$ and $\beta_{i_b}^*$ in *Hybrid 5* are both independently and uniformly chosen random strings. Thus, the distributions output by this hybrid for $b = 0$ and $b = 1$ are identical, and therefore even an unbounded adversary could have no advantage in distinguishing them i.e. $\boldsymbol{Adv}_{\mathcal{A}}^{Hyb\,5} = 0$. $\square$

Hence, $\boldsymbol{Adv}_{\mathcal{A}}^{Anony} = \boldsymbol{Adv}_{\mathcal{A}}^{Hyb\,0} - \boldsymbol{Adv}_{\mathcal{A}}^{Hyb\,5} \leq \mathsf{negl}(\lambda)$. $\hspace{3cm}\square$

Theorem 3. *The ARS scheme of Sect. 4 is traceable and traceably sound if the primitives of Sect. 3 satisfy their respective security properties.*

Proof. In order to prove this theorem it is enough to prove that the proof system between the opener (prover) and the judge (verifier) is a *non-interactive zero knowledge* (NIZK) proof system where *traceability* and *tracing soundness* respectively correspond to *completeness* and *soundness* properties of NIZK.

Lemma 14 ([19]). *The NIZK proof system between* ARS.Open *and* ARS.Judge *with common reference string* crs = (NIZKprove, NIZKverify) *is secure if i$\mathcal{O}$ is secure, F_3 is a secure PPRF, and f_2 is a secure one-way function.*

6 Conclusion

In this paper we proposed a new construction for accountable ring signature (ARS). This is the first indistinguishability obfuscation-based ARS scheme. Also, the signature-size of our scheme is constant, and does not vary with the size of the ring. We have provided proof of security in the standard model. Our construction can be viewed as a new application of *puncturable programming* and *hidden sparse trigger* techniques introduced by Sahai and Waters. We leave the issue of constructing fully unforgeable, fully anonymous, constant size ARS scheme in standard model as an open problem.

Acknowledgement. First author is supported by Tata Consultancy Services (TCS) research fellowship. We thank anonymous reviewers for their constructive comments.

References

1. Barak, B., Goldreich, O., Impagliazzo, R., Rudich, S., Sahai, A., Vadhan, S., Yang, K.: On the (im)possibility of obfuscating programs. In: Kilian, J. (ed.) CRYPTO 2001. LNCS, vol. 2139, pp. 1–18. Springer, Heidelberg (2001). https://doi.org/10.1007/3-540-44647-8_1
2. Boneh, D., Ishai, Y., Sahai, A., Wu, D.J.: Lattice-based SNARGs and their application to more efficient obfuscation. In: Coron, J.-S., Nielsen, J.B. (eds.) EUROCRYPT 2017, Part III. LNCS, vol. 10212, pp. 247–277. Springer, Cham (2017). https://doi.org/10.1007/978-3-319-56617-7_9
3. Boneh, D., Waters, B.: Constrained pseudorandom functions and their applications. In: Sako, K., Sarkar, P. (eds.) ASIACRYPT 2013, Part II. LNCS, vol. 8270, pp. 280–300. Springer, Heidelberg (2013). https://doi.org/10.1007/978-3-642-42045-0_15
4. Bootle, J., Cerulli, A., Chaidos, P., Ghadafi, E., Groth, J., Petit, C.: Short accountable ring signatures based on DDH. In: Pernul, G., Ryan, P.Y.A., Weippl, E. (eds.) ESORICS 2015, Part I. LNCS, vol. 9326, pp. 243–265. Springer, Cham (2015). https://doi.org/10.1007/978-3-319-24174-6_13
5. Bose, P., Das, D., Rangan, C.P.: Constant size ring signature without random oracle. In: Foo, E., Stebila, D. (eds.) ACISP 2015. LNCS, vol. 9144, pp. 230–247. Springer, Cham (2015). https://doi.org/10.1007/978-3-319-19962-7_14
6. Boyle, E., Goldwasser, S., Ivan, I.: Functional signatures and pseudorandom functions. In: Krawczyk, H. (ed.) PKC 2014. LNCS, vol. 8383, pp. 501–519. Springer, Heidelberg (2014). https://doi.org/10.1007/978-3-642-54631-0_29
7. Camenisch, J., Haralambiev, K., Kohlweiss, M., Lapon, J., Naessens, V.: Structure preserving CCA secure encryption and applications. In: Lee, D.H., Wang, X. (eds.) ASIACRYPT 2011. LNCS, vol. 7073, pp. 89–106. Springer, Heidelberg (2011). https://doi.org/10.1007/978-3-642-25385-0_5
8. Camenisch, J.: Efficient and generalized group signatures. In: Fumy, W. (ed.) EUROCRYPT 1997. LNCS, vol. 1233, pp. 465–479. Springer, Heidelberg (1997). https://doi.org/10.1007/3-540-69053-0_32
9. Chaum, D., van Heyst, E.: Group signatures. In: Davies, D.W. (ed.) EUROCRYPT 1991. LNCS, vol. 547, pp. 257–265. Springer, Heidelberg (1991). https://doi.org/10.1007/3-540-46416-6_22
10. Coron, J.-S., Lepoint, T., Tibouchi, M.: Practical multilinear maps over the integers. In: Canetti, R., Garay, J.A. (eds.) CRYPTO 2013, Part I. LNCS, vol. 8042, pp. 476–493. Springer, Heidelberg (2013). https://doi.org/10.1007/978-3-642-40041-4_26
11. Garg, S., Gentry, C., Halevi, S., Raykova, M., Sahai, A., Waters, B.: Candidate indistinguishability obfuscation and functional encryption for all circuits. In: 2013 IEEE 54th Annual Symposium on Foundations of Computer Science (FOCS), pp. 40–49. IEEE (2013)
12. Goldreich, O.: Foundations of Cryptography: Volume 2, Basic Applications. Cambridge University Press, New York (2009)
13. Goldreich, O., Goldwasser, S., Micali, S.: How to construct random functions. J. ACM (JACM) **33**(4), 792–807 (1986)
14. Goldwasser, S., Micali, S.: Probabilistic encryption & how to play mental poker keeping secret all partial information. In: Proceedings of the Fourteenth Annual ACM Symposium on Theory of Computing, pp. 365–377. ACM (1982)

15. Groth, J., Kohlweiss, M.: One-out-of-many proofs: or how to leak a secret and spend a coin. In: Oswald, E., Fischlin, M. (eds.) EUROCRYPT 2015, Part II. LNCS, vol. 9057, pp. 253–280. Springer, Heidelberg (2015). https://doi.org/10. 1007/978-3-662-46803-6_9

16. Kiayias, A., Papadopoulos, S., Triandopoulos, N., Zacharias, T.: Delegatable pseudorandom functions and applications. In: Proceedings of the 2013 ACM SIGSAC Conference on Computer & Communications Security, pp. 669–684. ACM (2013)

17. Lai, R.W.F., Zhang, T., Chow, S.S.M., Schröder, D.: Efficient sanitizable signatures without random oracles. In: Askoxylakis, I., Ioannidis, S., Katsikas, S., Meadows, C. (eds.) ESORICS 2016, Part I. LNCS, vol. 9878, pp. 363–380. Springer, Cham (2016). https://doi.org/10.1007/978-3-319-45744-4_18

18. Rivest, R.L., Shamir, A., Tauman, Y.: How to leak a secret. In: Boyd, C. (ed.) ASIACRYPT 2001. LNCS, vol. 2248, pp. 552–565. Springer, Heidelberg (2001). https://doi.org/10.1007/3-540-45682-1_32

19. Sahai, A., Waters, B.: How to use indistinguishability obfuscation: deniable encryption, and more. In: Proceedings of the Forty-sixth Annual ACM Symposium on Theory of Computing, pp. 475–484. ACM (2014)

20. Xu, S., Yung, M.: Accountable ring signatures: a smart card approach. In: Quisquater, J.J., Paradinas, P., Deswarte, Y., El Kalam, A.A. (eds.) Smart Card Research and Advanced Applications VI. IFIP AICT, vol. 153. Springer, Boston (2004). https://doi.org/10.1007/1-4020-8147-2_18

A Universal Designated Multi-Verifier Transitive Signature Scheme

Fei Zhu[1], Yuexin Zhang[2], Chao Lin[1], Wei Wu[1(✉)], and Ru Meng[3]

[1] Fujian Provincial Key Laboratory of Network Security and Cryptology,
School of Mathematics and Informatics, Fujian Normal University, Fuzhou, China
`feizcscoding@gmail.com, lcfjnu@foxmail.com, weiwu81@gmail.com`
[2] School of Software and Electrical Engineering, Swinburne University of Technology,
Hawthorn, VIC 3122, Australia
`zhangyuexin87@hotmail.com`
[3] School of Computer Science, Shaanxi Normal University, Xi'an, China
`mengru@snnu.edu.cn`

Abstract. A Universal Designated Verifier Transitive Signature (UDVTS) scheme is designed for the graph-based big data system. Specifically, it allows a transitive signature holder to convince the designated verifier with a transitive signature. Nevertheless, existing UDVTS schemes cannot be directly employed in the scenarios when multi-verifier are involved. Thus, in this paper, we extend the notion to the Universal Designated Multi-Verifier Transitive Signature (UDMVTS) scheme. Namely, our new scheme allows a transitive signature holder to designate the signature to multi-verifier. Furthermore, for the proposed scheme, we formalize its security notions and prove its security in the random oracle model. We also analyse the performance of our scheme to demonstrate its efficiency.

Keywords: Universal designated multi-verifier signature
Transitive signature · Privacy

1 Introduction

The era of "Data-Intensive Scientific" [11] has already arrived. Specifically, in the era, data come in all scales and shapes, and it creates the so-called big data system. In such a complex big data system, it should be guaranteed that data are collected and transmitted securely, and they are accessed only by authorized parties. Namely, security is one of the most crucial factors when designing the big data system.

In certain scenarios, such as administrative domains, certificate chains in PKI and military chains of command, a signer (we say Alice) needs to publish a graph (which is used to represent the structures) in an authenticated manner. Absolutely, she can create a digital signature for each entity at any time. However, she may feel thwarted, especially when the domain grows dynamically or when the components are pretty large.

© Springer International Publishing AG, part of Springer Nature 2018
X. Chen et al. (Eds.): Inscrypt 2017, LNCS 10726, pp. 180–195, 2018.
https://doi.org/10.1007/978-3-319-75160-3_12

At first sight, it looks as if Alice's needs can be satisfied by the Transitive Signatures, which was suggested by Micali and Rivest in [21]. Recall that in such a graph data system, entities and their relationships are represented by nodes and edges respectively. More precisely, an edge (i, j) implies that nodes i and j are located in the same administrative domain. Compared with a classical digital signature, a transitive signature achieves two types of properties. One is transitive, which means that given two different signatures σ_{ij} and σ_{jk}, one can easily derive a signature σ_{ik} for composed edge (i, k) from the adjacent edges (i, j) and (j, k) using the public key only. Technically, σ_{ik} is indistinguishable from the signature produced by the legitimate signer. This property allows a signer to sign a graph by signing its transitive reduction. In other words, it supports public edge composition, which can reduce the signing complexity and helps to protect the path privacy between edges. The other property is unforgeability. Namely, it should be difficult for an adaptive chosen-message and probabilistic polynomial time adversary to forge a valid signature for an edge (which is out of the transitive closure), even after the adversary adaptively queried signatures of a great many vertices and edges.

Consider the scenario that Alice authenticates the graph, where vertices represent computers and an undirected edge (i, j) implies that computers i, j are in the same administrative domain. Suppose Bob's computer and Cindy's computer are in the same domain. In that way, Bob could get a transitive signature from Alice, which can be used to prove such relationship between them. If Bob wants to send the signature to an external client (we say Dave) to state their relationship, and for privacy concerns, he does not want Dave to convince any other third party about the truth of this signature. In this scenario, the Universal Designated Verifier Transitive Signature scheme, as presented by Hou *et al.* in [13], can be employed to fulfill the requirements. However, the scheme presented in [13] cannot be employed when multi-verifier are involved. Motivated by this observation, we now propose a Universal Designated Multi-Verifier Transitive Signature (UDMVTS) scheme.

1.1   Related Work

In this section, we review these closely related work, including the transitive signature schemes and the universal designated multi-verifier signature schemes.

Transitive Signatures. Micali and Rivest [21] presented two transitive signature schemes DLTS and RSATS-1 based on the discrete logarithm problem and RSA assumption respectively. The former is provably secure under adaptive chosen-message attacks while the latter is only secure against the non-adaptive chosen-message attacks. They also classified the schemes into Directed Transitive Signature (DTS) schemes and Undirected Transitive Signature (UTS) schemes according to whether the graph is directed or not. Subsequent to their work, many new undirected transitive signature schemes rely on the difficulty of certain assumptions have been proposed, such as [4,5,10,28], and all of them are provably secure in their security models. It is worth mentioning that Bellare and Neven removed the node certificates which are employed in the above schemes. To achieve this, in [5], a node i's public label is served as the output of a hash function applied to i and it is used when constructing RSATS-2, FactTS-2 and GapTS-2. Additionally, the new designs are proved secure in the random oracle model.

The above transitive signature schemes can only suitable for solving the problem of authenticating undirected graphs. How to construct a practical directed transitive signature scheme was still an open problem since being envisioned by Micali and Rivest [21]. Actually, theoretical analysis in [12] shows that the general directed transitive signature schemes may be infeasible to construct. The reason is that the edge signatures form a special Abelian trapdoor group, and the existence of it is still unknown.

The research of the directed transitive signature schemes was restricted to a special directed graph, like directed tree. In 2007, Yi [31] provided a scheme RSADTS based on the difficulty of RSA-inversion problem in the standard model. Also for directed trees, in 2008, Neven [23] presented a more efficient construction from any standard signature scheme, without using any RSA-related assumptions [4,9]. However, [23,31] have two weaknesses. Firstly, the size of composed signature increases linearly with the number of recursive applications of composition. Secondly, the creating history of composed edge is not hidden properly. To address these issues, in 2009, Xu [30] raised a scheme DTTS, which has a constant signature size and it achieves privacy preserving property. Additionally, it is provably secure in the standard model. Later, Camacho and Hevia [7] exploited a late-model collision-resistant hashing and came up with a novel practical scheme DTTS. To our knowledge, this is the most effective scheme so far.

Universal Designated Multi-Verifier Signatures. The concept of Universal Designated Verifier Signature (UDVS), was defined by Steinfeld *et al.* [26] in 2003, is a significant tool to address the privacy issues associated with dissemination of signatures. In this notation, a signature holder (not necessarily the signer) can transfer the signature into a UDVS by using the verifier's public key. And only the designated verifier can verify the signature. Specifically, the designated verifier can be convinced if the above verification succeeds, because he is the only one who can forge the signature. This operation can prevent the abuse of spreading the signature. Subsequently, Steinfeld *et al.* also demonstrated other UDVS schemes in [27], which were extended by using standard RSA/Schnorr signature schemes. Additionally, the security of them is based on the random oracle model.

In 2005, on the basis of Boneh *et al.*'s BLS [6] scheme, Willy *et al.* [29] proposed a variant of UDVS scheme, coined as universal designated multi-verifier signature scheme, i.e., the UDVMS, which allows a signature holder to convince the designated multi-verifier with a signature. Later, new constructions were proposed in the standard model, such as [15,17,22,25,34]. In 2009,

Shahandashti and Safavi-Naini [24] introduced a generic construction for UDVS scheme from a large class of signature schemes. Moreover, [1,8,14,16,18,19,33], for instance, new constructions with additional properties have been further pursued by many researchers according to the various demands.

In 2015, Hou *et al.* [13] integrated the notions of UDVS and transitive signatures, and proposed the first universal designated verifier transitive signature (UDVTS) scheme. This kind of signature scheme is suitable for a lot of applications where the users' privacy and transitive signature are required. Their construction is designed on the basis of Bellare *et al.*'s GapTS-2 [5] and Steinfeld *et al.*'s DVSBM [26]. In 2017, Lin *et al.* [20] combined Bellare *et al.*'s RSATS-2 [5] and Steinfeld *et al.*'s RSAUDVS [27], and presented another UDVTS scheme. Specifically, both the above two UDVTS schemes could be proven secure in the random oracle model.

1.2   Our Contributions

In this study, we point out the drawback of literature [13] and firstly introduce the notion of Universal Designated Multi-Verifier Transitive Signature (UDMVTS) schemes. Our new notion possesses an additional property that it allows a transitive signature holder to designate the signature to any desired multi-verifier by using their public keys. Then we present an efficient construction to meet our model. Additionally, we prove the security of the new design in the random oracle model. According to the security analysis, our UDMVTS scheme can securely address privacy concerns in the aforementioned scenario. We also analyse the performance of our scheme to demonstrate its efficiency.

1.3   Organization

The remainder of the paper is organized as follows. Section 2 briefly reviews several basic concepts on preliminaries deployed in our scheme. Section 3 is dedicated to providing formal definitions of UDMVTS and relevant security models. In Sect. 4, we present our proposed UDMVTS scheme and give its security/performance analysis. Lastly, we conclude in Sect. 5.

2   Preliminaries

In this section, we introduce some fundamental backgrounds used in this paper.

2.1   Notations

We specify fairly standard notations in Table 1.

Table 1. Notations

Notation	Description
$\mathbb{Z}_q$	The definite field of the prime order q
$\mathbb{N}$	The set of positive integers
$f(\cdot) : \mathbb{N} \to \mathbb{R}$	A negligible function
$\mathcal{PPT}$	The abbreviation of probabilistic polynomial time
$x \xleftarrow{\$} \mathcal{R}$	x is sampled from the set $\mathcal{R}$ randomly
$H(\cdot) : \mathbb{N} \to \mathcal{Z}_N^*$	A MapToPoint[a] hash function

[a] As introduced in [32], this kind of hash is an admissible encoding function called MapToPoint. Additionally, it is used by many conventional cryptographic schemes from pairings, such as [3,5,6,13,26].

2.2 Graphs

This paper considers $G = (V, E)$ as an undirected graph with vertexes set V and edges set E. We implicitly assume that $V = \{1, 2, ..., n\}$ and (i, j) represent an edge in G.

- Transitive closure: $\widetilde{G} = (V, \widetilde{E})$ such that there is a path from i to j in G, where $(i, j) \in \widetilde{E}$.
- Transitive reduction: $G^* = (V, E^*)$ such that it has the minimum subset of edges with the same transitive closure as G, where $(i, j) \in E^*$.

2.3 Admissible Bilinear Mapping

Let $\mathbb{G}$, $\mathbb{G}_T$ be two multiplicative cyclic groups of prime order q, and $\mathbb{G}$ is generated by g. The efficient mapping $\hat{e} : \mathbb{G} \times \mathbb{G} \to \mathbb{G}_T$ has the following properties:

- Bilinearity: for all $\alpha, \beta \xleftarrow{\$} \mathbb{Z}_q^*$, $\hat{e}(g^\alpha, g^\beta) = \hat{e}(g, g)^{\alpha\beta}$.
- Non-degeneracy: $\hat{e}(g, g) \neq 1$.
- Computability: for all $\alpha, \beta \xleftarrow{\$} \mathbb{Z}_q^*$, $\hat{e}(g^\alpha, g^\beta)$ can be efficiently computed.

2.4 Complexity Problems

- Computational Diffie-Hellman (CDH) problem: Given g, g^α, $g^\beta \in \mathbb{G}$ for unknown randomly chosen $\alpha, \beta \xleftarrow{\$} \mathbb{Z}_q^*$, compute $g^{\alpha\beta} \in \mathbb{G}$.
- Bilinear Diffie-Hellman (BDH) problem: Given g, g^α, g^β, $g^\gamma \in \mathbb{G}$ for unknown randomly chosen $\alpha, \beta, \gamma \xleftarrow{\$} \mathbb{Z}_q^*$, compute $\hat{e}(g, g)^{\alpha\beta\gamma} \in \mathbb{G}_T$.

Since massive hard mathematic problems and their assumptions are extensively deployed in analyzing and proving the security of cryptographic schemes, similar to [13], we will formally define a security assumption about the above problems (as shown below):

Definition 1 *(one-more* BDH *problem). Let $x, y \xleftarrow{\$} \mathbb{Z}_q$ and let $y_x = g^x$, $y_y = g^y$. A $\mathcal{PPT}$ algorithm $\mathcal{F}$ is given $(\hat{e}, \mathbb{G}, \mathbb{G}_T, g, q, y_x, y_y)$ and has access to a Challenge oracle $\mathcal{O}_{\mathcal{CH}}(\cdot)$ and a* CDH *oracle $\mathcal{O}_{\mathcal{CDH}}(\cdot)$.*

- *$\mathcal{O}_{\mathcal{CH}}(\cdot)$: when invoked, it returns a random challenge point from $\mathbb{G}$.*
- *$\mathcal{O}_{\mathcal{CDH}}(\cdot)$: given a random point $h_i \xleftarrow{\$} \mathbb{G}$ as input, outputs $(h_i)^x \in \mathbb{G}$.*

The one-more BDH *problem is that the algorithm $\mathcal{F}$ inverts* BDH *solutions $\hat{e}(g, h_i)^{xy} \in \mathbb{G}_T$ with all s points $h_i \xleftarrow{\$} \mathbb{G}$ output by the Challenge oracle, but using strictly fewer than s queries to the* CDH *oracle.*

Defining the advantage of $\mathcal{F}$ in its attack to be $\mathsf{Adv}_{\mathcal{F}}^{om-\mathsf{BDH}}(k)$. We say one-more BDH *problem is hard if for any $\mathcal{PPT}$ algorithm $\mathcal{F}$ there exists a negligible function $f(\cdot)$ such that the success probability of $\mathcal{F}$ is $\Pr[\mathsf{Adv}_{\mathcal{F}}^{om-\mathsf{BDH}}(k) = 1] \leq f(\cdot)$.*

3 Description of Universal Designated Multi-Verifier Transitive Signature

In what follows, we devote to providing the formal definition of Universal Designated Multi-Verifier Transitive Signature (UDMVTS) schemes.

3.1 Outline of UDMVTS

A UDMVTS scheme comprises the following nine polynomial-time algorithms: UDMVTS = (SS, SKGe, VKGe, TSig, TVf, Comp, DS, $\widehat{\mathsf{DS}}$, DV).

- $cp \leftarrow$ SS (1^k) is the **System Setup algorithm** which, on input a security parameter 1^k, outputs a string consisting of common public parameters cp.
- $(sks, pks) \leftarrow$ SKGe (cp) is the **Signer Key Generation algorithm** which, on input a common parameter cp, outputs a private/public key pair (sks, pks) for the signer.
- $(skv_l, pkv_l) \leftarrow$ VKGe (cp) is the **Verifier Key Generation algorithm** which, on input a common parameter cp, outputs n private/public key pairs $(skv_l, pkv_l), l = 1, \ldots, n$, for n verifiers.
- $\sigma_{ij} \leftarrow$ TSig (sks, i, j) is the **Transitive Signing algorithm** which, on input signer's private key sks, nodes $i, j \in \mathbb{N}$, outputs a signature σ_{ij} for edge (i, j).
- $d \in \{0, 1\} \leftarrow$ TVf (pks, i, j, σ_{ij}) is the **Verification algorithm** which, on input signer's public key pks, nodes $i, j \in \mathbb{N}$ and a candidate transitive signature σ_{ij}, outputs the verification result $d \in \{0, 1\}$.
- $\{\sigma_{ik}, \perp\} \leftarrow$ Comp $(pks, i, j, k, \sigma_{ij}, \sigma_{jk})$ is the **Composition algorithm** which, on input signer's public key pks, nodes $i, j, k \in \mathbb{N}$ and corresponding transitive signatures σ_{ij}, σ_{jk}, outputs either a composed signature σ_{ik} of edge (i, k) or $\perp$ to indicate failure.
- $\sigma_{DV} \leftarrow$ DS $(pks, pkv_1, \ldots, pkv_n, i, j, \sigma_{ij})$ is the **Signature Holder's Designation algorithm** which, on input signer's public key pks, verifiers' public key $pkv_1, \ldots, pkv_n$, nodes $i, j \in \mathbb{N}$ and an edge signature σ_{ij}, outputs a designated multi-verifier signature σ_{DV}.

- $\sigma_{\widehat{DV}} \leftarrow \widehat{DS} (pks, skv_1, \ldots, skv_n, i, j, \sigma_{ij})$ is the **Designated Verifier's Designation algorithm** which, on input signer's public key pks, verifiers' private key $skv_1, \ldots, skv_n$, nodes $i, j \in \mathbb{N}$, outputs a designated multi-verifier signature $\sigma_{\widehat{DV}}$.
- $d \in \{0,1\} \leftarrow \mathsf{DV} (pks, skv_1, \ldots, skv_n, i, j, \sigma_{DV})$ is the **Designated Verifying algorithm** which, on input signer's public key pks, verifiers' private key $skv_1, \ldots, skv_n$, nodes $i, j \in \mathbb{N}$, a designated multi-verifier signature σ_{DV}, returns the verification result $d \in \{0,1\}$.

3.2   Completeness

Concerning these algorithms, we require that the UDMVTS scheme should satisfy the following probability equation:

$\Pr[\mathsf{DV}(pks, skv_1, \ldots, skv_n, i, j, \sigma_{DV})] = 1$ where $cp \leftarrow \mathsf{SS}(1^k)$, $(sks, pks) \leftarrow \mathsf{SKGe}(cp)$, $(skv_l, pkv_l) \leftarrow \mathsf{VKGe}(cp, n)$, $\sigma_{ij} \leftarrow \mathsf{TSig}(sks, i, j)$, $1 \leftarrow \mathsf{TVf}(pks, i, j, \sigma_{ij})$, $\sigma_{ik} \leftarrow \mathsf{Comp}(pks, i, j, k, \sigma_{ij}, \sigma_{jk})$, $\sigma_{DV} \leftarrow \mathsf{DS}(pks, pkv_1, \ldots, pkv_n, i, j, \sigma_{ij})$, $\sigma_{\widehat{DV}} \leftarrow \widehat{\mathsf{DS}}(pks, skv_1, \ldots, skv_n, i, j)$.

3.3   Security Notions

This section defines security requirements, such as completeness, unforgeability and privacy, which should be met by the UDMVTS scheme.

Unforgeability. With regard to unforgeability [13,20], there are two kinds of properties in UDMVTS. The first property refers that even an adaptive chosen-message attacker cannot forge a valid transitive signature, i.e., the transitive signature unforgeability (we say TVf-Unforgeability). The second property refers that anyone cannot forge a designated multi-verifier signature if there does not exist a valid transitive signature previously. Thus, no one can convince any one of the designated verifiers that he has a valid transitive signature, i.e., the designated multi-verifier signature unforgeability (we say DV-Unforgeability). Hou *et al.* [13] have pointed out that DV-Unforgeability signifies TVf-Unforgeability in UDVTS scheme. Although we extend *one-verifier* to *multi-verifier*, it still makes no difference to the TSig algorithm. Thus, we only consider DV-Unforgeability.

Specifically, the following simulation essentially captures a scenario that if no $\mathcal{PPT}$ adversary $\mathcal{A}$ has a non-negligible advantage against a challenger $\mathcal{C}$, we can assert that the UDMVTS scheme is secure against existential forgery on adaptively chosen message attacks.

We denote by $\mathsf{Adv}^{uf-cma}_{\mathcal{A}, \mathrm{UDMVTS}}(k)$ the advantage of $\mathcal{A}$ in winning the simulation. At the beginning of the simulation, $\mathcal{C}$ utilizes SS algorithm to generate the common parameters cp, and runs SKGe algorithm to obtain signer's private/public key pair (pks, sks). Meanwhile, he runs VKGe algorithm to generate all the verifiers' private/public key pairs (skv_l, pkv_l), $l = 1, \ldots, n$. Afterwards $\mathcal{C}$ sends $(cp, pks, pkv_l, \ldots, pkv_n)$ to the adversary $\mathcal{A}$. The TSig algorithm serves as a transitive signature oracle controlled by $\mathcal{C}$, $\mathcal{A}$ can query it to get the signature σ_{ij} on the edge (i, j) of his choice. Furthermore, $\mathcal{A}$ also has access to other queries provided by the algorithms (which supplied by the scheme) whenever it is needed.

At the end of simulation, $\mathcal{A}$ forges a valid UDMVTS signature $\sigma_{DV_{i'j'}}$ on edge (i', j') that has never submitted as one of signing queries, for all the designated verifiers' public keys. Among the rest, we assert that $\mathcal{A}$ wins the simulation if $\sigma_{DV_{i'j'}}$ is a valid forgery with additional needed, i.e., (i', j') is not in the transitive closure of the graph G formed by all $\mathcal{A}$'s TSig queries.

Definition 2. *The proposed* UDMVTS *scheme is unforgeable under chosen-message and public key attacks if* $\mathsf{Adv}^{uf-cma}_{\mathcal{A},\mathrm{UDMVTS}}$ *is negligible for any* $\mathcal{PPT}$ *adversary* $\mathcal{A}$.

Privacy. For a UDMVTS scheme, the basic privacy requirements are include: (1) the privacy of transitive signatures, and (2) non-transferability privacy of UDMVTS.

The former has been systematically discussed by Micali and Rivest in [21]. That is, for same edge, it is impossible for an adaptive chosen-message attacker to distinguish a valid signature from the original signer or the composition algorithm. This implies that the composition algorithm can operate properly even if the previously given signatures were generated by composition algorithm. As described above, the concept *multi-verifier* still makes no difference to the TSig algorithm. Thus, we only focus on another key issue, which requires a UDMVTS scheme to be non-transferable.

The non-transferability enables all designated verifiers to generate an indistinguishable signature from the one that ought to be generated by the signature holder.

Definition 3. *The proposed* UDMVTS *scheme is non-transferable, even if an adaptive chosen-message attacker cannot distinguish a valid designated signature from the signature holder or the designated verifier.*

4   Our Proposed UDMVTS Scheme

We now combine Hou *et al.*'s [13] work with Willy *et al.*'s [29] work to obtain a secure Universal Designated Multi-Verifier Transitive Signature (UDMVTS) scheme in this section. Later, we provide its formal security/performance analysis.

4.1   Our Concrete Scheme

Our UDMVTS scheme is trivially defined as follows:

- SS: Let $\mathbb{G}$ is a q-order multiplicative subgroup of $\mathbb{Z}^*_q$ generated by g. Note a bilinear mapping $\hat{e}$: $\mathbb{G} \times \mathbb{G} \to \mathbb{G}_T$ named $\hat{e}$. Then the common parameters are $cp = (\hat{e}, g)$.

- SKGe: Given cp, chooses $xs \xleftarrow{\$} \mathbb{Z}_p^*$ and computes $ys = g^{xs}$. The private/public key pair for the signer is (xs, ys).
- VKGe: Given cp and n as the number of the verifiers, chooses $xv_l \xleftarrow{\$} \mathbb{Z}_p^*$ for $l = 1, \ldots, n$ and computes $yv_l = g^{xv_l}$. The private/public key pair for verifier i is (xv_l, yv_l).
- TSig: Given the signer's private key xs and nodes $i, j \in \mathbb{N}$, computes $\sigma_{ij} = (h_i h_j^{-1})^{xs} \in \mathbb{G}$, where $h_i = H(i)$ and $h_j = H(j)$. (w.l.o.g., $i < j$, one can swap them if it is not the case.)
- TVf: Given the signer's public key ys, nodes $i, j \in \mathbb{N}$ and a candidate signature σ_{ij}, checks whether $\hat{e}(g, \sigma_{ij}) = \hat{e}(ys, h_i h_j^{-1})$, where $h_i = H(i)$ and $h_j = H(j)$. Then, the algorithm outputs the verification result 1 iff the equation holds and 0 for not.
- Comp: Given nodes $i, j, k \in \mathbb{N}$ and corresponding transitive signatures σ_{ij}, σ_{jk}, if $\mathsf{TVf}(ys, \sigma_{ij}, i, j) = 0$ or $\mathsf{TVf}(ys, \sigma_{jk}, j, k) = 0$ then returns $\perp$ as an indication of failure. Otherwise, returns $\sigma_{ik} = \sigma_{ij} \cdot \sigma_{jk}$ for the edge (i, k). (w.l.o.g., $i < j < k$, one can swap them if it is not the case.)
- DS: Given the signer's public key ys, verifiers' public key yv_l for $l = 1, \ldots, n$, nodes $i, j \in \mathbb{N}$ and an edge signature σ_{ij}, computes $\sigma_{DV} = \hat{e}(\prod_{l=1}^n yv_l, \sigma_{ij})$. Then, the algorithm outputs the designated multi-verifier signature σ_{DV} for the edge.
- $\widehat{\mathsf{DS}}$: Given the signer's public key ys and verifiers' private key xv_l for $l = 1, \ldots, n$, nodes $i, j \in \mathbb{N}$, each verifier individually computes $\sigma_{\widehat{DV}l} = (h_i h_j^{-1})^{xv_l}$ for $l = 1, \ldots, n$, where $h_i = H(i)$ and $h_j = H(j)$.
- DV: Given the signer's public key ys and verifiers' private key xv_l for $l = 1, \ldots, n$, and nodes $i, j \in \mathbb{N}$, and a set of designated verifier edge signature $\sigma_{\widehat{DV}l}$ for $l = 1, \ldots, n$, the algorithm outputs 1 iff $\sigma_{DV} = \prod_{l=1}^n \hat{e}(\sigma_{\widehat{DV}l}, ys)$ hold, otherwise outputs 0.

4.2   Completeness

Note that if $h_i = H(i)$, $h_j = H(j)$, $h_k = H(k)$, $\sigma_{ij} \overset{def}{=} (h_i h_j^{-1})^{xs}$, $\sigma_{ik} \overset{def}{=} \sigma_{ij} \cdot \sigma_{jk} = (h_i h_j^{-1})^{xs} \cdot (h_j h_k^{-1})^{xs} = (h_i h_k^{-1})^{xs}$, $\sigma_{DV} \overset{def}{=} \hat{e}(\prod_{l=1}^n yv_l, \sigma_{ij})$, $\sigma_{\widehat{DV}l} \overset{def}{=} \hat{e}(ys^{xv_l}, h_i h_j^{-1})$ for $l = 1, \ldots, n$. Then the completeness of the UDMVTS scheme is justified as follows:

$$\hat{e}(g, \sigma_{ij}) = \hat{e}(g, (h_i h_j^{-1})^{xs}) = \hat{e}(g, h_i h_j^{-1})^{xs}$$
$$= \hat{e}(g^{xs}, h_i h_j^{-1}) = \hat{e}(ys, h_i h_j^{-1}).$$
$$\hat{e}(g, \sigma_{ik}) = \hat{e}(g, (h_i h_k j^{-1})^{xs}) = \hat{e}(g, h_i h_k^{-1})^{xs}$$
$$= \hat{e}(g^{xs}, h_i h_k^{-1}) = \hat{e}(ys, h_i h_k^{-1}).$$

$$\prod_{l=1}^{n} \hat{e}(\sigma_{\widehat{DV}_l}, ys) = \prod_{l=1}^{n} \hat{e}((h_i h_j^{-1})^{xv_l}, g^{xs})$$

$$= \hat{e}((h_i h_j^{-1})^{xs}, \prod_{l=1}^{n} g^{xv_l})$$

$$= \hat{e}(\sigma_{ij}, \prod_{l=1}^{n} g^{xv_l}) = \hat{e}(\prod_{l=1}^{n} g^{xv_l}, \sigma_{ij})$$

$$= \hat{e}(\prod_{l=1}^{n} yv_l, \sigma_{ij}).$$

4.3 Security Analysis

By the following theorems, this section presents the rigorous security analysis of our UDMVTS scheme.

Theorem 1 (Unforgeability). *In the random oracle model, our UDMVTS scheme is existentially unforgeable against an adaptive chosen-message and public key PPT attacker $\mathcal{A}$ under the one-more BDH assumption. Assuming that there exists an adversary $\mathcal{A}$ who has an advantage $\mathsf{Adv}_{\mathcal{A},\mathrm{UDMVTS}}^{uf-cma}(k)$ in attacking UDMVTS, then there exists a PPT adversary $\mathcal{F}$ that solves the one-more BDH problem in a mapping $\hat{e}$ with an advantage $\mathsf{Adv}_{\mathcal{F}}^{om-\mathsf{BDH}}(k)$ at least:*

$$\mathsf{Adv}_{\mathcal{F}}^{om-\mathsf{BDH}}(k) \geq \mathsf{Adv}_{\mathcal{A},\mathrm{UDMVTS}}^{uf-cma}(k).$$

Proof. The similar simulation can be gathered by Bellare and Neven's security model [5], but contain some twists to make it meet our own. An adversary $\mathcal{F}$ is given access to a challenge oracle $\mathcal{O}_{\mathcal{CH}}(\cdot)$ and a CDH oracle $\mathcal{O}_{\mathcal{CDH}}(\cdot)$. Besides, $\mathcal{F}$ maintains state information (V, Δ), where V is the set of all queried nodes and $\Delta : V \times V \to \mathbb{G}$ is a function storing known edge signatures. The simulation is operated as the following.

At the beginning, $\mathcal{F}$ creates the common parameters cp and provides it to $\mathcal{A}$. Meanwhile, $\mathcal{F}$ chooses $xs, xv_l \xleftarrow{\$} \mathbb{Z}_p^*$ and computes $ys = g^{xs}$, $yv_l = g^{xv_l}$, $l = 1, \ldots, n$ (suppose n is the number of verifiers). Then, $\mathcal{F}$ sends the signer's public key ys, verifiers' public key $yv_1, \ldots, yv_n$ to $\mathcal{A}$. Recall that the main idea of constructing an adversary $\mathcal{F}$ to solve the one-more BDH problem is to compute all n points output by $\mathcal{O}_{\mathcal{CH}}(\cdot)$, using strictly less than n queries to $\mathcal{O}_{\mathcal{CDH}}(\cdot)$. By doing so, $\mathcal{F}$ can response all Hash queries and TSig queries of the adaptive adversary $\mathcal{A}$ as follows:

- Whenever $\mathcal{A}$ carries out a Hash query on any vertex i of his choice, $\mathcal{F}$ checks the set V. If i has not been created, $\mathcal{F}$ invokes the $\mathcal{O}_{\mathcal{CH}}(\cdot)$ to obtain $H(i)$. He then adds i to V, records $H(i)$ and sends it to $\mathcal{A}$ simultaneously.
- Each time when answering $\mathcal{A}$'s TSig query on a signature for edge (i, j), $\mathcal{F}$ checks whether a signature on (i, j) can be obtained by composing previously signed edges. If he cannot, i.e., (i, j) is not in the transitive closure of the graph

produced by $\mathcal{A}$' queries previously, he then utilizes $\mathcal{O}_{\mathcal{CDH}}(\cdot)$ to compute the edge signature $\sigma_{ij} = \Delta(i,j) = \mathcal{O}_{\mathcal{CDH}}(H(i)H(j)^{-1})$ (w.l.o.g., $i < j$, one can swap them if it is not the case.) Then, $\mathcal{F}$ adds $\Delta(i,j)$ to Δ and sends σ_{ij} to $\mathcal{A}$. Absolutely, we suppose the vertexes i, j are already in the set V, otherwise $\mathcal{F}$ can do as the same as Hash query first.

Apparently, a simulator can successfully respond to all the queries. And eventually $\mathcal{A}$ generates a forgery $\sigma_{DV_{i'j'}}$ as a designated multi-verifier transitive signature for edge (i', j'). Assume that $\mathcal{A}$ has been made the Hash queries on i' and j' (and hence that $i', j' \in V$.), otherwise $\mathcal{F}$ can ask the Hash oracle itself after $\mathcal{A}$ outputs his forgery. (w.l.o.g., $i' < j'$, one can swap them if it is not the case.) Let $G = (V, E)$ be the graph formed by $\mathcal{A}$'s TSig queries, and let $\widetilde{G} = (V, \widetilde{E})$ be its transitive closure. If $\sigma_{DV_{i'j'}}$ is not a valid forgery, implying that $\sigma_{DV_{i'j'}} \neq \prod_{l=1}^{n} \hat{e}(\sigma_{\widetilde{DV}_{l\,i'j'}}, ys)$ or $(i', j') \in \widetilde{E}$ (remark as the event E_1.), then $\mathcal{F}$ fails the simulation and aborts. And if (i', j') has been queried during the simulation (remark as the event E_2.), $\mathcal{F}$ reports failure and terminates. Otherwise, we then show how $\mathcal{F}$ solves BDH problems for all challenges in the following:

- $\mathcal{F}$ firstly divides $\widetilde{G}$ into r disjoint subgraphs $V_t \subset V$ for $t = 1, 2, \ldots, r$. Assuming that $V_{t'}$ is the subgraph containing node i', but not containing node j'.
- For every $v \in V_{t'} \backslash \{i'\}$, $\mathcal{F}$ computes $\hat{e}(\prod_{l=1}^{n} yv_l, \sigma_{vi'})$ and $\sigma_{DV_{vi'}} \cdot \sigma_{DV_{i'}}$ as σ_{DV_v} and σ_{DV_v} respectively, and computes $\sigma_{DV_{i'j'}} \cdot \sigma_{DV_{j'}}$ as $\sigma_{DV_{i'}}$ simultaneously.
- For all $t = 1, 2, \ldots, c$, $t \neq t'$, $\mathcal{F}$ chooses a reference node $r_t \in V_t$ in each subgraph, invokes the $\mathcal{O}_{\mathcal{CDH}}(\cdot)$ to get σ_{r_t}, computes $\hat{e}(\prod_{l=1}^{n} yv_l, \sigma_{r_t})$ as $\sigma_{DV_{r_t}}$. And then, for all $v \in V_t \backslash \{r_t\}$, $\mathcal{F}$ computes $\hat{e}(\prod_{l=1}^{n} yv_l, \sigma_{vr_t})$, $\sigma_{DV_{vr_t}} \cdot \sigma_{DV_{r_t}}$ as $\sigma_{DV_{vr_t}}$ and σ_{DV_v} respectively, alone the way, $\mathcal{F}$ solves the BDH solutions of all nodes in V_t.

Remember that for each $i \in V$, $\mathcal{F}$ can do so by invoking the challenge oracle $\mathcal{O}_{\mathcal{CH}}(\cdot)$ to obtain the hash $H(i)$, so he can spontaneously output BDH problems for all $i \in V$.

We now prove the claim by induction in details. Firstly, we would have to check the quantity of CDH queries. As we can see, $\mathcal{F}$ needs to invoke the $\mathcal{O}_{\mathcal{CDH}}(\cdot)$ at most once to response every signature query from $\mathcal{A}$, so the component $V_{t'}$ needs $|V_{t'}| - 1$ queries. Because it did not need the query to get $\mathcal{O}_{\mathcal{CDH}}(H(v))$. Simultaneously, for each component V_t, $t \neq t'$, $\mathcal{F}_{\mathcal{B}}$ needs $|V_t|$ CDH queries. In conclusion, by using $(|V_{t'}| - 1) + \sum_{t \neq t'} |V_t| = |V| - 1$ CDH queries, $\mathcal{F}$ solves $|V|$ BDH solutions and hence perfectly performs the simulation.

Now the probability that $\mathcal{F}$ does not abort during the simulation should be analyzed. $\mathcal{F}$ will not abort if both the events E_1 and E_2 are not happen. Accordingly we have

$$\mathsf{Adv}^{uf-cma}_{\mathcal{A},\mathrm{UDMVTS}}(k) = \Pr[\Pr^{uf-cma}_{\mathcal{A},\mathrm{UDMVTS}}(k) = 1 \wedge \bar{E}_1 \wedge \bar{E}_2]$$
$$\leq \mathsf{Adv}^{om-BDH}_{\mathcal{F}}(k).$$

Therefore, we complete the proof. $\square$

Theorem 2 (Non-Transferability). *The proposed* UDMVTS *scheme achieves the non-transferability.*

Proof. Since each verifier can always receive the transitive signature σ_{ij} and the designated multi-verifier signature σ_{DV} from the signature holder, the non-transferability of our UDMVTS scheme is perfectly achieved in privacy protection. More precisely, suppose n designated verifiers collude to obtain a signature on nodes $i, j \in \mathbb{N}$, they can simulate σ_{DV} by obtaining an indistinguishable signature $\sigma_{\widehat{DV}}$ as follows:

- Each of them signs the transitive signature σ_{ij} as $\sigma_{\widehat{DV}_l} = (h_i h_j^{-1})^{x v_l}$ for $l = 1, \ldots, n$, where $h_i = H(i)$ and $h_j = H(j)$. Observing that since $ys, H(i), H(j)$ are publicly available, any verifier can generate such a signature using his own private key sk_l. Afterwards, they publish it to the other verifiers.
- Upon receiving $\sigma_{\widehat{DV}_l}$, each of them verifies whether the $\sigma_{\widehat{DV}_l}, l = 1, \ldots, n$ are valid. if not, then reports failure and terminates. Otherwise, computes $\sigma_{\widehat{DV}} = \prod_{l=1}^{n} \hat{e}(\sigma_{\widehat{DV}_l}, ys)$.

Obviously, note that $\sigma_{\widehat{DV}}$ is indistinguishable from σ_{DV}, any third party will be incapable of convincing it that should have been produced by a signature holder, even if these verifiers provide their own private keys. □

4.4 Performance Analysis

In order to facilitate the analysis of our UDMVTS scheme, we start by defining the following notations. Assume that the bit length is $|G|$ in $\mathbb{G}$, and it is $|G_T|$ in $\mathbb{G}_T$. Abbreviations used are: "TSlen." for a transitive signature's length and "DSlen." for a designated multi-verifier signature's length. The notations used to denote the time costs in algorithms are: "T_E" for computing an exponentiation operation; "T_{Mt}" for mapping one point operation; "T_I" for computing one inversion operation; "T_{Mu}" for computing one modular multiplication; "T_P" for computing one pairing operation; "T_A" for computing one modular addition.

The computational cost for the System Setup algorithm SS, the Signer Key Generation algorithm SKGe and the Verifier Key Generation algorithm VKGe are significantly small and thus omitted. Take the algorithm TSig as an example, when running the Transitive Signing algorithm, we need two hash operation to obtain h_i, h_j, one inversion operation to compute h_j^{-1}, one modular multiplication operation to compute $h_i h_j^{-1}$ and one modular addition operation to generate $\sigma_{ij} = (h_i h_j^{-1})^{xs}$. We then list the result of the proposed scheme with n verifiers in Table 2.

From Table 2 we can see that, in our efficient construction, the signature is only $|G_T|$ bits. Thus, though our UDMVTS is designed for multi-verifier scenarios, the signature length of our scheme is independent from the number of verifiers. Furthermore, recall that the most time-consuming operations are the pairing operations, and our scheme only needs one pairing operation when generating a designated multi-verifier signature.

Table 2. Computational costs and signature lengths

TSig cost	Tvf cost	Comp cost	TSlen.
$2T_{Mt} + T_I + T_{Mu} + T_E$	$2T_{Mt} + T_I + T_{Mu} + 2T_P$	T_{Mu}	$|G|$
DS cost	$\widehat{DS}$ cost	DV cost	DSlen.
$(n-1)T_{Mu} + T_P$	$n(2T_{Mt} + T_I + T_{Mu} + T_E)$	$(n-1)T_P$	$|G_T|$

Furthermore, we also evaluate the performance of our UDMVTS scheme by analyzing the time cost of its sub algorithms. The UDMVTS model is implemented by using C language. The simulation ran on Intel Core $i5$ processors with 3.2 GHz and $4G$ RAM on Ubuntu 10.10. We use the PBC library (version $0.5.12$)[1] to conduct our simulation experiments. In our simulation, we assume that there are 100 verifiers. And we do not test the time cost of the Composition algorithm Comp for the reason that this algorithm is one simple multiply computation. Specifically, the simulation was run more than one hundred times, and Table 3 shows the time costs.

Table 3. Time costs (in s)

Operation	SKGe	VKGe	Tsig	Tvf	DS	$\widehat{DS}$	DV
Max time	0.003910	0.519995	0.024029	0.015475	0.012913	0.395942	0.399361
Min time	0.003670	0.491002	0.015171	0.006696	0.004953	0.361157	0.364484
Average time	0.003803	0.504790	0.018411	0.010744	0.008092	0.379986	0.383732

In Table 3, numbers are used to count the time costs when the sub algorithms in our proposed scheme were ran. For example, the Verifier Key Generation algorithm VKGe is the most time-costing algorithm among these sub algorithms. From Table 3 we can see that, the max time cost is 0.519995 s, the min time cost is 0.491002 s, and the average time cost is 0.504790 s. Namely, the time cost of our algorithm is extremely low.

Summarizing the above discussions, our proposed scheme achieves good performance and can be efficiently implemented.

5   Conclusions

In this paper, we formalize the concept of Universal Designated Multi-Verifier Transitive Signature (UDMVTS) in order to achieve authentication in the graph-based big data system. Specifically, we present its design model and provide detailed discussions on the requirements for achieving security and privacy. Based on the one-more BDH problem, our new scheme is proved secure in the random

---

[1] http://crypto.stanford.edu/pbc/download.html.

oracle model. Additionally, we implement our scheme to demonstrate its efficiency. Nevertheless, similar to Hou *et al.*'s UDVTS scheme, our construction also needs a special MapToPoint hash function. However, it is a probabilistic algorithm. As a promising future research work, we plan to design new schemes by using general cryptographic hash functions such as SHA-512 or MD6. Furthermore, we will try to design a UDMVTS scheme in the standard model.

Acknowledgment. This work is supported by National Natural Science Foundation of China (61472083, 61402110, 61771140), Program for New Century Excellent Talents in Fujian University (JA14067), Distinguished Young Scholars Fund of Fujian (2016J06013), and Fujian Normal University Innovative Research Team (NO. IRTL1207).

References

1. Baek, J., Safavi-Naini, R., Susilo, W.: Universal designated verifier signature proof (or how to efficiently prove knowledge of a signature). In: Roy, B. (ed.) ASIACRYPT 2005. LNCS, vol. 3788, pp. 644–661. Springer, Heidelberg (2005). https://doi.org/10.1007/11593447_35
2. Bao, F., Deng, R., Zhou, J. (eds.): PKC 2004. LNCS, vol. 2947. Springer, Heidelberg (2004). https://doi.org/10.1007/978-3-540-24632-9
3. Barreto, P.S.L.M., Kim, H.Y.: Fast hashing onto elliptic curves over fields of characteristic 3. IACR Cryptology ePrint Archive 2001, 98 (2001). http://eprint.iacr.org/2001/098
4. Bellare, M., Neven, G.: Transitive signatures based on factoring and RSA. In: Zheng, Y. (ed.) ASIACRYPT 2002. LNCS, vol. 2501, pp. 397–414. Springer, Heidelberg (2002). https://doi.org/10.1007/3-540-36178-2_25
5. Bellare, M., Neven, G.: Transitive signatures: new schemes and proofs. IEEE Trans. Inf. Theor. **51**(6), 2133–2151 (2005)
6. Boneh, D., Lynn, B., Shacham, H.: Short signatures from the weil pairing. In: Boyd, C. (ed.) ASIACRYPT 2001. LNCS, vol. 2248, pp. 514–532. Springer, Heidelberg (2001). https://doi.org/10.1007/3-540-45682-1_30
7. Camacho, P., Hevia, A.: Short transitive signatures for directed trees. In: Dunkelman, O. (ed.) CT-RSA 2012. LNCS, vol. 7178, pp. 35–50. Springer, Heidelberg (2012). https://doi.org/10.1007/978-3-642-27954-6_3
8. Chen, X., Chen, G., Zhang, F., Wei, B., Mu, Y.: Identity-based universal designated verifier signature proof system. Int. J. Netw. Secur. **8**(1), 52–58 (2009)
9. Goldwasser, S., Micali, S., Rivest, R.L.: A digital signature scheme secure against adaptive chosen-message attacks. SIAM J. Comput. **17**(2), 281–308 (1988)
10. Gong, Z., Huang, Z., Qiu, W., Chen, K.: Transitive signature scheme from LFSR. J. Inf. Sci. Eng. **26**(1), 131–143 (2010)
11. Gray, J.: Jim Gray on eScience: a transformed scientific method. In: Based on the transcript of a talk given by Jim Gray to the NRC-CSTB in Mountain View, CA, 11 January 2007. The Fourth Paradigm: Data-Intensive Scientific Discovery (2009)
12. Hohenberger, S.: The cryptographic impact of groups with infeasible inversion. Master's thesis. MIT (2003)
13. Hou, S., Huang, X., Liu, J.K., Li, J., Xu, L.: Universal designated verifier transitive signatures for graph-based big data. Inf. Sci. **318**, 144–156 (2015)

14. Huang, X., Susilo, W., Mu, Y., Wu, W.: Universal designated verifier signature without delegatability. In: Ning, P., Qing, S., Li, N. (eds.) ICICS 2006. LNCS, vol. 4307, pp. 479–498. Springer, Heidelberg (2006). https://doi.org/10.1007/11935308_34

15. Huang, X., Susilo, W., Mu, Y., Wu, W.: Secure universal designated verifier signature without random oracles. Int. J. Inf. Secur. **7**(3), 171–183 (2008)

16. Huang, X., Susilo, W., Mu, Y., Zhang, F.: Restricted universal designated verifier signature. In: Ma, J., Jin, H., Yang, L.T., Tsai, J.J.-P. (eds.) UIC 2006. LNCS, vol. 4159, pp. 874–882. Springer, Heidelberg (2006). https://doi.org/10.1007/11833529_89

17. Laguillaumie, F., Libert, B., Quisquater, J.-J.: Universal designated verifier signatures without random oracles or non-black box assumptions. In: De Prisco, R., Yung, M. (eds.) SCN 2006. LNCS, vol. 4116, pp. 63–77. Springer, Heidelberg (2006). https://doi.org/10.1007/11832072_5

18. Laguillaumie, F., Vergnaud, D.: On the soundness of restricted universal designated verifier signatures and dedicated signatures. In: Garay, J.A., Lenstra, A.K., Mambo, M., Peralta, R. (eds.) ISC 2007. LNCS, vol. 4779, pp. 175–188. Springer, Heidelberg (2007). https://doi.org/10.1007/978-3-540-75496-1_12

19. Li, J., Wang, Y.: Universal designated verifier ring signature (proof) without random oracles. In: Zhou, X., et al. (eds.) EUC 2006. LNCS, vol. 4097, pp. 332–341. Springer, Heidelberg (2006). https://doi.org/10.1007/11807964_34

20. Lin, C., Wu, W., Huang, X., Xu, L.: A new universal designated verifier transitive signature scheme for big graph data. J. Comput. Syst. Sci. **83**(1), 73–83 (2017)

21. Micali, S., Rivest, R.L.: Transitive signature schemes. In: Preneel, B. (ed.) CT-RSA 2002. LNCS, vol. 2271, pp. 236–243. Springer, Heidelberg (2002). https://doi.org/10.1007/3-540-45760-7_16

22. Ming, Y., Wang, Y.: Universal designated multi verifier signature scheme without random oracles. Wuhan Univ. J. Nat. Sci. **13**(6), 685–691 (2008)

23. Neven, G.: A simple transitive signature scheme for directed trees. Theoret. Comput. Sci. **396**(1–3), 277–282 (2008)

24. Shahandashti, S.F., Safavi-Naini, R.: Generic constructions for universal designated-verifier signatures and identity-based signatures from standard signatures. IET Inf. Secur. **3**(4), 152–176 (2009)

25. Shailaja, G., Kumar, K.P., Saxena, A.: Universal designated multi verifier signature without random oracles. In: Mohanty, S.P., Sahoo, A. (eds.) ICIT, pp. 168–171. IEEE Computer Society (2006)

26. Steinfeld, R., Bull, L., Wang, H., Pieprzyk, J.: Universal designated-verifier signatures. In: Laih, C.-S. (ed.) ASIACRYPT 2003. LNCS, vol. 2894, pp. 523–542. Springer, Heidelberg (2003). https://doi.org/10.1007/978-3-540-40061-5_33

27. Steinfeld, R., Wang, H., Pieprzyk, J.: Efficient extension of standard Schnorr/RSA signatures into universal designated-verifier signatures. In: Bao et al. [2], pp. 86–100

28. Wang, L., Cao, Z., Zheng, S., Huang, X., Yang, Y.: Transitive signatures from braid groups. In: Srinathan, K., Rangan, C.P., Yung, M. (eds.) INDOCRYPT 2007. LNCS, vol. 4859, pp. 183–196. Springer, Heidelberg (2007). https://doi.org/10.1007/978-3-540-77026-8_14

29. Willy, C.Y.N., Susilo, W., Mu, Y.: Universal designated multi verifier signature schemes. In: Proceedings of the 11th International Conference on Parallel and Distributed Systems, vol. 2, pp. 305–309. IEEE (2005)

30. Xu, J.: On directed transitive signature. IACR Cryptology ePrint Archive 2009, 209 (2009). http://eprint.iacr.org/2009/209

31. Yi, X.: Directed transitive signature scheme. In: Abe, M. (ed.) CT-RSA 2007. LNCS, vol. 4377, pp. 129–144. Springer, Heidelberg (2006). https://doi.org/10.1007/11967668_9

32. Zhang, F., Safavi-Naini, R., Susilo, W.: An efficient signature scheme from bilinear pairings and its applications. In: Bao et al. [2], pp. 277–290

33. Zhang, F., Susilo, W., Mu, Y., Chen, X.: Identity-based universal designated verifier signatures. In: Enokido, T., Yan, L., Xiao, B., Kim, D., Dai, Y., Yang, L.T. (eds.) EUC 2005. LNCS, vol. 3823, pp. 825–834. Springer, Heidelberg (2005). https://doi.org/10.1007/11596042_85

34. Zhang, R., Furukawa, J., Imai, H.: Short signature and universal designated verifier signature without random oracles. In: Ioannidis, J., Keromytis, A., Yung, M. (eds.) ACNS 2005. LNCS, vol. 3531, pp. 483–498. Springer, Heidelberg (2005). https://doi.org/10.1007/11496137_33

Cryptanalysis and Improvement of a Strongly Unforgeable Identity-Based Signature Scheme

Xiaodong Yang[1,2]([✉]) [iD], Ping Yang[2], Faying An[2], Shudong Li[3,4],
Caifen Wang[2], and Dengguo Feng[1]

[1] State Key Laboratory of Cryptology, Beijing 100878, China
y2008880163.com
[2] College of Computer Science and Engineering, Northwest Normal University,
Lanzhou 730070, China
[3] Cyberspace Institute of Advanced Technology, Guangzhou University,
Guangzhou 510006, China
[4] College of Mathematics and Information Science,
Shandong Technology and Business University, Yantai 264005, China

Abstract. Recently, Tsai et al. constructed an efficient identity-based signature (IBS) scheme and claimed that it was strongly unforgeable in the standard model. Unfortunately, we find that their scheme is insecure. By giving concrete attack, we show that their scheme does not meet the requirement of strong unforgeability. Meanwhile, we demonstrate that there are serious flaws in their security proof. The simulator cannot correctly answer the signing query in the security model. Furthermore, we propose an improved strongly unforgeable IBS scheme without random oracles. Compared with other strongly unforgeable IBS schemes in the standard model, our scheme is more efficient in terms of computation cost and signature size.

Keywords: Identity-based signature · Standard model
Strong unforgeability · Bilinear map

1 Introduction

Identity-based cryptosystems [1] allow a user to make use of email address and other unique identity information as its public key. At the same time, a trustworthy entity called private key generator (PKG) computes a corresponding private key from the public key of the user. As a result, the identity-based cryptosystem simplifies key management mechanism of traditional public-key cryptography. Under the framework of the identity-based encryption (IBE) scheme of Boneh and Franklin [2], lots of identity-based signature (IBS) schemes were presented in [3–6], and the random oracle model was used to prove the security of these schemes. However, a provably secure cryptography scheme in the random oracle model might be insecure in reality since the concrete hash function cannot implement the ideal random oracle. Therefore, it is too important to construct

© Springer International Publishing AG, part of Springer Nature 2018
X. Chen et al. (Eds.): Inscrypt 2017, LNCS 10726, pp. 196–208, 2018.
https://doi.org/10.1007/978-3-319-75160-3_13

efficient IBS schemes in the standard model without utilizing random oracles [7]. Based on Waters's IBE scheme [8], Paterson and Schuldt [9] presented an IBS scheme without random oracles. Later, many IBS schemes have been proposed in the standard model [10–13].

Most existing IBS schemes without random oracles only cover existential unforgeability, in which an adversary cannot forge signatures on messages not signed before. Since some existentially unforgeable IBS schemes are malleable [9,10], a previous valid signature on a message can easily be modified into a new legal signature on the same massage. In order to prevent the tampering of existing signatures, a notion called strong unforgeability has been introduced [14,15]. Strong unforgeability ensures that an attacker is unable to forge signatures for previously signed messages, and strongly unforgeable IBS schemes are used to construct chosen-ciphertext secure IBE schemes, identity-based group signature schemes, and so on. Nevertheless, the strongly unforgeable schemes [16–18] with applying transformation techniques lack efficiency in signature size and computation cost. In the standard model, Sato et al. [19] adopted a direct construction to present a strongly unforgeable IBS scheme, but it is inefficient since their scheme needs five signature parameters and six pairing operations for the signature verification phase. Recently, Kwon [20] put forward an efficient IBS scheme in the standard model and proved its strong unforgeability under the computational Diffie-Hellman (CDH) assumption, but Lee et al. [21] found that the security of the Know's scheme was not associated with the CDH assumption. Tsai et al. [22] also constructed a strongly unforgeable IBS scheme, and demonstrated that it was provably secure in the standard model. Unfortunately, our observation manifests that their scheme is insecure.

In this paper, we first give one kind of attack against Tsai et al.'s IBS scheme [22] to show that their scheme does not possess strong unforgeability. Next, we find that there are some flaws in the Tsai et al.'s security proof. The simulator of Tsai et al.'s security argument is unable to output a correct signature to respond to the adversary's signature query. Furthermore, we construct an improved IBS scheme that is strongly unforgeable in the standard model. In addition, the analysis results indicate that our scheme has higher performance and security.

2   Preliminaries

2.1   Bilinear Pairings

Suppose that two multiplicative cyclic groups G_1 and G_2 have the same prime order p, and g is a generator of G_1. Pick two random values $a, b \in Z_p$, and define a map $e : G_1 \times G_1 \to G_2$ to be a bilinear pairing if e meets the following conditions [23]:

(1) Bilinear: $e(g^a, g^b) = e(g, g)^{ab} = e(g^b, g^a)$.
(2) Non-degenerate: $e(g, g) \neq 1$.
(3) Computable: $e(g^a, g^b)$ can be computed efficiently.

2.2   Complexity Assumption

Given $(g, g^a, g^b) \in G_1^3$ for unknown $a, b \in Z_p$ and a generator g selected randomly in G_1, the CDH problem is to output g^{ab}.

Definition 1. *We say that the CDH assumption holds if no probabilistic polynomial-time (PPT) algorithm can solve the CDH problem in G_1 with a non-negligible probability [24].*

3   Revisiting the Tsai et al.'s IBS Scheme

3.1   The Original IBS Scheme

Tsai et al.'s IBS scheme [22] consists of the following four algorithms: **Setup**, **Extract**, **Sign** and **Verify**. Assume that all user's identities are bit strings of length m and all messages to be signed are bit strings of length n. In practice, by using two collision-resistant hash functions $H_1 : \{0,1\}^* \to \{0,1\}^m$ and $H_2 : \{0,1\}^* \to \{0,1\}^n$, a more flexible scheme can be constructed to achieve that messages and identities are strings of arbitrary lengths.

Setup: Taking as input a security parameter $\lambda \in Z$, the PKG selects two cyclic groups G_1, G_2 of prime order p, a generator g in G_1, and a bilinear pairing $e : G_1 \times G_1 \to G_2$. The PKG also selects $m + n + 4$ random elements $g_2, u', u_1, \cdots, u_m, v, w', w_1, \cdots, w_n$ in G_1 and a collision-resistant hash function $H : \{0,1\}^* \to \{0,1\}^n$. Furthermore, the PKG randomly picks $\alpha \in Z_p$, and computes $g_1 = g^\alpha$ and $msk = g_2^\alpha$. Finally, the PKG broadcasts system parameters $cp = (G_1, G_2, p, e, g, g_1, g_2, u', u_1, \cdots, u_m, v, w', w_1, \cdots, w_n, H)$ and stores the master key msk secretly.

Extract: For an identity $ID = (v_1, ..., v_m) \in \{0,1\}^m$, the PKG randomly selects $r_v \in Z_p$ and computes the user's private key $d_{ID} = (d_1, d_2) = (g_2^\alpha(u' \prod_{i=1}^{m} u_i^{v_i})^{r_v}, g^{r_v})$. Then, the PKG secretly transmits d_{ID} to the user ID.

Sign: On input a message $M = (M_1, ..., M_n) \in \{0,1\}^n$ and a private key $d_{ID} = (d_1, d_2)$ of an identity $ID = (v_1, ..., v_m) \in \{0,1\}^m$, it randomly selects $r_m \in Z_p$ and computes $h = H(M, g^{r_m})$. Next, it uses $d_{ID} = (d_1, d_2)$ to compute $\sigma = (\sigma_1, \sigma_2, \sigma_3) = (d_1^h(w' \prod_{j=1}^{n} w_j^{M_j})^{r_m}, d_2^h, g^{r_m})$ as a signature of ID on M.

Verify: On input a signer's identity $ID = (v_1, ..., v_m) \in \{0,1\}^m$, a message $M = (M_1, ..., M_n) \in \{0,1\}^n$ and a signature $\sigma = (\sigma_1, \sigma_2, \sigma_3)$, a verifier computes $h = H(M, \sigma_3)$ and checks the following verification equation:

$$e(\sigma_1, g) = e(g_2, g_1)^h e(u' \prod_{i=1}^{m} u_i^{v_i}, \sigma_2) e(w' \prod_{j=1}^{n} w_j^{M_j}, \sigma_3)$$

If the above equation holds, accept σ. Otherwise, reject σ.

3.2   The Security Proof

In this subsection, we briefly review the security proof for Tsai et al.'s IBS scheme [22].

Assume that there exists an adversary $\mathcal{A}$ who makes q_E key extraction queries and q_S signing queries, and $\mathcal{A}$ forges a valid signature for Tsai et al.'s IBS scheme with a non-negligible probability. Then, a simulator $\mathcal{B}$ is given a random instance (g, g^a, g^b) of the CDH problem in G_1. By using $\mathcal{A}$'s forgery, $\mathcal{B}$ can solve the CDH problem. To output g^{ab}, $\mathcal{B}$ interacts with $\mathcal{A}$ as follows.

Setup: Let $l_v = 2(q_S + q_E)$ and $l_m = 2q_S$. $\mathcal{B}$ first picks two random values $k_v \in \{1, ..., m\}$ and $k_m \in \{1, ..., n\}$. Then, $\mathcal{B}$ picks a collision-resistant hash function $H : \{0,1\}^* \rightarrow \{0,1\}^n$ and randomly selects $x', x_1, ..., x_m \in Z_{l_v}$ and $y', y_1, ..., y_m \in Z_p$. For each queried identity $ID = (v_1, ..., v_m) \in \{0,1\}^m$, we define two functions $F(ID) = -l_v k_v + x' + \sum_{i=1}^{m} x_i v_i$ and $J(ID) = y' + \sum_{i=1}^{m} y_i v_i$. Also, $\mathcal{B}$ randomly selects $c', c_1, ..., c_n \in Z_{l_m}$ and $t', t_1, ..., t_n \in Z_p$. For each queried message $M = (M_1, ..., M_n) \in \{0,1\}^n$, we similarly define functions $K(M) = -l_m k_m + c' + \sum_{j=1}^{n} c_j M_j$ and $L(M) = t' + \sum_{j=1}^{n} t_j M_j$. $\mathcal{B}$ sets $g_1 = g^a$, $g_2 = g^b$, $u' = g_2^{l_v k_v + x'} g^{y'}$, $u_i = g_2^{x_i} g^{y_i} (1 \leq i \leq m)$, $w' = g_2^{l_m k_m + c'} g^{t'}$ and $w_j = g_2^{c_j} g^{t_j} (1 \leq j \leq n)$. Finally, $\mathcal{B}$ sends $cp = (G_1, G_2, p, e, g, g_1, g_2, u', u_1, ..., u_m, w', w_1, ..., w_n, H)$ to $\mathcal{A}$.

Queries: $\mathcal{A}$ is allowed to adaptively access the following oracles which are built by the simulator $\mathcal{B}$.

Key extraction oracle: On receiving a private key query for an identity $ID = (v_1, ..., v_m) \in \{0,1\}^m$, $\mathcal{B}$ computes $F(ID)$ and $J(ID)$. Note that $l_v(m+1) < p$ and $x', x_1, ..., x_m \in Z_{l_v}$, we have $0 < l_v k_v < p$ and $0 < x' + \sum_{i=1}^{m} x_i v_i < p$. Then, it is easy to see that $F(ID) = 0 \bmod p$ such that $F(ID) = 0 \bmod l_v$. Furthermore, if $F(ID) = 0 \bmod l_v$, there exists a unique value $0 \leq k_v \leq m$ such that $F(ID) = 0 \bmod p$. On the other hand, $F(ID) \neq 0 \bmod l_v$ implies $F(ID) \neq 0 \bmod p$. To make the analysis of the simulation easier, if $F(ID) = 0 \bmod l_v$, $\mathcal{B}$ aborts. Otherwise, $\mathcal{B}$ randomly selects $r_v \in Z_p$, and returns the corresponding private key $d_{ID} = (d_1, d_2) = ((g_1)^{\frac{-J(ID)}{F(ID)}} (u' \prod_{i=1}^{m} u_i^{v_i})^{r_v}, g_1^{\frac{-1}{F(ID)}} g^{r_v})$ to the adversary $\mathcal{A}$.

Signing oracle: On receiving a signature query on a message $M = (M_1, ..., M_n) \in \{0,1\}^n$ for an identity $ID = (v_1, ..., v_m) \in \{0,1\}^m$, $\mathcal{B}$ computes $F(ID)$ and considers the following two cases:

- If $F(ID) \neq 0 \bmod l_v$, $\mathcal{B}$ first runs the key extraction oracle to get the ID's private key d_{ID}. Next, $\mathcal{B}$ executes the algorithm **Sign** to generate a signature σ on M, and transmits σ to $\mathcal{A}$.

– If $F(ID) = 0 \bmod l_v$, $\mathcal{B}$ first computes $K(M)$ and $L(M)$. By similar arguments, given the assumption $l_m(n+1) < p$, $K(M) = 0 \bmod l_m$ implies $K(M) = 0 \bmod p$. If $K(M) = 0 \bmod l_m$, $\mathcal{B}$ terminates the simulation. If $K(M) \neq 0 \bmod l_m$, $\mathcal{B}$ randomly picks two elements r_v, r_m from Z_p, and computes $h = H(M, g^{r_m})$. Finally, $\mathcal{B}$ returns a signature $\sigma = (\sigma_1, \sigma_2, \sigma_3)$ on M to $\mathcal{A}$, where $\sigma_1 = ((u' \prod_{i=1}^{m} u_i^{v_i})^{r_v})^h (g_1)^{\frac{-L(M)h}{K(M)}} (w' \prod_{j=1}^{n} w_j^{M_j})^{r_m}$, $\sigma_2 = (g^{r_v})^h$ and $\sigma_3 = (g_1)^{\frac{-h}{K(M)}} g^{r_m}$.

Note that the simulator $\mathcal{B}$ cannot generate a legal signature with $F(ID) = 0 \bmod l_v$ and $K(M) = 0 \bmod l_m$.

Forgery: $\mathcal{A}$ outputs a forged signature $\sigma^* = (\sigma_1^*, \sigma_2^*, \sigma_3^*)$ for identity ID^* on message M^*, where ID^* was not queried in key extraction oracle and σ^* was not produced by the signing oracle. If $F(ID^*) = 0 \bmod p$ and $K(M^*) = 0 \bmod p$, $\mathcal{B}$ computes $h^* = H(M^*, \sigma_3^*)$ and the CDH value $g^{ab} = \left(\frac{\sigma_1^*}{(\sigma_2^*)^{J(ID^*)}(\sigma_3^*)^{L(M^*)}} \right)^{\frac{1}{h^*}}$. Otherwise, $\mathcal{B}$ declares failure.

4   Cryptanalysis of Tsai et al.'s IBS Scheme

4.1   Attack Against Tsai et al.'s IBS Scheme

Tsai et al. [22] demonstrated that their IBS scheme was strongly unforgeable against adaptive chosen-message attacks. However, in this subsection, we describe an attack on Tsai et al.'s IBS scheme to show that this scheme does not cover strong unforgeability. If an adversary $\mathcal{A}$ obtains a legal signature on any message, $\mathcal{A}$ always succeeds in forging a new valid signature on the same message as follows.

(1) $\mathcal{A}$ randomly picks a message $M = (M_1, ..., M_n) \in \{0,1\}^n$ and an identity $ID = (v_1, ..., v_m) \in \{0,1\}^m$, and then requests a signing query of ID on M. Let $\sigma' = (\sigma_1', \sigma_2', \sigma_3')$ be the corresponding output of the signing oracle on (ID, M), where $\sigma_1' = g_2^{\alpha h}(u' \prod_{i=1}^{m} u_i^{v_i})^{r_v h}(w' \prod_{j=1}^{n} w_j^{M_j})^{r_m}$, $\sigma_2' = g^{r_v h}$, $\sigma_3' = g^{r_m}$ and $h = H(M, \sigma_3')$.

(2) $\mathcal{A}$ chooses a random value $r_v' \in Z_p$, and computes $\sigma_1 = \sigma_1'(u' \prod_{i=1}^{m} u_i^{v_i})^{r_v'}$, $\sigma_2 = \sigma_2' g^{r_v'}$ and $\sigma_3 = \sigma_3'$.

(3) $\mathcal{A}$ outputs $\sigma = (\sigma_1, \sigma_2, \sigma_3)$ as a new ID's signature on M.

It is clear that the forged signature $\sigma = (\sigma_1, \sigma_2, \sigma_3)$ is valid since σ satisfies the following verification equation.

$$e(\sigma_1, g) = e(\sigma_1'(u'\prod_{i=1}^{m} u_i^{v_i})^{r_v'}, g)$$

$$= e(g_2^{ah}(u'\prod_{i=1}^{m} u_i^{v_i})^{r_v}h(w'\prod_{j=1}^{n} w_j^{M_j})^{r_m}(u'\prod_{i=1}^{m} u_i^{v_i})^{r_v'}, g)$$

$$= e(g_2, g_1)^h e(u'\prod_{i=1}^{m} u_i^{v_i}, g^{r_v h + r_v'})e(w'\prod_{j=1}^{n} w_j^{M_j}, g^{r_m})$$

$$= e(g_2, g_1)^h e(u'\prod_{i=1}^{m} u_i^{v_i}, \sigma_2)e(w'\prod_{j=1}^{n} w_j^{M_j}, \sigma_3),$$

where $h = H(M, \sigma_3) = H(M, \sigma_3')$.

Note that $\mathcal{A}$ does not request the private key query with ID, and σ is not produced by the signing oracle. Moreover, $\mathcal{A}$ is a legal adversary since the signature query for ID on M is allowed in the security model of strongly unforgeable IBS [22]. Therefore, Tsai et al.'s IBS scheme is not strongly unforgeable in the standard model. The problem of this scheme is that the original signature σ' and the forged signature σ have the same hash value h. Consequently, the adversary can easily re-randomize previous signatures to produce new signatures on same messages.

4.2 Analysis of Tsai et al.'s Security Proof

We have reviewed the security proof of Tsai et al.'s IBS scheme [22] in Sect. 3.2. Nevertheless, we find that there are some flaws in their security proof. The detailed analysis is described below.

If an adversary $\mathcal{A}$ requests a signature query on a message $M = (M_1, ..., M_n) \in \{0, 1\}^n$ for an identity $ID = (v_1, ..., v_m) \in \{0, 1\}^m$, the simulator $\mathcal{B}$ of Tsai et al.'s security proof [22] randomly selects r_v, r_m from Z_p and computes $h = H(M, g^{r_m})$. If $F(ID) = 0 \bmod l_v$ and $K(M) \neq 0 \bmod l_m$, $\mathcal{B}$ outputs a signature $\sigma = (\sigma_1, \sigma_2, \sigma_3)$ on M, where

$$\sigma_1 = ((u'\prod_{i=1}^{m} u_i^{v_i})^{r_v})^h (g_1)^{\frac{-L(M)h}{K(M)}}(w'\prod_{j=1}^{n} w_j^{M_j})^{r_m}$$

$$= g_2^{ah}(u'\prod_{i=1}^{m} u_i^{v_i})^{r_v h}(w'\prod_{j=1}^{n} w_j^{M_j})^{r_m - \frac{ah}{K(M)}}$$

$$= g_2^{ah}(u'\prod_{i=1}^{m} u_i^{v_i})^{r_v h}(w'\prod_{j=1}^{n} w_j^{M_j})^{r_m'},$$

$\sigma_2 = g^{r_v h}$ and $\sigma_3 = (g_1)^{\frac{-h}{K(M)}}g^{r_m} = g^{r_m'}$.

Note that the hash function value h in σ_1 is computed by M and g^{r_m}, just as $h = H(M, g^{r_m})$. According to the algorithm **Verify** of Tsai et al.'s IBS scheme described in Sect. 3.1, the hash function value h' in verification equation is computed by M and σ_3, just as $h^{'} = H(M, \sigma_3)$. Since $r_m^{'} = r_m - \frac{ah}{K(M)}$, we have $r_m^{'} \neq r_m$. Then, we can obtain $h^{'} = H(M, \sigma_3) = H(M, g^{r_m^{'}}) \neq H(M, g^{r_m}) = h$ since H is a collision-resistant hash function. Hence, we have

$$e(\sigma_1, g) = e(g_2, g_1)^h e(u^{'} \prod_{i=1}^{m} u_i{}^{v_i}, \sigma_2) e(w^{'} \prod_{j=1}^{n} w_j{}^{M_j}, \sigma_3)$$

$$\neq e(g_2, g_1)^{h^{'}} e(u^{'} \prod_{i=1}^{m} u_i{}^{v_i}, \sigma_2) e(w^{'} \prod_{j=1}^{n} w_j{}^{M_j}, \sigma_3).$$

From the above description, we know that the simulated signature $\sigma = (\sigma_1, \sigma_2, \sigma_3)$ generated by the simulator $\mathcal{B}$ does not satisfy the verification equation of the algorithm **Verify** in Tsai et al.'s IBS scheme [22]. Therefore, the simulator $\mathcal{B}$ cannot correctly answer the signing query defined in the security model, and thus the adversary $\mathcal{A}$ cannot obtain any valid signature for $F(ID) = 0 \bmod l_v$ and $K(M) \neq 0 \bmod l_m$. Furthermore, based on the wrong simulation, it is not correct that the security of Tsai et al.'s IBS scheme is directly reduced to the hardness of the CDH assumption by using the advantage of the adversary against their scheme.

5 Improved Strongly Unforgeable IBS Scheme

5.1 Construction

Here, we propose an efficient IBS scheme motivated from Tsai et al.'s IBS [22] and Paterson and Schuldt's IBS [9], and our scheme is proven to be strongly unforgeable under the CDH assumption in the standard model. The details of our IBS scheme are described below:

The algorithms **Setup** and **Extract** are identical to those of Tsai et al.'s IBS scheme described in Sect. 3.1. Note that the user can re-randomize the private key d_{ID} after receiving d_{ID} sent by the PKG.

Sign: Given a message $M = (M_1, ..., M_n) \in \{0, 1\}^n$ and a private key $d_{ID} = (d_1, d_2)$ of an identity $ID = (v_1, ..., v_m) \in \{0, 1\}^m$, the signer with ID randomly chooses $r_m \in Z_p$. Next, the signer computes $h = H(M, ID, d_2, g^{r_m})$ and $\sigma = (\sigma_1, \sigma_2, \sigma_3) = (d_1((w^{'} \prod_{j=1}^{n} w_j{}^{M_j}) v^h)^{r_m}, d_2, g^{r_m})$ as a signature of ID on M.

Verify: Given a signer's identity $ID = (v_1, ..., v_m) \in \{0, 1\}^m$, a message $M = (M_1, ..., M_n) \in \{0, 1\}^n$ and a signature $\sigma = (\sigma_1, \sigma_2, \sigma_3)$, it computes $h = H(M, ID, \sigma_2, \sigma_3)$ and checks the following verification equation:

$$e(\sigma_1, g) = e(g_2, g_1) e(u^{'} \prod_{i=1}^{m} u_i{}^{v_i}, \sigma_2) e((w^{'} \prod_{j=1}^{n} w_j{}^{M_j}) v^h, \sigma_3).$$

If this equation holds, accept σ. Otherwise, reject σ.

Note that it is difficult to re-randomize our ID-based signature to compute a new valid signature on the same message without the signer's private key. It is because that the hash value $h = H(M, ID, \sigma_2, \sigma_3)$ is hashed by a message M, an identity ID, the element d_2 for randomness of a signer's private key $d_{ID} = (d_1, d_2)$ issued by the PKG and the element σ_3 for randomness of a message M chosen by the signer. Owing to the collision resistance of H, the adversary cannot find two different inputs (r_1, r_2) such that $H(r_1) = H(r_2)$. Hence, our improved scheme could withstand the attack presented in Sect. 4.1.

5.2   Proof of Security

Our proof approach is similar to the Tsai et al.'s proof method [22], but we modify the signing oracle to avoid the flaws given in Sect. 4.2.

Theorem 1. *In the standard model, our IBS scheme is strongly unforgeable against adaptive chosen-message attacks under the CDH assumption.*

Proof. Assume that an adversary $\mathcal{A}$ forges a valid signature for our IBS scheme with a non-negligible probability after making q_E extraction queries and q_S signing queries. Then, we can construct a simulator $\mathcal{B}$ to break the CDH assumption by using $\mathcal{A}$. Let (g, g^a, g^b), where $a, b \in Z_p$, be a random instance of the CDH problem in G_1, which is given to $\mathcal{B}$ as input. The simulator $\mathcal{B}$ performs the following simulation operations, and the goal of $\mathcal{B}$ is to output g^{ab}.

Setup: $\mathcal{B}$ randomly chooses $d \in Z_p$ and computes $v = g^d$. Then, $\mathcal{B}$ sets other public parameters $(G_1, G_2, p, e, g, g_1, g_2, u', u_1, ..., u_m, w', w_1, ..., w_n, H)$ described in Sect. 3.2. Note that $\mathcal{B}$ assigns $g_1 = g^a$ and $g_2 = g^b$. For any identity $ID = (v_1, ..., v_m) \in \{0, 1\}^m$ and message $M = (M_1, ..., M_n) \in \{0, 1\}^n$, we define four functions $F(ID)$, $J(ID)$, $K(M)$ and $L(M)$ in a similar way as in Sect. 3.2. Accordingly, we have two equations $u' \prod_{i=1}^{m} u_i^{v_i} = g_2^{F(ID)} g^{J(ID)}$ and $w' \prod_{j=1}^{n} w_j^{M_j} = g_2^{K(M)} g^{L(M)}$.

Queries: $\mathcal{B}$ simulates the following oracles to answer $\mathcal{A}$ for each query.

Key extraction oracle: On receiving a private key query for an identity $ID = (v_1, ..., v_m) \in \{0, 1\}^m$, $\mathcal{B}$ computes $F(ID)$ and $J(ID)$. If $F(ID) = 0 \bmod l_v$, $\mathcal{B}$ terminates the simulation. If $F(ID) \neq 0 \bmod l_v$, $\mathcal{B}$ randomly selects $r_v \in Z_p$. Next, $\mathcal{B}$ computes $d_1 = (g_1)^{\frac{-J(ID)}{F(ID)}} (u' \prod_{i=1}^{m} u_i^{v_i})^{r_v}$ and $d_2 = g_1^{\frac{-1}{F(ID)}} g^{r_v}$, and then sends the ID's private key $d_{ID} = (d_1, d_2)$ to $\mathcal{A}$.

Signing oracle: On receiving a signature query on message $M = (M_1, ..., M_n) \in \{0, 1\}^n$ for an identity $ID = (v_1, ..., v_m) \in \{0, 1\}^m$, $\mathcal{B}$ computes $F(ID)$ and considers two cases:

– If $F(ID) \neq 0 \bmod l_v$, $\mathcal{B}$ first runs the key extraction oracle to get the ID's private key d_{ID}. Next, $\mathcal{B}$ executes the algorithm **Sign** to generate a signature σ on M, and transmits σ to $\mathcal{A}$.

– If $F(ID) = 0 \bmod l_v$, $\mathcal{B}$ first computes $K(M)$ and $L(M)$. If $K(M) = 0 \bmod l_m$, $\mathcal{B}$ terminates the simulation. Otherwise, $\mathcal{B}$ selects a random value $r_v \in Z_p$. Next, $\mathcal{B}$ randomly selects $r_m \in Z_p$, and then computes $\sigma_3 = (g_1)^{\frac{-1}{K(M)}} g^{r_m}$, $\sigma_2 = g^{r_v}$, $h = H(M, ID, \sigma_2, \sigma_3)$ and

$$\sigma_1 = (u' \prod_{i=1}^{m} u_i^{v_i})^{r_v} (g_1)^{\frac{-L(M)-hd}{K(M)}} ((w' \prod_{j=1}^{n} w_j^{M_j}) v^h)^{r_m}.$$

Finally, $\mathcal{B}$ creates a signature $\sigma = (\sigma_1, \sigma_2, \sigma_3)$ on M, and returns σ to $\mathcal{A}$.

For $r'_m = r_m - \frac{a}{K(M)}$, we have that

$$\sigma_1 = (u' \prod_{i=1}^{m} u_i^{v_i})^{r_v} (g_1)^{\frac{-L(M)-hd}{K(M)}} ((w' \prod_{j=1}^{n} w_j^{M_j}) v^h)^{r_m}$$

$$= (g_2^a)(g_2^{-a})(u' \prod_{i=1}^{m} u_i^{v_i})^{r_v} (g^a)^{\frac{-L(M)-hd}{K(M)}} ((w' \prod_{j=1}^{n} w_j^{M_j}) v^h)^{r_m}$$

$$= (g_2^a)(u' \prod_{i=1}^{m} u_i^{v_i})^{r_v} (g_2^{K(M)} g^{L(M)} g^{hd})^{\frac{-a}{K(M)}} ((w' \prod_{j=1}^{n} w_j^{M_j}) v^h)^{r_m}$$

$$= (g_2^a)(u' \prod_{i=1}^{m} u_i^{v_i})^{r_v} ((w' \prod_{j=1}^{n} w_j^{M_j}) v^h)^{\frac{-a}{K(M)}} ((w' \prod_{j=1}^{n} w_j^{M_j}) v^h)^{r_m}$$

$$= (g_2^a)(u' \prod_{i=1}^{m} u_i^{v_i})^{r_v} ((w' \prod_{j=1}^{n} w_j^{M_j}) v^h)^{r_m - \frac{a}{K(M)}}$$

$$= (g_2^a)(u' \prod_{i=1}^{m} u_i^{v_i})^{r_v} ((w' \prod_{j=1}^{n} w_j^{M_j}) v^h)^{r'_m},$$

$\sigma_3 = (g_1)^{\frac{-1}{K(M)}} g^{r_m} = (g^a)^{\frac{-1}{K(M)}} g^{r_m} = g^{r_m - \frac{a}{K(M)}} = g^{r'_m}$, $\sigma_2 = g^{r_v}$ and $h = H(M, ID, \sigma_2, \sigma_3) = H(M, ID, g^{r_v}, g^{r'_m})$.

In the following, we show that the above signature $\sigma = (\sigma_1, \sigma_2, \sigma_3)$ generated by the simulator is correct since σ satisfies the verification equation of the algorithm **Verify** in our scheme.

$$e(\sigma_1, g) = e((g_2^a)(u' \prod_{i=1}^{m} u_i^{v_i})^{r_v} ((w' \prod_{j=1}^{n} w_j^{M_j}) v^h)^{r'_m}, g)$$

$$= e(g_2, g_1) e(u' \prod_{i=1}^{m} u_i^{v_i}, \sigma_2) e((w' \prod_{j=1}^{n} w_j^{M_j}) v^h, \sigma_3).$$

From the prospective of the adversary $\mathcal{A}$, the above relation shows that the signature produced by the simulator is indistinguishable to the real signature computed by the signer.

Forgery: Suppose that $\mathcal{A}$ outputs a forged signature $\sigma^* = (\sigma_1^*, \sigma_2^*, \sigma_3^*)$ of an identity $ID^* = (v_1^*, ..., v_m^*) \in \{0,1\}^m$ on a message $M^* = (M_1^*, ..., M_n^*) \in \{0,1\}^n$, where ID^* was not queried in key extraction oracle and σ^* was not produced by the signing oracle. If $F(ID^*) \neq 0 \bmod p$ or $K(M^*) \neq 0 \bmod p$, $\mathcal{B}$ declares failure. Otherwise, $\mathcal{B}$ computes $h^* = H(M^*, ID^*, \sigma_2^*, \sigma_3^*)$ and outputs g^{ab} as a solution to the CDH instance (g, g^a, g^b) by calculating

$$
\frac{\sigma_1^*}{(\sigma_2^*)^{J(ID^*)}(\sigma_3^*)^{L(M^*)}(\sigma_3^*)^{h^* d}}
$$

$$
= \frac{(g_2^a)(u' \prod_{i=1}^{m} u_i^{v_i^*})^{r_v^*}((w' \prod_{j=1}^{n} w_j^{M_j^*})v^{h^*})^{r_m^*}}{(g^{r_v^*})^{J(ID^*)}(g^{r_m^*})^{L(M^*)}(g^{r_m^*})^{h^* d}}
$$

$$
= \frac{(g_2^a)(g_2^{F(ID^*)}g^{J(ID^*)})^{r_v^*}(g_2^{K(M^*)}g^{L(M^*)}(g^d)^{h^*})^{r_m^*}}{(g^{r_v^*})^{J(ID^*)}(g^{r_m^*})^{L(M^*)}(g^{r_m^*})^{h^* d}}
$$

$$
= g_2^a \quad (\text{since} \quad F(ID^*) = K(M^*) = 0 \bmod p)
$$

$$
= g^{ab}.
$$

We now analyze the probability that $\mathcal{B}$ does not abort in the above simulation. In the phase of key extraction queries, $\mathcal{B}$ aborts if $F(ID) = 0 \bmod l_v$ for a queried identity ID. In the phase of signing queries, $\mathcal{B}$ aborts if $F(ID) = 0 \bmod l_v$ and $K(M) = 0 \bmod l_m$ for a queried message M and an identity ID. In the forgery phase, $\mathcal{B}$ aborts if $F(ID^*) \neq 0 \bmod p$ or $K(M^*) \neq 0 \bmod p$ for a forged signature σ^* of an identity ID^* on a message M^*. From the analysis of Paterson and Schuldt [9], we know that the probability of $F(ID) \neq 0 \bmod l_v$ and $K(M) \neq 0 \bmod l_m$ is at least $\frac{1}{2}$, and the probability of $F(ID^*) = 0 \bmod p$ and $K(M^*) = 0 \bmod p$ is at least $\frac{1}{2(q_E + q_S)(m+1)}$ and $\frac{1}{2q_S(n+1)}$, respectively. Hence, the probability of $\mathcal{B}$ succeeding in accomplishing the above simulation is at least $\frac{1}{2} \cdot \frac{1}{2} \cdot \frac{1}{2(q_E + q_S)(m+1)} \cdot \frac{1}{2q_S(n+1)} = \frac{1}{16 q_S(q_E + q_S)(m+1)(n+1)}$ (see the proof of Paterson and Schuldt [9] for detailed probability analysis). If $\mathcal{A}$ breaks the strong unforgeability of our IBS scheme with probability ε, $\mathcal{B}$ solves the CDH problem with probability at least $\frac{\varepsilon}{16 q_S(q_E + q_S)(m+1)(n+1)}$.

5.3   Comparison

We compare our IBS scheme with the existing strongly unforgeable IBS schemes in the standard model [19,20,22] from performance and security. The results are shown in Table 1, in which m, n and $|G_1|$ denote the bit length of an identity, a message and an element in G_1, respectively. Also, E and P represent an exponentiation and a pairing operation respectively. We ignore the multiplication and hash operations since they are more efficient when compared with pairing and exponentiation operations [25]. Additionally, the *Size* column shows that the signature contains the number of group elements in G_1. The *Security* column specifies whether the scheme is secure.

Table 1. Comparison of strongly unforgeable IBS scheme

Scheme	Size	Sign	Verify	Security
Sato et al.'s scheme [19]	$5\|G_1\|$	$3E$	$6P$	Secure
Kwon's scheme [20]	$3\|G_1\|$	$3E$	$3P + E$	Insecure
Tsai et al.'s scheme [22]	$3\|G_1\|$	$4E$	$3P + E$	Insecure
Our scheme	$3\|G_1\|$	$3E$	$3P + E$	Secure

Considering computation complexity, the value $e(g_2, g_1)$ can be pre-computed. For the computation cost in both stages of signing and verifying, our scheme is more efficient than Sato et al.'s IBS [19] and Tsai et al.'s IBS [22]. Our scheme has the same signature size and computation cost as the Know's scheme [20], However, the Tsai et al.'s scheme as well as the Know's scheme have some security flaws in their security proofs. As seen from Table 1, our scheme has better performance in terms of signature size and computation cost since it only needs three signature parameters and three exponentiation operations for the signature generation phase. Meanwhile, our scheme covers strong unforgeability, and it is provably secure in the standard model.

We also evaluate the performance of signature generation algorithm of our scheme and Tsai et al.'s scheme. The experiment is conducted in the hardware environment with an Intel core i7-6500 processor (2.5 GHz), 8 GB RAM and 512G hard disk space. The corresponding simulation algorithm using C programming language is implemented on the Windows 10 operating system and the PBC-0.47-VC library. In addition, we select the prime p with the length of 160 bits. To generate a valid signature on any message, our scheme needs 3 exponential operations, while Tsai et al.'s scheme requires 4 exponential operations. As shown in Fig. 1, the time overhead of our scheme is lower than that of Tsai et al.'s scheme. Obviously, our scheme has higher efficiency.

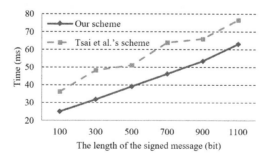

Fig. 1. Comparison of signing time for messages of different lengths

6   Conclusion

In this paper, we give cryptanalysis of a strongly unforgeable IBS scheme proposed by Tsai et al. [22]. We have demonstrated one kind of concrete attack against it. We also shows that the Tsai et al.'s security proof is seriously flawed. Therefore, Tsai et al.'s IBS scheme does not meet the security requirements of strong unforgeability. To solve these security problems, we construct an improved strongly unforgeable IBS scheme without random oracles.

Acknowledgements. This work was partially supported by the National Natural Science Foundation of China (61662069, 61672020, 61472433, 61702309), China Postdoctoral Science Foundation (2017M610817, 2013M542560, 2015T81129), Natural Science Foundation of Gansu Province of China (145RJDA325, 1506RJZA130), Research Fund of Higher Education of Gansu Province (2014-A011), Science and Technology Project of Lanzhou City of China (2013-4-22), Foundation for Excellent Young Teachers by Northwest Normal University (NWNU-LKQN-14-7), Shandong Province Higher Educational Science and Technology Program (No. J16LN61).

References

1. Shamir, A.: Identity-based cryptosystems and signature schemes. In: Blakley, G.R., Chaum, D. (eds.) CRYPTO 1984. LNCS, vol. 196, pp. 47–53. Springer, Heidelberg (1985). https://doi.org/10.1007/3-540-39568-7_5
2. Boneh, D., Franklin, M.: Identity-based encryption from the weil pairing. SIAM J. Comput. **32**(3), 586–615 (2003)
3. Paterson, K.G.: ID-based signatures from pairings on elliptic curves. Electron. Lett. **38**(18), 1025–1026 (2002)
4. Yi, X.: An identity-based signature scheme from the weil pairing. IEEE Commun. Lett. **7**(2), 76–78 (2003)
5. Tseng, Y.M., Wu, T.Y., Wu, J.D.: An efficient and provably secure ID-based signature scheme with batch verifications. Int. J. Innovative Comput. Inf. Control **5**(11), 3911–3922 (2009)
6. Shim, K.A.: An ID-based aggregate signature scheme with constant pairing computations. J. Syst. Softw. **83**(10), 1873–1880 (2010)
7. Sun, X., Li, J., Chen, G.: Identity-based verifiably committed signature scheme without random oracles. J. Shanghai Jiaotong Univ. (Science) **13**(1), 110–115 (2008)
8. Waters, B.: Efficient identity-based encryption without random oracles. In: Cramer, R. (ed.) EUROCRYPT 2005. LNCS, vol. 3494, pp. 114–127. Springer, Heidelberg (2005). https://doi.org/10.1007/11426639_7
9. Paterson, K.G., Schuldt, J.C.N.: Efficient identity-based signatures secure in the standard model. In: Batten, L.M., Safavi-Naini, R. (eds.) ACISP 2006. LNCS, vol. 4058, pp. 207–222. Springer, Heidelberg (2006). https://doi.org/10.1007/11780656_18
10. Narayan, S., Parampalli, U.: Efficient identity-based signatures in the standard model. IET Inf. Secur. **2**(4), 108–118 (2008)
11. Li, F., Gao, W., Wang, G., et al.: Efficient identity-based threshold signature scheme from bilinear pairings in standard model. Int. J. Internet Protocol Technol. **8**(2/3), 107–115 (2014)

12. Sahu, R.A., Padhye, S.: Provable secure identity-based multi-proxy signature scheme. Int. J. Commun. Syst. **28**(3), 497–512 (2015)
13. Hu, X.M., Wang, J., Xu, H.J., et al.: An improved efficient identity-based proxy signature in the standard model. Int. J. Comput. Math. **94**(1), 22–38 (2017)
14. Boneh, D., Shen, E., Waters, B.: Strongly unforgeable signatures based on computational Diffie-Hellman. In: Yung, M., Dodis, Y., Kiayias, A., Malkin, T. (eds.) PKC 2006. LNCS, vol. 3958, pp. 229–240. Springer, Heidelberg (2006). https://doi.org/10.1007/11745853_15
15. Hung, Y.H., Tsai, T.T., Tseng, Y.M., et al.: Strongly secure revocable ID-based Signature without random oracles. Inf. Technol. Control **43**(3), 264–276 (2014)
16. Galindo, D., Herranz, J., Kiltz, E.: On the generic construction of identity-based signatures with additional properties. In: Lai, X., Chen, K. (eds.) ASIACRYPT 2006. LNCS, vol. 4284, pp. 178–193. Springer, Heidelberg (2006). https://doi.org/10.1007/11935230_12
17. Steinfeld, R., Pieprzyk, J., Wang, H.: How to strengthen any weakly unforgeable signature into a strongly unforgeable signature. In: Abe, M. (ed.) CT-RSA 2007. LNCS, vol. 4377, pp. 357–371. Springer, Heidelberg (2006). https://doi.org/10.1007/11967668_23
18. Huang, Q., Wong, D.S., Li, J., et al.: Generic transformation from weakly to strongly unforgeable signatures. J. Comput. Sci. Technol. **23**(2), 240–252 (2008)
19. Sato, C., Okamoto, T., Okamoto, E.: Strongly unforgeable ID-based signatures without random oracles. Int. J. Appl. Cryptogr. **2**(1), 35–45 (2010)
20. Kwon, S.: An identity-based strongly unforgeable signature without random oracles from bilinear pairings. Inf. Sci. **276**, 1–9 (2014)
21. Lee, K., Lee, D.H.: Security analysis of an identity-based strongly unforgeable signature scheme. Inf. Sci. **286**, 29–34 (2014)
22. Tsai, T.T., Tseng, Y.M., Huang, S.S.: Efficient strongly unforgeable ID-based signature without random oracles. Informatica **25**(3), 505–521 (2014)
23. Zhang, L., Wu, Q., Qin, B.: Identity-based verifiably encrypted signatures without random oracles. In: Pieprzyk, J., Zhang, F. (eds.) ProvSec 2009. LNCS, vol. 5848, pp. 76–89. Springer, Heidelberg (2009). https://doi.org/10.1007/978-3-642-04642-1_8
24. Wei, J., Hu, X., Liu, W.: Traceable attribute-based signcryption. Secur. Commun. Netw. **7**(12), 2302–2317 (2014)
25. Cheng, L., Wen, Q., Jin, Z., et al.: Cryptanalysis and improvement of a certificateless aggregate signature scheme. Inf. Sci. **295**, 337–346 (2015)

Encryption

Parallel Long Messages Encryption Scheme Based on Certificateless Cryptosystem for Big Data

Xuguang Wu, Yiliang Han$^{(\boxtimes)}$, Minqing Zhang, and Shuaishuai Zhu

Engineering University of China Armed Police Force, Xian, China
`wxguang0210@gmail.com, hanyil@163.com`

Abstract. In big data environment, the quantity of generated and stored data is huge, and the size is larger than before. A general solution to encrypt large messages is to adopt the hybrid encryption method, that is, one uses an asymmtric cryptosystem to encrypt the symmetric key, and needs a symmetric cryptosystem to encrypt the real message. To eliminate this requirement for an additional cryptosystem, a parallel long message encryption scheme based on certificateless cryptosystem is proposed, which eliminates the needs for public key certificates, and avoids the key escrow problem. In combination with parallel computer hardware, we further improve the performance. The simulation results show that it can make full use of CPU resources and has high efficiency advantages. In the random oracle model, the presented scheme is secure in a One-Way Encryption (OWE) model.

Keywords: Certificateless cryptosystem · Parallel encryption
Long messages encryption · Big data

1 Introduction

In recent years, with the rapid development Internet, Internet of Things, cloud computing, and social networks, big data processing has become crucial to most enterprise and government applications [1]. The emergence of large data changes the way people work and business mode of operation, and even cause a fundamental change in scientific research model. Moreover, the transiting, storing and using of big data faces lots of threats [2]. The collected data often contains private information about individuals or corporate secrets that would cause great harm if they fell into the wrong hands. Criminal groups are creating underground markets where one can buy and sell stolen personal information [3]. This potential for harm is clearly demonstrated by many recent, highly publicized cyber attacks against commercial and government targets [4], costing these organizations millions of dollars and causing serious damage to the individuals and institutions affected.

Security tools especially modern cryptography are needed to protect collecting and handling of big data to allow applications to reap the benefits of big

© Springer International Publishing AG, part of Springer Nature 2018
X. Chen et al. (Eds.): Inscrypt 2017, LNCS 10726, pp. 211–222, 2018.
https://doi.org/10.1007/978-3-319-75160-3_14

data analysis without the risk of such catastrophic attacks. In 1976, Diffie and Hellman of Stanford University published the famous paper "New Directions in Cryptography" [5], and proposed the concept of public key cryptography, that is, using different encryption key and decryption key, so that deriving a decryption key from a known encryption key is computationally infeasible. Compared with traditional cryptosystems, the public key cryptosystem not only solves the complicated problem of key distributions, but also adapts to the demand of society development for digital signature. However, it has the shortcomings of high computational complexity and low efficiency. Moreover, public key cryptography schemes are mostly based on one-way trap-gate permutation, such as RSA [6], ElGamal [7] and Waters encryption scheme [8], etc. But asymmtric cryptosystem has a lower performance, especially in a big data environment, where the data is not only in a wide variety, but also the amount of is also very large. As a result, when encrypting a long message in big data, such as a multimedia file in social networks, it is no longer feasible to simply call the public key algorithm without modification. A general solution for big data encryption is to adopt the encryption method of KEM-DEM [9], which is (1) using public key cryptography to encrypt the key of symmetric cryptosystem, called Key Encapsulation Mechanism (KEM); (2) using a symmetric cryptosystem to encrypt long messages, called Data Encapsulation Mechanism (DEM). This scheme solves the problem of efficiency in long message encryption, as a result of the symmetric encryption scheme with fast operation speed.

In order to eliminate the need for an additional cryptosystem of the hybrid KEM-DEM encryption method, long message encryption scheme based public key cryptography is designed and proposed, which is an alternative to KEM-DEM and achieves a high performance when encrypting long length messages. Hwang et al. proposed the first long message public-key cryptography scheme based on ElGamal [10]. Zhong [11] proposed a new long message encryption scheme, and gives a formal proof of security. In the same year, Wang et al. [12] pointed out the security problem of Hwang's scheme, and further improved it. In 2012, Chang et al. [13] pointed out that paper [10] does not satisfy the security of chosen plaintext attack and proposed a long message encryption scheme which satisfies the selected ciphertext attack security. Based on Elliptic Curve Discrete Logarithm Problem (ECDLP), Debasish et al. [14] proposed an elliptic curve long message encryption scheme. Yang and Liu [15] give two schemes of threshold public key cryptosystem for encrypting long messages. It can be said that the above schemes are based on public key infrastructure (PKI, Public Key Infrastructure) of the public key cryptosystem construction, and it requires a global Certification Authority (CA, Certification Authority) to issue public and private key for every user. The complex certificate management problem, which includes the distribution, storage, revocation and verification of certificates, is brought. These certificate management problems consume a great deal of computational and communication resources, especially in bandwidth-constrained network environments.

To reduce the complexity of key management issues, Liu et al. [16] proposed a long message encryption scheme based on IBE (Identity Based Encryption System), and proved the security under random oracle model. In the identity based public key cryptosystem, the user's public key is its own public identity, such as e-mail address, ID number, IP address and so on, and the private key is issued from the trusted third-party key generation center (PKG, Private Key Generator). Therefore, Liu's algorithm eliminates the certificate requirements, but there is a key escrow problem, that PKG can decrypt any cipher text, and then can forge the user's signature.

Towards long message encryption problem in big data environment, this paper proposes a parallel long message encryption scheme based on certificate encryption cryptosystem. The parallel encryption scheme does not only divide the long message into short pieces and process the short one parallel, just like the symmetric cipher block encryption mode ECB [17], but also the idea of parallel computing is optimized in the algorithm level improving the efficiency. In special cases, it is even more efficient than AES algorithm. And the presented scheme is based on the certificateless public key cryptosystem [18], so it neither needs public key certificate nor has the key escrow problem.

Rest of the paper is organized as follows: in Sect. 2, we introduce the concept of certificateless public key cryptosystem and the idea of parallel cryptography. Section 3 introduces our proposed parallel long messages encryption scheme based on certificateless cryptosystem. Section 4 proves the security. Simulation results show that the proposed scheme has high efficiency. The security analysis is given in Sect. 5. Finally, we conclude the paper in Sect. 6.

2   Relate Works

2.1   Certificateless Public Key Cryptosystem

In 2003, Al-Riyami and Paterson [18] proposed the concept of certificateless public key cryptosystem. There is a trusted third-party key generation center KGC, which owns the master key of the system. Its role is to calculate the partial private key of a user according to its identity. Then it transmits the partial private key to the user securely. After receiving the key, a user calculates the complete private key with its own randomly selected secret value. According to the secret value, identity and system parameters, the user calculate its public key. In such a system, KGC cannot know the private key of any user, thus effectively overcoming the key escrow problem in the identity based cryptosystem. Compared with traditional public key cryptosystems, the certificateless public key cryptosystem does not need the public key certificate. It can be said that the certificateless public key cryptosystem not only combines the advantages of the two cryptosystems well, but also overcomes their shortcomings to a certain extent.

Certificateless public key cryptosystem includes seven algorithms: system establishment, partial private key generation, secret value generation, private key generation, public key generation, encryption and decryption.

System establishment: The algorithm is executed by KGC, which generates the master key and keeps it private. Then it generates and exposes the system parameters *params*.

Partial private key generation: This algorithm inputs the system parameters *params*, the master key and the identity $ID_A \in \{0, 1\}^*$ of user A, and returns a partial private key D_A. Normally, KGC runs the algorithm and passes the private key D_A to user A over a secure channel.

Secret value generation: The algorithm inputs system parameters *params* and user's identity ID_A, and output the user's secret value x_A.

Private key generation: This algorithm enters the system parameters *params*, partial private key of user A D_A and secret value x_A, which is used to convert D_A to the private key S_A of user A, and as the output of the algorithm.

Public key generation: The algorithm inputs system parameters *params* and the secret value x_A of user A to construct the output public key P_A of the user. In normal circumstances, the private key generation algorithm and public key generation algorithm are implemented by the user A. The secret value x_A is used by both algorithms.

Encryption: The system input system parameters *params*, plaintext message $m \in M$, public key P_A and user's identity ID_A, the success of encryption is output to the ciphertext $c \in C$. On the contrary, the encryption failure is output.

Decryption: The algorithm inputs system parameters *params*, ciphertext $c \in C$ and private key S_A, the successful decryption of the output message is $m \in M$. Otherwise, the decryption failure is output.

2.2   Parallel Encryption

Multi-core parallel processing technology only depending on hardware, is unable to play full parallel processing effects, so we need to combine parallel software. Running a parallel computing program, a multi-core CPU allocates the computing tasks to different cores and performs computations simultaneously to achieve spatial and temporal parallelism. It is unlike single-core CPU multi-thread computing, where threads are still executed on a single-core CPU.

Cipher algorithm parallelism is lager behind parallel programming in software, most of current algorithms are still following the serial idea. These algorithms executes in serial sequence in line with people's thinking habits, which is easy to be understood. However, the efficiency cannot be further improved. Pieprzyk and Pointcheval [19] proposed a parallel signcryption scheme, in which a single operation achieves both encryption and signature, while the time does not exceed a single encryption or signature operation. Han et al. [20] proposed a parallel multi-receiver signcryption scheme framework, and a parallel multi-receiver signcryption scheme for multi-user multicast communication environment. Through experiments and comparative analysis, they conclude that multi-core parallel technology can significantly improve the calculation efficiency.

Zhang et al. [21] proposed a re-encryption scheme for parallel multi-receivers in wireless environment, and adopted the public key certificate encryption scheme to solve the problem of secure multicast.

3 Parallel Long Messages Encryption Scheme Based on Certificateless Cryptosystem

In this section, we propose a parallel long messages encryption scheme based on certificateless cryptosystem, which consists of the following algorithms.

System establishment: Choose two groups G_1, G_2 with order q, then we have a bilinear pairing $e : G_1 \times G_1 \to G_2$, where the generator is $P \in G_1$. Select a random number $s \in Z_q{}^*$ as the master key, and compute $P_0 = s \cdot P$. Pick two hash functions $H_1 : \{0,1\}^* \to G_1^*$ and $H_2 : G_1^* \to \{0,1\}^n$, where n is the length of plain texts. KGC publishes the system parameter $params = < G_1, G_2, e, n, P, P_0, H_1, H_2 >$, and keeps $s \in Z_q^*$ secret. Then the plain text message space is $M \in \{0,1\}^{tn}$, while the cipher text message space is $C \in G_1 \times \{0,1\}^{tn}$.

Partial private key generation: With $ID_A \in \{0,1\}^*$, the KGC computes $Q_A = H_1(ID_A) \in G_1^*$, and outputs the partial key $D_A = s \cdot Q_A \in G_1^*$.

Secret value generation: User A selects the random number $x_A \in Z_q^*$ as its secret value.

Private key generation: With the partial key D_A and secret value x_A, the user A computes $S_A = x_A \cdot D_A = x_A \cdot s \cdot Q_A \in G_1^*$.

Public key generation: With the secret value x_A, user A constructs the output public key $P_A = < X_A, Y_A >$, where $X_A = x_A \cdot P$ and $Y_A = x_A \cdot P_0 = x_A \cdot s \cdot P$.

Encryption: In order to encrypt the long message M, a user can divide it into t pieces n-bit partial plain texts $(m_1, m_2, \cdots, m_t)$. Then the t pieces are assigned to $\lceil t/N \rceil$ groups, where N represents the number of CPU cores. The calculations are shown as follows:

(1) Very the equation $e(X_A, P_0) = e(Y_A, P)$, where $X_A, Y_A \in G_1^*$. If the equation does not hold, then the output is $\perp$, and it will interrupt the encryption process.
(2) Compute $Q_A = H_1(ID_A) \in G_1^*$.
(3) Pick $r \in Z_q^*$ randomly, and calculate $U = r \cdot P$ and $W = e(Q_A, Y_A^r)$.
(4) Parallel computation:

```
// Para begin
for each i ∈ [0, N − 1] do
    for each j ∈ [0, ⌈t/N − 1⌉] do
        calculate V_{j+i*N} = m_i ⊕ H_2(W, i);
        publish C_{j+i*N} =< U, V_{j+i*N} >;
    end for
end for
// Para end
```

(5) With $C_{j+i*N} =< U, V_{j+i*N} >$, output the aggregation $C =< C_1, C_2, \cdots,$
    $C_t >=< U, V_1, V_2, \cdots, V_t >$.

Decryption: To decrypt the cipher text $C =< U, V_1, V_2, \cdots, V_t >$, user A performs the calculation as follows:

(1) Calculate common value $w = e(S_A, U)$.
(2) Parallel computation:

   // Para begin
   for each $i \in [0, N-1]$ **do**
      for each $j \in [0, \lceil t/N - 1 \rceil]$ **do**
         Calculate $V_{j+i*N} \oplus H_2(w, j + i * N) = m_{j+i*N}$;
      end for
   end for
   // Para end

(3) Aggregate m_{j+i*N} to M.

4   Experiment

In order to test the performance of the algorithm, this paper has carried on the simulation. We choose the ThinkPad computer X201i as the hardware running platform, whose configuration parameters are Inter® Core I5 CPU 2.53 GHz, 2.00 GB RAM, and 120G solid hard disk. The software is running on Ubuntu 14.04 operating system with GCC 4.6 [22]. The PBC cryptographic library [23] is used to program the cryptographic algorithm. The elliptic curve group is 64 Bytes, and the single plain text is 32 Bytes. The hash algorithm is chosen as Sha256, implemented by the hash library Mhash [24], which is a free (under GNU Lesser GPL) library providing a uniform interface to many hash algorithms.

Experiment I. In this experiment, we compared the computation time of CPU to encrypt the same long messages under the proposed scheme SmartParallel and other public key encryption schemes. There are three types schemes: (1) executing the message encryption in serial sequence, called Serial scheme, (2) simply paralleling processing to the Serial algorithm, called SimpleParallel, (3) optimization of parallel in the algorithm level, called SmartParallel. The plaintext length is 500 KB, 1024 KB and 2048 KB. Experimental results are shown in Table 1. As can be seen from Table 1, this algorithm significantly saves CPU time and increases the computational efficiency.

Experiment II. In this experiment, we compared the CPU computing time of encrypting the same long messages between SmartParallel scheme and AES-128 algorithm. The AES-128 algorithm encrypts 128 bits' plain text per time in the ECB model. From the Fig. 1, we can see that the decryption time of Smart-Parallel scheme is less than AES-128; when the length of message is shorter than 40 KB, our SmartParallel scheme spends more time than AES-128. While, SmartParallel scheme becomes more and more efficient with the increasement

Table 1. Computation time of CPU (ms)

		Length of messages		
		500 KB	1024 KB	2048 KB
Serial	Encryption	41385.8	84931.2	169201.2
	Decryption	23781.3	49509.3	98121.6
SimpleParallel	Encryption	19525	39344.4	79470.5
	Decryption	13610.1	29119.5	57086.9
SmartParallel	Encryption	34.98	56.42	97.32
	Decryption	23.71	46.29	91.89

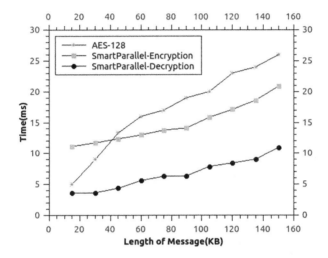

Fig. 1. Computation time of CPU (ms) with increasement of length of message

of cipher text length. Because the common random parameters in both Encryption and Decryption are computed once and reused in parallel, it saves a lot of computations. And the computations of Hash and XOR are executed in different CPU cores at same time, it further saves time.

Experiment III. In this experiment, we compared the CPU utilization of encrypting and decrypting the same long message between Serial scheme and SimpleParallel scheme. Through this experiment, we want to show that parallel computing can make good use of CPU resources. The results are shown in Figs. 2 and 3. Inter Core I5 CPU is the dual core CPU, through virtualization technology it can become to 4-core CPU. Serial scheme only effectively uses parts of CPU cores, while the CPU utilization rate is still very low. It can be seen that only one CPU core's utilization reaches 100% at one time and lasts a short period. SimpleParallel scheme makes 4-core CPU running at full capacity

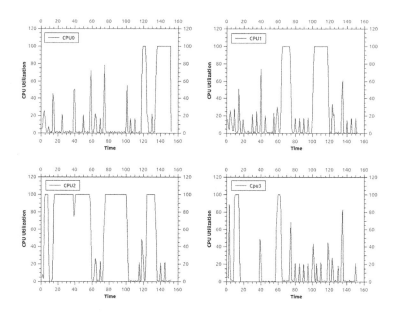

Fig. 2. CPU utilization with the time consume, when encrypting and decrypting long message by *Serial* scheme

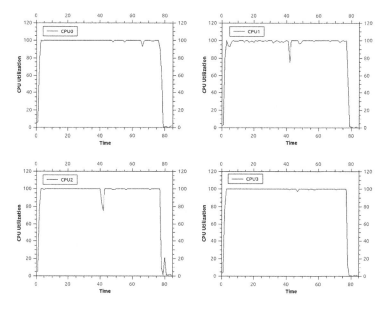

Fig. 3. CPU utilization with the time consume, when encrypting and decrypting long message by *SimpleParallel* scheme

simultaneously, whose utilization is almost 100% for all the lifecycle. Compared with SimpleParallel scheme, SmartParallel scheme is parallel optimized with randomness reusing, and therefore the computation time is further reduced.

5   Security Analysis

In the security analysis of certificateless encryption, there are two types adversaries Type I and Type II [18]. Such an adversary of Type I does not have access to master-key, but it may request public keys and replace public keys with values of its choice, extract partial private and private keys queries. While an adversary of Type II does have access to master-key, but may not replace public keys of entities. And it has the ability to compute partial private keys for itself, given master-key and also request public keys, make private key extraction queries.

In the random oracle model, the presented scheme is secure in a One-Way Encryption (OWE) model [25]. The security analysis process is as follows:

Lemma 1: The parallel certificateless encryption scheme is Type-I secure if H_1 and H_2 are random oracles, and GBDHP assumption [26] holds.

Proof:
Setup: The challenger $\mathbb{C}$ chooses $s \in Z_q^*$ as the master key, computes $P_0 = s \cdot P$, then sends the system parameter $< G_1, G_2, e, n, P, P_0, H_1, H_2 >$ to the Type-I adversary $\mathbb{A}_1$. H_1, H_2 are random oracles.

Phase1:

- Partial-Key-Extract: On receiving a query ID_i to H_1:
  (1) If $< ID_i, e >$ exists in H_1 List, return e as answer.
  (2) Otherwise, pick $e \in G_1^*$ at random, add $< H_1(ID_i), e >$ to H_1 List and return e as answer.
- Private-Key-Extract-Query: Upon receiving a query for a private key of an identity ID_i, $\mathbb{C}$ accesses its database to locate the corresponding entry. If necessary, $\mathbb{C}$ chooses $r_i \in Z_p$ at random, and computes $S_{ID_i} = e \cdot r_i = x_{ID_i} \cdot D_A = x_{ID_i} \cdot s \cdot Q_A \in G_1^*$.
- Public-Key-Request: Upon receiving a query for a public key of an identity ID_i, $\mathbb{C}$ accesses the database DB to locate the corresponding entry. If the corresponding entry does not exist, $\mathbb{C}$ performs first **Secret-value- generation** to choose $r \in Z_q^*$ at random, then the **Public-key-generation** algorithm to obtain secret-public keys $PK_{ID_i} = < X_{ID_i, Y_{ID_i}} >$. Then $\mathbb{C}$ stores the pair keys in its database and returns the public key.
- Replace-Public-Key-Query: When the adversary requests to replace the current public key PK_{ID_i} of an identity ID_i with a new and valid public key PK'_{ID_i} chosen by him, $\mathbb{C}$ accesses its database DB to locate PK_{ID_i}, and replace it with the new public key PK'_{ID_i}. If PK_{ID_i} does not exist, $\mathbb{C}$ directly sets $PK'_{ID_i} = PK_{ID_i}$.
- Challenge: After a polynomial bounded number of queries, $\mathbb{A}_1$ chooses a distinct identity PK_{ID_S} on which he wishes to challenge. $\mathbb{A}_1$ has not asked a private key extraction query with identity ID_S in Phase 1. $\mathbb{C}$ selects the plaintext M_r and splits it into n pieces. Then he picks $r_r \in Z_q^*$ randomly, computes $W = e(Q_{ID_S}, Y_{ID_S}^{r_r})$, and sets $< U = r_r \cdot P, V_i >$, where $V_i = m_j \oplus H_2(W, j), j \in (1, \cdots, n)$.

Phase2: $\mathbb{A}_1$ continues to probe the challenger in Phase 2 with the same type of queries made in Phase 1. The only constraint is that ID_i should never been queried.

Analysis: Assumes H_1 and H_2 are random oracles. The scheme security can be reduced to the difficulty of computing the value $e(Q_{ID_S}, x_{ID_i} \cdot s \cdot P)^{r_r}$. $\mathbb{A}_1$ does not know the master key, but might replace Y_{ID_S} by a new value Y'_{ID_S}. With the input $< P, Q_{ID_S} = a \cdot P, U = r_r \cdot P, P_0 = s \cdot P, V_i = m_i \oplus H_2(W, i) >$, it can be reduced to the GBDHP with solution $Y'_{ID_S}, e(P, Y'_{ID_S})^{a \cdot s \cdot r_r}$.

Lemma 2: The parallel certificateless encryption scheme is Type-II secure if H_1, H_2 are random oracles, and BDHP assumption [27] holds.

Proof:
Setup: The Type II adversary $\mathbb{A}_2$ randomly chooses $\alpha \in Z_q^*$ as the master key, and computes $P_0 = \alpha \cdot P$. Other public parameters are identical to those of Lemma 1. Then $\mathbb{A}_2$ sends the system parameters and the master secret key α to $\mathbb{C}$.

Phase1:

- Private-Key-Extract-Query: When $\mathbb{A}_2$ make a private key query on an identity ID_i, $\mathbb{C}$ accesses its database to locate the corresponding entry (ID_i, sk_{ID_i}). If necessary, $\mathbb{C}$ chooses $r_i \in Z_p$ at random, and computes $S_{ID_i} = x_{ID_i} \cdot r_i \in G_1^*$.
- Public-Key-Request: When $\mathbb{A}_2$ make a public key query on identity ID_i, if the database contains the corresponding entry, $\mathbb{C}$ returns PK_{ID_i}. Otherwise, $\mathbb{C}$ runs algorithm to obtain secret-public keys (sk_{ID_i}, PK_{ID_i}), where $PK_{ID_i} =< X_{ID_i}, Y_{ID_i} >$. Then $\mathbb{C}$ stores the pair keys in its database and returns the public key.
- Challenge: After a polynomial bounded number of queries, $\mathbb{A}_2$ chooses a distinct identity PK_{ID_S} on which he wishes to challenge. $\mathbb{A}_2$ cannot replace public key and extract the private key at any point in Phase 1. $\mathbb{C}$ selects the plain text M_r and splits it into n pieces. Then he picks $r_r \in Z_q^*$ randomly, computes $W = e(Q_{ID_S}, Y_{ID_S}^{r_r})$, and sets $< U = r_r \cdot P, V_i >$, where $V_i = H_2(W, j), j \in (1, \cdots, n)$.

Phase2: $\mathbb{A}_2$ continues to probe the challenger in Phase 2 with the same type of queries made in Phase 1. The only constraint is that ID_i should never been queried.

Analysis: Assumes H_1 and H_2 are random oracles. The scheme security can be reduced to the difficulty of computing the value $e(Q_{ID_S}, x_{ID_i} \cdot s \cdot P)^{r_r}$. $\mathbb{A}_2$ has the master key, but do not know x_{ID_i}. With the input $< P, Q_{ID_S}, U, x_{ID_i}, V_i = m_i \oplus H_2(W, i) >$, it can be reduced to the BDHP.

6   Conclusion

In view of the long message encryption problem, a parallel long message encryption scheme is proposed, which is based on certificateless encryption cryptosystem, eliminating the needs for public key certificates and avoiding the key escrow problem. It is suitful to used in big data environment. We think that it is a useful attempt to design cryptographic scheme combining computer hardware, which is usually ignored by cryptography scholars. In the next work, we will improve the security of the parallel long message encryption scheme.

Acknowledgments. This work is supported by National Cryptology Develepment Foundation of China (No: MMJJ20170112), National Nature Science Foundation of China under grant 61572521, Natural Science Basic Research Plan in Shaanxi Province of China (2015JM6353), and Basic Research Plan of Engineering College of the Chinese Armed Police Force (WJY201523, WJY201613).

References

1. Hashem, I.A.T., Chang, V., Anuar, N.B., Adewole, K., Yaqoob, I., Gani, A., Ahmed, E., Chiroma, H.: The role of big data in smart city. Int. J. Inf. Manage. **36**(5), 748–758 (2016)
2. Hamlin, A., Schear, N., Shen, E., Varia, M., Yakoubov, S., Yerukhimovich, A.: Cryptography for big data security. Big Data Storage Sharing Secur. **3S**, 241–288 (2016)
3. Gurrin, C., Smeaton, A.F., Doherty, A.R.: Lifelogging: personal big data. Found. Trends Inf. Retr. **8**(1), 1–125 (2014)
4. Chen, M., Mao, S., Liu, Y.: Big data: a survey. Mob. Netw. Appl. **19**(2), 171–209 (2014)
5. Diffie, W., Hellman, M.: New directions in cryptography. IEEE Trans. Inf. Theory **22**(6), 644–654 (1976)
6. Rivest, R.L., Shamir, A., Adleman, L.: A method for obtaining digital signatures and public-key cryptosystems. Commun. ACM **26**(1), 96–99 (1983)
7. ElGamal, T.: A public key cryptosystem and a signature scheme based on discrete logarithms. IEEE Trans. Inf. Theory **31**(4), 469–472 (1985)
8. Waters, B.: Efficient identity-based encryption without random oracles. Eurocrypt **3494**, 114–127 (2005)
9. Giri, D., Barua, P., Srivastava, P.D., Jana, B.: A cryptosystem for encryption and decryption of long confidential messages. In: Bandyopadhyay, S.K., Adi, W., Kim, T., Xiao, Y. (eds.) ISA 2010. CCIS, vol. 76, pp. 86–96. Springer, Heidelberg (2010). https://doi.org/10.1007/978-3-642-13365-7_9
10. Hwang, M.S., Chang, C.C., Hwang, K.F.: An ElGamal-like cryptosystem for enciphering large messages. IEEE Trans. Knowl. Data Eng. **14**(2), 445–446 (2002)
11. Zhong, S.: An efficient and secure cryptosystem for encrypting long messages. Fundam. Informaticae **71**(4), 493–497 (2006)
12. Wang, M.N., Yen, S.M., Wu, C.D., Lin, C.T.: Cryptanalysis on an Elgamal-like cryptosystem for encrypting large messages. In: Proceedings of the 6th WSEAS International Conference on Applied Informatics and Communications, pp. 418–422 (2006)
13. Chang, T.Y., Hwang, M.S., Yang, W.P.: Cryptanalysis on an improved version of ElGamal-like public-key encryption scheme for encrypting large messages. Informatica **23**(4), 537–562 (2012)
14. Jena, D., Panigrahy, S.K., Jena, S.K.: A novel and efficient cryptosystem for long message encryption. In: Proceedings of International Conference on Industrial and Information Systems, p. 79 (2009)

15. Yang, G., Liu, J.: Threshold public key cryptosystem for encrypting long messages. J. Comput. Inf. Syst. **11**(2), 671–681 (2015)
16. Liu, J., Zhong, S., Han, L., Yao, H.: An identity-based cryptosystem for encrypting long messages. Int. J. Innov. Comput. Inf. Control **7**(6), 3295–3301 (2011)
17. Ferguson, N., Schneier, B., Kohno, T.: Cryptography engineering: design principles and practical applications (2011)
18. Al-Riyami, S.S., Paterson, K.G.: Certificateless public key cryptography. Asiacrypt **2894**, 452–473 (2003)
19. Pieprzyk, J., Pointcheval, D.: Parallel authentication and public-key encryption. In: Safavi-Naini, R., Seberry, J. (eds.) ACISP 2003. LNCS, vol. 2727, pp. 387–401. Springer, Heidelberg (2003). https://doi.org/10.1007/3-540-45067-X_33
20. Han, Y., Gui, X., Wu, X.: Parallel multi-recipient signcryption for imbalanced wireless networks. Int. J. Innov. Comput. Inf. Control **6**(8), 3621–3630 (2010)
21. Zhang, M., Wu, X., Han, Y., Guo, Y.: Secure group communication based on distributed parallel ID-based proxy re-encryption. In: 2013 32nd Chinese Control Conference (CCC), pp. 6364–6367 (2013)
22. GCC. http://gcc.gnu.org/
23. Lynn, B.: PBC: the pairing-based cryptography library (2011). https://crypto.stanford.edu/pbc/
24. Mavroyanopoulos, N., Schumann, S.: Mhash library. http://mhash.sourceforge.net/
25. Di Crescenzo, G., Ostrovsky, R., Rajagopalan, S.: Conditional oblivious transfer and timed-release encryption. EuroCrypt **99**, 74–89 (1999)
26. Barbosa, M., Farshim, P.: Certificateless signcryption. In: Proceedings of the 2008 ACM Symposium on Information, Computer and Communications Security. ACM (2008)
27. Boneh, D., Franklin, M.: Identity-based encryption from the weil pairing. In: Kilian, J. (ed.) CRYPTO 2001. LNCS, vol. 2139, pp. 213–229. Springer, Heidelberg (2001). https://doi.org/10.1007/3-540-44647-8_13

Constant Decryption-Cost Non-monotonic Ciphertext Policy Attribute-Based Encryption with Reduced Secret Key Size (and Dynamic Attributes)

Geng Wang[✉], Xiao Zhang, and Yanmei Li

Science and Technology on Information Assurance Laboratory,
Beijing 100072, People's Republic of China
cnpkw@126.com

Abstract. Attribute-based encryption, especially ciphertext policy attribute based encryption (CP-ABE), is a standard method for achieving access control using cryptography. The access control policy is determined by access structure in a CP-ABE scheme. If negative permission is required in the access control model, which is a quite common setting, then non-monotonic access structures must be allowed in the CP-ABE scheme.

In 2011, Chen et al. proposed a CP-ABE scheme with non-monotonic access structures that has constant decryption cost. However, it requires a secret key size linear to the number of total attributes, which is hard to implement when the resources are limited for *both* computation and storage. In this paper, we improve this scheme to get a CP-ABE scheme where access structure is non-monotonic AND-gate, while the secret key size is only linear to the number of attributes held by a user, without increasing the decryption cost. This scheme will be useful if the total attributes are much more than attributes for each user. Our scheme is provably secure for selective CPA-security under the decision n-BDHE assumption. We also show that our scheme can be naturally extended to supporting attribute addition and revocation, where the attribute set of each user can be updated dynamically, without any complicated proxy re-encryption or decryption procedure.

Keywords: Attribute-based encryption · Short secret-key
Non-monotonic access structure · Attribute revocation

1 Introduction

Access control, which is used to protect information from unauthorized users, is an important part in achieving information security. However, sometimes the information provider may think the server as vulnerable, and wish to protect the information by encryption, while keeping the access control policy, and avoiding

© Springer International Publishing AG, part of Springer Nature 2018
X. Chen et al. (Eds.): Inscrypt 2017, LNCS 10726, pp. 223–241, 2018.
https://doi.org/10.1007/978-3-319-75160-3_15

complex key distribution matters. Attribute-based encryption (ABE) brought by Sahai and Waters [29] in 2005 is a cryptographic primitive to achieve such a goal, and many ABE schemes [8,9,12,16,18,24,27,28,30] have been proposed since then.

There are two basic types of attribute-based encryption, called key policy attribute-based encryption (KP-ABE) [18] and ciphertext policy attribute-based encryption (CP-ABE) [8]. The former embeds the access policy in the decryption key, while the ciphertext is related to a set of attributes; the latter does the opposite, the access policy is embedded in the ciphertext, and attributes are related to the decryption key, held by the users. We can see that CP-ABE is more flexible for access control model, as attributes can be naturally assigned to a user, such as roles in the role-based access control model.

There are many types of access control policy in practical use, but many of them can be modeled by a combination of positive and negative permissions. Here, positive permission means that a group of users are allowed to access a piece of information, and negative permission means that a group of users are denied of access. Negative permission is often useful, if not necessary. For example, a teaching assistant is allowed to gain access to the score server, but if the teaching assistant is also a student, she/he will not be able to gain such access. This can easily be done by adding a negative permission on the score server for students. Another example is that all employees in a company are voting for some groups, say, financial department, excluding the financial staff themselves. So one can put a negative permission on financial department, without tediously setting positive permission on all other departments.

In a CP-ABE scheme, the access policy is called *access structure*, and access policy with negative permission is often called *non-monotonic access structure*. A type of non-monotonic access structure is called non-monotonic AND-gate, which is represented by the conjunction of a set of either positive or negative attributes. We can see that it naturally represents the concept of positive and negative permissions. The first CP-ABE scheme with non-monotonic AND-gate is presented by Cheung and Newport [12], and their scheme is also proved secure in the standard model instead of the more unreliable generic group model.

Most ABE schemes are based on bilinear groups, which is usually implemented by pairing on elliptic curves, and such operation is quite time consuming. In order to make ABE to be practical on devices where resources are limited, researchers had made great efforts on decreasing the computation cost and communication cost for users. In 2011, Chen, Zhang and Feng [13] presented a CP-ABE scheme with constant-size ciphertext and constant computation-cost. Also, the access structure of their scheme is non-monotonic AND-gate, which is quite expressive as we have claimed. However, followed by the scheme of Cheung and Newport, in their scheme for each user, a key element is given for *every* attribute, no matter the user holds it or not. If a user does not hold an attribute, then a key related to the negative attribute is given to the user.

There are quite a few disadvantages for such treatment. The first obvious one is the storage cost for users. In many applications, the total number of attributes is much greater than the number of attributes held by a user. For example, there are more than 100 different nationalities in the world, while a person can usually have only one or two. It is unwise that each user must store 100+ pieces of private key, instead of two or three, especially when such information is stored in a RFID card, which has very limited memory capacity. Another disadvantage is the lack of flexibility. If a new attribute is added, even it is totally unrelated to the user, its secret key must be updated, which brings unnecessary trouble.

In this paper, we present a CP-ABE scheme, where its secret key size for each user is only related to the number of attribute the user holds, rather than the number of total attributes, in the cost of a no-longer-constant ciphertext size. What more important is that we remain its computation-cost for decryption. And although we have a larger ciphertext, we show that the communication cost is not increased. So our scheme is more useful under cases where all kinds of resources, computation, communication or *storage* are limited for decryption, but not for encryption.

1.1 Dynamic Settings

User revocation has been a tricky task even at the beginning of public key cryptography. There has been a lot of discussion under the context of identity-based encryption (IBE), such as [7], but revocation under ABE is more complex to handle, since the problem is not simply deactivating the secret key of a user, but may only deactivate part of the attributes held by the user. Also, other than subtraction, there may also addition, which gives a user more attributes.

Follow the concept in role-based access control model, we use the term *dynamic* for the scenarios which attribute set of a user can change over time. Note that it is common to use a dynamic model in the real world access control. For example, an employee moves from department A to department B, the administrator may deactivate the attribute of department A, and active the attribute of department B for the user, while other attributes remain unchanged, such as the role of engineer, manager, etc.

There are currently a few solutions for the attribute revocation problem, but they usually require some kind of proxy re-encryption to update on each involved attribute and ciphertext, which sometimes too complicate to use. However, we will see that in our scheme, the dynamic settings can be applied naturally, with only an update on a version number, that is, nothing related to attributes need to be updated. This will bring great convenience on the key management. Also, in our scheme, the update of ciphertext on version number can do either by the encrypter or the server, without publishing any re-encryption key.

1.2 Related Works

After Sahai and Waters brought the idea of ABE [29], the first ABE scheme was proposed by Goyal et al. [18] which is a key-policy scheme. The first ciphertext-policy scheme is constructed by Bethencourt et al. [8]. Schemes of non-monotonic access structures are given independently by Ostrovsky et al. [27] and Cheung and Newport [12], which are KP-APE and CP-ABE, respectively.

For more efficient CP-ABE schemes, CP-ABE with constant ciphertext size and constant decryption cost has been proposed in [14,19], while their access structures allow only threshold gates. In 2011, Chen, Zhang and Feng [13] proposed the first CP-ABE scheme with constant ciphertext size and constant computation cost which allows non-monotonic access structure, that our scheme is mainly based on, which is followed by other works including [10,36]. There are also discussions on reducing the secret key size. In [17,26], the authors proposed CP-ABE schemes with constant secret key size, but the access structure only allow monotonic AND-gates, which is not quite expressive.

In 2014, Attrapadung [1] proposed a generic framework called *pair encoding scheme* which can be used to prove full security for ABE schemes, and there are many following works constructing ABE schemes with short ciphertexts or short secret keys using pair encoding. Among them, [2,3,5] also presented CP-ABE schemes with short secret keys. The technique used in these schemes are quite different from ours. However, these schemes cover only monotonic access structures, and although they also have constant decryption cost, they are not as efficient as our scheme, for example, the scheme presented in [5] uses 8 pairing operations in decryption while we only use 2. In [4], the authors presented a non-monotonic CP-ABE scheme with large universe and constant size ciphertexts, but it is unclear whether it can be modified to get a scheme with short secret keys.

The problem of attribute revocation has been discussed in [20–22,31,35]. These schemes use a proxy to re-encrypt or decrypt the ciphertext. Also there are some discussion on multi-authorized attribute revocation, such as in [33,34]. It is also remarkable, that a revocation system with small private keys for broadcast encryption was proposed in [25], and the authors extended their scheme to ABE settings. However, the ABE version inherits neither the revocation nor the small private keys.

1.3   Organization

This paper is organized as follows: in Sect. 2, we introduce some basic notions which are useful in our discussion. In Sect. 3, we present our CP-ABE scheme, and prove its security. In Sect. 4, we extend our scheme into the dynamic settings. Finally in Sect. 5, we draw the conclusion.

2   Preliminaries

2.1   Bilinear Maps

Let $\mathbb{G}$ and $\mathbb{G}_T$ be two multiplicative cyclic groups of prime order p. Let g be a generator of $\mathbb{G}$, and e be a mapping $e : \mathbb{G} \times \mathbb{G} \to \mathbb{G}_T$. e is called a bilinear map, if it satisfies the following properties:

(1) Bilinearity: for all $u, v \in \mathbb{G}$ and $a, b \in \mathbb{Z}_p$, $e(u^a, v^b) = e(u, v)^{ab}$.
(2) Non-degeneracy: $e(g, g) \neq 1$.

We say that $\mathbb{G}$ is a bilinear group, if the group operation in $\mathbb{G}$ and the bilinear map $e : \mathbb{G} \times \mathbb{G} \rightarrow \mathbb{G}_T$ are both efficiently computable.

2.2   Hardness Assumption

We assume the hardness of the decision n-BDHE problem, which is already shown to be hard in the generic group model [6].

Definition 1. *Let $\mathbb{G}$ be a bilinear group of prime order p, and g, h are two independent generators of $\mathbb{G}$. Denote $\overrightarrow{y}_{g,\theta,n} = (g_1, g_2, ..., g_n, g_{n+2}, ..., g_{2n}) \in \mathbb{G}^{2n-1}$, where $g_i = g^{\theta^i}$ for some unknown $\theta \in \mathbb{Z}_p^*$. An algorithm $\mathcal{B}$ that outputs $\nu \in \{0, 1\}$ has advantage ϵ in solving the n-BDHE problem, if:*

$$|Pr[\mathcal{B}(g, h, \overrightarrow{y}_{g,\theta,n}, e(g_{n+1}, h)) = 1] - Pr[\mathcal{B}(g, h, \overrightarrow{y}_{g,\theta,n}, R) = 1]| \geq \epsilon$$

where the probability is over the random choice of g, h in $\mathbb{G}$, the random choice $\theta \in \mathbb{Z}_p^*$, the random choice of $R \in \mathbb{G}_T$, and the random bits consumed by $\mathcal{B}$. We say that the decision n-BDHE assumption holds in $\mathbb{G}$ if no polynomial algorithm has advantage for some non-negligible ϵ in solving the n-BDHE problem.

3   Non-monotonic CP-ABE with Short Secret Key

We first briefly take an overview on our scheme. Since we give private key elements only on the attributes that the user holds, we must solve the problem that the user hides some of its attributes with ill intent. We solve this by bringing on a "check code", which is related to all attributes held by the user. Without the check code and all its related attributes, the user is unable to decrypt correctly. However, there are also some attributes that are unrelated to the access structure, so we need additional information to get rid of these attributes. We put these information in the encrypted ciphertext, so in the cost of an increasing ciphertext size, we successfully reduce the secret key size.

3.1   Syntactic Definition

First, we specify what we mean by access structure in our scheme. Intuitively speaking, an access structure $\mathbb{W}$ is a boolean formula that returns either 0 or 1 for a given attribute set S. We say that S satisfies $\mathbb{W}$, written as $S \vDash \mathbb{W}$, if and only if $\mathbb{W}$ returns 1 on S. We focus on the non-monotonic AND formulas, which is the conjunction between literals (i or $\neg i$). We also write an access structure $\mathbb{W} = \{X, Y, Z\}$, X, Y, Z are set of attributes, for a set S, $S \vDash \mathbb{W}$ if and only of $X \subseteq S$ and $Y \cap S = \emptyset$. We call X the set of positive attributes, Y the set of negative attributes, and Z is the set of all other attributes we do not care.

In addition, we require that $X \neq \emptyset$, which means that there is at least one positive attribute. Note that if there is not, one could always add a dummy attribute to meet our requirement.

A ciphertext policy attribute-based encryption (CP-ABE) system consists of four algorithms, called **Setup, Encrypt, KeyGen, Decrypt**, which is defined as:

Setup$(1^\lambda, U)$. The setup algorithm takes as input a security parameter 1^λ and an attribute universe U. It outputs a public key PK and a master key MK. The master key is kept secret, and used to generate secret keys for users. The public key is published for encryption.

KeyGen(MK, S). The key generation algorithm takes as input the master key MK and a set of attributes $S \subseteq U$, which is the attribute set held by a user. It returns a secret key SK related to S for the user.

Encrypt$(PK, M, \mathbb{W})$. The encryption algorithm takes as input the public key, a message M, and an access structure $\mathbb{W}$. It returns a ciphertext CT. CT could be decrypted successfully by a secret key SK, if and only if SK is generated by an attribute set S that S satisfies $\mathbb{W}$.

Decrypt(PK, CT, SK). The decryption algorithm takes as input the public key PK, a ciphertext CT, which contains an access structure $\mathbb{W}$, and a secret key SK which is a private key for an attribute set S. It returns the message M if and only if S satisfies $\mathbb{W}$.

3.2   Security Definition

We now describe the selective CPA security model for CP-ABE schemes by a selective CPA security game. In the selective security settings, the adversary will first chooses an access structure $\mathbb{W}^*$ to be challenged, and then asks for private keys with any attribute sets S providing $S \not\models \mathbb{W}^*$, before and after the challenger produces the ciphertext for either M_0 or M_1, where M_0 and M_1 are chosen by the adversary. The adversary wins the game only if it can successfully determine which message, M_0 or M_1, that the ciphertext is encrypted for. The formal definition of CPA-CP-ABE game is as follows.

Init. The adversary chooses the challenge access structure $\mathbb{W}^*$ and gives it to the challenger.

Setup. The challenger runs the **Setup** algorithm and gives the adversary PK.

Phase 1. The adversary submits an attribute set S for a **KeyGen** query. If $S \not\models \mathbb{W}^*$, the challenger answers with a secret key SK for S. These queries can be repeated adaptively.

Challenge. The adversary submits two messages M_0 and M_1 of equal length. The challenger chooses a random bit $\mu \in \{0, 1\}$, and encrypts M_μ under $\mathbb{W}^*$. The encrypted ciphertext CT^* is returned to the adversary.

Phase 2. The adversary repeats **Phase 1** to get more secret keys.

Guess. The adversary outputs a guess μ' for μ.

The advantage of an adversary Adv in the CPA-CP-ABE game is defined by $|Pr[\mu' = \mu] - 1/2|$.

Definition 2. *A ciphertext policy attribute-based encryption scheme is selective CPA secure, if for all polynomial time adversaries, there is at most a negligible advantage in the CPA-CP-ABE game.*

Also, it has been pointed out that a CPA-secure scheme can be converted into a CCA-secure one use the general Fujisaki-Okamoto transformation [15], or some other techniques [11,23,32]. We will not do this in our paper.

3.3 Our Construction

In this section, we present our CP-ABE scheme. For the sake of simplicity, we suppose that our attribute universe $U = \{1, 2, ..., n\}$, n is the total number of attributes. Also we use lower case Greek letters, lower case Latin letters, upper case Latin letters for elements in $\mathbb{Z}_p^*, \mathbb{G}, \mathbb{G}_T$.

Setup$(1^\lambda, U)$. The setup algorithm use the security parameter 1^λ to determine a group $\mathbb{Z}_p^*$ and a cyclic group $\mathbb{G}$ with prime order p. Let g be a generator of $\mathbb{G}$. The algorithm randomly choose $\alpha_1, ..., \alpha_n, \beta_1, ..., \beta_n \in \mathbb{Z}_p^*$ and $h_1, ..., h_n \in \mathbb{G}$. For $i = 1, ..., n$, compute $a_i = g^{\alpha_i}$, $b_i = g^{\beta_i}$, and $H_i = e(g, h_i)$. The public key is $PK := \{\langle a_i, b_i, H_i \rangle | i = 1, ..., n\}$, and the master secret key is $MK := \{\langle \alpha_i, \beta_i, h_i \rangle | i = 1, ..., n\}$.

Encrypt$(PK, M, \mathbb{W})$. Take as input the public key PK, a message M, and an access structure $\mathbb{W} = \{X, Y, Z\}$, where X, Y, Z are positive, negative, and non-relative attributes in $\mathbb{W}$. Choose a random element $\kappa \in \mathbb{Z}_p^*$, and calculate $k = g^\kappa$, $C = M(\prod_{i \in X} H_i)^\kappa$, $u = (\prod_{i \in X} a_i)^\kappa$. Furthermore, for each attribute $i \in Z$, set $z_i = (b_i)^{-\kappa}$. The ciphertext is $CT := \langle \mathbb{W}, k, u, C, \{z_i | i \in Z\} \rangle$.

KeyGen(MK, S). Take as input the master secret key MK and the attribute set S for user. Choose a random element $v \in \mathbb{G}$, for each $i \in S$, calculate $p_i = h_i v^{-\alpha_i - \beta_i}$. Calculate $t = v^{\sum_{i \in S} \beta_i}$. The secret key is $SK := \langle v, \{p_i | i \in S\}, t \rangle$.

Decrypt(PK, CT, SK). The decryption algorithm first check if $S \vDash \mathbb{W}$, if not, returns $\bot$. Otherwise, set $u' = u \prod_{i \in S \cap Z} z_i$. Note that this step can be done by the server to reduce the communication cost. Then, the decryption algorithm calculates $p = t \prod_{i \in X} p_i$, and calculates $M = C/(e(v, u')e(p, k))$, outputs M as the plaintext.

Correctness of the scheme. We have:

$$e(v, u')e(p, k)$$

$$= e(v, (\prod_{i \in X} a_i \prod_{i \in S \cap Z} (b_i)^{-1})^\kappa) e(v^{\sum_{i \in S} \beta_i} \prod_{i \in X} h_i v^{-\alpha_i - \beta_i}, g^\kappa)$$

$$= e(v, (g^{\sum_{i \in X} \alpha_i + \sum_{i \in S \cap Z} -\beta_i})^\kappa)) e(v^{\sum_{i \in S} \beta_i}, g^\kappa) e(\prod_{i \in X} h_i, g^\kappa) e(v^{\sum_{i \in X} -\alpha_i - \beta_i}, g^\kappa)$$

$$= e(v, g^\kappa)^{\sum_{i \in X} \alpha_i - \sum_{i \in S \cap Z} \beta_i + \sum_{i \in S} \beta_i - \sum_{i \in X} (\alpha_i + \beta_i)} e(\prod_{i \in X} h_i, g^\kappa).$$

If $X \subseteq S$ and $S \cap Y = \emptyset$, then $X \cup (S \cap Z) = S$. So the first factor has its exponent 0, and can be omitted. Then, we have:

$$C/e(v, u')e(p, k) = M(\prod_{i \in X} H_i)^\kappa / e(\prod_{i \in X} h_i, g^\kappa)$$

$$= M(\prod_{i \in X} e(h_i, g))^\kappa / e(\prod_{i \in X} h_i, g^\kappa)$$

$$= M.$$

Theorem 1. *The scheme we defined above is selective CPA-secure if the decision n-BDHE assumption holds in $\mathbb{G}$.*

3.4   Security Proof

In this section, we prove the selective CPA-security of our scheme in the standard model under the decision n-BDHE assumption.

Suppose that an adversary Adv can win the CPA-CP-ABE game with a non-negligible advantage ϵ for a CP-ABE scheme describe above with n attributes. We can construct a simulator Sim that uses an n-BDHE challenge to simulate all the requested service for adversary Adv. Then Sim uses output of Adv to solve the decision n-BDHE assumption.

Let $\nu \in \{0, 1\}$ be selected at random. Sim takes as input a random n-BDHE challenge $(g, h, \overrightarrow{y}_{g,\theta,n}, R)$, where $\overrightarrow{y}_{g,\theta,n} = (g_1, g_2, ..., g_n, g_{n+2}, ..., g_{2n}) \in \mathbb{G}^{2n-1}$, $R = e(g_{n+1}, h)$ if $\nu = 1$, and R is a random element in $\mathbb{G}_T$ if $\nu = 0$.

The simulator Sim now plays the role of challenger in the CPA-CP-ABE game, and interacts with Adv as follows.

Init. During the initial phase, Sim receives a challenge gate $\mathbb{W}^* = \{X, Y, Z\}$ specified by the adversary Adv. We denote the attribute index in $X \cup Y$ by $\{i_1, ..., i_m\}$, m is the total attributes involved in the challenged access structure $\mathbb{W}^*$. Without loss of generality, we can assume that $i_1 \in X$ is positive in $\mathbb{W}^*$, since we assumed that there is at least one positive attribute.

Setup. The simulator Sim chooses at random $\alpha_i', \beta_i', \sigma_i' \in \mathbb{Z}_p^*$ for $i = 1, ..., n$.

(a) For $i \in X$, $i \neq i_1$, Sim constructs public key elements (a_i, b_i, H_i) by:

$$a_i = g^{\alpha_i'} g_{n+1-i}^{-1}, b_i = g^{\beta_i'} g_{n+1-i}, H_i = e(g, g)^{\sigma_i'}.$$

(b) For $i \in Y$, Sim constructs public key elements (a_i, b_i, H_i) by:

$$a_i = g^{\alpha_i'} g_{n+1-i}, b_i = g^{\beta_i'} g_{n+1-i}^{-1}, H_i = e(g, g)^{\sigma_i'}.$$

(c) For $i = i_1$, Sim constructs public key elements $(a_{i_1}, b_{i_1}, H_{i_1})$ by:

$$a_{i_1} = g^{\alpha_{i_1}'} \prod_{k \in X - \{i_1\}} g_{n+1-k}, b_{i_1} = g^{\beta_{i_1}'} \prod_{k \in Y \cup \{i_1\}} g_{n+1-k}, H_{i_1} = e(g, g)^{\sigma_{i_1}'} e(g_1, g_n).$$

(d) For other i that $i \notin \{i_1, ..., i_m\}$, Sim constructs public key elements (a_i, b_i, H_i) by: $a_i = g^{\alpha'_i}, b_i = g^{\beta'_i}, H_i = e(g, g)^{\sigma'_i}$.

Phase 1. Adv can submit a few sets $S \subseteq \{1, 2, ..., n\}$ in a secret key query, where $S \not\vdash W^*$, and Sim constructs a set of secret key elements for Adv. Then, there must exists an i', either $i' \in X$ and $i' \notin S$, or $i' \in Y$ and $i' \in S$. If $i_1 \notin S$, we choose $i' = i_1$, otherwise we choose a random i' satisfy the condition above. Sim randomly selects $\tau \in Z_p^*$, and let $v = g_{i'} g^\tau$. Each secret key elements is computed below:

(a) For $i \in S \cap (X \cup Y)$ and $i \neq i_1$, compute: $p_i = g^{\sigma'_i} (g_{i'} g^\tau)^{-\alpha'_i - \beta'_i}$.
    Correctness of this key element: In setup phase, $a_i b_i = g^{\alpha'_i} g^{\beta'_i}$. So $\alpha_i + \beta_i$ should be $\alpha'_i + \beta'_i$, and $H_i = e(g, g)^{\sigma'_i}$, so $h_i = g^{\sigma'_i}$. Then: $p_i = h_i v^{-\alpha_i - \beta_i} = g^{\sigma'_i} (g_{i'} g^\tau)^{-\alpha'_i - \beta'_i}$.

(b) If $i_1 \in S$, compute:

$$p_{i_1} = g^{\sigma_{i_1}} (g_{i'})^{-\alpha_{i_1} - \beta_{i_1}} \Big(\prod_{k \in X \cup Y - \{i'\}} g_{n+1-k+i'}^{-1} \Big) (a_{i_1} b_{i_1})^{-\tau}.$$

Correctness of this key element:

In setup phase, $a_{i_1} b_{i_1} = g^{\alpha'_{i_1}} g^{\beta'_{i_1}} \prod_{k \in X \cup Y} g_{n+1-k}$. So $\alpha_{i_1} + \beta_{i_1} = \alpha'_{i_1} + \beta'_{i_1} + \sum_{k \in X \cup Y} \theta^{n+1-k}$. $H_{i_1} = e(g, g)^{\sigma'_{i_1}} e(g_1, g_n)$, so $h_{i_1} = g^{\sigma'_{i_1}} g^{\theta^{n+1}}$. Then:

$$
\begin{aligned}
p_{i_1} &= h_{i_1} v^{-\alpha_{i_1} - \beta_{i_1}} \\
&= g^{\sigma'_{i_1}} g^{\theta^{n+1}} (g_{i'} g^\tau)^{-\alpha'_{i_1} - \beta'_{i_1} - \sum_{k \in X \cup Y} \theta^{n+1-k}} \\
&= g^{\sigma'_{i_1}} g^{\theta^{n+1}} g_{i'}^{-\alpha'_{i_1} - \beta'_{i_1} - \sum_{k \in X \cup Y} \theta^{n+1-k}} (g^\tau)^{-\alpha'_{i_1} - \beta'_{i_1} - \sum_{k \in X \cup Y} \theta^{n+1-k}} \\
&= g^{\sigma'_{i_1}} g^{\theta^{n+1}} g_{i'}^{-\alpha'_{i_1} - \beta'_{i_1}} g_{i'}^{-\sum_{k \in X \cup Y} \theta^{n+1-k}} (g^{-\alpha'_{i_1} - \beta'_{i_1} - \sum_{k \in X \cup Y} \theta^{n+1-k}})^\tau \\
&= g^{\sigma'_{i_1}} g_{i'}^{-\alpha'_{i_1} - \beta'_{i_1}} (g^{\theta^{n+1}} \prod_{k \in X \cup Y} g_{n+1-k+i'}^{-1}) (a_{i_1} b_{i_1})^{-\tau} \\
&= g^{\sigma'_{i_1}} g_{i'}^{-\alpha'_{i_1} - \beta'_{i_1}} \Big(\prod_{k \in X \cup Y - \{i'\}} g_{n+1-k+i'}^{-1} \Big) (a_{i_1} b_{i_1})^{-\tau}.
\end{aligned}
$$

(c) For $i \in S - (X \cup Y)$, compute: $p_i = g^{\sigma'_i} (g_{i'} g^\tau)^{-\alpha'_i - \beta'_i}$.
    Correctness of this key element: From setup phase, it is obvious that $\alpha_i = \alpha'_i$, $\beta_i = \beta'_i$, and $h_i = g^{\sigma'_i}$. Then: $p_i = h_i v^{-\alpha_i - \beta_i} = g^{\sigma'_i} (g_{i'} g^\tau)^{-\alpha'_i - \beta'_i}$.

(d) If $i_1 \in S$, compute: $t = (\prod_{k \in (X \cap S) \cup (Y - S)} g_{n+1-k+i'})(g_{i'})^{\sum_{k \in S} \beta'_k} (\prod_{k \in S} b_k)^\tau$.

We can see that by our assumption, $i' \notin (X \cap S) \cup (Y - S)$, so that g_{n+1} does not occur, and the key could be correctly generated.
If $i_1 \notin S$, set $i' = i_1$ and compute: $t = (\prod_{k \in S \cap X} g_{n+1-k+i'} \prod_{k \in S \cap Y} g_{n+1-k+i'}^{-1})(g_{i'})^{\sum_{k \in S} \beta'_k} (\prod_{k \in S} b_k)^\tau$.
Since $i' = i_1 \notin S$, g_{n+1} does not occur either.

Correctness of this key element: In setup phase, the public key element b is computed as: $b_{i_1} = g^{\beta'_{i_1}} g_{n+1-i_1} \prod_{k\in Y} g_{n+1-k}$; for $i \in X$ and $i \neq i_1$, $b_i = g^{\beta'_i} g_{n+1-i}$; for $i \in Y$, $b_i = g^{\beta'_i} g^{-1}_{n+1-i}$; and for $i \notin X \cup Y$, $b_i = g^{\beta'_i}$.

So, $\beta_{i_1} = \beta'_{i_1} + \theta^{n+1-i_1} + \sum_{k\in Y} \theta^{n+i-k}$, for $i \neq i_1$ and $i \in X$, $\beta_i = \beta'_i + \theta^{n+1-i}$, for $i \in Y$, $\beta_i = \beta'_i - \theta^{n+1-i}$, and for $i \notin X \cup Y$, $\beta_i = \beta'_i$.

Then, if $i_1 \in S$:

$$
\begin{aligned}
t &= v^{\sum_{i\in S} \beta_i} = (g_{i'} g^\tau)^{\sum_{i\in S} \beta_i} \\
&= g_{i'}^{\sum_{i\in S} \beta_i} (g^\tau)^{\sum_{i\in S} \beta_i} \\
&= g_{i'}^{\sum_{i\in S} \beta'_i + \sum_{i\in X\cap S} \theta^{n+1-i} - \sum_{i\in Y\cap S} \theta^{n+1-i} + \sum_{k\in Y} \theta^{n+1-k}} (g^{\sum_{i\in S} \beta_i})^\tau \\
&= g_{i'}^{\sum_{i\in S} \beta'_i} g_{i'}^{\sum_{k\in (X\cap S)\cup(Y-S)} \theta^{n+1-k}} (g^{\sum_{i\in S} \beta_i})^\tau \\
&= g_{i'}^{\sum_{k\in S} \beta'_k} \left(\prod_{k\in (X\cap S)\cup(Y-S)} g_{i'}^{\theta^{n+1-k}} \right) \left(\prod_{k\in S} b_k \right)^\tau \\
&= g_{i'}^{\sum_{k\in S} \beta'_k} \left(\prod_{k\in (X\cap S)\cup(Y-S)} g_{n+1-k+i'} \right) \left(\prod_{k\in S} b_k \right)^\tau.
\end{aligned}
$$

If $i_1 \notin S$:

$$
\begin{aligned}
t &= v^{\sum_{i\in S} \beta_i} = (g_{i'} g^\tau)^{\sum_{i\in S} \beta_i} \\
&= g_{i'}^{\sum_{i\in S} \beta_i} (g^\tau)^{\sum_{i\in S} \beta_i} \\
&= g_{i'}^{\sum_{i\in S} \beta'_i + \sum_{i\in X\cap S} \theta^{n+1-i} - \sum_{i\in Y\cap S} \theta^{n+1-i}} (g^{\sum_{i\in S} \beta_i})^\tau \\
&= g_{i'}^{\sum_{i\in S} \beta'_i} g_{i'}^{\sum_{k\in X\cap S} \theta^{n+1-k}} g_{i'}^{-\sum_{k\in Y\cap S} \theta^{n+1-k}} (g^{\sum_{i\in S} \beta_i})^\tau \\
&= g_{i'}^{\sum_{k\in S} \beta'_k} \left(\prod_{k\in X\cap S} g_{i'}^{\theta^{n+1-k}} \right) \left(\prod_{k\in Y\cap S} g_{i'}^{-\theta^{n+1-k}} \right) \left(\prod_{k\in S} b_k \right)^\tau \\
&= g_{i'}^{\sum_{k\in S} \beta'_k} \left(\prod_{k\in X\cap S} g_{n+1-k+i'} \right) \left(\prod_{k\in Y\cap S} g^{-1}_{n+1-k+i'} \right) \left(\prod_{k\in S} b_k \right)^\tau.
\end{aligned}
$$

Sim returns to Adv with the secret key $SK = \langle v, \{p_i | i \in S\}, t\rangle$.

Challenge. Adv runs the CPA-CP-ABE game under the public encryption key, submit two messages M_0 and M_1 of equal length. Sim choose $\mu \in \{0, 1\}$ at random. Then Sim computes the challenge ciphertext:

$$
CT^* = (\mathbb{W}^*, k = h, u = h^{\sum_{i\in X} \alpha'_i}, C = M_\mu Re(g, h)^{\sum_{i\in X} \sigma'_i}, \{z_i = h^{\beta_i} | i \in Z\}).
$$

Correctness of the ciphertext when $R = e(g_{n+1}, h)$: Suppose that $h = g^\kappa$.

$$
\begin{aligned}
\prod_{i\in X} a_i &= a_{i_1} \prod_{i\in X-\{i_1\}} a_i = g^{\alpha'_{i_1}} \prod_{k\in X-\{i_1\}} g_{n+1-k} \prod_{i\in X-\{i_1\}} (g^{\alpha'_i} g^{-1}_{n+1-i}) \\
&= g^{\alpha'_{i_1}} \prod_{i\in X-\{i_1\}} g^{\alpha'_i} = \prod_{i\in X} g^{\alpha'_i}.
\end{aligned}
$$

So we have: $u = (\prod_{i \in X} a_i)^\kappa = (g^{\sum_{i \in X} \alpha'_i})^\kappa = (g^\kappa)^{\sum_{i \in X} \alpha'_i} = h^{\sum_{i \in X} \alpha'_i}$.
For C, we have:

$$\prod_{i \in X} H_i = e(g,g)^{\sigma'_{i_1}} e(g_1, g_n) \prod_{i \in X - \{i_1\}} e(g,g)^{\sigma'_i}$$

$$= e(g_{n+1}, g) \prod_{i \in X} e(g,g)^{\sigma'_i} = e(g_{n+1} \prod_{i \in X} g^{\sigma'_i}, g).$$

So:

$$(\prod_{i \in X} H_i)^\kappa = e(g_{n+1} \prod_{i \in X} g^{\sigma'_i}, g)^\kappa = e(g_{n+1} \prod_{i \in X} g^{\sigma'_i}, g^\kappa)$$

$$= e(g_{n+1} \prod_{i \in X} g^{\sigma'_i}, h) = e(g_{n+1}, h) e(\prod_{i \in X} g^{\sigma'_i}, h)$$

$$= e(g_{n+1}, h) e(g, h)^{\sum_{i \in X} \sigma'_i}.$$

So $C = M_\mu R e(g,h)^{\sum_{i \in X} \sigma'_i}$.
For z_i, each $i \notin X \cup Y$, $b_i = g^{\beta'_i}$. So $z_i = b_i^\kappa = (g^{\beta'_i})^\kappa = h^{\beta'_i}$.
So we have that the challenge ciphertext is a valid encryption of M_μ whenever $R = e(g_{n+1}, h)$, on the other hand, when R is uniform and independent in $\mathbb{G}_T$, the challenge ciphertext C^* is independent of μ in Adv's view.

Phase 2. The same as Phase 1.

Guess. If Adv outputs a correct guess after the Challenge phase, Sim answers 1 in the n-BDHE game to guess that $R = e(g_{n+1}, h)$. Otherwise, Sim answers 0 to indicate that it believes R is a random group element in $\mathbb{G}_T$. If $R = e(g_{n+1}, h)$, then CT^* is a valid ciphertext, the simulator Sim gives a perfect simulation so we have that:
$Pr[Sim(g, h, \overrightarrow{y}_{g,\theta,n}, e(g_{n+1}, h)) = 1] = 1/2 + \epsilon$
If R is a random group element, the message M_μ is completely hidden from the adversary, and we have
$Pr[Sim(g, h, \overrightarrow{y}_{g,\theta,n}, R) = 1] = 1/2$
Therefore, Sim can play the decisional n-BDHE game with non-negligible advantage. This concludes our proof.

3.5   Performance Analysis

In this section, we simply discuss the performance of our scheme. We write n as the number of total attributes, m as the number of attributes involved in the access structure, and r the number of attributes the user holds.

In **Setup**, we use $2n$ exponentiations and n pairings, but since it will only be run once, it does not really matter. In each **KeyGen**, we use $r + 1$ exponentiations, and the secret key size is $(r+2)|\mathbb{G}|$. In each **Encrypt**, we use $n - m + 3$ exponentiations, and the ciphertext length is $|\mathbb{G}_T| + (n - m + 2)|\mathbb{G}|$. In each **Decrypt**, we use 2 pairings and no exponentiations. We point out that the calculation of $u' = u \prod_{i \in S \cap Z} z_i$ can be done by the server, so while decryption, the client only needs to download k, u', C rather than the whole ciphertext from the server. Although the ciphertext might be large, the communication cost (size of transmitted data) for decryption is only $|\mathbb{G}_T| + 2|\mathbb{G}|$.

Compared to the CZF scheme [13], our computation cost and communication cost for decryption are the same. Our **KeyGen** use only $r + 1$ exponentiations and has an $(r+2)|\mathbb{G}|$ size of secret key, which could be much more smaller than n exponentiations and $(n + 1)|\mathbb{G}|$ secret key size in the CZF scheme. However, our ciphertext has a larger size and more computation cost. Although our scheme will not show any advantage when user for decrypt has sufficient memory, but as we have shown, our scheme can be useful when the storage for decryption is limited, especially when it is done by a lightweight device. Also, if there is more users than data, which the computation cost for **KeyGen** instead of **Encrypt** is concerned, our scheme can also show its advantages.

4   Dynamic Non-monotonic CP-ABE

Our scheme can be naturally extended to the dynamic settings, which allows attribute update for users. Since we have a check code in the secret key of each user, each time the secret key is updated, it suffices to change the check code only, while keeping the other secret key elements. By such technique, we can do attribute revocation without an explicit revocation algorithm.

We use a version number for the dynamic settings, while each time the version number adds by 1 when there is a user update. All keys and ciphertexts have a version number, and the user can successfully decrypt the ciphertext only if its secret key has the same version number as the ciphertext.

4.1   Syntactic Definition

A dynamic ciphertext policy attribute-based encryption (d-CP-ABE) system consists of 7 algorithms, called **Setup, Encrypt, KeyGen, ReKey, ReEncrypt, ReKeyGen, Decrypt**, which is defined as:

Setup$(1^\lambda, U)$. The setup algorithm takes as input a security parameter 1^λ and an attribute universe U. It outputs a public key PK and a master key MK. The master key is kept secret, and used to generate secret key for users. The public key is published for encryption. The version number of this master key and public key is set to 1.

Encrypt$(PK, M, \mathbb{W})$. The encryption algorithm takes as input the public key, a message M, and an access structure $\mathbb{W}$. It returns a ciphertext CT, with the same version number as PK. CT could be decrypted successfully by a secret key SK, if and only if SK is generated by an attribute set S that S satisfies $\mathbb{W}$, and SK has the same version number with CT. Also, some additional information κ will not be published, but used for the **ReEncrypt** algorithm.

KeyGen(MK, S). The key generation algorithm takes as input the master key MK and a set of attributes $S \subseteq U$, which is the attribute set held by a user.

It returns a secret key SK related to S for the user, and its version number is set to the same as the master key.

ReKey(U^*, MK). This algorithm takes as input the master key MK and a set of attributes U^* which is needed for attribute update, and outputs the new master key MK' and new public key PK'. The version number is added by 1. (We will show that in our scheme, U^* is not needed.)

ReEncrypt(PK, PK', CT, κ). This algorithm takes as input the old public key PK which version number is the same as CT, the new public key PK', the ciphertext CT, and the additional information κ. It outputs a new ciphertext CT' which version number is the same as PK'. This algorithm can also be written as **ReEncrypt**(MK, MK', CT), when the previous encrypter is not available, and the re-encryption must be done by the server. (Since the server uses the master key, κ is not needed.)

ReKeyGen(MK, SK, S'). This algorithm takes as input the master key, the old secret key SK, a set of attributes S' as the attribute change for the user. If the secret key for the user has not been generated yet at the current version number, it returns to user an updated secret key SK' which version number is the same as MK.

Decrypt(PK, CT, SK). The decryption algorithm takes as input the public key PK, a ciphertext CT, which contains an access structure $\mathbb{W}$, and a secret key SK which is a private key for an attribute set S. It returns the message M if and only if S satisfies $\mathbb{W}$ and CT has the same version number with PK and SK.

4.2 Security Definition

We now describe the selective CPA security model for dynamic CP-ABE schemes by a selective dynamic CPA security game, which is similar to the non-dynamic settings. The formal definition of d-CPA-CP-ABE game is as follows.

Init. The adversary chooses the challenge access structure $\mathbb{W}^*$ and a version number ver^*, gives them to the challenger.

Setup. The challenger runs the **Setup** algorithm, and **ReKey** repeatedly, until the version number reaches ver^*, and gives the adversary all ver^* versions of PK.

Phase 1. The adversary submits an attribute set S for a **KeyGen** query, and query **ReKeyGen** repeatedly until the version number reaches ver^* to get all ver^* versions of SK. The only restriction is that $S \not\models \mathbb{W}^*$ at version number ver^*. These queries can be repeated adaptively.

Challenge. The adversary submits two messages M_0 and M_1 of equal length. The challenger choose a random bit $\mu \in \{0, 1\}$, and encrypt M_μ under $\mathbb{W}^*$ and version number ver^*. The encrypted ciphertext CT^* is returned to the adversary.

Phase 2. The adversary repeats **Phase 1** to get more secret keys.

Guess. The adversary outputs a guess μ' for μ.

The advantage of an adversary Adv in the d-CPA-CP-ABE game is defined by $|Pr[\mu' = \mu] - 1/2|$.

Definition 3. *A dynamic ciphertext policy attribute-based encryption scheme is selective CPA secure, if for all polynomial time adversaries, there is at most a negligible advantage in the d-CPA-CP-ABE game.*

4.3   Our Construction

In this section, we present our dynamic CP-ABE scheme. This scheme is based on our non-dynamic scheme in Sect. 3.3. We have an element related to the version number in both the public key and the master key, and only one element in each user secret key and ciphertext is related to the version number. So the computational cost is low for each secret key/ciphertext update. By merging the version number into the "check code" t, the user must provide exactly the attributes it has at the current version number, so it cannot use any revoked attributes even there is no explicit attribute revocation.

Setup$(1^\lambda, U)$. The setup algorithm uses the security parameter 1^λ to determine a group $\mathbb{Z}_p^*$ and a cyclic group $\mathbb{G}$ with prime order p. Let g be a generator of $\mathbb{G}$. The algorithm randomly chooses $\alpha_1, ..., \alpha_n, \beta_1, ..., \beta_n, \omega \in \mathbb{Z}_p^*$ and $h_1, ..., h_n \in \mathbb{G}$. For $i = 1, ..., n$, compute $a_i = g^{\alpha_i}$, $b_i = g^{\beta_i}$, $w = g^\omega$ and $H_i = e(g, h_i)$. The public key is $PK := \langle \{\langle a_i, b_i, H_i \rangle | i = 1, ..., n\}, w \rangle$, and the master secret key is $MK := \langle \{\langle \alpha_i, \beta_i, h_i \rangle | i = 1, ..., n\}, \omega \rangle$.

Encrypt$(PK, M, \mathbb{W})$. Take as input the public key PK, a message M, and an access structure $\mathbb{W} = \{X, Y, Z\}$, where X, Y, Z are positive, negative, and non-relative attributes in $\mathbb{W}$. Choose a random element $\kappa \in \mathbb{Z}_p^*$, and calculate $k = g^\kappa$, $C = M(\prod_{i \in X} H_i)^\kappa$, $u = (w \prod_{i \in X} a_i)^\kappa$. Furthermore, for each attribute $i \in Z$, set $z_i = (b_i)^{-\kappa}$. The ciphertext is $CT := \langle \mathbb{W}, k, u, C, \{z_i | i \in Z\} \rangle$, and κ is kept for **ReEncrypt**.

KeyGen(MK, S). Take as input the master key MK and the attribute set S for user. Choose a random element $v \in \mathbb{G}$, for each $i \in S$, calculate $p_i = h_i v^{-\alpha_i - \beta_i}$. Calculate $t = v^{\sum_{i \in S} \beta_i - \omega}$. The secret key is $SK := \langle v, \{p_i | i \in S\}, t \rangle$.

ReKey. Generate a new random number $\omega' \in \mathbb{Z}_p^*$, and calculate $w' = g^{\omega'}$. Replace the master key element ω in MK by ω' and the public key element w in PK by w' to get the new master key MK' and new public key PK'. Return MK' and PK'.

ReEncrypt(PK, PK', CT, κ). Suppose the last key element of PK is w and for PK' is w'. Then update the ciphertext element u into $u(w'/w)^\kappa$, while other ciphertext elements remain the same. Note that this can also be done by the server holding MK by calculating $uk^{\omega' - \omega}$, when the encrypter is not currently available.

ReKeyGen(MK, SK, S'). Let S be the attribute set of SK. For each attribute $i \in S' - S$, generate new key element $p_i = h_i v^{-\alpha_i - \beta_i}$. Calculate $t = v^{\sum_{i \in S'} \beta_i - \omega}$, and return them to the user. (Note that we do not need explicit revocation for attributes in $S - S'$.)

Decrypt(PK, CT, SK). The decryption algorithm first check if $S \models \mathbb{W}$ and there is a same version number between CT and SK. If not, returns $\perp$. Otherwise, set $u' = u \prod_{i \in S \cap Z} z_i$. Note that this step can be done by the server to reduce the communication cost. Then, the decryption algorithm calculates $p = t \prod_{i \in X} p_i$, and calculates $M = C/(e(v, u')e(p, k))$, outputs M as the plaintext.

Correctness of the scheme:

$$e(v, u')e(p, k)$$

$$= e(v, (w \prod_{i \in X} a_i \prod_{i \in S \cap Z} (b_i)^{-1})^\kappa) e(v^{\sum_{i \in S} \beta_i - \omega'} \prod_{i \in X} h_i v^{-\alpha_i - \beta_i}, g^\kappa)$$

$$= e(v, (g^{\omega + \sum_{i \in X} \alpha_i + \sum_{i \in S \cap Z} -\beta_i)\kappa}))$$

$$e(v^{\sum_{i \in S} \beta_i - \omega'}, g^\kappa) e(\prod_{i \in X} h_i, g^\kappa) e(v^{\sum_{i \in X} -\alpha_i - \beta_i}, g^\kappa)$$

$$= e(v, g^\kappa)^{\omega + \sum_{i \in X} \alpha_i - \sum_{i \in S \cap Z} \beta_i + \sum_{i \in S} \beta_i - \omega' - \sum_{i \in X} (\alpha_i + \beta_i)} e(\prod_{i \in X} h_i, g^\kappa)$$

If $X \subseteq S$ and $S \cap Y = \emptyset$, then $X \cup (S \cap Z) = S$. So when $\omega = \omega'$, which means the version numbers of CT and SK are the same, the first term has its exponent 0, and can be omitted. Then, we have:

$$C/e(v, u')e(p, k) = M(\prod_{i \in X} H_i)^\kappa / e(\prod_{i \in X} h_i, g^\kappa)$$

$$= M(\prod_{i \in X} e(h_i, g))^\kappa / e(\prod_{i \in X} h_i, g^\kappa) = M.$$

Theorem 2. *The scheme we defined above is selective CPA-secure under generic group model.*

Security proof of this scheme will be given in the appendix.

5   Conclusion and Future Work

In this paper, we present a CP-ABE scheme with non-monotonic AND-gate access structure, which has constant decryption cost and reduced secret key size. Each decryption only requires two pairing and no exponential computation, the communication cost is also constant, and the secret key size is linear only to the attributes held by the user. By reducing the secret key size while retaining the low decryption cost, our scheme is useful at the situations where all resources including storage is limited for user decryption. Also, we show that our scheme can be naturally extended into a dynamic scheme, which allows dynamic update on user attributes, with a very low computational overhead.

There are more things we can do to extend our scheme. Although non-monotonic AND-gate access structures are expressive enough, it is still an interesting problem to extend our scheme into more expressive non-monotonic boolean functions as its access structures. Also our security definition is under the selective model, we can extend it to gain full security. Note that the security proof of our dynamic scheme is only under generic group model, so it will be an important work to prove it under standard model.

Acknowledgement. This work is partially supported by Foundation of Science and Technology on Information Assurance Laboratory under Grant 6142112010202.

A Security for Our Dynamic Scheme

We shall give the security proof sketch for our dynamic scheme under generic group model. Since the challenged ciphertext $C = M_\mu(\prod_{i \in X} H_i)^\kappa$, $\mu \in \{0, 1\}$, it suffices to show that given all information in the d-CPA-CP-ABE game for Adv, $(\prod_{i \in X} H_i)^\kappa$ is uniformly random under Adv's view.

We first list all elements of $\mathbb{G}$ and $\mathbb{G}_T$ that Adv can get from the game, and then analyse the elements Adv can get from the oracles assumed by the generic group model. In the d-CPA-CP-ABE game, Adv is provided with (for each w at version number ver, we distinguish it among others by w_{ver}):

In $\mathbb{G}$: $a_1, ..., a_n, b_1, ..., b_n, w_1, ..., w_{ver^*}$,
$\{(v)_k, (p_1)_k, ..., (p_n)_k, (t_1)_k, ..., (t_{ver^*})_k | (v)_k ... \text{ is } v, ... \text{ in the } k\text{-th query}\}$, $k, u, \{z_i | i \in Z\}$.

In $\mathbb{G}_T$: $H_1, ..., H_n$.

We can write each element by its exponent. For the given generator g of $\mathbb{G}$, we write an element in $\mathbb{G}$, g^i by i, and an element in $\mathbb{G}_T$, $e(g, g)^j$ by j. We need to include some new variables. We suppose that $h_i = g^{\sigma_i}$, so $H_i = e(g, g)^{\sigma_i}$. Also we suppose that $(v)_k = g^{v_k}$.

Then, we write the elements Adv can get by:

In $\mathbb{G}$: $1, \alpha_1, ..., \alpha_n, \beta_1, ..., \beta_n, \omega_1, ..., \omega_{ver^*}$,
$\{v_k, \{\sigma_i - v_k(\alpha_i + \beta_i) | i = 1, ..., n\}, \{v_k(\sum_{i \in S_j} \beta_i - \omega_j) | j = 1, ..., ver^*\} | k = 1, ..., s\}$, s is the number of total private key queries,
$\kappa, \kappa(\omega_{ver^*} + \sum_{i \in X} \alpha_i), \{\kappa\beta_i | i \in Z\}$.

In $\mathbb{G}_T$: $\sigma_1, ..., \sigma_n$. Also the required $(\prod_{i \in X} H_i)^\kappa$ could be written by $\kappa \sum_{i \in X} \sigma_i$.

The only restriction on them is that either $X \nsubseteq S_{ver^*}$ or $Y \cap S_{ver^*} \neq \emptyset$.

There are two kinds of oracles in the generic group model: one can call the mapping oracle e which maps two elements a, b in $\mathbb{G}$ into element ab in $\mathbb{G}_T$, and the group operation oracle which maps two elements a, b in $\mathbb{G}$ (or $\mathbb{G}_T$) into an element $a + b$ in $\mathbb{G}$ (or $\mathbb{G}_T$, respectively). We show that it is unable for Adv to get $\kappa \sum_{i \in X} \sigma_i$ by these oracles. Without loss of generality, we assume that the mapping oracle is always called between two elements which Adv gotten from the game.

For each $i \in X$, in order to get the term $\kappa\sigma_i$, Adv must at least call the mapping oracle once with two elements, one has the term κ and the other has the term σ_i. The only element contains term κ is κ itself, and the element contains the term σ_i is $\sigma_i - v_k(\alpha_i + \beta_i)$ for any k. Then, Adv must further get the term $\kappa v_k \alpha_i$ and $\kappa v_k \beta_i$ from other mapping oracle calls. The former can only be get from $(\kappa(\omega_{ver^*} + \sum_{i \in X} \alpha_i), v_k)$, while the latter can only be get from $(\kappa, v_k(\sum_{i \in S_j} \beta_i - \omega_j))$ for some j. Here, j must be equal to ver^*, or there is no way for Adv to get rid of the term $\kappa v_k \omega_j$.

Note that although Adv can the term $\kappa\sigma_i$ for each $i \in X$ from $(\kappa, \sigma_i - v_k(\alpha_i+\beta_i))$ with different ks, but the additional terms in $(\kappa(\omega_{ver^*}+\sum_{i \in X}\alpha_i), v_k)$ requires that all k to be same.

We sum up all the oracle calls Adv must make, and get:

$$\sum_{i \in X} \kappa(\sigma_i - v_k(\alpha_i + \beta_i)) + \kappa v_k(\omega_{ver^*} + \sum_{i \in X} \alpha_i) + \kappa v_k\left(\sum_{i \in S_{ver^*}} \beta_i - \omega_{ver^*}\right)$$

$$= \sum_{i \in X} \kappa\sigma_i + \sum_{i \in S_{ver^*}-X} \kappa v_k\beta_i - \sum_{i \in X-S_{ver^*}} \kappa v_k\beta_i.$$

The additional term of $\kappa v_k\beta_i$ must be gotten from mapping queries different from those above, which means that only from $(\kappa\beta_i, v_k)$ where $i \in Z$. So we must have that $(S_{ver^*} - X) \cup (X - S_{ver^*}) \subseteq Z$. But $X \cap Z = \emptyset$, so we must have $X - S_{ver^*} = \emptyset$, which means $S_{ver^*} \subseteq X$. Also, $Y \cap Z = \emptyset$, so $(S_{ver^*} - X) \cap Y = \emptyset$. But $X \cap Y = \emptyset$, so $S_{ver^*} \cap Y = \emptyset$. This contradicts with our requirement for S_{ver^*}.

The only chance for Adv to get the hidden elements, is that there happen to be two same elements in two different oracle calls. Suppose that the number of total oracle calls is q, so the advantage for Adv cannot be greater than $O(q^2/p)$.

References

1. Attrapadung, N.: Dual system encryption via doubly selective security: framework, fully secure functional encryption for regular languages, and more. In: Nguyen, P.Q., Oswald, E. (eds.) EUROCRYPT 2014. LNCS, vol. 8441, pp. 557–577. Springer, Heidelberg (2014). https://doi.org/10.1007/978-3-642-55220-5_31
2. Attrapadung, N.: Dual System Encryption Framework in Prime-Order Groups. IACR Cryptology ePrint Archive 2015 (2015). 390
3. Attrapadung, N., Hanaoka, G., Matsumoto, T., Teruya, T., Yamada, S.: Attribute based encryption with direct efficiency tradeoff. In: Manulis, M., Sadeghi, A.-R., Schneider, S. (eds.) ACNS 2016. LNCS, vol. 9696, pp. 249–266. Springer, Cham (2016). https://doi.org/10.1007/978-3-319-39555-5_14
4. Attrapadung, N., Hanaoka, G., Yamada, S.: Conversions among several classes of predicate encryption and applications to ABE with various compactness tradeoffs. In: Iwata, T., Cheon, J.H. (eds.) ASIACRYPT 2015. LNCS, vol. 9452, pp. 575–601. Springer, Heidelberg (2015). https://doi.org/10.1007/978-3-662-48797-6_24
5. Attrapadung, N., Yamada, S.: Duality in ABE: converting attribute based encryption for dual predicate and dual policy via computational encodings. In: Nyberg, K. (ed.) CT-RSA 2015. LNCS, vol. 9048, pp. 87–105. Springer, Cham (2015). https://doi.org/10.1007/978-3-319-16715-2_5
6. Boneh, D., Boyen, X., Goh, E.-J.: Hierarchical identity based encryption with constant size ciphertext. In: Cramer, R. (ed.) EUROCRYPT 2005. LNCS, vol. 3494, pp. 440–456. Springer, Heidelberg (2005). https://doi.org/10.1007/11426639_26
7. Boldyreva, A., Goyal, V., Kumar, V.: Identity-based encryption with efficient revocation. In: Proceedings of the 15th ACM Conference on Computer and Communications Security, pp. 417–426. ACM (2008)

8. Bethencourt, J., Sahai, A., Waters, B.: Ciphertext-policy attribute-based encryption. In: IEEE Symposium on Security and Privacy, SP 2007, pp. 321–334. IEEE (2007)

9. Chase, M.: Multi-authority attribute based encryption. In: Vadhan, S.P. (ed.) TCC 2007. LNCS, vol. 4392, pp. 515–534. Springer, Heidelberg (2007). https://doi.org/10.1007/978-3-540-70936-7_28

10. Chen, C., Chen, J., Lim, H.W., Zhang, Z., Feng, D., Ling, S., Wang, H.: Fully secure attribute-based systems with short ciphertexts/signatures and threshold access structures. In: Dawson, E. (ed.) CT-RSA 2013. LNCS, vol. 7779, pp. 50–67. Springer, Heidelberg (2013). https://doi.org/10.1007/978-3-642-36095-4_4

11. Canetti, R., Halevi, S., Katz, J.: Chosen-ciphertext security from identity-based encryption. In: Cachin, C., Camenisch, J.L. (eds.) EUROCRYPT 2004. LNCS, vol. 3027, pp. 207–222. Springer, Heidelberg (2004). https://doi.org/10.1007/978-3-540-24676-3_13

12. Cheung, L., Newport, C.: Provably secure ciphertext policy ABE. In: Proceedings of the 14th ACM Conference on Computer and Communications Security, pp. 456–465. ACM (2007)

13. Chen, C., Zhang, Z., Feng, D.: Efficient ciphertext policy attribute-based encryption with constant-size ciphertext and constant computation-cost. In: Boyen, X., Chen, X. (eds.) ProvSec 2011. LNCS, vol. 6980, pp. 84–101. Springer, Heidelberg (2011). https://doi.org/10.1007/978-3-642-24316-5_8

14. Emura, K., Miyaji, A., Nomura, A., Omote, K., Soshi, M.: A ciphertext-policy attribute-based encryption scheme with constant ciphertext length. In: Bao, F., Li, H., Wang, G. (eds.) ISPEC 2009. LNCS, vol. 5451, pp. 13–23. Springer, Heidelberg (2009). https://doi.org/10.1007/978-3-642-00843-6_2

15. Fujisaki, E., Okamoto, T.: How to enhance the security of public-key encryption at minimum cost. In: Imai, H., Zheng, Y. (eds.) PKC 1999. LNCS, vol. 1560, pp. 53–68. Springer, Heidelberg (1999). https://doi.org/10.1007/3-540-49162-7_5

16. Goyal, V., Jain, A., Pandey, O., Sahai, A.: Bounded ciphertext policy attribute based encryption. In: Aceto, L., Damgård, I., Goldberg, L.A., Halldórsson, M.M., Ingólfsdóttir, A., Walukiewicz, I. (eds.) ICALP 2008. LNCS, vol. 5126, pp. 579–591. Springer, Heidelberg (2008). https://doi.org/10.1007/978-3-540-70583-3_47

17. Guo, F., Mu, Y., Susilo, W., et al.: CP-ABE with constant-size keys for lightweight devices. IEEE Trans. Inf. Forensics Secur. 9(5), 763–771 (2014)

18. Goyal, V., Pandey, O., Sahai, A., et al.: Attribute-based encryption for fine-grained access control of encrypted data. In: Proceedings of the 13th ACM Conference on Computer and Communications Security, pp. 89–98. ACM (2006)

19. Herranz, J., Laguillaumie, F., Ràfols, C.: Constant size ciphertexts in threshold attribute-based encryption. In: Nguyen, P.Q., Pointcheval, D. (eds.) PKC 2010. LNCS, vol. 6056, pp. 19–34. Springer, Heidelberg (2010). https://doi.org/10.1007/978-3-642-13013-7_2

20. Hur, J., Noh, D.K.: Attribute-based access control with efficient revocation in data outsourcing systems. IEEE Trans. Parallel Distrib. Syst. 22(7), 1214–1221 (2011)

21. Ibraimi, L., Petkovic, M., Nikova, S., Hartel, P., Jonker, W.: Mediated ciphertext-policy attribute-based encryption and its application. In: Youm, H.Y., Yung, M. (eds.) WISA 2009. LNCS, vol. 5932, pp. 309–323. Springer, Heidelberg (2009). https://doi.org/10.1007/978-3-642-10838-9_23

22. Jahid, S., Mittal, P., Borisov, N.: EASiER: encryption-based access control in social networks with efficient revocation. In: Proceedings of the 6th ACM Symposium on Information, Computer and Communications Security, pp. 411–415. ACM (2011)

23. Lai, J., Deng, R.H., Liu, S., Kou, W.: Efficient CCA-secure PKE from identity-based techniques. In: Pieprzyk, J. (ed.) CT-RSA 2010. LNCS, vol. 5985, pp. 132–147. Springer, Heidelberg (2010). https://doi.org/10.1007/978-3-642-11925-5_10

24. Lewko, A., Okamoto, T., Sahai, A., Takashima, K., Waters, B.: Fully secure functional encryption: attribute-based encryption and (Hierarchical) inner product encryption. In: Gilbert, H. (ed.) EUROCRYPT 2010. LNCS, vol. 6110, pp. 62–91. Springer, Heidelberg (2010). https://doi.org/10.1007/978-3-642-13190-5_4

25. Lewko, A., Sahai, A., Waters, B.: Revocation systems with very small private keys. In: IEEE Symposium on Security and Privacy (SP), pp. 273–285. IEEE (2010)

26. Odelu, V., Das, A.K., Rao, Y.S., et al.: Pairing-based CP-ABE with constant-size ciphertexts and secret keys for cloud environment. Comput. Stan. Interfaces (2016)

27. Ostrovsky, R., Sahai, A., Waters, B.: Attribute-based encryption with non-monotonic access structures. In: Proceedings of the 14th ACM Conference on Computer and Communications Security, pp. 195–203. ACM (2007)

28. Rouselakis, Y., Waters, B.: Practical constructions and new proof methods for large universe attribute-based encryption. In: Proceedings of the 2013 ACM SIGSAC Conference on Computer and Communications Security, pp. 463–474. ACM (2013)

29. Sahai, A., Waters, B.: Fuzzy identity-based encryption. In: Cramer, R. (ed.) EUROCRYPT 2005. LNCS, vol. 3494, pp. 457–473. Springer, Heidelberg (2005). https://doi.org/10.1007/11426639_27

30. Waters, B.: Ciphertext-policy attribute-based encryption: an expressive, efficient, and provably secure realization. In: Catalano, D., Fazio, N., Gennaro, R., Nicolosi, A. (eds.) PKC 2011. LNCS, vol. 6571, pp. 53–70. Springer, Heidelberg (2011). https://doi.org/10.1007/978-3-642-19379-8_4

31. Wang, G., Liu, Q., Wu, J.: Hierarchical attribute-based encryption for fine-grained access control in cloud storage services. In: Proceedings of the 17th ACM Conference on Computer and Communications Security, pp. 735–737. ACM (2010)

32. Yamada, S., Attrapadung, N., Hanaoka, G., Kunihiro, N.: Generic constructions for chosen-ciphertext secure attribute based encryption. In: Catalano, D., Fazio, N., Gennaro, R., Nicolosi, A. (eds.) PKC 2011. LNCS, vol. 6571, pp. 71–89. Springer, Heidelberg (2011). https://doi.org/10.1007/978-3-642-19379-8_5

33. Yang, K., Jia, X.: DAC-MACS: effective data access control for multi-authority cloud storage systems. In: Security for Cloud Storage Systems, pp. 59–83. Springer, New York (2014). https://doi.org/10.1007/978-1-4614-7873-7_4

34. Yang, K., Jia, X., Ren, K.: Attribute-based fine-grained access control with efficient revocation in cloud storage systems. In: Proceedings of the 8th ACM SIGSAC Symposium on Information, Computer and Communications Security, pp. 523–528. ACM (2013)

35. Yu, S., Wang, C., Ren, K., et al.: Attribute based data sharing with attribute revocation. In: Proceedings of the 5th ACM Symposium on Information, Computer and Communications Security, pp. 261–270. ACM (2010)

36. Zhang, Y., Zheng, D., Chen, X., Li, J., Li, H.: Computationally efficient ciphertext-policy attribute-based encryption with constant-size ciphertexts. In: Chow, S.S.M., Liu, J.K., Hui, L.C.K., Yiu, S.M. (eds.) ProvSec 2014. LNCS, vol. 8782, pp. 259–273. Springer, Cham (2014). https://doi.org/10.1007/978-3-319-12475-9_18

Fully Homomorphic Encryption Scheme Based on Public Key Compression and Batch Processing

Liquan Chen[1(✉)], Ming Lim[2], and Muyang Wang[1]

[1] School of Information Science and Engineering, Southeast University,
Nanjing 210096, China
{Lqchen,muyangwang}@seu.edu.cn
[2] Center of Supply Chain and Operations Management, Coventry University,
Coventry CV1 5FB, UK
ac2912@coventry.ac.uk

Abstract. Fully homomorphic encryption is a type of encryption technique that allows arbitrary complex operations to be performed on the ciphertext, thus generating an encrypted result that, when decrypted, matches the results of those operations performed on the plaintext. The DGHV scheme over the integers is one of the key schemes in fully homomorphic encryption research field, but the incredible size of the public key and the low computational efficiency are the main challenges. Based on the original DGHV encryption structure and parameters' design, the idea of batch processing was introduced in this paper. With the combination of the quadratic parameter-based public key compression mechanism, a complete public key compression and batch processing fully homomorphic encryption (PKCB-FHE) scheme was presented. Like those in the original DGHV scheme, the parameter restriction of the proposed scheme was presented. Further analysis and simulation of the proposed scheme indicate that the required storage space of the public key is immensely reduced and that the overall length of the public key is compressed. Furthermore, the total processing time of the proposed scheme is reduced, which makes it much more efficient than those existing schemes.

Keywords: Fully homomorphic encryption · Public key compression
Batch technology · Quadratic parameter

1 Introduction

Fully homomorphic encryption (FHE) allows us to perform arbitrarily complex operations over encrypted data without performing decryption, which is equivalent to the same operations over plaintext. This method is proposed to provide security for cloud computing and multi-party computations now [1, 2] and is also used in other security fields with public key technology [3, 4]. The idea of FHE was first proposed by Rivest et al. in 1978 [5]. Since then, many researchers have suggested a variety of solutions to improve or realize the FHE function. However, all of those existing schemes were designed and studied at the theory and model levels; the load of computation and the

© Springer International Publishing AG, part of Springer Nature 2018
X. Chen et al. (Eds.): Inscrypt 2017, LNCS 10726, pp. 242–259, 2018.
https://doi.org/10.1007/978-3-319-75160-3_16

algorithm's complexity were still high. In 2009, Gentry et al. proposed a new FHE scheme (Gentry scheme) based on an ideal lattice technique [6, 7]. However, due to the overload of the public key size and low operational efficiency, the Gentry scheme could not be used in practical application. In 2010, van Dijk and Gentry et al. proposed an FHE scheme (DGHV scheme) based on integer arithmetic [8]. It has higher efficiency and has attracted much attention from the researchers in this field.

The DGHV scheme inherits the framework of the lattice technique and has improved in computation complexity, but it still faces two problems: the length of the public key size is as long as $\tilde{O}(\lambda^8)$, and the efficiency in encryption processing is low. The plaintext space of the DGHV scheme is $\{0, 1\}$. It can only process one bit in one encryption processing. Currently, those existing FHE schemes [9–14], including the schemes based on lattice and integers, always use a one-bit plaintext processing method to perform encryption and decryption. Their processing efficiency is at a low level.

To solve the first problem mentioned above, the basic improvement idea is to compress the public key storage space [15]. Coron et al. proposed a public key compression scheme with quadratic form (QF-PKC scheme) in 2011 [16]. It generates the public key parameters from a subprime parametric quadratic form that replaces the original linear form used before. In this way, the size of the public key is compressed from $\tilde{O}(\lambda^8)$ to $\tilde{O}(\lambda^{6.5})$. In 2012, Coron et al. proposed another public key compression scheme with correction (C-PKC scheme) [17], and the size of the public key was compressed to $\tilde{O}(\lambda^5)$. Chen et al. also proposed a fully homomorphic encryption scheme with a smaller public key size [18]. However, this scheme is based on the LWE assumption.

To improve the efficiency of the encryption processing, Smart and Vercauteren proposed a new scheme based on the Gentry scheme to improve the frame structure and realized the fully homomorphic encryption with specific design [19]. This scheme can support the single instruction multiple data (SIMD) style of operation and can support plaintext parallel processing by adjusting some parameters. From there, Smart and Vercauteren further discussed how to adjust the specific parameters [8] and established a new FHE scheme [20]. The above SIMD operations are always used in those FHE schemes based on the learning with errors (LWE). However, the SIMD technique used in the FHE scheme based on the integers has not been reported yet.

Here, with the consideration of the challenges and problems presented above, a new FHE scheme based on public key compression and batch processing (PKCB-FHE scheme) is proposed. In this scheme, the technology of public key parametric quadratic with correction is used to reduce the size of the public key, and the idea of batch processing is used to improve the encryption efficiency. A somewhat homomorphic encryption scheme based on public key compression and batch processing (PKCB-SWHE scheme) is first presented. Then, using a squashing and bootstrapping technique, the PKCB-SWHE scheme is improved to a fully homomorphic PKCB-FHE scheme. Further, the parameter restriction of the proposed PKCB-FHE scheme is presented, and its correctness and semantic security are proven as well. Finally, the analysis and empirical results show the superiority in public key size and processing efficiency of the proposed PKCB-FHE scheme.

The remainder of this paper is organized as follows: Sect. 2 presents the notations and knowledge of batch processing. Section 3 describes the proposed PKCB-SWHE scheme and its parameter constraints. Then, the proposed PKCB-FHE scheme is presented in Sect. 4. The correctness and semantic security of the PKCB-FHE scheme are proved in Sect. 5, and the empirical studies and data analyses are presented in Sect. 6. Finally, the conclusions are presented.

2 Preliminary Knowledge

Notation. Here we give the definition of the parameters. For a real number z, the rounding of z up, down and to the nearest integer are set as $\lceil z \rceil$, $\lfloor z \rfloor$ and $\lfloor z \rceil$ as in [8]. These operations are conducted on the $(z, z+1], (z-1, z]$ and $(z-1/2, z+1/2]$ semi-closed intervals. For a real number z and an integer p, $q_p(z)$ and $r_p(z)$ represent the quotient and remainder of z divided by p. $q_p(z) \stackrel{\text{def}}{=} \lfloor z/p \rceil$ and $r_p(z) \stackrel{\text{def}}{=} z - q_p(z) \cdot p$, in which $r_p(z) \in (-p/2, p/2]$ and $r_p(z) = z \bmod p$.

The scheme parameters. Given the security parameter λ, n is the precision of the squash coefficient y, and the following parameters are defined: γ is the bit-length of the subset public key elements $x'_{i,j}$, τ is the number of public key elements, 2β is the number of the subset public key elements, where $\beta = \sqrt{\tau}$. α is the bit-length of the noise $b_{i,j}$ used for encryption to increase the randomness of the ciphertext. ρ' is the bit-length of the noise r used for encryption.

l is the number of the private key $\left(p_j\right)_{0 \le j < l}$, which is decided by the length of plaintext vector $\mathbf{m} \in \{0, 1\}^l$. This number also limits the vector length of batch processing noise $\mathbf{b}'$ in the encryption process. $\eta = \tilde{O}(\lambda^2) \ge \alpha' + \rho' + 1 + \log_2(l)$. The length of the private key is $\left(p_j\right)_{0 \le j < l}$, which is used to meet the requirements of the proper data decompression. $\gamma = \tilde{O}(\lambda^5)$ is the length of the public key parameter x_0. It can limit the range of q_0, and further limit the length of parameters x_i. $\rho = \lambda$, which is the range of noise $r'_{i,j}$ when x'_i is generated. At the same time, it also can indicate the range of Π_i, when $w'_{i,j}$ is generated.

$(\rho' - 1)$ is the range of the noise $r_{i,j}$ when x_i is generated, where $\rho' = \tilde{O}(\lambda^2) \ge p + \lambda$ is defined to ensure the semantic security. α' is the vector length of the noise $\mathbf{b}'$ in batch processing, where $\alpha' \ge \alpha + \lambda$ is used to ensure the semantic security. θ is the Hamming weight of the discrete subset elements.

The batch processing in the proposed PKCB-FHE scheme will use the technology of SIMD [20]. It includes (1) the mechanism that packages plaintexts $m_0, \ldots, m_{l-1}$ into a ciphertext and (2) the technique that realizes homomorphic addition and multiplication operation in parallel.

For a better understanding of the design idea of the proposed PKCB-FHE scheme, the following three problems are presented: (1) How can we keep the encryption structure of FHE scheme with the batch processing technique? (2) How can we meet the requirement of the plaintext tank structure? (3) How can we ensure the semantic security of the proposed scheme?

2.1   Batch Processing Technology

Based on a set of prime $p_0, \ldots, p_{l-1}$, the plaintexts m_i are encrypted into a ciphertext c. According to the Chinese remainder theorem (CRT) [21], the packaged cipher structure is constructed as the following equation:

$$c = q \cdot \prod_{i=0}^{l-1} p_i + CRT_{p_0,\ldots,p_{l-1}}(2r_0 + m_0, \ldots, 2r_{l-1} + m_{l-1}) \tag{1}$$

Here, $CRT_{p_0,\ldots,p_{l-1}}(a_0, \ldots, a_{l-1})$ indicates that for a single integer $u < \prod_{i=0}^{l-1} p_i$, there is u mod $p_i = a_i$. The decryption process is that for all $0 \leq i < l$, a bit vector **m** can by calculated through $m_i = [c \bmod p_i]_2$ equation. The operation c mod p_i is the same as the operation in the original DGHV scheme.

2.2   Encryption Design Satisfies the Plaintext Tank Structure

The plaintext tank structure for SIMD on a different modulus is designed based on the CRT technique. This specific technique is to adopt some parameters in the original DGHV scheme and then encrypt the plaintext vector $\mathbf{m} = (m_0, \ldots, m_{l-1})$ at the same time. The ciphertext structure is shown as follows:

$$c = \left[\sum_{i=0}^{l-1} m_i \cdot x_i' + 2r + 2 \sum_{i \in S} x_i \right]_{x_0} \tag{2}$$

We can get $\left[c \bmod p_j \right]_2 = m_j$, which meets the requirements of FHE. The parameters used here are defined as follows:

(1) x_i', which satisfies $x_i' \bmod p_j = \delta_{i,j} + 2r_{i,j}'$ to provide batch processing structure for the plaintext, where $\delta_{i,j} = \begin{cases} 1, & i = j \\ 0, & i \neq j \end{cases}$, $r_{i,j}'$ is the noise value for x_i'.

(2) Public key parameter x_i, which satisfies $x_i \bmod p_j = r_{i,j}$ to provide the necessary encryption structure for ciphertext, where $r_{i,j}$ is the noise value for x_i.

(3) The noise for the ciphertext is 2r, which provides noise for the ciphertext to ensure security.

2.3   Semantic Security of FHE with Batch Processing

The semantic security of the original DGHV scheme is based on the subset of the remaining hash lemma. Meanwhile, the parameter 2r is added to the ciphertext to further randomize the ciphertext. However, for each p_i in FHE with batch processing, there is 2r mod p_i to be inserted into the ciphertext, and it is necessary to redesign the noise structure of the ciphertext to ensure the independent of each noise variable for semantic security.

The FHE scheme proposed by Cheon et al. has been proved to be semantic secure [21]. When the DGHV scheme is combined with batch processing, the ciphertext structure is

$$c = \left[m + 2b \cdot \Pi + 2 \sum\nolimits_{i \in S} x_i \right]_{x_0} \tag{3}$$

where b is a large random integer and $\Pi \bmod p = \mathfrak{w}$. We can also expand the random multiplier Π to l subsets to fulfil the equation $\Pi_i \bmod p_i = \mathfrak{w}_i$. Moreover, it is shown that the random noise variable is basically independently distributed after mod p_i processing, which is required for semantic security proof.

3   The Proposed PKCB-SWHE Scheme

Here, the somewhat homomorphic encryption scheme based on a public key compression and batch processing (PKCB-SWHE) scheme is introduced. This encryption scheme combines the public key compression technique based on public key parametric quadratic with correction and the batch processing technique. They are presented in the following subsections.

3.1   The Public Key Parametric Quadratic with Correction

The public key compression technique is a solution for reducing the length of the public key in the FHE scheme. Here, we propose a parametric quadratic public key compression technique with correction. This method decreases the length of the public key parameter by integrating the auxiliary parameters with correction. These parameters are selected based on the assumption of the approximate greatest common divisor without noise.

The main steps of the proposed public key compression scheme are as follows: the public key parameter is first transformed into subset parameters; then, the subset parameters are modified into auxiliary parameters with correction. Here, we can get τ public key parameters using the equation $x_{i,j} = x'_{i,0} \cdot x'_{j,1}$ $1 \leq i, j \leq \beta$. Thus, the transformation of the public key parameter into the subset parameter $x'_{i,b}$ can be realized. Finally, the subset parameter $x'_{i,b}$ with a length of γ-bits is transformed into a random number with a length of γ bits and a differential value of small length. The random number is called the auxiliary parameter; the differential value is called the parameter with correction. The auxiliary parameter $\chi_{i,b}$ is generated by the random number generator f based on the seed "se". To ensure that x_0 can effectively limit the length of the ciphertext while not causing a noise problem, it is necessary to take $\chi_{i,b} \in [0, x_0)^{2\beta}$ in order to ensure that all of the subset parameters $x'_{i,0}$ are lower than x_0. Moreover, the parameters with correction will be generated by $\chi_{i,b}$, according to the structure $\langle \chi_{i,b} \rangle_p + \xi_{i,b} \cdot p - r_{i,b}$, where $r_{i,b} \leftarrow \mathbb{Z} \bigcap (-2^\rho, 2^\rho)$ and $\xi_{i,b} \leftarrow \mathbb{Z} \bigcap [0, 2^{\lambda+\eta}/p)$. This structure can ensure that the generated public key parameter still maintains the secure feature of the original DGHV scheme.

3.2   The DGHV SWHE Scheme with Batch Processing

Here, we present the DGHV SWHE scheme with batch processing. It includes the following main parts:

The structure of ciphertext. The ciphertext noise parameter is chosen as the random integer vector $\mathbf{b} = (b_i)_{0 \leq i < \tau} \in (-2^\alpha, 2^\alpha)^\tau$. The batch processing noise parameter is chosen as the random integer vector $\mathbf{b}^{(W)} = \left(b_i^{(W)}\right)_{0 \leq i < l} \in \left(-2^{\alpha^{(W)}}, 2^{\alpha^{(W)}}\right)^l$. Then, the ciphertext is derived as

$$c = \left[\sum_{i=0}^{l-1} m_i \cdot \mathfrak{B}_i + \sum_{i=0}^{l-1} b_i^{(W)} \cdot \mathcal{W}_i + \sum_{i=0}^{\tau-1} b_i \cdot x_i\right]_{x_0} \tag{4}$$

The encryption parameters. The encryption parameters in Eq. (4) are described as follows: $\mathfrak{B}_i$ is the plaintext parameter of batch processing. It is used to provide batch structure for ciphertext. $\mathcal{W}_i$ is the noise for the ciphertext. It is used to add a large p multiple index for the noise in ciphertext in order to realize the noise variable distribution, which is similar to a uniform distribution. x_i is the main parameter of the public key. It is needed for providing the encryption structure for the ciphertext, which is similar to the role of public key parameter in the original DGHV scheme. The generation of different parameters is shown in Table 1.

Table 1. Generation of each encryption parameters in DGHV SWHE scheme with batch processing

$0 \leq i < \tau$	$x_i \bmod p_j = 2r_{i,j}$	$r_{i,j} \leftarrow \mathbb{Z} \cap \left(-2^{\rho'-1}, 2^{\rho'-1}\right)$
$0 \leq i < l$	$\mathfrak{B}_i \bmod p_j = 2r_{i,j}^{(\mathfrak{B})} + \delta_{i,j}$	$\mathfrak{B}_i \leftarrow \mathbb{Z} \cap (-2^\rho, 2^\rho)$
$0 \leq i < l$	$\mathcal{W}_i \bmod p_j = 2w_{i,j} + \delta_{i,j} \cdot 2^{\rho'+1}$	$w_{i,j} \leftarrow \mathbb{Z} \cap (-2^\rho, 2^\rho)$

The generation of the other parameters in Table 1 is presented below. The random prime set $p_0, \ldots, p_{l-1}$ is generated, while p_i is η-bits as the private key. Here, $\pi = p_0 \cdot \ldots \cdot p_{l-1}$. $q_0 \leftarrow \mathbb{Z} \cap [0, 2^\gamma/\pi)$ is selected, which is a 2^{λ^2}-rough integer (an integer that does not contain any element factors less than b is called a b-rough integer). The public key without noise is computed as $x_0 = q_0 \cdot \pi$.

The process of decryption. The decryption for the ciphertext c is achieved through computing the equation $m_j \leftarrow [c]_{p_j} \bmod 2$. It is also presented as $m_j \leftarrow \left[c - \lfloor c/p_j \rceil\right]_2 = (c \bmod 2) \oplus \left(\lfloor c/p_j \rceil \bmod 2\right)$.

3.3   The Proposed PKCB-SWHE Scheme

Combining the batch processing and the technique of public key parametric quadratic with correction, the PKCB-SWHE scheme works as the following steps.

(1) Key generation module PKCBS.KeyGen$\left(1^{\lambda}\right)$

In this module, a random prime set $p_0, \ldots, p_{l-1}$ and a parameter $q_0 \leftarrow \mathbb{Z} \cap [0, 2^{\gamma}/\pi)$ are generated. Here, q_0 is a 2^{λ^2}-rough integer which is calculated by multiplying the 2^{λ^2}-bits prime numbers. The public key module $x_0 = q_0 \cdot \pi$ is also generated.

Pseudo-random initialization $(0 \leq i < \beta, 0 \leq j < \beta^{(l)}, b \in \{0, 1\})$ is processed. A pseudo-random generator f_x with a random seed se_x is initialized while the public key auxiliary parameter $\chi_{i,b} \in [0, x_0)^{2\beta}$, plaintext auxiliary parameter $\chi_{j,b}^{(\mathfrak{B})} \in [0, x_0)^{2\beta^{(l)}}$, and the ciphertext noise auxiliary parameter $\chi_{j,b}^{(\mathcal{W})} \in [0, x_0)^{2\beta^{(l)}}$ integer sets are generated.

Then, the public key main parameter $(0 \leq i < \beta, 0 \leq k < l)$ is designed. The public key main parameter with correction is calculated as $\varphi_{i,b} = \left[\chi_{i,b}\right]_\pi + \xi_{i,b} \cdot \pi - \mathrm{CRT}_{p_0, \ldots, p_{l-1}}\left(2r_{i,b,0}, \ldots, 2r_{i,b,(l-1)}\right)$. Here, $r_{i,b,k} \leftarrow \mathbb{Z} \cap \left(-2^{\rho'-1}, 2^{\rho'-1}\right)$ and $\xi_{i,b} \leftarrow \mathbb{Z} \cap \left[0, 2^{\lambda + \log_2(l) + l \cdot \eta}/\pi\right)$.

The plaintext parameter with batch processing $(0 \leq i < \beta^{(l)})$ is generated. The plaintext parameter with correction and batch processing is generated as $\varphi_{i,b}^{(\mathfrak{B})} = \left[\chi_{i,b}^{(\mathfrak{B})}\right]_\pi + \xi_{i,b}^{(\mathfrak{B})} \cdot \pi - \mathrm{CRT}_{p_0, \ldots, p_{l-1}}\left(2r_{i,b,0}^{(\mathfrak{B})} + \delta_{i,b,0}, \ldots, 2r_{i,b,(l-1)}^{(\mathfrak{B})} + \delta_{i,b,(l-1)}\right)$, Here, $r_{i,b,k}' \leftarrow \mathbb{Z} \cap (-2^\rho, 2^\rho)$ and $\xi_{i,b}' \leftarrow \mathbb{Z} \cap \left[0, 2^{\lambda + \log_2(l) + l \cdot \eta}/\pi\right)$.

The ciphertext noise parameters $(0 \leq i < \beta^{(l)}, 0 \leq k < l, b \in \{0, 1\})$ are also generated. The ciphertext noise with correction is computed as $\varphi_{i,b}^{(\mathcal{W})} = \left[\chi_{i,b}^{(\mathcal{W})}\right]_\pi + \xi_{i,b}^{(\mathcal{W})} \cdot \pi - \mathrm{CRT}_{p_0, \ldots, p_{l-1}}\left(2w_{i,b,0} + \delta_{i,b,0} \cdot 2^{\rho'+1}, \ldots, 2w_{i,b,(l-1)} + \delta_{i,b,(l-1)} \cdot 2^{\rho'+1}\right)$. Here, $w_{i,b,0} \leftarrow \mathbb{Z} \cap (-2^\rho, 2^\rho)$ and $\xi_{i,b}^{(\mathcal{W})} \leftarrow \mathbb{Z} \cap \left[0, 2^{\lambda + \log_2(l) + l \cdot \eta}/\pi\right)$.

Finally, the keys are generated. The public key is $pk = \langle x_0, se_x, \varphi_{i,b}, \varphi_{j,b}^{\mathcal{W}}, \varphi_{j,b}^{(\mathfrak{B})} \rangle$, and the secret key is $sk = p_0, \ldots, p_{l-1}$.

(2) Encryption module PKCBS.Encrypt$\left(pk, m \in \{0, 1\}^l\right)$

Here, $f_x(se_x)$ is used to generate the integer set $\chi_{i,b}, \chi_{j,b}^{(\mathfrak{B})}, \chi_{j,b}^{(\mathcal{W})}$. The main parameter of the subset public key $x_{i,b}' = \chi_{i,b} - \varphi_{i,b}$ is computed. Here, the subset plaintext parameter with batch processing is $\mathfrak{B}_{i,b}' = \chi_{i,b}^{(\mathfrak{B})} - \varphi_{i,b}^{(\mathfrak{B})}$, the subset ciphertext noise parameter $\mathcal{W}_{i,b}' = \chi_{i,b}^{(\mathcal{W})} - \varphi_{i,b}^{(\mathcal{W})}$.

The encryption noise parameter is $(b_{i,j})_{0 \leq i,j < \beta} \in [0, 2^\alpha)^\tau$, and the noise parameter with batch processing $\mathbf{b}' = \left(b_{i,j}'\right)_{0 \leq i,j < \beta^{(l)}} \in [0, 2^{\alpha'})^l$. The main parameter of public key $(x_{i,j})_{0 \leq i,j < \beta} = x_{i,0}' \cdot x_{j,1}'$, the plaintext parameter with batch processing $(\mathfrak{B}_{i,j})_{0 \leq i,j < \beta^{(l)}} = \mathfrak{B}_{i,0}' \cdot \mathfrak{B}_{i,1}'$, while the ciphertext noise parameter $(\mathcal{W}_{i,j})_{0 \leq i,j < \beta^{(l)}} = \mathcal{W}_{i,0}' \cdot \mathcal{W}_{i,1}'$. Then, the ciphertext $c = \left[\sum_{1 \leq i,j < \beta^{(l)}} m_i \cdot \mathfrak{B}_{i,j} + \sum_{1 \leq i,j < \beta^{(l)}} b_{i,j}' \cdot \mathcal{W}_{i,j} + \sum_{1 \leq i,j < \beta} b_{i,j} \cdot x_{i,j}\right]_{x_0}$ is gained.

(3) The data processing module PKCBS.Evaluate$(\mathbf{pk}, \mathcal{C}_\varepsilon, \mathbf{c_1}, \dots, \mathbf{c_t})$

This module is used to evaluate the fully homomorphic capability. For processing circuit $\mathcal{C}_\varepsilon$ with input t, all of the ciphertext $c_1, \dots, c_t$ are used as input. Then, through the addition gate and multiplication gate, which are used in processing of the function $\mathcal{C}_\varepsilon$, the following processing is achieved: PKCB.Add$(\mathbf{pk}, c_1, c_2)$: $c_1 + c_2 \bmod x_0$; PKCB.Mult$(\mathbf{pk}, c_1, c_2)$: $c_1 \cdot c_2 \bmod x_0$.

(4) Decryption module PKCBS.Decrypt$(\mathbf{sk}, \mathbf{c})$

Finally, the operation $m_j \leftarrow [c]_{p_j} \bmod 2$ for c is used to do the decryption processing,

$$m_j \leftarrow \left[c - \left\lfloor c/p_j \right\rceil \right]_2 = (c \bmod 2) \oplus \left(\left\lceil c/p_j \right\rceil \bmod 2 \right). \text{ Then, the decrypted plaintext}$$

vector m $= (m_0, \dots, m_{l-1})$ is output.

4 The Proposed PKCB-FHE Scheme

The purpose of the full homomorphism of the SWHE scheme is to tackle the problem of cross-border noise in the SWHE-style scheme. Methods of noise management mainly include re-encryption, decryption circuit compression, modulus conversion, and dimension reduction. Through certain structural adjustment, these methods can be applied to different FHE schemes. Here, we adopt the framework of the Gentry's "squash" technique [8] and the bootstrapping technique to expand the PKCB-SWHE scheme into a fully homomorphic encryption scheme based on public key compression and batch processing (PKCB-FHE scheme).

4.1 Squash Decryption Circuit for Batch Processing

To realize the bootstrapping function in the FHE encryption step, the re-encryption preparation work is performed first. The preparation includes converting the decryption circuit into a lower order circuit to meet the requirements of "permissive circuit". Specifically, the decryption process is changed in order to express the decryption circuit in a way which links to the private key with low polynome.

The method to the squash decryption circuit has been given in the original DGHV scheme, but it cannot directly meet the requirements of the batch processing scheme with public key compression. Thus, certain adjustments for a "permissive circuit" have to be considered. The literature [21] has expanded the discrete subset $\mathbf{s} = (s_0, \dots, s_{\Theta-1})$ of the squash circuit in the original DGHV scheme to l similar functions of discrete subsets according to the requirement of batch processing, and the subset is used to filter the random numbers $\mathbf{y} = \{y_0, \dots, y_{\Theta-1}\}$. After the expansion, the existing discrete subset $S_j \subset [0, \Theta - 1]$ for $0 \leq j < l$, its hamming weight is θ. Moreover, it fulfils the equation $\sum_{i \in S_j} y_j \simeq 1/p_j \bmod 2$.

4.2   Design of PKCB-FHE Scheme

There are three main problems to be solved in the fully homomorphic transformation of the PKCB-SWHE scheme: (1) new re-encryption parameters should be generated to meet the parallel re-encryption; (2) new batch replacement structure should be designed for re-encryption; (3) the length of the new re-encryption parameters should be reduced. The following steps show the processing of the fully homomorphic transformation of the PKCB-SWHE scheme. Then the PKCB-FHE scheme is obtained.

Design of the parallel re-encryption parameters. The l number of private key $s_{j,i}$ is used for encryption to generate the ciphertext $\sigma_i = \text{Encrypt}(s_{0,i}, \ldots, s_{l-1,i})$. This gives the cryptograph f the same l-bits plaintext vector, but the noise will be smaller. In this way, the re-encryption is a parallel process for l plaintext troughs, and its complexity is equivalent to the single re-encryption process in DGHV scheme. Based on the above processing, a homomorphic encryption scheme is obtained.

Design of the batch replacement structure fit for re-encryption. The practical solution is to use a Beneš replacement network [22] as in [21] to design the batch replacement structure fit for re-encryption. It is sufficient to achieve the $\pm 2^i$-bits cycle shift when $2\log_2(l)$ replacement parameters are added to the public key. Every replacement can be completed within $2\log(l) - 1$ steps. Each step contains up to two shift operations and two screening operations. Regarding the screening operations, c_1 and c_2 are used to construct a ciphertext, while the l number of the included plaintext slots were extracted from c_1 and c_2. It is easier to realize the screening operations when two multiplication operations and one addition operation are used. However, this method has a limited impact on the public key, which can cause a plurality of $2\log_2(l) \cdot \Theta \cdot \gamma$ bits, while any replacement will be completed in a maximum length of $(2\log_2(l) - 1) \cdot 6$ re-encryption operations.

4.3   The Whole PKCB-FHE Scheme

Based on the PKCB-SWHE scheme and the squashing and bootstrapping techniques, the whole PKCB-FHE scheme is obtained. The flow chart of the whole PKCB-FHE scheme is shown in Fig. 1, and the processing steps are presented in detail as follows:

(1) Re-encryption initialization module PKCB.KeyGen(1^λ)

- **The initialization of related parameters $\left(0 \leq i < \beta^{(l)}, 0 \leq j < l\right)$**

    The private key $\mathbf{sk}^* = (p_0, \ldots, p_{l-1})$ and the public key $\mathbf{pk}^*$ are generated, which is the same as the processing in the PKCB-SWHE scheme. $x_{p_j} \leftarrow \left\lfloor 2^\kappa / p_j \right\rceil$. $2\beta'$ number of β_Θ-bits random re-encryption private key auxiliary parameter vector $\mathcal{S}_{i,b} = \left(\mathcal{S}_{i,b,0}, \ldots, \mathcal{S}_{i,b,\lceil\sqrt{\Theta}\rceil-1}\right)$ are selected, and the hamming weight is β_θ, $b \in \{0, 1\}$, $\beta_\theta^2 = \theta$.

The re-encryption private key $s_{i,k} = S_{i,1,k} \cdot S_{j,0,k}$, $0 \leq k < \beta_\Theta$ is generated by $S_{i,b}$ to constitute re-encryption private key vector $\mathbf{s}_j = (s_{j,0}, \ldots, s_{j,\Theta-1})$. The vector elements dispersed in θ slots, and each slot contains $B = \Theta/\theta$ elements.

After the above selection, a random seed se_u is used to initialize a pseudo-random generator f_u, and some integers are generated with $f_u(se_u)$: Θ integer squash parameters $u_k \in [0, 2^{\kappa+1})$, where $0 \leq k < \Theta$. For all j, there is $x_{p_j} = \sum_{k=0}^{\Theta-1} s_{j,k} \cdot u_k \mod 2^{\kappa+1}$.

Squeeze coefficient $y_k = u_k/2^\kappa$ is calculated. Each y_i is an integer less than 2, and the precision is κ-bits. Then, the squash coefficient sets $\mathbf{y} = (y_0, \ldots, y_{\Theta-1})$ are obtained, which can satisfy the equation $\left| \left[\sum_{k=0}^{\Theta-1} s_{j,k} \cdot y_k - \frac{1}{p_j} \right]_2 \right| < 2^{-\kappa}$.

- **The initialization of re-encryption parameter** $(0 \leq i < \beta_\Theta, b \in \{0, 1\})$

After the above selection, a pseudo-random generator f_σ is initialized using a random seed se_σ, and an integer set $f_\sigma(se_\sigma)$ is also generated. Here, the re-encryption auxiliary parameter $\chi_{i,b}^\sigma \in [0, 2^\gamma)^{2\beta_\Theta}$.

The re-encryption parameter with correction is calculated as $\varphi_{i,b}^{(\sigma)} = \left[\chi_{i,b}^{(\sigma)} \right]_\pi + \xi_{i,b}^{(\sigma)} \cdot$

$\pi - CRT_{p_0, \ldots, p_{l-1}} \left(2r_{i,b,0}^{(\sigma)} + S_{i,b,0}, \ldots, 2r_{i,b,(l-1)}^{(\sigma)} + S_{i,b,(l-1)} \right)$, where $r_{i,b,j}^{(\sigma)} \leftarrow \mathbb{Z} \cap (-2^p,$

$2^p)$, $\xi_{i,b}^{(\sigma)} \leftarrow \mathbb{Z} \cap [0, 2^{\lambda + \log_2(l) + l \cdot \eta}/\pi)$.

- **Output the keys**

Private key $\mathbf{sk} = (\mathbf{s}_0, \ldots, \mathbf{s}_{l-1})$, where $\mathbf{s}_j = (s_{j,0}, \ldots, s_{j,\Theta-1})$, $0 \leq j < l$. Public key $\mathbf{pk} = (\mathbf{pk}^*, \mathbf{pk}_\sigma, \mathbf{y})$, where $\mathbf{pk}^* = x_0, se_x, \varphi_{i,b}, \varphi_{j,b}^{(W)}, \varphi_{j,b}^{(\mathcal{B})}$, $\mathbf{pk}_\sigma = \left(se_\sigma, \varphi_{k,b}^\sigma \right)$, $0 \leq i < \beta$, $0 \leq j < \beta^{(l)}$, $0 \leq k < \beta_\Theta$, $b \in \{0, 1\}$.

(2) Encryption module PKCB.Encrypt$\left(\mathbf{pk}, \mathbf{m}_i \in \{0, 1\}^l \right)$

The encryption module of PKCB-FHE scheme works the same as that in the PKCB-SWHE scheme.

(3) Ciphertext expansion module PKCB.Expand$(\mathbf{pk}, \mathbf{c})$

The expanded ciphertext $z_i = \lfloor c \cdot y_i \rceil_2$ is calculated, where $0 \leq i < \Theta$ and the precision of z_i keeps $n = \lceil \log_2(\theta) \rceil$. The expanded ciphertext vector $\mathbf{z} = (z_i)_{i=0,\ldots,\Theta-1}$ is obtained.

(4) Re-encryption module PKCB.Recrypt$(\mathbf{pk}, \mathbf{c}, \mathbf{z})$

The set of integers re-encryption auxiliary parameter $(0 \leq i < \beta_\Theta, b \in \{0, 1\})$ is restored with $f_\sigma(se_\sigma)$. The secondary re-encryption parameter $\sigma_{i,b}' = \chi_{i,b}^\sigma - \varphi_{k,b}^\sigma$ is calculated. For all $0 \leq i < \beta_\Theta$, $b \in \{0, 1\}$, the re-encryption parameter $\sigma_i = \left(\sigma_{i,0}' \cdot \sigma_{i,1}' \right)_{0 \leq i < \beta_\Theta}$ is computed, and the expansion ciphertext group is homomorphic decrypted by parameter $\boldsymbol{\sigma}$. Finally, the new ciphertext c^* is output.

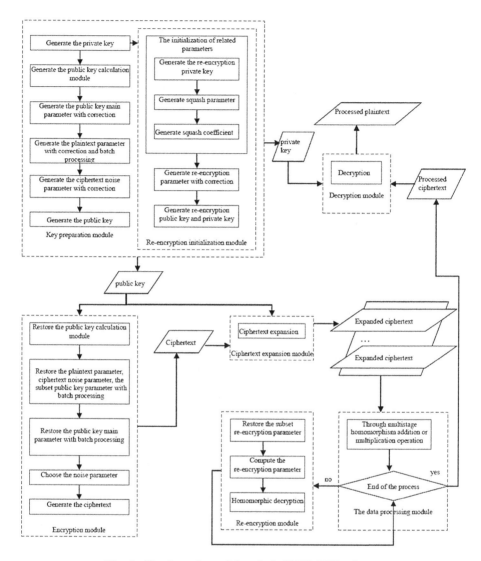

Fig. 1. The flow chart of the whole PKCB-FHE scheme

(5) Decryption module PKCB.Decrypt(sk, c, z)

We compute the plaintext $m_j \leftarrow \left[\left\lfloor \sum_{i=0}^{\Theta-1} s_{j,i} \cdot z_i \right\rceil \right]_2 \oplus (c \bmod 2)$, where $0 \le j < l$. The plaintext vector $\mathbf{m} = (m_0, \ldots, m_{l-1})$ is obtained. The plaintext vector is output.

4.4   Parameter Restriction

The main parameters used in the proposed PKCB-FHE scheme are presented as follows. Here, λ is the safety parameter and n is the precision of the squash coefficient y.

- $\Theta = \tilde{O}(\lambda^3)$: It is the number of discrete subset elements.
- $\gamma = \omega(\eta^2 \log \lambda)$: It is used to prevent the attacks of the approximate greatest common divisor problem based on lattice.
- $\kappa = \gamma + 2$: The parameter indicates the value accuracy of y_i, namely, the length of u_i.
- $\beta_\Theta = \tilde{O}(\lambda^{1.5})$: $\beta_\Theta^2 = \Theta$. It indicates the number of secondary re-encryption parameter.
- $\beta_\theta = \tilde{O}(\lambda^{0.5})$: $\beta_\theta^2 = \theta$. This is the Hamming weight of the bootstrap key auxiliary parameter vector.
- $\beta^{(l)}$: $\left(\beta^{(l)}\right)^2 = l$. It represents the number of secondary batch processing plaintext parameters.
- $\eta \geq \rho \cdot \Theta(\lambda \log^2 \lambda)$. It is used to meet the requirements of homomorphic operation, thus achieving the evaluation of the "squashed decryption circuit".
- $\alpha \cdot \tau \geq \gamma + \lambda$ and $\tau \geq l \cdot (\rho' + 2) + \lambda$. It is used to fulfil the Hash Lemma.

In this paper, two specific parameters are provided: $\theta = 16$, $n = 4$. They can meet the requirements of the PKCB-FHE scheme to generate subset parameters in every level. Furthermore, this method takes the efficiency requirement into consideration.

5 Security Proof of the Proposed Scheme

5.1 Correctness of the PKCB-SWHE Scheme

(1) Permissive circuit

Similar to the cases in [8], we define a permissive circuit C_ε. It means that for any $i \geq 1$ and integer set of the input whose absolute value is less than $l^i 2^{i(\alpha' + \rho' + 2)}$, the absolute value of the output in C_ε circuit will not exceed $2^{i(\eta - 3 - n)}$, where $n = \log_2(\lambda + 1)$.

Lemma 1. The PKCB-SWHE scheme is correct for C_ε.

(2) The definition of correctness

The definition of correctness in [8] can be adjusted to fit in the circumstance with batch processing. Considering that there is a homomorphic public key encryption algorithm $\mathcal{E}$, it has the assessment algorithm *Evaluate*, the public key pk and the modulo-2 algebra circuit C which has t number of inputs, and t number of ciphertexts c_i, another ciphertext c is regarded as output. If we define the algorithm to be $\mathcal{E} = (\text{KeyGen}, \text{Encrypt}, \text{Decrypt}, \text{Evaluate})$, the input includes t-input circuit C, random key pair (sk, pk) generated by $\text{KeyGen}(\lambda)$, any t plaintexts vector $m_1, \ldots, m_t$ which is l-bits and the arbitrary ciphertext $C = (c_1, \ldots, c_t)$, we can get $c_i \leftarrow \text{Encrypt}(pk, m_i)$. That is, the equation $\text{Decrypt}(sk, \text{Evaluate}(pk, C, \mathbf{C})) = (C(m_1[0], \ldots, m_t[0]), \ldots, C(m_1[l-1], \ldots, m_t[l-1]))$ is gained. Consequently, the algorithm $\mathcal{E}$ is considered correct for C.

(3) The verification of the correctness

We use the following steps to prove the Lemma 1 in order to prove that the PCKB-SWHE scheme proposed in Sect. 3.2 is correct.

Proof. Given ciphertext c, which is the output from PKCB.Encrypt(pk, **m**), there is an integer vector $\mathbf{b} = (b_{i,j})_{0 \le i,j < \beta} \in [0, 2^\alpha)^\tau$ and $\mathbf{b}' = (b'_{i,j})_{0 \le i,j < \beta^{(l)}} \in [0, 2^{\alpha'})^l$ to meet the requirement $c = \left[\sum_{1 \le i,j < \beta^{(l)}} m_i \cdot \mathcal{B}_{i,j} + \sum_{1 \le i,j < \beta^{(l)}} b'_{i,j} \cdot \mathcal{W}_{i,j} + \sum_{1 \le i,j < \beta} b_{i,j} \cdot x_{i,j} \right]_{x_0}$.
For each $j = 0, \ldots, l - 1$, the relation can be written as

$$\left| c \bmod p_j \right| \le l \cdot 2^{\rho+1} + \tau \cdot 2^{\alpha+\rho'+1} + l \cdot 2^{\alpha'+\rho'+1} \le l \cdot 2^{\alpha'+\rho'+2} \tag{5}$$

Suppose that C is a permissive circuit with t inputs, C' is the corresponding operation of the circuit which is based on integer rather than on modulo 2. Given $c_i \leftarrow$ PKCB.Encrypt(pk, $\mathbf{m_i}$), then for each $j = 0, \ldots, l - 1$, there is

$$c \bmod p_j = C'(c_1, \ldots, c_t) \bmod p_j = C'\left(c_1 \bmod p_j, \ldots, c_t \bmod p_j \right) \bmod p_j \tag{6}$$

Based on the definition of the permissive circuit and Eq. (5), we can get $\left| C'\left(c_1 \bmod p_j, \ldots, c_t \bmod p_j \right) \bmod p_j \right| \le 2^{\eta-4} \le p_j/8$, so $\left| C'\left(c_1 \bmod p_j, \ldots, c_t \bmod p_j \right) \bmod p_j \right| = C'\left(c_1 \bmod p_j, \ldots, c_t \bmod p_j \right)$, then the Eq. (6) can be change to $c \bmod p_j = C'\left(c_1 \bmod p_j, \ldots, c_t \bmod p_j \right)$, namely, $\left[c \bmod p_j \right]_2 = \left[C'\left(\left[c_1 \bmod p_j \right]_2, \ldots, \left[c_t \bmod p_j \right]_2 \right) \right]_2 = C(\mathbf{m}_1[j], \ldots, \mathbf{m}_t[j])$. The correctness is proved.

5.2   Semantic Security Proof of the PKCB-FHE Scheme

From an ideal perspective, we hope to prove the PKCB-FHE scheme based on the same security assumption with the original DGHV scheme. That is to say, the approximate greatest common divisor (GCD) problem with the unsolvable assumption is semantically secure. However, in actual design, a slightly stronger assumption, namely, the approximate GCD of non-noise, was introduced.

(1) Proof strategy

The semantic security proof of the PKCB-FHE scheme is based on the following strategy. First, it is necessary to prove that the scheme is semantically secure under a new assumption, which is referred to as "non-noise l policy decision approximate GCD assumption". Then, we hope to prove the non-noise l policy decision approximate GCD problem is difficult if the non-noise approximate GCD problem is difficult. Finally, we can standardize the semantic security of the PKCB-FHE scheme to non-noise approximate GCD problem.

(2) Non-noise l policy decision approximate GCD problem and its security proof

Theorem 1. Based on the random oracle model, the PKCB-FHE scheme satisfies the semantic security under the non-noise approximate GCD assumption.

Given the integer q_0 and $p_0, \ldots, p_{l-1}$, the oracle $O_{q_0,(p_i)_{0 \leq i < l}}(\mathbf{v})$ is defined as follows. Vector $\mathbf{v} \in \mathbb{Z}^l$ is defined as input, $x = \mathrm{CRT}_{q_0,(p_i)_{0 \leq i < l}}(q, 2r_0 + v_0, \ldots, 2r_{l-1} + v_{l-1})$ is output, where $q \leftarrow [0, q_0)$, $r_i \leftarrow (-2^\rho, 2^\rho)$. $\mathrm{CRT}_{q_0,(p_i)_{0 \leq i < l}}(q, a_0, \ldots, a_{l-1})$ represents the integer u smaller than $x_0 = q_0 \cdot \prod_{i=0}^{l-1} p_i$, and for all $0 \leq i < l$, u satisfies the condition $u \equiv q[q_0]$ and $u \equiv a_i[p_i]$. Regarding an output $O_{q_0,(p_i)}(\mathbf{v})$ that corresponds to the ciphertext of the plaintext, the related component v_i of vector $\mathbf{v}$ can be any integer.

Definition 1. The non-noise l policy decision approximate GCD problem (hereinafter referred to as EF-l-dec-AGCD problem) is defined as follows: The random integers $p_0, \ldots, p_{l-1}$ with η-bits are chosen to make sure their product is π. Then, a non-square random 2^{λ^2}-rough number $q_0 \leftarrow \mathbb{Z} \cap [0, 2^\gamma/\pi)$ and a random bit b are selected. The conditions $\mathbf{v_0} = (0, \ldots, 0)$ and $\mathbf{v_1} \leftarrow \{0, 1\}^l$ are defined, $x_0 = q_0 \cdot p_0 \cdots p_{l-1}$, $z = O_{q_0,(p_i)}(\mathbf{v_b})$ and the input of oracle is $O_{q_0,(p_i)}$, then b is predicted.

Lemma 2. Based on the random oracle model, the PKCB-FHE scheme is semantically secure under the assumption of the EF-l-dec-AGCD unsolvable problem.

(3) Proof of Lemma 2

The proof process is presented as follows: in an attacking case, the adversary receives the public key first. Then, two l-bits messages $\mathbf{m}_0$ and $\mathbf{m}_1$ are generated. The challenger will send back the encrypted ciphertext $\mathbf{m}_b$ in terms of the random bit b. The adversary will generate predicted result b'. If $b' = b$, then the attack is successful.

6 Performance Analysis

We analyse and compare the performance of the proposed PKCB-FHE in terms of the size of subset public key and storage space, the complexity of public key computation and the efficiency of scheme processing.

The length of the subset public key parameter is the actual size of the public key parameters to be stored. The public key is formed by a shorter length of subset public key parameters through a certain computation, and this subset public key parameter also retains the basic algebraic structure of the public key and reverts back to the public key easily. The subset public key parameters are the public key parameters we actually need to store.

6.1 The Length of Subset Public Key and the Storage Space

Based on the parameters' restrictions described in Sect. 4.4, the performance analysis of the PKCB-FHE scheme is presented. Theoretically, we compare the lengths of the subset public key parameter as follows. Here, $p\mathfrak{k}$ represents the subset public key parameter.

The subset public key parameter of the PKCB-FHE scheme is $\mathfrak{pf} = (\text{se}, \delta_{1,0}, \delta_{1,1}, \ldots \delta_{\beta,0}, \delta_{\beta,1})$, $1 \le i \le \beta$, $b \in \{0, 1\}$. Its overall size is $\text{sizeof}(\text{se}) + \text{numof}(\delta_{i,b}) \cdot \text{sizeof}(\delta_{i,b}) = 2\beta \cdot (\eta + \lambda) = \tilde{O}(\lambda^{3.5})$. However, the public key parameter of the original DGHV scheme is $\mathfrak{pf} = (x_1, \ldots, x_\tau)$, $1 \le i \le \tau$, its size is $\text{numof}(x_i) \cdot \text{sizeof}(x_i) = \tau \cdot \gamma = \tilde{O}(\lambda^8)$. Moreover, the subset public key parameter of QF-PKC scheme [16] is $\mathfrak{pf} = (x_{1,0}, x_{1,1}, \ldots x_{\beta,0}, x_{\beta,1})$, and its size is $\text{numof}(x_{i,b}) \cdot \text{sizeof}(x_{i,b}) = (2\beta + 1) \cdot \gamma = \tilde{O}(\lambda^{6.5})$. The subset public key parameter of C-PKC scheme [17] is $\mathfrak{pf} = (\text{se}, \delta_1, \ldots, \delta_\tau)$, and its size is $\text{sizeof}(\text{se}) + \text{numof}(\delta_i) \cdot \text{sizeof}(\delta_i) = \tau \cdot (\eta + \lambda) = \tilde{O}(\lambda^5)$. Table 2 lists the lengths of the subset public key parameter of these four schemes.

Table 2. Comparison of the public key size

	PKCB-FHE	DGHV [8]	QF-PKC [16]	C-PKC [17]
The size of the subset public key parameter	$\tilde{O}(\lambda^{3.5})$	$\tilde{O}(\lambda^8)$	$\tilde{O}(\lambda^{6.5})$	$\tilde{O}(\lambda^5)$

In addition, based on the FHE code package provided by Coron et al. and the python's Sage 4.7.2 algebra libraries [16], we test and analyse the performance and the length of the public key. Table 3 shows the public key size comparison of the QF-PKC and C-PKC schemes and the proposed PKCB-FHE scheme when the security parameter is in a larger scale such as $\lambda = 72$.

Table 3. Parameters selection and the public key size

$\lambda = 72$	ρ	η	$\gamma \times 10^{-6}$	β	α	$\mathfrak{pf}$
QF-PKC [16]	39	2652	19.00	88	/	797.5 MB
C-PKC [17]	71	2698	19.35	2556	7659	5.77 MB
PKCB-FHE	71	2840	19.00	88	2556	135.7 KB

As shown in Table 3, the proposed PKCB-FHE scheme, which uses the quadratic variable and combines the offset processing technique when generating the public key, can reduce the size of the public key by an order of 10^3 compared to the other schemes. For instance, when $\lambda = 72$, the storage space needed for the PKCB-FHE scheme is 135.7 KB. This is far less than the 797.5 MB of the QF-PKC scheme and 5.77 MB of the C-PKC scheme.

6.2   The Length of the Public Key

The public key of the PKCB-FHE scheme is generated by the public key generation module $\mathbf{pk} = (\mathbf{pk}^*, \mathbf{pk}_\sigma, \mathbf{y})$, with $\mathbf{pk}^* = \langle x_0, \text{se}_x, \varphi_{i,b}, \varphi_{j,b}^{(\mathcal{W})}, \varphi_{j,b}^{(\mathcal{B})} \rangle$, $\mathbf{pk}_\sigma = \left(\text{se}_\sigma, \varphi_{k,b}^\sigma \right)$, $\mathbf{y} = (y_0, \ldots, y_{\Theta-1})$, $0 \le i < \beta$, $0 \le j < \beta^{(l)}$, $0 \le k < \beta_\Theta$, $b \in \{0, 1\}$.

We calculate the length of the public key in the PKCB-FHE scheme as
$\text{sizeof}(se) + \text{sizeof}(x_0) + \text{numof}(\varphi_{I,b}) \cdot \text{sizeof}(\varphi_{I,b}) + \text{numof}(\varphi_{j,b}^{(W)}) \cdot$
$\text{sizeof}(\varphi_{j,b}^{(W)}) + \text{numof}(\varphi_{j,b}^{(\mathcal{B})}) \cdot \text{sizeof}(\varphi_{j,b}^{(\mathcal{B})}) + \text{numof}(\varphi_{k,b}^{\sigma}) \cdot \text{sizeof}(\varphi_{k,b}^{\sigma}) +$
$\text{sizeof}(\mathbf{y}) = \gamma + 2\beta \cdot (\eta + \lambda) + 4\beta^{(l)} \cdot (\lambda + \log_2(l) + l \cdot \eta) + \kappa \cdot \Theta = \tilde{O}(\lambda^5) +$
$\tilde{O}(\lambda^{3.5}) + (4\sqrt{l} + \Theta) \cdot (\lambda + \log_2(l) + l \cdot \tilde{O}(\lambda^2)) = \tilde{O}(\lambda^5) + l^{1.5} \cdot \tilde{O}(\lambda^2) + l \cdot$
$\tilde{O}(\lambda^5) + \tilde{O}(\lambda^8)$ when l bits of plaintext is processed.

The public key of the C-PKC scheme is $\mathbf{pk} = (se, x_0, \delta_1, \ldots, \delta_\tau, \mathbf{y})$, $1 \leq i \leq \tau$. The
length of the public key is computed as $l \cdot (\text{sizeof}(x_0) + \text{sizeof}(se) + \text{numof}(\delta_i) \cdot$
$\text{sizeof}(\delta_i) + \text{sizeof}(\mathbf{y})) = l \cdot (\tau \cdot (\eta + \lambda) + \kappa \cdot \Theta) = l \cdot \tilde{O}(\lambda^8)$ with respect to l bits
plaintext processing.

The public key of the original DGHV scheme is $\mathbf{pk} = (x_0, x_1, \ldots, x_\tau, \mathbf{y})$, $0 \leq i \leq \tau$.
The length of its public key is $l \cdot (\text{numof}(x_i) \cdot \text{sizeof}(x_i) + \text{sizeof}(\mathbf{y})) =$
$l \cdot (\tau \cdot \gamma + \kappa \cdot \Theta) = l \cdot \tilde{O}(\lambda^8)$ when l bits plaintext are processed.

If we define $l \ll \tilde{O}(\lambda^3)$, the specific security parameters are $\lambda = 72$ and $l = 512$.
Thus, $l^{1.5} \cdot \tilde{O}(\lambda^2) \ll l \cdot \tilde{O}(\lambda^5) \ll \tilde{O}(\lambda^8)$. Then, the total length of the public key in the
PKCB-FHE scheme is $\tilde{O}(\lambda^8)$ when l bits of plaintext are processed at the same time.

Table 4 presents the different public key lengths in the three different schemes
when l bits of plaintext are processed. The proposed PKCB-FHE scheme has the
minimum public key length. It is less than that of the original DGHV scheme and the
C-PKC scheme when l bits of plaintext are processed.

Table 4. Comparison of the length of public key when l bits plaintext is processed

	PKCB-FHE	DGHV [8]	C-PKC [17]
The length of public key	$\tilde{O}(\lambda^8)$	$l \cdot \tilde{O}(\lambda^8)$	$l \cdot \tilde{O}(\lambda^8)$

6.3 Processing Time

In the key generation phase of the PKCB-FHE scheme, primary generation of q_0
prolongs the key generation time. However, due to the introduction of batch processing, the prolonged time to generate q_0 in the PKCB-FHE scheme is less than the
time to process the l-bit plaintexts in the QFC-PKC scheme. Thus, additional time cost
is not incurred in the PKCB-FHE scheme.

The key generation phase of the PKCB-FHE scheme also includes the generation of
the public key primary parameter with correction, batch processing plaintext parameter,
and ciphertext noise parameter with correction. However, that time only increases
approximately three times relative to the correction parameter generation time of the
QFC-PKC scheme. Moreover, if we use the QFC-PKC scheme to process l bits of
plaintext instead of the batch processing technology, the time of the correction
parameter generation will be l times longer than the time of the PKCB-FHE scheme.

The recovery rate of public parameters in the encryption module of PKCB-FHE scheme is faster than that in the C-PKC scheme. In the PKCB-FHE scheme, the total overhead times required to recover the public primary parameter, batch processing plaintext parameter, and ciphertext noise parameters are less than the overhead time required to process l bits of plaintext in non-batch processing C-PKC scheme.

7   Conclusions

The design ideas and technical framework of the FHE scheme with public key compression and batch processing are proposed in this paper. We present the structural design of FHE with semantic security firstly, and then propose the PKCB-SWHE scheme based on a parametric quadratic public key with correction and batch processing techniques. Furthermore, the squash decryption circuit based on batch processing is analysed. Together with the bootstrapping technique and in-depth replacement operation, the PKCB-SWHE scheme is extended to a fully homomorphic encryption scheme (PKCB-FHE scheme). With the determined parameter limitation, the PKCB-FHE scheme is proved to be semantically secure based on the non-noise approximate GCD assumption and the random oracle model. Finally, the size of the storage public key, the length of public key and the scheme efficiency are analysed. Compared to those existing schemes such as the DGHV scheme, the C-PKC scheme and the QF-PKC scheme, the proposed PKCB-FHE scheme has the advantage of a smaller public key size and higher processing efficiency.

Acknowledgements. This work was supported in part by the European Commission Marie Curie IRSES project "AdvIOT" and the national Natural Science Foundation of China (NSFC) under grant No.61372103.

References

1. Kaosar, M.G., Paulet, R., Yi, X.: Fully homomorphic encryption based two-party association rule mining. Data Knowl. Eng. **76**, 1–15 (2012)
2. Yan, H., Li, J., Han, J.: A novel efficient remote data possession checking protocol in cloud storage. IEEE Trans. Inf. Forensics Secur. **12**(1), 78–88 (2017)
3. Wang, W., Hu, Y., Chen, L., Huang, X.: Exploring the feasibility of fully homomorphic encryption. IEEE Trans. Comput. **64**(3), 698–706 (2015)
4. Cheon, J.H., Kim, J.: A hybrid scheme of public-key encryption and somewhat homomorphic encryption. IEEE Trans. Inf. Forensics Secur. **10**(5), 1208–1212 (2015)
5. Rivest, R., Adleman, L., Dertouzos, M.: On data banks and privacy homomorphisms. Found. Secur. Comput. **4**(11), 169–180 (1978)
6. Gentry, C.: A Fully Homomorphic Encryption Scheme. Stanford University, Stanford (2009)
7. Gentry, C.: Fully homomorphic encryption using ideal lattices. In: Proceedings of the 41st Annual ACM Symposium on Theory of Computing, New York, vol. 9, pp. 169–178 (2009)
8. van Dijk, M., Gentry, C., Halevi, S., Vaikuntanathan, V.: Fully homomorphic encryption over the integers. In: Gilbert, H. (ed.) EUROCRYPT 2010. LNCS, vol. 6110, pp. 24–43. Springer, Heidelberg (2010). https://doi.org/10.1007/978-3-642-13190-5_2

9. Stehlé, D., Steinfeld, R.: Faster fully homomorphic encryption. In: Abe, M. (ed.) ASIACRYPT 2010. LNCS, vol. 6477, pp. 377–394. Springer, Heidelberg (2010). https://doi.org/10.1007/978-3-642-17373-8_22

10. Brakerski, Z.: Fully homomorphic encryption without modulus switching from classical GapSVP. In: Safavi-Naini, R., Canetti, R. (eds.) CRYPTO 2012. LNCS, vol. 7417, pp. 868–886. Springer, Heidelberg (2012). https://doi.org/10.1007/978-3-642-32009-5_50

11. Brakerski, Z., Vaikuntanathan, V.: Efficient fully homomorphic encryption from (Standard) LWE. In: Proceedings of IEEE 52nd Annual Symposium on Foundations of Computer Science (FOCS), pp. 97–106 (2011)

12. Brakerski, Z., Gentry, C., Vaikuntanathan, V.: Fully homomorphic encryption without bootstrapping. In: Proceedings of the 3rd Innovations in Theoretical Computer Science Conference (ITCS), pp. 309–325 (2012)

13. Zhang, X., Xu, C., Jin, C.: Efficient fully homomorphic encryption from RLWE with an extension to a threshold encryption scheme. Future Gener. Comput. Syst. **36**, 180–186 (2014)

14. Plantard, T., Susilo, W., Zhang, Z.: Fully homomorphic encryption using hidden ideal lattice. IEEE Trans. Inf. Forensics Secur. **8**(12), 2127–2137 (2013)

15. Coron, J.S., Naccached, D., Tibouchi, M.: Optimization of fully homomorphic encryption. IACR Cryptology ePrint Archive, pp. 440–458 (2011)

16. Coron, J.-S., Mandal, A., Naccache, D., Tibouchi, M.: Fully homomorphic encryption over the integers with shorter public keys. In: Rogaway, P. (ed.) CRYPTO 2011. LNCS, vol. 6841, pp. 487–504. Springer, Heidelberg (2011). https://doi.org/10.1007/978-3-642-22792-9_28

17. Coron, J.-S., Naccache, D., Tibouchi, M.: Public key compression and modulus switching for fully homomorphic encryption over the integers. In: Pointcheval, D., Johansson, T. (eds.) EUROCRYPT 2012. LNCS, vol. 7237, pp. 446–464. Springer, Heidelberg (2012). https://doi.org/10.1007/978-3-642-29011-4_27

18. Chen, Z., Wang, J., Zhang, Z., Song, X.: A fully homomorphic encryption scheme with better key size. China Commun. **28**(4), 82–92 (2014)

19. Smart, N.P., Vercauteren, F.: Fully homomorphic encryption with relatively small key and ciphertext sizes. In: Nguyen, P.Q., Pointcheval, D. (eds.) PKC 2010. LNCS, vol. 6056, pp. 420–443. Springer, Heidelberg (2010). https://doi.org/10.1007/978-3-642-13013-7_25

20. Smart, N.P., Vercauteren, F.: Fully homomorphic SIMD operations. Des. Codes Crypt. **71**(1), 57–81 (2014)

21. Cheon, J.H., Coron, J.-S., Kim, J., Lee, M.S., Lepoint, T., Tibouchi, M., Yun, A.: Batch fully homomorphic encryption over the integers. In: Johansson, T., Nguyen, P.Q. (eds.) EUROCRYPT 2013. LNCS, vol. 7881, pp. 315–335. Springer, Heidelberg (2013). https://doi.org/10.1007/978-3-642-38348-9_20

22. Beneš, V.E.: Optimal rearrangeable multistage connecting networks. Bell Syst. Tech. J. **43**(4), 1641–1656 (2013)

Leveled FHE with Matrix Message Space

Biao Wang[1,2], Xueqing Wang[1,2], and Rui Xue[1,2(✉)]

[1] State Key Laboratory of Information Security, Institute of Information
Engineering, Chinese Academy of Sciences, Beijing 100093, China
{wangbiao,wangxueqing,xuerui}@iie.ac.cn
[2] School of Cyber Security, University of Chinese Academy of Sciences,
Beijing 100049, China

Abstract. Up to now, almost all fully homomorphic encryption (FHE)
schemes can only encrypt bit or vector. In PKC 2015, Hiromasa et al.
[12] constructed the only leveled FHE scheme that encrypts matrices
and supports homomorphic matrix addition and multiplication. But the
ciphertext size of their scheme is somewhat large and the security of their
scheme depends on some special kind of circular security assumption.

We propose a leveled FHE scheme that encrypts matrices and supports homomorphic matrix addition, multiplication and Hadamard product. It can be viewed as matrix-packed FHE, and has much smaller
ciphertext size. Its security is only based on LWE assumption. In particular, the advantages of our scheme are:

1. Supporting homomorphic matrix Hadamard product. All entries in
   plaintext matrices can be viewed as plaintext slots. While the scheme
   in [12] doesn't support this homomorphic operation and only the
   diagonal entries of plaintext matrix can be viewed as plaintext slots.
2. Small ciphertext size. For a plaintext matrix $M \in \{0,1\}^{r \times r}$, the size
   of ciphertext matrix is $r \times (n+r)$, in contrast to $(n+r) \times (n+r) \lceil \log q \rceil$
   in [12].
3. Standard assumption. The security is based on LWE assumption
   merely, while the security of scheme in [12] depends additionally on
   some special kind of circular security assumption.

As Brakerski's work [3] in CRYPTO 2012, our scheme can be improved
in efficiency by using ring-LWE (RLWE).

Keywords: Fully homomorphic encryption · LWE · Matrix · Packing

1 Introduction

Fully Homomorphic Encryption (FHE) is one of the holy grails of modern cryptography. For short, a FHE scheme is an encryption scheme that allows anyone
to perform arbitrary computations on encrypted data using only public information. With this fascinating feature, FHE has many theoretical and practical
applications, a typical one of which is outsourcing computation to untrusted
entities without compromising one's privacy.

© Springer International Publishing AG, part of Springer Nature 2018
X. Chen et al. (Eds.): Inscrypt 2017, LNCS 10726, pp. 260–277, 2018.
https://doi.org/10.1007/978-3-319-75160-3_17

FHE was first introduced by Rivest, Adleman and Dertouzos [17] in 1978. But the first candidate scheme, Gentry's groundbreaking work in 2009 [7,8], came thirty years later. While Gentry's work is a major breakthrough, it is far from efficient in the practical point of view. Since 2009, a lot of designs [2–6,9–11,19,20] have been proposed towards more efficient FHE schemes. Among these works, ciphertext packing is one of the main techniques [2,10,20]. With ciphertext packing, we can pack multiple messages into one ciphertext and apply Single Instruction Multiple Data (SIMD) homomorphic operations to many encrypted messages. Smart and Vercauteren [20], for the first time, introduced a polynomial-CRT based ciphertext-packing technique, which partitions the message space of the Gentry's FHE scheme [7] into a vector of plaintext slots. By employing this technique, Gentry, Halevi, and Smart [10] presented a FHE scheme with poly-log overhead. Based on [10], Gentry et al. [9] obtained a bootstrapping method that works in time quasi-linear in the security parameter. Brakerski, Gentry and Halevi [1] proposed a new design of LWE-based FHE by employing another packing technique [15] to pack messages in [2,3,5].

After that, Hiromasa et al. [12] constructed a matrix variant of GSW-FHE [11]. Their scheme can be viewed as packed FHE in the sense that plaintext slots in packed FHE correspond to diagonal entries of plaintext matrices in their matrix GSW-FHE scheme.

1.1 Related Work

In PKC 2015, Hiromasa et al. [12] constructed the only leveled fully homomorphic encryption scheme that encrypts matrices and supports homomorphic matrix addition and multiplication. Their scheme is a matrix variant of GSW-FHE [11]. Recall that in GSW-FHE, a ciphertext of a plaintext $m \in \{0,1\}$ is a matrix $C \in \mathbb{Z}_q^{(n+1) \times N}$ such that $sC = m \cdot sG + e$ for a secret key vector $s \in \mathbb{Z}_q^{n+1}$, a small noise vector $e \in \mathbb{Z}^N$, and a fixed matrix $G \in \mathbb{Z}_q^{(n+1) \times N}$. The idea of Hiromasa et al. in constructing matrix FHE is to extend the above equation to the form of $SC = MSG + E$ where $S \in \mathbb{Z}_q^{r \times (n+r)}$ is a secret key matrix, $C \in \mathbb{Z}_q^{(n+r) \times N}$ is a ciphertext matrix, $M \in \{0,1\}^{r \times r}$ is a plaintext matrix, $G \in \mathbb{Z}_q^{(n+r) \times N}$ is a fixed matrix, and $E \in \mathbb{Z}^{r \times N}$ is a small noise matrix. In their construction, homomorphic matrix addition is just matrix addition. Homomorphic matrix multiplication is computed by $C_1 G^{-1}(C_2)$ where for a matrix $C \in \mathbb{Z}_q^{(n+r) \times N}$, $G^{-1}(C)$ is the function that outputs a matrix $X \in \mathbb{Z}^{N \times N}$ with subgaussian parameter $O(1)$ such that $GX = C$. To satisfy the equation $SC = MSG + E$, their construction (symmetric scheme) uses ciphertexts of the following form:

$$BR + \begin{pmatrix} MS \\ 0 \end{pmatrix} G$$

where $S = (I_r \parallel S') \in \mathbb{Z}_q^{r \times (n+r)}$ is a secret key matrix, $B \in \mathbb{Z}_q^{(n+r) \times N}$ is the LWE matrix such that $SB = E$, and $R \in \{0,1\}^{N \times N}$ is a random matrix.

The method similar to [18] is used to translate the resulting symmetric scheme to the asymmetric one. In particular, they published symmetric encryption ciphertexts $P_{(i,j)}$ of $M_{(i,j)}$ for all $i, j \in [r]$ where $M_{(i,j)}$ is the matrix with 1 in the (i,j)-th entry and 0 in the others. A ciphertext (asymmetric scheme) C for a plaintext matrix M is computed by summing all $P_{(i,j)}$ for which the (i,j)-th entry of M is equal to 1, and adding BR to randomize the result.

When the matrix FHE scheme in [12] is viewed as packed FHE, only the diagonal entries of plaintext matrix can be viewed as plaintext slots. This property is not so well since there are many "invalid" entries in plaintext matrix. Additionally, a special kind of circular security assumption is needed to prove the security of asymmetric construction in [12].

1.2   Our Results

We propose a leveled FHE scheme that encrypts matrices and supports homomorphic matrix addition, multiplication and Hadamard product. (We also call this scheme "matrix FHE" for short)

Our scheme has the following features:

1. **Hadamard Product**. Our scheme supports homomorphic matrix Hadamard product. As a result, all entries in plaintext matrix can be viewed as plaintext slots. While the scheme in [12] doesn't support this homomorphic operation and only the diagonal entries of plaintext matrix can be viewed as plaintext slots.
2. **Short Ciphertext**. For a plaintext matrix $M \in \{0,1\}^{r \times r}$, the size of ciphertext matrix is $r \times (n+r)$, which is much smaller than $(n+r) \times (n+r)\lceil \log q \rceil$ in [12].
3. **Standard assumption**. The security of our scheme is based on LWE assumption merely, while the security of scheme in [12] depends additionally on some special kind of circular security assumption.

We believe that ciphertext extension of our matrix FHE scheme is optimal for LWE-based FHE schemes that encrypt matrices with plaintext space $\{0,1\}^{r \times r}$. The reason is as follows. When encrypting matrices in $\{0,1\}^{r \times r}$, the secret key matrix must have the form of $S = \left(\begin{smallmatrix} I_r \\ S' \end{smallmatrix} \right) \in \mathbb{Z}_q^{(n+r) \times r}$ or $S = (I_r \parallel S') \in \mathbb{Z}_q^{r \times (n+r)}$. Recall that the decryption function for every LWE-based cryptosystem requires inner product operation between ciphertext vector and secret key vector, thus the ciphertext matrix for LWE-based FHE schemes that encrypt matrices must have a size of at least $r(n+r)$. A similar analysis also holds for our matrix FHE scheme in RLWE case.

1.3   Our Techniques

Our starting point is the Brakerski's FHE scheme [3]. Recall that in [3], the ciphertext of a plaintext $m \in \{0,1\}$ is a vector $c \in \mathbb{Z}_q^{n+1}$ such that $[\langle c, s \rangle]_q = \lfloor q/2 \rfloor \cdot m + e'$ for a secret key vector $s \in \mathbb{Z}_q^{n+1}$ and a small noise $e' \in \mathbb{Z}$.

To generalize this scheme to a matrix FHE scheme, we first turn ciphertext vector $\boldsymbol{c}$ and secret key vector $\boldsymbol{s}$ to the form of matrix, and convert $e' \in \mathbb{Z}$ to $\boldsymbol{E}' \in \mathbb{Z}^{r \times r}$. In this case, the corresponding plaintext space becomes $\{0,1\}^{r \times r}$. More details are given in the following sections.

Note that our goal is to achieve the generalized relationship $[\boldsymbol{CS}]_q = \lfloor q/2 \rfloor \cdot \boldsymbol{M} + \boldsymbol{E}'$. For this purpose, let $\boldsymbol{S} = \left(\begin{smallmatrix} I_r \\ \boldsymbol{S}' \end{smallmatrix} \right) \in \mathbb{Z}_q^{(n+r) \times r}$ where $\boldsymbol{S}' \leftarrow \mathbb{Z}_q^{n \times r}$, $\boldsymbol{A}$ be a LWE matrix and $\boldsymbol{E}$ denote the corresponding noise matrix such that $\boldsymbol{AS} = \boldsymbol{E}$ (take $\boldsymbol{A} = ((\boldsymbol{E} + \boldsymbol{A}'\boldsymbol{S}') \| - \boldsymbol{A}') \in \mathbb{Z}_q^{m \times (n+r)}$ where $\boldsymbol{A}' \leftarrow \mathbb{Z}_q^{m \times n}$), and sample $\boldsymbol{R} \leftarrow \{0,1\}^{r \times m}$. Set $\boldsymbol{C} = \lfloor q/2 \rfloor \cdot (\boldsymbol{M} \| \boldsymbol{0}_{r \times n}) + \boldsymbol{RA}$, then we have $[\boldsymbol{CS}]_q = \lfloor q/2 \rfloor \cdot \boldsymbol{M} + \boldsymbol{E}'$ where $\boldsymbol{E}' = \boldsymbol{RE}$.

We consider three types of homomorphic operations (matrix addition, matrix multiplication and matrix Hadamard product), and the corresponding plaintext operations are all under mod 2.

It is easy to see that homomorphic matrix addition is just matrix addition, since

$$[(\boldsymbol{C}_1 + \boldsymbol{C}_2)\boldsymbol{S}]_q = \lfloor q/2 \rfloor \cdot [\boldsymbol{M}_1 + \boldsymbol{M}_2]_2 + (\boldsymbol{E}_1 + \boldsymbol{E}_2).$$

Homomorphic matrix multiplication is somewhat intractable. Let $\boldsymbol{C}_1$ and $\boldsymbol{C}_2$ be two ciphertexts which satisfy $[\boldsymbol{C}_1\boldsymbol{S}]_q = \lfloor q/2 \rfloor \cdot \boldsymbol{M}_1 + \boldsymbol{E}_1$ and $[\boldsymbol{C}_2\boldsymbol{S}]_q = \lfloor q/2 \rfloor \cdot \boldsymbol{M}_2 + \boldsymbol{E}_2$ respectively. By multiplying the above two equations, we have $[\lfloor \frac{2}{q}(\boldsymbol{C}_1\boldsymbol{S}) \cdot (\boldsymbol{C}_2\boldsymbol{S})]]_q = \lfloor \frac{q}{2} \rfloor \cdot [\boldsymbol{M}_1\boldsymbol{M}_2]_2 + \boldsymbol{E}'$. In order to obtain the resulting ciphertext of homomorphic multiplication, $\boldsymbol{C}_1\boldsymbol{S}\boldsymbol{C}_2\boldsymbol{S}$ must have the form of $\boldsymbol{C}^*\boldsymbol{S}^*$, i.e., the product of ciphertext matrix and secret key matrix. Fortunately, this requirement can be satisfied by setting $\boldsymbol{C}^* = (\boldsymbol{C}_1 \otimes \boldsymbol{C}_2)_{sr}$, $\boldsymbol{S}^* = (\boldsymbol{S} \otimes' \boldsymbol{S})_{sc}$. Here, the operators $\otimes$, $\otimes'$, sr and sc are defined at the beginning of Sect. 2.

Homomorphic matrix Hadamard product is implemented in the way similar to homomorphic matrix multiplication. In particular, let $\boldsymbol{C}_1$ and $\boldsymbol{C}_2$ be two ciphertexts which satisfy $[\boldsymbol{C}_1\boldsymbol{S}]_q = \lfloor q/2 \rfloor \cdot \boldsymbol{M}_1 + \boldsymbol{E}_1$ and $[\boldsymbol{C}_2\boldsymbol{S}]_q = \lfloor q/2 \rfloor \cdot \boldsymbol{M}_2 + \boldsymbol{E}_2$ respectively. By executing Hadamard product (denoted by $\circ$) of the above two equations, we have $[\lfloor \frac{2}{q}(\boldsymbol{C}_1\boldsymbol{S}) \circ (\boldsymbol{C}_2\boldsymbol{S})]]_q = \lfloor \frac{q}{2} \rfloor \cdot [\boldsymbol{M}_1 \circ \boldsymbol{M}_2]_2 + \boldsymbol{E}''$. Setting $\boldsymbol{C}^* = (\boldsymbol{C}_1 \otimes \boldsymbol{C}_2)_{er}$, $\boldsymbol{S}^* = (\boldsymbol{S} \otimes \boldsymbol{S})_{ec}$ in which the operators er and ec are also defined at the beginning of Sect. 2, we have $[\lfloor \frac{2}{q}\boldsymbol{C}^*\boldsymbol{S}^*]]_q = \lfloor \frac{q}{2} \rfloor \cdot [\boldsymbol{M}_1 \circ \boldsymbol{M}_2]_2 + \boldsymbol{E}''$.

Up to now, we obtain the resulting ciphertexts of homomorphic operations which involve Kronecker product of matrices and some other operations. These operations enlarge the size of ciphertext matrices and secret key matrices. To solve this problem, we generalize the key switching technology in [2] and transform the resulting ciphertexts to ciphertexts of the same size as the original ciphertexts.

1.4   Organization

The rest of the paper is organized as follows. In Sect. 2, we first define some basic notation and notion, and then introduce some preliminaries on the LWE assumption, homomorphic encryption and bootstrapping. Section 3 presents our construction of (leveled) matrix FHE scheme. In Sect. 4, we discuss the homomorphic properties and security of our matrix FHE scheme. Section 5 analyzes our matrix FHE scheme from the aspect of packing.

2   Preliminaries

Let λ denote a security parameter. When we speak of a negligible function $\text{negl}(\lambda)$, we mean a function that is asymptotically bounded from above by the reciprocal of all polynomials in λ.

Basic Notation. A bold upper-case letter like M will denote a matrix, and M_{ij} is the element in its i-th row and j-th column. The transpose of matrix M is denoted as M^T, meanwhile the i-th row and i-th column of M is denoted, respectively, by M_i and M_i^T. Matrix I_r is the $r \times r$-dimension identity matrix, and $0_{m \times n}$ is the $m \times n$-dimension zero matrix. $A \| B$ will be the concatenation of A and B.

The ℓ_1 norm $\ell_1(M)$ of the matrix M means the maximum ℓ_1 norm of columns in matrix M and the ℓ_∞ norm $\|M\|_\infty$ of the matrix M means the maximum ℓ_∞ norm of columns in matrix M.

We assume that vectors are in column form and are denoted by bold lower-case letters, e.g., x. The inner product between two vectors is denoted by $\langle x, y \rangle$.

For a nonnegative integer n, we let $[n] = \{1, 2, ..., n\}$. We denote the set of integers by $\mathbb{Z}$, and for an integer q, we let $\mathbb{Z}_q$ denote the quotient ring of integers modulo q. For $a \in \mathbb{Z}$, we use the notation $[a]_q$ to refer to $a \bmod q$, with coefficients reduced into the range $(-q/2, q/2]$.

Basic Notion. Hadamard product of matrices (also known as entrywise product): For two matrices A and B of the same dimension $m \times n$, the Hadamard product $A \circ B$ is the matrix of the same dimension with elements given by $(A \circ B)_{ij} = A_{ij} \cdot B_{ij}$. Note that the Hadamard product is associative and distributive, and unlike the matrix product it is also commutative.

Kronecker product of matrices and its variant: For two matrices $A_{m \times n}$ and $B_{s \times t}$, the Kronecker product (also known as tensor product) $(A \otimes B)_{ms \times nt}$ and its variant $(A \otimes' B)_{ms \times nt}$ are defined, respectively, as following:

$$A \otimes B := \begin{pmatrix} a_{11}B & \cdots & a_{1n}B \\ \vdots & \ddots & \vdots \\ a_{m1}B & \cdots & a_{mn}B \end{pmatrix}$$

$$A \otimes' B := \begin{pmatrix} a_{11}B_1^T & \cdots & a_{1n}B_1^T & \cdots & a_{11}B_t^T & \cdots & a_{1n}B_t^T \\ \vdots & & \vdots & \ddots & & & \vdots \\ a_{m1}B_1^T & \cdots & a_{mn}B_1^T & \cdots & a_{m1}B_t^T & \cdots & a_{mn}B_t^T \end{pmatrix}$$

Stretching by row (or by column): For a matrix C of dimension $m^2 \times n^2$, the stretching matrix C_{sr} of C by row is the matrix of dimension $m \times mn^2$ whose

i-th row is composed of the concatenation of $((i-1)m+1)$-th to im-th row of C. The stretching matrix C_{sc} of C by column is defined in the similar way, which is a matrix of dimension $m^2 n \times n$.

Extracting by row (resp. column): For a matrix C of dimension $m^2 \times n^2$, the result of extracting by row (resp. column) of matrix C is the matrix C_{er} (resp. C_{ec}) of dimension $m \times n^2$ (resp. $m^2 \times n$) whose i-th row (resp. column) is the $((i-1)m+i)$-th row (resp. the $((i-1)n+i)$-th column) of C.

2.1 The Learning with Errors (LWE) Problem

The learning with errors (LWE) problem was introduced by Regev [16] as a generalization of learning parity with noise.

Definition 1 (LWE). *For security parameter λ, let $n = n(\lambda)$ be an integer dimension, $q = q(\lambda) \geq 2$ an integer, and $\chi = \chi(\lambda)$ a distribution over $\mathbb{Z}$. The $\mathrm{LWE}_{n,q,\chi}$ problem is to distinguish the following distribution $(\boldsymbol{a}_i, b_i)$ from the uniform distribution over $\mathbb{Z}_q^{n+1}$: to draw $\boldsymbol{a}_i$, $\boldsymbol{s} \leftarrow \mathbb{Z}_q^n$ uniformly at random and $e_i \leftarrow \chi$, and set $b_i = \langle \boldsymbol{a}_i, \boldsymbol{s} \rangle + e_i \bmod q$.*

The $\mathrm{LWE}_{n,q,\chi}$ assumption is that the $\mathrm{LWE}_{n,q,\chi}$ problem is infeasible.

There are known quantum ([16]) and classical ([14]) reductions between $\mathrm{LWE}_{n,q,\chi}$ and worst-case lattice problems as follows. These reductions take χ as a discrete Gaussian distribution, which is statistically indistinguishable from a B-bounded distribution for an appropriate B [3].

Definition 2 (B-bounded distributions). *A distribution ensemble $\{\chi_n\}_{n \in \mathbb{N}}$, supported over the integers, is called B-bounded if $\Pr_{e \leftarrow \chi_n}[|e| > B] = \mathrm{negl}(n)$.*

Theorem 1 ([3,14,16]). *Let $q = q(n) \in \mathbb{N}$ be either a prime power or a product of small (size $\mathrm{poly}(n)$) co-prime numbers, and let $B \geq \omega(\log n) \cdot \sqrt{n}$. Then there exists an efficiently sampleable B-bounded distribution χ such that if there is an efficient algorithm that solves the (average-case) $\mathrm{LWE}_{n,q,\chi}$ problem:*

- *There is an efficient quantum algorithm that solves $\mathrm{GapSVP}_{\widetilde{O}(n \cdot q/B)}$ (and $\mathrm{SIVP}_{\widetilde{O}(n \cdot q/B)}$) on any n-dimensional lattice.*
- *If in addition $q \geq \widetilde{O}(2^{n/2})$, then there is an efficient classical algorithm for $\mathrm{GapSVP}_{\widetilde{O}(n \cdot q/B)}$ on any n-dimensional lattice.*

In both cases, if one also considers distinguishers with sub-polynomial advantage, then we require $B \geq \widetilde{O}(n)$ and the resulting approximation factor is slightly larger $\widetilde{O}(n^{1.5} \cdot q/B)$.

Recall that GapSVP_γ is the promise problem to distinguish between the case in which the lattice has a vector shorter than $r \in \mathbb{Q}$, and the case in which all the lattice vectors are longer than $\gamma \cdot r$. SIVP_γ is the problem to find the set of short linearly independent vectors in a lattice. The best known algorithms for GapSVP_γ ([13]) require time $2^{\widetilde{\Omega}(n/\log \gamma)}$ (n is the dimension of the lattice).

2.2 Homomorphic Encryption and Bootstrapping

We now describe the definition of homomorphic encryption and introduce Gentry's bootstrapping theorem. Our definitions are mostly taken from [3,12]. Let $\mathcal{M}$ and $\mathcal{C}$ be the message and ciphertext space. A homomorphic encryption scheme HE consists of the following four algorithms:

- KeyGen(1^λ): output a public encryption key pk, a secret decryption key sk, and a public evaluation key evk.
- Enc(pk, m): using the public key pk, encrypt a plaintext $m \in \mathcal{M}$ into a ciphertext $c \in \mathcal{C}$.
- Dec(sk, c): using the secret key sk, recover the message $m \in \mathcal{M}$ from the ciphertext c.
- Eval($evk, f, c_1, ..., c_t$): using the evaluation key evk, output a ciphertext $c_f \in \mathcal{C}$ by applying the function $f : \mathcal{M}^t \to \mathcal{M}$ to $c_1, ..., c_t$.

A homomorphic encryption scheme is said to be secure if it is semantically secure (the adversary is given both pk and evk). Homomorphism w.r.t depth-bounded circuits and full homomorphism are defined as:

Definition 3 (L-homomorphism). *A scheme* HE *is said to be L-homomorphic if for any depth L arithmetic circuit f and any set of inputs $m_1, ..., m_t$, it holds that*

$$\Pr[\mathsf{Dec}(sk, \mathsf{Eval}(evk, f, c_1, ..., c_t)) \neq f(m_1, ..., m_t)] = \mathrm{negl}(\lambda),$$

where $(pk, sk, evk) \leftarrow$ KeyGen (1^λ) and $c_i = $ Enc(pk, m_i).

Definition 4 (Compactness and full homomorphism). *A homomorphic encryption scheme is compact if its ciphertexts are independent of the evaluated function. A compact scheme is fully homomorphic if it is L-homomorphic for any polynomial L. The scheme is leveled fully homomorphic if it takes 1^L as additional input in key generation.*

Theorem 2 (Bootstrapping [7,8]). *If there is an L-homomorphic scheme whose decryption circuit depth is less than L, then there exists a leveled fully homomorphic encryption scheme.*

Furthermore, if the aforementioned L-homomorphic scheme is also weak circular secure (remains secure even against an adversary who gets encryptions of the bits of the secret key), then there exists a fully homomorphic encryption scheme.

3   Construction of (Leveled) Matrix FHE

In this section, we present our leveled matrix FHE construction. First, we give the matrix variant of Regev's basic public-key encryption scheme [16], and then we extend the key-switching methodology of [2] to matrix form. At last, the final matrix FHE construction is presented.

3.1   Matrix Variant of Regev's Encryption Scheme

To begin with, we present a basic LWE-based encryption scheme without considering homomorphic operations. Let λ be the security parameter, an integer modulus $q = q(\lambda)$, an integer dimension $n = n(\lambda)$ and a noise distribution $\chi = \chi(\lambda)$ over $\mathbb{Z}$, which are chosen following LWE assumption. Let r be the dimension of plaintext matrices and the plaintext space $\mathcal{M} = \{0,1\}^{r \times r}$. The ciphertext space is $\mathcal{C} = \mathbb{Z}_q^{r \times (n+r)}$. Let $m = \lceil (2n+1) \log q \rceil$ and $params = (n, q, m, \chi, r)$.

- E.SecretKeyGen($params$): Sample $\boldsymbol{S'} \leftarrow \mathbb{Z}_q^{n \times r}$ and output $sk = \boldsymbol{S} = \left(\begin{smallmatrix} I_r \\ \boldsymbol{S'} \end{smallmatrix} \right) \in \mathbb{Z}_q^{(n+r) \times r}$.
- E.PublicKeyGen($params, sk$): Sample $\boldsymbol{A'} \leftarrow \mathbb{Z}_q^{m \times n}$, $\boldsymbol{E} \leftarrow \chi^{m \times r}$. Set $\boldsymbol{A} = ((\boldsymbol{E} + \boldsymbol{A'S'})\| - \boldsymbol{A'}) \in \mathbb{Z}_q^{m \times (n+r)}$(Observe that $\boldsymbol{AS} = \boldsymbol{E}$). Output $pk = \boldsymbol{A}$.
- E.Enc($params, pk, \boldsymbol{M}$): To encrypt a message $\boldsymbol{M} \in \{0,1\}^{r \times r}$, pad $\boldsymbol{M}$ to obtain $\boldsymbol{M'} = (\boldsymbol{M} \| \boldsymbol{0}_{r \times n}) \in \mathbb{Z}_2^{r \times (n+r)}$. Sample $\boldsymbol{R} \leftarrow \{0,1\}^{r \times m}$ and output the ciphertext $\boldsymbol{C} = \lfloor \frac{q}{2} \rfloor \boldsymbol{M'} + \boldsymbol{RA} \in \mathbb{Z}_q^{r \times (n+r)}$.
- E.Dec($sk, \boldsymbol{C}$): Output $\boldsymbol{M} = [\lfloor \frac{2}{q} \cdot [\boldsymbol{CS}]_q \rceil]_2$.

Now we consider correctness and security of the above basic scheme.

Correctness: Note that $\boldsymbol{AS} = \boldsymbol{E}$, then $[\boldsymbol{CS}]_q = [(\lfloor \frac{q}{2} \rfloor \boldsymbol{M'} + \boldsymbol{RA})\boldsymbol{S}]_q = [\lfloor \frac{q}{2} \rfloor \boldsymbol{M'S} + \boldsymbol{RAS}]_q = [\lfloor \frac{q}{2} \rfloor \boldsymbol{M} + \boldsymbol{RE}]_q$. We define the noise of ciphertext $\boldsymbol{C}$ under key $\boldsymbol{S}$ as $\|\boldsymbol{RE}\|_\infty$, assume that $|\chi| \leq B$, then $\|\boldsymbol{RE}\|_\infty \leq mB$. Decryption succeeds as long as the magnitude of the noise stays smaller than $\lfloor q/2 \rfloor /2$.

Security: It is obvious that the security of our basic encryption scheme directly holds by the LWE assumption. We capture the security of our scheme in the following lemma, which is used to prove the security of Regev's encryption scheme [16] and GSW-FHE scheme [11].

Lemma 1 ([11,16]). *Let* $params = (n, q, m, \chi, r)$ *be such that the* $\mathrm{LWE}_{n,q,\chi}$ *assumption holds. Then, for* $m = \lceil (2n+1) \log q \rceil$ *and* $\boldsymbol{A}, \boldsymbol{R}$ *generated as above, the joint distribution* $(\boldsymbol{A}, \boldsymbol{RA})$ *is computationally indistinguishable from uniform over* $\mathbb{Z}_q^{m \times (n+r)} \times \mathbb{Z}_q^{r \times (n+r)}$.

Now we consider homomorphic operations roughly.

- For homomorphic matrix addition $\boldsymbol{C}_1 \oplus \boldsymbol{C}_2$: Output $\boldsymbol{C}_{add} := [\boldsymbol{C}_1 + \boldsymbol{C}_2]_q$. The correctness is obvious.
- For homomorphic matrix multiplication $\boldsymbol{C}_1 \odot \boldsymbol{C}_2$: We have $\lfloor \frac{2}{q}(\boldsymbol{C}_1 \boldsymbol{S}) \cdot (\boldsymbol{C}_2 \boldsymbol{S}) \rceil = \lfloor \frac{q}{2} \rfloor \boldsymbol{M}_1 \boldsymbol{M}_2 + \boldsymbol{E'} \bmod q$, then $\boldsymbol{M}_1 \boldsymbol{M}_2 = [\lfloor \frac{2}{q} \cdot \lfloor \frac{2}{q}(\boldsymbol{C}_1 \boldsymbol{S}) \cdot (\boldsymbol{C}_2 \boldsymbol{S}) \rceil \rceil]_2$. Set $\boldsymbol{C}^* = (\boldsymbol{C}_1 \otimes \boldsymbol{C}_2)_{sr}$, $\boldsymbol{S}^* = (\boldsymbol{S} \otimes' \boldsymbol{S})_{sc}$. We claim that $(\boldsymbol{C}_1 \boldsymbol{S}) \cdot (\boldsymbol{C}_2 \boldsymbol{S}) = \boldsymbol{C}^* \boldsymbol{S}^* \pmod{q}$.

– For homomorphic matrix Hadamard product $C_1 \bullet C_2$: We have $\lfloor \frac{2}{q}(C_1 S) \circ$
  $(C_2 S)\rceil = \lfloor \frac{q}{2} \rfloor M_1 \circ M_2 + E''$ mod q, then $M_1 \circ M_2 = [[\frac{2}{q} \cdot \lfloor \frac{2}{q}(C_1 S) \circ$
  $(C_2 S)\rceil]]_2$. Set $C^* = (C_1 \otimes C_2)_{er}$, $S^* = (S \otimes S)_{ec}$. We claim that $(C_1 S) \circ$
  $(C_2 S) = C^* S^*$ (mod q).

We capture the correctness of the above two claims in the following theorem.

Theorem 3. *For $A, B \in \mathbb{Z}_q^{m \times n}, S \in \mathbb{Z}_q^{n \times m}$, we have*

$$(AS) \cdot (BS) = (A \otimes B)_{sr}(S \otimes' S)_{sc} \pmod{q},$$
$$(AS) \circ (BS) = (A \otimes B)_{er}(S \otimes S)_{ec} \pmod{q}.$$

The proof of Theorem 3 is in Appendix A.

Up to now, we have showed that ciphertexts of homomorphic matrix multiplication $C_1 \odot C_2$ (resp. ciphertexts of homomorphic matrix Hadamard product $C_1 \bullet C_2$) are related to $(C_1 \otimes C_2)_{sr}$ (resp. $(C_1 \otimes C_2)_{er}$).

3.2 Matrix Variant of Key Switching

From the above section, we know that $M_1 M_2$ (and $M_1 \circ M_2$) can be obtained by C^*. But the dimensions of C^* and S^* are much larger than those of the initial ciphertexts and secret keys, which are $r \times r(n+r)^2$ and $r(n+r)^2 \times r$ (resp. $r \times (n+r)^2$ and $(n+r)^2 \times r$ for $M_1 \circ M_2$). We can extend the key switching technology in [2] to transform C^* and S^* to a ciphertext and a secret key of the same dimensions as the original ciphertext and secret key. The extended key switching technology will use two subroutines as follows.

– BitDecomp($A \in \mathbb{Z}_q^{m \times n}, r$) decomposes the rows of A into its bit representation. Namely, write $A_i = \sum_{j=0}^{\lceil \log q \rceil - 1} 2^j \cdot a_{i,j}$, where all of the vectors $a_{i,j}$ are in $\mathbb{Z}_2^n$, and output $(a_{i,0}, a_{i,1}, ..., a_{i,\lceil \log q \rceil - 1}) \in \mathbb{Z}_2^{n \times \lceil \log q \rceil}$ to replace A_i for $i = 1$ to m. Similarly, BitDecomp($A \in \mathbb{Z}_q^{m \times n}, c$) decomposes the columns of A into its bit representation.
– Powersof2($A \in \mathbb{Z}_q^{m \times n}, r$) outputs the vector $(A_i, 2A_i, ..., 2^{\lceil \log q \rceil - 1} A_i)$ to replace A_i for $i = 1$ to m. Similarly, Powersof2($A \in \mathbb{Z}_q^{m \times n}, c$) applies the same operation to the columns of A.

Lemma 2. *For matrices $C \in \mathbb{Z}_q^{m \times n}$, $S \in \mathbb{Z}_q^{n \times m}$, we have*

BitDecomp(C, r) $\cdot$ Powersof2(S, c) $= CS$ (mod q) $=$ Powersof2(C, r) $\cdot$ BitDecomp(S, c).

Proof. For $i = 1$ to m, $j = 1$ to n, $\langle$BitDecomp(C_i), Powersof2(S_j^T)$\rangle = \sum_{k=0}^{\lceil \log q \rceil - 1}$ $\langle c_{i,k}, 2^k S_j^T \rangle = \sum_{k=0}^{\lceil \log q \rceil - 1} \langle 2^k c_{i,k}, S_j^T \rangle = \langle \sum_{k=0}^{\lceil \log q \rceil - 1} 2^k c_{i,k}, S_j^T \rangle = \langle C_i, S_j^T \rangle$. Similarly, the second equation is correct.

The extended key switching includes two procedures: the first one is SwitchKey-Gen($\boldsymbol{S}_a \in \mathbb{Z}_q^{n' \times r}, \boldsymbol{S}_b \in \mathbb{Z}_q^{(n+r) \times r}$), which takes as input the two secret key matrices, the corresponding dimensions $n' \times r, (n+r) \times r$, and the modulus q (we often omit the parameters n', n, r, q when they are clear in the context), and outputs the switching key $t_{\boldsymbol{S}_a \to \boldsymbol{S}_b}$; and the second one is SwitchKey($t_{\boldsymbol{S}_a \to \boldsymbol{S}_b}$, $\boldsymbol{C}_a$) that takes as input the switching key $t_{\boldsymbol{S}_a \to \boldsymbol{S}_b}$ and a ciphertext $\boldsymbol{C}_a$ encrypted under $\boldsymbol{S}_a$, and outputs a new ciphertext $\boldsymbol{C}_b$ that encrypts the same message under the secret key $\boldsymbol{S}_b$.

- SwitchKeyGen($\boldsymbol{S}_a \in \mathbb{Z}_q^{n' \times r}, \boldsymbol{S}_b \in \mathbb{Z}_q^{(n+r) \times r}$):
  1. Run $\boldsymbol{A} \leftarrow$ E.PublicKeyGen($\boldsymbol{S}_b, m$) for $m = n' \cdot \lceil \log q \rceil$
  2. Set $\boldsymbol{B} = \boldsymbol{A} + (\text{Powersof2}(\boldsymbol{S}_a, c) \| \boldsymbol{0}_{m \times n})$, output $t_{\boldsymbol{S}_a \to \boldsymbol{S}_b} = \boldsymbol{B} \in \mathbb{Z}_q^{m \times (n+r)}$
- SwitchKey($t_{\boldsymbol{S}_a \to \boldsymbol{S}_b}$, $\boldsymbol{C}_a \in \mathbb{Z}_q^{r \times n'}$): Output $\boldsymbol{C}_b = \text{BitDecomp}(\boldsymbol{C}_a, r) \cdot \boldsymbol{B} \in \mathbb{Z}_q^{r \times (n+r)}$

The correctness of the extended key switching procedure is showed in the following lemma. It ensures that the key switching procedure is meaningful, in the sense that it preserves the correctness of decryption under the new key.

Lemma 3 (Correctness). *Let* $\boldsymbol{S}_a, \boldsymbol{S}_b, \boldsymbol{C}_a, \boldsymbol{C}_b, q, n, n', r$ *be the parameters as defined above, and* $\boldsymbol{A}\boldsymbol{S}_b = \boldsymbol{E}_b$, *we have* $\boldsymbol{C}_b \boldsymbol{S}_b = \text{BitDecomp}(\boldsymbol{C}_a, r) \cdot \boldsymbol{E}_b + \boldsymbol{C}_a \boldsymbol{S}_a$ *(mod q).*

Proof

$$
\begin{aligned}
\boldsymbol{C}_b \boldsymbol{S}_b &= \text{BitDecomp}(\boldsymbol{C}_a, r) \cdot \boldsymbol{B} \cdot \boldsymbol{S}_b \\
&= \text{BitDecomp}(\boldsymbol{C}_a, r) \cdot (\boldsymbol{A} + (\text{Powersof2}(\boldsymbol{S}_a, c) \| \boldsymbol{0}_{m \times n})) \cdot \boldsymbol{S}_b \\
&= \text{BitDecomp}(\boldsymbol{C}_a, r) \cdot (\boldsymbol{E}_b + \text{Powersof2}(\boldsymbol{S}_a, c)) \\
&= \text{BitDecomp}(\boldsymbol{C}_a, r) \cdot \boldsymbol{E}_b + \text{BitDecomp}(\boldsymbol{C}_a, r) \cdot \text{Powersof2}(\boldsymbol{S}_a, c) \\
&= \text{BitDecomp}(\boldsymbol{C}_a, r) \cdot \boldsymbol{E}_b + \boldsymbol{C}_a \boldsymbol{S}_a \pmod{q}.
\end{aligned}
$$

Note that the ℓ_∞ norm of the product of $\text{BitDecomp}(\boldsymbol{C}_a, r)$ and $\boldsymbol{E}_b$ is small, since the former is a bit-matrix and the ℓ_∞ norm of the latter is small. Therefore, $\boldsymbol{C}_b$ is a valid encryption of $\boldsymbol{M}$ under key $\boldsymbol{S}_b$, with noise expanded by a small additive factor.

Now we consider the security of the key switching procedure. By the properties of the basic scheme, we know that the matrix $\boldsymbol{A} \leftarrow$ E.PublicKeyGen($\boldsymbol{S}_b, m$) is computationally indistinguishable from uniform, so the switch key generated by SwitchKeyGen($\boldsymbol{S}_a, \boldsymbol{S}_b$) is computationally indistinguishable from uniform for any $\boldsymbol{S}_a \in \mathbb{Z}_q^{n' \times r}$ and $\boldsymbol{S}_b \in \mathbb{Z}_q^{(n+r) \times r} \leftarrow$ E.SecretKeyGen($params$). We capture the security of the key switching procedure in the following lemma.

Lemma 4 (Security). *For any* $\boldsymbol{S}_a \in \mathbb{Z}_q^{n' \times r}$ *and* $\boldsymbol{S}_b \leftarrow$ E.SecretKeyGen *($params$), the following two distributions are computationally indistinguishable:*

1. $\{(\boldsymbol{A}, t_{\boldsymbol{S}_a \to \boldsymbol{S}_b})\ :\ \boldsymbol{A}\ \leftarrow\ \mathsf{E.PublicKeyGen}(\boldsymbol{S}_b, m), t_{\boldsymbol{S}_a \to \boldsymbol{S}_b}\ \leftarrow\ \mathsf{SwitchKeyGen}$
   $(\boldsymbol{S}_a, \boldsymbol{S}_b)\}$
2. $\{(\boldsymbol{A}, t_{\boldsymbol{S}_a \to \boldsymbol{S}_b}) : \boldsymbol{A} \leftarrow \mathbb{Z}_q^{m \times (n+r)}, t_{\boldsymbol{S}_a \to \boldsymbol{S}_b} \leftarrow \mathbb{Z}_q^{m \times (n+r)}\}$

where the randomness is over the choice of $\boldsymbol{S}_b \in \mathbb{Z}_q^{(n+r) \times r}$, and the coins of E.PublicKeyGen *and* SwitchKeyGen.

3.3   Leveled FHE with Matrix Message Space

In this Section, we present the (leveled) matrix FHE scheme using components described in the previous subsections. In the scheme, we will use a parameter L to indicate the number of levels of arithmetic circuit that our scheme is capable of evaluating.

<u>Our Leveled Matrix FHE Scheme</u>

– FHE.KeyGen(L, n, q, r):
   1. Sample $\boldsymbol{S}_0, ..., \boldsymbol{S}_L \leftarrow$ E.SecretKeyGen(n, q, r). Compute a public key for the first one: $\boldsymbol{A}_0 \leftarrow$ E.PublicKeyGen($\boldsymbol{S}_0$).
   2. For all $k \in [L]$, set $\boldsymbol{S}_{k-1}^0 = (\mathsf{BitDecomp}(\boldsymbol{S}_{k-1}, c) \otimes' \mathsf{BitDecomp}(\boldsymbol{S}_{k-1}, c))_{sc} \in \{0,1\}^{r(n+r)^2 \lceil \log q \rceil^2 \times r}$ for FHE.Mult and
      $\boldsymbol{S}_{k-1}^1 = (\mathsf{BitDecomp}(\boldsymbol{S}_{k-1}, c) \otimes \mathsf{BitDecomp}(\boldsymbol{S}_{k-1}, c))_{ec} \in \{0,1\}^{(n+r)^2 \lceil \log q \rceil^2 \times r}$ for FHE.pMult.
   3. For all $k \in [L]$, run $t_{\boldsymbol{S}_{k-1}^0 \to \boldsymbol{S}_k} \leftarrow$ SwitchKeyGen($\boldsymbol{S}_{k-1}^0, \boldsymbol{S}_k$) and
      $t_{\boldsymbol{S}_{k-1}^1 \to \boldsymbol{S}_k} \leftarrow$ SwitchKeyGen($\boldsymbol{S}_{k-1}^1, \boldsymbol{S}_k$).
   Output $pk = \boldsymbol{A}_0$, $evk = \{t_{\boldsymbol{S}_{k-1}^b \to \boldsymbol{S}_k}\}_{k \in [L], b \in \{0,1\}}$, $sk = \boldsymbol{S}_L$.
– FHE.Enc($pk, \boldsymbol{M}$): To encrypt $\boldsymbol{M} \in \{0,1\}^{r \times r}$, run $\boldsymbol{C} \leftarrow$ E.Enc($pk, \boldsymbol{M}$).
– FHE.Dec($sk, \boldsymbol{C}$): Assume w.l.o.g that $\boldsymbol{C}$ is a ciphertext that corresponds to $\boldsymbol{S}_L (= sk)$, run $\boldsymbol{M} =$ E.Dec($\boldsymbol{S}_L, \boldsymbol{C}$).
– FHE.Add($evk, \boldsymbol{C}_1, \boldsymbol{C}_2$): Assume w.l.o.g that the two ciphertexts are under the same secret key $\boldsymbol{S}_{k-1}$. First compute

$$\widetilde{\boldsymbol{C}}_{add} := (\mathsf{Powersof2}(\boldsymbol{C}_1 + \boldsymbol{C}_2, r) \otimes \mathsf{Powersof2}(I_r \| \boldsymbol{0}_{r \times n}, r))_{sr} \in \mathbb{Z}_q^{r \times r(n+r)^2 \lceil \log q \rceil^2}$$

then output

$$\boldsymbol{C}_{add} := \mathsf{SwitchKey}(t_{\boldsymbol{S}_{k-1}^0 \to \boldsymbol{S}_k}, \widetilde{\boldsymbol{C}}_{add}) \in \mathbb{Z}_q^{r \times (n+r)}$$

– FHE.Mult($evk, \boldsymbol{C}_1, \boldsymbol{C}_2$): Assume w.l.o.g that the two ciphertexts are under the same secret key $\boldsymbol{S}_{k-1}$. First compute

$$\widetilde{\boldsymbol{C}}_{mult} := \lfloor \frac{2}{q} \cdot (\mathsf{Powersof2}(\boldsymbol{C}_1, r) \otimes \mathsf{Powersof2}(\boldsymbol{C}_2, r))_{sr} \rceil \in \mathbb{Z}_q^{r \times r(n+r)^2 \lceil \log q \rceil^2}$$

then output

$$C_{mult} := \mathsf{SwitchKey}(t_{S^0_{k-1} \to S_k}, \widetilde{C}_{mult}) \in \mathbb{Z}_q^{r \times (n+r)}$$

– FHE.pMult(evk, C_1, C_2): Assume w.l.o.g that the two ciphertexts are under the same secret key S_{k-1}. First compute

$$\widetilde{C}_{pmult} := \lfloor \frac{2}{q} \cdot (\mathsf{Powersof2}(C_1, r) \otimes \mathsf{Powersof2}(C_2, r))_{er} \rceil \in \mathbb{Z}_q^{r \times (n+r)^2 \lceil \log q \rceil^2}$$

then output

$$C_{pmult} := \mathsf{SwitchKey}(t_{S^1_{k-1} \to S_k}, \widetilde{C}_{mult}) \in \mathbb{Z}_q^{r \times (n+r)}$$

Remark 1. Our scheme has two kinds of switching keys for FHE.pMult and FHE.Mult respectively, since their intermediate ciphertexts have different keys.

4   Homomorphic Properties and Security

In this Section, we show the homomorphic properties and security of our matrix FHE scheme.

4.1   Homomorphic Properties

Lemma 5. *Let $q, n, |\chi| \le B, L, r$ be parameters for FHE and (pk, evk, sk) be the corresponding keys. Let C_1, C_2 be such that*

$$C_1 S_{k-1} = \lfloor \frac{q}{2} \rfloor M_1 + E_1 \pmod q$$
$$C_2 S_{k-1} = \lfloor \frac{q}{2} \rfloor M_2 + E_2 \pmod q,$$

with $|E_1|_\infty, |E_2|_\infty \le E < \lfloor \frac{q}{2} \rfloor / 2$. Consider ciphertexts C_{add} :=FHE.Add(evk, C_1, C_2), C_{mult} :=FHE.Mult(evk, C_1, C_2), C_{pmult} :=FHE.pMult(evk, C_1, C_2). Then

$$C_{add} S_k = \lfloor \frac{q}{2} \rfloor (M_1 + M_2) + E_{add} \pmod q$$
$$C_{mult} S_k = \lfloor \frac{q}{2} \rfloor (M_1 M_2) + E_{mult} \pmod q,$$
$$C_{pmult} S_k = \lfloor \frac{q}{2} \rfloor (M_1 \circ M_2) + E_{pmult} \pmod q,$$

where

$$|E_{add}|_\infty, |E_{mult}|_\infty, |E_{pmult}|_\infty \le O(n \log q) \cdot \max\{E, (n \log^2 q) \cdot B\}$$

Proof. Since the error growth in homomorphic multiplication is the largest, we only need to analyze homomorphic multiplication.

According to the correctness of the extended key switching procedure, we have

$$\boldsymbol{C}_{mult}\boldsymbol{S}_k = \widetilde{\boldsymbol{C}}_{mult}\boldsymbol{S}_{k-1}^0 + \mathsf{BitDecomp}(\widetilde{\boldsymbol{C}}_{mult}, r) \cdot \boldsymbol{E}_{k-1:k} \pmod{q}$$

where $\boldsymbol{E}_{k-1:k} \sim \chi^{r(n+r)^2\lceil\log q\rceil^3 \times r}$

Define $\Delta_1 = \mathsf{BitDecomp}(\widetilde{\boldsymbol{C}}_{mult}, r) \cdot \boldsymbol{E}_{k-1:k}$, then $\|\Delta_1\|_\infty \leq r(n+r)^2\lceil\log q\rceil^3 \cdot B = O(n^2\log^3 q) \cdot B$ (since r is not dependent on n and q, we can view r as a constant).

Now we consider $\widetilde{\boldsymbol{C}}_{mult}\boldsymbol{S}_{k-1}^0$:

$$\widetilde{\boldsymbol{C}}_{mult}\boldsymbol{S}_{k-1}^0 = \lfloor\frac{2}{q} \cdot (\mathsf{Powersof2}(\boldsymbol{C}_1, r) \otimes \mathsf{Powersof2}(\boldsymbol{C}_2, r))_{sr}\rceil$$
$$\cdot(\mathsf{BitDecomp}(\boldsymbol{S}_{k-1}, c) \otimes' \mathsf{BitDecomp}(\boldsymbol{S}_{k-1}, c))_{sc} \pmod{q}$$

Define $\Delta_2 = (\lfloor\frac{2}{q}\cdot(\mathsf{Powersof2}(\boldsymbol{C}_1, r)\otimes\mathsf{Powersof2}(\boldsymbol{C}_2, r))_{sr}\rceil - \frac{2}{q}\cdot(\mathsf{Powersof2}(\boldsymbol{C}_1, r)\otimes\mathsf{Powersof2}(\boldsymbol{C}_2, r))_{sr}) \cdot \boldsymbol{S}_{k-1}^0$, then $\|\Delta_2\|_\infty \leq r(n+r)^2\lceil\log q\rceil^2 = O(n^2\log^2 q)$.

Using Theorem 3, we have

$$\widetilde{\boldsymbol{C}}_{mult}\boldsymbol{S}_{k-1}^0 - \Delta_2 = \frac{2}{q} \cdot (\mathsf{Powersof2}(\boldsymbol{C}_1, r) \cdot \mathsf{BitDecomp}(\boldsymbol{S}_{k-1}, c))$$
$$\cdot(\mathsf{Powersof2}(\boldsymbol{C}_2, r) \cdot \mathsf{BitDecomp}(\boldsymbol{S}_{k-1}, c)) \qquad (1)$$

According to the assumption of the lemma and using Lemma 2, there exist $\boldsymbol{K}_1, \boldsymbol{K}_2 \in \mathbb{Z}^{r\times r}$ such that

$$\mathsf{Powersof2}(\boldsymbol{C}_i, r) \cdot \mathsf{BitDecomp}(\boldsymbol{S}_{k-1}, c) = \lfloor\frac{q}{2}\rfloor\boldsymbol{M}_i + \boldsymbol{E}_i + q \cdot \boldsymbol{K}_i, \quad i = 1, 2 \qquad (2)$$

Plugging Eq. (2) into Eq. (1), we get

$$\widetilde{\boldsymbol{C}}_{mult}\boldsymbol{S}_{k-1}^0 - \Delta_2 = \frac{2}{q} \cdot (\lfloor\frac{q}{2}\rfloor\boldsymbol{M}_1 + \boldsymbol{E}_1 + q \cdot \boldsymbol{K}_1) \cdot (\lfloor\frac{q}{2}\rfloor\boldsymbol{M}_2 + \boldsymbol{E}_2 + q \cdot \boldsymbol{K}_2)$$
$$= \lfloor\frac{q}{2}\rfloor\boldsymbol{M}_1\boldsymbol{M}_2 + \Delta_3 + q \cdot (\boldsymbol{K}_1\boldsymbol{M}_2 + \boldsymbol{M}_1\boldsymbol{K}_2 + 2\boldsymbol{K}_1\boldsymbol{K}_2)$$

where $\Delta_3 \triangleq 2\boldsymbol{K}_1\boldsymbol{E}_2 + 2\boldsymbol{E}_1\boldsymbol{K}_2 + \boldsymbol{E}_1\boldsymbol{M}_2 + \boldsymbol{M}_1\boldsymbol{E}_2 + 2/q \cdot \boldsymbol{E}_1\boldsymbol{E}_2$ (For simplicity, we only consider the case q is even, but the following results also hold for the case q is odd).

Since

$$\|\boldsymbol{K}_i\|_\infty = 1/q \cdot \|\mathsf{Powersof2}(\boldsymbol{C}_i, r) \cdot \mathsf{BitDecomp}(\boldsymbol{S}_{k-1}, c) - \lfloor\frac{q}{2}\rfloor\boldsymbol{M}_i - \boldsymbol{E}_i\|_\infty$$
$$\leq 1/q \cdot \|\mathsf{Powersof2}(\boldsymbol{C}_i, r) \cdot \mathsf{BitDecomp}(\boldsymbol{S}_{k-1}, c)\|_\infty + 1$$
$$\leq (n+r)\log q + 1 = O(n\log q)$$

We have

$$\|\varDelta_3\|_\infty \leq 2 \cdot 2rO(n\log q)E + 2rE + 2/q \cdot rE^2 = O(n\log q) \cdot E.$$

Putting all the above together, we obtain that

$$\boldsymbol{C}_{mult}\boldsymbol{S}_k = \lfloor\frac{q}{2}\rfloor\boldsymbol{M_1}\boldsymbol{M_2} + \varDelta_1 + \varDelta_2 + \varDelta_3 \pmod{q}$$

where $\|\boldsymbol{E}_{mult}\|_\infty = \|\varDelta_1 + \varDelta_2 + \varDelta_3\|_\infty \leq O(n\log q) \cdot E + O(n^2\log^3 q) \cdot B$.
    Above all, the lemma holds.

From the above lemma, we can immediately obtain the following theorem, which demonstrates the homomorphic properties of our scheme.

Theorem 4. *The scheme* FHE *with parameters* $n, q, |\chi| \leq B, L, r$ *for which*

$$q/B \geq (O(n\log q))^{L+O(1)}$$

is L-homomorphic.

That is to say, our scheme is a leveled FHE scheme with matrix message space.

4.2   Security

The security of our matrix FHE scheme follows by a standard hybrid argument from the security of the basic scheme E described in Sect. 3.1. At a high level, the view of a CPA adversary for our scheme is very similar to that for the basic scheme E, except that our adversary can also see the switching keys. However, the switching keys are composed of a set of outputs of the SwitchKeyGen algorithm which are indistinguishable from uniform by Lemma 4. A formal proof is very similar to that of [2, Theorem 5] (full version), and please refer to [2] for more details.

Theorem 5. *Let* n, q, χ *be some parameters such that* $\mathrm{LWE}_{n,q,\chi}$ *holds, L be polynomially bounded, and* $m = \lceil(2n+1)\log q\rceil$. *Then our matrix* FHE *scheme is CPA secure under* $\mathrm{LWE}_{n,q,\chi}$ *assumption.*

5   Viewed as Packed FHE

Our matrix FHE scheme can be viewed as vector-packed FHE in [1]. Plaintext slots of vector-packed FHE in [1] correspond to diagonal entries (or entries in one column) of plaintext matrices in our matrix FHE scheme. We can correctly compute homomorphic slot-wise addition and multiplication by FHE.Add and FHE.Mult (or FHE.pMult). In some applications ofvector-packed FHE such as [9], the operation of permuting plaintext slots is required. When diagonal entries are viewed as plaintext slots, we can implement the operation of permuting plaintext slots as same as that in [12] by FHE.Mult. To be more specific, this operation can be implemented by multiplying the encryptions of a permutation matrix and its transpose from left and right respectively. When entries in one column are viewed as plaintext slots, we can implement this operation more succinctly by multiplying the encryption of a permutation matrix from left merely. And in this case, we can implement the same permutation on r sets of plaintext slots at one time. The algorithm of permuting plaintext slots is as follows:

– SlotSwitchKG(pk, σ): Given a public key $pk = A \in \mathbb{Z}_q^{m \times (n+r)}$ and a permutation σ on r elements, let $P_\sigma \in \{0, 1\}^{r \times r}$ be a permutation matrix corresponding to σ, and generate

$$W_\sigma \leftarrow \mathsf{E.Enc}(pk, P_\sigma).$$

Output the switch key $sw_\sigma := W_\sigma$.

– SlotSwitch(sw_σ, C): Take as input a switch key sw_σ and a ciphertext C, output

$$C_\sigma \leftarrow W_\sigma \odot C.$$

In addition, our matrix FHE can also be viewed as matrix-packed FHE (in contrast to vector-packed FHE in [1]), which means that every entry of plaintext matrices can be viewed as one plaintext slot. In this case homomorphic slot-wise addition and multiplication exactly correspond to FHE.Add and FHE.pMult. This is one of our advantages over the scheme in [12].

Acknowledgment. This work is supported by National Natural Science Foundation of China (No. 61402471, 61472414, 61602061, 61772514).

A   Proof of Theorem 3

Theorem 3. For $A, B \in \mathbb{Z}_q^{m \times n}, S \in \mathbb{Z}_q^{n \times m}$, we have

$$(AS) \cdot (BS) = (A \otimes B)_{sr}(S \otimes' S)_{sc} \pmod{q},$$
$$(AS) \circ (BS) = (A \otimes B)_{er}(S \otimes S)_{ec} \pmod{q}.$$

Proof. For the conciseness of description, we let $m = n = 2$. One can easily check that the proof can be generalized to any positive integer m and n. Assume $A = \begin{pmatrix} a_{11} & a_{12} \\ a_{21} & a_{22} \end{pmatrix}, B = \begin{pmatrix} b_{11} & b_{12} \\ b_{21} & b_{22} \end{pmatrix}, S = \begin{pmatrix} s_{11} & s_{12} \\ s_{21} & s_{22} \end{pmatrix}$, then we have

$$AS = \begin{pmatrix} a_{11}s_{11} + a_{12}s_{21} & a_{11}s_{12} + a_{12}s_{22} \\ a_{21}s_{11} + a_{22}s_{21} & a_{21}s_{12} + a_{22}s_{22} \end{pmatrix}$$

$$BS = \begin{pmatrix} b_{11}s_{11} + b_{12}s_{21} & b_{11}s_{12} + b_{12}s_{22} \\ b_{21}s_{11} + b_{22}s_{21} & b_{21}s_{12} + b_{22}s_{22} \end{pmatrix}$$

$$A \otimes B = \begin{pmatrix} a_{11}b_{11} & a_{11}b_{12} & a_{12}b_{11} & a_{12}b_{12} \\ a_{11}b_{21} & a_{11}b_{22} & a_{12}b_{21} & a_{12}b_{22} \\ a_{21}b_{11} & a_{21}b_{12} & a_{22}b_{11} & a_{22}b_{12} \\ a_{21}b_{21} & a_{21}b_{22} & a_{22}b_{21} & a_{22}b_{22} \end{pmatrix}$$

$$S \otimes S = \begin{pmatrix} s_{11}s_{11} & s_{11}s_{12} & s_{12}s_{11} & s_{12}s_{12} \\ s_{11}s_{21} & s_{11}s_{22} & s_{12}s_{21} & s_{12}s_{22} \\ s_{21}s_{11} & s_{21}s_{12} & s_{22}s_{11} & s_{22}s_{12} \\ s_{21}s_{21} & s_{21}s_{22} & s_{22}s_{21} & s_{22}s_{22} \end{pmatrix}$$

$$S \otimes' S = \begin{pmatrix} s_{11}s_{11} & s_{12}s_{11} & s_{11}s_{12} & s_{12}s_{12} \\ s_{11}s_{21} & s_{12}s_{21} & s_{11}s_{22} & s_{12}s_{22} \\ s_{21}s_{11} & s_{22}s_{11} & s_{21}s_{12} & s_{22}s_{12} \\ s_{21}s_{21} & s_{22}s_{21} & s_{21}s_{22} & s_{22}s_{22} \end{pmatrix}$$

$$(A \otimes B)_{sr} = \begin{pmatrix} a_{11}b_{11} & a_{11}b_{12} & a_{12}b_{11} & a_{12}b_{12} & a_{11}b_{21} & a_{11}b_{22} & a_{12}b_{21} & a_{12}b_{22} \\ a_{21}b_{11} & a_{21}b_{12} & a_{22}b_{11} & a_{22}b_{12} & a_{21}b_{21} & a_{21}b_{22} & a_{22}b_{21} & a_{22}b_{22} \end{pmatrix}$$

$$(S \otimes' S)_{sc} = \begin{pmatrix} s_{11}s_{11} & s_{11}s_{12} \\ s_{11}s_{21} & s_{11}s_{22} \\ s_{21}s_{11} & s_{21}s_{12} \\ s_{21}s_{21} & s_{21}s_{22} \\ s_{12}s_{11} & s_{12}s_{12} \\ s_{12}s_{21} & s_{12}s_{22} \\ s_{22}s_{11} & s_{22}s_{12} \\ s_{22}s_{21} & s_{22}s_{22} \end{pmatrix}$$

$((A \otimes B)_{sr}(S \otimes' S)_{sc})_{11} = a_{11}b_{11}s_{11}s_{11} + a_{11}b_{12}s_{11}s_{21} + a_{12}b_{11}s_{21}s_{11} + a_{12}b_{12}s_{21}s_{21} + a_{11}b_{21}s_{12}s_{11} + a_{11}b_{22}s_{12}s_{21} + a_{12}b_{21}s_{22}s_{11} + a_{12}b_{22}s_{22}s_{21} = (ASBS)_{11}$

For other entries of $ASBS$, one can check the correctness in the same way. For the second equation, we have

$$(A \otimes B)_{er} = \begin{pmatrix} a_{11}b_{11} & a_{11}b_{12} & a_{12}b_{11} & a_{12}b_{12} \\ a_{21}b_{21} & a_{21}b_{22} & a_{22}b_{21} & a_{22}b_{22} \end{pmatrix}, (S \otimes S)_{ec} = \begin{pmatrix} s_{11}s_{11} & s_{12}s_{12} \\ s_{11}s_{21} & s_{12}s_{22} \\ s_{21}s_{11} & s_{22}s_{12} \\ s_{21}s_{21} & s_{22}s_{22} \end{pmatrix}$$

then $((A \otimes B)_{er}(S \otimes S)_{ec})_{11} = a_{11}b_{11}s_{11}s_{11} + a_{11}b_{12}s_{11}s_{21} + a_{12}b_{11}s_{21}s_{11} + a_{12}b_{12}s_{21}s_{21} = ((AS) \circ (BS))_{11}$

For other entries of $(AS) \circ (BS)$, one can check the correctness in the same way.

References

1. Brakerski, Z., Gentry, C., Halevi, S.: Packed ciphertexts in LWE-based homomorphic encryption. In: Kurosawa, K., Hanaoka, G. (eds.) PKC 2013. LNCS, vol. 7778, pp. 1–13. Springer, Heidelberg (2013). https://doi.org/10.1007/978-3-642-36362-7_1

2. Brakerski, Z., Gentry, C., Vaikuntanathan, V.: (Leveled) fully homomorphic encryption without bootstrapping. In: ITCS, pp. 309–325 (2012), Full Version, http://people.csail.mit.edu/vinodv/6892-Fall2013/BGV.pdf

3. Brakerski, Z.: Fully homomorphic encryption without modulus switching from classical GapSVP. In: Safavi-Naini, R., Canetti, R. (eds.) CRYPTO 2012. LNCS, vol. 7417, pp. 868–886. Springer, Heidelberg (2012). https://doi.org/10.1007/978-3-642-32009-5_50

4. Brakerski, Z., Vaikuntanathan, V.: Fully homomorphic encryption from ring-LWE and security for key dependent messages. In: Rogaway, P. (ed.) CRYPTO 2011. LNCS, vol. 6841, pp. 505–524. Springer, Heidelberg (2011). https://doi.org/10.1007/978-3-642-22792-9_29

5. Brakerski, Z., Vaikuntanathan, V.: Efficient fully homomorphic encryption from (standard) LWE. In: FOCS, pp. 97–106 (2011)

6. van Dijk, M., Gentry, C., Halevi, S., Vaikuntanathan, V.: Fully homomorphic encryption over the integers. In: Gilbert, H. (ed.) EUROCRYPT 2010. LNCS, vol. 6110, pp. 24–43. Springer, Heidelberg (2010). https://doi.org/10.1007/978-3-642-13190-5_2

7. Gentry, C.: A Fully Homomorphic Encryption Scheme. PhD thesis. Stanford University (2009). http://crypto.stanford.edu/craig

8. Gentry, C.: Fully homomorphic encryption using ideal lattices. In: STOC, pp. 169–178 (2009)

9. Gentry, C., Halevi, S., Smart, N.P.: Better bootstrapping in fully homomorphic encryption. In: Fischlin, M., Buchmann, J., Manulis, M. (eds.) PKC 2012. LNCS, vol. 7293, pp. 1–16. Springer, Heidelberg (2012). https://doi.org/10.1007/978-3-642-30057-8_1

10. Gentry, C., Halevi, S., Smart, N.P.: Fully homomorphic encryption with polylog overhead. In: Pointcheval, D., Johansson, T. (eds.) EUROCRYPT 2012. LNCS, vol. 7237, pp. 465–482. Springer, Heidelberg (2012). https://doi.org/10.1007/978-3-642-29011-4_28

11. Gentry, C., Sahai, A., Waters, B.: Homomorphic encryption from learning with errors: conceptually-simpler, asymptotically-faster, attribute-based. In: Canetti, R., Garay, J.A. (eds.) CRYPTO 2013. LNCS, vol. 8042, pp. 75–92. Springer, Heidelberg (2013). https://doi.org/10.1007/978-3-642-40041-4_5

12. Hiromasa, R., Abe, M., Okamoto, T.: Packing messages and optimizing bootstrapping in GSW-FHE. In: Katz, J. (ed.) PKC 2015. LNCS, vol. 9020, pp. 699–715. Springer, Heidelberg (2015). https://doi.org/10.1007/978-3-662-46447-2_31

13. Micciancio, D., Voulgaris, P.: A deterministic single exponential time algorithm for most lattice problems based on voronoi cell computations. In: Schulman, L.J. (ed.) STOC, pp. 351–358. ACM (2010)

14. Peikert, C.: Public-key cryptosystems from the worst-case shortest vector problem. In: STOC, pp. 333–342. ACM (2009)

15. Peikert, C., Vaikuntanathan, V., Waters, B.: A framework for efficient and composable oblivious transfer. In: Wagner, D. (ed.) CRYPTO 2008. LNCS, vol. 5157, pp. 554–571. Springer, Heidelberg (2008). https://doi.org/10.1007/978-3-540-85174-5_31

16. Regev, O.: On lattices, learning with errors, random linear codes, and cryptography. In: Gabow, H.N., Fagin, R. (eds.) STOC, pp. 84–93. ACM, New York (2005)
17. Rivest, R., Adleman, L., Dertouzos, M.: On data banks and privacy homomorphisms. In: Foundations of Secure Computation, pp. 169–180 (1978)
18. Rothblum, R.: Homomorphic encryption: from private-key to public-key. In: Ishai, Y. (ed.) TCC 2011. LNCS, vol. 6597, pp. 219–234. Springer, Heidelberg (2011). https://doi.org/10.1007/978-3-642-19571-6_14
19. Smart, N.P., Vercauteren, F.: Fully homomorphic encryption with relatively small key and ciphertext sizes. In: Nguyen, P.Q., Pointcheval, D. (eds.) PKC 2010. LNCS, vol. 6056, pp. 420–443. Springer, Heidelberg (2010). https://doi.org/10.1007/978-3-642-13013-7_25
20. Smart, N.P., Vercauteren, F.: Fully homomorphic SIMD operations. Des. Codes Crypt. **71**(1), 57–81 (2014)

Predicate Fully Homomorphic Encryption: Achieving Fine-Grained Access Control over Manipulable Ciphertext

Hanwen Feng[1,2], Jianwei Liu[1], Qianhong Wu[1(✉)], and Weiran Liu[1]

[1] School of Electronic and Information Engineering, Beihang University,
Beijing, China
{feng_hanwen,liujianwei,qianhong.wu}@buaa.edu.cn,
liuweiran900217@gmail.com
[2] State Key Laboratory of Cryptology, P.O. Box 5159, Beijing 100878, China

Abstract. With the popularity of cloud computing, there is an increasing demand for enforcing access control over outsourced files and performing versatile operations on encrypted data. To meet this demand, a novel primitive called predicate fully homomorphic encryption (PFHE) is introduced and modeled in this work, which can provide the security guarantee that neither cloud computing server nor invalid cloud users can acquire any extra information about the processed data, while the server can still process the data correctly. We give a generic construction for PFHE, from any predicate key encapsulation mechanism (PKEM) and any LWE-based multi-key fully homomorphic encryption (MFHE). Compared with previously proposed generic construction for attribute-based fully homomorphic encryption (ABFHE), which can naturally be extended to one for PFHE, our construction has advantages in both time for encryption and space for encrypted data storage. In addition, our construction can achieve CCA1-secure. Thus it directly implies approaches for CCA1-secure FHE, CCA1-secure PFHE and CCA1-secure MFHE. The latter two have not been touched in previous work. In addition, we give a conversion which results a CCA1-secure PFHE scheme from a CPA-secure one, drawing on the techniques for CCA2-secure PE schemes.

Keywords: Fine-grained access control · Cloud computing security
Fully homomorphic encryption · Predicate encryption

1 Introduction

Cloud computing has been regarded as the trend of IT industry, as promoting the fully utilization of information technology resources and providing quality service. While bringing convenience to users, cloud computing services inherently incur potential risks for user privacy, i.e. computing security and storage security must be considered on an equal basis, while the former is not necessary in cloud storage services. Since computing and data sharing [22] are two of the most

© Springer International Publishing AG, part of Springer Nature 2018
X. Chen et al. (Eds.): Inscrypt 2017, LNCS 10726, pp. 278–298, 2018.
https://doi.org/10.1007/978-3-319-75160-3_18

important services, one typical scenario of cloud computing may be as follows: a data owner outsources his data items to cloud computing server and defines a fine-grained access control policy for them. He wishes that the data items can be processed by the server correctly and only valid users can get access to them.

Fully homomorphic encryption (FHE) [15,30] is a technique which can provide computing security, while similar techniques only provide security for some specific operations [12]. It enables an untrusted cloud server without decryption ability to evaluate any function on cipheretexts through a carefully designed algorithm, and result a cipheretext which is a valid encryption of the value of function evaluated on the corresponding plaintexts. A data owner can use this technique to encrypt his data and outsource the cipheretexts to the server. But enforcing fine-grained access control cannot be achieved through this technique.

Recall what we use to achieve fine-grained access control without homomorphism property such as in cloud storage services. There are numerous researches on this topic, achieving various functionalities (e.g. hierarchies [10,11] and revocation [35]) for various scenes (e.g. web-based cloud computing [25]). And the main cryptographic tools can be summarized as predicate encryption (PE) [19,24]. It's a public key encryption paradigm that enables users to encrypt data according to pre-defined boolean predicate, to realize fine-grained access control mechanisms. This primitive was introduced by Katz, Sahai and Waters in 2008 [24], and then described as an important subclass of functional encryption (FE) by Boneh et al. in 2011 [4]. PE is an integrated primitive, and many previously proposed primitives, such as identity-based encryption (IBE) [2], attribute-based encryption (ABE) [20] and broadcast encryption(BE) [33], can be viewed as instances of PE. Existing PE schemes are based on various assumptions, and most of them can be classified into the following two types, pairing-based ones [14] and lattice-based ones [28]. While lattice-based PE schemes provide post-quantum security and can support any policy computable by general circuits at cost of large computation overhead [3,18,19], pairing-based PE schemes serve as practical tools for nowadays application.

It raises the question that how to achieve fine-grained access control and homomorphic property simultaneously. This is essentially what the primitive, predicate fully homomorphic encryption (PFHE), aims to solve. As an integrated primitive, PFHE subsumes identity-based fully homomorphic encryption (IBFHE) [13,17] and attribute-based fully homomorphic encryption (ABFHE) [5] et al. primitives. Considering the constructed methods of existing PFHE schemes, they can be classified into two types.

First, most of existing PFHE schemes follows Gentry et al.'s work [17]. Gentry, Sahai and Waters proposed a simple FHE scheme based on learning with errors (LWE) problem [29], the encrypted data of which could be computed without any auxiliary information, while it was necessary in all previous FHE schemes [6,15]. Cashing in on the simplicity, Gentry et al. constructed a compiler, which complied an IBE(ABE) scheme that satisfies certain conditions to an IBFHE (ABFHE) scheme. They showed that almost all LWE-based IBE schemes [9,16] satisfied these conditions, and gave a concrete IBFHE scheme from [16]. The basic idea behind this compiler is embedding encryptions of zeros of the IBE scheme to the ciphertexts of resulting IBFHE scheme. Therefore when ciphertext of the original IBE scheme is a high dimensional lattice vector, the ciphertext of the resulted IBFHE scheme would be a matrix or a vector with the same dimension, at least. This is an inherently barrier to make the resulting schemes practical. Another serious problem is the case that how to evaluate a function on encrypted data items associated with different attributes (i.e., across-attribute computing) is unclear. Multi-key IBFHE was introduced by Clear and McGoldrick in 2015 [13], where across-identity computing could be executed at the cost of that the length ciphertext grew quadratically with the number of identities involved. But multi-key ABFHE has not been realized via this constructing method. Brakerski et al. introduced a weaker primitive called target homomorphic attribute-based encryption (THABE) [5], where key attributes of the target decryptors should be known to the evaluator. The length of ciphertexts also grew quadratically with the number of the target attributes. Since the target attributes are necessary for the evaluator, THABE cannot handle many scenarios where the attributes of receivers need to be kept secret.

Second, advances of MFHE can yield a generic construction for PFHE [5]. Using MFHE and ABE as black box tools, the construction can be expressed in a simple way:to encrypt a bit, the encryptor generate a key pair, public key pk and secret key sk, of MFHE, then encrypt this bit under pk and encrypt sk under the ABE scheme. A valid decryptor can recover sk from the ABE ciphertext using his own secret key, then get the encrypted bit. Homomorphic operations can be performed under the MFHE manner, no matter whether cipheretexts involved are associated with the same attribute. The generic construction is not compact. Asymptotically, as mentioned in [5], the ciphertext size grows(quadratically or linearly) with the number of input ciphertexts, while ciphertext's size of THABE scheme is only relevant to the number of target attributes; practically, because of the large size of sk in MFHE schemes, the size of ABE ciphertext which encrypt the secret key would be extremely large.

1.1   Our Results

As far as we know, the primitive PFHE has not been formally introduced in previous work, although a weaker primitive ABFHE was introduced in 2013 [17]. As discussed in [19], privacy of attributes can be provided by PFHE but not by ABFHE, thus PFHE is necessary when attributes are sensitive. The reason why PFHE hasn't emerged is that there is no candidate predicate encryption for Gentry et al.'s compiler, but it won't be a barrier for generic construction of PFHE. In this work, we formally introduced the definition and security model of PFHE, and gave a generic construction for it, which can be summarised in the following result.

Starting from a PKEM scheme and any one of existing LWE-based MFHE schemes, there is a PFHE scheme constructed as a hybrid encryption, in which,

- Efficiency: public parameters of the KEM scheme and ones of the MFHE scheme are generated independently, only determined by the security parameter. And to encrypt each bit, the KEM scheme only needs to work for once.
- Functionality: across-attribute computing can be executed at cost of almost nothing.
- Security: the PFHE scheme still remains indistinguishability against an adversary with almost the same ability as the one in the PKEM scheme, except that all quires to decryption oracle should be completed before the challenge phase(if adversary is allowed to make decryption quires in the PKEM scheme).
- Extension: A PFHE scheme with CPA security can be converted into one with CCA1 security, through a generic conversion.

Concretely, we proposed a new generic construction for PFHE, given a PKEM scheme which is commonly used to construct a PE scheme or constructed from a PE scheme, and any one of existing LWE-based MFHE schemes. Compared to the previous generic construction, in our work, the length of additional information(that is, ABE ciphertexts for secret decryption key, in previous work), for a valid decryptor to recover the secret decryption key in MFHE scheme, is independent from the length of the key. Current PE schemes are based on bilinear pairings or lattice, while the former one based PE schemes are much more efficient than the latter one based. It is necessary to construct from a pairing-based KEM for efficiency consideration. To do this, firstly, we carefully designed a transfer map from a random element in the target group of a pairing mechanism, to a high dimensional lattice vector, which can serve as a secret decryption key in MFHE schemes. The transfer map is built upon standard tools, which are randomness extractor [31] and pseudorandom generator [23]. Then we showed that with this transfer map, any one of existing LWE-based MFHE schemes can work as a data encapsulation mechanism(DEM) for all pairing-based PKEM. Thus we can get a PFHE scheme via constructing a hybrid encryption given the KEM and DEM. For a lattice-based PKEM, existing LWE-based MFHE schemes can work as DEM easily.We also note that across-attribute computing can be executed at cost of almost nothing, except that different predicate secret keys may need to be collected together to decrypt the resulting ciphertext.

We analyzed the security of the proposed construction and got an interesting result, that is, we can use the construction to get a CCA1-secure PFHE scheme, where attackers can make queries to the decryption oracle before challenge. Thus, a CCA1-secure IBFHE scheme is available, then a CCA1-secure (traditional) FHE scheme can be constructed by encapsulating the public parameters and an id as a public key. Furthermore, our PFHE supports across-attribute computing. Thus we can get a CCA1-secure MFHE by choosing a proper predicate relation. We note that CCA2 security, in which an adversary can make decryption queries after submitting the challenge plaintexts, is impossible for a FHE scheme [15]. And how to construct a FHE scheme with CCA1 security under standard assumptions was an open problem until Canetti et al. implied that a CCA1-secure FHE scheme could be transformed from a multi-key IBFHE scheme [8]. However, the question how to arrive at a CCA1-secure PFHE or

MFHE scheme still remains difficult so far. The basic idea behind our work is that, when a scheme works as a DEM rather than an encryption scheme, it can enjoy higher security, due to the encryption key used only for once in DEM. We showed that when a PFHE is a hybrid encryption scheme of a CCA-ecure KEM and a CPA-secure MFHE scheme(working as DEM), it enjoys CCA1 security. One more step, we showed that a CPA-secure PFHE scheme with verifiability [34], which guarantees that two secret decryption key associated with different attributes have the same output in decryption if they are in the same predicate relation with attribute of the ciphertext, can be converted into one with CCA1 security.

1.2   Related Work

Multi-key FHE is a special type of FHE which allows evaluating function on ciphertexts under different keys. The first MFHE scheme was proposed by López-Alt, Tromer and Vaikuntanthan in 2012 [26]. Their scheme was built upon provable secure NTRU encryption [21,32], and had a limitation that the maximal number of participants in a computation had to be known at the time a key was generated. Then the limitation was removed by Clear and McGoldrick in 2015 [13]. They used IBFHE proposed in [17] as a stepping stone, and constructed an MFHE scheme based on LWE problem. Their construction worked in single-hop setting, where all participants were determined when a computation started. And the size of cipheretext grew quadratically with the number of participants involved. Mukherjee and Wichs simplified this scheme [27]. In 2016, Brakerski and Perlman proposed a fully dynamic MFHE scheme [7], where the number of operations and participants is unbounded, and ciphertexts from new participants are allowed to join a computation. [7] also enjoyed a short ciphertext, which only grew linearly with the number of participants. However, Since bootstrapping technique [15] is used to realize homomorphic operation, the per gate computation of Brakerski et al.'s scheme is larger than previous ones.

While there are only limited ways to construct CCA-secure FHE schemes under standard assumptions, the question how to arrive CCA-secure PE schemes has various answers. Canetti et al. proposed an approach which was widely used to construct a CCA2-secure public key encryption scheme [1]. Their approach can be employed to convert CPA-secure IBE to CCA2-secure PKE, and convert CPA-secure hierarchical IBE to CCA2-secure IBE, and obtain CCA2-secure ABE from CPA-secure ABE. Yamada et al. further showed that any CPA-secure PE scheme with verifiability can be converted into CCA2-secure PE scheme [34].

2   Preliminaries

2.1   Notations

We use λ to denote the security parameter, use bold-face capital letters, such as A, to denote matrices, and use bold-face small letters, such as v, to denote vectors. $\mathbb{Z}$ denotes the set of integers, and $\mathbb{N}$ denotes the set of positive integers, as usual. And $\mathbb{Z}_q$ denotes $\mathbb{Z}/q\mathbb{Z}$, where $q \in \mathbb{N}$. We use $\mathbb{Z}_q^n$ to denote the $n \in \mathbb{N}$ dimensional space over $\mathbb{Z}_q$. For a vector $\boldsymbol{v} \in \mathbb{Z}_q^n$, $\boldsymbol{v}[i]$ denotes its ith component, where $i \in [n]$ and $[n]$ denotes the set $\{1, 2, \ldots, n\}$. We also use $\mathbb{F}_q$ to denote a finite field, whose order is $q \in \mathbb{N}$.

We take the standard definitions for some useful concepts of statistics and computational theory, such as *statistical distance, negligible funciton, statistically indistinguishable and computationally indistinguishable*. And we use negl(λ) to denote a negligible function in λ, use $x \leftarrow \mathbb{A}$ to denote choosing a uniform random element in $\mathbb{A}$ and $x \leftarrow X$ to denote choosing a random sample from distribution X.

2.2   Predicate Key Encapsulation Mechanism

Let $P_t = \{K_t \times E_t \rightarrow \{0, 1\} | t \in \mathbb{N}\}$ be a predicate family, where t denotes the dimension of the predicate P_t, K_t denotes the key attribute space and E_t denotes the ciphertext attribute family. For such a P_t, a PKEM scheme consists of the following four algorithms:

- PKEM.Setup(λ, t): a randomized algorithm which outputs a public parameter pp and a master key msk.
- PKEM.KeyGen(pp, msk, x): a randomized algorithm which outputs a secret key sk_x associated with the key attribute x.
- PKEM.Encaps(pp, y): a randomized algorithm which outputs a ciphertext c_y and a key k.
- PKEM.Decaps(pp, c_y, y, sk_x, x): an algorithm which outputs the key k.

Definition 1 (Correctness of PKEM). *A PKEM scheme is correct if for all* $(msk, pp) \leftarrow PKEM.Setup(\lambda, t)$, *all* $sk_x \leftarrow PKEM.KeyGen(pp, msk, x)$ *and all* $(c_y, k) \leftarrow PKEM.Encaps(pp, y)$, *it holds that*

1. *if* $P_t(x, y) = 1$, $PKEM.Decaps(pp, c_y, y, sk_x, x) = k$;
2. *otherwise* $PKEM.Decaps(pp, c_y, y, sk_x, x) = \bot$.

For $b \in \{0, 1\}$, we define that $PKEM.Encaps_b(pp, y)$ equals (c_y, k) for $b = 0$ and equals (c_y, k_R) for $b = 1$, where $(c_y, k) \leftarrow PKEM.Encaps(pp, y)$ and k_R is a random element.

Definition 2 (Security of PKEM). *A PKEM scheme* $\Pi = (PKEM.Setup, PKEM.KeyGen, PKEM.Encaps, PKEM.Decaps)$ *is chosen ciphertext secure if for all valid PPT adversary* $\mathcal{A}$,

$$Adv_{\mathcal{A}}^{PKEM}(\lambda) = |Pr[\mathcal{A}^{PKEM.KeyGen(pp,msk,\cdot), PKEM.Encaps_0(pp,\cdot)}(pp, \lambda, t) = 1]$$
$$- Pr[\mathcal{A}^{PKEM.KeyGen(pp,msk,\cdot), PKEM.Encaps_1(pp,\cdot)}(pp, \lambda, t)] = 1| \le negl(\lambda)$$

The probabilities are taken over the randomness of PKEM.Setup, PKEM.KeyGen, PKEM.Encaps and $\mathcal{A}$. *And* $\mathcal{A}$ *cannot take her challenge attribute as input for PKEM.KeyGen($pp, msk, \cdot$).*

Some Details About Pairing-Based PKEM Schemes. Pairing-based PKEM schemes constitute the main part of all current PKEM schemes. In this section, we will give some common properties of pairing-based PKEM schemes, other than the standard definition of PKEM given above.

We conclude these properties in the following lemma.

Lemma 1. *Almost all pairing-based PKEM schemes comply with the properties below.*

1. *The key k encapsulated is uniformly distributed in a certain group called target group, denoted G_T.*
2. *G_T is finite multiplicative cyclic group with order $r \in \mathbb{N}$, where r is about equal to the module of the pairing-based PKEM scheme, denoted q.*
3. *G_T consists of all r-th roots of unity, which are all in an extension field $\mathbb{F}_{q^\alpha}$, where $\alpha \in \mathbb{N}$, a small integer usually less than 12, is the embedding parameter of the underlying elliptic curve.*

2.3   Multi-key Fully Homomorphic Encryption

Since a common matrix which is shared by all participants is necessary in all existing LWE-based MFHE schemes, the setup algorithm should be described separately to generate the common matrix for all participants in context. Thus a MFHE scheme consists of five PPT algorithms {MFHE.Setup, MFHE.KeyGen, MFHE.Enc, MFHE.Dec, MFHE.Eval} as follows:

- MFHE.Setup(λ): a randomized algorithm which outputs the public parameter *params* of the system.
- MFHE.KeyGen(λ): a randomized algorithm which outputs a public encryption key pk along with a secret decryption key sk.
- $\varphi \leftarrow$ MFHE.Enc(pk, μ): a randomized algorithm which outputs a ciphertext φ.
- MFHE.Dec($(sk_1, \ldots, sk_N), \varphi$): an algorithm which outputs the decryption result μ.
- MFHE.Eval($f, (\varphi_1, \ldots, \varphi_\ell), (pk_1, \ldots, pk_N)$): an algorithm which outputs the evaluating result φ^f.

Definition 3 (Correctness of MFHE). *An MFHE scheme is correct if for all params $\leftarrow$ MFHE.Setup(λ), $(pk_i, sk_i) \leftarrow$ MFHE.KeyGen(params), $\varphi_j \leftarrow$ MFHE.Enc(pk_i, μ_j), where for simplicity we set $SK := \{sk_1, \ldots, sk_N\}$ and $PK := \{pk_i, \ldots, pk_N\}$, It holds that*

$$Pr[MFHE.Dec(SK, MFHE.Eval(f, (\varphi_1, \ldots, \varphi_\ell), PK) \neq f(\mu_1, \ldots, \mu_\ell)] = negl(\lambda)$$

Security. The security definition of an MFHE scheme is the basic IND-CPA security.

Key Generation Algorithm in LWE-Based MFHE Schemes. All existing LWE-based MFHE schemes share some common properties in their key generation algorithms other than the standard definition of MFHE scheme. Due to their important roles in our construction, we state them below.

Lemma 2. *The key generation algorithm in any LWE-based MFHE scheme can be deconstructed into two steps:*

1. *The secret decryption key is generated by selecting a uniform random vector in $\mathbb{Z}_p^n$: $sk \leftarrow \mathbb{Z}_p^n$, where $\log p$ and n are all polynomial in the security parameter λ.*
2. *The public encryption key is generated by a subalgorithm, denoted Gp, whose inputs are sk and the public parametrization params: $pk \leftarrow Gp(params, sk)$.*

We note that if an MFHE scheme satisfy the property described in Lemma 2, it can work as a symmetric encryption scheme, thus a DEM scheme if its secret key can be encapsulated in some KEM scheme.

2.4 Randomness Extractor and Pseudorandom Generator

Randomness Extractor. Before introducing the formal definition of randomness extractor, we give the definition for *min-entropy*.

Definition 4 (min-entropy). *Let X be a distribution over a set $\mathbb{A}$. The min-entropy of X, denoted by $H_\infty(X)$, is*

$$H_\infty(X) = min_{\eta \in \mathbb{A}}\{\log \frac{1}{Pr[X = \eta]}\}.$$

if $H_\infty(X) \geq \theta$, we call X a $\theta - source$.

Now we introduce a subclass of randomness extractor, *strong extractor*.

Definition 5 (strong extractor). *A function $Ext: \{0,1\}^u \times \{0,1\}^v \rightarrow \{0,1\}^w$ is a strong (θ, ε)-extractor if for every θ-source X on $\{0,1\}^u$, $(U_v, Ext(X, U_v))$ is ε-close to (U_v, U_w), where U_v denotes uniform distribution on $\{0,1\}^v$, so does U_w.*

Such a strong extractor exists.

Theorem 1. *For every $u, \theta \in \mathbb{N}$ and $\varepsilon > 0$ there exists a strong (θ, ε)-extractor $Ext: \{0,1\}^u \times \{0,1\}^v \rightarrow \{0,1\}^w$ with $w = \theta - 2\log(\frac{1}{\varepsilon}) - O(1)$ and $v = \log(u - \theta) + 2\log(\frac{1}{\varepsilon} + O(1))$.*

Pseudorandom Generator. A pseudorandom generator G is an efficient, deterministic algorithm for transforming a short, uniform string called *seed* into a longer pseudorandom one. A formal definition follows.

Definition 6 (Pseudorandom generator). *A deterministic polynomial-time algorithm* $G : \{0,1\}^s \rightarrow \{0,1\}^{g(s)}$, *where g is a polynomial function, is a* pseudorandom generator *if the following conditions hold:*

1. *Expansion: For every s it holds that $g(s) > s$.*
2. *Pseudorandom: For any PPT algorithm D, it holds that*

$$|Pr[D(G(\sigma)) = 1] - Pr[D(\gamma) = 1]| = negl(s).$$

where the first probability is taken over uniform choice of $\sigma \in \{0,1\}^s$ and the randomness of D, and the second probability is taken over uniform choice of $\gamma \in \{0,1\}^{g(s)}$ and the randomness of D.

g is the expansion factor of G.

We take note that, the existence of pseudorandom generator is a weaker assumption, than the existence of one-way function and public key cryptosystem.

3   Predicate Fully Homomorphic Encryption

3.1   Definition

For a predicate family $P_t = \{K_t \times E_t \rightarrow \{0,1\}|t \in \mathbb{N}\}$, defined as in Subsect. 2.2, a PFHE scheme consists the following five algorithms.

- PFHE.Setup(λ, t): a randomize algorithm which outputs a public parametrization pp of the system and a master secret key msk.
- PFHE.KeyGen(pp, msk, x): a randomize algorithm which outputs a secret decryption key sk_x associated with x.
- PFHE.Enc(pp, y, μ): a randomized algorithm which outputs a ciphertext ϕ_y associated with y.
- PFHE.Dec(pp, y, ϕ_y, x, sk_x). an algorithm which outputs the decryption result μ.
- PFHE.Eval($f, \{\phi_{y,1}, \ldots, \phi_{y,\ell}\}, pp$). an algorithm which outputs the evaluating result ϕ_y^f.

If across-attribute computing is allowed in a PFHE scheme, we call this scheme a *multi-attribute PFHE* scheme, definition of which can be derived by slightly modifying the definition above.

Definition 7 (multi-attribute PFHE). *A multi-attribute PFHE scheme consists of five algorithms: {MPFHE.Setup, MPFHE.KeyGen, MPFHE.Enc, MPFH- E.Dec, MPFHE.Eval }, and most of them can be described in the same way as a PFHE scheme, while MPFHE.Dec and MPFHE.Eval are defined as follows.*

- *MPFHE.Dec($pp, \{y_j\}_{j\in[N]}, \phi', \{x_k\}_{k\in[M]}, \{sk_{x_k}\}_{k\in[M]}$) : an algorithm which outputs a decryption result μ.*

– $MPFHE.Eval(f, \{\phi_{y_j,i}\}_{i\in[\ell],j\in[N]}, pp)$: *an algorithm which outputs the evaluating result* ϕ^f.

Definition 8 (Correctness of PFHE). *A PFHE scheme is correct, if for all* $(pp, msk) \leftarrow PFHE.Setup(\lambda, t)$, $sk_x \leftarrow PFHE.KeyGen(pp, msk, x)$, $\phi_{y,i} \leftarrow PFHE.Enc(pp, y, \mu_i)$ *where* $i \in [\ell]$, *it holds that:*

1. *if* $P_t(x, y) = 1$, $PFHE.Dec(pp, y, PFHE.Eval(f, \{\phi_{y,1}, \ldots, \phi_{y,\ell}\}, pp), x, sk_x) = f(\mu_1, \ldots, \mu_\ell)$.
2. *if* $P_t(x, y) = 0$, $PFHE.Dec(pp, y, PFHE.Eval(f, \{\phi_{y,1}, \ldots, \phi_{y,\ell}\}, pp), x, sk_x) = \bot$.

3.2 Security of PFHE

The chosen ciphertext security (precisely, CCA1 security) of PFHE is defined through the following interactive game played between an adversary $\mathcal{A}$ and a challenger $\mathcal{C}$.

1. **Setup.** $\mathcal{C}$ draws $(pp, msk) \leftarrow PFHE.Setup(\lambda, t)$ and gives pp to $\mathcal{A}$.
2. **Phase 1.** $\mathcal{A}$ does the following two things.
   (a) Adaptively sends ϕ_y along with y and some x, and gets μ from $\mathcal{C}$.
   (b) Adaptively sends x, and gets sk_x from $\mathcal{C}$.
3. **Challenge.** $\mathcal{A}$ claims y^* and two challenge messages μ_0 and μ_1, with the restriction that $P_t(x, y^*) = 0$ for all x been quired in **Phase 1** and $|\mu_0| = |\mu_1|$. $\mathcal{C}$ choose $b \in \{0, 1\}$, and runs $\phi_{y^*,b} \leftarrow PFHE.Enc(pp, y^*, \mu_b)$ and returns it to $\mathcal{A}$.
4. **Phase 2.** $\mathcal{A}$ can adaptively submit secret key quires for x with the restriction that $P_t(x, y^*) = 0$.
5. **Guess.** $\mathcal{A}$ outputs a bit $b' \in \{0, 1\}$ and wins the game only if $b = b'$.

The advantage of $\mathcal{A}$ in the game above with security parameter λ is defined as $Adv_{\mathcal{A}}^{PFHE}(\lambda) = |Pr[b' = 1|b = 1] - Pr[b' = 1|b = 0]|$. We say the PKEM is chosen ciphertext secure if and only if $Adv_{\mathcal{A}}^{PFHE}(\lambda) = negl(\lambda)$.

4 Our Construction

In this section, we build a transfer map Tr from an encapsulated key k in a PKEM scheme to a secret decryption sk in an LWE-based MFHE scheme. Then we give a generic construction for a PFHE scheme, basing on a PKEM scheme, an LWE-based MFHE scheme and the transfer map Tr.

4.1 The Transfer Map

Generally, a hybrid encryption scheme can function only if the key of the DEM scheme can be encapsulated in the KEM scheme. However, the key of a symmetric LWE-based MFHE scheme, which we have shown in Sect. 2.3 and will work as a basic block for our PFHE scheme in this section, is a high dimensional vector over $\mathbb{Z}_p$, thus the PKEM scheme need to work with a huge parameter or work for a large amount of times to encapsulate the vector in a direct way, both resulting inefficiency.

We want to build a transfer map Tr to assist the KEM scheme for an MFHE scheme. Concretely, if the space of k is denoted by $\mathcal{K}$, and the space of sk is denoted by $\mathcal{SK}$, the transfer map

$$Tr : \mathcal{K} \to \mathcal{SK}$$
$$k \to sk.$$

should satisfy the following property.

Property 1 (Pseudorandomness). For any PPT algorithm D, it holds that

$$|\Pr[D(Tr(k)) = 1] - \Pr[D(sk) = 1]| = \mathrm{negl}(\lambda)$$

where $k \leftarrow \mathcal{K}$, $sk \leftarrow \mathrm{MFHE.KeyGen}(\lambda)$.

Due to the difference between the key space $\mathcal{K}$ of pairing-based PKEM and the one of lattice-based PKEM, we will build the transfer map for them separately.

Transfer Map for Lattice-Based PKEM. A lattice-based PKEM scheme can be constructed from Lattice-based PE schemes through standard techniques, the k produced by encapsulation algorithm of which is uniformly distributed over $\{0,1\}^\lambda$.

If there exists a pseudorandom generator G

$$G : \{0,1\}^\lambda \to \{0,1\}^{n\lceil \log p \rceil}.$$

We can build the transfer map

$$Tr_0 : \mathcal{K} \to \mathcal{SK},$$

as required, through the following steps:

1. Take as input the key k, run $bsk \leftarrow G(k)$.
2. Divide bsk into n equal-length parts, denoted bsk_i, $i \in [n]$.
3. Let $\boldsymbol{g} = \{1, 2, \ldots, 2^{\lceil \log q \rceil - 1}\}$, and regard bsk_i as a vector over $\{0,1\}$, for $i \in [n]$. Compute
$$sk[i] \leftarrow < \boldsymbol{g}, bsk_i >$$
then get a vector $\boldsymbol{sk} \in \mathbb{Z}_p^n$.
4. Output $sk = \boldsymbol{sk}$.

We note that in an LWE-based MFHE scheme, sk is uniformly distributed over $\mathbb{Z}_p^n$. The *pseudorandomness* of the map Tr_0 can be easily proved.

Transfer Map for Pairing-Based PKEM. Some basic knowledge of pairing-based PKEM schemes is introduced in Sect. 2.2. According to Lemma 1, k encapsulated in a pairing-based PKEM scheme is uniformly distributed over a group G_T, which is a finite multiplicative cyclic group in an extension filed $\mathbb{F}_{q^\alpha}$. Trivially, there is an isomorphic map, denoted by $BitDecomp$, from elements of the filed $\mathbb{F}_{q^\alpha}$ to strings in $\{0,1\}^{\alpha \lceil \log q \rceil}$. Elements of G_T can also be mapped to strings in $\{0,1\}^{\alpha \lceil \log q \rceil}$. However, $BitDecomp(k)$ where $k \leftarrow G_T$ is not uniformly distributed on $\{0,1\}^{\alpha \lceil \log q \rceil}$, thus it cannot work as a seed for a pseudorandom generator. Actually, its distribution is X, where

$$Pr(X = \eta) = \begin{cases} \dfrac{1}{r}, & \eta = BitDecomp(k), \text{for some } k \leftarrow G_T \\ 0, & \text{Otherwise.} \end{cases}$$

The min-entropy of X is

$$H_\infty(X) = min_{\eta \in \{0,1\}^{\alpha \lceil \log q \rceil}} \{ \frac{1}{\Pr(X = \eta)} \} = \log r.$$

To get a seed for pseudorandom generator, we use a strong (θ, ε)-extractor,

$$\text{Ext} : \{0,1\}^u \times \{0,1\}^v \to \{0,1\}^w$$

where $\theta = \log r$, $\varepsilon = negl(\theta)$, $u = \alpha \lceil \log q \rceil$, $w = \theta - 2\log(\frac{1}{\varepsilon}) - O(1)$ and $v = \log(u - \theta) + 2\log(\frac{1}{\varepsilon} + O(1))$. From Theorem 2, such a extractor exists. Then we assume there is a pseudorandom generator $G : \{0,1\}^s \to \{0,1\}^{g(s)}$ with $s = w$ and $g(s) = n\lceil \log p \rceil$.

Now, the transfer map Tr_1 for pairing-based KEM schemes can be built as follows.

1. Take as inputs the key k, run $\eta \leftarrow BitDecomp(k)$;
2. Choose $a \leftarrow \{0,1\}^v$, run $\sigma \leftarrow \text{Ext}(\eta, a)$.
3. Run $bsk \leftarrow G(\sigma)$.
4. Get $\boldsymbol{sk} \in \mathbb{Z}_p^n$ through the method described in Tr_0.
5. Output $sk = \boldsymbol{sk}$ and a.

We note the form of map Tr_1 has been slightly modified as

$$Tr_1 : U_v \times \mathcal{K} \to U_v \times \mathcal{SK}$$
$$(a, k) \to (a, sk).$$

The *pseudorandomness* of Tr_1 can be easily proved. And the *consistency* of Tr_1 can be shown through a standard argument. We refer the reader to the full version for details.

4.2 Our Scheme

Starting from a PKEM scheme {PKEM.Setup, PKEM.KeyGen, PKEM.Encaps PKEM.Decaps } w.r.t a predicate family $P_t = \{K_t \times E_t \to \{0,1\}\}$, an LWE-based MFHE scheme {MFHE.Setup, MFHE.KeyGen, MFHE.Enc, MFHE.Dec,

MFHE.Eval} with an additional subalgorithm $pk \leftarrow Gp(params, sk)$ in MFHE. K-eyGen algorithm, both of which are defined in Sect. 2, and a transfer map $Tr : U \times \mathcal{K} \rightarrow U \times \mathcal{SK}$ as defined in Subsect. 4.1, we build a PFHE scheme w.r.t. the same predicate family, which consists five algorithms PFHE.Setup, PFHE.KeyGen, PFHE.Enc, PFHE.Dec, PFHE.Eval.

Concretely, the five algorithms are defined as following.

PFHE.Setup(λ, t). Do the followings:

1. Run $(\overline{pp}, msk) \leftarrow$ PKEM.Setup(λ, t).
2. Run $params \leftarrow$ MFHE.Setup(λ).
3. Return $pp = \{\overline{pp}, params\}$ and msk.

PFHE.KeyGen(pp, msk, x). Do the followings:

1. Parse $pp = \{\overline{pp}, params\}$
2. Run $sk_x \leftarrow$ PKEM.KeyGen$(\overline{pp}, msk, x)$.
3. Return sk_x.

PFHE.Enc(pp, y, μ). Do the followings:

1. Parse $pp = \{\overline{pp}, params\}$.
2. Run $(c_y, k) \leftarrow$ PKEM.Encaps$(\overline{pp}, y)$.
3. Select $a \leftarrow U$, Run $(a, sk) \leftarrow Tr(a, k)$.
4. Run $pk \leftarrow Gp(params, sk)$.
5. Run $\varphi \leftarrow$ MFHE.Enc(pk, μ).
6. Return $\phi_y = \{c_y, pk, a, \varphi\}$.

PFHE.Dec$(pp, y, \phi'_y, x, sk_x)$. ϕ'_y is a fresh ciphertext or an evaluated ciphertext, which consists of $\{c_{y,i}, pk_i, a_i\}_{i \in [\ell]}$ for some $\ell \in \mathbb{N}$, and φ'. Do the followings:

1. Parse $pp = \{\overline{pp}, params\}$ and $\phi_y = \{\{c_{y,i}, pk_i, a_i\}_{i \in [\ell]}, \varphi'\}$.
2. For $i \in [\ell]$, do the followings:
   – Run $k_i \leftarrow$ PKEM.Decaps$(\overline{pp}, c_{y,i}, y, sk_x, x)$.
   – Run $sk_i \leftarrow Tr(a_i, k_i)$.
3. Run $\mu \leftarrow$ MFHE.Dec$(\{sk_1, \ldots, sk_\ell\}, \varphi')$ and return it.

PFHE.Eval$(pp, \{\phi_{y,1}, \ldots, \phi_{y,\ell}\}, pp)$. Do the followings.

1. Parse $pp = \{\overline{pp}, params\}$ and $\phi_{y,i} = \{c_{y,i}, pk_i, a_i, \varphi_i\}$.
2. Run $\varphi^f \leftarrow$ MFHE.Eval$(f, \{\varphi_i\}_{i \in [\ell]}, \{pk_i\}_{i \in [\ell]})$.
3. Return $\phi_y^f = \{\{c_{y,i}, pk_i, a_i\}_{i \in [\ell]}, \varphi^f\}$

Correctness

Lemma 3. *If both the underlying PKEM scheme and MFHE scheme are correct, the scheme PFHE is correct.*

We refer the reader to the full version for the proof.

Multi-attribute PFHE

Lemma 4. *The PFHE scheme inherently supports across-attribute computing. Any function can be evaluated on ciphertext $\{\phi_{y_j,i}\}_{i\in[\ell],j\in[N]}$, (we assume there is no pair (i, j_1), if (i, j_2) exists and $j_i \neq j_2$.) where $\phi_{y_j,i} \leftarrow PFHE.Enc(pp, y_j, \mu_i)$. And the ciphertext output by the evaluated function can be decrypted correctly with the assistance of a secret key set $\{sk_{x_k}\}_{k\in[M]}$, in which for every y_j, there is a x_k s.t. $P_t(x_k, y_j) = 1$.*

Proof. We will build two algorithms MFPHE.Eval and MFPHE.Dec, as defined in Definition 7, to complete this proof.

MPFHE.Eval$(f, \{\phi_{y_j,i}\}_{i\in[\ell],j\in[N]}, pp)$. Do the followings.

1. Parse $\phi_{y_j,i} = \{c_{y_j,i}, pk_i, a_i, \varphi_i\}$ for $i \in [\ell], j \in [N]$, $pp = \{\overline{pp}, params\}$.
2. Run $\varphi^f \leftarrow$ MFHE.Eval$(f, \{pk_i\}_{i\in[\ell]}, \{\varphi_i\}_{i\in[\ell]})$.
3. Return $\phi_Y^f = \{\{c_{y_j,i}, pk_i, a_i\}_{i\in[\ell],j\in[N]}, \varphi^f\}$, where Y denotes the set $\{y_j\}_{j\in[N]}$.

MPFHE.Dec$(pp, \{y_j\}_{j\in[N]}, \phi', \{x_k\}_{k\in[M]}, \{sk_{x_k}\}_{k\in[M]})$. Do the followings.

1. Parse $\phi' = \{\{c_{y_j,i}, pk_i, a_i\}_{i\in[\ell],j\in[N]}, \varphi'\}$ and $pp = \{\overline{pp}, params\}$.
2. For every $i \in [\ell]$, do the followings.
   - For the $c_{y_j,i}$, find $k \in [M]$ s.t $P_t(x_k, y_j) = 1$, then run

$$k_i \leftarrow \text{PKEM.Decaps}(\overline{pp}, c_{y_j,i}, y_j, sk_{x_k}, x_k).$$

   - Run $sk_i \leftarrow Tr(a_i, k_i)$.
3. Run $\mu \leftarrow$ MFHE.Dec$(\{sk_i\}_{i\in[\ell]}, \varphi')$.
4. Return μ.

4.3 Security Analysis, PFHE with Chosen Ciphertext Security

In this subsection, we will prove the following theorem.

Theorem 2. *If the underlying PKEM scheme is indistinguishability under chosen attribute attack and chosen ciphertext attack(IND-CAA-CCA) secure, and the MFHE scheme is indistinguishability under chosen plaintext attack(IND-CPA) secure, the proposed PFHE scheme is IND-CAA-CCA1 secure.*

Proof. The proof proceeds in a sequence of games where the first game is identical to the game defined in Sect. 3.2. And in the last game of the sequence, the adversary has advantage zero. These games are built as follow.

Game 0. This game is identical to the game in Sect. 3.2.
Game 1. This game is built by slightly modifying **Game 0** in *Challenge* phase. Concretely,

1. $\mathcal{A}$ submits μ_0 and μ_1 with $|\mu_0| = |\mu_1|$, and the challenge attribute y^*.
2. $\mathcal{C}_1$ Runs $(c_{y^*}, k^*) \leftarrow$ PKEM.Encaps$(\overline{pp}, y^*)$.
3. $\mathcal{C}_1$ selects $k_R^* \leftarrow \mathcal{K}$, $a^* \leftarrow U$, then run $sk^* \leftarrow Tr(a^*, k_R^*)$.

4. $\mathcal{C}_1$ Runs $pk^* \leftarrow Gp(params, sk^*)$, then chooses a bit $b \in \{0,1\}$, runs $\varphi^* \leftarrow$ MFHE.Enc(pk^*, μ_b).
5. $\mathcal{C}_1$ returns $\phi^* = \{c_{y^*}, pk^*, a^*, \varphi^*\}$ as the challenge response.

Game 2. This game is built by slightly modifying Game 1 in *challenge* phase. Concretely,

1. $\mathcal{A}$ submits her challenge attribute y^* and plaintexts μ_0 and μ_1 as in Game 1.
2. $\mathcal{C}_2$ runs $(c_{y^*}, k^*) \leftarrow$ PKEM.Encaps$(\overline{pp}, y^*)$.
3. $\mathcal{C}_2$ runs $(pk^*, sk^*) \leftarrow$ MFHE.KeyGen$(params)$ and $a^* \leftarrow U$.
4. $\mathcal{C}_2$ selects $b \in \{0,1\}$, draws $\varphi_b \leftarrow$ MFHE.Enc(pk, μ_b).
5. $\mathcal{C}_2$ returns $\phi_{y^*,b} = \{c_{y^*}, pk^*, a^*, \varphi_b\}$.

Game 3. Similarly to previous constructions, this game is built by slightly modifying **Game 2** in *Challenge* phase, where

1. $\mathcal{A}$ submits her challenge attribute y^* and plaintexts μ_0 and μ_1 as in Game 1.
2. $\mathcal{C}_3$ runs $(c_{y^*}, k^*) \leftarrow$ PKEM.Encaps$(\overline{pp}, y^*)$.
3. $\mathcal{C}_3$ runs $(pk^*, sk^*) \leftarrow$ MFHE.KeyGen$(params)$ and $a^* \leftarrow U$.
4. $\mathcal{C}_3$ selects a random element in the ciphertext space of the MFHE scheme, which denoted by $\mathcal{CT}$: $\varphi^* \leftarrow \mathcal{CT}$.
5. $\mathcal{C}_3$ returns $\phi_{y^*,b} = \{c_{y^*}, pk^*, a^*, \varphi^*\}$.

Since the challenge ciphertext is always a fresh random element in the ciphertext space, $\mathcal{A}$'s advantage in this game is zero.

Lemmas 5, 6 and 7 show that there is no PPT adversary can distinguish between these games with non-negligible advantage. Thus, no PPT adversary can win the original game with non-negligible advantage. We refer the reader to the full version for the proof of Lemmas 5, 6 and 7.

Lemma 5. *Any PPT adversary cannot distinguish between Game 0 and Game 1 with non-negligible advantage, if the underlying PKEM scheme is IND-CAA-CCA secure.*

Lemma 6. *Any PPT adversary cannot distinguish between Game 1 and Game 2, if the underlying transfer map Tr has the property of* pseudorandomness.

Lemma 7. *Any PPT adversary cannot distinguish between Game 2 and Game 3 with non-negligible advantage, if the underlying MFHE scheme is IND-CPA secure.*

5   CCA1-secure PFHE from Various Way

In this section, we will provide a different way to approach CCA1 security. Our starting point is the work of Yamada et al. [34], who showed that any CPA-secure PE scheme with verifiability could be converted into a CCA2-secure one. There are two main obstacles in applying the method in a direct way. First, in [34], every valid ciphertext includes a one time signature. But with homomorphism, we can obtain a valid ciphertext through evaluating on some valid ciphertexts and the resulting cipheretext doesn't have a valid signature. In our method, one time signature is removed, at cost of that adaptive decryption quires are forbidden, which is inherently forbidden by homomorphism. Second, in their method, in each encryption, a message is encrypted under an original attribute and a randomly chosen attribute. Thus computation designed under one certain attribute is essentially performed on different attributes. We exploit the property that across-attribute computation can be performed at cost of almost nothing, to overcome the second problem.

5.1   Related Definitions

The following three definitions are from [34].

Definition 9 (Verifiability). *A PE scheme Π = {PE.Setup, PE.KeyGen, PE.E- nc, PE.Dec}w.r.t. $P_t = \{K_t \times E_t \to \{0,1\}|t \in \mathbb{N}\}$ is said to have verifiablity if that for all $\lambda, t \in \mathbb{N}$, $(pp, msk) \leftarrow PE.Setup(\lambda, t)$, $x, x' \in K_t$, $y \in E_t$, the following holds:*

If $sk_x \leftarrow PE.KeyGen(pp, msk, x)$, $sk_{x'} \leftarrow PE.KeyGen(pp, msk, x')$ and $P_t(x, y) = P_t(x', y)$, then for all ciphertext ϕ, $PE.Dec(pp, \phi, y, sk_x, x) = PE.Dec(pp, \phi, y, sk_{x'}, x')$ holds.

Definition 10 (OR-compatibility). *A predicate family $P_t = \{K_t \times E_t \to \{0,1\}|t \in \mathbb{N}\}$ is OR-compatible, if for all $d, t \in \mathbb{N}$, there are maps OR: $K_t \times K_d \to K_{t+d}$ and $ET : K_t \to K_{t+d}$, $ED : K_d \to K_{t+d}$, such that for all $x_1 \in K_t$, $x_2 \in K_d$ and $y_1 \in E_t$, $y_2 \in E_d$ it holds that*

$$P_{t+d}(ET(x_1), OR(y_1, y_2)) = P_t(x_1, y_1);$$

$$P_{t+d}(ED(x_2), OR(y_1, y_2)) = P_d(x_2, y_2).$$

Definition 11 (Equality test). *For a predicate family $P_t = \{K_t \times E_t \to \{0,1\}|t \in \mathbb{N}\}$ and a set V, we say that P_t can perform equality test over V by using dimension d, if there are maps $\tau : V \to K_d$ and $\pi : V \to E_d$ s.t. for all $e, e' \in V$, we have $P_d(\tau(e), \pi(e)) = 1$ and $P_d(\tau(e), \pi(e')) = 0$.*

5.2   Conversion from CPA-secure MPFHE

Let Π = (MPFHE.Setup, MPFHE.KeyGen, MPFHE.Enc, MPFHE.Dec, MPFHE.Eval) be a CPA-secure MPFHE scheme w.r.t. predicate family $P_t = \{K_t \times E_t \to \{0,1\}|t \in \mathbb{N}\}$ (such a scheme can be derived from a PFHE scheme easily via Lemma 4.) We assume that Π has verifiability, OR-compatibility and it can perform equality test on one set U, We can construct a PFHE scheme with CCA1 security w.r.t P_t. The resulting scheme Π' ={CCA.Setup, CCA.KeyGen, CCA.Enc, CCA.Dec, CCA.Eval} is constructed as follows.

CCA.Setup(λ, t). Run $(pp, msk) \leftarrow$ MPFHE.Setup$(\lambda, t + d)$, where d is dimension for equality test.
CCA.KeyGen(pp, msk, x). Run $sk'_{ET(x)} \leftarrow$ MPFHE.KeyGen$(pp, msk, ET(x))$, then let $sk_x = sk'_{ET(x)}$.
CCA.Enc(pp, y, μ). Choose a random element $e \leftarrow V$, then run $\phi'_{OR(y, \pi(e))} \leftarrow$ MPFHE.Enc$(pp, \mu, OR(y, \pi(e)))$, return $\phi_y = (\phi'_{OR(y, \pi(e))}, e)$.
CCA.Dec(pp, ϕ_y, y, sk_x, x).

- Parse $\phi_y = \{\{e_i\}_{i \in [\ell]}, \phi_Y\}$, Y denotes the set $\{OR(y, \pi(e_i))\}_{i \in [\ell]}$ and ℓ is some integer.
- Run $\mu \leftarrow$ MPFHE.Dec$(pp, \phi'_Y, Y, sk'_{ET(x)}, x)$.
- Return μ.

CCA.Eval$(f, \{\phi_{y,i}\}_{i \in [\ell]}, pp)$.

- For $i \in [\ell]$, parse $\phi_{y,i} = (\phi'_{OR(y, ED(e_i)), i}, e_i)$.
- Then run $\phi'^f_y \leftarrow$ MPFHE.Eval$(f, \{\phi'_{OR(y, ED(e_i)), i}\}_{i \in [\ell]}, pp)$.
- Finally return $\phi^f_y = \{\phi'^f_y, \{e_i\}_{i \in [\ell]}\}$.

Correctness. According to Definition 10, we have

$$P_{t+d}(ET(x), OR(Y, \pi(e))) = P_t(x, y).$$

Thus the correctness of Π' follows directly from correctness Π.

5.3 Security Analysis for the Conversion

In this section, we will prove the following theorem.

Theorem 3. *If Π is CPA-secure MPFHE w.r.t. predicate family P_t, then Π' is CCA1-secure PFHE w.r.t. the same predicate family.*

Proof. If an adversary $\mathcal{A}$ can win the CCA1 security game for Π' with non-negligible advantage, we can invoke it to construct an adversary $\mathcal{B}$ which breaks CPA security of Π. $\mathcal{B}$ is defined as follows:

Setup: The challenger $\mathcal{C}$ runs $(pp, msk) \leftarrow$ MPFHE.Setup$(\lambda, t + d)$, then gives pp to $\mathcal{B}$. $\mathcal{B}$ gives pp to $\mathcal{A}$ and selects $e^* \in V$.
Phase 1:

- **Game for Π:**
    1. For every secret key quire for some attribute $x \in K_{t+d}$ from $\mathcal{B}$, $\mathcal{C}$ runs $sk'_x \leftarrow$ MPFHE.KeyGen(pp, msk, x) and returns it.
- **Game for Π':**
    1. For every secret key quire for some attribute $x \in K_t$ from $\mathcal{A}$, $\mathcal{B}$ computes $ET(x) \in K_{t+d}$, then submits $ET(x)$ to $\mathcal{C}$ as his own quire. $\mathcal{B}$ receives $sk'_{ET(x)}$ from $\mathcal{C}$, and lets $sk_x = sk'_{ET(x)}$ then returns it to $\mathcal{A}$.

2. For every decryption quire for some ciphertext ϕ_y along with its attribute $y \in E_t$ and some key attribute $x \in K_t$, $\mathcal{B}$ checks whether $P_t(x, y) = 1$, if not, returns $\bot$. Otherwise, he parses it $\phi_y = (\{e_i\}_{i \in [\ell]}, \phi'_Y)$ where $Y = \{OR(y, \pi(e_i))\}_{i \in [\ell]}$. Then $\mathcal{B}$ checks whether $e_i = e^*$ holds for all $i \in [\ell]$. If it holds for some i, $\mathcal{B}$ aborts. Otherwise, he computes $ED(\tau(e_i))$ for all $i \in [\ell]$, and submits them to $\mathcal{C}$ as his own secret quires. Once he receives $\{sk'_{ED(\tau(e_i))}\}_{i \in [\ell]}$, runs $\mu \leftarrow$ MPFHE.Dec$(pp, Y, \phi'_Y, \{ED(\tau(e_i))\}_{i \in [\ell]}, \{sk'_{ED(\tau(e_i))}\}_{i \in [\ell]})$ and returns it to $\mathcal{A}$.

Challenge:

- **Game for Π:** For the challenge attribute $y^* \in E_{t+d}$, which satisfies that $P_t(x, y^*) \neq 0$ for all x quired in **Phase 1**, and two equal-length challenge messages μ_0 and μ_1 from $\mathcal{B}$, $\mathcal{C}$ selects a bit $b \in \{0, 1\}$, then $\phi'_{y^*, b} \leftarrow$ MPFHE.Enc(pp, y, μ_b) and returns it to $\mathcal{B}$.
- **Game for Π':** For the challenge attribute $y^* \in E_t$, which satisfies that $P_t(x, y^*) \neq 0$ for all x quired in **Phase 1**, two equal-length challenge plaintexts μ_0 and μ_1 from $\mathcal{A}$, $\mathcal{B}$ computes $OR(y^*, \pi(e^*))$ and submits it and the two challenge messages to $\mathcal{C}$ as his own challenge attribute and plaintexts. He receives the response ciphertext $\phi'_{OR(y^*, \pi(e^*)), b}$, and returns $(\phi'_{OR(y^*, \pi(e^*)), b}, e^*)$ to $\mathcal{A}$.

Phase 2: $\mathcal{A}$ can make secret key quires for x with restriction that $P_t(x, y^*) = 0$, and $\mathcal{B}$ can respond these quires in the same way as in **Phase 1**.

Guess: Finally, $\mathcal{A}$ outputs $b' \in \{0, 1\}$ as her guess for b. And $\mathcal{B}$ takes b' as his own guess.

What remains to show is that the simulation of the decryption oracle is correct. Since

$$P_{t+d}(ET(x), OR(y, \pi(e))) = P_{t+d}(ED(\tau(e)), OR(y, \pi(e))) = 1,$$

if $P_t(x, y) = 1$, by the verifiability, we have that

$$\text{MPFHE.Dec}(pp, \phi'_Y, Y, \{sk'_{ED(\tau(e_i))}\}_{i \in [\ell]}, \{ED(\tau(e_i))\}_{i \in [\ell]}) =$$
$$\text{MPFHE.Dec}(pp, \phi'_Y, Y, sk'_{ET(x)}, ET(x)),$$

if $P_t(x, y) = 1$.

Thus, the simulation is perfect if $\mathcal{B}$ doesn't abort. And since $\mathcal{A}$ doesn't know e^* until getting the response ciphertext, the probability which $\mathcal{B}$ aborts with is equal to $1 - (1 - \frac{1}{|V|})^Q$, where Q is the number of secret key quires that $\mathcal{A}$ makes in **Phase 1**. Thus, the probability is a negligible function of λ. We completed the proof.

6 Conclusion

We provide a general framework for constructing PFHE from PKEM and LWE-based MFHE. The resulting PFHE schemes have relatively small ciphertext size and support across-attribute computing. We prove that if the underlying PKEM is CCA-secure and MFHE is CPA-secure, the resulting PFHE is CCA1 secure. Thus we proposed a generic construction for CCA1-secure PFHE, which subsumes CCA1-secure FHE and CCA1-secure MFHE. We further give a conversion from CPA-secure PFHE to CCA1-secure PFHE, assuming the underlying CPA-secure PFHE has verifiability.

Acknowledgment. Qianhong Wu is the corresponding author. This paper is supported by the National Key Research and Development Program of China through project 2017YFB0802505, the Natural Science Foundation of China through projects 61772538, 61672083, 61370190, 61532021, 61472429 and 61402029, and by the National Cryptography Development Fund through project MMJJ20170106.

References

1. Boneh, D., Canetti, R., Halevi, S., Katz, J.: Chosen-ciphertext security from identity-based encryption. SIAM J. Comput. **36**(5), 1301–1328 (2007)
2. Boneh, D., Franklin, M.: Identity-based encryption from the weil pairing. In: Kilian, J. (ed.) CRYPTO 2001. LNCS, vol. 2139, pp. 213–229. Springer, Heidelberg (2001). https://doi.org/10.1007/3-540-44647-8_13
3. Boneh, D., Gentry, C., Gorbunov, S., Halevi, S., Nikolaenko, V., Segev, G., Vaikuntanathan, V., Vinayagamurthy, D.: Fully key-homomorphic encryption, arithmetic circuit ABE and compact garbled circuits. In: Nguyen, P.Q., Oswald, E. (eds.) EUROCRYPT 2014. LNCS, vol. 8441, pp. 533–556. Springer, Heidelberg (2014). https://doi.org/10.1007/978-3-642-55220-5_30
4. Boneh, D., Sahai, A., Waters, B.: Functional encryption: definitions and challenges. In: Ishai, Y. (ed.) TCC 2011. LNCS, vol. 6597, pp. 253–273. Springer, Heidelberg (2011). https://doi.org/10.1007/978-3-642-19571-6_16
5. Brakerski, Z., Cash, D., Tsabary, R., Wee, H.: Targeted homomorphic attribute-based encryption. In: Hirt, M., Smith, A. (eds.) TCC 2016. LNCS, vol. 9986, pp. 330–360. Springer, Heidelberg (2016). https://doi.org/10.1007/978-3-662-53644-5_13
6. Brakerski, Z., Gentry, C., Vaikuntanathan, V.: (Leveled) fully homomorphic encryption without bootstrapping. In: ITCS, pp. 309–325 (2012)
7. Brakerski, Z., Perlman, R.: Lattice-based fully dynamic multi-key FHE with short ciphertexts. In: Robshaw, M., Katz, J. (eds.) CRYPTO 2016. LNCS, vol. 9814, pp. 190–213. Springer, Heidelberg (2016). https://doi.org/10.1007/978-3-662-53018-4_8
8. Canetti, R., Raghuraman, S., Richelson, S., Vaikuntanathan, V.: Chosen-ciphertext secure fully homomorphic encryption. In: Fehr, S. (ed.) PKC 2017. LNCS, vol. 10175, pp. 213–240. Springer, Heidelberg (2017). https://doi.org/10.1007/978-3-662-54388-7_8
9. Cash, D., Hofheinz, D., Kiltz, E., Peikert, C.: Bonsai trees, or how to delegate a lattice basis. In: Gilbert, H. (ed.) EUROCRYPT 2010. LNCS, vol. 6110, pp. 523–552. Springer, Heidelberg (2010). https://doi.org/10.1007/978-3-642-13190-5_27

10. Castiglione, A., Santis, A.D., Masucci, B., Palmieri, F., Castiglione, A., Huang, X.: Cryptographic hierarchical access control for dynamic structures. IEEE Trans. Inf. Forensics Secur. **11**(10), 2349–2364 (2016)

11. Castiglione, A., Santis, A.D., Masucci, B., Palmieri, F., Castiglione, A., Li, J., Huang, X.: Hierarchical and shared access control. IEEE Trans. Inf. Forensics Secur. **11**(4), 850–865 (2016)

12. Chen, X., Li, J., Ma, J., Tang, Q., Lou, W.: New algorithms for secure outsourcing of modular exponentiations. IEEE Trans. Parallel Distrib. Syst. **25**(9), 2386–2396 (2014)

13. Clear, M., McGoldrick, C.: Multi-identity and multi-key leveled FHE from learning with errors. In: Gennaro, R., Robshaw, M. (eds.) CRYPTO 2015. LNCS, vol. 9216, pp. 630–656. Springer, Heidelberg (2015). https://doi.org/10.1007/978-3-662-48000-7_31

14. Galbraith, S.D., Paterson, K.G., Smart, N.P.: Pairings for cryptographers. Discrete Appl. Math. **156**(16), 3113–3121 (2008)

15. Gentry, C.: Fully homomorphic encryption using ideal lattices. In: STOC, pp. 169–178 (2009)

16. Gentry, C., Peikert, C., Vaikuntanathan, V.: Trapdoors for hard lattices and new cryptographic constructions. In: STOC, pp. 197–206 (2008)

17. Gentry, C., Sahai, A., Waters, B.: Homomorphic encryption from learning with errors: conceptually-simpler, asymptotically-faster, attribute-based. In: Canetti, R., Garay, J.A. (eds.) CRYPTO 2013. LNCS, vol. 8042, pp. 75–92. Springer, Heidelberg (2013). https://doi.org/10.1007/978-3-642-40041-4_5

18. Gorbunov, S., Vaikuntanathan, V., Wee, H.: Attribute-based encryption for circuits. In: STOC, pp. 545–554 (2013)

19. Gorbunov, S., Vaikuntanathan, V., Wee, H.: Predicate encryption for circuits from LWE. In: Gennaro, R., Robshaw, M. (eds.) CRYPTO 2015. LNCS, vol. 9216, pp. 503–523. Springer, Heidelberg (2015). https://doi.org/10.1007/978-3-662-48000-7_25

20. Goyal, V., Pandey, O., Sahai, A., Waters, B.: Attribute-based encryption for fine-grained access control of encrypted data. In: CCS, pp. 89–98 (2006)

21. Hoffstein, J., Pipher, J., Silverman, J.H.: NTRU: a ring-based public key cryptosystem. In: Buhler, J.P. (ed.) ANTS 1998. LNCS, vol. 1423, pp. 267–288. Springer, Heidelberg (1998). https://doi.org/10.1007/BFb0054868

22. Huang, X., Liu, J.K., Tang, S., Xiang, Y., Liang, K., Xu, L., Zhou, J.: Cost-effective authentic and anonymous data sharing with forward security. IEEE Trans. Comput. **64**(4), 971–983 (2015)

23. Indyk, P.: Stable distributions, pseudorandom generators, embeddings and data stream computation. In: FOCS, pp. 189–197 (2000)

24. Katz, J., Sahai, A., Waters, B.: Predicate encryption supporting disjunctions, polynomial equations, and inner products. In: Smart, N. (ed.) EUROCRYPT 2008. LNCS, vol. 4965, pp. 146–162. Springer, Heidelberg (2008). https://doi.org/10.1007/978-3-540-78967-3_9

25. Liu, J.K., Au, M.H., Huang, X., Lu, R., Li, J.: Fine-grained two-factor access control for web-based cloud computing services. IEEE Trans. Inf. Forensics Secur. **11**(3), 484–497 (2016)

26. López-Alt, A., Tromer, E., Vaikuntanathan, V.: On-the-fly multiparty computation on the cloud via multikey fully homomorphic encryption. In: STOC, pp. 1219–1234 (2012)

27. Mukherjee, P., Wichs, D.: Two round multiparty computation via multi-key FHE. In: Fischlin, M., Coron, J.-S. (eds.) EUROCRYPT 2016. LNCS, vol. 9666, pp. 735–763. Springer, Heidelberg (2016). https://doi.org/10.1007/978-3-662-49896-5_26

28. Peikert, C.: A decade of lattice cryptography. Found. Trends Theor. Comput. Sci. **10**(4), 283–424 (2016)

29. Regev, O.: On lattices, learning with errors, random linear codes, and cryptography. In: STOC, pp. 84–93 (2005)

30. Rivest, R.L., Adleman, L., Dertouzos, M.L.: On data banks and privacy homomorphisms. Found. Secur. Comput. **4**(11), 169–180 (1978)

31. Shaltiel, R.: An introduction to randomness extractors. In: Aceto, L., Henzinger, M., Sgall, J. (eds.) ICALP 2011. LNCS, vol. 6756, pp. 21–41. Springer, Heidelberg (2011). https://doi.org/10.1007/978-3-642-22012-8_2

32. Stehlé, D., Steinfeld, R.: Making NTRU as secure as worst-case problems over ideal lattices. In: Paterson, K.G. (ed.) EUROCRYPT 2011. LNCS, vol. 6632, pp. 27–47. Springer, Heidelberg (2011). https://doi.org/10.1007/978-3-642-20465-4_4

33. Wu, Q., Qin, B., Zhang, L., Domingo-Ferrer, J., Farràs, O., Manjón, J.A.: Contributory broadcast encryption with efficient encryption and short ciphertexts. IEEE Trans. Comput. **65**(2), 466–479 (2016)

34. Yamada, S., Attrapadung, N., Santoso, B., Schuldt, J.C.N., Hanaoka, G., Kunihiro, N.: Verifiable predicate encryption and applications to CCA security and anonymous predicate authentication. In: Fischlin, M., Buchmann, J., Manulis, M. (eds.) PKC 2012. LNCS, vol. 7293, pp. 243–261. Springer, Heidelberg (2012). https://doi.org/10.1007/978-3-642-30057-8_15

35. Yang, K., Jia, X., Ren, K.: Attribute-based fine-grained access control with efficient revocation in cloud storage systems. In: ASIACCS, pp. 523–528 (2013)

Cryptanalysis and Attack

NativeSpeaker: Identifying Crypto Misuses in Android Native Code Libraries

Qing Wang, Juanru Li, Yuanyuan Zhang[✉], Hui Wang, Yikun Hu,
Bodong Li, and Dawu Gu

Shanghai Jiao Tong University, Shanghai, China
yyjess@sjtu.edu.cn

Abstract. The use of native code (ARM binary code) libraries in Android apps greatly promotes the execution performance of frequently used algorithms. Nonetheless, it increases the complexity of app assessment since the binary code analysis is often sophisticated and time-consuming. As a result, many defects still exist in native code libraries and potentially threat the security of users. To assess the native code libraries, current researches mainly focus on the API invoking correctness and less dive into the details of code. Hence, flaws may hide in internal implementation when the analysis of API does not discover them effectively.

The assessment of native code requires a more detailed code comprehension process to pinpoint flaws. In response, we design and implement NATIVESPEAKER, an Android native code analysis system to assess native code libraries. NATIVESPEAKER provides not only the capability of recognizing certain pattern related to security flaws, but also the functionality of discovering and comparing native code libraries among a large-scale collection of apps from non-official Android markets. With the help of NATIVESPEAKER, we analyzed 20,353 dynamic libraries (.so) collected from 20,000 apps in non-official Android markets. Particularly, our assessment focuses on searching crypto misuse related insecure code pattern in those libraries. The analyzing results show even for those most frequently used (top 1%) native code libraries, one third of them contain at least one misuse. Furthermore, our observation indicates the misuse of crypto is often related to insecure data communication: about 25% most frequently used native code libraries suffer from this flaw. Our conducted analysis revealed the necessity of in-depth security assessment against popular native code libraries, and proved the effectiveness of the designed NATIVESPEAKER system.

1 Introduction

Android apps are typically written in Java. However, the limitations of Java such as memory management and performance drives many Android apps to contain

This work was partially supported by the Key Program of National Natural Science Foundation of China (Grants No. U1636217), the Major Project of the National Key Research Project (Grants No. 2016YFB0801200), and the Technology Project of Shanghai Science and Technology Commission under Grants No. 15511103002.

X. Chen et al. (Eds.): Inscrypt 2017, LNCS 10726, pp. 301–320, 2018.
https://doi.org/10.1007/978-3-319-75160-3_19

components implemented in native code. Android provides Native Development Kit (NDK) to support native development in C/C++, and the app supports a hybrid execution mode that allows a seamless switch between Java code and native code. In a hybrid execution, native code is largely compiled as the form of shared library (.so file) and its exported functions are invoked by Java code via a Java Native Interface (JNI). Since native code achieves better performance and flexible data manipulation, it is especially suitable for implementing data encoding/decoding (e.g., crypto transformation) and raw socket communication.

Although more efficient, native code can be more harmful compared to Java code. Despite the common memory corruption vulnerabilities, high level security flaws are also contained in native code especially some third-party libraries. Even though security assessment of native code libraries is essential, flaws are more difficult to be discovered. The audit of Android native code is sophisticated for two main reasons: First, the binary code is hard to be understood since the compilation process removes a large amount of symbol information from the source code. Without such symbol information the binary code contains little semantics and the comprehension of low-level disassembling code is also time-consuming. Second, on Android platform the lack of fine-grained dynamic analysis tools (e.g., code instrumentation) restricts analysts from collecting runtime data to supplement the code comprehension. Therefore, an improper designed logic in a native code library often requires an in-depth analysis to be excavated.

Among all security flaws, crypto misuses in Android apps ia a major security issue of Android app security [6,20]. Although the security community has proposed utilities to detect crypto misuse in Android apps, the designed technique is mainly effective when analyzing Java code of Android apps. Our observation in recent app development reveals that apps tend to use native code version of crypto implementations in shared libraries rather than that of Java code version to fulfil the crypto operations. The main consideration is that crypto in native code is efficient and is not easily analyzed by reverse engineers. To assess crypto misuse in native code of apps, existing native code analysis techniques [8,12,14,19,22] are generally not domain-specific and thus are less effective.

To further improve the status quo and address the crypto misuse analysis issue against native code of Android app, in this paper we propose a native code analysis to help identifying typical crypto misuse patterns in Android native code libraries practically. Our approach firstly utilizes several heuristics to identify third-party libraries and then locate crypto functions in their native code. By summarizing typical implementation features of crypto in native code, our approach is able to locate two common patterns of crypto functions in native code. After the locating of crypto functions, we further detect relevant crypto misuse through checking obsolete algorithms and incorrect parameters. In particular, we design and implement NATIVESPEAKER, an automated native code analysis tool for crypto misuse identification. NATIVESPEAKER is able to analyze common Android third-party libraries and find certain crypto misuses such as predicable key generation and incorrect parameters for crypto APIs. Our evaluation is based on a corpus of 20,000 Android apps, which contain 20,353 instances of native shared library files (.so). Our analysis demonstrated that the occurrence of crypto and crypto-like functions are very popular in those files, and 21 potential crypto misuses were reported by our analysis. Furthermore, a fine-grained pattern-matching assessment on 310 frequently used native code shared libraries was conducted to find insecure communication issue. The results showed that NATIVESPEAKER is effective to find complicated crypto misuse cases among a huge amount of dynamic libraries, and revealed that communication with broken encryption routine is common in many shared libraries.

The main contributions of this paper are the followings:

- We achieve a large-scale security assessment of apps in those non-official Android markets. We collected 20,000 popular Android apps and extracted all 20,353 native code libraries used. Then we deduplicate them using semantic similarity comparison to reduce the number of targets to be analyzed. This native code library dataset reflects the common features of how functions are used in native code of Android apps.
- We propose a practical and lightweight analysis approach to find crypto misuses in native code. The approach combines static taint analysis and natural language processing of function names to locate crypto functions. Then, the crypto misuses are then found by searching typical patterns summarized from empirical studies.
- We made use of NATIVESPEAKER to search a sophisticated insecure behavior—raw socket data communication without encryption. We find several flaws in real-world implementations of native code which lead to the broken of communication protection. Compared with previous studies, our system provides not only capability of large-scale assessment on native code libraries in a reasonable time, but also find internal implementation vulnerabilities caused by cryptographic misuse.

2 Excavating Semantic Information of Native Code Library

The usage rate of native code libraries in Android app is rapidly increasing [7]. And it also brings many challenges to Android native code analysis especially large-scale analysis. An important aspect often ignored by existing analyses is how to utilize intrinsic characteristics and semantic information (e.g., functionality of the library, feature of exported interfaces) of native code libraries. In this section, We firstly list typical encountered challenges of native code library analysis. Then, we illustrate common features in native code libraries that can be leveraged to help retrieve more semantic information through an empirical study of libraries in current Android apps.

2.1 Challenges of Native Code Analysis

To conduct effective and scalable security analysis against widely used Android native code libraries, several restrictions should be taken into account. Typically,

Android app contains native code in the form of shared library (.*so* file). Java code and native code communicates with each other through Android provided JNI interface: functions in native code can be invoked from Java layer through JNI interfaces and vice versa. To develop native code libraries, the Android Native Development Kit (NDK), a companion tool to the Android SDK, is used to help developers build performance-critical portions of apps in native code. It provides headers and libraries that allow developers to build activities, handle user input, use hardware sensors, and access application resources by programming in C or C++. As a result, almost all Android native code libraries are written in C or C++ and are compiled using the NDK. To analyze them, binary code reverse engineering techniques such as disassembling and decompilation are necessary. However, since the inherent complexity of binary code analysis [17], understanding those libraries in native code form is not easy.

Moreover, native code libraries are often provided to achieve low latency or run computationally intensive applications, such as games or physics simulations, and to reuse existing C/C++ libraries. Thus most Android native code libraries are implemented by third-party developers to fulfil some universal algorithms (e.g., crypto algorithm). App developers often directly integrate an Android native code library and invoke its functions through exported interfaces without knowing the implementation details. Due to the lack of source code, security assessment of those libraries are often ignored. Although this is convenient for app developers, potential security flaws in Android native code libraries may be introduced to a wide variety of apps.

Although static code analysis is often harnessed to help assess the security of Android app on a grand scale, finding security flaws, especially sophisticated logic vulnerabilities related to high-level functionality (e.g., data protection), is generally restricted to Java code with rich semantics. Existing flaw detection approaches (e.g., crypto misuse detection) strongly rely on static patterns of code to find vulnerability. Native code, due to the lack of symbol information, does not contain enough static pattern and thus a simple pattern matching approach is inadequate to effectively locate its contained flaws.

Dynamic analysis can collect more runtime information of an app and complements static analysis. Ideally, fine-grained dynamic analysis combined with static analysis is expected to generate precise analysis results and find security flaws. However, due to the considerable analysis time, dynamic analysis system such as the one proposed by Afonso *et al.* [7] to obtain a comprehensive characterization of native code usage in real world applications are often not applicable to large-scale security analysis. Such issue also exists among other heavyweight program analysis techniques (e.g., symbolic execution). Since not only the amount of native code libraries but also the code size of each library have increased to a considerable scale, a more lightweight analysis should be introduced.

Table 1. Word indicators and their related functionalities

Word Indicator	Occurrence	Functionality
xml	11200	file parsing
png	10572	picture processing
pthread	6828	thread controlling
curl	6220	network
ssl	5260	crypto
http	3860	network
crypto	3553	crypto
evp	3546	crypto
x509	4182	crypto
mutex	2859	thread control

2.2 Extracing Semantics in Native Code Library

To obtain an in-depth understanding of how native code libraries are used, the first step we conduct is an analysis based on the interface name of a library. As a shared library, Android native libraries usually export numbers of functions as interfaces, and the name of those interfaces are generally not obfuscated. Thus, we utilize those interfaces in exported table as an important source of information to classify libraries. We utilize a simple natural language processing approach to analyze those interface names: an N-gram algorithm [3,18] is used to extract the English word sequences in interface names. After splitting an interface name into different units as a sequence, we can deduce relevant functionalities of the interface according to specific word indicator. For instance, if an interface name contains the word "x509", it is possibly related to cryptographic certificate operating and can be considered as possessing the functionality of "crypto".

We collected 20,000 apps and extracted 20,353 native code libraries (details are illustrated in Sect. 4.1). A part of the analyzing results with related deduction rules are listed in Table 1. In further, we choose 180 frequently used samples to conduct a manual investigation. The following observations of native code libraries are summarized through this manual investigation:

Code reuse: Developers transplant existing open source C/C++ projects to Android platform. In our investigation, the portion of native libraries including code migrated form open source project reaches 41.1% (74 of 180). Most commonly used open source projects are *bspatch*, *base64encoder*, *stagefright_honeycomb*, and *tnet*. The reuse of existing open source code allows analysts to utilize code similarity comparison techniques [16] to obtain more semantic informations.

Native API invoking: The invoking of certain API indicates the specific behavior of the program. Unlike Java code, native code can conveniently invoke many low-level system API such as *fork*. This helps analyze the behaviors of thread/memory management, process controlling, and network communication. In our investigation, we found there are 25.5% (46 of 180) of the analyzed libraries invoke at least one API related to the mention behaviors. Through monitoring such API invoking, we can better understand the library.

Crypto functions: An important trend of app protection is that developers tend to implement security related function in native code instead of in Java code. Java code is easily decompiled and most of the function logic can be recovered even if the method-renaming obfuscation has been broadly used by many Android apps. In contrast with Java code, native code is more difficult to be comprehended. Thus many developers tend to hide the critical function in native code. However, to protect secret, standard crypto functions are often adopted. The domain knowledge of cryptography can be leveraged as a supplementary semantic information. If the crypto functions can be identified, the relevant semantics can hugely assist the understanding of the behavior of the app.

There are two ways for an interface to fulfil cryptographic function: one is to take advantage of the Java Cryptography Architecture (JCA) which provides cryptographic services and specifies the developers how to invoke Cryptographic APIs on Android platform. The other is to implement a cryptographic function in native code directly. For those two cases, our investigation indicates more than 28.8% (52 of 180) of native library included at least one crypto encryption function.

3   NativeSpeaker

In this section we describe the design and architecture the proposed NATIVES-PEAKER native code security analysis system. The workflow of NATIVESPEAKER system is depicted in Fig. 1. It first extracts all native code shared libraries from apps, then dedpulicates native libraries from same code base via semantic similarity comparison techniques. After the deduplication, a dataset of native code libraries is built. Then, NATIVESPEAKER conducts both *Java cryptography architecture (JCA) interface analysis* and *bitwise operation analysis* to locate crypto functions in those libraries. If crypto functions are found in a native code shared library, they are further checked with a crypto algorithm analysis–checking the use of obsolete crypto algorithms and incorrect crypto parameters. With the entire analysis workflow, NATIVESPEAKER could help analyst pinpoint typical crypto misuses such as insecure encryption mode and non-random crypto key in native code.

In the following, we detailed the process of how NATIVESPEAKER analyzes native code and search crypto misuses.

3.1   Preprocessing of Native Library

Obtaining ARM Binary Code. Notice that different Android devices use different CPUs, which in turn support different instruction sets. To adapt multiple

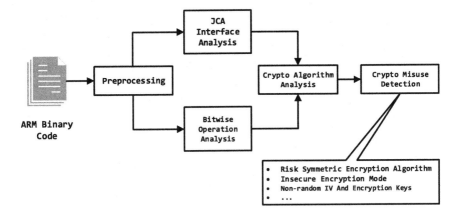

Fig. 1. Workflow of NATIVESPEAKER

architectures of mobile devices, Android NDK supports a variety of Application Binary Interfaces (ABIs) such as *armeabi, armeabi-v7a, arm64-v8a, x86, x86_64, mips, mips64*, etc. As a result, a released APK usually contains multiple native code shared libraries with same functions. To simplify the analysis work, in our analysis only the most frequently used ARM version is analyzed if multiple libraries with same functions are extracted.

We mainly searched for ARM binary files in app's `/lib` and `/lib/armabi(XXX)` directories, which are the default directories for developers to store their native shared libraries. In addition, we found in those directories not all files are with the same file extension (.so). As a result, we further checked the file header to find those files started with ELF magic number (`7f454c46010101`). Thus even if some apps change a regular shared library's file extension to others such as *.xml* and *.dex* to hide it, our analysis would not miss it.

Native Library Deduplication. Among the 20,353 extracted native code files, many of them are the same (or adjacent versions) libraries integrated by different APKs. If duplicated libraries can be excluded, the amount of analysis could be reduced significantly.

However, it is not trivial to find duplicated libraries. We argue that using simple rules such as judging with file hashing is inadequate to deduplicate similar libraries. In general, two classes of similarity are considered:

(i) **Libraries compiled by different developers:** If two apps are developed by different developers, the integrated native code libraries are possibly compiled using different compilation kits but are from the same code base. In this case, libraries are slightly different and file hashing is not able to handle this similarity.

(ii) **Libraries of different versions:** Same libraries with adjacent code versions (e.g., ver 0.9 and 1.0) in our analysis are considered as similar ones. The product iteration often update its integrated native libraries, but the contents of the updated libraries are often similar to the old ones.

To cluster similar native code libraries efficient, we utilized an interface-based analysis to judge whether two libraries are similar. Given a native code library, four attributes can be considered to judge its provenance: **a** file hash, **b** file names, **c** author signature, and **d** interface names. The first three attributes of one particular native code library change frequently, but the names of exported interfaces are often consistent. Hence we make use of them to cluster similar libraries. First, we generate for each library files an *interface set*, which contains all function names extracted from its export table. Then, two sets from different libraries are compared if both sets contain at least 20 function names. We consider two files as the same library of different versions if a high overlap percentages (90+%) of their interface sets is found, and only choose the latest one as our analysis target.

In addition, even if two libraries are similar according to our analysis, we still tend to analyze them respectively if each library is integrated by more than 10 different apps. In this case, we believe that these frequently used libraries affect enough apps and should be meticulously checked.

3.2   Crypto Function Recognition

The problem of crypto function identification in binary programs of desktop platform has been studied previously for different motivations. But implementation of the cryptographic function in mobile platform is different from other platforms. In Android apps, crypto function can be implemented through two styles: Java style and Native style. To implement in the Java style, native code invokes crypto APIs in Java layer through JNI; to implement in the Native style, the crypto algorithms are directly developed using C/C++ and then compiled into native functions. In the following, we present the identification of crypto function implemented in each style, respectively.

Java Style Crypto Identification. JNI allows native code to interact with the Java code to perform actions such as calling Java methods and modifying Java fields. Developers can complete cryptographic functions by leveraging the Java Cryptography Architecture (JCA) which provides cryptographic services and specifies the developers how to invoke Cryptographic APIs on Android platform. The JCA uses a provider-based architecture and contains a set of APIs for various purposes, such as encryption, key generation and management, certificate validation, etc. We first use a concrete example to illustrate how native code invoke JCA API in Java layer. As Listing 1.1 shows, the entire sample involves the phase of initializing crypto key (Line 1–5), choosing crypto algorithm (Line 6–13), initializing IV, and executing encryption routine.

Regulated by the invoke convention of JNI, before executing each method, the following procedures are essentially employed: First, the code utilizes the *FindClass* method to search the class containing methods to be invoked. The, the *GetMethodID* is used to find the ID of specific method. Finally, a *CallObjectMethod* is invoked to conduct the concrete encryption (i.e., DES encryption).

```
 0
 1  ...
 2  KeyClass = env->FindClass("javax/crypto/spec/SecretKeySpec");
 3  KeyInitMethodId = env->GetMethodID(KeyClass,
 4              "<init>",
 5              "([BLjava/lang/String;)V");
 6  KeyObj = env->NewObject(KeyClass, KeyInitMethod, key);
 7  CipherClass = env->FindClass("javax/crypto/Cipher");
 8  CipherInstance = env->CallStaticByteMethod(CipherClass,
 9              "getInstance",
10              "(Ljava/lang/String;)Ljavax/crypto/Cipher;",
11              env->NewStringUTF("DES"));
12  DesInstance = env->CallStaticObjectMethod(CipherClass,
13              CipherInstance,
14              env->NewStringUTF("DES/CBC/PKCS5Padding"));
15  ...
16  /*IV initialization here*/
17  ...
18  DofinalMethod = env->GetMethodID(CipherClass,
19              "doFinal",
20              "([B)[B");
21  result = env->CallObjectMethod(DesInstance,
22              DofinalMethod,
23              msg);
```

Listing 1.1. Java crypto example

The identification of Java style crypto function in native code mainly relies on the string information of the JNI parameters. We can capture the JNI parameters involved and locate relevant JNI invoking. A total number of 230 frequently invoked JNI methods (see Appendix A) are monitored to collect such parameters. After obtaining the information, we further build a (*method, parameter*) tuple. For all tested functions in native code, we collect tuples from them and analyze them. If we find a (*FindClass, javax/crypto/Cipher*) tuple, the host function is expected to invoke crypto APIs in Java layer. Moreover, we can further analyze collected tuples to help recover the information of used crypto algorithms and operation modes. For instance, if a (*NewStringUTF, AES/CBC/P-KCS5Padding*) tuple is found close to the (*FindClass, javax/crypto/Cipher*) tuple, it implies that *java.crypto.cipher* executes an AES-CBC encryption/decryption operation.

Although there are many crypto schemes such as *javax.crypto.Cipher, Bouncy Castle*, and *Spongy Castle* that support crypto operations in Java layer. Our analysis only observed the situation of native code utilizing *javax.crypto.Cipher*. Therefore, we only focus on the situation of using *javax.crypto.Cipher*. If any other crypto providers are involved, similar patterns can also be included.

Native Style Crypto Identification. Compared to Java style crypto code, native style crypto functions possess less features. They are generally implemented in C or C++ and are compiled to assemble code. The identification of such crypto functions is actually a procedure of understanding certain semantics in ARM binary code. In this case, there is no general standard of cryptographic cipher coding template. Recent researches have proposed multiple techniques on identifying crypto primitives. To meet our requirement, a technique to identify both symmetric cryptography and public key cryptography with an acceptable overhead is expected. We compared these approaches and choose the approach proposed by Caballero *et al.* [10], which utilizes a heuristic detection to locate potential crypto functions. The intuition of this approach is that the substantive characteristics of cryptographic and encoding methods, a high proportion of bitwise operations is necessary, and for ordinary methods, bitwise operation would hardly be used. This standard then leads to a efficient static analysis that could identify both symmetric and asymmetric crypto algorithms.

In detail, we re-implement Caballero approach through the following five steps: First, we statically disassemble the native code library to label every function. Second, each function is divided into several basic blocks and for each basic block, the number of contained instructions and that of bitwise instructions are counted. Third, the ratio of bitwise instructions to all instructions in one basic block is calculated. If the ratio exceeds a particular threshold (e.g., 55%), this basic block is considered as a potential crypto related block. Fourth, if a function contains more than one crypto related block, this function is labeled as a potential crypto/encoding function. Finally, we conduct a function-name heuristic filtering to determine whether a potential crypto/encoding function is actually a crypto function. Since most crypto functions are exported by the native libraries to provide particular functionalities, we believe a function without exporting information (i.e., interface name) is unnecessary to be analyzed. We directly remove those potential crypto/encoding functions without an exported function name. Then, we make use of words segmentation technique to handle function name as the following sample shows:

```
VP8Lencodestream => VP 8L encode stream
```

An N-gram algorithm [3,17] is employed to all function names collected. If the words set of one function contains names of mainstream crypto ciphers such as AES, DES, DESede, blowfish, RC4, RSA, etc., this function is labeled as a crypto function. To improve the efficiency, we optimize the adopted N-gram algorithm through carefully choosing the data set used. The words segmentation data set we used is a subset of Linguistic Data Consortium data set. Our data set contains 333,000 unigram words, and 250,000 bigram phrases.

Notice that using simple string matching or regular expression matching could accelerate the analysis, this however causes false positive. Take the function of *VP8Lencodestream* as an example, it contains a "des" substring but is actually not a crypto function.

3.3 Cryptographic Misuse Detection

cryptographic misuse can be very diverse and complex. In this paper, we only focus on crypto misuse of symmetric encryption algorithms. We refer to flaw model from the study of Shuai et al. [20] and mainly concern about two kinds of misuses: the **misuse of crypto algorithm** and the **misuse of crypto parameters**. The misuse of crypto algorithm include the case of using obsolete crypto algorithms such as DES and MD5. This kind of misuse often leads to brute-force attacks. The misuse of crypto parameters include the case of using non-random key material and the case of using improper mode. Using non-random key or IV directly leads to a weak or broken cryptosystem, while the improperly used mode such as ECB significantly weakens the security of adopted cryptosystem.

To detect crypto misuse in native code, we employ a series of analyzing strategies as follows:

Non-random Key Material. Using a non-random cryptographic key material to deduce crypto key, or directly using hard-coded cryptographic keys for encryption, is a severe and critical mistake of crypto engineering practice. However, this situation is still popular due to the reasons of ignorant developers or misunderstanding of cryptography. Also, using a non-random Initialization Vector (IV) with Cipher Block Chaining (CBC) Mode causes algorithms to be vulnerable to dictionary attacks. If an attacker knows the IV before he specifies the next plaintext, he can check his guess about plaintext of some blocks that was encrypted with the same key before. In order to find such misuse of key or IV in native cryptographic functions, we proposed a simple data dependency analysis approach. In List 3.3, the example demonstrates how developer uses a fixed string as key of the *DES_Encrypt_string* function. Through checking the function name (*DES_Encrypt_string*, an exported function in native library), the use of certain crypto algorithms has been located. Then, we follow a simple rule that all parameters of a symmetric crypto function should be dynamically generated. We conduct a simple intra-procedural data flow analysis only focus on the caller of the crypto function. If a parameter of crypto function is generated without involving of the caller function's parameters (i.e., random information from outside) or system APIs, a potential warning of key misuse is reported. Then we can conduct a manual verification to assure the misuse.

```
1 msg = JNIEnv::GetStringUTFChars(*env, msg_input, 0);
2 msg_len = strlen(msg);
3 DES_Encrypt_string(msg, msg_len + 1, "akazwlan", &output);
4 base64_encode(output, &base64_output, out_len);
5 result = JNIEnv::NewStringUTF(*env, (const char *)&base_output);
```

Improper Encryption Mode. The use of vulnerable modes such as Electronic Code Book (ECB) in symmetric encryption is common. For Java style crypto functions, we can obtain the encryption mode through analyzing its JNI parameters, searching certain string related to vulnerable mode (e.g., AES/ECB) and pinpoint typical misuses. However, this approach is not effective when analyzing native style crypto function if the implementation does not regulate the format encryption mode. We address this through a heuristic detection: we observe that most ECB encryptions are implemented within a loop to handle long messages. In this loop, the message parameter is directly handled by the encryption routine instead of firstly masked by the IV. Hence, we first identify the encryption routine with its name, and then check the caller function to search whether the encryption routine is invoked in a loop. If so, the parameter of the routine is checked with any related exclusive-or operation to find potential IV. The missing of IV implies a misuse of ECB mode.

Obsolete Algorithms. To find obsolete crypto algorithms, the major source of information is the function name. Notice that we can obtain the name of crypto algorithms from JNI parameters and the exported interfaces, which indicates for both Java style and Native style crypto functions the used obsolete algorithms can be searched. Although this method is straight-forward, it is effective to find typical misuse of crypto algorithm.

4   Evaluation

4.1   Dataset

To evaluate whether NATIVESPEAKER is able to analyze native code third-party libraries and find crypto misuses, we build an app dataset with a corpus of 20,000 popular Android apps downloaded from *myapp*, the largest non-official Android APP market. We unpacked apps and extract 20,353 ARM native code shared library files. We observed due to the strict regulation of Google Play market, many popular apps are not uploaded. Instead, users are often leaded to website of third-party non-official Android app markets to download them. Moreover, some apps are published to both Google Play market and non-official Android app markets, and the released versions for non-official Android app markets are usually different from that for Google Play market (e.g., add some functionalities not allowed by Google Play market). As a result, we choose to build the dataset through collecting apps from a non-official Android app market to cover more apps in use and find more potential flaws.

In our dataset, the chosen apps possess the following features: (1) Each app had been downloaded by at least 30,000 times. The top 12.5% (2,496 in 20,000) own more than one million users. (2) The category of these apps are various including shopping, gaming, news, traveling, social contacting, etc., Therefore, if any flaw is found in those apps, its influence is significant and it is expected to infect a huge amount of users.

4.2   Native Code Analysis

After unpacking the 20,000 apps and extracting the 20,353 native code shared libraries, we further disassembled those binaries to collect more information. We leveraged the *objdump* utility and the state-of-the-art disassembler IDA Pro (version 6.95) to analyze the collected native code. We found that among all shared libraries, there were 279 malformat files containing no export function information. Our manual analysis revealed that those files adopt native code packing protection. In addition, there are 13 files adopting anti-analysis protection and IDA Pro is not able to disassemble them. Since the code protection issues are outside the scope of this paper, we choose to ignored those failure cases.

Table 2. The 15 most popular Android native code libraries

Library Name	Occurrence	Functional Description
liblocSDK6a.so	2721	Geographic Information System service
ibbspatch.so	2017	Incremental updating
libunity.so	1205	Game engine
libmono.so	1203	Game engine
libweibosdkcore.so	1156	Social networking services
libtpnsSecurity.so	1064	Security service
libtpnsWatchdog.so	972	Security service
libmain.so	935	Game engine
libgame.so	882	Game engine
libBugly.so	847	Crash information service
libcocklogic.so	766	Task restart service
libidentifyapp.so	750	Security service
libcasdkjni.so	743	In-app payment service
libgetuiext.so	665	Push service
libcocos2dcpp.so	636	Game engine

For the rest files, we then conducted a library deduplication process. According to this analysis, 5,970 unique libraries were finally determined. For those library files, we build their profiles including interface information, disassembled code, and call graph through an automated analysis with IDAPython [2].

The deduplication of library significantly reduces the amount of analysis. As Table 2 shows, each of the 15 most popular third party native code libraries is frequently integrated by at least 600 different apps. In this situation, deduplicating those repeatedly used library files saves unnecessary expenses.

4.3   Cryptographic Algorithm Recognition

We run our analysis on a HP Z840 machine, with an Intel Xeon E5-2643 v3, 12-thread processor. We use twelve threads to run analysis task concurrently, the average time to analyze an so file is 4.75 s, and the cost of two function name filtering is 30 ms, which is within an acceptable range.

The function boundary and disassembled code was generated by IDA, even though identificating function boundary and resulting disassembled code with IDA Pro is not perfect, it is sufficient in our scenario.

Java Crypto. In Java cryptographic function recognition, we select Java class "javax.crypto.Cipher" as identification symbol. There are more than one way to implement Java cryptographic function in Java code, such as Bouncy Castle [1] and Spongy Castle [5], and these crypto libraries can be used in native code theoretically. But in our research, we didn't find any Java cryptographic method implemented without the "javax.crypto.Cipher" class. In this Java class, developers could realize cryptographic ciphers such as AES (CBC), AES (ECB), DES (CBC), DES (ECB), DESede (CBC) and DESede (ECB), all these ciphers could be recognized by our work.

In the experiment, we found a total of 47 libraries using JNI interface to invoke Java's cryptographic algorithms, there are 122 such cryptographic algorithms. The result is shown in Table 3. It is noteworthy that DES act as a cryptographic algorithm that obey best practice principles are still used widely, besides in the commonly used cryptos, blowfish, RC2, 3DES are all outdated cryptographic algorithms.

Table 3. Java cryptographic function occurrence

Java cryptographic function	Encrypt mode	Occurrence
AES	CBC	5
AES	ECB	10
AES	CFB	15
AES	None	30
DES	ECB	7
DES	None	12
DESede	CBC	4
DESede	ECB	14
DESede	None	2
RSA	ECB	23

Among these cryptographic algorithms, AES has the highest usage, but most of the scenarios that use AES are to encrypt a short string of information such as a string, so the encryption mode is not used. In addition to AES, DES and 3DES algorithm are frequently used as well. We also found 23 cases using RSA for encryption.

Native Crypto. In our experiment, the lowest value of the instruction number in Native Style Crypto Identification is 20, we select the threshold as 50% and get 100,218 encryption/encoding functions. After our first function name filtering, the remaining number of encryption/encoding functions with function names is 42,897, and it reduces to 13,642 after the word segmentation. The result is illustrated as Fig. 2.

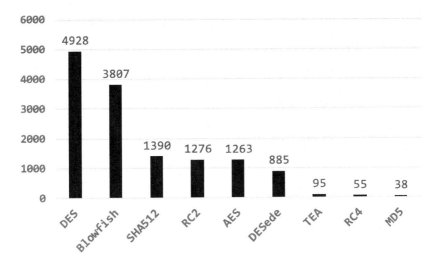

Fig. 2. Native crypto result

Obviously, the Java cryptographic functions usage is significantly less than the native cryptos. Reason for this may be the coding complexity for Java code in native programming environment is very high, and there are many mature cryptographic functions written in C language, such as OpenSSL project [4] (Table 4).

Table 4. Cryptographic misuse in top 60 libraries

Function misuse		Encryption mode misuse	Parameter misuse		
DES	MD5	ECB	Key	IV	Hard-coded Key
3	2	8	16	11	4

4.4 Cryptographic Misuse Detection

In order to analyze the misuse of cryptographic algorithms implemented in native code, we performed manual analysis to 60 most frequently used shared libraries implementing cryptographic algorithms. 21 of the 60 libraries misused cryptographic algorithms, 3 of them used the obsolete DES algorithm, 16 of them adopted the insecure ECB mode for encryption, we also found 27 cases using predictable keys or IVs, 4 of them hard-coded the cryptographic key.

Case Studies. We describe a typical example which results in insecure communication in real world to illustrate the dangers of crypto misuse. In this example, native libraries implement cryptographic algorithms incorrectly by using non-random keys or IVs. We identify such libraries in three steps:

(i) We collect the shared libraries using communication-related APIs (e.g. socket, send, sendto) in our dataset.
(ii) We analyze whether the collected libraries use non-random keys or IVs when implementing cryptographic algorithms.
(iii) We further identify the shared libraries using such insecure cryptographic algorithms to encrypt communication traffic.

We found a total of 13 native libraries existing such problems. We attempted to decrypt the traffic sent out from these shared libraries, and successfully restored the encrypted traffic from eight native libraries, the traffic content included audio, video, program running information and so on. We couldn't trigger sockets in the remaining five cases, these libraries contained associated code but provided no relevant call interface. We think these code may be deprecated or remains to be further developed in the future, and does not affect the correctness of our identification scheme.

"*Libanychatcore.so*" is extracted from the *Anychat* SDK, it is used for transmitting audio and video, and the content is encrypted by AES. We find it hard-codes secret keys when analyzing its implementation of cryptographic algorithms, the hard-coded key is "*BaiRuiTech.Love*". This shared library is found to be used by a popular stock app named *DaZhiHui*, which has been downloaded more than 30 million times. We are able to decrypt the network traffic from *DaZhiHui* with the extracted hard-coded key.

"*Libgwallet.so*" is a shared library used for in-app payments in games released by *GLU Mobile*. It synchronizes data with the server, and the communication is encrypted by AES. In our analysis, we find it uses a hard-coded key "*3A046BB89F76AC7CBA488348FE64959C*" and a fixed IV "*Glu Mobile Games*" for encryption. This shared library is used by more than 10 game apps for synchronizing payment data with their servers. We conduct traffic analysis to these apps and decrypt their payment data successfully.

5   Related Work

– **Android Native Code:** The sandboxing mechanisms can be feasible and useful in restraining privileged API invoking from native code. NativeGuard [21] is a framework isolates native libraries into a non-privileged application, dangerous behavior of native code would be restricted by the sandbox mechanism. Afonso et al. [7] complement the sandboxing mechanisms and generate a native code sandboxing policy to block malicious behavior in realworld applications. While for security flaws like crypto misuses in native code, the sandboxing mechanisms are less effective.
Virtual machine based dynamic analysis platform provide a feasible way to track information flow and implement dynamic taint analysis. Henderson et al. [14] proposed a virtual machine based binary analysis framework, it provides whole-system dynamic binary analytical ability, for Android platform, they include DroidScope [22] as an extension. DroidScope is a dynamic analysis platform based on the Android Emulator, it reconstructs both the OS-level

and Java-level semantics simultaneously and seamlessly, and enable dynamic instrumentation of both the Dalvik bytecode as well as native instructions. NDroid [19] tracks information flows cross the boundary between Java code and native code and the information flows within native codes.

All these dynamic analysis platform may incur 5 times overhead at least, meanwhile, these techniques are not domain-specific and thus are less effective for assessing crypto misuse in native code of apps.

- **Cryptographic Misuse in Android Applications:** A number of efforts have been made to investigate the cryptographic misuse problem in Android Java code. Shao et al. [20] build the cryptographic misuse vulnerability model in Android Java code, they conclude the main classes of cryptographic misuse are misuse of cryptographic algorithm, mismanagement of crypto keys and use inappropriate encryption mode. Egele et al. [11] made an empirical study of the cryptographic misuse. They adopt a light-weight static analysis approach to find cryptographic misuse in real word Android applications, but their work only targets Dalvik bytecode, therefor, applications that invoke misused cryptographic primitives from native code cannot be assessed out. Enck et al. [13] design and execute a horizontal study of Android applications, from a vulnerability perspective, they found that many developers fail to take necessary security precautions. Our analysis complements all these research efforts by performing an in-depth analysis focused on native code.

- **Third Party Library Detection:** Backes et al. [9] proposes a Android Java library detection technique that is resilient against common code obfuscations and capable of pinpointing the exact library version used in apps. Li et al. [15] collect a set of 1,113 libraries supporting common functionality and 240 libraries for advertisement from 1.5 million Android applications, they investigated several aspects of these libraries, including their popularity and their proportion in Android app code. Li et al. [16] utilizes the internal code dependencies of an Android app to detect and classify library candidates, their method is based on feature hashing and can better handle code obfuscation.

A   Appendix

Monitored JNI Functions

AllocObject	CallStaticBooleanMethod	GetDoubleArrayRegion	NewObjectA
CallBooleanMethod	CallStaticBooleanMethodA	GetDoubleField	NewObjectArray
CallBooleanMethodA	CallStaticBooleanMethodV	GetFieldID	NewObjectV
AllocObject	CallStaticBooleanMethod	GetDoubleArrayRegion	NewObjectA
CallBooleanMethod	CallStaticBooleanMethodA	GetDoubleField	NewObjectArray
CallBooleanMethodA	CallStaticBooleanMethodV	GetFieldID	NewObjectV
CallBooleanMethodV	CallStaticByteMethod	GetFloatArrayElements	NewShortArray
CallByteMethod	CallStaticByteMethodA	GetFloatArrayRegion	NewString
CallByteMethodA	CallStaticByteMethodV	GetFloatField	NewStringUTF
CallByteMethodV	CallStaticCharMethod	GetIntArrayElements	NewWeakGlobalRef
CallCharMethod	CallStaticCharMethodA	GetIntArrayRegion	PopLocalFrame
CallCharMethodA	CallStaticCharMethodV	GetIntField	PushLocalFrame
CallCharMethodV	CallStaticDoubleMethod	GetJavaVM	RegisterNatives
CallDoubleMethod	CallStaticDoubleMethodA	GetLongArrayElements	ReleaseBooleanArrayElements
CallDoubleMethodA	CallStaticDoubleMethodV	GetLongArrayRegion	ReleaseByteArrayElements
CallDoubleMethodV	CallStaticFloatMethod	GetLongField	ReleaseCharArrayElements
CallFloatMethod	CallStaticFloatMethodA	GetMethodArgs	ReleaseDoubleArrayElements
CallFloatMethodA	CallStaticFloatMethodV	GetMethodID	ReleaseFloatArrayElements
CallFloatMethodV	CallStaticIntMethod	GetObjectArrayElement	ReleaseIntArrayElements
CallIntMethod	CallStaticIntMethodA	GetObjectClass	ReleaseLongArrayElements
CallIntMethodA	CallStaticIntMethodV	GetObjectField	ReleasePrimitiveArrayCritical
CallIntMethodV	CallStaticLongMethod	GetPrimitiveArrayCritical	ReleaseShortArrayElements
CallLongMethod	CallStaticLongMethodA	GetShortArrayElements	ReleaseStringChars
CallLongMethodA	CallStaticLongMethodV	GetShortArrayRegion	ReleaseStringCritical
CallLongMethodV	CallStaticObjectMethod	GetShortField	ReleaseStringUTFChars
CallNonvirtualBooleanMethod	CallStaticObjectMethodA	GetStaticBooleanField	reserved1
CallNonvirtualBooleanMethodA	CallStaticObjectMethodV	GetStaticByteField	reserved2
CallNonvirtualBooleanMethodV	CallStaticShortMethod	GetStaticCharField	reserved3
CallNonvirtualByteMethod	CallStaticShortMethodA	GetStaticDoubleField	SetBooleanArrayRegion
CallNonvirtualByteMethodA	CallStaticShortMethodV	GetStaticFieldID	SetBooleanField
CallNonvirtualByteMethodV	CallStaticVoidMethod	GetStaticFloatField	SetByteArrayRegion
CallNonvirtualCharMethod	CallStaticVoidMethodA	GetStaticIntField	SetByteField
CallNonvirtualCharMethodA	CallStaticVoidMethodV	GetStaticLongField	SetCharArrayRegion
CallNonvirtualCharMethodV	CallVoidMethod	GetStaticMethodID	SetCharField
CallNonvirtualDoubleMethod	CallVoidMethodA	GetStaticObjectField	SetDoubleArrayRegion
CallNonvirtualDoubleMethodA	CallVoidMethodV	GetStaticShortField	SetDoubleField
CallNonvirtualDoubleMethodV	DefineClass	GetStringChars	SetFloatArrayRegion
CallNonvirtualFloatMethod	DeleteGlobalRef	GetStringCritical	SetFloatField
CallNonvirtualFloatMethodA	DeleteLocalRef	GetStringLength	SetIntArrayRegion
CallNonvirtualFloatMethodV	DeleteWeakGlobalRef	GetStringRegion	SetIntField
CallNonvirtualIntMethod	EnsureLocalCapacity	GetStringUTFChars	SetLongArrayRegion
CallNonvirtualIntMethodA	ExceptionCheck	GetStringUTFLength	SetLongField
CallNonvirtualIntMethodV	ExceptionClear	GetStringUTFRegion	SetObjectArrayElement
CallNonvirtualLongMethod	ExceptionDescribe	GetSuperclass	SetObjectField
CallNonvirtualLongMethodA	ExceptionOccurred	GetVersion	SetShortArrayRegion
CallNonvirtualLongMethodV	FatalError	IsAssignableFrom	SetShortField
CallNonvirtualObjectMethod	FindClass	IsInstanceOf	SetStaticBooleanField
CallNonvirtualObjectMethodA	FromReflectedField	IsSameObject	SetStaticByteField
CallNonvirtualObjectMethodV	FromReflectedMethod	MonitorEnter	SetStaticCharField
CallNonvirtualShortMethod	GetArrayLength	MonitorExit	SetStaticDoubleField
CallNonvirtualShortMethodA	GetBooleanArrayElements	NewBooleanArray	SetStaticFloatField
CallNonvirtualShortMethodV	GetBooleanArrayRegion	NewByteArray	SetStaticIntField
CallNonvirtualVoidMethod	GetBooleanField	NewCharArray	SetStaticLongField
CallNonvirtualVoidMethodA	GetByteArrayElements	NewDirectByteBuffer	SetStaticObjectField
CallNonvirtualVoidMethodV	GetByteArrayRegion	NewDoubleArray	SetStaticShortField
CallObjectMethod	GetByteField	NewFloatArray	Throw
CallObjectMethodA	GetCharArrayElements	NewGlobalRef	ThrowNew
CallObjectMethodV	GetCharArrayRegion	NewIntArray	ToReflectedField
CallShortMethod	GetCharField	NewLocalRef	UnregisterNatives
CallShortMethodA	GetDirectBufferAddress	NewLongArray	
CallShortMethodV	GetDoubleArrayElements	NewObject	

References

1. Bouncy castle. https://www.bouncycastle.org/
2. Ida-python. https://github.com/idapython/src/
3. N-gram wiki. https://en.wikipedia.org/wiki/N-gram
4. openssl project. https://www.openssl.org/
5. Spongy castle. https://rtyley.github.io/spongycastle/
6. Acar, Y., Backes, M., Bugiel, S., Fahl, S., McDaniel, P., Smith, M.: SoK: lessons learned from android security research for appified software platforms. In: 2016 IEEE Symposium on Security and Privacy (SP), pp. 433–451. IEEE (2016)
7. Afonso, V.M., de Geus, P.L., Bianchi, A., Fratantonio, Y., Kruegel, C., Vigna, G., Doupé, A., Polino, M.: Going native: using a large-scale analysis of android apps to create a practical native-code sandboxing policy. In: NDSS (2016)
8. Arzt, S., Rasthofer, S., Fritz, C., Bodden, E., Bartel, A., Klein, J., Le Traon, Y., Octeau, D., McDaniel, P.: FlowDroid: precise context, flow, field, object-sensitive and lifecycle-aware taint analysis for android apps. Acm Sigplan Not. **49**(6), 259–269 (2014)
9. Backes, M., Bugiel, S., Derr, E.: Reliable third-party library detection in android and its security applications. In: Proceedings of the 2016 ACM SIGSAC Conference on Computer and Communications Security, pp. 356–367. ACM (2016)
10. Caballero, J., Poosankam, P., Kreibich, C., Song, D.: Dispatcher: enabling active botnet infiltration using automatic protocol reverse-engineering. In: Proceedings of the 16th ACM Conference on Computer and Communications Security, pp. 621–634. ACM (2009)
11. Egele, M., Brumley, D., Fratantonio, Y., Kruegel, C.: An empirical study of cryptographic misuse in android applications. In: Proceedings of the 2013 ACM SIGSAC Conference on Computer & Communications Security, pp. 73–84. ACM (2013)
12. Enck, W., Gilbert, P., Han, S., Tendulkar, V., Chun, B.-G., Cox, L.P., Jung, J., McDaniel, P., Sheth, A.N.: Taintdroid: an information-flow tracking system for realtime privacy monitoring on smartphones. ACM Trans. Comput. Syst. (TOCS) **32**(2), 5 (2014)
13. Enck, W., Octeau, D., McDaniel, P.D., Chaudhuri, S.: A study of android application security. In: USENIX Security Symposium, vol. 2, p. 2 (2011)
14. Henderson, A., Prakash, A., Yan, L.K., Hu, X., Wang, X., Zhou, R., Yin, H.: Make it work, make it right, make it fast: building a platform-neutral whole-system dynamic binary analysis platform. In: Proceedings of the 2014 International Symposium on Software Testing and Analysis, pp. 248–258. ACM (2014)
15. Li, L., Bissyandé, T.F., Klein, J., Le Traon, Y.: An investigation into the use of common libraries in android apps. In: 2016 IEEE 23rd International Conference on Software Analysis, Evolution, and Reengineering (SANER), vol. 1, pp. 403–414. IEEE (2016)
16. Li, M., Wang, W., Wang, P., Wang, S., Wu, D., Liu, J., Xue, R., Huo, W.: LibD: scalable and precise third-party library detection in android markets. In: Proceedings of the 39th International Conference on Software Engineering, pp. 335–346. IEEE Press (2017)
17. Meng, X., Miller, B.P.: Binary code is not easy. In: Proceedings of the 25th International Symposium on Software Testing and Analysis, pp. 24–35. ACM (2016)

18. Mochihashi, D., Yamada, T., Ueda, N.: Bayesian unsupervised word segmentation with nested Pitman-Yor language modeling. In: Proceedings of the Joint Conference of the 47th Annual Meeting of the ACL and the 4th International Joint Conference on Natural Language Processing of the AFNLP: vol. 1, pp. 100–108. Association for Computational Linguistics (2009)

19. Qian, C., Luo, X., Shao, Y., Chan, A.T.: On tracking information flows through JNI in android applications. In: 2014 44th Annual IEEE/IFIP International Conference on Dependable Systems and Networks (DSN), pp. 180–191. IEEE (2014)

20. Shuai, S., Guowei, D., Tao, G., Tianchang, Y., Chenjie, S.: Modelling analysis and auto-detection of cryptographic misuse in android applications. In: 2014 IEEE 12th International Conference on Dependable, Autonomic and Secure Computing (DASC), pp. 75–80. IEEE (2014)

21. Sun, M., Tan, G.: NativeGuard: protecting android applications from third-party native libraries. In: Proceedings of the 2014 ACM Conference on Security and Privacy in Wireless & Mobile Networks, pp. 165–176. ACM (2014)

22. Yan, L.-K., Yin, H.: DroidScope: seamlessly reconstructing the OS and Dalvik semantic views for dynamic android malware analysis. In: USENIX Security Symposium, pp. 569–584 (2012)

A Game-Based Framework Towards Cyber-Attacks on State Estimation in ICSs

Cong Chen[1,2]($\boxtimes$), Dongdai Lin[1], Wei Zhang[1,2], and Xiaojun Zhou[1,2]

[1] State Key Laboratory of Information Security, Institute of Information Engineering, Chinese Academy of Sciences, Beijing 100093, China
`chencong253@126.com`

[2] School of Cyber Security, University of Chinese Academy of Sciences, Beijing 100090, China

Abstract. The security issue on remote state estimation process against false data injection (FDI) attacks in Industrial Control Systems (ICSs) is considered in this paper. To be practically, it is more reasonable to assume whether or not a meter measurement could be compromised by an adversary does depend on the defense budget deployed on it by the system defender. Based on this premise, this paper focuses on designing the defense budget strategy to protect state estimation process in ICSs against FDI attacks by applying a game-based framework. With resource-constraints for both the defender and the attacker side, the decision making process of how to deploy the defending budget for defenders and how to launch attacks on the meters for an attacker are investigated. A game-based framework is formulated and it has been proved that the Nash equilibrium is existed. For practical computation convenience, an on-line updating algorithm is proposed. What's more, the simulation of the game-based framework described in this paper is demonstrated to verify its validity and efficiency. The experimental results have shown that the game-based framework could improve performance of the decision making and estimation process and mitigate the impact of the FDI attack. This may provide a novel and feasible perspective to protect the state estimation process and improve the intrusion tolerance in ICSs.

Keywords: Industrial Control System (ICS)
Critical infrastructures · False data injection (FDI) · State estimation
Intrusion tolerance · Game theory · Nash equilibrium

1 Introduction

Nowadays, Industrial Control Systems (ICSs) are ubiquitously applied in various industrial processes, especially national critical infrastructures including power system, chemical production, oil and gas refining and water desalination [1,2]. ICSs are computer-controlled systems that manage industrial processes automatically in the physical world. These systems include DCS, SCADA, PLC,

© Springer International Publishing AG, part of Springer Nature 2018
X. Chen et al. (Eds.): Inscrypt 2017, LNCS 10726, pp. 321–341, 2018.
https://doi.org/10.1007/978-3-319-75160-3_20

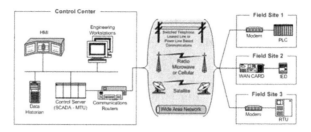

Fig. 1. Structure of SCADA.

and devices such as remote telemetry units (RTU), smart meters, and intelligent field instruments including remotely programmable valves and intelligent electronic relays [2,3]. Figure 1 shows the classical structure of the SCADA system in [3].

As the key drivers of sensing, monitoring, control and management, ICSs are the hearts of critical infrastructure. While sharing basic constructs with Information Technology (IT) business systems, ICSs are technically, administratively, and functionally more complex and unique than IT systems. Hence, the well-behaved security methods in IT system, such as authentication, access control, and message integrity, appear inadequate for a satisfactory protection of ICSs. Indeed, these methods do not exploit the compatibility of the measurements with the underlying physical process or control mechanism, and they are therefore ineffective against insider attacks targeting the physical dynamics [4].

In the past, ICSs were mainly conceived as isolated systems. While, with the ever growing demand of both highly ubiquitous computing services and location-independent access to Information and Communication Technology (ICT) resources, they are increasingly rely on remote operations via local area networks or the Internet, which are enabled by software with limited security protections. As a consequence, ICSs are inviting targets for adversaries attempting to disable critical infrastructure through cyber attacks. Generally, cyber attacks on such systems can lead to devastating effects on the functionality of national critical infrastructures, which has been demonstrated by the Stuxnet attack in 2011 [1,5–7]. Therefore, the security of ICS has attracted considerable interest from both academic and industrial communities in the past few years.

Challenges include data integrity, data veracity and trustworthiness, and resources availability are at hand. The general approach has been to study the effect of specific attacks against particular systems. For instance, deception and DoS attacks against a networked control system are defined in [8]. Deception attacks compromise the integrity of control packets of measurements, while DoS attacks refer to compromising the availability of resources by, for instance, jamming the communication channels. As specific deception attacks in the context of static estimators, FDI attacks against static estimators are introduced in [9]. Stealthy deception attacks against SCADA system are introduced in [10]. Reply attacks and covert attacks effect are introduced in [11] and [12] respectively.

Meanwhile, Clark et al. in [1] proposed a proactive defense framework against Stuxnet. Commands from the system operator to the PLC are authenticated using a randomized set of cryptographic keys. It leverages cryptographic analysis and control and game-theoretic methods to quantify the impact of malicious commands on the performance of physical plants. The works in [13,14] studies the timing of DoS in the term of attackers. Considering the vulnerabilities of stale data in ICSs, properly timed DoS could drive the system to unsafe states. By attacking sensor and controller signals, the attacker can manipulate the process at will. Similarly, Zhang et al. in [15] studies the optimal DoS attack schedule in Wireless Networked Control System (WNCS). The optimal jamming attack schedule is proposed to maximize the attack effect on the control system. Also, centralized and distributed monitors are proposed to detect and identify the attacks in Cyber-physical Systems(CPSs) by Pasqualetti et al. in [16]. Krotofil et al. in [17] investigated a set of lightweight real-time algorithms for spoofing sensor signals directly at the micro-controller of the field device. The data veracity detection takes the form of plausibility and consistency checks with the help of the correlation entropy in a cluster of related sensors.

As a kind of deception attack on state estimation process [9], common countermeasures to defend against FDI attacks could be securing measurements physically or monitoring states directly by sensors such as phasor measurement units (PMUs). For example, Bobba et al. proposed to detect FDI by protecting a strategically selected set of measurements and states in [18]. When there is no verifiable state variable, it is necessary and sufficient to secure a set of basic measurements to detect attacks. Kim and Poor in [19] proposed a fast greedy algorithm to select a subset of meter measurements to protect against FDI attacks. Jia et al. studied the impacts of malicious data attacks on real-time price of electrical market operations [20]. The chance that the adversary can make profit by intelligently manipulating values of meter measurements is analyzed.

However, these works have focused only on one side, i.e., the attacker or the defender. If attackers have knowledge of system parameters, then both defender and attacker may involve in an interactive decision making process. To handle these issues, some researchers proposed the game-theory approach [21–25]. In the similar fashion, Li et al. in [26,27] introduced zero-sum game approaches to optimize the decision process both for the jamming attacker and the remote state estimator in CPSs. The existence of optimal solutions, i.e., Nash equilibrium of zero-sum game, in the decision process is proved and illustrated in numerical examples.

Whatever, to secure system against the FDI attack by protecting some measurements and/or states, the aforementioned works are launched based on the very strong assume that some meter measurements can be absolutely protected from being compromised no matter how powerful the adversary is. While to be more practically, it is more reasonable to assume that whether a measurements could be compromised mainly depends on the defense budget deployed on the meters. From this perspective, the defense budget strategy to guarantee that the adversary cannot modify any set of state variables has been investigated in this paper.

As far as we know, there isn't any relative researches on the game-based framework for defense budgets decision making process in deception attacks in ICSs. This paper presents an analytical gamep-based defense budget approach to analyzing and mitigating the efficacy of FDI attack on remote state estimation in ICSs. In this context, our work will provide insightful guidance on protecting power systems from cyber attacks in real practice. The main contributions can be summarized as follows:

(1) The interaction behaviors between the system defender and a rational attacker is investigated, and the game-based defense budget strategy to improve the tolerance against FDI attacks on state estimation process of power systems is proposed for the first time as far as we known. The game framework is based on the problem of commons model.
(2) A practical updating algorithm is proposed to apply the game based framework into the practical situation and update the framework on-line.
(3) Simulation of the proposed game-based framework is executed. Experimental results shown verification of the validity and efficacy of this framework to the FDI attack.

The reminder of the paper is organized as follows. In Sect. 2, we present the system model and problem formulation. In Sect. 3, we introduce some basic properties of system performance under arbitrary false data injection attacks under game-theory framework. We construct game-theory framework and analyze the system performance. In Sect. 4, dynamic update of the game theory action is illustrated. In Sect. 5, numerical examples are shown to demonstrate the effectiveness of the proposed framework. Finally, Sect. 6 concludes this paper.

Notations: Z denotes the set of all integers and N the set of all positive integers. R is the set of all real numbers. R^n is the n-dimensional Euclidean space. S_+^n (and S_{++}^n) is the set of n by n positive semi-definite matrices (and positive definite matrices). When $X \in S_+^n$ (and S_{++}^n), we write $X \geq 0$ (and $X > 0$). $X \geq Y$ if $X - Y \in S_+^n$. The notation $P[\cdot]$ refers to probability and $E[\cdot]$ to expectation. $T!$ stands for the factorial of T. We write C_M^T for $\binom{T}{M} = T!/(M!(T-M)!)$.

2   Problem Analysis

2.1   System Model

In this paper, a general discrete linear time-invariant (LTI) control process is considered, which is described by the following state space expression:

$$\begin{aligned} x_{k+1} &= Ax_k + w_k \\ y_k &= Cx_k + v_k \end{aligned} \tag{1}$$

where $k \in N, x_k \in R^{n_x}$ is the control process state vector at time k, $y_k \in R^{n_y}$ is the measurement taken by the meter, $w_k \in R^{n_x}$ and $v_k \in R^{n_y}$ are zero-mean i.i.d. Gaussian noises with $E[w_k w_j'] = \delta_{kj} Q (Q \geq 0)$. $E[v_k v_j'] = \delta_{kj} R (R > 0)$, $E[w_k v_j'] = 0, \forall j, k \in N$. The pair (A, C) is assumed to be observable and $(A, Q^{1/2})$ controllable.

2.2   State Estimation

The aim of state estimation process is to find an estimate $\hat{x}$ of state variables x that is the best fit of the meter measurements z. The meter measurements z are related to the state variables x as: $z = h(x) + e$, where $h(x) = [h_1(x), h_2(x), ..., h_m(x)]^T$ and $h_i(x)$ is a nonlinear function of x. States $x = [x_1, x_2, ..., x_n]^T$ could be estimated according to measurements $z = [z_1, z_2, ..., z_m]^T$, and $e = [e_1, e_2, ..., e_m]^T$ is the independent random measurement errors (noises). In DC (linearized) state estimator model [28]:

$$z = \mathbf{H}x + e \tag{2}$$

where $\mathbf{H} = [h_{ij}]_{m \times n}$ is the measurement Jacobian matrix. The measurement residual is the difference between the observed measurements z and its estimate $\hat{z}$ as $r = z - \hat{z} = z - \mathbf{H}\hat{x}$. The weighted least-squares (WLS) criterion problem is to find an estimate $\hat{x}$ that minimizes the performance index $J(\hat{x})$ [29]. Besides WLS, some other statistical estimation criteria, such as maximum likelihood criterion and minimum variance criterion, are commonly used in state estimation [30].

Let $\hat{x}_k$ and P_k as the Minimum Mean-Squares Error (MMSE) estimate of the state x_k and the corresponding estimation error covariance as:

$$\hat{x}_k = E[x_k | y_1, y_2, ..., y_k] \tag{3}$$

$$P_k = E[(x_k - \hat{x}_k)(x_k - \hat{x}_k)' | y_1, y_2, ..., y_k] \tag{4}$$

As it is well-known that under suitable conditions the estimation error covariance of the Kalman filter converges to a unique value $\overline{P}$ [31] from any initial condition (similar assumptions as in [32,33]), simplify the subsequent discussion by setting $P_k = \overline{P}, k \geq 1$.

2.3   Defender Model

Consider a system with a set $N_s = \{1, 2, ..., n\}$ of state variables and a set $M = \{1, 2, ..., m\}$ of meter measurements. It is natural to secure meter measurements physically to defend system against the FDI attacks by countermeasures such as guards, video monitoring, temper-proof communication systems and etc.

In this paper, the defense budget problem will be considered. The defense budget for m meter measurements $\mathbf{b} = [b_1, b_2, ..., b_m]^T$ denotes the defender's budget allocation vector. With the knowledge of the system configuration, the system defender allocates the defense budget for m meter measurements by making decision of applying certain strategy under the budget constrain B. here, $b_i (i \in M)$ denotes the defense budget for the i-th meter measurement z_i. For example, $b_i = 0$ means no defense deployed on z_i, otherwise, it means the defense budget amount on z_i.

2.4   Attack Model

In this paper, a kind of special deception attack - FDI attack is considered. With a certain degree of knowledge of the system configuration and the historical transportation on the communication channel, the attacker can systematically generate and inject malicious measurements which will mislead the state estimation process without being detected by any of the exiting techniques for BDD(Bad Data Detection) as shown in [9].

Let z_a represent the measurements vector that may contain malicious data as $z_a = z + a$, where $z = (z_1, ..., z_m)^T$ is the vector of original measurements and $a = (a_1, ..., a_m)^T$ is the malicious data(i.e. attack vector) added to the original measurements. The non-zero a_i means that the i-th measurements is compromised and then the original measurements z_i is replaced by a phoney value z_i^a. The attacker can choose any non-zero arbitrary vector as the attack vector a, and then construct the malicious measurements $z_i^a = z_i + a$.

In general, a is likely to be identified by BDD if it is unstructured. However, it has been found that some well-structured attack vectors can systematically bypass the existing residual-based BDD mechanism in [9]. For example, $a = Hc$, where $c = [c_1, c_2, ..., c_n]^T$ is an arbitrary nonzero vector. Then the well-constructed attack measurements $z_i^a = z + a$ could eventually bypass the BDD.

Let $\hat{x}_k^a$ and $\hat{x}_k$ denote the estimate of x_k using the malicious measurements z_a and the original measurements z, respectively. $\hat{x}_k^a$ can be represented as $\hat{x}_k + c$, where c is a non-zero vector of length n. Note that c reflects the estimation error injected by the attacker. The attack model can be described as:

$$\hat{x}_k = \begin{cases} \hat{x}_k^a, & \text{if an attack vector arrives} \\ \hat{x}_k, & \text{otherwise} \end{cases} \tag{5}$$

where $\hat{x}_k^a$ is an attack on k-th state variable at the moment k with the attacking vector z_a. And under the defender's strategy **b**, the attack cost k_i denotes the payment to compromise the meter measurement z_i successfully for an attacker. k_i is assumed to be a function of the defense budget b_i:

$$k_i = f_i(b_i), i \in M. \tag{6}$$

Here, the cost function $f_i(\cdot)$ should be formulated as a monotonic function. Thus the k_i will rise correspondingly when b_i increases and it means that it will be more difficult to compromise z_i. and $\mathbf{k} = [k_1, k_2, ..., k_m]^T$ denotes the attacker's attack cost vector.

Distinctly, the state variables with less defense budget will be much more easier to be compromised. A rational attacker will choose the state variable with the least cost (i.e., the easiest target) to attack. A successful attack is to modify (i.e., to introduce arbitrary errors into) at least one state variable without being detected. The attacker's objective is to launch an attack with the least cost, whose capability includes:

(1) A certain degree of the knowledge of the network topology and configuration of the ICS (such as power system), i.e., the **H** matrix;
(2) The ability to access any set of meter measurements simultaneously, which may or may not be compromised depending on the defender's protection budget.

2.5 Communication Channel

As described in [9], typical FDI attack can inject malicious measurements into the data transportation process between components(i.e. meters and remote state estimator) in networked control systems i.e. ICSs) without being detected by BDD, which may mislead the state estimation process and drive the system into undesired situations.

In practice, for both meters and attackers, energy and source constraint is a natural concern(B for defender and R for attacker), which affects the state estimation performance and attacking policies. Within a given time horizon $t \in [0, T](t \in N, T \in N)$, let $B'_t(B'_t \leq B)$ denotes the total defense budget on meters at time t in the system, and $R'_{t_a}(t'_a \in [0, T], R'_{t_a} \leq R)$ denotes the attacker's cost on compromising or manipulating measurements at each attacking time t_a.

Thus, the meter's defense budget strategy on each meter at each sampling time is denoted as $\boldsymbol{\psi}_t$.

$$\boldsymbol{\psi}_t \triangleq \{\psi_1, \psi_2, ..., \psi_m\} \tag{7}$$

where, $\psi_i \in [0, 1]$ denotes the measurement z_i of i-th meter is secured or not, and $\psi_i = 1$ means that meter measurement z_i is secured, otherwise no defense is deployed on it. Then the corresponding total budget B'_t is obtained by applying the defense budget b_i:

$$B'_t = \sum_{i=1}^{m} b_i \psi_i, \ \forall i \in M \tag{8}$$

Consequently, we have the following constraint:

$$B'_t \leq B, \ \forall i \in M \tag{9}$$

Similarly, the attacker's attacking cost strategy is denoted as:

$$\boldsymbol{\theta}_t \triangleq \{\theta_1, \theta_2, ..., \theta_m\} \tag{10}$$

where $\theta_i = 1$ for $i = 1, 2, ..., m$ means that the attacker launches a FDI attack on the i-th measurement z_i, otherwise there is no attack on it. Applying the attack cost function (6), the corresponding total cost is:

$$R'_{t_a} = \sum_{i=1}^{m} \theta_i k_i, \ \forall i \in M \tag{11}$$

The associated constraint is

$$R'_{t_a} \leq R, \ \forall i \in M \tag{12}$$

In practical communication systems, packet dropouts may occur due to different reasons, including signal degradation, channel fading and channel congestion. However, we assume the communication dropout probability of data packet from the meter arrives at the remote estimator is 0 all the time to simplify the simulation and explanation.

2.6   Main Problem

Here, the defense budget problem will be discussed considering the constrains on both sides of attacking and defense. According to the perspective aforementioned, the attacker can only increase the probability of launching a successful attack, but cannot further cut down the least attack cost. Thus, the defender's best strategy against any attacker's strategy is to raise up the least attack cost as much as possible, under the constraint of the total defense budget B. Then, we could discuss and deduce the primal problem as following.

To ease the illustration, an H^* matrix is constructed from the H matrix firstly as:

$$h_{ij}^* = \begin{cases} 0, & \text{if } h_{ij} = 0; \\ 1, & \text{otherwise.} \end{cases} \quad \forall i \in M, \forall j \in N_s. \tag{13}$$

Takea 5-bus power system as example shown in Fig. 2 and applying (13), the $\mathbf{H}^*$ matrix is constructed as p(14), where the j-th column of $\mathbf{H}^*$ is denoted by $\mathbf{h}_j^* \in \mathbb{R}^{m \times 1}$, e.g. $\mathbf{h}_1^* = [1, 1, 0, 0, 1, 1]^T$.

$$\mathbf{H}^* = \begin{bmatrix} 1 & 0 & 0 & 0 \\ 1 & 0 & 1 & 0 \\ 0 & 1 & 0 & 1 \\ 0 & 0 & 1 & 1 \\ 1 & 1 & 0 & 0 \\ 1 & 0 & 1 & 1 \end{bmatrix} \tag{14}$$

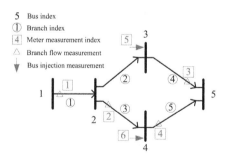

Fig. 2. Example of a fully measured 5-bus power system [34].

Considering the FDI attack scene, an attacker could launch the attack successfully to modify the state x_j (where $j \in N_s$) without being detected. Then, the cost $r(j)$ should be:

$$r(j) = h_j^{*T}k = \sum_{i=1}^{m} h_{ij}^* k_i, \quad \forall j \in N_s \tag{15}$$

Applying (14), $r(j)$ in (15) could be translated as:

$$r(j) = \sum_{i=1}^{m} h_{ij}^* k_i \psi_i, \quad \forall j \in N_s \tag{16}$$

Let $i = 1$ in the system in Fig. 2 as example, and $r(1) = k_1 + k_2 + k_5 + k_6$ according to ψ. That is to say, if the attacker wants to modify the state variable x_1, the total attack cost is the sum of costs to compromise the meter measurements z_1, z_2, z_5, and z_6. It is obvious that a rational attacker will reasonably choose the state with the least cost as the easiest target to attack. The corresponding strategy could be formulated according to [34] as:

$$\min_{j \in N} \quad r(j) \tag{17}$$

The interaction between the defender and attacker can be viewed as a two-player zero-sum game. The defender plays first deploying the protection strategy $\mathbf{b}$ and φ, then followed by the attacker's decision with strategy θ to launch a FDI attack on the easiest target x_j without being detected by BDD.

It is practically reasonable to assume that the defender makes decision without any information about attacker's strategy beforehand, while the attacker may possess knowledge of both the system deployment and the defender's strategy partially or totally in advance. In real situation, it is true the attacker will try his best to collect such information as much as possible during the preparation period. The more information being grasped by the attacker, the much higher probability of launching a successful attack. It is the worst situation corresponding to the attacker with full knowledge of the defender's strategy, i.e., given $\mathbf{b}$, the worst situation can be achieved by solving (17).

As mentioned above, the attacker can only increase the risk probability of launching a successful attack with more information, but cannot further cut down the least attack cost. Thus, the defender's best strategy against any attacker's strategy is to raise up the least attack cost as much as possible under the constraint of the total defense budget B. Thus, we have a primal problem:

$$\max_{\mathbf{b} \geq 0} \min_{j \in N} r(j)$$

$$s.t. \begin{cases} \sum_{i=1}^{m} b_i \leq B; \\ \min_{i \in N} r(j) \geq R; \end{cases} \tag{18}$$

The defender intends to deploy the defense budget as low as possible while guaranteeing that the attacker cannot modify any set of state variables (i.e., even the least attack cost needed is still greater than the attacker's limited resource R). In this context, the primal problem (18) can be transformed into an equivalent problem:

$$\min_{b \geq 0} \sum_{i-1}^{m} b_i$$

$$s.t. \begin{cases} \sum_{i=1}^{m} b_i \leq B; \\ \min_{i \in N} r(j) \geq R; \end{cases} \tag{19}$$

The detailed transformation process is elaborated in [34]. For (18), the constraint can be rewritten into the equivalent form of: $r(j) \geq R, \forall j \in N$. The linear approximation to the cost function (6) is: $\mathbf{k} = \mathbf{f} \cdot \mathbf{b}$, where f is the cost function Jacobian vector.

Besides, recall the constraint (15), and thus we have: $r = [h_1^*, h_2^*, ..., h_n^*]^T \mathbf{k} = \mathbf{H}^{*T} \mathbf{k} = \mathbf{H}^{*T} (\mathbf{f} \cdot \mathbf{b})$. Thus the cost function of this problem is obtained as:

$$J(\cdot) = \mathbf{H}^{*T} \mathbf{k} = \mathbf{H}^{*T} (\mathbf{f} \cdot \mathbf{b}) \tag{20}$$

In term of the defender's side, the goal of the decision maker is to minimize the cost function as $J(\cdot)$ in (20). Oppositely, the goal of the attacker is to maximize it. So, the objective function of defender's side is:

$$J_D(\boldsymbol{\psi_t}) \triangleq J(\boldsymbol{\psi}) \tag{21}$$

while the objective function of attacker's side is obtained as:

$$J_A(\boldsymbol{\theta_t}) \triangleq -J(\boldsymbol{\theta}) \tag{22}$$

The optimal strategies for both sides subjecting to the constrains described in (9) and (12) are the solutions of the optimal problems based on the above analysis. Since getting more measurements always benefits for improving the performance for both sides, it is not difficult to show that the optimal strategies for both sides remain the same if (9) and (12) are changed to $\sum_{i=1}^{m} = B$ and $\sum_{i=1}^{m} = R$, respectively.

Thus, the optimization problem is obtained as to get $\boldsymbol{\varphi_t}$ (i.e. the $\boldsymbol{\varphi}$ at the sampling time t) following for the defender's side in overall:

$$\max_{\boldsymbol{\psi_t}} \quad J_D(\boldsymbol{\psi_t})$$

$$s.t. \begin{cases} \sum_{i=1}^{m} b_i \psi_i = B \\ r(j) = R, \quad \forall j \in N_s. \end{cases} \tag{23}$$

And the optimization problem to get b_i for each single meter is:

$$\max_{\boldsymbol{\psi_t}} \quad b_i$$

$$s.t. \sum_{i=1}^{m} b_i \psi_i = B \tag{24}$$

In the similar fashion, for the attacker's side (where $\boldsymbol{\theta_t}$ is the corresponding $\boldsymbol{\theta}$ at the sampling time t):

$$\max_{\boldsymbol{\theta_t}} \quad J_A(\boldsymbol{\theta_t})$$

$$s.t. \sum_{i=1}^{m} k_i \theta_i = R \tag{25}$$

Take the system in Fig. 2 as an example. For ease of illustration, consider the simplest case where $f = R = 1$. By means of LP, we may obtain the solution 1 as shown$[0, 0.2, 0.4, 0, 0.6, 0.8]$. The solution satisfying defense budget needed is solution 2 as $[0, 1, 1, 0, 0, 0]$, which means that if the defender strategically deploys $[0, 0.2, 0.4, 0, 0.6, 0.8]$ on the set $[z_2, z_3, z_5, z_6]$ of meter measurements, it is guaranteed that the attacker cannot modify any set of state variables. Although LP can efficiently get a solution for (21), however, since $\mathbf{H}^{*T}$ is the $n \times m$ matrix and $n < m$ is the typical case, (21) is an under determined linear system, and theoretically has infinitely many solutions. For example, the solution $[0, 1, 1, 0, 0, 0]$ is also a solution. Nevertheless, from the practical point of view, we may consider that the solution 2 would be better than the solution 1 since the total attack cost that the attacker could modify all state variables, i.e., $\sum_{j=1}^{n} = r(j)$, under the former defense strategy is larger. From the above observation, we consider additional framework applying Game theory to let the solution make more sense in practical applications.

3   The Game-Based Framework

Here, the deciding process of defense strategy for system defender is focused. The decision making processes between the defender side and the attacker side consist a zero-sum game which is investigated in details in [29]. Among that, the deciding process of defense strategy for system defender are demonstrated in a game-theory framework based on the optimization problems, i.e. (17) and (18).

3.1   The Problem of Commons-Based Framework

Since at lease Hume in 1739, political philosophers and economists have understood that if citizens respond only to private incentives, public goods will be under-provided and public resources over-utilized. Hardin's much cited paper in 1968 brought the problem to the attention of non-economists and developed a model of the problem of commons trying to solve the profits disputes of single or commons [35].

In the problem of commons model, the two sides make choices separately and then the final decision maker picks one of the offers as the settlement according to the prime problem. In the conventional problem of commons model, in contrast, the decision maker is free to impose any strategy as the settlement. Here, the former model would be taken into consideration.

So far, relative researches in the existing literature mainly focus on only one side with energy or resource constraint under strong assumption. In this paper, for the case with constraints for both sides, i.e., $B'_t \leq B$, and $R'_{t_a} \leq R$, and loose assumption, those cannot be used. For there may be kinds of different strategies for both sides, the problem will be investigated from a game-theoretic view adopting the following definitions according to the problem of commons model:

(1) *Players:* there are two players: the system controller in overall and meters deployed in the field.
(2) *Strategy space:* ψ and $\mathbf{b}$ for the both sides.
(3) *Payoff function:* $J_M(\psi\mathbf{b})$ for the defender and $J_S(\psi\mathbf{b})$ for the attacker.

Here, defense budget decision process includes the system controller in overall and each single meter deployed in the field with the same strategy space. In the problem of commons model, there is an decision maker in the final decision making process for defender. For the situation in this paper, the decision maker is the algorithm on defender located between the local meter measurement and the remote state estimate calculated upon the historical data. Decision maker of system defender determines the final budget strategy by evaluate the strategies to benefit both the single meter and the system controller. The local measurement is obtained by meters and transferred to remote estimator through the communication channel. The defender runs the algorithm during the decision making process in this game model. The output of the decision making process will be accepted as the system's defense budget allocation to ensured the well-function of remote state estimator, according to which the controllers regulate and control the whole system's performance.

3.2 Existence of Nash Equilibrium

To analyze the Nash equilibrium of the game between the defender and the attacker, the number of all the pure strategies need to be taken into account first. For the defender, the number of pure strategies P is $P = C_m^0 + C_m^1 + \ldots + C_m^m = 2^m$, which are denoted as $\psi^{prue}(1), \psi^{prue}(2),\ldots, \psi^{prue}(P)$. Correspondingly, the mixed strategies can be denoted as: $\psi^{mixed}(\pi_1, \pi_2, ..., \pi_P) = \psi^{prue}(p)$ with $\pi_p, p = 1, 2, ..., P$, where π_p is the probability of $\psi_S^{prue}(p)$ and $\sum_{p=1}^P \pi_p = 1, \pi_p \in [0, 1]$.

In the same way, for the attacker, the number of pure strategies Q is $Q = 2^m$, which are denoted as $\theta^{prue}(1), \theta^{prue}(2),\ldots, \theta^{prue}(Q)$. Correspondingly, the mixed strategies can be denoted as: $\theta^{mixed}(\pi_1, \pi_2, ..., \pi_Q) = \theta^{prue}(q)$ with π_q, $q = 1, 2, ..., Q$, where π_q is the probability of $\theta^{prue}(q)$ and $\sum_{q=1}^Q \pi_q = 1, \pi_q \in [0, 1]$.

Different combinations of π_p and π_q constitute different mixed strategies for both the defender and attacker, respectively. Obviously, the number of pure strategies is finite, thus there are infinitely many mixed strategies for each side.

It has been proved in [32] that a Nash equilibrium exists for the considered two-player zero-sum game between the defender and the attacker. The optimal strategies for each side are denoted as $\psi^\star$ and $\theta^\star$, respectively. By giving the optimal strategy $\theta^\star$ chosen by the attacker, the optimal strategy for the defender is the one that maximizes its objective function $J_D(\boldsymbol{\psi})$ as described in (17), i.e., $J_D(\boldsymbol{\psi}^*|\boldsymbol{\theta}^*) \geq J_D(\boldsymbol{\psi}|\boldsymbol{\theta}^*), \forall \boldsymbol{\psi}$. For the attacker, a similar conclusion is obtained, i.e., $J_A(\boldsymbol{\theta}^*|\boldsymbol{\psi}^*) \geq J_A(\boldsymbol{\theta}|\boldsymbol{\psi}^*), \forall \boldsymbol{\theta}$. Since the payoff functions are objective functions for each sides respectively, the optimal strategies for the defender and the attacker constitute a Nash equilibrium of this game naturally.

In the same way, applying the above conclusion in the similar decision process of the two-player zero-sum game (similar as in [34]) - the problem of commons between the system controller in overall and meters deployed in the field, the Nash equilibrium of this game could be obtained in the same way, which is the final budget strategy deployed in the system defense scheme.

4   Update of the Game Theory Action

All decisions investigated in the last section are made before the initial time of the system. Anyway, it can just be regarded as a off-line schedule. In some practical situations, both sides may be able to monitor the performance of the opponent and renew their choices at each sampling time according to the practical status of the system. For example, after the attacker launches a attack, the system server may be able to detect and identify the abnormality and inform the responsive meters about that [16], through which the meters and the state estimation process could do some corresponding adjustment, i.e. abandon the abnormal measurements, increase the budget of the objective meters or refuse the relevant suspicious meter's data uploading mission by applying the SDN technology. The attacker can also detect whether its data packet is accepted or not based on the system status. Thus, though each side can not be sure about the next decision of their opponent, they can still make prediction on the opponent's future action through the system's present performance and therefore narrow the scope of their opponent's action sets.

Algorithm 1. Game updating for both sides

Initialize: $\overline{x}_S$, $\boldsymbol{\psi}^*(T, M, N, \Phi_0)$ and $\boldsymbol{\theta}^*(T, M, N, \Phi_0)$.

1. Game begins with the initial parameters;
2. **for** $t = 1 : T$ **do**
3. Solve for $\boldsymbol{\psi}^*(T, M, N_s, \Phi_0)$ and $\boldsymbol{\theta}^*(T, M, N_s, \Phi_0)$;
4. Employ the actions of $\boldsymbol{\psi}^*(T, M, N_s, \Phi_0)$ and $\boldsymbol{\theta}^*(T, M, N_s, \Phi_0)$ designed for the first time step for the new game as the action of the current time step t;
5. Each side observe and compute the action taken by the opponent at time step t;
6. Calculating $\hat{x}, \Phi_0, P_t$;
7. Estimating and adjusting for both attacker and defender;
8. $T = T - 1$;
9. $\Phi_0 = E[P_t]$;
10. **end for**

For this situation, based on the observation on the past action of opponents, a new game with new constrains will be considered at each time step for both defender and attacker during the whole time horizon T. And simultaneously, the defender's decision should be updated by applying the game mentioned above. At each time step, let $\psi^\star(T, B, R, \Phi_0)$ denotes the optimal mixed strategies for the defender and $\theta^\star(T, B, R, \Phi_0)$ denotes the one for attacker with parameters of T, B, R, Φ_0. Here, T is the time-horizon, B and R are the resource constraints of defender and attacker respectively, and Φ_0 is the initial estimation error covariance. $\psi^\star(T, B, R, \Phi_0)$ and $\theta^\star(T, B, R, \Phi_0)$ is calculated at each time and contain the action sequences for the whole time-horizon of each side respectively, but only the first step will be kept as the initial parameters of the next time step at both sides. Moving to the next step, the parameters, e.g., system outcomes, constraints and time-horizon, are updated, and thus the game, decision and constraints will be renewed continually.

To update the whole decision making process at each time step for each sides in the game-based framework, a recursive algorithm is proposed here as showing in Algorithm 1. Following this algorithm, both sides are able to involve in the decision making process of the game-based framework described above with the time varying, resource constrains and any initial states.

5 Example and Simulation Analysis

To verify the validity and efficiency of this game-based decision making framework, simulation experiment is employed in this section.

Consider a LTI control process with Gaussian White Noise which is similar to function (3) as following:

$$
\begin{aligned}
x_{t+1} &= Ax_t + Bu_t + w_t \\
y_t &= Cx_t + Du_t + v_t
\end{aligned}
\tag{26}
$$

with parameters as following:

$$
A = \begin{bmatrix} 1.1269 & -0.4940 & 0.1129 & 0.1129 \\ 1.0000 & 0 & 0 & 0 \\ 1.0000 & 0 & -0.0055 & 0 \\ 0 & 1.0000 & 0 & 0 \end{bmatrix} \quad B = \begin{bmatrix} -0.3832 \\ 0.5919 \\ -0.5919 \\ 0.5191 \end{bmatrix}
\tag{27}
$$

$$
C = \begin{bmatrix} 1 & 0 & 0 & 0 \end{bmatrix} \quad D = \begin{bmatrix} 0 \\ 0 \\ 0 \\ 0 \end{bmatrix} \quad \mathbf{H}^* = \begin{bmatrix} 1 & 0 & 0 & 0 \\ 1 & 0 & 1 & 0 \\ 0 & 1 & 0 & 1 \\ 0 & 0 & 1 & 1 \\ 1 & 1 & 0 & 0 \\ 1 & 0 & 1 & 1 \end{bmatrix}
\tag{28}
$$

where $t \in [0, T]$ is the time horizon, $u_t = sin(t/5)$ is the input of the system, $w_k \in R^{n_x}$ and $v_k \in R^{n_y}$ are Gaussian White Noise with error covariances of

$Q = 1$ and $R = 1$ respectively, which are generated and mixed into the system by the function $awgn(\cdot)$ provided by MATLAB. In this system, the number of system state is $n = 4$, the number of system output is 1, the number of meter is $m = 6$, the constrains for both sides are $B = 2$ and $R = 5$, respectively. The state estimator is standard Kalman filters as described in [31]. Calculating the ranks of the pair (A, C) and $(A, Q^{1/2})$ respectively, it is confirmed that the system is controllable and observable. The attacker takes an action with FDI on state estimate represented by a sequence of random values generated by an algorithm based on the observation of the system's previous states.

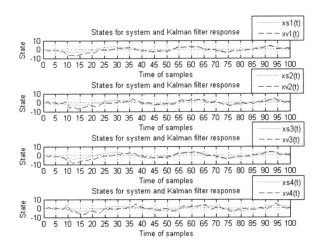

Fig. 3. States of the original system with and without Noise.

In the following simulation results, the attack moments denoted as set kat where $kat = [8, 24, 31, 42, 92]$, and the responding sequence number of attacked meters are denoted as Ka where $Ka(kat) = [1, 3, 1, 3, 2]$. Actually, both kat and Ka are generated randomly in the practical.

Figure 3 shows the comparison of the states time-changing curve during the time window T of the original system with and without Gaussian White Noise w_k and v_k. Here in Fig. 3, the states of original system without noise are x_s. x_v are the states of the same system with Gaussian White Noises.

Figure 4 shows the comparison of the outputs time-changing curve during the time window T of the system with and without Gaussian White Noise w_k and v_k in the same situation described in Fig. 3. Here in Fig. 4, the output of original system without noise is y_s. y_v is the output of the same system with Gaussian White Noises.

According to Figs. 3 and 4, it is easy to figure that the system performance, i.e., x_s and y_s, driven by the input u_t can follow the expected performance well, i.e., x_v and y_v.

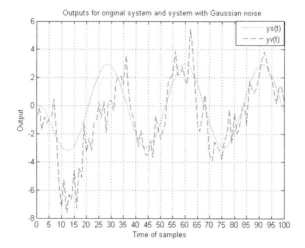

Fig. 4. Outputs of the original system with and without Noise.

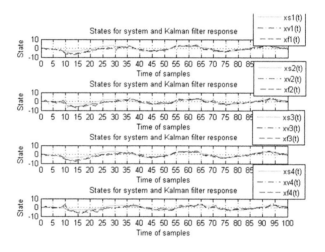

Fig. 5. States of the above system with state estimator.

Figure 5 shows the comparison of the states time-changing curve during the time window T of the above system with the state estimator. Here, the xs and xv are states same as they are described in Fig. 3, and xf are the results of the state estimation process.

Figure 6 shows the comparison of the outputs time-changing curve during the time window T of the above system with the state estimator in the same situation described in Fig. 5. Here in Fig. 6, ys and yv are outputs same as they are described in Fig. 4, and ye is the output of the system with state estimation.

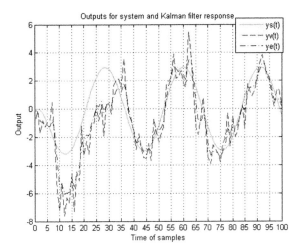

Fig. 6. Outputs of the above system with state estimator.

Figure 7 shows the comparison of the states time-changing curve during the time window T of the above system and the same system applying the game framework under the FDI attack on the state estimate. Here, the original system state are xs, the states of system with attack obtained from the communication channel are xle. $\bar{x}$ are the expected states of the system, which equal to states estimate xre as shown in the Fig. 7. Xcf are the results of the arbitration process of the game-based decision making framework in the same situation, which are accepted by the remote state estimator and then employed by the controllers of the system into the adjustment of the control process. It is obvious that the states estimation process could be improved and the attacking effectiveness could be relieved by applying the game framework into the system described above.

According to the simulation experiment and corresponding to the time steps $kat = [8, 24, 31, 42, 92]$, the state estimate received by remote state estimation generated by local estimation process are as shown by green line in Fig. 7: $xre = [-3.8599, -0.7811, -3.0910, 0.0208, -1.7168]$; the expected state estimate $\bar{x} = x_{le}$ received by calculation based on previous measurement are as shown by red line in Fig. 7: $xle = [-0.7228, -0.8879, 0.5121, -2.4349, 1.7364]$; the final state outputs of the game process is denoted as x_{cf}, which are shown by blue line in Fig. 7: $xcf = [-0.7228, -0.8879, 0.5121, -2.4349, 1.7364]$.

In addition, Fig. 8 shows the overall estimation error covariance trace time-varying during the time window T. The system state estimation error covariance trace Trs of the steady state is $Trs = 3.3301$. While in the same time window T, the corresponding remote state estimation error covariance trace at the attack moments $kat = [8, 24, 31, 42, 92]$ are denoted as $Trat$, where $Trat(kat) = [3.3412, 3.4113, 3.5090, 3.3993, 3.3604]$. It is easy to figure that the traces of the attack moments are higher than the stable state without any attack.

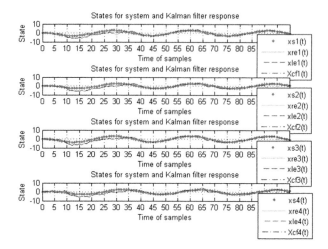

Fig. 7. States of the above system under attack.

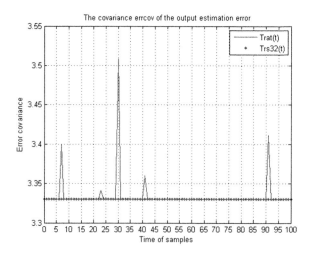

Fig. 8. Estimation error covariances of the above system under attacker.

Obviously, by applying the problem of commons game-based framework in the decision making process for the system defender during the state estimation process, performance of the same system under FDI, i.e., xle, has been improved and the attack impact on the remote state estimate has been relieved, i.e., Xcf.

6   Conclusion

As the former ICSs with poor security mechanism are feeble and vulnerable to malicious attacks from internal and external network, the ICSs need to be able to relieved the attack impact of certain degree to keep plants work as normal and to minimize the loss and harm caused by the malicious action.

A situation where a game-based framework under a kind of special deception attack - FDI attack on the remote estimator in the ICSs has been studied in this paper. The behavior of a rational attacker is studied, the interaction between the defender and attacker is investigated, and the decision making process of defense budget strategy deployed on meters against FDI attacks is solved. The problem of commons-based decision making framework is applied to improve the system control performance and mitigate the FDI attack impact.

To verify the validity and efficiency of this game-based decision making framework, simulation of a discrete LTI control process applying the framework has been done in MATLAB. Experimental results have shown that the problem of commons game-based decision making framework could relief the impact of the false data injection attack on the remote state estimator and improve the state estimation performance of the ICSs.

In our future work, the other game frameworks applying into the similar attacking situations aiming at the ICSs would be our interests.

Acknowledgement. The authors would like to thank anonymous reviewers for considerate and helpful comments. The work described in this paper is supported by National Natural Science Foundation of China (61379139) and the "Strategic Priority Research Program" of the Chinese Academy of Sciences, Grant No. XDA06010701.

References

1. Clark, A., Zhu, Q., Poovendran, R., Başar, T.: An impact-aware defense against stuxnet. In: 2013 American Control Conference, pp. 4140–4147. IEEE (2013)
2. Cheminod, M., Durante, L., Valenzano, A.: Review of security issues in industrial networks. IEEE Trans. Industr. Inf. **9**(1), 277–293 (2013)
3. Stouffer, K., Falco, J., Scarfone, K.: Guide to industrial control systems (ICS) security. NIST Spec. Publ. **800**(82), 16–16 (2011)
4. Slay, J., Miller, M.: Lessons learned from the maroochy water breach. In: Goetz, E., Shenoi, S. (eds.) ICCIP 2007. IIFIP, vol. 253, pp. 73–82. Springer, Boston, MA (2008). https://doi.org/10.1007/978-0-387-75462-8_6
5. Byres, E., Ginter, A., Langill, J.: How stuxnet spreads-a study of infection paths in best practice systems. Tofino Security, White paper (2011)
6. Falliere, N., Murchu, L.O., Chien, E.: W32. Stuxnet Dossier. White paper, Symantec Corp., Security Response, 5, 6 (2011)
7. Albright, D., Brannan, P., Walrond, C.: Did Stuxnet Take Out 1,000 Centrifuges at the Natanz Enrichment Plant? Institute for Science and International Security (2010)
8. Amin, S., Cárdenas, A.A., Sastry, S.S.: Safe and secure networked control systems under denial-of-service attacks. In: Majumdar, R., Tabuada, P. (eds.) HSCC 2009. LNCS, vol. 5469, pp. 31–45. Springer, Heidelberg (2009). https://doi.org/10.1007/978-3-642-00602-9_3
9. Liu, Y., Ning, P., Reiter, M.K.: False data injection attacks against state estimation in electric power grids. ACM Trans. Inf. Syst. Secur. (TISSEC) **14**(1), 13 (2011)

10. Teixeira, A., Amin, S., Sandberg, H., Johansson, K.H., Sastry, S.S.: Cyber security analysis of state estimators in electric power systems. In: 49th IEEE Conference on Decision and Control (CDC), pp. 5991–5998. IEEE (2010)
11. Mo, Y., Sinopoli, B.: Secure control against replay attacks. In: 47th Annual Allerton Conference on Communication, Control, and Computing, Allerton 2009, pp. 911–918. IEEE (2009)
12. Smith, R.S.: A decoupled feedback structure for covertly appropriating networked control systems. IFAC Proc. Volumes **44**(1), 90–95 (2011)
13. Krotofil, M., Cárdenas, Á.A.: Is this a good time?: Deciding when to launch attacks on process control systems. In: Proceedings of the 3rd International Conference on High Confidence Networked Systems, pp. 65–66. ACM (2014)
14. Krotofil, M., Cardenas, A., Larsen, J., Gollmann, D.: Vulnerabilities of cyber-physical systems to stale data–determining the optimal time to launch attacks. Int. J. Crit. Infrastruct. Prot. **7**(4), 213–232 (2014)
15. Zhang, H., Cheng, P., Shi, L., Chen, J.: Optimal dos attack scheduling in wireless networked control system. IEEE Trans. Control Syst. Technol. **24**(3), 843–852 (2016)
16. Pasqualetti, F., Dörfler, F., Bullo, F.: Attack detection and identification in cyber-physical systems. IEEE Trans. Autom. Control **58**(11), 2715–2729 (2013)
17. Krotofil, M., Larsen, J., Gollmann, D.: The process matters: ensuring data veracity in cyber-physical systems. In: Proceedings of the 10th ACM Symposium on Information, Computer and Communications Security, pp. 133–144. ACM (2015)
18. Bobba, R.B., Rogers, K.M., Wang, Q., Khurana, H., Nahrstedt, K., Overbye, T.J.: Detecting false data injection attacks on DC state estimation. In: Preprints of the First Workshop on Secure Control Systems, CPSWEEK, vol. 2010 (2010)
19. Kim, T.T., Vincent Poor, H.: Strategic protection against data injection attacks on power grids. IEEE Trans. Smart Grid **2**(2), 326–333 (2011)
20. Jia, L., Thomas, R.J., Tong, L.: Impacts of malicious data on real-time price of electricity market operations. In: 2012 45th Hawaii International Conference on System Science (HICSS), pp. 1907–1914. IEEE (2012)
21. Bhattacharya, S., Başar, T.: Game-theoretic analysis of an aerial jamming attack on a UAV communication network. In: Proceedings of the 2010 American Control Conference, pp. 818–823. IEEE (2010)
22. Roy, S., Ellis, C., Shiva, S., Dasgupta, D., Shandilya, V., Wu, Q.: A survey of game theory as applied to network security. In: 2010 43rd Hawaii International Conference on System Sciences (HICSS), pp. 1–10. IEEE (2010)
23. Kashyap, A., Basar, T., Srikant, R.: Correlated jamming on mimo Gaussian fading channels. IEEE Trans. Inf. Theory **50**(9), 2119–2123 (2004)
24. Gupta, A., Langbort, C., Başar, T.: Optimal control in the presence of an intelligent jammer with limited actions. In: 49th IEEE Conference on Decision and Control (CDC), pp. 1096–1101. IEEE (2010)
25. Agah, A., Das, S.K., Basu, K.: A game theory based approach for security in wireless sensor networks. In: 2004 IEEE International Conference on Performance, Computing, and Communications, pp. 259–263. IEEE (2004)
26. Li, Y., Shi, L., Cheng, P., Chen, J., Quevedo, D.E.: Jamming attacks on remote state estimation in cyber-physical systems: a game-theoretic approach. IEEE Trans. Autom. Control **60**(10), 2831–2836 (2015)
27. Li, Y., Quevedo, D.E., Dey, S., Shi, L.: Sinr-based DoS attack on remote state estimation: a game-theoretic approach (2016)

28. Ekneligoda, N.C., Weaver, W.W.: A game theoretic bus selection method for loads in multibus DC power systems. IEEE Trans. Industr. Electron. **61**(4), 1669–1678 (2014)
29. Chen, C., Lin, D.: Cyber-attacks on remote state estimation in industrial control system: a game-based framework. In: Chen, K., Lin, D., Yung, M. (eds.) Inscrypt 2016. LNCS, vol. 10143, pp. 431–450. Springer, Cham (2017). https://doi.org/10.1007/978-3-319-54705-3_27
30. Wood, A.J., Wollenberg, B.F.: Power Generation, Operation, and Control. Wiley, New York (2012)
31. Anderson, B.D.O., Moore, J.B.: Optimal filtering. Courier Corporation (2012)
32. Li, Y., Shi, L., Cheng, P., Chen, J., Quevedo, D.E.: Jamming attack on cyber-physical systems: a game-theoretic approach. In: 2013 IEEE 3rd Annual International Conference on Cyber Technology in Automation, Control and Intelligent Systems (CYBER), pp. 252–257. IEEE (2013)
33. Shi, L., Epstein, M., Murray, R.M.: Kalman filtering over a packet-dropping network: a probabilistic perspective. IEEE Trans. Autom. Control **55**(3), 594–604 (2010)
34. Deng, R., Xiao, G., Rongxing, L.: Defending against false data injection attacks on power system state estimation. IEEE Trans. Industr. Inf. **13**(1), 198–207 (2017)
35. Gibbons, R.: A Primer in Game Theory. Harvester Wheatsheaf, New York (1992)

Cryptanalysis of Acorn in Nonce-Reuse Setting

Xiaojuan Zhang[1,2(✉)] and Dongdai Lin[1]

[1] State Key Laboratory of Information Security, Institute of Information Engineering, Chinese Academy of Sciences, Beijing, China
{zhangxiaojuan,ddlin}@iie.ac.cn
[2] School of Cyber Security, University of Chinese Academy of Sciences, Beijing, China

Abstract. Acorn is a third-round candidate of the CAESAR competition. It is a lightweight authenticated stream cipher. In this paper, we show how to recover the initial state of Acorn when one pair of Key and IV is used to encrypt three messages. Our method contains two main steps: (1) gathering different states; (2) retrieving linear equations. At the first step, we demonstrate how to gather the relation between states when two different plaintexts are encrypted under the same nonce. And at the second step, we exploit how to retrieve a system of linear equations with respect to the initial state, and how to recover the initial state from this system of equations. We apply this method to both Acorn v2 and Acorn v3. The time complexity to recover the initial state of Acorn v2 is $2^{78}c$, where c is the time complexity of solving linear equations. It is lower than that of the previous methods. For Acorn v3, we can recover the initial state with the time complexity of $2^{120.6}c$, lower than that of the exhaustion attack. We also apply it on shrunk ciphers with similar structure and properties of Acorn v2 and Acorn v3 to prove the validity of our method. This paper is the first time to analyze Acorn v3 when a nonce is reused and our method provides some insights into the diffusion ability of such stream ciphers.

Keywords: CAESAR · Authenticated cipher · Stream cipher
Acorn · State recovery attack

1 Introduction

The ongoing CAESAR competition is to find authenticated encryption schemes that offer advantages over AES-GCM and are suitable for widespread adoption. In total, 57 candidates were submitted to the CAESAR competition, and after the challenge of two rounds, 15 submissions have been selected for the third round. Acorn containing three versions [1–3] is one of them. It is a lightweight authenticated encryption cipher, submitted by Hongjun Wu. Actually, it is a simple binary feedback shift register based cipher, containing a 293-bit internal state. Its goals are to protect up to 2^{64} bits of associated data and up to 2^{64} bits

© Springer International Publishing AG, part of Springer Nature 2018
X. Chen et al. (Eds.): Inscrypt 2017, LNCS 10726, pp. 342–361, 2018.
https://doi.org/10.1007/978-3-319-75160-3_21

of plaintext, and to generate up to 128-bit authentication tag by using a 128-bit secret key and a 128-bit initialization vector (IV).

There are some attacks against Acorn. Liu et al. showed the slid properties of Acorn v1 which could be used to propose a state recovery attack using a guess-and-determine technique and a differential-algebraic method [4]. An attack, described by Chaigneau et al., allowed instant key recovery when nonce was reused to encrypt a small amount of chosen plaintexts [5]. Salam et al. investigated cube attacks on Acorn v1 and v2 of 477 initialization rounds [6]. Under the assumption that the key is known, Salam et al. developed an attack to find collisions [7]. Lafitte et al. claimed that they proposed practical attacks to recover the state and the key [8]. Keeping the key and IV unchanged, and modifying the associated data, Josh et al. claimed that the associate data dose not affect any keystream bits if the size of the associated data is small [9]. Roy et al. gave some results on ACORN in [10], one of which was that they found a probabilistic linear relation between plaintext bits and ciphertext bits with probability $\frac{1}{2} + \frac{1}{2^{350}}$. However, the bias was too small to be tested. The comments on the analysis of Acorn were given in google groups of cryptographic competitions by the designer [11].

In [4], the authors gave a key recovery attack on Acorn v1 when a nonce is reused twice. In [5], it showed that the state of Acorn v2 could be recovered when a nonce is reused more than four times, which can also be seen as a key recovery attack on Acorn v1. Furthermore, the attack in [5] is practical when a nonce is reused more than six times. However, they do not consider how to recover the state when a nonce is reused three times.

In this paper, we show that why the cipher is insecure when one nonce is used to protect three messages, and apply it to Acorn v2 and v3. Our results do not contradict the security claims made by the designer, but provide some insights into the diffusion ability of such stream ciphers. The main contributions can be summarized as follows.

1. For Acorn v2, we show that when a nonce is reused three times, the state can be recovered with the time complexity of $2^{78}c$, where c is the time complexity of solving linear equations. The idea of our method is based on the fact that choosing new different plaintexts, we can get more linear equations to decrease the time complexity. Although our attack is not practical, the time complexity is lower than that of [5] when a nonce is reused one more times.
2. For Acorn v3, we can recover the initial state of Acorn v3 when a nonce is reused three times. The time complexity is $2^{120.6}c$, which is lower than that of the exhaustion attack. This is the first time to analyze Acorn v3 when nonce is reused, and the methods in [5] are not applicable to Acorn v3. The results of Acorn in nonce-reuse setting are listed in Table 1.
3. We also apply our method on the toy ciphers designed with the similar structure and properties of Acorn v2 and Acorn v3 to prove the validity of our method.

The rest of this paper is organized as follows. Section 2 is a brief description of Acorn v2 and v3. Section 3 is the state recovery attack on Acorn v2, and Sect. 4 is on Acorn v3. The paper concludes in Sect. 5. The attacks described in this

Table 1. Summary of results

Nonce reuse-time	The complexity
2 (Acorn v1)	$c \cdot 2^{110.23}$ [4]
3 (Acorn v2)	$c \cdot 2^{78}$ Sect. 3
4 (Acorn v2)	$c \cdot 2^{109}$ [5]
5 (Acorn v2)	$c \cdot 2^{67}$ [5]
6 (Acorn v2)	$c \cdot 2^{25}$ [5]
7 (Acorn v2)	c [5]
3 (Acorn v3)	$c \cdot 2^{120.6}$ Sect. 4

paper are sharing the same associated data field which we will not emphasize later.

2   Brief Descriptions of Acorn v2 and v3

In this section, we briefly recall Acorn v2 and v3. One can refer to [1,3] for more details. Considering our attack does not involve the procedures of the initialization, the process of associated data and the finalization, here we do not intend to introduce them, and just restate briefly the encryption procedure.

Denote by $s = (s_0, s_1, ..., s_{292})$ the initial state before the keystream bits are generated, and let p be a plaintext. There are three functions used in the encryption procedure of Acorn v2, the feedback function $f(s, p)$, the state update function $F(s, p)$ and the filter function $g(s)$. The keystream bit z is generated by the filter function $g(s)$, defined as

$$g(s) = s_{12} \oplus s_{154} \oplus s_{61}s_{235} \oplus s_{193}s_{235} \oplus s_{61}s_{193}.$$

The feedback bit is generated by the boolean function $f(s, p)$, expressed as

$$f(s, p) = 1 \oplus s_0 \oplus s_{66} \oplus s_{107} \oplus s_{196} \oplus s_{23}s_{160} \oplus s_{23}s_{244} \oplus$$

$$s_{160}s_{244} \oplus s_{66}s_{230} \oplus s_{111}s_{230} \oplus p.$$

Intermediate variables $x = (x_0, \cdots, x_{292})$ are introduced below

$$x_{289} = s_{289} \oplus s_{235} \oplus s_{230},$$
$$x_{230} = s_{230} \oplus s_{196} \oplus s_{193},$$
$$x_{193} = s_{193} \oplus s_{160} \oplus s_{154},$$
$$x_{154} = s_{154} \oplus s_{111} \oplus s_{107},$$
$$x_{107} = s_{107} \oplus s_{66} \oplus s_{61},$$
$$x_{61} = s_{61} \oplus s_{23} \oplus s_0,$$
$$x_i = s_i, 0 \le i \le 292 \text{ and } i \notin \{61, 107, 154, 193, 230, 289\}.$$

Then the state update function $F(s,p)$ can be described as

$$s_{292} = f(x,p),$$

$$s_i = x_{i+1}, \ 0 \le i \le 291.$$

It is easy to check that the state update function $F(s,p)$ is invertible with respect to s when p is given.

At step i of the encryption, one plaintext bit p_i is injected into the state, and c_i is obtained by p_i XOR z_i. The pseudo-code of the generation of the ciphertext is given as follow.

$l \leftarrow$ the bit length of the plaintext
for i from 0 to $l-1$ **do**
    $z_i = g(x)$
    $c_i = z_i \oplus p_i$
    $s = F(s,p_i)$
end for

In Acorn v3, the only tweak is that a part of the feedback function is moved to the filter function. More precisely, the feedback function and the filter function used in Acorn v3 are

$$g(s) = s_{12} \oplus s_{154} \oplus s_{61}s_{235} \oplus s_{193}s_{235} \oplus s_{61}s_{193} \oplus s_{66} \oplus s_{66}s_{230} \oplus s_{111}s_{230},$$

and

$$f(s,p) = 1 \oplus s_0 \oplus s_{107} \oplus s_{196} \oplus s_{23}s_{160} \oplus s_{23}s_{244} \oplus s_{160}s_{244} \oplus p.$$

3   The State Recovery Attack on Acorn V2

Before introducing our state recovery attack on Acorn v2, we first recall the state recovery attacks proposed in [5].

In [5], it shows that if a nonce is reused more than four times, i.e. a pair of Key and IV is used to encrypt more than four messages, the initial state can be recovered. The main idea is to gather enough linear equations with respect to the initial state, by exploiting the fact that the keystream may be affected by a plaintext after a few rounds. Once the initial state is recovered, the full key can be recovered by stepping the cipher backward in Acorn v1, while in Acorn v2 and v3, it is difficult to recover the full key, but we can make forgery attacks on them.

Let $s^t = (s_0^t, s_1^t, ..., s_{292}^t)$ be a state at time t, and let P and P' be two different plaintexts. Time t starts from 0 in the encryption process, and the initial state before the encryption process is $s^0 = (s_0^0, s_1^0, ..., s_{292}^0)$. A key observation of [5] is that when P and P' are different in the first bit, the first different keystream bits are z_{58} and z'_{58}. Then, two linear equations can be obtained, that is,

$$z'_{58} \oplus z_{58} = x_{61} \oplus x_{193},$$

$$z_{58} = (z'_{58} \oplus z_{58})(x_{235} \oplus x_{193}) \oplus x_{12} \oplus x_{154} \oplus x_{193},$$

where x_i $(1 \leq i \leq 293)$ is the internal variables at time $t = 58$. If the difference in plaintexts P and P'_i is introduced in the i'th bit, where $0 \leq i' \leq 57$, the first different keystream bits are z_{58+i} and z'_{58+i}. So, by choosing 58 plaintexts, 116 linear equations with respect to the state at time $t = 58$ can be obtained. In order to get more linear equations, extra unknowns y_k satisfying

$$y_k = s_{292}^{58+k}$$

are introduced, where $1 \leq k \leq 177$. Utilizing the same method, one can obtain other 354 linear equations.

However, the above attack needs to encrypt 236 messages under the same nonce. If differences in plaintexts P and P' satisfy

$$P' = P \oplus (\overbrace{0, \cdots, 0,}^{l} \overbrace{1, \cdots, 1,}^{42} \overbrace{0, \cdots, 0,}^{j}),$$

where l and j are integers, and l is a multiple of 42, then 84 linear equations can be obtained. That is because the 42 bits differences in P and P' can affect the keystream z_t independently, in the case of $l + 58 \leq t \leq l + 99$. If n different P' are chosen, then $n + 1$ messages of length at least $42n + 58$ bits are used, and $84n$ linear equations with $235 + 42n$ unknowns are obtained. The expected number of solutions is $2^{235-42n}$, which is lower than 2^{128} when $n \geq 3$. For more details one can refer to [5].

In this paper, we give a method to recover the initial state when $n = 2$. The main idea is to choose different plaintexts P and P' to retrieve more linear equations in order to recover the initial state. In our attack, there are two main steps: (1) gathering different states; (2) retrieving linear equations.

3.1 Gathering Different States

As discussed in [3], if one bit difference is injected into a state by modifying a plaintext, the state will be changed in many bits after a few rounds. In this section, we will first give some new observations, which are useful to retrieve more linear equations. Then we show how the states change when different P and $P'(P'')$ are injected, where

$$P' = P \oplus (\overbrace{1, \cdots, 1,}^{49} \overbrace{0, \cdots, 0,}^{39} \overbrace{0, \cdots, 0,}^{146}),$$

$$P'' = P \oplus (\overbrace{0, \cdots, 0,}^{88} \overbrace{1, \cdots, 1,}^{49} \overbrace{0, \cdots, 0,}^{97}).$$

At last, we give some experimental confirmation of the result on a shrunk cipher with the similar structure and properties of ACORN v2.

New Observations. For simplicity, we rewrite the filter function and the feedback function of Acorn v2 used in encryption phase as

$$g(x) = x_{12} \oplus x_{154} \oplus x_{61}x_{193} \oplus x_{61}x_{235} \oplus x_{235}x_{193}$$

and

$$f(x, p) = 1 \oplus p \oplus x_0 \oplus x_{66} \oplus x_{107} \oplus x_{23}x_{160} \oplus x_{23}x_{244}\oplus$$
$$x_{244}x_{160} \oplus x_{230}(x_{111} \oplus x_{66}) \oplus x_{196},$$

where $x = (x_0, x_1, \cdots, x_{291}, x_{292})$ are internal variables. Let $x' = (x'_0, x'_1, \cdots, x'_{291}, x'_{292})$. Then we can obtain the following observations.

Og1: If $x'_i = x_i$ and $x'_{235} = x_{235} \oplus 1$, where $i \in \{12, 61, 154, 193\}$, then we can get

$$g(x') \oplus g(x) = x_{61} \oplus x_{193},$$

$$g(x) = (g(x') \oplus g(x))(x_{235} \oplus x_{193}) \oplus x_{12} \oplus x_{154} \oplus x_{193}.$$

Og2: If $x'_i = x_i$ and $x'_{193} = x_{193} \oplus 1$, where $i \in \{12, 61, 154, 235\}$, then we have

$$g(x') \oplus g(x) = x_{61} \oplus x_{235},$$

$$g(x) = (g(x') \oplus g(x))(x_{235} \oplus x_{193}) \oplus x_{12} \oplus x_{154} \oplus x_{235}.$$

Og3: If $x'_i = x_i$, $x'_{193} = x_{193} \oplus 1$ and $x'_{235} = x_{235} \oplus 1$, where $i \in \{12, 61, 154\}$, then

$$g(x') \oplus g(x) = x_{193} \oplus x_{235} \oplus 1,$$

$$g(x) = (g(x') \oplus g(x))(x_{61} \oplus x_{193}) \oplus x_{12} \oplus x_{154} \oplus x_{61}.$$

Og4: If $x'_i = x_i$, $x'_{193} = x_{193} \oplus 1$ and $x'_{154} = x_{154} \oplus 1$, where $i \in \{12, 61, 235\}$, then we obtain that

$$g(x') \oplus g(x) = x_{61} \oplus x_{235} \oplus 1,$$

$$g(x) = (g(x') \oplus g(x))(x_{61} \oplus x_{193}) \oplus x_{12} \oplus x_{154} \oplus x_{193}.$$

Og5: If $x'_i = x_i$ and $x'_j = x_j \oplus 1$, where $i \in \{12, 61\}$ and $j \in \{154, 193, 235\}$, then we get

$$g(x') \oplus g(x) = x_{193} \oplus x_{235},$$

$$g(x) = (g(x') \oplus g(x))(x_{61} \oplus x_{193}) \oplus x_{12} \oplus x_{154} \oplus x_{193}.$$

Of6: If $x'_i = x_i$ and $x'_{244} = x_{244} \oplus 1$, where $i \in \{0, 23, 66, 107, 111, 160, 196, 230\}$, then the difference in $f(x')$ and $f(x)$ is

$$f(x') \oplus f(x) = x_{23} \oplus x_{160}.$$

Of7: If $x'_i = x_i$ and $x'_j = x_j \oplus 1$, where $i \in \{0, 23, 66, 107, 111, 160, 196\}$ and $j \in \{230, 244\}$, then the difference in $f(x')$ and $f(x)$ is

$$f(x') \oplus f(x) = x_{23} \oplus x_{160} \oplus x_{111} \oplus x_{66}.$$

States Classification. Let $s^t = (s_0^t, s_1^t, ..., s_{292}^t)$ and $c^t = (c_0^t, c_1^t, ..., c_{292}^t)$ be the states at time t when P and P' are, respectively, encrypted under the same nonce. It is easy to see that $s^0 = c^0$, and the corresponding keystreams are the same until the 58th bits are outputted. When $58 \leq t \leq 145$, the difference in the keystreams at time t is caused by the difference in s_i^t and c_i^t, where $i \in \{235, 193, 160, 154\}$. Next, we will present how the differences transfer in these locations. For simplicity, we will use some symbols as follows to represent $c^t = (c_0^t, c_1^t, ..., c_{292}^t)$ which is an intermediate step.

- s_i^t in c^t means $c_i^t = s_i^t$,
- $\underline{s_i^t}$ in c^t means $c_i^t = \underline{s_i^t} = s_i^t \oplus 1$,
- $\underbrace{s_i^t}$ in c^t means $c_i^t = \underbrace{s_i^t} = s_i^t \oplus a_{i+t-341}$,
- $\underbrace{\underline{s_i^t}}$ in c^t means $c_i^t = \underbrace{\underline{s_i^t}} = s_i^t \oplus a_{i+t-341} \oplus 1$,

where $a_{i+t-341}$ are internal variables. According to differential properties, we divide c^t into 4 cases, where $1 \leq t \leq 145$. The symbol Δs_i^t denotes a bit difference in s_i^t and c_i^t.

Case 1. $1 \leq t \leq 49$

As P and P' are different in the first 49 bits, we can get

$$c^t = (s_0^t, s_1^t, ..., s_{292-t}^t, \underline{s_{292-t+1}^t}, ..., \underline{s_{292-t+t}^t}),$$

where $\underline{s_i^t} = s_i^t \oplus 1$ and $293 - t \leq i \leq 292$.

Case 2. $50 \leq t \leq 63$

If $50 \leq t \leq 63$, then the value of Δs_{292}^t is determined by the difference in s_{244}^{t-1} and c_{244}^{t-1}. According to Of6, we can obtain

$$\Delta s_{292}^t = s_{292}^t \oplus c_{292}^t = s_{23}^{t-1} \oplus s_{160}^{t-1}.$$

c^t can be represented as

$$c^t = (s_0^t, s_1^t, ..., s_{292-t}^t, \underline{s_{292-t+1}^t}, ..., s_{292-t+49}^t, \underbrace{s_{292-t+50}^t}, ..., s_{292}^t),$$

where $\underbrace{s_i^t} = s_i^t \oplus a_{i-341+t}$ and $a_{t-49} = \Delta s_{292}^t$, where $342 - t \leq i \leq 292$. The linear transformation would also cause some differences when $59 \leq t \leq 63$, which is

$$\Delta s_{288}^t = \Delta s_{289}^{t-1} \oplus \Delta s_{235}^{t-1} \oplus \Delta s_{230}^{t-1} = a_{t-53} \oplus 1.$$

The right representation of c^t is

$$c^t = (s_0^t, s_1^t, ..., s_{292-t}^t, \underline{s_{292-t+1}^t}, ..., s_{292-t+49}^t,$$

$$\underbrace{s_{292-t+50}^t}, ..., \underbrace{s_{292-t+55}^t}, ..., s_{292-t+59}^t, ..., s_{292}^t),$$

where $s_i^t = a_{i-341+t} \oplus 1$ and $347 - t \le i \le 351 - t$.

Case 3. $64 \le t \le 88$

If $64 \le t \le 88$, then the value of Δs_{292}^t is determined by the values of s_{230}^{t-1}, c_{230}^{t-1}, s_{244}^{t-1} and c_{244}^{t-1}. According to Of7, we have

$$\Delta s_{292}^t = s_{292}^t \oplus c_{292}^t = s_{23}^{t-1} \oplus s_{160}^{t-1} \oplus s_{111}^{t-1} \oplus s_{66}^{t-1}.$$

The linear transformation does not cause any difference, since

$$\Delta s_{288}^t = \Delta s_{289}^{t-1} \oplus \Delta s_{235}^{t-1} \oplus \Delta s_{230}^{t-1} = a_{t-53}.$$

Let $a_{t-49} = \Delta s_{292}^t$. c^t can be represented as

$$c^t = (s_0^t, s_1^t, ..., s_{292-t}^t, \underbrace{s_{292-t+1}^t, ..., s_{292-t+49}^t,}$$

$$\underbrace{s_{292-t+50}^t, ..., s_{292-t+55}^t, ..., s_{292-t+59}^t, ..., s_{292}^t}).$$

Case 4. $89 \le t \le 145$

When $89 \le t \le 145$, s_{292}^t and c_{292}^t will not be used in our attack. Hence, we only need to analyze the differences caused by the linear transformation. For simplicity, we only give the representation of c^t at some critical time t. For other internal states c^t, they can be obtained by shifting.

1. If $98 \le t \le 100$, then

$$\Delta s_{229}^t = \Delta s_{230}^{t-1} \oplus \Delta s_{196}^{t-1} \oplus \Delta s_{193}^{t-1} = 0.$$

The representation of c^{100} is

$$c^{100} = (s_0^{100}, ..., s_{192}^{100}, \underbrace{s_{193}^{100}, ..., s_{226}^{100}}, s_{227}^{100}, ..., s_{229}^{100},$$

$$s_{230}^{100}, ..., s_{241}^{100}, \underbrace{s_{242}^{100}, ..., s_{247}^{100}, ..., s_{251}^{100}}, ..., s_{280}^{100}, c_{281}^{100}, ...).$$

2. If $132 \le t \le 133$, then there is

$$\Delta s_{229}^t = \Delta s_{230}^{t-1} \oplus \Delta s_{196}^{t-1} \oplus \Delta s_{193}^{t-1} = a_{t-112} \oplus 1.$$

c^{133} can be write as

$$c^{133} = (s_0^{133}, ..., s_{159}^{133}, \underbrace{s_{160}^{133}, ..., s_{193}^{133}}, ..., s_{197}^{133}, ..., s_{208}^{133},$$

$$s_{209}^{133}, ..., s_{214}^{133}, ..., s_{218}^{133}, ..., \underbrace{s_{228}^{133}, s_{229}^{133}}, ..., s_{247}^{133}, c_{248}^{133}, ...).$$

3. For $t = 134$, the differences caused by the linear transformation are

$$\Delta s_{229}^{134} = \Delta s_{230}^{133} \oplus \Delta s_{196}^{133} \oplus \Delta s_{193}^{133} = a_{22} \oplus 1,$$

$$\Delta s_{192}^{134} = \Delta s_{193}^{133} \oplus \Delta s_{160}^{133} \oplus \Delta s_{154}^{133} = 0,$$

and c^{134} is

$$c^{134} = (s_0^{134}, ..., s_{158}^{134}, \underbrace{s_{159}^{134}, ..., s_{191}^{134}}, s_{192}^{134}, ..., s_{196}^{134}, ..., s_{207}^{134},$$

$$s_{208}^{134}, ..., s_{213}^{134}, ..., s_{217}^{134}, ..., \underbrace{s_{227}^{134}, ..., s_{229}^{134}}, ..., s_{246}^{134}, c_{247}^{133}, ...).$$

4. If $135 \leq t \leq 137$, then

$$\Delta s^t_{192} = \Delta s^{t-1}_{193} \oplus \Delta s^{t-1}_{160} \oplus \Delta s^{t-1}_{154} = 1,$$

$$\Delta s^t_{229} = \Delta s^{t-1}_{230} \oplus \Delta s^{t-1}_{196} \oplus \Delta s^{t-1}_{193} = a_{t-112} \oplus 1.$$

The representation of c^{137} is

$$c^{137} = (s^{137}_0, ..., s^{137}_{155}, s^{137}_{156}, ..., s^{137}_{188}, s^{137}_{189}, s^{137}_{190}, ..., s^{137}_{204},$$

$$s^{137}_{205}, ..., s^{137}_{210}, ..., s^{137}_{214}, ..., s^{137}_{224}, ..., s^{137}_{229}, ..., s^{137}_{243}, c^{137}_{244}, ...).$$

5. If $138 \leq t \leq 139$, then

$$\Delta s^t_{192} = \Delta s^{t-1}_{193} \oplus \Delta s^{t-1}_{160} \oplus \Delta s^{t-1}_{154} = 0,$$

$$\Delta s^t_{229} = \Delta s^{t-1}_{230} \oplus \Delta s^{t-1}_{196} \oplus \Delta s^{t-1}_{193} = a_{t-112}.$$

Thus, c^{139} can be represented as

$$c^{139} = (s^{139}_0, ..., s^{139}_{153}, s^{139}_{154}, ..., s^{139}_{186}, s^{139}_{187}, s^{139}_{188}, s^{139}_{189}, s^{139}_{190}, s^{139}_{191}, s^{139}_{192},$$

$$s^{139}_{193}, ..., s^{139}_{202}, s^{139}_{203}, ..., s^{139}_{208}, ..., s^{139}_{212}, ..., s^{139}_{222}, ..., s^{139}_{227}, ..., s^{139}_{241}, c^{139}_{242}, ...).$$

6. If $140 \leq t \leq 145$, then there is not any new difference introduced by the linear transformation. Hence, the representation of c^{145} is

$$c^{145} = (s^{145}_0, ..., s^{145}_{147}, s^{145}_{148}, ..., s^{145}_{180}, s^{145}_{181}, s^{145}_{182}, s^{145}_{183}, s^{145}_{184}, s^{145}_{185}, s^{145}_{186},$$

$$s^{145}_{187}, ..., s^{145}_{196}, s^{145}_{197}, ..., s^{145}_{202}, ..., s^{145}_{206}, ..., s^{145}_{216}, ..., s^{145}_{221}, ..., s^{145}_{235}, c^{145}_{236}, ...).$$

Once the differences in states s^t and c^t are known, where $0 \leq t \leq 145$, it needs to retrieve enough linear equations in order to recover the initial states.

3.2   Retrieving Linear Equations

Since the first difference in keystream appears at time $t = 58$ when P and P' are encrypted under the same nonce, we choose

$$s^{58}_0, s^{58}_1, ..., s^{58}_{292}, s^{59}_{292}, s^{60}_{292}, ..., s^{88}_{292}$$

and a_i with $0 < i < 40$, as variables in our equations. Once the internal state at time $t = 58$ is recovered, it is easy to obtain the initial state at time $t = 0$ by stepping the cipher backward. In this section we will show how to retrieve linear equations.

1. If $58 \leq t \leq 99$, then the differences in s^t and c^t satisfy conditions of Og1, and the two linear equations become

$$z_t' \oplus z_t = x_{61}^t \oplus x_{193}^t,$$

$$z_t = (z_t' \oplus z_t)(x_{235}^t \oplus x_{193}^t) \oplus x_{12}^t \oplus x_{154}^t \oplus x_{193}^t.$$

2. If $100 \leq t \leq 106$, then the differences in s^t and c^t satisfy conditions of Og3, and the two linear equations are

$$z_t' \oplus z_t = x_{235}^t \oplus x_{193}^t \oplus 1,$$

$$z_t = (z_t' \oplus z_t)(x_{61}^t \oplus x_{193}^t) \oplus x_{12}^t \oplus x_{154}^t \oplus x_{61}^t.$$

3. If $107 \leq t \leq 138$, then the linear equations are determined by the value of a_{t-106}.

(a) $107 \leq t \leq 111$, $117 \leq t \leq 132$, $134 \leq t \leq 136$

If $a_{t-106} = 1$, then according to Og3, the linear equations are

$$z_t' \oplus z_t = x_{235}^t \oplus x_{193}^t \oplus 1,$$

$$z_t = (z_t' \oplus z_t)(x_{61}^t \oplus x_{193}^t) \oplus x_{12}^t \oplus x_{154}^t \oplus x_{61}^t.$$

If $a_{t-106} = 0$, then the conditions of Og2 are satisfied and the equations are

$$z_t' \oplus z_t = x_{235}^t \oplus x_{61}^t,$$

$$z_t = (z_t' \oplus z_t)(x_{61}^t \oplus x_{193}^t) \oplus x_{12}^t \oplus x_{154}^t \oplus x_{61}^t.$$

Put them together, the linear equations can be expressed as

$$z_t' \oplus z_t = (a_{t-106} \oplus 1)(x_{235}^t \oplus x_{61}^t) \oplus a_{t-106}(x_{235}^t \oplus x_{193}^t \oplus 1),$$

$$z_t = (z_t' \oplus z_t)(x_{61}^t \oplus x_{193}^t) \oplus x_{12}^t \oplus x_{154}^t \oplus x_{61}^t.$$

(b) $112 \leq t \leq 116$

The linear equations are

$$z_t' \oplus z_t = a_{t-106}(x_{235}^t \oplus x_{61}^t) \oplus (a_{t-106} \oplus 1)(x_{235}^t \oplus x_{193}^t \oplus 1),$$

$$z_t = (z_t' \oplus z_t)(x_{61}^t \oplus x_{193}^t) \oplus x_{12}^t \oplus x_{154}^t \oplus x_{61}^t.$$

(c) $t = 133$, 137, 138 if $a_{t-106} = 0$, then the two outputted bits are the same. So there is not any linear equations. If $a_{t-106} = 1$, then the conditions of Og1 are satisfied, and the linear equations are

$$z_t' \oplus z_t = x_{61}^t \oplus x_{193}^t,$$

$$z_t = (z_t' \oplus z_t)(x_{235}^t \oplus x_{193}^t) \oplus x_{12}^t \oplus x_{154}^t \oplus x_{193}^t.$$

4. $139 \leq t \leq 145$

The equations are also determined by the value of a_{t-106}. They are

$$z'_t \oplus z_t = a_{t-106}(x^t_{235} \oplus x^t_{193}) \oplus (a_{t-106} \oplus 1)(x^t_{235} \oplus x^t_{61} \oplus 1),$$

$$z_t = (z'_t \oplus z_t)(x^t_{61} \oplus x^t_{193}) \oplus x^t_{12} \oplus x^t_{154} \oplus x^t_{193}.$$

By guessing 39 values of a_i, where $1 \leq i \leq 39$, we can get at least 209 linear equations with respect to 323 unknowns

$$s^{58}_0, s^{58}_1, ..., s^{58}_{292}, s^{59}_{292}, s^{60}_{292}, ..., s^{88}_{292},$$

when P and P' are encrypted under the same nonce.

With the same method, we can encrypt P and P'' under the same nonce. The first different states are at time $t = 88$, while the first different keystream bits are at time $t = 146$. By guessing another 39 values of the internal variables, we can get at least 209 linear equations with respect to 323 variables

$$s^{146}_0, s^{146}_1, ..., s^{146}_{291}, s^{146}_{292}, s^{147}_{292}, ..., s^{176}_{292}.$$

By guessing 78 bits, we can get at least 418 linear equations over 411 variables

$$s^{58}_0, s^{58}_1, ..., s^{58}_{292}, s^{59}_{292}, ..., s^{176}_{292},$$

as s^{146}_i ($0 \leq i \leq 234$), can be represent as the linear combination of s^{58}_j, where $0 \leq j \leq 292$. As the linear transformation is applied, there may be few linearly dependent equations. If the linearly independent equations are not enough, we can use the feedback functions which are regarded as variables in our method. By solving the linear equations, we can recover the state at time $t = 58$ with the time complexity of $2^{78}c$, where c is the complexity of solving linear equations. Once the state s^{58} is recovered, the initial state at time $t = 0$ can be easily obtained by stepping the cipher backward. We give some experimental confirmation of the result on a shrunk cipher with similar structure and properties of ACORN v2.

3.3   Implementation and Verification

To prove the validity of our method, we apply it on a shrunk cipher A with similar structure and properties. More specifically, we built a small stream cipher according to the design principles used for Acorn but with a small state of 20 bits. We then implemented our attack to recover the initial state.

Denote by $s = (s_0, s_1, ..., s_{19})$ the initial state of the toy cipher and p the plaintext. The feedback function $f(s, p)$ is defined as

$$f(s, p) = s_0 \oplus s_3 \oplus s_{10} \oplus s_2 s_7 \oplus s_2 s_{16} \oplus s_7 s_{16} \oplus s_{14} s_4 \oplus s_{14} s_3 \oplus p.$$

Introduce intermediate variables x_i ($1 \le i \le 20$):

$$x_{20} = f(s, p),$$
$$x_{14} = s_{14} \oplus s_{11} \oplus s_{10},$$
$$x_{10} = s_{10} \oplus s_8 \oplus s_7,$$
$$x_7 = s_7 \oplus s_2 \oplus s_0,$$

$$x_i = s_i \text{ for } 1 \le i \le 19 \text{ and } i \notin \{7, 10, 14\}.$$

Then the state update function $F(s, p)$ can be described as

$$s_i = x_{i+1} \text{ for } 0 \le i \le 19.$$

The filter function $g(s)$ is used to derive a keystream z and defined as

$$g(s) = s_1 \oplus s_{15}s_{10} \oplus s_{15}s_5 \oplus s_{10}s_5.$$

The encryption procedure of the toy cipher is the same as that of ACORN v2. Here, we choose

$$P' = P \oplus (\overbrace{1, \cdots, 1}^{4}, \overbrace{0, \cdots, 0}^{4}, \overbrace{0, 0, \cdots, 0}^{\cdots})$$

and

$$P'' = P \oplus (\overbrace{0, \cdots, 0}^{8}, \overbrace{1, \cdots, 1}^{4}, \overbrace{0, \cdots, 0}^{\cdots}).$$

States Classification. Let $s^t = (s_0^t, s_1^t, ..., s_{292}^t)$ and $c^t = (c_0^t, c_1^t, ..., c_{292}^t)$ be the states at time t when P and P' are, respectively, encrypted under the same nonce. It is easy to see that $s^0 = c^0$, and the corresponding keystreams are the same until the 5th bits are outputted. $c^t (t > 0)$ can be represented as follows.

Case 1. $1 \le t \le 4$

$$c^t = (s_0^t, s_1^t, ..., s_{19-t}^t, \underline{s_{19-t+1}^t}, ..., \underline{s_{19}^t}),$$

where $\underline{s_i^t} = s_i^t \oplus 1$ and $20 - t \le i \le 19$.

Case 2. $5 \le t \le 8$

$$c^t = (s_0^t, s_1^t, ..., s_{19-t}^t, \underline{s_{19-t+1}^t}, ..., \underline{s_{19-t+4}^t}, \underbrace{s_{19-t+5}^t, ..., s_{19}^t}),$$

where $\underbrace{s_i^t} = s_i^t \oplus a_{i-24+t} (24 - t \le i \le 19)$ and $a_{t-5} = \Delta s_{19}^t$.

Case 3. $t = 9$

$$c^9 = (s_0^9, s_1^9, ..., s_{10}^9, \underline{s_{11}^9}, ..., \underline{s_{14}^9}, \underbrace{s_{15}^9, ..., s_{18}^9}, c_{19}^9).$$

Case 4. $10 \le t \le 12$

$$c^t = (s_0^t, s_1^t, ..., s_{19-t}^t, \underline{s_{19-t+1}^t}, ..., \underline{s_{19-t+3}^t}, \underbrace{s_{19-t+4}^t, s_{19-t+5}^t, ..., s_{19-t+8}^t}, c_{19-t+9}^t, ..., c_{19}^t).$$

Retrieving Linear Equations. In this section we will show how to retrieve linear equations. Since the first difference in keystream appears at time $t = 5$ when P and P' are encrypted under the same nonce, we choose

$$s_0^5, s_1^5, ..., s_{19}^5, s_{19}^6, s_{19}^7, s_{19}^8$$

and a_i with $0 \leq i \leq 3$, as variables in our equations. Once the internal state at time $t = 5$ is recovered, it is easy to obtain the initial state at time $t = 0$ by stepping the cipher backward.

1. When $5 \leq t \leq 8$, the two linear equations are

$$z_t' \oplus z_t = s_5^t \oplus s_{10}^t,$$

$$z_t = (z_t' \oplus z_t)(s_{15}^t \oplus s_5^t) \oplus s_1^t \oplus s_5^t.$$

2. When $t = 9$, the differences in z^t and $z^{t'}$ depend on $a_0 = \Delta s_{15}^9$. If $a_0 = 0$, there isn't any linear equations. Otherwise, the two linear equations are

$$z_9' \oplus z_9 = s_5^9 \oplus s_{10}^9,$$

$$z_9 = (z_9' \oplus z_9)(s_{15}^9 \oplus s_5^9) \oplus s_1^9 \oplus s_5^9.$$

3. When $10 \leq t \leq 12$, the differences in z^t and $z^{t'}$ depend on $a_{t-9} = \Delta s_{15}^t$ and Δs_{10}^t. If $a_{t-9} = 0$, the two linear equations are

$$z_t' \oplus z_t = s_{15}^t \oplus s_5^t,$$

$$z_t = (z_t' \oplus z_t)(s_{10}^t \oplus s_5^t) \oplus s_1^t \oplus s_5^t.$$

Otherwise, the two linear equations are

$$z_t' \oplus z_t = s_{15}^t \oplus s_{10}^t \oplus 1,$$

$$z_t = (z_t' \oplus z_t)(s_{10}^t \oplus s_5^t) \oplus s_1^t.$$

By guessing 4 values a_i, where $0 \leq i \leq 3$, we can get at least 18 linear equations with respect to 23 unknowns

$$s_0^5, s_1^5, ..., s_{19}^5, s_{19}^6, s_{19}^7, s_{19}^8,$$

when P and P' are encrypted under the same nonce.

With the same method, we can encrypt P and P'' under the same nonce. The first different states are at time $t = 8$, while the first different keystream bits are at time $t = 12$. By guessing another 4 values of the internal variables, we can get at least 18 linear equations with respect to 23 variables:

$$s_0^{12}, s_1^{12}, ..., s_{19}^{12}, s_{19}^{13}, s_{19}^{14}, s_{19}^{15}.$$

Putting them together, we can get at least 36 linear equations over 26 variables:

$$s_0^5, s_1^5, ..., s_{19}^5, s_{19}^6, ..., s_{19}^{15}$$

by guessing 6 bits. All the equations are listed in Appendix A, see Table 3.

By solving the linear equations, we can recover the state at time $t = 5$ with the time complexity of $2^8 c$, where c is the complexity of solving linear equations. Then by stepping the cipher backward, the initial state at time $t = 0$ can be easily obtained.

In Acorn v1, the secret key can be obtained using the knowledge that the internal state is initially 0. In Acorn v2, although the tweaks are provided to protect the secret key, we can make a forgery attack on it. In Acorn v3, the functions used are changed so that there are quite a few differences.

4   The State Recovery Attack on Acorn V3

In this section, we will apply the above method to Acorn v3, which also contains two steps: (1) gathering different states; (2) retrieving linear equations.

4.1   Gathering Different States

For Acorn v3, we choose the same plaintexts P, P' and P''. We find that the state cases are the same as those of Acorn v2, except that the expression of a_{t-49} which is

$$a_{t-49} = \Delta s_{292}^t = s_{292}^t \oplus c_{292}^t = s_{23}^{t-1} \oplus s_{160}^{t-1},$$

where $64 \leq t \leq 88$.

As the feedback function and the filter function used in Acorn v3 are different from Acorn v2, which are

$$g(s) = s_{12} \oplus s_{154} \oplus s_{61}s_{235} \oplus s_{193}s_{235} \oplus s_{61}s_{193} \oplus s_{66} \oplus s_{66}s_{230} \oplus s_{111}s_{230},$$

and

$$f(s, p) = 1 \oplus s_0 \oplus s_{107} \oplus s_{196} \oplus s_{23}s_{160} \oplus s_{23}s_{244} \oplus s_{160}s_{244} \oplus p.$$

The observations are different as well.

Observation 1. *Let x_i be the linear transformation of s_i, where $0 \leq i \leq 292$. If there is at least one i satisfying $x_i' = x_i \oplus 1$ and others are satisfying $x_i' = x_i$, where*

$$i \in \{61, 66, 111, 193, 230, 235\},$$

and the degrees of $x_{12}' \oplus x_{12}$ and $x_{54}' \oplus x_{54}$ are less than 2, $g(x') \oplus g(x)$ can determine one linear equation. Moreover, if $x_i' = x_i$ and $x_{244}' = x_{244} \oplus 1$, where $i \in \{0, 23, 107, 160, 196\}$, then the difference in $f(x')$ and $f(x)$ is

$$f(x') \oplus f(x) = x_{23} \oplus x_{160}.$$

4.2 Retrieving Linear Equations

When P and $P^{'}$ are encrypted under the same nonce, the differences in states s^t and c^t can be obtained according to Sect. 4.1, where $0 \leq t \leq 145$. According to Observation 1, we can retrieve at least 124 linear equations with respect to 323 variables, which are

$$s_0^{58}, s_1^{58}, ..., s_{292}^{58}, s_{292}^{59}, s_{292}^{60}, ..., s_{292}^{88},$$

by guessing 39 values of a_t $(1 \leq t \leq 39)$. The concrete process is the same as that of Acorn v2. With the same methods, we can encrypt $P^{''}$. In total, we can get at least 248 linear equations with respect to 411 variables by guessing 78 values. It seems that the number of equations is not enough to recover the initial state.

However, in our equations, the feedback variables s_{292}^t can be expressed as some simple quadratic equations, where the quadratic part is of the form

$$(s_{23}^{t-1} \oplus s_{160}^{t-1})s_{244}^{t-1} \oplus s_{23}^{t-1}s_{160}^{t-1}$$

with $59 \leq t \leq 88$. What's more, the guessed variables a_t satisfy

$$a_t = s_{23}^{t+48} \oplus s_{160}^{t+48},$$

where $1 \leq t \leq 39$. So by guessing the values of a_t, where $10 \leq t \leq 39$, we can retrieve other 29 linear equations from the feedback functions, see Example 1 below.

Example 1. *For a_{10} and s_{292}^{59}, if $a_{10} = 0$, then*

$$s_{23}^{58} \oplus s_{160}^{58} = 0$$

and

$$(s_{23}^{t-1} \oplus s_{160}^{t-1})s_{244}^{t-1} \oplus s_{23}^{t-1}s_{160}^{t-1} = s_{23}^{58},$$
$$s_{292}^{59} = 1 \oplus s_0^{58} \oplus s_{107}^{58} \oplus s_{66}^{58} \oplus s_{61}^{58} \oplus s_{196}^{58} \oplus s_{23}^{58}$$

Otherwise,

$$s_{23}^{58} = s_{160}^{58} \oplus 1$$

and

$$(s_{23}^{t-1} \oplus s_{160}^{t-1})s_{244}^{t-1} \oplus s_{23}^{t-1}s_{160}^{t-1} = s_{244}^{58},$$
$$s_{292}^{59} = 1 \oplus s_0^{58} \oplus s_{107}^{58} \oplus s_{66}^{58} \oplus s_{61}^{58} \oplus s_{196}^{58} \oplus s_{244}^{58}$$

When $P^{''}$ is encrypted, we can retrieve other 39 linear equations like this, rather than 29, because variables $s_{292}^{79}, ..., s_{292}^{88}$ can be used when $P^{''}$ is encrypted, which is helpful to retrieve other 10 linear equations.

The following lemma is a useful observation of the functions used in Acorn which we can use to retrieve more linear equations.

Lemma 1. *Let*

$$y = x_i x_j \oplus x_i x_k \oplus x_j x_k, \tag{1}$$

where x_i, x_j and x_k are linear functions of the initial state. Then

$$\Pr[y = x_i] = \frac{3}{4}$$

and

$$\Pr[y = x_j = x_k | y \neq x_i] = 1.$$

If there are n_1 equations of the form (1), *then the time complexity to linearize the quadric equations is* $(\frac{4}{3})^{n_1}$.

The quadratic part of the feedback function used in Acorn v3 is of the form (1) which can be linearized with Lemma 1. We can retrieve other 49 linear equations with the time complexity of $2^{20.1}$. According to the filter function used in Acorn v3, we can also retrieve other 176 quadratic equations, each of which can lead to 1 bit information by guessing one bit.

As shown above, the linear equations are enough to recover the state at time $t = 58$. Hence, the initial state at time $t = 0$ can be easily obtained by stepping the cipher backward. The time complexity is

$$2^{78} \times 2^{20.1} \times 2^{22.5} \times c = 2^{120.6}c,$$

where c is the time complexity of solving linear equations. It is lower than that of the exhaustion attack.

4.3   Implementation and Verification

We also apply it on a shrunk cipher B with similar structure and properties of Acorn v3. The shrunk cipher B is different from A in the feedback function $f(s, p)$ and filter function $g(s)$. More specifically, these two functions are

$$f(s, p) = s_0 \oplus s_{10} \oplus s_2 s_7 \oplus s_2 s_{16} \oplus s_7 s_{16} \oplus p$$

$$g(s) = s_1 \oplus s_3 \oplus s_{15} s_{10} \oplus s_{15} s_5 \oplus s_{10} s_5 \oplus s_{14} s_4 \oplus s_{14} s_3.$$

The encryption procedures of the two toy ciphers are the same.

Here, we choose the same P, P', P'' as that in the shrunk cipher A. The equations obtained are listed in listed in Appendix A, see Table 2. In total, we can get 32 linear equations with respect to 27 variables by guessing 8 bits. We can recover the state at time $t = 5$ with the time complexity of $2^8 c$, where c is the complexity of solving linear equations. Then by stepping the cipher backward, the initial state at time $t = 0$ can be easily obtained.

5   Conclusion

In our attack, we have shown that when a nonce is reused three times, we could recover the state of Acorn v2 and v3. On Acorn v2, the time complexity is $2^{78}c$, where c is the complexity of solving linear equations. The idea of our analysis is based on the fact that choosing new different plaintexts, we can get more linear equations which can decrease the time complexity of guessing. Although our attack is not practical, the time complexity is lower than that of Chaignean et al. when a nonce is reused one more times. We analyzed why Acorn v2 is insecure when one nonce is used to protect more than one messages. We also recovered the initial state of Acorn v3 when a nonce is reused three times, in which Chaignean's method is not applicable. The time complexity of our result is lower than that of the exhaustion attack.

Acknowledgment. The authors would like to thank anonymous reviewers for considerate and helpful comments. This work is supported by National Natural Science Foundation of China (Grant No. 61379139) and the "Strategic Priority Research Program" of the Chinese Academy of Sciences (Grant No. XDA06010701).

A Appendix

Table 2. Equations of the toy cipher A

	P and P' are encrypted		
time	guessing	equation	
$t=5$	none	$z_5' \oplus z_5 = s_5^5 \oplus s_{10}^5$	$z_5 = (z_5' \oplus z_5)(s_{15}^5 \oplus s_5^5) \oplus s_1^5 \oplus s_5^5$
$t=6$	none	$z_6' \oplus z_6 = s_5^6 \oplus s_{10}^6$	$z_6 = (z_6' \oplus z_6)(s_{15}^6 \oplus s_5^6) \oplus s_1^6 \oplus s_5^6$
$t=7$	none	$z_7' \oplus z_7 = s_5^7 \oplus s_{10}^7$	$z_7 = (z_7' \oplus z_7)(s_{15}^7 \oplus s_5^7) \oplus s_1^7 \oplus s_5^7$
$t=8$	none	$z_8' \oplus z_8 = s_5^8 \oplus s_{10}^8$	$z_8 = (z_8' \oplus z_8)(s_{15}^8 \oplus s_5^8) \oplus s_1^8 \oplus s_5^8$
$t=9$	$a_0 = 0$	none	
	$a_0 = 1$	$z_9' \oplus z_9 = s_5^9 \oplus s_{10}^9$	$z_9 = (z_9' \oplus z_9)(s_{15}^9 \oplus s_5^9) \oplus s_1^9 \oplus s_5^9$
$t=10$	$a_1 = 0$	$z_{10}' \oplus z_{10} = s_{15}^{10} \oplus s_5^{10}$	$z_{10} = (z_{10}' \oplus z_{10})(s_{10}^{10} \oplus s_5^{10}) \oplus s_1^{10} \oplus s_5^{10}$
	$a_1 = 1$	$z_{10}' \oplus z_{10} = s_{15}^{10} \oplus s_{10}^{10} \oplus 1$	$z_{10} = (z_{10}' \oplus z_{10})(s_{10}^{10} \oplus s_5^{10}) \oplus s_1^{10}$
$t=11$	$a_2 = 0$	$z_{11}' \oplus z_{11} = s_{15}^{11} \oplus s_5^{11}$	$z_{11} = (z_{11}' \oplus z_{11})(s_{10}^{11} \oplus s_5^{11}) \oplus s_1^{11} \oplus s_5^{11}$
	$a_2 = 1$	$z_{11}' \oplus z_{11} = s_{15}^{11} \oplus s_{11}^{11} \oplus 1$	$z_{11} = (z_{11}' \oplus z_{11})(s_{10}^{11} \oplus s_5^{11}) \oplus s_1^{11}$
$t=12$	$a_3 = 0$	$z_{12}' \oplus z_{12} = s_{15}^{12} \oplus s_5^{12}$	$z_{12} = (z_{12}' \oplus z_{12})(s_{10}^{12} \oplus s_5^{12}) \oplus s_1^{12} \oplus s_5^{12}$
	$a_3 = 1$	$z_{12}' \oplus z_{12} = s_{15}^{12} \oplus s_{11}^{12} \oplus 1$	$z_{12} = (z_{12}' \oplus z_{12})(s_{10}^{12} \oplus s_5^{12}) \oplus s_1^{12}$
$a_0 = s_2^4 \oplus s_7^4$ $a_1 = s_2^5 \oplus s_7^5$ $a_2 = s_2^6 \oplus s_7^6 \oplus s_4^6 \oplus s_3^6$ $a_3 = s_2^7 \oplus s_7^7 \oplus s_4^7 \oplus s_3^7$			

	P and P'' are encrypted		
time	guessing	equation	
$t=12$	none	$z_{12}' \oplus z_{12} = s_5^{12} \oplus s_{10}^{12}$	$z_{12} = (z_{12}' \oplus z_{12})(s_{15}^{12} \oplus s_5^{12}) \oplus s_1^{12} \oplus s_5^{12}$
$t=13$	none	$z_{13}' \oplus z_{13} = s_5^{13} \oplus s_{10}^{13}$	$z_{13} = (z_{13}' \oplus z_{13})(s_{15}^{13} \oplus s_5^{13}) \oplus s_1^{13} \oplus s_5^{13}$
$t=14$	none	$z_{14}' \oplus z_{14} = s_5^{14} \oplus s_{10}^{14}$	$z_{14} = (z_{14}' \oplus z_{14})(s_{15}^{14} \oplus s_5^{14}) \oplus s_1^{14} \oplus s_5^{14}$
$t=15$	none	$z_{15}' \oplus z_{15} = s_5^{15} \oplus s_{10}^{15}$	$z_{15} = (z_{15}' \oplus z_{15})(s_{15}^{15} \oplus s_5^{15}) \oplus s_1^{15} \oplus s_5^{15}$
$t=16$	$a_0 = 0$	none	
	$a_0 = 1$	$z_{16}' \oplus z_{16} = s_5^{16} \oplus s_{10}^{16}$	$z_{16} = (z_{16}' \oplus z_{16})(s_{15}^{16} \oplus s_5^{16}) \oplus s_1^{16} \oplus s_5^{16}$
$t=17$	$a_0 = 0$	$z_{17}' \oplus z_{17} = s_{15}^{17} \oplus s_5^{17}$	$z_{17} = (z_{17}' \oplus z_{17})(s_{10}^{17} \oplus s_5^{17}) \oplus s_1^{17} \oplus s_5^{17}$
	$a_0 = 1$	$z_{17}' \oplus z_{17} = s_{15}^{17} \oplus s_{10}^{17} \oplus 1$	$z_{17} = (z_{17}' \oplus z_{17})(s_{10}^{17} \oplus s_5^{17}) \oplus s_1^{17}$
$t=18$	$a_0 = 0$	$z_{18}' \oplus z_{18} = s_{15}^{18} \oplus s_5^{18}$	$z_{18} = (z_{18}' \oplus z_{18})(s_{10}^{18} \oplus s_5^{18}) \oplus s_1^{18} \oplus s_5^{18}$
	$a_0 = 1$	$z_{18}' \oplus z_{18} = s_{15}^{18} \oplus s_{11}^{18} \oplus 1$	$z_{18} = (z_{18}' \oplus z_{18})(s_{10}^{18} \oplus s_5^{18}) \oplus s_1^{18}$
$t=19$	$a_0 = 0$	$z_{19}' \oplus z_{19} = s_{15}^{19} \oplus s_5^{19}$	$z_{19} = (z_{19}' \oplus z_{19})(s_{10}^{19} \oplus s_5^{19}) \oplus s_1^{19} \oplus s_5^{19}$
	$a_0 = 1$	$z_{19}' \oplus z_{19} = s_{15}^{19} \oplus s_{11}^{19} \oplus 1$	$z_{19} = (z_{19}' \oplus z_{19})(s_{10}^{19} \oplus s_5^{19}) \oplus s_1^{19}$
$a_0 = s_2^9 \oplus s_7^9$ $a_1 = s_2^{10} \oplus s_7^{10}$ $a_2 = s_2^{11} \oplus s_7^{11} \oplus s_4^{11} \oplus s_3^{11}$ $a_3 = s_2^{12} \oplus s_7^{12} \oplus s_4^{12} \oplus s_3^{12}$			

Conversion between variables
$s_2^4 = s_1^5$ $s_1^6 = s_2^7$ $s_1^7 = s_3^5$ $s_1^8 = s_4^5$ $s_1^9 = s_5^5$ $s_1^{10} = s_5^6 = s_6^5$ $s_1^{11} = s_5^7 = s_7^5 \oplus s_2^5 \oplus s_2^5$
$\cdots$
$s_{11}^{19} = s_{19}^{11} \oplus s_{19}^8 \oplus s_{19}^7 \oplus s_{19}^5 \oplus s_{17}^5 \oplus s_{16}^5 \oplus s_{15}^5 \oplus s_{14}^5 \oplus s_{10}^5$ $\cdots$ $s_{15}^{19} = s_{19}^{15}$

Table 3. Equations of the toy cipher B

P and P' are encrypted		
time	guessing	equation
$t=5$	none	$z_5' \oplus z_5 = s_5^5 \oplus s_{10}^5$
$t=6$	none	$z_6' \oplus z_6 = s_5^6 \oplus s_{10}^6 \oplus s_4^6 \oplus s_3^6$
$t=7$	none	$z_7' \oplus z_7 = s_5^7 \oplus s_{10}^7 \oplus s_4^7 \oplus s_3^7$
$t=8$	none	$z_8' \oplus z_8 = s_5^8 \oplus s_{10}^8 \oplus s_4^8 \oplus s_3^8$
$t=9$	$a_0=0$	$z_9' \oplus z_9 = s_4^9 \oplus s_3^9$
	$a_0=1$	$z_9' \oplus z_9 = s_5^9 \oplus s_{10}^9 \oplus s_4^9 \oplus s_3^9$
$t=10$	a_0, a_1	$z_{10}' \oplus z_{10} = a_1(s_{15}^{10} \oplus s_{10}^{10} \oplus 1) \oplus (a_1 \oplus 1)(s_{15}^{10} \oplus s_5^{10}) \oplus a_0(s_4^{10} \oplus s_3^{10})$
$t=11$	a_1, a_2	$z_{11}' \oplus z_{11} = a_2(s_{15}^{11} \oplus s_{10}^{11} \oplus 1) \oplus (a_2 \oplus 1)(s_{15}^{11} \oplus s_5^{11}) \oplus a_1(s_4^{11} \oplus s_3^{11})$
$t=12$	a_2, a_3	$z_{12}' \oplus z_{12} = a_3(s_{15}^{12} \oplus s_{10}^{12} \oplus 1) \oplus (a_3 \oplus 1)(s_{15}^{12} \oplus s_5^{12}) \oplus a_2(s_4^{12} \oplus s_3^{12})$
	$a_0 = \oplus s_0^4 \oplus s_7^4$	$a_1 = s_0^5 \oplus s_7^5 \quad a_2 = s_0^6 \oplus s_7^6 \quad a_3 = s_2^0 \oplus s_7^7$
Feedback functions		
$s_{19}^5 = s_0^4 \oplus s_{10}^4 \oplus (a_0 \oplus 1)s_2^4 \oplus a_0 s_{16}^5 \quad s_{19}^6 = s_0^5 \oplus s_{10}^5 \oplus (a_1 \oplus 1)s_2^5 \oplus a_1 s_{16}^5$		
$s_{19}^7 = s_0^6 \oplus s_{10}^6 \oplus (a_2 \oplus 1)s_2^6 \oplus a_2 s_{16}^6 \quad s_{19}^8 = s_0^7 \oplus s_{10}^7 \oplus (a_3 \oplus 1)s_2^7 \oplus a_3 s_{16}^7$		
P and P'' are encrypted		
time	guessing	equation
$t=12$	none	$z_{12}' \oplus z_{12} = s_5^{12} \oplus s_{10}^{12}$
$t=13$	none	$z_{13}' \oplus z_{13} = s_5^{13} \oplus s_{10}^{13} \oplus s_4^{13} \oplus s_3^{13}$
$t=14$	none	$z_{14}' \oplus z_{14} = s_5^{14} \oplus s_{10}^{14} \oplus s_4^{14} \oplus s_3^{14}$
$t=15$	none	$z_{15}' \oplus z_{15} = s_5^{15} \oplus s_{10}^{15} \oplus s_4^{15} \oplus s_3^{15}$
$t=16$	$a_0=0$	$z_{16}' \oplus z_{16} = s_4^{16} \oplus s_3^{16}$
	$a_0=1$	$z_{16}' \oplus z_{16} = s_5^{16} \oplus s_{10}^{16} \oplus s_4^{16} \oplus s_3^{16}$
$t=17$	a_0, a_1	$z_{17}' \oplus z_{17} = a_1(s_{15}^{17} \oplus s_{10}^{17} \oplus 1) \oplus (a_1 \oplus 1)(s_{15}^{17} \oplus s_5^{17}) \oplus a_0(s_4^{17} \oplus s_3^{17})$
$t=18$	a_1, a_2	$z_{18}' \oplus z_{18} = a_2(s_{15}^{18} \oplus s_{10}^{18} \oplus 1) \oplus (a_2 \oplus 1)(s_{15}^{18} \oplus s_5^{18}) \oplus a_1(s_4^{18} \oplus s_3^{18})$
$t=19$	a_2, a_3	$z_{19}' \oplus z_{19} = a_3(s_{15}^{19} \oplus s_{10}^{19} \oplus 1) \oplus (a_3 \oplus 1)(s_{15}^{19} \oplus s_5^{19}) \oplus a_2(s_4^{19} \oplus s_3^{19})$
	$a_0 = s_0^{11} \oplus s_7^{11}$	$a_1 = s_0^{12} \oplus s_7^{12} \quad a_2 = s_0^{13} \oplus s_7^{13} \quad a_3 = s_0^{14} \oplus s_7^{14}$
Feedback functions		
$s_{19}^{12} = s_0^{11} \oplus s_{10}^{11} \oplus (a_0 \oplus 1)s_2^{11} \oplus a_0 s_{16}^{11} \quad s_{19}^{13} = s_0^{12} \oplus s_{10}^{12} \oplus (a_1 \oplus 1)s_2^{12} \oplus a_1 s_{16}^{12}$		
$s_{19}^{14} = s_0^{13} \oplus s_{10}^{13} \oplus (a_2 \oplus 1)s_2^{13} \oplus a_2 s_{16}^{13} \quad s_{19}^{15} = s_0^{14} \oplus s_{10}^{14} \oplus (a_3 \oplus 1)s_2^{14} \oplus a_3 s_{16}^{14}$		
Conversion between variables		
$s_2^4 = s_1^5 \quad s_1^6 = s_2^5 \quad s_1^7 = s_3^5 \quad s_1^8 = s_4^5 \quad s_1^9 = s_5^5 \quad s_1^{10} = s_5^6 = s_5^5 \quad s_1^{11} = s_5^7 = s_7^5 \oplus s_2^5 \oplus s_2^5$		
$\cdots$		
$s_{11}^{19} = s_{19}^{11} \oplus s_{19}^8 \oplus s_{19}^7 \oplus s_{19}^5 \oplus s_{17}^5 \oplus s_{16}^5 \oplus s_{15}^5 \oplus s_{14}^5 \oplus s_{10}^5 \quad \cdots \quad s_{15}^{19} = s_{19}^{15}$		

References

1. Wu, H.: Acorn: a lightweight authenticated cipher (v3) (2016). http://competitions.cr.yp.to/round3/Acornv3.pdf
2. Wu, H.: Acorn: a lightweight authenticated cipher (v1) (2014). http://competitions.cr.yp.to/round1/Acornv1.pdf
3. Wu, H.: Acorn: a lightweight authenticated cipher (v2) (2015). http://competitions.cr.yp.to/round2/Acornv2.pdf
4. Liu, M., Lin, D.: Cryptanalysis of Lightweight Authenticated Cipher ACORN. Posed on the crypto-competition mailing list (2014)

5. Chaigneau, C., Fuhr, T., Gilbert, H.: Full Key-recovery on Acorn in Nonce-reuse and Decryption-misuse settings. Posed on the crypto-competition mailing list (2015)

6. Salam, M.I., Bartlett, H., Dawson, E., Pieprzyk, J., Simpson, L., Wong, K.K.-H.: Investigating cube attacks on the authenticated encryption stream cipher Acorn. In: Batten, L., Li, G. (eds.) ATIS 2016. CCIS, vol. 651, pp. 15–26. Springer, Singapore (2016). https://doi.org/10.1007/978-981-10-2741-3_2

7. Salam, M.I., Wong, K.K.-H., Bartlett, H., Simpson, L., Dawson, E., Pieprzyk, J.: Finding state collisions in the authenticated encryption stream cipher Acorn. In: Proceedings of the Australasian Computer Science Week Multiconference, p. 36. ACM (2016)

8. Lafitte, F., Lerman, L., Markowitch, O., Heule, D.V.: SAT-based cryptanalysis of Acorn. IACR Cryptology ePrint Archive, 521 (2016)

9. Josh, R.J., Sarkar, S.: Some observations on Acorn v1 and Trivia-SC. In: Lightweight Cryptography Workshop, NIST, USA, pp. 20–21 (2015)

10. Roy, D., Mukhopadhyay, S.: Some results on ACORN. IACR cryptology ePrint report 1132 (2016)

11. https://groups.google.com/forum/#!forum/crypto-competitions/dzzNcybqFP4

An Improved Method to Unveil Malware's Hidden Behavior

Qiang Li[1,2], Yunan Zhang[1,2(✉)], Liya Su[1,2], Yang Wu[1], Xinjian Ma[1,2], and Zeming Yang[1,2]

[1] Institute of Information Engineering, Chinese Academy of Sciences, Beijing, China
{liqiang7,zhangyunan,suliya,wuyang,maxinjian,yangzeming}@iie.ac.cn
[2] School of Cyber Security, University of Chinese Academy of Sciences, Beijing, China

Abstract. Sandbox technique is widely used in automated malware analysis. However, it can only see one path during its analysis. This is fatal when meeting the targeted malware. The challenge is how to unleash the hidden behaviors of targeted malware. Many works have been done to mitigate this problem. However, these solutions either use limited and fixed sandbox environments or introduce time and space consuming multi-path exploration. To address this problem, this paper proposes a new hybrid dynamic analysis scheme by applying function summary based symbolic execution of malware. Specifically, by providing Windows APIs' summary stub and using unicorn CPU emulator, we can effectively extract malware's hidden behavior which are not shown in sandbox environment. Without the usage of full system emulation, our approach achieve much higher speed than existing schemes. We have implemented a prototype system, and evaluated it with typical real-world malware samples. The experiment results show that our system can effectively and efficiently extract malware's hidden behavior.

Keywords: Dynamic malware analysis · Function summary
Symbolic execution

1 Introduction

In the past few years, we have witnessed a new evolution of malware attacks from blindly or randomly attacking all of the Internet machines to targeting only specific systems, with a great deal of diversity among the victims, including government, military, business, education and civil society networks [3]. Through querying the victim environment, such as the version of the operating system, the keyboard layout, or the existence of vulnerable software, malware can precisely determine whether it infects the targeted machine or not. Such query-then-infect pattern has been widely employed by emerging malware attacks. As one representative example, advanced persistent threat (APT), a unique category of targeted attacks that sets its goal at a particular individual or organization, are consistently increasing and they have caused massive damage [38]. According to

© Springer International Publishing AG, part of Springer Nature 2018
X. Chen et al. (Eds.): Inscrypt 2017, LNCS 10726, pp. 362–382, 2018.
https://doi.org/10.1007/978-3-319-75160-3_22

an annual report from Symantec Inc., in 2011 targeted malware has a steady uptrend of over 70%, increasing since 2010 [38], such overgrowth has never been slow down, especially for the growth of malware binaries involved in targeted attacks in 2012 [37].

To defeat such massive intrusions, one critical challenge for malware analysis is how to effectively and efficiently expose these environment-sensitive behaviors and in further derive the specification of environments, especially when we have to handle a large volume of malware corpus everyday. Moreover, in the context of defeating targeted attacks, deriving the malware targeted environment is an indispensable analysis step. If we can derive the environment conditions that trigger malware's malicious behavior, we can promptly send out alerts or patches to the systems that satisfy these conditions.

In this paper, we focus on environment-targeted malware, i.e., malware that contains query-then-infect features. To analyze such malware and extract the specification of their targeted environment, we have to refactor our existing malware analysis infrastructure, especially for dynamic malware analysis. Because of the limitation of static analysis [27], dynamic malware analysis is recognized as one of the most effective solutions for exposing malicious behaviors [26,27]. However, existing dynamic analysis technique are not effective and efficient enough, and, as mentioned, we are facing two new challenges: First, we need highly efficient techniques to handle a great number of environment-targeted malware samples collected every day. Second, we require the analysis environment to be more adaptive to each individual sample since malware may only exhibit its malicious intent in its targeted environment.

As such, in this paper we attempt to fill the aforementioned gaps. Specifically, we present a novel dynamic analysis scheme, for agile and effective malware targeted environment analysis. To serve as an efficient tool for malware analysts, our system is able to proactively capture malwares environment-sensitive behaviors in progressive running, dynamically determine the malwares possible targeted environments, and online switch its system environment adaptively for further analysis.

The key idea is that by using function summary technique, we improved the analysis speed of symbolic execution. The major technical contributions of this paper are as follows:

- We propose one method to evaluate the execution progress of the sample in sandbox analysis. This method is used to pre-select the potential targeted malware.
- We propose one function summary based method to conduct symbolic execution of the sample. This method leaves the whole system emulation away, can save space and time consumptions.
- We implement the prototype of the system and conduct experiment. The experiment result proves that our method is very effective.

The rest of this paper is organized as follows. We discuss backgrounds and related work in Sect. 2. In Sect. 3, we present the pre-selection method.

We describe our function summary based symbolic execution method in Sect. 4. Section 5 shows the experiment results. In Sect. 6 we discuss our work and Sect. 7 concludes this paper.

2   Background and Related Work

2.1   Problem Statement

The focal point of this paper is on a set of malware families, namely environment targeted malware. In our context, we adopt the same definition of environment in related work [25], i.e., we define an environment as a system configuration, such as the version of operating system, system language, and the existence of certain system objects, such as file, registry and devices. Environment-targeted malware families commonly contain some customized environment check logic to identify their targeted victims. Such logic can thus naturally lead us to find out the malwares targeted running environment. For example, Stuxnet [44], an infamous targeted malware family, embeds a logic to detection PLC device so as to infect machines that connect to PLC control devices. Banking Trojans, such as Zeus [40], only steal information from users who have designated bank accounts. Other well-known examples include Flame [43], Conficker [31] and Duqu [20].

As a result, different from the traditional malware analysis, which mainly focuses on malwares behaviors, environment-targeted malware analysis has to answer the following two questions: (1) Given a random malware binary, can we tell whether this sample is used for environment-targeted attacks? (2) If so, what is its targeted victim or targeted running environment? Consequently, the goal of our work is to design techniques that can (1) identify possible targeted malware; (2) unveil targeted malwares environment sensitive behaviors; and (3) provide environment information to describe malwares targeted victims.

Threat Model. In our method, we deal with the Windows platform targeted malware, to constrain the scale of the problem we make the following assumptions.

- Malware would be trigged by querying Windows API functions, like: GetSystemTime, GetOSVersion, recv et al. to determine system time, operation system version or receive command from network.
- Malware always operate system resources after they were triggered like: create file, create registry key, log keyboard, send data through network, etc.
  In this model, adversary has the following capabilities:
- They use static obfuscation technique.
- They can detect whether they were executed in sandbox environment.
- They can display disguised behavior once they found they were analyzed in sandbox.

2.2   Related Work

Here we discuss the previous related work on targeted malware. We summary all these works into two categories as follows.

Research on Multi-path Analysis. Exposing malicious behaviors has been extensively discussed in existing research [1,11,22,25,33,49]. [49] proposed one forced path exploration scheme. Instead of providing semantics information for a paths trigger condition, the technique was designed for brute-force exhausting path space only. Latter, X-Force [30] has improved this approach by designing a crash-free engine. To provide the semantics of the trigger, Brumley et al. [5] proposed an approach that uses taint analysis and symbolic execution to derive the condition of malwares hidden behavior. In [23], Hasten is proposed to automatically identify malwares stalling code and deviate the execution from it. In [24], Kolbitsch et al. propose a multipath execution scheme for Java-script-based malware. Other research [9,46] proposed techniques to enforce execution of different malware functionalities.

One important work in this domain [26] introduced a snapshot based approach which could be applied to expose malwares environment-sensitive behaviors. However, this approach is not efficient for large-scale analysis of environment-targeted malware: it is typically very expensive and it may provide too much unwanted information. This approach essentially requires to run the malware multiple times to explore different paths. After each path exploration, we need to rewind to a previous point, deduce the trigger condition of branches and explore unobserved paths by providing a different set of input, or sometimes enforce the executing of branches in a brute-force way. Obviously this kind of frequent forward execution and then rolling back is very resource-consuming, thus making it not very scalable to be applied for analyzing a large volume of malware samples collected each day. Last but not least, the possible path explosion problem [26] is another problem faced by this approach.

Research on Malwares Environment-Sensitive Behaviors. The other way to deal with targeted malware starts with the detection of malware's environment sensitive behaviors [1,8,25,28,32,33]. These studies fall into three categories: (1) Identify malwares anti-debugging and anti-virtualization logic [1,8]; (2) Discovering malwares different behaviors in different system configurations [25]; (3) Discovering behaviors in network-contained environment [18]. The main idea in these studies is to provide possible target environments before applying the traditional dynamic analysis. The possible target environment could be a running environment without debuggers [8], introspection tools [1], or patched vulnerabilities involved.

In a recent representative study [21], the authors provided several statically-configured environments to detect malwares environment sensitive behaviors. While efficient (not carrying the overhead of multi-path exploration), this approach is not effective (we cannot predict and enumerate all possible target environments in advance). In particular, in the case of targeted malware, we often are not able to predict malwares targeted environments.

We summarize the pros and cons of previous research in Table 1. We analyze these techniques from several aspects: Completeness, Resource Consumption, Analysis Speed, Assisting Techniques.

Table 1. Summaries of existing techniques

Approach category	I	II
Representative work	[26]	[21]
Completeness	High	Low
Resource consumption	High	Low
Analysis speed	Slow	Fast
Assisting techniques	Symbolic execution	
	Taint analysis	Trace comparison
	Execution snapshot	

Table 2. Techniques in existing multi-path exploring system

Related work	Based platform	Technique type	Year
[4]	QEMU	PC emulator	2007
[26]	QEMU	PC emulator	2007
[7]	Pin	Binary instrumentation	2012
[50]	DynamoRIO	Binary instrumentation	2014

As illustrated, the first category of solution, such as [5,26], has theoretically full completeness but with high resource consumption. It requires the execution to periodically store execution context and roll back analysis after one-round exploration, thus very slow. Meanwhile, it requires some assisting techniques, such as symbolic execution which is slow.

For the second category, such as [1,25], it has less resource consumption and fast analysis speed. However, all the environments require manual expertise knowledge and need to be configured statically beforehand. More importantly, it is incomplete, limited to these fixed number of preconfigured environments, and has a low analysis coverage.

Further more, for current multi-path exploring system, they usually use a live operating system environment, and using Emulator or Binary Instrumentation technique to retrieve sample's instruction, as shown in Table 2. This way induce a heavy analysis environment.

3   Pre-selection of Potential Targeted Malware

We propose a system for agile and effective malware analysis. Our system contains two parts, pre-selection and symbolic execution as shown in Fig. 1. Pre-selection is used to select the potential targeted malware through static function call graph and dynamic analysis information. Symbolic Execution is used to find malware which is improved by the computing summaries and focus on

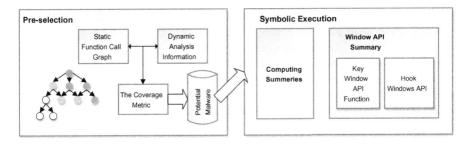

Fig. 1. System overview

windows API. Below we give the detail of pre-selection and in the Part 4 we will talk more about the symbolic execution.

Sandbox technique is widely used in automated malware analysis [2,11,47], it retrieves program dynamic behavior by monitoring the program or operating system. However, dynamic execution only show limited behaviors of the sample as shown in previous work [13]. If the program shows nothing special during executing, the detection system would tag it as legitimate [33]. Based on this idea, we extract function call graph from programs as their reference functionality. By comparing dynamic analysis information with its function call graph, we calculate one metric to measure the coverage of its functionality. When this metric is too small, it is determined to have the potential hidden behavior.

3.1 Function Call Graph

Function call graph (short for FCG) is a directed graph that represents calling relationships between functions in program. In this graph, each node represented a function, each edge (f_1, f_2) indicates function f_1 calls function f_2. A recursive function would create a cycle in the graph. Function call graph could be retrieved in static or runtime form. Since runtime FCG only represent the part code of being executed, it was not suitable in this condition. We focus on static form. Static FCG was created by conducting static analysis of program, the program could be in source code or binary code form. As for malware, source code were almost impossible to get, the problem remains deal with binary code in static form. For binary code, it should be disassembled first and identify functions, then calling relations between functions were analyzed. With the function name and their calling relations the function call graph was created.

We use the concise function call graph to estimate program functionality. When represent program functionality, we only concern Windows API functions, since they are the only way to operate system resources. From this point, the paths which we concerned in the graph have the following characteristics:

- They start with "main" function.
- They end with system functions.

Since malware could use obfuscation in their code, we should de-obfuscate malware if they were packed. Since it is impossible to deal with all obfuscate technique [27], we only deal with currently known types of packer. After that, IDA Pro is used to generate the disassembly file of the sample, then FCG is created by analyzing the disassembly file. In practice, we found the initial output of IDA Pro catch many "noises", which make it not suitable to analysis directly. The "noises" include: start up function (like: CRTStartUp), runtime environment initial function (like: initenv), system exception handler function (like: SEH), system secure guard function (like: security_cookie), etc. The typical function call graph of a Visual C++ compiled program is shown in Fig. 2.

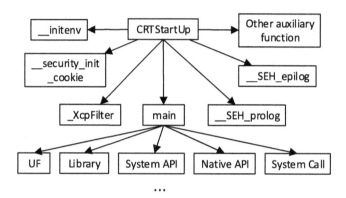

Fig. 2. Typical function call graph of Visual C++ compiled program

Those "noise" functions has nothing to do with the functionality of user program, they only deal with exception caused by program or prevent program being exploited. Since different program may have different "noise", they would introduce errors on coverage metric, so they should be eliminated. To do that, we create a dummy program (with no actual user function), extract function call graph and subtract it from sample function call graph.

3.2   Dynamic Information

Dynamic information could be retrieved through many tools. We used an open source sandbox [10] to fulfill this work. Cuckoo is an open source automated malware analysis system. It could monitor native API calls as well as API calls. Dynamic information output by the sandbox was commonly described as a sequence of operate, like in Fig. 3.

Since arguments and return value could only be determined at run time, they are not suitable for merging with static information. We use only the API call name to represent the dynamic execution process and organize them as a sequence.

```
{
    "category": "file",
    "api": "GetFileType",
    "return_value": 0,
    "arguments": {
        "file_handle": "0x00000000"
    }
    [...]
}
```

Fig. 3. Dynamic API calling information output by cuckoo

3.3   Map Dynamic Information on FCG

There is a problem that function name could miss match between dynamic information and FCG since dynamic information could come from different API levels. Functions in FCG and dynamic information could mismatch just like Table 3.

Table 3. Function mismatch between dynamic information and FCG

Function in dynamic info	Function in FCG
NtDelayExecution	Sleep
NtCreateFile	CreateFileW
NtReadFile	OpenFileW

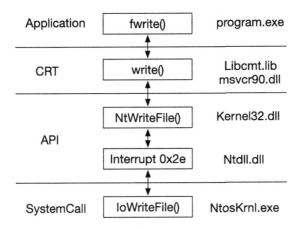

Fig. 4. The different levels of Windows API

There was three levels of API in Windows: normal API, native API and system call. The relationship between them is shown in Fig. 4. Developers were suggested to use normal APIs to create their own program, but they could also directly use native API and system calls. However, the native API and system call are poorly documented. Through analyzing Windows system library files, it is found that high level APIs are wrappers of low level APIs. For example, kernel32.dll map normal APIs to native APIs, while ntdll.dll map native APIs to system calls. The relationship between different levels of APIs can be analyzed through FCG of system DLLs. We could provide an adaption to bridge the gap between different levels of system functions. This adaption map one high level system function to several low level system functions.

Simply, the adaption was a map table like Table 4. When comparing dynamic information with FCG, high level functions could be mapped to lower level functions so they could be calculated on the same level.

Table 4. Function map relations between different level

High level functions	Low level functions
OpenFile	NtOpenFile, NtQueryAttributesFile
CreateFileA	CreateFileW, NtQueryInformationFile, NtCreateFile, NtClose

In order to bridge dynamic information and static information, we analysed 7 Windows system dynamic link libraries, including 4,191 functions as shown in Table 5.

Table 5. Analysed windows system libraries for function map

Order	Library name	Number of functions
1	advapi32.dll	676
2	iphlpapi.dll	155
3	kernel32.dll	953
4	ntdll.dll	1315
5	user32.dll	733
6	wininet.dll	242
7	ws2_32.dll	117
8	Total	4191

Using these adaption tables, we can draw dynamic APIs on FCG to fulfill the map process of dynamic to FCG. The process is shown in Fig. 5.

The coverage of graph. When dynamic is mapped on FCG, we get a sub graph of FCG. So the coverage can be measured by metric in graph theory.

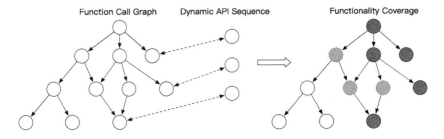

Fig. 5. Map dynamic information on FCG

Let G_{static} denote the function graph and $G_{dynamic}$ denote the mapped dynamic graph. The coverage metrics can be represented as follows:

$$G_{static} = (V_{static}, E_{static}) \tag{1}$$

$$G_{dynamic} = (V_{dynamic}, E_{dynamic}) \tag{2}$$

$$V_{dynamic} \subseteq V_{static}, E_{dynamic} \subseteq E_{static} \tag{3}$$

$$Coverage_{node} = \frac{|V_{dynamic}|}{|V_{static}|} \tag{4}$$

According to graph theory, this the value of this metric $Coverage_{node}$ is between 0 and 1, the higher the value the less uncovered behavior, thus less likely be a potential targeted malware.

4 Function Summary Based Symbolic Execution

4.1 Definition of Function Summary

As in [16], a function summary ϕ_f for a function f is defined as a formula of propositional logic whose propositions are constraints expressed in theory T. ϕ_f can be computed by successive iterations and defined as a disjunction of formulas ϕ_w of the form $\phi_w = pre_w \wedge post_w$, where pre_w is a conjunction of constraints on the inputs of f while $post_w$ is a conjunction of constraints on the outputs of f. ϕ_w can be computed from the path constraint corresponding to the execution path w. An input to a function f is any address (memory location) that can be read by f in some of its execution, while an output of f is any address that can be written by f in some of its executions and later read by P after f returns.

Preconditions in function summaries are expressed in terms of constraints on function inputs instead of program inputs in order to avoid duplication of identical summaries in equivalent but different calling contexts. For instance, in the following program the summary for the function is positive could be $(x > 0 \wedge ret = 1) \vee (x \leqslant 0 \wedge ret = 0)$ (if T includes linear arithmetic) where ret denotes the value returned by the function. This summary is expressed in terms of the function input x, independently of specific calling contexts which may map x to different program inputs like y and z in this example.

```
1   int is_positive(int x)
2   {
3       if (x > 0) return 1;
4       return 0;
5   }
6   void test(int y, int z)
7   {
8       int a,b;
9       a = is_positive(y);
10      b = is_positive(z);
11      if (a && b)
12      {
13          ...
14      }
15  }
```

4.2   Computing Summaries

If the execution of the function terminates on a return statement, a postcondition $post_w$ can be computed by taking the conjunction of constraints associated with memory locations $m \in Write(f, \vec{I}_f, w)$ written during the execution of f during the last execution w generated from a context (set of inputs) $\vec{I}_f$. Precisely, we have

$$post_w = \bigwedge_{m \in Write(f, \vec{I}_f, w)} (m = evaluate\_symbolic(m, \mathcal{M}, \mathcal{S})) \tag{5}$$

Otherwise, if the function terminates on a halt or abort statement, we define $post_w = false$ to record this in the summary for possible later use in the calling context, as described later.

A summary for the execution path w in f is then $\phi_w = pre_w \wedge post_w$. The process is repeated for other DART-exercised paths w in f, and the overall summary for f is defined as $\phi_f = \vee_w \phi_w$.

By default, the above procedure can always be used to compute function summaries path by path. But more advanced techniques, such as automatically-inferred loop invariants, could also be used. Note that pre_w can always be approximated by false (the strongest precondition) while $post_w$ can always be approximated by true (the weakest postcondition) without compromising the correctness of summaries, and that any technique for generating provably correct weaker preconditions or stronger post conditions can be used to improve precision.

Loops in program could induce an infinite execution path for symbolic execution. This would make symbolic execution unusable if the loop is not been processed properly. Usually, loop is bypassed using timeout and iteration limit.

```
1   void loop()
2   {
3        int sum = 0;
4        int N = get_input();
5        while (N > 0)
6        {
7             sum = sum+N;
8             N = get_input();
9        }
10  }
```

Symbolic execution of code containing loops or recursion may result in an infinite number of paths if the termination condition for the loop or recursion is symbolic. For example, the code in Fig. 3 has an infinite number of execution paths, where each execution path is either a sequence of an arbitrary number of trues followed by a false or a sequence of infinite number of trues. The symbolic path constraint of a path with a sequence of n trues followed by a false is:

$$\left(\bigwedge_{i \in [1,n]} N_i > 0 \right) \wedge (N_{n+1} \leqslant 0) \tag{6}$$

Where each N_i is a fresh symbolic value, and the symbolic state at the end of the execution is $\{N \mapsto N_{n+1}, sum \mapsto \sum_{i \in [1,n]} N_i\}$. In practice, one needs to put a limit on the search, e.g., a timeout, or a limit on the number of paths, loop iterations, or exploration depth.

In order to speed up the symbolic execution process of malware sample, we calculate function summary for external functions in Windows PE file.

4.3   Automatic Generate Windows API Summary

Most of Windows APIs are exposed through Dynamic Linked Library files. In actual development, developers can use these API functions through static import and dynamic load. When using static import, APIs get involved would be inserted into Import Address Table (short for IAT) in the head of PE file.

Windows has thousands of APIs, we only care about the security related ones. For most of these APIs, we do not need to actually execute the original code, instead we could use a faked API stub to summary its behavior. Note that Windows API functions use the calling convention called: stdcall, in which the callee is respond for release parameters space on stack. During the implementation of Windows API summary, we have to detect the parameters number and type, and release its space on stack manually. For the release process, we firstly copy return address to the address of the first parameter, then increase the value of ESP register value according to the number of parameters, finally call the ret instruction to set return address to EIP.

4.4 Emulate Key Windows API Function

Since symbolic execution process is really expensive, so we optimized the function summary to reduce unnecessary symbolic value. For example, in the initial process of Visual C++ compiled program, $GetSystemTimeAsFileTime$ is called to calculate the cookie for stack guard. We could simply mark the return value to be symbolic, it would not affect the correctness of the execution process. However, these symbolic value would spread to other place and cause further more symbolic value. So we read current time and write the concrete value to the return value address. Some other Windows API, for example the return value of $VirtualFree$, $CloseHandle$ actually dose not have effective on the sample's execution, so we just set the return value as s sign of success.

4.5 Static and Dynamic Hook Windows API

Windows API can be used in two ways: get from IAT or dynamic determine the address of API function use $GetProcAddress$. Since we do not have Windows operating system environment, in order to hook Windows API functions, we firstly get PE executable file's IAT, and hook functions in IAT. Be noticed that, some functions in IAT are imported by ordinal, so we have to maintain a relationship between function's name and ordinal. To hook APIs which are dynamically loaded, we hooked $GetProcAddress$ function, then implement the hook of been queried APIs inside $GetProcAddress$'s summary. Through parsing arguments of $GetProcAddress$, we determine the name of called APIs, then assign the address of its corresponding function summary.

5 Evaluation

The experiments include two steps work: first we calculate the coverage metric of every sample, and select out samples with potential hidden behaviors, then the selected samples are analysed by our function summary based symbolic execution system.

5.1 Experiment Dataset

Our test dataset consists of 63, 653malware samples, collected from online malware repositories [14]. This dataset is randomly collected without any preselection involved. We scan these malware using a free virus classification tool [17] and classify them into 417 distinct malware families. Analyzing the classification result, we further categorize these malware seven classes: Trojan, Backdoor, Virus, Worm, Adware, Packed and Others. The statistics about our dataset is shown in Table 6.

Table 6. Experiment dataset category classification by Kaspersky

Category	Number of samples	Percent
Trojan	37281	58.59%
Backdoor	9641	15.15%
Virus	6428	10.10%
AdWare	3214	5.05%
Worm	1928	3.03%
Packed	1286	2.02%
Others	3857	6.06%
Total	63635	100%

5.2 Experiment Setup

In our experiment setting, all our experiments are conducted in a machine with Intel Core i7-2600 3.40 GHz processor and 8GB memory. In the pre-select process, FCG was extracted by IDA Pro using a customized script, dynamic information was extracted by Cuckoo with a customized script to parse its output file. We implement our analysis system on an open source multi-architecture binary analysis platform: angr [41]. This platform has the capability to perform dynamic symbolic execution (like Mayhem [7], KLEE [6], etc.) and some static analyses on binaries. Since angr is not developed to perform malware analysis, it can not directly run the whole Windows malware sample. We implemented a Windows operating system environment to support the execution of malware sample. Since we are focusing on Windows platform, we developed API functions summary for Windows. Note that, it should be easy to develop such a environment to support another operating system environment. So this would not be a limitation of our method. We implement 1579 Windows API function summaries in total, including automatically generated 1428 functions and 151 manually generated functions.

5.3 Experiments on General Malware Corpus

Before the symbolic execution, we use the coverage metric to pre-select potential targeted malware. The experiment result of coverage is in Fig. 6.

We conduct the following experiments to evaluate our system with selected 1185 samples whose coverage is less than 25% (this is based on our experiment experience).

Measurement of Effectiveness. First, we study the effectiveness of our approach in terms of the code coverage in analysis. To measure that, we first collect a baseline trace by naturally running each malware sample in our virtual environment for 5 min. Then we use our symbolic execution system to analysis the sample again to get multiple paths. To evaluate the multiple paths

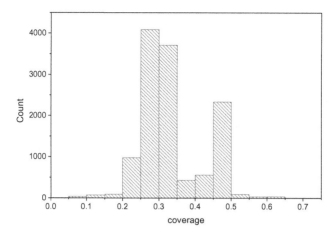

Fig. 6. Pre-selection experiment result

effects, we measure the relative increase in the number of Windows API functions between the base run and symbolic run. The distribution of increased APIs among all malware samples is shown in Fig. 4. As seen in Fig. 4, over 500 malware samples exhibit over 50% more APIs in the new run. It shows that our system can expose more malwares environment-sensitive behaviors. From the result, we also find that over 10% Adware/Spyware exhibits 100% more behaviors. It may imply that Spyware is more sensitive to the running environment compared with other malware categories. This is reasonable because Spyware normally exhibits its malicious behavior after it collects enough information about the infected user. This further proves the usefulness of our system. Examining the quantitative results of other categories, it is evident that our system can efficiently discover malwares environment-sensitive functionalities.

Comparison with Related Work. The last set of our experiment is to compare the effectiveness and efficiency of our method with other approaches. We configure four virtual environments according to the descriptions in related work [25]. We test malware samples in all four environments and choose the best one as the result. Then we randomly select 100 malware samples from each category of malware and collect the traces generated by our method, and Related Work II. When collecting each execution path trace, we terminate the analysis if no further Windows API are observed for 3 min, or if it reaches maximum analysis time which we set as 5 min. We use the following metrics for the comparison:

- Increased APIs. For each of three approaches, we pick the longest trace during any single run to compare with the normal run. For each approach, we record the percentage of malware samples whose increased APIs belonging to different level, as is shown in Table 7.

Table 7. Relative increase of code coverage

Relative increase	Number of samples
0%–10%	145
10%–50%	489
50%–100%	255
> 100%	296

- Time Consumption. According to the description in [26], average analysis time consumes around 30 min. Considering their approach needs to maintain the context of the whole system for branches and sub-branches, and needs to restore the snapshot of the whole system after each analysis. The memory/disk overhead should be very high. While in our method, we only need to maintain context for the sample, and because the usage of function summary, the sample would not affect the analysis environment, so we do not need to keep the snapshot of the analysis environment. Those make our method more lightweight. Hence, for analysis speed, our system also outperforms the compared solution.

Finally, we compare the total time to complete analysis for our method and Related Work on two typical samples: MyDoom [15] and NetSky [29]. The result is summarized in Table 8. As we can see, our system is faster than Related Work.

Table 8. Performance comparison with existing techniques

Program	Related work [5]	Our system
MyDoom	28 min	20 min
NetSky	9 min	5 min
Perfect keylogger	2 min	less than 1 min
TFN	21 min	10 min

In summary, it is evident that our approach has better performance regarding the trade-off of effectiveness and efficiency. We believe the main reason that other solutions have a higher overhead or lower effectiveness is because they are not designed to analyze malwares targeted environment. In other words, our approach is more proactive and dynamic to achieve the goal of targeted malware analysis.

5.4 Case Studies

Next, we study some cases in our analysis. We list several environment targets which may trigger malwares activities.

System Time. NetSky is a Win32 worm that spreads via email. It is known to have time triggered functionality, however different variants trigger at different times. For example, the C variant is triggered on February 26, 2004 between 6 am and 9 am [12]. The D variant is triggered on March 2, 2004, when the hour is between 6 am and 8 am [19].

Targeted Location. For Conficker A, we successful captures the system call GetKeyboardLayout and automatically extracts malwares intention of not infecting the system with Ukrainian keyboard [31]. For some variants of Bifrost [34], we find they query the system language to check whether the running OS is Chinese system or not, which is their targeted victim environment. For these cases, our system can intelligently change the query result of APIs, such as GetKeyboardLayout, to make malware believe they are running in their targeted machine/location.

User Credentials. We found several malware samples target at user credentials to conduct their malicious activities. For example, we found that Neloweg [39] will access registry at Microsoft\Internet Account Manager\Accounts key, which stores users outlook credentials. Similar examples also include Koobface [35], which targets at users facebook credentials. Our system successfully captures these malicious intentions by providing fake credentials/file/registry to malware and allowing the malware to continue execution. While the malwares further execution may fail because our system may not provide the exact correct content of the credential, our method can still provide enough targeted environment information to malware analysts.

System Invariants. In our test, we extracted one mutex from Sality [36] whose name is uxJLpe1m. In the report, we found that the existence of such mutex may disable Salitys execution. This turns out to be some common logic for a set of malware to prevent multiple infections. Similar logic has also been found in Zeus [40] and Conficker [31]. This information is useful, potentially for malware prevention, as discussed in [48].

6 Discussion

Exposing malicious behaviors of environment-targeted malware is a challenging research task for the whole malware defense community. As a new step towards systematic environment-targeted malware analysis, our solution is not perfect and not targeting to completely solve the problem. We now discuss limitations/evasions below.

Efficient of our System. One limitation of our approach is that it is implemented in Python. The core of symbolic execution is base on python emulation of VEX. So, it is slow for programs with a lot of concrete instructions compared to relaterd work. However, this limitation can be overcome by using a dedicated CPU emulator (like unicorn [42]).

Possible Problems of Instruction Emulation. In experiment, we found that angr sometimes would complain that it does not support certain merely used instruction. Since angr is not an CPU emulator, it dose not try to implement the emulation of all the instructions. As we discussed previously, this could be enhanced by the aids of a dedicated CPU emulator.

Environment-Uniqueness Malware. A recent study [33] discussed a novel anti-analysis technique, which applies environment primitives as the decryption key for the malware binary. In the real world, flashback [45] malware has exhibited similar interesting attributes. To the best of our knowledge, there is no research or tool can automatically analyze such kind of malware. Even though our approach cannot provide correct analysis environment for the captured sample, we believe our analysis can still discover more information than traditional automatic analysis techniques. For example, our approach can detect malwares query for system environment and deduce what are likely environment elements that compose the decryption key. We leave the analysis of such malware to our future work.

7   Conclusions

In this paper, we propose one method to evaluate the cover degree of sandbox analysis in order to pre-select the potential targeted malware. Then we presented a new dynamic analysis system to facilitate targeted malware analysis by efficiently and effectively exposing its targeted environments. To achieve our goal, we design several new dynamic analysis techniques based on function summary, such as automatic generate function summary for Windows API, automatically hook Windows API statically and dynamically, to solve the technical challenges faced by targeted malware analysis. In the evaluation, we show that our scheme can work on a large real-world malware corpus and achieve a better performance trade-off compared with existing approaches. While not perfect, we believe this is a right step towards an interesting new topic, which needs further research from the community.

Acknowledgments. This work was partially supported by The National Key Research and Development Program of China (2016YFB0801004 and 2016YFB0801604).

References

1. Balzarotti, D., Cova, M., Karlberger, C., Kruegel, C., Kirda, E., Vigna, G.: Efficient detection of split personalities in malware. In: NDSS 2010, 17th Annual Network and Distributed System Security Symposium, February 2010
2. Bayer, U., Moser, A., Kruegel, C., Kirda, E.: Dynamic analysis of malicious code. J. Comput. Virol. **2**(1), 67–77 (2006)
3. Bilge, L., Dumitras, T.: Before we knew it: an empirical study of zero-day attacks in the real world. In: Proceedings of the 2012 ACM Conference on Computer and Communications Security, CCS 2012, pp. 833–844. ACM, New York (2012)

4. Brumley, D., Hartwig, C., Kang, M.G., Liang, Z., Newsome, J., Poosankam, P., Song, D., Yin, H.: Bitscope: automatically dissecting malicious binaries. Technical report, In CMU-CS-07-133 (2007)
5. Brumley, D., Hartwig, C., Liang, Z., Newsome, J., Song, D., Yin, H.: Automatically identifying trigger-based behavior in malware. In: Lee, W., Wang, C., Dagon, D. (eds.) Botnet Detection. Advances in Information Security, vol. 36. Springer, Boston (2008). https://doi.org/10.1007/978-0-387-68768-1_4
6. Cadar, C., Dunbar, D., Engler, D.R.: Klee: unassisted and automatic generation of high-coverage tests for complex systems programs. In: OSDI, vol. 8, pp. 209–224 (2008)
7. Cha, S.K., Avgerinos, T., Rebert, A., Brumley, D.: Unleashing mayhem on binary code. In: Proceedings of the 2012 IEEE Symposium on Security and Privacy, SP 2012, pp. 380–394. IEEE Computer Society, Washington, DC (2012)
8. Chen, X., Andersen, J., Mao, Z.M., Bailey, M., Nazario, J.: Towards an understanding of anti-virtualization and anti-debugging behavior in modern malware. In: 2008 IEEE International Conference on Dependable Systems and Networks With FTCS and DCC (DSN), pp. 177–186, June 2008
9. Comparetti, P.M., Salvaneschi, G., Kirda, E., Kolbitsch, C., Kruegel, C., Zanero, S.: Identifying dormant functionality in malware programs. In: 2010 IEEE Symposium on Security and Privacy (SP), pp. 61–76. IEEE (2010)
10. Cuckoo: Automated malware analysis - cuckoo sandbox (2016). http://www.cuckoosandbox.org/
11. Dinaburg, A., Royal, P., Sharif, M., Lee, W.: Ether: malware analysis via hardware virtualization extensions. In: CCS 2008, pp. 51–62. ACM (2008)
12. Ferrie, T.L.: Win32.netsky.c. https://www.symantec.com/security_response/writeup.jsp?docid=2004-022417-4628-99
13. Fleck, D., Tokhtabayev, A., Alarif, A., Stavrou, A., Nykodym, T.: Pytrigger: a system to trigger & extract user-activated malware behavior. In: 2013 Eighth International Conference on Availability, Reliability and Security (ARES), pp. 92–101. IEEE (2013)
14. GeorgiaTech: Open malware (2016). http://www.offensivecomputing.net/
15. Gettis, S.: W32.mydoom.b@mm. https://www.symantec.com/security_response/writeup.jsp?docid=2004-022011-2447-99
16. Godefroid, P.: Compositional dynamic test generation. In: Proceedings of the 34th Annual ACM SIGPLAN-SIGACT Symposium on Principles of Programming Languages, POPL 2007, pp. 47–54. ACM, New York (2007)
17. Google: Virustotal (2016). https://www.virustotal.com/
18. Graziano, M., Leita, C., Balzarotti, D.: Towards network containment in malware analysis systems. In: Proceedings of the 28th Annual Computer Security Applications Conference, ACSAC 2012, pp. 339–348. ACM, New York (2012)
19. Hindocha, N.: Win32.netsky.d. https://www.symantec.com/security_response/writeup.jsp?docid=2004-030110-0232-99
20. Kaspersky: Duqu (2016). http://www.kaspersky.com/about/press/major_malware_outbreaks/duqu
21. Kirat, D., Vigna, G., Kruegel, C.: Barecloud: bare-metal analysis-based evasive malware detection. In: Proceedings of the 23rd USENIX conference on Security Symposium (SEC 2014), pp. 287–301. USENIX Association, Berkeley (2014)
22. Kolbitsch, C., Comparetti, P.M., Kruegel, C., Kirda, E., Zhou, X., Wang, X.: Effective and efficient malware detection at the end host. In: Proceedings of the 18th Conference on USENIX Security Symposium, SSYM 2009, pp. 351–366. USENIX Association, Berkeley (2009)

23. Kolbitsch, C., Kirda, E., Kruegel, C.: The power of procrastination: detection and mitigation of execution-stalling malicious code (2011)
24. Kolbitsch, C., Livshits, B., Zorn, B., Seifert, C.: Rozzle: de-cloaking internet malware. In: Proceedings of the 2012 IEEE Symposium on Security and Privacy, SP 212, pp. 443–457. IEEE Computer Society, Washington, DC (2012)
25. Lindorfer, M., Kolbitsch, C., Milani Comparetti, P.: Detecting environment-sensitive malware. In: Sommer, R., Balzarotti, D., Maier, G. (eds.) RAID 2011. LNCS, vol. 6961, pp. 338–357. Springer, Heidelberg (2011). https://doi.org/10. 1007/978-3-642-23644-0_18
26. Moser, A., Kruegel, C., Kirda, E.: Exploring multiple execution paths for malware analysis. In: IEEE Symposium on Security and Privacy, SP 2007, pp. 231–245 (2007)
27. Moser, A., Kruegel, C., Kirda, E.: Limits of static analysis for malware detection. In: Twenty-Third Annual Computer Security Applications Conference, ACSAC 2007, pp. 421–430 (2007)
28. Nappa, A., Xu, Z., Rafique, M.Z., Caballero, J., Gu, G.: Cyberprobe: towards internet-scale active detection of malicious servers. In: Proceedings of the 2014 Network and Distributed System Security Symposium (NDSS 2014), pp. 1–15 (2014)
29. NetSky (2016). https://en.wikipedia.org/wiki/Netsky_(computer_worm)
30. Peng, F., Deng, Z., Zhang, X., Xu, D., Lin, Z., Su, Z.: X-force: force-executing binary programs for security applications. In: Proceedings of the 23rd USENIX Conference on Security Symposium, SEC 2014, pp. 829–844. USENIX Association, Berkeley (2014)
31. Porras, P., Saïdi, H., Yegneswaran, V.: A foray into conficker's logic and rendezvous points. In: Proceedings of the 2nd USENIX Conference on Large-scale Exploits and Emergent Threats: Botnets, Spyware, Worms, and More, LEET 2009, p. 7. USENIX Association, Berkeley (2009)
32. Shin, S., Xu, Z., Gu, G.: Effort: efficient and effective bot malware detection. In: 2012 Proceedings IEEE INFOCOM, pp. 2846–2850, March 2012
33. Song, C., Royal, P., Lee, W.: Impeding automated malware analysis with environmentsensitive malware. In: USENIX Workshop on Hot Topics in Security (2012)
34. Symantec: Bifrost (2016). http://www.symantec.com/security_response/writeup. jsp?docid=2004-101214-5358-99
35. Symantec: Koobface (2016). http://www.symantec.com/security_response/ writeup.jsp?docid=2008-080315-0217-99&tabid=2
36. Symantec: Sality (2016). http://www.symantec.com/security_response/writeup. jsp?docid=2006-011714-3948-99
37. Symantec: Symantec intelligence quarterly (2016). http://www.symantec.com/ threatreport/quarterly.jsp
38. Symantec: Triage analysis of targeted attacks (2016). http://www.symantec.com/ threatreport/topic.jsp?id=malicious_code_trend
39. Symantec: Trojan.neloweg (2016). http://www.symantec.com/security_response/ writeup.jsp?docid=2012-020609-4221-99
40. Symantec: Zeus Trojan Horse (2016). http://www.symantec.com/security_ response/writeup.jsp?docid=2010-011016-3514-99
41. UCSB: Angr (2016). https://github.com/angr/angr
42. Unicorn: The ultimate CPU emulator (2016). http://www.unicorn-engine.org/
43. Wikipedia: Flame (2016). http://en.wikipedia.org/wiki/Flame_malware
44. Wikipedia: Stuxnet (2016). http://en.wikipedia.org/wiki/Stuxnet
45. Wikipedia: Trojan backdoor.flashback (2016). http://en.wikipedia.org/wiki/ Trojan_BackDoor.Flashback

46. Wilhelm, J., Chiueh, T.C.: A forced sampled execution approach to kernel rootkit identification (2007)
47. Willems, C., Holz, T., Freiling, F.: Toward automated dynamic malware analysis using cwsandbox. IEEE Secur. Privacy **5**(2), 32–39 (2007)
48. Xu, Z., Zhang, J., Gu, G., Lin, Z.: Autovac: automatically extracting system resource constraints and generating vaccines for malware immunization. In: 2013 IEEE 33rd International Conference on Distributed Computing Systems (ICDCS), pp. 112–123, July 2013
49. Xu, Z., Chen, L., Gu, G., Kruegel, C.: Peerpress: utilizing enemies' P2P strength against them. In: Proceedings of the 2012 ACM Conference on Computer and Communications Security, CCS 212, pp. 581–592. ACM, New York (2012)
50. Xu, Z., Zhang, J., Gu, G., Lin, Z.: GOLDENEYE: efficiently and effectively unveiling malware's targeted environment. In: Stavrou, A., Bos, H., Portokalidis, G. (eds.) RAID 2014. LNCS, vol. 8688, pp. 22–45. Springer, Cham (2014). https://doi.org/10.1007/978-3-319-11379-1_2

BotTokenizer: Exploring Network Tokens of HTTP-Based Botnet Using Malicious Network Traces

Biao Qi[1,2], Zhixin Shi[1(✉)], Yan Wang[1,2], Jizhi Wang[1,2], Qiwen Wang[1,2], and Jianguo Jiang[1]

[1] Institute of Information Engineering, Chinese Academy of Sciences, Beijing, China
{qibiao,shizhixin,wangyan,wangjizhi,wangqiwen,jiangjianguo}@iie.ac.cn
[2] School of Cyber Security, University of Chinese Academy of Sciences, Beijing, China

Abstract. Nowadays, malicious software and especially botnets leverage HTTP protocol as their communication and command (C&C) channels to connect to the attackers and control compromised clients. Due to its large popularity and facility across firewall, the malicious traffic can blend with legitimate traffic and remains undetected. While network signature-based detection systems and models show extraordinary advantages, such as high detection efficiency and accuracy, their scalability and automatization still need to be improved.

In this work, we present BotTokenizer, a novel network signature-based detection system that aims to detect malicious HTTP C&C traffic. BotTokenizer automatically learns recognizable network tokens from known HTTP C&C communications from different botnet families by using words segmentation technologies. In essence, BotTokenizer implements a coarse-grained network signature generation prototype only relying on Uniform Resource Locators (URLs) in HTTP requests. Our evaluation results demonstrate that BotTokenizer performs very well on identifying HTTP-based botnets with an acceptable classification errors.

Keywords: HTTP-based botnet detection · Network tokens
Words segmentation

1 Introduction

Botnets, collections of malware-infected machines (i.e. bots) that are remotely controlled by an attacker, known as botherder, pose a huge threat to cyber security. A botherder operates a botnet to carry out his or her nefarious tasks, such as stealing personal credentials, spreading new malware, distributing spam emails, performing click-frauds, launching distributed denial of service (DDoS) attacks through a command and control (C&C) channel. Botnets uses a large varity of network protocols as their carrying protocols to implement the C&C channels ranging from earlier IRC to current HTTP and Peer-to-Peer (P2P) for a decentralized C&C topology [5].

© Springer International Publishing AG, part of Springer Nature 2018
X. Chen et al. (Eds.): Inscrypt 2017, LNCS 10726, pp. 383–403, 2018.
https://doi.org/10.1007/978-3-319-75160-3_23

To ease the threats caused by botnets, research community have proposed numerous techniques to detect botnets or bots at the network perimeter. Lots of detection systems focus on C&C channels due to they can be regarded as crucial bridges that connect botherders to bots. From this perspective, interesting works are represented by [7,8,14,24,30,32]. Although these approaches may detect infections at the first place, existing malware or its family categories will probably not be discovered. Another direction to detect botnets is aims to identify common spatio-temporal behaviors of bots that come from the same botnet, for instance, infected hosts communicate with the same C&C server at the same time, this leads to the emergence of same or similar group behaviors within a range number of hosts. In this regard, [4,6,23,37–39] can be important literatures. However, the above-mentioned approaches will be less effective when detection systems come across small size botnets or botnets generates only a small fraction of overall traffic (e.g., an attacker may control only a small number of bots to fulfill his goals) and hides their malicious activities and C&C communications in legitimate-looking traffic (e.g., HTTP requests) [36].

Nowadays, Many botnets leverage HTTP protocol as their C&C channels due to its large popularity and facility across firewall, hence, malicious activities and communications can easily mingle with benign HTTP traffic and sneak out of traditional IDS systems [27].

Network signatures are widely used by network administrators to detect malware families and polymorphic variants, the usual approach to produce network signatures for each malware family is that firstly collect a flood of unlabeled malware by using trapping tools, such as honeynet [28], and then generate malicious traffic in a controlled environment, lastly cluster all malware into families and produce network signatures in each cluster [2,17,20–22,35]. It is reasonable to do this, because malware or bot spontaneously searches and communicates with the C&C server, vicious activities and behaviors can be discovered in the generated C&C traffic, some keywords or fields and even their aggregations become unique and distinguishing network signatures between various malware families. For matching the network signatures handily, security experts accomplish the final signatures in the syntax used by popular signature-matching IDSes, such as Snort[1] or Suricata[2]. While network signature-based detection approaches show extraordinary advantages, such as high detection efficiency and efficacy, their scalability and automatization still need to be improved.

In this work, we propose BotTokenizer, a novel network signature-based detection tool that given a number of HTTP-based botnet network traces obtained by executing HTTP-based malware samples automatically produces a set of network tokens for each HTTP-based botnet family. More specifically, BotTokenizer gets recognizable string sequences for each HTTP request from different HTTP-based botnet network traces using TF-IDF words segmentation technologies [15]. The set of obtained string sequences can be used to identify malicious HTTP C&C traffic. Compared to the prior literatures [2,17],

---

[1] Snort. http://www.snort.org/.
[2] Suricata. http://suricata-ids.org/.

BotTokenizer essentially offers a coarse-grained but very fast version of automatic network signature generation system framework.

In summary, our main contributions are as follows:

- We propose a novel HTTP-based C&C traffic detection tool called BotTokenizer that automatically discover recognizable network tokens that capture HTTP-based botnet C&C messages. To best of our knowledge, BotTokenizer is the first system to use word segmentation technologies to automatically generate network tokens for botnet C&C traffic.
- We implemented a prototype version of BotTokenizer, and apply it on actual public traffic datasets, the effectiveness of BotTokenizer was validated with several machine learning methods. Experimental results demonstrate that the proposed approach can detect botnet traffic traces successfully with very low classification errors.
- Last but not the least, we offer a new coarse-grained but very automatic ideas to network signature generation.

The remainder of this work is organized as follows. We introduce the background of network signature generation of botnet C&C traffic in Sect. 2. We present our system design and implementation in Sect. 3. We analyze collected datasets and conduct numerous experiments to validate the proposed approach in Sect. 4. We present our experimental results and analysis in Sect. 5. We survey related work in Sect. 6 and finally conclude the paper in Sect. 7.

2   Background

About a decade ago, a number of automated malware analysis tools such as ANUBIS[3], CWSandbox[4] were developed by anti-malware companies to process a large quantity of malware samples. These tools usually execute a malware sample in a controlled environment and produce reports that summarize the program's actions. Afterwards, security experts determine the type and severity of the threat that the malware constitutes based on the reports. Also, this information is valuable to cluster analyzed malware samples into families with similar behaviors and create consistent malware families signatures. However, these host-based detection systems can be very effective only if they are deployed at every end host [12,16,34]. Network signatures, by contrast, are easier to deploy than host-based detection systems, only requiring a signature-matching IDS at a vantage network point.

2.1   Network Behavioral Signatures

There have been extensive prior works proposed by security researchers to automatically generate different types of network behavioral signatures to identify

---

[3] ANUBIS. http://www.assiste.com/Anubis.html/.
[4] CWSandbox. http://www.cwsandbox.org/.

worm traffic [11,13,18,25,29]. Afterwards, a number of research efforts are also made to automatically generate network signatures to identify malware and cluster it to the corresponding family [19,20,22].

It's consistent with many security experts that a powerful behavioral signature is network traffic because the vast majority of malware families require network communication with an attacker and receive commands to perpetrate various cyber crimes. And botnets are even more so. Figure 1 shows an example that malware (i.e. bot) communicates with the attacker. The black solid line represents legitimate hosts visit the Internet sites, but a malware-infected computer connects to the ill-disposed attacker is denoted by the red dashed line.

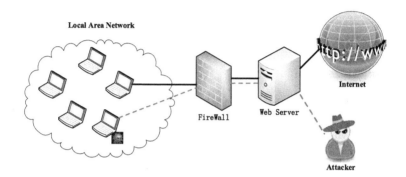

Fig. 1. An example that malware communicates with the attacker (Color figure online)

According to the above description, network traffic are strong evidence of cybercrimes, important clues can also be found in malicious network traces. Hence, some crucial network fields, keywords in the network traffic and their aggregations of course become discriminating network signatures for each malware family. Note that code reuse and polymorphism are common practices in malware [10], this naturally results in many different malware instances sharing a C&C protocol, and the same C&C protocol will reappear in network traffic, even when control servers owned by different attackers use different C&C domains and IPs. As a result, on some level, network signatures can detect and mitigate practical nefarious network threats caused by malware. Concretely, in most situations, network signature-based detection approaches can identify known network attacks caused by the malware we have come across and unknown network attacks caused by the malware variants or polymorphism from the known malware family.

Unfortunately, it is difficult to automatically generate network signatures for lots of malware, despite the fact that we know network signatures have unique advantages according to our above discussions. The main reason we consider is that a large number of malware, especially, botnets may use multiple C&C protocols to perform attacker's commands, hence, valuable information that constitutes network signatures should be elaborately sought from large and complex network traffic.

2.2   Network Tokens of HTTP-Based Botnet C&C Traffic

The main difference between botnets and other malware is that botnets have particular C&C centers to manipulate a collection of compromised machines (bots). A malicious bot is self-propagating malware designed to infect a host and connect back to a central server or servers that acts as a C&C server for the entire botnet. With a botnet, attackers can launch broad-based, remote-control, flood-type attacks against their target(s). In addition to the worm-like ability to self-propagate, bots can include the ability to provide computing resources, log keystrokes, gather passwords, capture and analyze packets, gather customer credit information, launch D(D)oS attacks, distribute spams, and open back doors on the infected host. From this perspective, the whole malicious traffic of botnets include not only communications between bots and C&C server, but also other information generated by launching various attacks. Therefore, network behavioral signatures for botnets are theoretically more complex and diverse than other types of malicious software. Practically, however, it is very difficult to capture the entire lifecycle of a botnet, this results in generating network behavioral signatures for botnets is much more challenging.

In order to simply the problem, our approach is as the same as some security researchers, we regard network connections initiated by malware as the main C&C traffic. Concretely, as for HTTP-based botnets, we refer to [2, 17, 35] and also regard malicious HTTP requests as HTTP C&C traffic. From an early detection perspective, successfully building HTTP requests for bots are a preliminary to launch an attack. Hence, if we discover some unusual signs or strange activities in HTTP requests, it can be believed that there exists malicious HTTP C&C traffic.

Botnets often use a specific set of parameter names and values that must be embedded in the URL for the C&C requests to avoid exposure of C&C server's true nature or functionalities to security researchers who may probe the server as part of an investigation. Also, some botnets exchanges information with the C&C server by first encrypting it, encoding it (e.g.,base-64), and then embedding it in the URL path [1]. Moreover, a redirected URL can also be embedded in URL path (e.g., phishing sites [9]).

In order to reduce false positives, BotProfiler [2] uses invariable keywords to replace regular expressions (e.g., $< str; length >$) proposed by ExecScent [17]. For example, for the URL in a HTTP request "GET /images/logos.gif?id=12345", ExecScent generates a control protocol template (CPT) "GET $/< str; 6 > / < str; 5 > . < str; 3 >? < str; 2 >=< int; 5 >$", but BotProfiler obtains a more concrete version "GET /**images**/$< str; 5 >$.**gif**?**id**=$< int; 5 >$". In the example, BotProfiler retains the invariability of substrings in HTTP requests such as "images", "gif". In our research, we indeed find that BotProfiler can successfully detect part of malicious HTTP C&C traffic, However, its automation and intelligence still need to be improved. Figure 2 shows two examples of different types of HTTP requests from the same botnet C&C traffic. We can clearly see that BotProfiler is easily able to generate a template to detect other HTTP C&C requests in Example (1), however,

```
GET /r.php?a=v&n=401   www.lddwj.com HTTP/1.1 Mozilla/4.0
GET /r.php?a=v&n=4459 www.lddwj.com HTTP/1.1 Mozilla/4.0
GET /r.php?a=v&n=4388 www.lddwj.com HTTP/1.1 Mozilla/4.0
GET /r.php?a=v&n=4386 www.lddwj.com HTTP/1.1 Mozilla/4.0
                          Example(1)
```

```
cmap.an.ace.advertising.com/ancm.ashx?appnexus_uid=0
cmap.dc.ace.advertising.com/dccm.ashx?id=E0
                          Example(2)
```

Fig. 2. Examples of HTTP C&C requests

BotProfiler intuitively fails to produce a suitable template using one URL to match the other one in Example (2).

Based on the above discussions and the following fact that there may be many different types of HTTP C&C requests from the same botnet, we wonder that how to automatically generate sound and comprehensive network behavioral signatures or templates, satisfying the low false positives and false negatives requirements. So in this paper, we propose a coarse-grained study prototype, called BotTokenizer, to explore network tokens of HTTP C&C traffic. To this end, we use word segmentation technology, concretely, term frequency inverse document frequency (TF-IDF) [26] to get tokens for each botnet C&C traffic just relying on URL information in HTTP requests.

3   System Design and Implementation

The primary goal of BotTokenizer is to automatically explore network tokens of HTTP-based C&C traffic generated by known malware. Figure 3 presents BotTokenizer system overview.

We start with a pool of botnet C&C traces generated by executing known malware in a controlled environment. First, we extract HTTP requests from every botnet trace, in this process, a large number of noise packets such as ARP-based packets, LLMNR-based packets are filtered and deleted. After the first step, we obtain pure HTTP C&C requests for each botnet. Second, in order to reduce false positives, we perform whitelist filtering, and the whitelist is provided by the top Alexa sites[5].

Third, we merge legitimate HTTP requests with botnet C&C requests obtained after the first two preprocessing steps, in this process, legitimate HTTP requests can be regarded as legitimate background traffic. And then we get network tokens for each botnet from the hybrid traffic. Note that the methods of getting network tokens can be various. For convenience, we chooses a comparatively

---

[5] Alexa. http://www.alexa.com/topsites/.

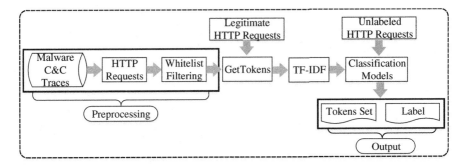

Fig. 3. BotTokenizer system overview

simple one, it is that each URL is segmented into substrings with symbols '/', '?', '.', '&', '-'. For example, for URL "nocomcom.com/temp/3425.exe?t=6.715029E-02", we will get the network tokens 'nocomcom', 'com', 'temp', '3425', 'exe', 't=6', '715029E', and '02'. In this way, each URL involved is segmented into a set of substrings (i.e., tokens). Fourth, we vectorize all HTTP requests by using TF-IDF word segmentation technology.

Intuitively, TF-IDF is the measure of a term's significance to a document given its significance among a set of document (or corpus). TF-IDF is often used in information retrieval such as automatic text classification, Automatic Keyphrase extraction [31]. Mathematically, we compute TF-IDF in the following way: TF-IDF = TF × IDF. TF represents term frequency, and IDF represents inverse document frequency. Without loss of generality, Given a document collection D, a word t, and an individual document $d \in D$, we calculate TF-IDF(t, d) = $tf(t, d) \times log(\frac{1+|D|}{1+df(t,d)})$, where $tf(t, d)$ equals the number of times t appears in d, $|D|$ is the number of the documents, $df(t, d)$ is the number of documents that contain term t.

In the fourth step, we can model different network behaviors between legitimate HTTP requests and malicious C&C HTTP requests. Besides, we can directly explore network tokens for each botnet in HTTP-based C&C traffic.

4   Experiments

4.1   Experiment Overview

To validate BotTokenizer, we conduct a comprehensive measurement study using several public botnet traces. Concretely, after the process of BotTokenizer, we utilize five machine learning algorithms to test the validity of BotTokenizer, namely Random Forest (RF), Linear Support Vector Machine (LinearSVM), Logistic Regression (LR), Decision Tree (DT), and Stochastic Gradient Descent (SGD). As a matter of fact, our approach is lightweight and very automatic multiple classification detection system. From the point of traceability, the intuition of our study is identical with many malware clustering and network signature generation literatures such as [20,22], what we are able to do is that given a HTTP request, we not only define whether it is legitimate or not, but also we can loosely infer which botnet it may be generated from.

4.2 Dataset

In this subsection, we describe the datasets used to evaluate BotTokenizer. We collected several botnet traces publicly provided by Malware Capture Facility Project (MCFP)[6] that is an effort from the Czech Technical University ATG Group for capturing, analyzing and publishing real and long-lived malware traffic. MCFP captured botnet traffic by executing real malware in controlled virtual environment for long periods of time. All published datasets include the pcap files of the botnet traffic and other files such as Binetflow files. For each botnet, a pcap file contains only botnet traffic, this facilitate us to extract valuable data. For more detailed information about the dataset, can refer to [3].

For convenience, BotTokenizer takes as input eight botnet network traces, summarized in Table 1. Both datasets contains a pcap file that only captures the botnet traffic.

After we preprocess each pcap file such as remove noise packets, filter whitelist, we obtain a number of HTTP requests for each botnet as shown in the third column in Table 1. The fourth column represents the output result of VirusTotal[7] about each malware that generates the botnet. Moreover, capture duration for each botnet is shown in the last column. Note that eight botnet datasets are generated by malware that come from different families, the behaviors between the eight botnets may be different. So eight datasets can represent eight different kinds of malicious botnet traffic, and the name of botnet is used as the label for each botnet. It's worth noting that we also need legitimate HTTP requests as background traffic. To this end, we captured legitimate campus traffic at edge link in our research institute for 72 h. After parsed and filtered the PCAP file, we obtained a total of about 60,000 HTTP requests. We can ensure that HTTP requests we collected are benign for two reasons, the one is the sites students and staffs visited are all legal, the other one is security protection measures have been taken to guard our network environment.

Generally speaking, an HTTP request includes request line, request header fields, and request body. The request line begins with a method token, followed by the request Uniform Resource Identifier (URI) and the protocol version (e.g., HTTP/1.1), and ending with CRLF. The request header fields allow the client to pass additional information about the request, and the client itself, to the server. These fields act as request modifiers. The request header fields contains several optional fields such as Accept-Encoding, Host. The request body is optionally a message-body. There is an HTTP request example as shown in Fig. 4. Note that a URL is a formatted text string used by Web browsers, email clients and other software to identify a network resource on the Internet, and usually consists

---

[6] MCFP. https://stratosphereips.org/category/dataset.html.
[7] VirusTotal. https://www.virustotal.com/.

of several components: a scheme (i.e., method token), a domain name or an IP address, a port (The port is the destination port of the web server running on this server. When no port is specified, standard ports are implied. HTTP uses TCP port 80 and HTTPS TCP port 443 as standard ports), a path and a fragment identifier. Actually, a URL is composed of HTTP(s), host, and request URI. For example, a URL is "HTTP://www.goole.com/hello.htm".

Fig. 4. Example of an HTTP request

Several related works such as [2, 20, 22] use HTTP request containing method token (e.g. GET), host, User-agent as the main components of network signatures. However, in our research, method token and User-agent have not much discrimination, for instance, all the method tokens are GET, and all the User agents are Mozilla/4.0 (compatible; MSIE5.01; Windows NT). So we only use URL without method token as our experimental data.

Table 1. Datasets used in the evaluation

Dataset	Pcap size	#HTTP requests	Type	Duration
Sality	395 MB	14,592	Win32.Sality.3	45 days
OpenCandy	417 MB	4,000	Win32.OpenCandy	26 days
Miuref	1.1 GB	12,496	Gen:Variant.Symmi	10 days
Virut	30 MB	1,025	Virus.Win32.Virut	11.63 h
Bunitu	222 MB	4,731	Trojan.Win32.Generic	24 days
Andromeda	69 MB	7,886	CRDF.Gen.Variant-Generic	3.8 days
BABO	54 MB	10,615	babo.exe	2.5 h
Neris	56 MB	3,133	Gen:Trojan.Heur.GZ	6.15 h

4.3   Performance Metric

In fact, we complete multi-classification tasks in our experimental setup. So the performance criteria for our scenario should be specified. It is reasonable that reflect the number of the correctly classified samples for each botnet in a given test data. To this end, we define detection rate (DR) as follows: $DR_i = \frac{NumTP_i}{TotalNum_i}$, where $NumTP_i$ represents the number of the correctly classified instances for ith class data, such as Sality botnet, and $TotalNum_i$ represents the total number samples in the test dataset for ith class data.

Besides, Complexity can be measured on different criteria such as memory consumption, computation time, or solution. In this case, we use computation time as the time complexity, which is an important metric for machine learning classifiers to measure their computational cost during the training phase.

Naturally, a high DR and a low time complexity are the desirable outcomes.

Algorithm 1. Extraction of tokens from HTTP requests with TF_IDF algorithm

Input:
H: set of HTTP requests;
α: the number of the token terms with top TF_IDF;
Output:
T: set of top α token terms for each botnet and legitimate traffic;
Execute:
1: **for** each http request $h \in H$ **do**
2:      $HW \Longleftarrow tokenizer(h)$
3:      **for** each $w \in HW$ **do**
4:          **if** w is not $\in BW$ **then**
5:              BW.append(w)
6:          **end if**
7:      **end for**
8: **end for**
9: **for** each $w \in BW$ **do**
10:      **for** each $h \in H$ **do**
11:          $tf = HW.counter(w)/len(HW)$
12:          $idf = log(|H|/|\{w \in h\}|)$
13:          $TF\_IDF = tf * idf$
14:      **end for**
15: **end for**
16: **for** each botnet or legitimate traffic **do**
17:      $L = sort(TF\_IDF)$
18:      Top_α=$L[0:\alpha]$
19:      **for** each $tf\_idf \in L$ **do**
20:          $Index = tf\_idf.get\_index()$
21:          $T.append(BW(Index[1])$
22:      **end for**
23: **end for**
24: **return** T

We implemented our framework in Python and utilized the popular scikit-learn machine learning framework[8] and libraries for our classification algorithms. To void bias, we partition our dataset into 10 random subsets, of which 1 is used for test and 9 others are used for training. This process is repeated 10 times, and we take the mean of 10 times test outcomes as the final result.

5   Experimental Results and Discussions

5.1   Experimental Results

Table 2 summarizes the classification results of five machine learning algorithms. As mentioned in Sect. 4, for each classification method, it is repeated for ten times so as to compute the average classification accuracy and standard deviation.

The evaluation of different methods is based on DR metric for each class. From Table 2, we can come into the conclusion that five machine learning algorithms perform very well on this multi-classification task except the Virut botnet class. Moreover, the performance of RF and LinearSVM are more better than other methods. Table 3 shows the computation time of each method. We think that LinearSVM offers major advantages over RF when we take into account classification accuracy and time complexity.

It is worthy to noting that the DRs of five classification algorithms are not good on the Virut botnet dataset. To make clear the causes of this, we plotted normalized confusion matrix of LinearSVM and RF. Confusion matrix is often used to evaluate the quality of the output of a classifier on the benchmark data set. The diagonal elements represent the number of points for which the predicted label is equal to the true label, while off-diagonal elements are those that are mislabeled by the classifier. The higher the diagonal values of the confusion matrix the better, indicating many correct predictions. The normalization version of confusion matrix is to scale elements in confusion matrix to the $[0,1]$

Table 2. Classification accuracy DR comparison on all data sets.(%)

Dataset	RF	LinearSVM	LR	DT	SGD
Sality	100.00 ± 0.01	100.00 ± 0.00	99.97 ± 0.03	99.96 ± 0.03	99.98 ± 0.00
OpenCandy	99.73 ± 0.22	99.62 ± 0.15	98.87 ± 0.71	99.72 ± 0.15	98.95 ± 0.53
Miuref	99.72 ± 0.13	99.70 ± 0.05	99.07 ± 0.13	99.73 ± 0.08	98.36 ± 0.51
Virut	90.33 ± 2.85	92.54 ± 0.47	84.93 ± 5.40	90.04 ± 1.22	81.69 ± 2.31
Bunitu	98.23 ± 0.32	99.99 ± 0.00	95.22 ± 0.72	97.92 ± 0.64	93.01 ± 1.82
Andromeda	100.00 ± 0.01	99.99 ± 0.00	99.96 ± 0.02	100.00 ± 0.02	99.99 ± 0.02
BABO	100.00 ± 0.01	99.99 ± 0.02	99.92 ± 0.04	99.98 ± 0.02	99.96 ± 0.02
Neris	99.76 ± 0.12	99.79 ± 0.08	99.26 ± 0.07	99.76 ± 0.16	97.55 ± 0.26
Legitimate	99.90 ± 0.06	99.96 ± 0.07	99.94 ± 0.02	99.68 ± 0.04	99.94 ± 0.06

---

[8] scikit-learn: http://scikit-learn.org/stable/.

Table 3. Classification time comparison on all data sets. (In seconds)

Methods	RF	LinearSVM	LR	DT	SGD
Time cost	41.65 ± 0.48	8.65 ± 0.15	11.63 ± 0.29	24.79 ± 0.55	4.31 ± 0.09

around rows, and can be evident in case of class imbalance to have a more visual interpretation of which class is being misclassified.

The results are reported in Figs. 5 and 6 respectively. From Figs. 5 and 6, we can see part of Virut samples are misclassified into the Legitimate class, the deep reason we consider is class imbalance between Virut botnet dataset and Legitimate traffic.

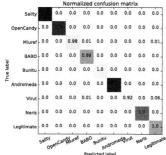
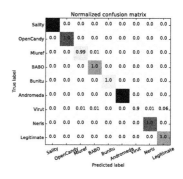

Fig. 5. Classification results of linearSVM

Fig. 6. Classification results of RF

Network Tokens. The goal of our work is to explore network tokens for each botnet, to this end, we get Top-5 TF-IDF token terms for all botnets and legitimate traffic with Algorithm 1 and thus set the parameter α to 5. Figure 7 shows histograms of network tokens for each botnet and legitimate traffic. In Fig. 7, such as subgraph (a), the vertical axis displays TF-IDF value of a network token, while the horizontal axis shows a network token. For the length of certain network tokens are too long to display, we keep the first four places of the terms, and the remaining is represented by two points. To clarify what exactly Top-5 network tokens for botnets and legitimate traffic in Fig. 7. We represent them in the form of set:

{Sality:33e4419e=1741194044,34064a11=872827409,3469bc3c=1860445604,34cd
289e=134335254,348a95e8=1875515736},
{OpenCandy:bench,swfobject,base64,es5,quant},
{Miuref:lzviz1ngc71umpidcdw10m5odp0qbcyuebd4+etd3puhtj97ukflajyr6z2z7p
b44bafbzz7og0w70neo0efx3frkijhd7b6ghhizse4wzr8emotqwizc9zv3j3b9tmimz+h
9zo2i1ajlmbhuizj8lc16g,daxpjyrqekmfyz7nbcfvswumy9qwozj5j44iekrsxywbnvvcd
sheur7odxz5iiawzvo9vbugswxtojczipst7lysyolgc6iwuqsbiffhqyx6onpqnh4x7vztht4

ebhjxdoe1ayvbpvr+fikomqev+dxcong,1n8zb6ra1dl+kx3kbsvrma66iq4h6y1jclldzv
w4o8tues+mlbhja1mhlbzd1xse8wd95uj72ugi6nk4go5ntsynjux7szy7byazkgbp4txn
37e5dxbtuhttrwyuc9jjeohehfoipw0chc52bqrd55hjt6psyj98+7ybaaff7ewim1bmvsn
mwrkbeukd0l4cpc1gsnfxuws,xenrg,tdsmanage},
{BABO:symcd,dat,wpad,sf,mfewtzbnmeswstajbgurdgmcgguabbs56bkhaoud%2bo
yl%2b0lhpg9jxyqm4gquf9nlp8ld7lvwmanzqzn6aq8zmtmcebkamst1nje4z6wrjdusf0
k%3d},
{Bunitu: whatismyipaddress,check2ip,cashunclaimed,rariab,airbnb},
{Andromeda :scomumpee,scerrims,vaccaplog,turerkila,orrereide},
{Virut:d1cec2b28022124c358459fee1f4290d,quant,897234kjdsf4523234,whatisrelat ion-
shipmarketing,thevinylsidinginstallation},
{Neris:podwine,wingpoker,sexoar,yahooadvertisingfeed,fallleague},
{Legitimate:cellphoneamp,eagemultimedia,hotelonthelake,jasoncosmo,kcpremier
apts}. From Fig. 7, we can easily find that network tokens for each botnet and
legitimate traffic are different from each other. Moreover, from this point, we
also know that our method is effective.

5.2 Comparison with State of the Art

Next we mainly compare our approach with a conventional system BotProfiler proposed by Chiba et al. [2]. BotProfiler generates detection templates by
aggregating HTTP requests with agglomerative hierarchical clustering and then
replacing substrings in HTTP requests with regular expressions in which Bot-
Profiler creatively reserves invariable keywords such as "images", "gif", "id".
Besides, BotProfiler profiles the rarity of each element in the templates to reduce
false positives. Finally, when BotProfiler matches generated templates with traffic in a deployed network, BotProfiler considers both the similarity between the
templates and HTTP requests and the rarities of elements of templates in the
deployment network. Note that BotProfiler is a closed-source detection system
and therefore we re-implement it based on the paper [2] and validate it with our
datasets. Meanwhile, our method and BotProfiler are compared in both data
utilization and detection performance.

5.2.1 Data Utilization

Data utilization in our work is defined as the adaptation of a detection algorithm to its network data. We can conclude that the more rigorous requirements
for data a method has, the lower its data utilization. According to BotProfiler,
its final cluster centroid is considered as a detection template, so the previous clustering results will directly affect the subsequent template generation. In
its clustering process, BotProfiler must calculate the similarity between HTTP
requests, and the similarity metric $Sim(h_a, h_b)$ between HTTP requests h_a and
h_b is defined by the following equation $Sim(h_a, h_b) = \frac{1}{n} \sum_{k=1}^{N} \sigma_k(h_a, h_b)$, where
σ_1 is the similarity between URL paths, σ_2 is the similarity between the combination of parameter names in the URL queries, σ_3 is the similarity between
the values in the URL queries, and σ_4 is the similarity between user agents. It

is well known that a generic URL is of the following form:

$scheme : [//[user[: password]@]host[: port]][/path][?query][\#fragment]$

where $password, port, query, fragment$ are optional items, so generally speaking, for optional query and fragment a URL could be three forms:

(α) $scheme : [//[user[: password]@]host[: port]][/path]$
(β) $scheme : [//[user[: password]@]host[: port]][/path][?query]$
(γ) $scheme : [//[user[: password]@]host[: port]][/path][?query][\#fragment]$

so when BotProfiler calculate the similarity between HTTP requests, only three scenarios satisfy the similarity metric, they are:

Scenario 1: all HTTP requests satisfy $\{h|h \in (\alpha)\}$
Scenario 2: all HTTP requests satisfy $\{h|h \in (\beta)\}$
Scenario 3: all HTTP requests satisfy $\{h|h \in (\gamma)\}$

and any other scenario such as the similarity between $h_a(h_a \in (\alpha))$ and $h_c(h_c \in (\gamma))$ can't be computed according to $Sim(h_a, h_b)$ for h_a lacks URL query. However, in reality, three forms of URL are very common and therefore the above circumstances will directly affect subsequent cluster results and template generation and finally contribute to high false negative rate. In short, BotProfiler is slightly confined to specific URL form and its data utilization is low.

Compared to BotProfiler, BotTokenizer extracts discriminating network tokens for each botnet by tokenizing HTTP request firstly and then calculating the TF-IDF value of each separated string, thus when BotTokenizer obtain network tokens, its operation is irrelevant to URL form, hence, the adaptation of BotTokenizer to HTTP requests is robust, its data utilization is comparatively high.

5.2.2   Detection Performance

In this subsection, we compare our approach with BotProfiler in detection performance. For BotProfiler is finally a binary classification, we also change our method to binary classification (such as one botnet traffic vs legitimate traffic). In this case, we still use the original botnet HTTP requests and 15000 legitimate HTTP requests as datasets, and we partition each dataset into 10 random subsets, of which 1 is used for test and 9 others are used for training. We need to redefine two performance metrics: detection rate (DR) and false positive rate (FPR). DR is defined by the ratio of correctly detected malicious HTTP requests, and FPR is defined by the ratio of falsely detected benign HTTP requests.

Results and findings. Figure 8 shows the detection results for BotProfiler and our approach. In particular, Fig. 8(a) and (b) show the detection rate and false positive rate for two systems respectively. According to Fig. 8(a), the detection rate of BotProfiler is lower than most machine learning algorithms on most datasets. This can be interpreted as the generated detection templates are not comprehensive enough, this results in more missing positive rate. Besides, for

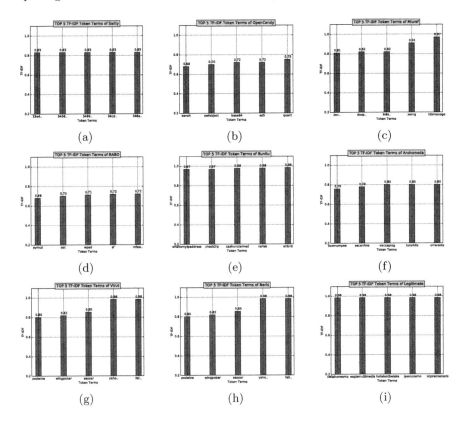

Fig. 7. Network tokens for each botnet and legitimate traffic

the reason that malicious infrastructures are more likely visited by malicious hosts and conversely legitimate sites are rarely visited by malicious devices, malicious stuff will show very rare in legitimate traffic, and thus BotProfiler doesn't perform much weaker than machine learning methods in false positive rate. This fact is illustrated by Fig. 8(b). In addition, the worst DR of Virut dataset reflects that the classification performance of machine learning methods will suffer from data imbalance. To summarize, despite BotProfiler won't be as affected by data distribution between classes (such as data imbalance) as machine learning methods are, on the whole, the whole detection performance of BotProfiler is not better than machine learning classification methods, and thus this verify the effectiveness of our method.

5.3   Discussions

Our experimental results demonstrate the feasibility and effectiveness of the proposed method in automatically detecting each type of malicious HTTP botnet C&C traffic in the case that several botnets and legitimate traffic coexist. We think it might be caused by the following reasons.

First, malicious HTTP requests are different from legitimate HTTP requests. In other words, the websites or contents visited by malware (especially, bots) are different from those visited by valid users. The differences here are reflected by tokens such as URL domains, paths, query-strings and so on. Thus, we are able to distinguish malicious HTTP requests from a great number of normal HTTP requests by using network tokens obtained by TF-IDF word segmentation technique. Second, each botnet should own different C&C channels, which result in bots from different botnet families will communicate with different C&C servers, and thus communications between different bots and C&C servers are distinguishing. In addition, from the view of TF-IDF word segmentation, an HTTP request can be regarded as a document, all tokens obtained by BotTokenizer constitute a very large bag of words, all HTTP requests are transformed into very high dimensional vectors after TF-IDF model, in binary classification (botnet vs benign traffic) scenario, for tokens of malicious traffic are rarely seen in benign traffic, the TF-IDF of these tokens are large in malicious traffic vector, and conversely they are small in corresponding benign traffic vector, thereby, vectors composed of TF-IDF of tokens are greatly distinguishing. By that analogy, it also works in multi-classification scenario, therefore it is easier to distinguish each class data for each class HTTP requests have different tokens.

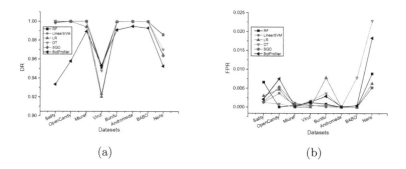

Fig. 8. Comparison between BotProfiler and our approach in DR and FPR

5.4   BotTokenizer Deployment Environment

Figure 9 presents BotTokenizer deployment overview. Practically, BotTokenizer can automatically explore network tokens for each botnet after data training. It is noted that the number of network tokens for each botnet can be adjusted as required. Moreover, each set of network tokens is labeled with the name of the botnet. Similarly to [17], once network tokens generated by BotTokenizer are deployed at the edge of a network, any HTTP traffic that contains network tokens is identified as C&C traffic. The domain names associated with the recognized traffic are then flagged as C&C domains, and attributed to the botnet family, and also blacklisted. Besides, hosts that made such malicious HTTP requests are labeled as infected hosts, and security experts make takedown efforts to such infected hosts.

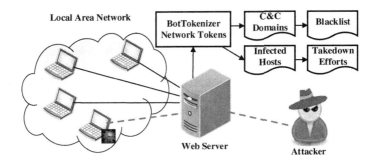

Fig. 9. BotTokenizer deployment overview

5.5 Limitations

Ideally, we make an attempt to get network tokens for each botnet from malicious C&C traffic obtained by running malware binaries in a contained environment. Unfortunately, we can't get all types of botnets since it is difficult to collect all kinds of malware binaries. From this point, the limitation of our work is similar to other literatures [2,17,20,35] that aim at malware clustering and network signature generation. In addition, we rely on both malicious C&C traffic and legitimate traffic, an attacker who gains knowledge of how BotTokenizer works may try to avoid detection by choosing requested fields that are common in legitimate HTTP requests as their C&C protocol components. Or attackers may also try to mislead the detector by injecting noise into the URL. This makes it more difficult to determine true C&C domains. Finally, our method may work very well on known botnets detection for it depends very closely on labelled training dataset, it may fail to identify unknown threats whose network tokens don't appear in training datasets.

6 Related Work

A number of prior works have been done on generating network signatures or templates as network-based countermeasures against botnet threat. Xie et al. proposed a spam signature generation system called AutoRE to detect spam botnets by building URL regular expression signatures from emails [33]. Wurzinger et al. proposed a system for automatically generating botnet C&C signatures. To this end, they leveraged the fact that after a bot received a command, the behavior of the bot changes considerably, and extracted possible commands in C&C traffic as network signatures [32]. Perdisci et al. proposed a network-level behavioral malware clustering and signature generation system that uses HTTP traffic using sending profiles and features on the HTTP method and URL. They generate network signatures that cover the HTTP request and the URL [20]. And then Nelms et al. improved [20] in their system called ExecScent, which used adaptive control templates to mine for new C&C domains [17]. Rafique and Caballero presented a system called FIRMA that produced network signatures for each of the network behaviors of a malware family, regardless of the type of traffic the malware uses [22]. Zand et al. introduced a botnet C&C signature extraction approach that finds frequent words in C&C communications with information theoretic rank criterion [35]. Daiki et al. proposed a system called BotProfiler that profiled variability of substrings in HTTP requests to generate C&C templates to detect malware-infected hosts [2]. Their system can cause fewer false positives than ExecScent.

The differences between our work and the above related works are mainly reflected in two aspects. First, we automatically explore network tokens for each botnet only relying on URL in HTTP requests. The network tokens are substrings like those in [35], however, we get tokens by dividing symbols such as '/', and then we use TF-IDF word segmentation technology to assign a score to each network token. This score represents the importance that the network token in the mixed network traffic. In general, network tokens for each botnet are different from each other, so network tokens can constitute distinguishing network signatures. Second, our approach are able to produce network tokens for several botnets after only one time data training, and the final discriminator for each botnet can consist of several network tokens.

7   Conclusion

In this paper, we proposed a system called BotTokenizer that automatically explores network tokens for each botnet to detect malicious HTTP C&C traffic. Network tokens for each botnet remain unchanged and discriminant for malicious codes and C&C reuse by an attacker. Based on this, we leverage TF-IDF word segmentation technology to assign a score to each token, and the tokens whose scores are bigger are more likely indicative of C&C traffic. In fact, we propose a new coarse-grained but very automatic framework to network signature generation. To validate BotTokenizer, five machine learning algorithms are used to multi-classification experiments on several public botnet traces and legitimate traffic. Our evaluation results revealed that our method can produce distinguishing network tokens for each botnet only relying on URL in HTTP requests. Lastly, we also described a limitation of our framework and the problems that remain to be solved regarding botnet threat detection.

References

1. Antonakakis, M., Demar, J., Stevens, K., Dagon, D.: Unveiling the network criminal infrastructure of TDSS/TDL4. Damballa Research Report 2012 (2012)
2. Chiba, D., Yagi, T., Akiyama, M., Aoki, K., Hariu, T., Goto, S.: BotProfiler: profiling variability of substrings in HTTP requests to detect malware-infected hosts. In: 2015 IEEE Trustcom/BigDataSE/ISPA, vol. 1, pp. 758–765. IEEE (2015)
3. Garcia, S., Grill, M., Stiborek, J., Zunino, A.: An empirical comparison of botnet detection methods. Comput. Secur. **45**, 100–123 (2014)

4. Goebel, J., Holz, T.: Rishi: identify bot contaminated hosts by IRC nickname evaluation. HotBots **7**, 8–8 (2007)

5. Goodman, N.: A survey of advances in botnet technologies. arXiv preprint arXiv:1702.01132 (2017)

6. Gu, G., Perdisci, R., Zhang, J., Lee, W., et al.: BotMiner: clustering analysis of network traffic for protocol-and structure-independent botnet detection. In: USENIX Security Symposium, vol. 5, pp. 139–154 (2008)

7. Gu, G., Yegneswaran, V., Porras, P., Stoll, J., Lee, W.: Active botnet probing to identify obscure command and control channels. In: Annual Computer Security Applications Conference, ACSAC 2009, pp. 241–253. IEEE (2009)

8. Gu, G., Zhang, J., Lee, W.: BotSniffer: detecting botnet command and control channels in network traffic (2008)

9. Han, X., Kheir, N., Balzarotti, D.: PhishEye: live monitoring of sandboxed phishing kits. In: Proceedings of the 2016 ACM SIGSAC Conference on Computer and Communications Security, pp. 1402–1413. ACM (2016)

10. Jang, J., Brumley, D., Venkataraman, S.: BitShred: feature hashing malware for scalable triage and semantic analysis. In: Proceedings of the 18th ACM Conference on Computer and Communications security, pp. 309–320. ACM (2011)

11. Kim, H.A., Karp, B.: Autograph: toward automated, distributed worm signature detection. In: USENIX Security Symposium, San Diego, CA, vol. 286 (2004)

12. Kirda, E., Kruegel, C., Banks, G., Vigna, G., Kemmerer, R.: Behavior-based spyware detection. In: USENIX Security, vol. 6 (2006)

13. Li, Z., Sanghi, M., Chen, Y., Kao, M.Y., Chavez, B.: Hamsa: fast signature generation for zero-day polymorphic worms with provable attack resilience. In: 2006 IEEE Symposium on Security and Privacy, pp. 15. IEEE (2006)

14. Lu, W., Rammidi, G., Ghorbani, A.A.: Clustering botnet communication traffic based on n-gram feature selection. Comput. Commun. **34**(3), 502–514 (2011)

15. Ma, J., Saul, L.K., Savage, S., Voelker, G.M.: Identifying suspicious URLs: an application of large-scale online learning. In: Proceedings of the 26th Annual International Conference on Machine Learning, pp. 681–688. ACM (2009)

16. Malan, D.J., Smith, M.D.: Host-based detection of worms through peer-to-peer cooperation. In: Proceedings of the 2005 ACM Workshop on Rapid Malcode, pp. 72–80. ACM (2005)

17. Nelms, T., Perdisci, R., Ahamad, M.: ExecScent: mining for new C&C domains in live networks with adaptive control protocol templates. In: USENIX Security, pp. 589–604 (2013)

18. Newsome, J., Karp, B., Song, D.: Polygraph: automatically generating signatures for polymorphic worms. In: 2005 IEEE Symposium on Security and Privacy, pp. 226–241. IEEE (2005)

19. Perdisci, R., Ariu, D., Giacinto, G.: Scalable fine-grained behavioral clustering of HTTP-based malware. Comput. Netw. **57**(2), 487–500 (2013)

20. Perdisci, R., Lee, W., Feamster, N.: Behavioral clustering of HTTP-based malware and signature generation using malicious network traces. In: NSDI, vol. 10, p. 14 (2010)

21. Perdisci, R., et al.: VAMO: towards a fully automated malware clustering validity analysis. In: Proceedings of the 28th Annual Computer Security Applications Conference, pp. 329–338. ACM (2012)

22. Rafique, M.Z., Caballero, J.: FIRMA: malware clustering and network signature generation with mixed network behaviors. In: Stolfo, S.J., Stavrou, A., Wright, C.V. (eds.) RAID 2013. LNCS, vol. 8145, pp. 144–163. Springer, Heidelberg (2013). https://doi.org/10.1007/978-3-642-41284-4_8

23. Saad, S., Traore, I., Ghorbani, A., Sayed, B., Zhao, D., Lu, W., Felix, J., Hakimian, P.: Detecting P2P botnets through network behavior analysis and machine learning. In: 2011 Ninth Annual International Conference on Privacy, Security and Trust (PST), pp. 174–180. IEEE (2011)
24. Sakib, M.N., Huang, C.T.: Using anomaly detection based techniques to detect HTTP-based botnet C&C traffic. In: 2016 IEEE International Conference on Communications (ICC), pp. 1–6. IEEE (2016)
25. Singh, S., Estan, C., Varghese, G., Savage, S.: Automated worm fingerprinting. In: OSDI, vol. 4, p. 4 (2004)
26. Small, S., Mason, J., Monrose, F., Provos, N., Stubblefield, A.: To catch a predator: a natural language approach for eliciting malicious payloads. In: USENIX Security Symposium, pp. 171–184 (2008)
27. Sourdis, I., Pnevmatikatos, D.: Fast, large-scale string match for a 10 Gbps FPGA-based network intrusion detection system. In: Y. K. Cheung, P., Constantinides, G.A. (eds.) FPL 2003. LNCS, vol. 2778, pp. 880–889. Springer, Heidelberg (2003). https://doi.org/10.1007/978-3-540-45234-8_85
28. Spitzner, L.: The honeynet project: trapping the hackers. IEEE Secur. Priv. **99**(2), 15–23 (2003)
29. Wang, K., Cretu, G., Stolfo, S.J.: Anomalous payload-based worm detection and signature generation. In: Valdes, A., Zamboni, D. (eds.) RAID 2005. LNCS, vol. 3858, pp. 227–246. Springer, Heidelberg (2006). https://doi.org/10.1007/11663812_12
30. Wang, X., Zheng, K., Niu, X., Wu, B., Wu, C.: Detection of command and control in advanced persistent threat based on independent access. In: 2016 IEEE International Conference on Communications (ICC), pp. 1–6. IEEE (2016)
31. Witten, I.H., Paynter, G.W., Frank, E., Gutwin, C., Nevill-Manning, C.G.: KEA: practical automatic keyphrase extraction. In: Proceedings of the Fourth ACM Conference on Digital Libraries, pp. 254–255. ACM (1999)
32. Wurzinger, P., Bilge, L., Holz, T., Goebel, J., Kruegel, C., Kirda, E.: Automatically generating models for botnet detection. In: Backes, M., Ning, P. (eds.) ESORICS 2009. LNCS, vol. 5789, pp. 232–249. Springer, Heidelberg (2009). https://doi.org/10.1007/978-3-642-04444-1_15
33. Xie, Y., Yu, F., Achan, K., Panigrahy, R., Hulten, G., Osipkov, I.: Spamming botnets: signatures and characteristics. ACM SIGCOMM Comput. Commun. Rev. **38**(4), 171–182 (2008)
34. Yin, H., Song, D., Egele, M., Kruegel, C., Kirda, E.: Panorama: capturing system-wide information flow for malware detection and analysis. In: Proceedings of the 14th ACM Conference on Computer and Communications Security, pp. 116–127. ACM (2007)
35. Zand, A., Vigna, G., Yan, X., Kruegel, C.: Extracting probable command and control signatures for detecting botnets. In: Proceedings of the 29th Annual ACM Symposium on Applied Computing, pp. 1657–1662. ACM (2014)
36. Zarras, A., Papadogiannakis, A., Gawlik, R., Holz, T.: Automated generation of models for fast and precise detection of http-based malware. In: 2014 Twelfth Annual International Conference on Privacy, Security and Trust (PST), pp. 249–256. IEEE (2014)
37. Zeidanloo, H.R., Manaf, A.B.A.: Botnet detection by monitoring similar communication patterns. arXiv preprint arXiv:1004.1232 (2010)

38. Zeng, Y., Hu, X., Shin, K.G.: Detection of botnets using combined host- and network-level information. In: 2010 IEEE/IFIP International Conference on Dependable Systems and Networks (DSN), pp. 291–300. IEEE (2010)
39. Zhang, J., Perdisci, R., Lee, W., Sarfraz, U., Luo, X.: Detecting stealthy P2P botnets using statistical traffic fingerprints. In: 2011 IEEE/IFIP 41st International Conference on Dependable Systems & Networks (DSN), pp. 121–132. IEEE (2011)

Improved Cryptanalysis of an ISO Standard Lightweight Block Cipher with Refined MILP Modelling

Jun Yin[1,3,4,5($\boxtimes$)], Chuyan Ma[6], Lijun Lyu[3,4,5], Jian Song[1], Guang Zeng[1], Chuangui Ma[7], and Fushan Wei[1,2]

[1] State Key Laboratory of Mathematical Engineering and Advanced Computing, Zhengzhou 450001, China
yinjun66888@163.com

[2] State Key Laboratory of Cryptology, P.O. Box 5159, Beijing 100878, China

[3] Institute of Information Engineering, Chinese Academy of Sciences, Beijing 100093, China

[4] Data Assurance and Communication Security Research Center, Chinese Academy of Sciences, Beijing 100093, China

[5] School of Cyber Security, University of Chinese Academy of Sciences, Beijing 100049, China

[6] National University of Defense Technology, Changsha 410073, China

[7] Army Aviation Institute, Beijing 101116, China

Abstract. Differential and linear cryptanalysis are two of the most effective attacks on block ciphers. Searching for (near) optimal differential or linear trails is not only useful for the security evaluation of block ciphers against these attacks, but also indispensable to the cryptanalysts who want to attack a cipher with these techniques. In recent years, searching for trails automatically with Mixed-Integer Linear Programming (MILP) gets a lot of attention. At first, Mouha *et al.* translated the problem of counting the minimum number of differentially active S-boxes into an MILP problem for word-oriented block ciphers. Subsequently, in Asiacrypt 2014, Sun *et al.* extended Mouha *et al.*'s method, and presented a technique which can find actual differential or linear characteristics of a block cipher in both the single-key and related-key models. In this paper, we refine the constraints of the 2-XOR operation in order to reduce the overall number of variables and constraints. Experimental results show that MILP models with the refined constraints can be solved more efficiently. We apply our method to HIGHT (an ISO standard), and we find differential (covering 11 rounds) or linear trails (covering 10 rounds) with higher probability or correlation. Moreover, we find so far the longest differential and linear distinguishers of HIGHT.

Keywords: Lightweight block cipher · Differential attack
Linear attack · HIGHT · MILP

© Springer International Publishing AG, part of Springer Nature 2018
X. Chen et al. (Eds.): Inscrypt 2017, LNCS 10726, pp. 404–426, 2018.
https://doi.org/10.1007/978-3-319-75160-3_24

1   Introduction

In recent years, with the rapid development of the Internet of Things (IoT), the application of micro computing equipment is more and more popular, such as RFID chips and wireless sensor networks. At the same time, how to ensure the security of information stored on or transmitted over such devices with constrained resources attracts more and more attention. Hence, the pursuit of efficient and secure lightweight block ciphers came into being. Researchers have put forward many lightweight block ciphers. Roughly speaking, those lightweight block cipher can be divided into two categories, one type based on small S-boxes, such as LBlock [1], PRESENT [2], SKINNY [3] and RECTANGLE [4]. Another type doesn't use S-boxes. Instead, they adopt the ARX construction, where modular addition, rotation and XOR are used. These operations are easy to implement in software, such as HIGHT [5], TEA [6], SPECK [7], Sparx [8] *etc.*

Differential cryptanalysis [9] and linear cryptanalysis [10] are two main attacks on symmetric-key ciphers. For these attacks, finding an optimal differential or linear trails are important to make an effective attack. Among the methods proposed in the literature on finding optimal differential and linear characteristics, automatic searching is a very popular one, which is relatively easy to implement. Matsui's branch and bound search algorithm is the classic methods for obtaining DES differential characteristics [11]. Recently, with the aim to raise the efficiency of it, Chen *et al.* proposed some variant methods [12,13]. In CT-RSA 2014, Biryukov and Velichkov extended Matsui's algorithm [14], they proposed a new automatic search tool to search for the differential characteristics of ARX ciphers by introducing the new concept of a partial difference distribution table (pDDT). In 2013, Mouha and Preneel proposed an automatic tool to search for the optimal differential characteristic for ARX ciphers Salsa20 [15], The main idea is to convert the problem of searching for differential characteristics to a Boolean satisfiability problem, which only involves writing out simple equations for every operation in the cipher, and applies an off-the-shelf SAT solver. In 2011, Mouha *et al.* translated the problem of counting the number of active S-boxes into an MILP problem which can be solved with MILP solvers [16]. Subsequently, In Asiacrypt 2014, Sun *et al.* extended Mouha *et al.*'s method, and presented methods for searching the differential or linear characteristics of bit-oriented block ciphers both the single-key and related-key models [17]. In FSE 2016, Fu *et al.* proposed an MILP-based tool for automatic search for differential and linear trails in ARX ciphers, through the properties of differential and linear characteristics for modular addition operation, and gave a systematic method to describe the differential and linear characteristics with some constraints [18]. In FSE 2017, automatic search was also conducted based on constraint programming, which was able to analyze ciphers with 8×8 S-boxes [19].

HIGHT [5] was introduced by Hong *et al.* in CHES 2006, which is an ISO standard lightweight block cipher [20]. The designers gave the differential and linear attack results, and found some 11-round differential characteristics with probability 2^{-58} and several 10-round linear approximations with correlation 2^{-26}.[1]

In this paper, we improve Sun *et al.*'s method for automatic search differential and linear trails based on the MILP model. We accurately describe the 2-XOR operations with new constraints. The new constraints can reduce the overall number of variables and constraints in MILP model, which can save the time for solving the MILP model. Subsequently, we apply our refined MILP model to the lightweight block cipher HIGHT. As a result, we not only search the better differential characteristic for 11-round HIGHT and linear approximation for 10-round HIGHT, but also find the optimal differential characteristic for 13-round HIGHT and linear approximation for 11-round HIGHT. These results are shown in Table 1. (The p and cor in the table, represent the probability of differential characteristic and the correlation of linear approximation respectively)

Table 1. Summary of differential characteristics and linear approximations for HIGHT

Differential characteristics			Linear approximations		
Rounds	log_2^p	Reference	Rounds	log_2^{cor}	Reference
11	−58	[5]	10	−26	[5]
11	−45	This paper	10	−25	This paper
12	−53	This paper	11	−31	This paper
13	−61	This paper	-	-	-

Organization. This paper is organised as follows. In Sect. 2, we introduce the related knowledge, and give a brief description of the HIGHT. In Sect. 3, we give a brief introduction the automatic search method of differential and linear trails based on MILP model. In Sect. 4, a refined MILP model is presented. As an application, we utilize the refined MILP model to search the differential and linear characteristics for HIGHT. Then in Sect. 5, the results of differential and linear cryptanalysis of HIGHT are given. Finally, we conclude the paper in Sect. 6.

2   Preliminaries

In this section, we introduce some notations and terms, and briefly describe the lightweight block cipher HIGHT.

---

[1] In [5], the 10-round linear approximation with $\varepsilon^2 = 2^{-54}$, ε is called bias. Correspondingly, converted into the 10-round linear approximation with correlation 2^{-26}.

2.1   Notations

The following notations are used in this paper:

- $(X_i^7\|X_i^6\|\cdots\|X_i^0)$: The 64-bit input of the i-th round is considered as con-catenations of 8 bytes X_i^j, $0 \le j \le 7$.
- $(SK_{4i+3}\|SK_{4i+2}\|SK_{4i+1}\|SK_{4i})$: The 32-bit subkey of the i-th round is con-sidered as concatenations of 4 bytes SK_{4i+j}, $0 \le j \le 3$.
- $\oplus$: Bitwise exclusive OR (XOR)
- $\boxplus$: Addition modulo 2^n
- $x \lll s$: Rotation of x to the left by s positions
- Δx: The XOR difference of x_1 and x_2, $x_1 \oplus x_2 = \Delta x$
- $x[i]$: The bit at position i of word x

2.2   Description of HIGHT

The HIGHT [5] is a lightweight block cipher, it was proposed in CHES 2006, and was adopted as an ISO standard cryptography. The HIGHT utilize an 8-branch Type-II generalized Feistel structure, 64-bit block size and 128-bit key size, consisting of 32 rounds with four parallel Feistel functions in each round. The round function of HIGHT is shown in Fig. 1.[2]

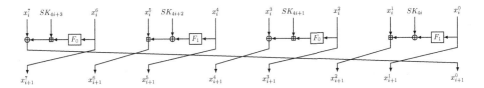

Fig. 1. The round function of HIGHT.

The round function transforms $(x_i^7\|x_i^6\|\cdots\|x_i^0)$ into $(x_{i+1}^7\|x_{i+1}^6\|\cdots\|x_{i+1}^0)$ as follows:

$$x_{i+1}^1 = x_i^0; x_{i+1}^3 = x_i^2; x_{i+1}^5 = x_i^4; x_{i+1}^7 = x_i^6;$$
$$x_{i+1}^0 = x_i^7 \oplus (F_0(x_i^6) \boxplus SK_{4i+3});$$
$$x_{i+1}^2 = x_i^1 \boxplus (F_1(x_i^0) \oplus SK_{4i+2});$$
$$x_{i+1}^4 = x_i^3 \oplus (F_0(x_i^2) \boxplus SK_{4i+1});$$
$$x_{i+1}^6 = x_i^5 \boxplus (F_1(x_i^4) \boxplus SK_{4i}).$$

The F_0 and F_1 used in the round function are defined as follows:

$$F_0(x) = (x <<< 1) \oplus (x <<< 2) \oplus (x <<< 7);$$
$$F_1(x) = (x <<< 3) \oplus (x <<< 4) \oplus (x <<< 6).$$

---

[2] The Figs. 1, 2 and 3 are generated by TikZ for Cryptographers, please refer to http://www.iacr.org/authors/tikz/.

The inner functions F_0 and F_1 provide bitwise diffusion. These functions can be regarded as linear transformations from $GF(2)^8$ to $GF(2)^8$. The two linear transformations selected in the design of the cipher have the best diffusion property.

In this paper, we only consider the single-key model. Therefore, we omit key schedule in this paper. For further details, please refer to [5].

2.3   Security Analysis Results of HIGHT

Since the HIGHT has been put forward, it has received a great deal of attention. At present, there are many cryptanalysis results of HIGHT, which also includes some cryptanalysis results given by the designer.

In [5], the designer gives the differential characteristic probability is 2^{-58} for the 11-round HIGHT, and gave the 13-round HIGHT differential attack result. At the same time, find the 10-round linear approximation with bias $\varepsilon = 2^{-27}$. By using the linear approximation, the designer proposed 13-round linear attack for HIGHT, in the attack process, it requires 2^{57} plaintexts with the success rate 96.7% to recover 36 bits of the subkeys. In addition, the designer use this 14-round impossible differential characteristic to attack 18-round HIGHT. This attack requires $2^{46.8}$ chosen-plaintexts and $2^{109.2}$ encryptions of 18-round HIGHT. Except for some of the above attacks, the designer also presented truncated differential cryptanalysis [21], boomerang attack [22], sliding attack [23] and related key attack [24] and so on. These results indicate that the HIGHT has sufficient security.

Lu et al. [25] gave the first impossible differential cryptanalysis result for 25-round HIGHT, this attack requires 2^{60} chosen-plaintexts and $2^{126.78}$ encryptions. At ACISP 2009, Özen et al. [26] applied the impossible differential technique to attack 26-round HIGHT with data complexity of 2^{61} plaintexts, and time complexity of about $2^{119.53}$ encryptions. Then, At AFRICACRYPT 2012, Chen et al. [27] presented the impossible differential attack on the 27-round HIGHT with data complexity of 2^{58}, and time complexity of about $2^{126.6}$ encryptions, which is smaller than exhaustive search. In 2016, Cui et al. [28] proposed MILP-based automatic tool to search all cases of 17-round impossible differentials that both hamming weights of input and output differences are one, They found 4 impossible differentials for 17-round HIGHT, which are the longest ones until now.

At ICISC 2010, Koo et al. [29] presented the first attack on the full HIGHT using related-key rectangle attack with $2^{123.169}$ encryptions, $2^{57.84}$ data, and 4 related keys.

In 2015, Igarashi et al. [30] gave 19-round HIGTH using meet-in-the-middle attack with Splice-and-Cut technique, the attacked with 2^8 bytes of memory, $2^8 + 2$ pairs of chosen plain and cipher texts, and $2^{120.7}$ times of the encryption operation.

3 MILP-Based Automatic Search for Differential and Linear Trails

In this section, we first introduce the mixed integer linear programming problem, Then we briefly describe the method of constructing constraint inequalities for every operation in the ARX ciphers.

3.1 Mixed Integer Linear Programming (MILP)

MILP: Assume $A \in R^{m \times n}$, $b \in R^m$ and $c_1, c_2, \cdots, c_n \in R^n$, find a vector $x = (x_1, x_2, \cdots, x_n)$, such that the linear function $c_1 x_1 + c_2 x_2 + \cdots + c_n x_n$ is minimized (or maximized) with respect to the linear constraint $Ax \leq b$.

The MILP problem is a kind of optimization problem, which aims at finding the optimal solution of the objective function under the constraints. This problem can be solved by a lot of commercial software, such as Gurobi [31], CPLEX [32], MAGMA [33], *etc.*

3.2 Differential Constraints for Different Operations

Suppose an ARX cipher is composed of the following three operations:

- Rotations
- XOR
- Modular addition

It is obvious that the differential constraint of rotations operation can be obtained, according to [17], the constraints on the XOR operation as follows.

Constraints for XOR Operation [17]. According to Sun *et al.*'s differential automatic search method. For XOR operation with input differences Δa, Δb and output difference Δc, the constraints are presented as follows:

$$\begin{cases} \Delta a + \Delta b + \Delta c \geq 2d_\oplus \\ d_\oplus \geq \Delta a, d_\oplus \geq \Delta b, d_\oplus \geq \Delta c \\ \Delta a + \Delta b + \Delta c \leq 2 \end{cases} \tag{1}$$

where $d_\oplus$ is a dummy variable.

Constraints for Modular Addition Operation [18]. Assume α, β and γ be n-bit XOR differences, α, β are the input differences for modular addition operation, and γ is the output difference. In [18], if $i = 0$, $\alpha[i] \oplus \beta[i] = \gamma[i]$, the constraints are shown in formula (1), if $i \in [1, n-1]$, Fu *et al.* proposed 13 inequalities in formula (2) to express it.

$$\begin{cases} \beta[i] - \gamma[i] + T(\alpha[i], \beta[i], \gamma[i]) \geq 0 \\ \alpha[i] - \beta[i] + T(\alpha[i], \beta[i], \gamma[i]) \geq 0 \\ -\alpha[i] + \gamma[i] + T(\alpha[i], \beta[i], \gamma[i]) \geq 0 \\ -\alpha[i] - \beta[i] - \gamma[i] - T(\alpha[i], \beta[i], \gamma[i]) \geq -3 \\ \alpha[i] + \beta[i] + \gamma[i] - T(\alpha[i], \beta[i], \gamma[i]) \geq 0 \\ -\beta[i] + \alpha[i+1] + \beta[i+1] + \gamma[i+1] + T(\alpha[i], \beta[i], \gamma[i]) \geq 0 \\ \beta[i] + \alpha[i+1] - \beta[i+1] + \gamma[i+1] + T(\alpha[i], \beta[i], \gamma[i]) \geq 0 \\ \beta[i] - \alpha[i+1] + \beta[i+1] + \gamma[i+1] + T(\alpha[i], \beta[i], \gamma[i]) \geq 0 \\ \beta[i] + \alpha[i+1] + \beta[i+1] - \gamma[i+1] + T(\alpha[i], \beta[i], \gamma[i]) \geq 0 \\ \gamma[i] - \alpha[i+1] - \beta[i+1] - \gamma[i+1] + T(\alpha[i], \beta[i], \gamma[i]) \geq -2 \\ -\beta[i] + \alpha[i+1] - \beta[i+1] - \gamma[i+1] + T(\alpha[i], \beta[i], \gamma[i]) \geq -2 \\ -\beta[i] - \alpha[i+1] + \beta[i+1] - \gamma[i+1] + T(\alpha[i], \beta[i], \gamma[i]) \geq -2 \\ -\beta[i] - \alpha[i+1] - \beta[i+1] + \gamma[i+1] + T(\alpha[i], \beta[i], \gamma[i]) \geq -2 \end{cases} \quad (2)$$

When $\alpha[i] = \beta[i] = \gamma[i]$, $T(\alpha[i], \beta[i], \gamma[i]) = 1$; otherwise, $T(\alpha[i], \beta[i], \gamma[i]) = 0$. the differential probability xdp^+ is calculated as follows:

$$xdp^+ = 2^{-\sum_{i=0}^{n-2} T(\alpha[i], \beta[i], \gamma[i])}$$

Fu *et al.* set the objective function for r-round differential MILP model as the $\sum_{j=0}^{r} \sum_{i=0}^{n-2} T(\alpha[i], \beta[i], \gamma[i])$.

3.3 Linear Constraints for Different Operations

In order to automatically search the linear trial of HIGHT, we must consider the propagation of the linear masks. Notice that the rotations is a simple bit permutation, we can give the corresponding linear constraints. Next, the constraints for the following operations can be given.

- XOR
- Branching
- Modular addition

Constraints for XOR Operation [34]. For XOR operation with input masks a, b and output mask c, include the following constraints:

$$a = b = c$$

Constraints for Branching Operation [34]. For branching operation with input mask a and output masks b, c, include the following constraints:

$$a \oplus b \oplus c = 0$$

Constraints for Modular Addition Operation [18]. Let modular addition operation with input masks $\wedge_\alpha, \wedge_\beta \in F_2^n$ and output mask $\Gamma \in F_2^n$. and

$\wedge_\alpha = (\wedge_\alpha[n-1], \cdots, \wedge_\alpha[0]), \wedge_\beta = (\wedge_\beta[n-1], \cdots, \wedge_\beta[0]), \Gamma = (\Gamma[n-1], \cdots, \Gamma[0])$. In [18], Fu et $al.$ utilize 8 linear inequalities to describe the possible transitions, as shown in formula (3).

$$
\begin{cases}
s_{i+1} - \Gamma[i] - \Lambda_\alpha[i] + \Lambda_\beta[i] + s_i \geq 0 \\
s_{i+1} + \Gamma[i] + \Lambda_\alpha[i] - \Lambda_\beta[i] - s_i \geq 0 \\
s_{i+1} + \Gamma[i] - \Lambda_\alpha[i] - \Lambda_\beta[i] + s_i \geq 0 \\
s_{i+1} - \Gamma[i] + \Lambda_\alpha[i] - \Lambda_\beta[i] + s_i \geq 0 \\
s_{i+1} + \Gamma[i] - \Lambda_\alpha[i] + \Lambda_\beta[i] - s_i \geq 0 \\
s_{i+1} - \Gamma[i] + \Lambda_\alpha[i] + \Lambda_\beta[i] - s_i \geq 0 \\
-s_{i+1} + \Gamma[i] + \Lambda_\alpha[i] + \Lambda_\beta[i] + s_i \geq 0 \\
s_{i+1} + \Gamma[i] + \Lambda_\alpha[i] + \Lambda_\beta[i] + s_i \leq 4
\end{cases}
\tag{3}
$$

The correlation of addition modulo 2^n ($cor_\boxplus$) can be computed as follows: $|cor_\boxplus(\Gamma, \wedge_\alpha, \wedge_\beta)| = 2^{-\sum_{i=1}^{n-1} s_i}$. Taking the above observation into account, Fu et $al.$ set the objective function for r-round linear MILP model as the $\sum_{j=1}^{r} \sum_{i=1}^{n-1} s_i$.

For more details, please refer to [18,35].

4   The Refined MILP Model and Application to HIGHT

In this section, we present our refined MILP model, and then we apply the refined MILP model to the lightweight block cipher HIGHT.

4.1   The Refined MILP Model

It is observed that the number of variables in the MILP model will affect the efficiency of the solver. By analyzing the differential propagation of XOR operation in detail, in Eurocrypt 2017, Sasaki et $al.$ gave the following constraints to model XOR operation.[3]

$$
\begin{cases}
\Delta a + \Delta b + \Delta c \leq 2 \\
\Delta a + \Delta b \geq \Delta c \\
\Delta a + \Delta c \geq \Delta b \\
\Delta b + \Delta c \geq \Delta a
\end{cases}
\tag{4}
$$

where Δa and Δb are the input differences of the XOR operation, and Δc is the output difference.

In the previous work, we obtain the five constraint inequalities by computing the H-representation of the convex hull for the four possible differential propagation modes. Now, we obtain the same four constraint inequalitys with the

---

[3] The constraints appear in the slide that Sasaki et $al.$ were reported in Eurocrypt 2017, please refer to https://eurocrypt2017.di.ens.fr/slides/A09-new-impossible-differential.pdf

greedy algorithm [17]. Compared with/the constraints given in the formula (1), the formula (4) not only introduces no new dummy variables, but also reduces one constraint, which can reduce 16 variables and constraints in just one round HIGHT. Similarly, in the modular addition and the branching operations, the XOR constraints also reduce a part of variables and constraints.

Next, we focus on the functions F_0 and F_1 for HIGHT.

$$F_0(x) = (x <<< 1) \oplus (x <<< 2) \oplus (x <<< 7)$$
$$F_1(x) = (x <<< 3) \oplus (x <<< 4) \oplus (x <<< 6)$$

For the 2-XOR operations in each function, in accordance with the above formula (4), we need to introduce an intermediate variable to generate the constraints, in this case, we convert constrained F_0 and F_1 to the following question.

Let $\Delta a \oplus \Delta b \oplus \Delta c = \Delta d$, where Δa, Δb and Δc are the input differences, and Δd is the output difference. The differential propagation of the 2-XOR operations is shown in Fig. 2.

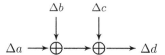

Fig. 2. The differential propagation of the 2-XOR operations.

By analyzing these possible differential patterns in detail, We give the constraints as shown in formula (5). These constraints can be clearly seen from Table 2, we can know the 8 constraints are chosen to describe the possible difference patterns for the 2 XOR operations. In Table 2, the constraint $\Delta a + \Delta b + \Delta c - \Delta d \geq 0$ can remove the impossible differential propagation mode $(0, 0, 0, 1)$. The eight constraint inequalities in formula (5) satisfy all possible input-output differential modes, and also exclude all impossible input-output differential modes.

$$\begin{cases} \Delta a + \Delta b + \Delta c - \Delta d \geq 0 \\ \Delta a + \Delta b + \Delta c - \Delta d \leq 2 \\ \Delta a + \Delta b + \Delta d - \Delta c \geq 0 \\ \Delta a + \Delta b + \Delta d - \Delta c \leq 2 \\ \Delta a + \Delta c + \Delta d - \Delta b \geq 0 \\ \Delta a + \Delta c + \Delta d - \Delta b \leq 2 \\ \Delta b + \Delta c + \Delta d - \Delta a \geq 0 \\ \Delta b + \Delta c + \Delta d - \Delta a \leq 2 \end{cases} \qquad (5)$$

By comparing the introduction of the intermediate variable with the constraint given by the formula (4), we can reduce 8 variables in each function by the formula (5). When solving the MILP model with Gurobi, it can save computing time and improve the solving efficiency through the above constraints.

Table 2. Remove all impossible differential propagations for the 2-XOR operations

Δa	Δb	Δc	Δd	Impossible
0	0	0	0	
0	0	0	1	✓ $\Delta a + \Delta b + \Delta c - \Delta d \geq 0$
0	0	1	0	✓ $\Delta a + \Delta b + \Delta d - \Delta c \geq 0$
0	0	1	1	
0	1	0	0	✓ $\Delta a + \Delta c + \Delta d - \Delta b \geq 0$
0	1	0	1	
0	1	1	0	
0	1	1	1	✓ $\Delta b + \Delta c + \Delta d - \Delta a \leq 2$
1	0	0	0	✓ $\Delta b + \Delta c + \Delta d - \Delta a \geq 0$
1	0	0	1	
1	0	1	0	
1	0	1	1	✓ $\Delta a + \Delta c + \Delta d - \Delta b \leq 2$
1	1	0	0	
1	1	0	1	✓ $\Delta a + \Delta b + \Delta d - \Delta c \leq 2$
1	1	1	0	✓ $\Delta a + \Delta b + \Delta c - \Delta d \leq 2$
1	1	1	1	

4.2   Construct the Refined MILP Model for HIGHT

According to the refined constraints given by the XOR and 2-XOR operations, we apply these refined constraints to the MILP model of the HIGHT. According to Sun et al. MILP-based automatic search technology and constraints for modular addition operation, it is easy to construct a refined MILP model for HIGHT.

Without loss of generality, we just give the method of constructing the differential MILP model for 1-round HIGHT in detail. The differential and linear MILP model for r-round HIGHT can be constructed in the similar way.

We assume that the new value introduced in modular addition operation $T(\alpha[i], \beta[i], \gamma[i])$ is denoted by $T_1^{(i)}$, where $i \in [0, 6]$. As there are 4 modular addition operations in a round of HIGHT, the number of the new values $T(\alpha[i], \beta[i], \gamma[i])$ is 28 in total, which is denoted by $(T_1^{(0)}, T_1^{(1)}, \cdots, T_1^{(27)})$. Then the sum of the values: $T_1^{(0)} + T_1^{(1)} + \cdots + T_1^{(27)}$ is chosen as the objective function to be minimized.

Assume that the 64-bit input difference for the 1-round HIGHT is denoted by $\Delta X = (\Delta X_0, \cdots, \Delta X_i, \cdots, \Delta X_7)$, where ΔX_i is an 8-bit differential variable, $\Delta X_i = (\Delta X_i^{(0)}, \Delta X_i^{(1)} \cdots, \Delta X_i^{(7)})$. According to the round function of HIGHT, the 1-8, 17-24, 33-40, 49-56 bits position of the outputs are the same as the corresponding differential positions of the inputs for 1-round HIGHT. So, the 64-bit output difference for the 1-round HIGHT is denoted by $\Delta Y = (\Delta X_1, \Delta Y_0, \Delta X_3, \Delta Y_1, \Delta X_5, \Delta Y_2, \Delta X_7, \Delta Y_3)$, where ΔY_i is an 8-bit differential variable, $\Delta Y_i = (\Delta Y_i^{(0)}, \Delta Y_i^{(1)} \cdots, \Delta Y_i^{(7)})$. At the same time, the output difference between the four F-functions from left to right is $\Delta Z = (\Delta Z_0, \Delta Z_1)$, ΔZ_i is an 8-bit differential variable, and $\Delta Z_i = (\Delta Z_i^{(0)}, \Delta Z_i^{(1)} \cdots, \Delta Z_i^{(7)})$. The difference between the output of the first and third addition modulo operation from left to right is $\Delta M = (\Delta M_0, \Delta M_1)$, where ΔM_i is an 8-bit differential variable, then $\Delta M_i = (\Delta M_i^0, \Delta M_i^{(1)}, \cdots, \Delta M_i^{(7)})$. These differential variables are shown in Fig. 3.

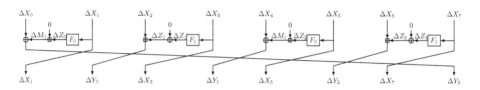

Fig. 3. Differential variable values for the 1-round MILP model of HIGHT.

Firstly, to make sure the non-zero input difference, we should add the constraint:

$$\Delta X_0^{(0)} + \Delta X_0^{(1)} + \cdots + \Delta X_0^{(63)} \geq 1$$

In the first generalized feistel structure, the input difference of the F-function is $(\Delta X_1^{(0)}, \Delta X_1^{(1)}, \cdots, \Delta X_1^{(7)})$, and the output difference is $(\Delta Z_0^{(0)}, \Delta Z_0^{(1)}, \cdots, \Delta Z_0^{(7)})$. When $j = 0$, according to formula (5), we can get the constraints as follows:

$$
\begin{cases}
\Delta X_1^{(1)} + \Delta X_1^{(2)} + \Delta X_1^{(7)} - \Delta Z_0^{(0)} \geq 0 \\
\Delta X_1^{(1)} + \Delta X_1^{(2)} + \Delta X_1^{(7)} - \Delta Z_0^{(0)} \leq 2 \\
\Delta X_1^{(1)} + \Delta X_1^{(2)} + \Delta Z_0^{(0)} - \Delta X_1^{(7)} \geq 0 \\
\Delta X_1^{(1)} + \Delta X_1^{(2)} + \Delta Z_0^{(0)} - \Delta X_1^{(7)} \leq 2 \\
\Delta X_1^{(1)} + \Delta X_1^{(7)} + \Delta Z_0^{(0)} - \Delta X_1^{(2)} \geq 0 \\
\Delta X_1^{(1)} + \Delta X_1^{(7)} + \Delta Z_0^{(0)} - \Delta X_1^{(2)} \leq 2 \\
\Delta X_1^{(2)} + \Delta X_1^{(7)} + \Delta Z_0^{(0)} - \Delta X_1^{(1)} \geq 0 \\
\Delta X_1^{(2)} + \Delta X_1^{(7)} + \Delta Z_0^{(0)} - \Delta X_1^{(1)} \leq 2
\end{cases}
\tag{6}
$$

when $j \geq 1$, the constraints can be obtained with the same method. So $8 \times 32 = 256$ linear constraints are proposed to describe the differential property for the 2 XOR operations in the 1-round HIGHT.

The constraints of the modular addition operations are introduced in the following text. In the first generalized feistel structure, the two input differences are $(\Delta Z_0^{(0)}, \Delta Z_0^{(1)}, \cdots, \Delta Z_0^{(7)})$ and 0, the output difference is $(\Delta M_0^{(0)}, \Delta M_0^{(1)}, \cdots, \Delta M_0^{(7)})$. When $j = 7$, the constraint is $\Delta Z_0^{(0)} - \Delta M_0^{(0)} = 0$, obviously. According to formula (2), when $j = 0$, the constraints are shown in formula (7).

$$\begin{cases} -\Delta M_0^{(1)} + T_0^{(0)} \geq 0 \\ \Delta Z_0^{(1)} + T_0^{(0)} \geq 0 \\ \Delta M_0^{(1)} - \Delta Z_0^{(1)} + T_0^{(0)} \geq 0 \\ -\Delta Z_0^{(1)} + \Delta M_0^{(1)} + T_0^{(0)} \leq 3 \\ -\Delta Z_0^{(1)} + \Delta M_0^{(1)} - T_0^{(0)} \geq 0 \\ \Delta Z_0^{(1)} + \Delta M_0^{(0)} + T_0^{(0)} \geq 0 \\ \Delta Z_0^{(0)} + \Delta M_0^{(0)} \geq 0 \\ -\Delta Z_0^{(0)} + \Delta M_0^{(0)} + T_0^{(0)} \geq 0 \\ \Delta Z_0^{(1)} + \Delta Z_0^{(0)} - \Delta M_0^{(0)} + T_0^{(0)} \geq 0 \\ \Delta M_0^{(1)} + \Delta Z_0^{(0)} - \Delta M_0^{(0)} + T_0^{(0)} \geq -2 \\ \Delta Z_0^{(0)} - \Delta M_0^{(0)} + T_0^{(0)} \geq -2 \\ -\Delta Z_0^{(0)} - \Delta M_0^{(0)} + T_0^{(0)} \geq -2 \\ \Delta M_0^{(0)} - \Delta Z_0^{(0)} + T_0^{(0)} \geq -2 \end{cases} \tag{7}$$

The constraints when $1 \leq j \leq 6$ can be calculated in the similar way. Therefore, we can produce $2 \times (7 \times 13 + 1) + 2 \times (7 \times 13 + 4) = 374$ constraints to represent the differential property for modular addition in the 1-round HIGHT.

Finally, we focus on the XOR operations, whose input difference are $(\Delta X_0^{(0)}, \Delta X_0^{(1)}, \cdots, \Delta X_0^{(7)})$ and $(\Delta M_0^{(0)}, \Delta M_0^{(1)}, \cdots, \Delta M_0^{(7)})$, and the output difference is $(\Delta Y_3^{(0)}, \Delta Y_3^{(1)}, \cdots, \Delta Y_3^{(7)})$. When $j = 0$, the constraints are shown in formula (8).

$$\begin{cases} \Delta X_0^{(0)} + \Delta M_0^{(0)} - \Delta Y_3^{(0)} \geq 0 \\ \Delta X_0^{(0)} + \Delta Y_3^{(0)} - \Delta M_0^{(0)} \geq 0 \\ \Delta Y_3^{(0)} + \Delta M_0^{(0)} - \Delta X_0^{(0)} \geq 0 \\ \Delta X_0^{(0)} + \Delta M_0^{(0)} + \Delta Y_3^{(0)} \leq 2 \end{cases} \tag{8}$$

The constraints when $j \geq 1$ can be calculated in the same way. Thus, we have $2 \times 4 \times 8 = 64$ constraints for XOR operation in the 1-round HIGHT. So far, we construct a complete MILP model for 1-round HIGHT. In total, we have $256 + 374 + 64 + 1 = 695$ constraints to exactly describe the difference $\Delta X \to \Delta Y$.

4.3   Comparison of Constraints and Variables in the MILP Model

In order to distinguish these two types of MILP models and also for the convenience of our statement in this paper, the MILP model without adding new constraints is named as the original MILP model, and the MILP model with new constraints is named as the refined MILP model.

We establish the differential MILP model for r-round HIGHT. The original MILP model have $776r + 2$ constraints and $214r + 65$ variables for the r-rounds HIGHT, but the refined MILP model for r-round HIGHT only needs $694r + 2$

constraints and $108r + 65$ variables, Figs. 4 and 5 show the number of constraints and variables in the original and the refined differential MILP model for the first 12-round HIGHT, respectively.

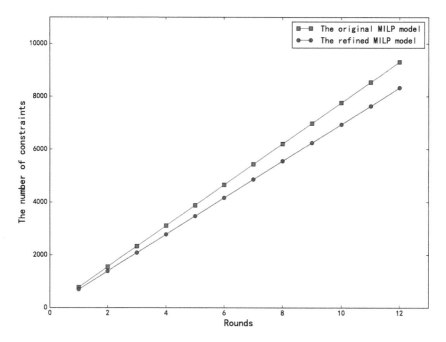

Fig. 4. Comparison of the number of constraints in the original and refined differential MILP model for the first 12-round.

As you can see in Fig. 4, the original model has 7762 constraints for the 10-round HIGHT, while the refined model has only 6942 constraints, in comparison, 820 constraint inequalities are reduced. In Fig. 5, the original model has 2205 variables for the 10-round HIGHT, while the refined model has only 1145 variables, 1060 variables are reduced.

We establish the linear MILP model for r-round HIGHT. The original MILP model have $736r + 2$ constraints and $288r + 65$ variables for the r-rounds HIGHT, but the refined MILP model for r-round HIGHT only needs $640r + 2$ constraints and $192r + 65$ variables, Figs. 6 and 7 show the number of constraints and variables in the original and the refined linear MILP model for the first 12-round HIGHT, respectively.

As you can see in Fig. 6, the original model has 7362 constraints for the 10-round HIGHT, while the refined model has only 6402 constraints, in comparison, 960 constraint inequalities are reduced. In Fig. 7, the original model has 2945 variables for the 10-round HIGHT, while the refined model has only 1985 variables, 960 variables are reduced.

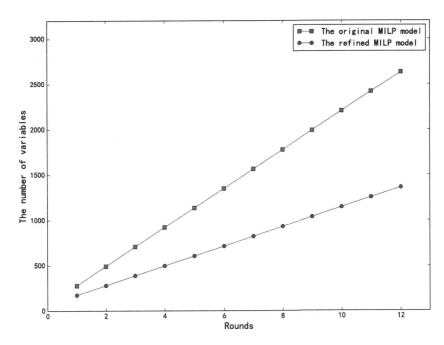

Fig. 5. Comparison of the number of variables in the original and refined differential MILP model for the first 12-round.

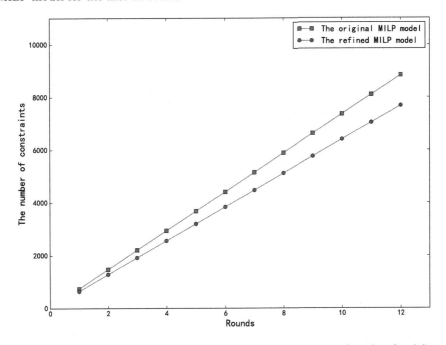

Fig. 6. Comparison of the number of constraints in the original and refined linear MILP model for the first 12-round.

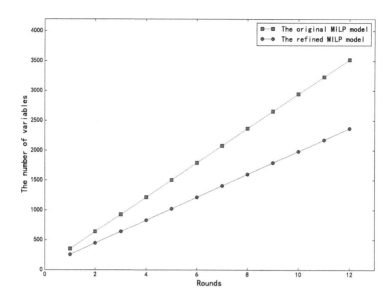

Fig. 7. Comparison of the number of variables in the original and refined linear MILP model for the first 12-round.

5   The Differential and Linear Cryptanalysis for HIGHT

Based on the refined MILP model, according to Sect. 4.2, we generate the differential and linear MILP models in "lp" format [14] for HIGHT through a small python program, and call Gurobi 7.0.2 to solve it. There are $108r + 65$ variables and $694r + 1$ constraints for r-round HIGHT in differential MILP model, and there are $192r + 65$ variables and $640r + 2$ constraints for r-round HIGHT in linear MILP model. The MILP model was solved on a server, the server configuration is shown in Table 3.

Table 3. Experimental environment for solving the MILP model

Item	Configuration
CPU	Intel Xeon E7-4820 v2
RAM	512 GB
OS	Windows Server 2008 R2 Enterprise
Software	Python3.5, Gurobi7.0.2

5.1   The Differential Cryptanalysis for HIGHT

By solving the refined differential MILP model, the probability of differential characteristic for reduced-round HIGHT obtained by our refined MILP model is listed in Table 4, the p_r denotes the probability of differential characteristic for the r-round HIGHT.

Table 4. The differential cryptanalysis results for refined MILP model application to HIGHT

Rounds	#variable	#constraint	$\log_2 p_r$	Timing(s)
1	173	695	0	1
2	281	1389	0	1
3	389	2083	-3	15
4	497	2777	-8	130
5	605	3471	-11	636
6	713	4165	-15	8362
7	821	4859	-19	18456
8	929	5553	-25	41251
9	1037	6247	-30	125565
10	1145	6941	-38	489785
11	1253	7635	-45	1012556
12	1361	8329	-53	1801255
13	1469	9023	-61	2518256
14	1577	9717	<-64	-

From Table 4 we know that the refined MILP model for the 11-round HIGTH consists of 1253 0-1 variables, 7635 constraints. The MILP model can be solved within 1012556 s and we find the better probability of differential characteristic for 11-round HIGHT is 2^{-45}. Note that the probability of the best single-key characteristic previously published covering 11-round is 2^{-58}. Furthermore, using the refined tool, we obtain the new single-key differential characteristics for HIGHT, which cover larger number of rounds. We obtain the 12- and 13-round single-key differential characteristics of HIGHT with probability 2^{-53} and 2^{-61}. For the 14-round HIGHT, the optimal probability for differential characteristic is less than 2^{-64}. The probability of success for an exhaustive search, thus, we concluded that the all-round HIGHT has a sufficient resistance to differential attacks.

Finally, the differential trails for 12 and 13-round HIGHT are listed in Tables 5 and 6, respectively.

In order to clarify that the refined MILP model can solve more efficiently, we establish the MILP model for the first 9-round HIGHT, and solve the MILP model in the same experimental environment. In less than 10 days, the original differential MILP model of the first 9-round HIGHT was solved, and the optimal differential probability is the same as Table 4. But the time expendure of round 1 to 9 is 1 s, 1 s, 123 s, 568 s, 2135 s, 21268 s, 48392 s, 124536 s and 671258 s, respectively. Figure 8 shows a comparison of the solve time between the original and the refined differential MILP model for the first 9-round HIGHT.

Table 5. Differential trail for 12-round HIGHT

Rounds	Difference	$\log_2 p_r$
0	00008227213AEA01	-0
1	000027A03A4E0100	-6
2	0000A0B84E010000	-6
3	0000B8C801000000	-4
4	0000C80100000000	-4
5	0000010000000000	-3
6	0001000000000000	-1
7	0100000000000082	-2
8	00000000009C8201	-3
9	000000039C7A0100	-8
10	00E803BC7A010000	-5
11	E800BCF801000002	-6
12	00B6F801009002E8	-5
Probability	2^{-53}	

Table 6. Differential trail for 13-round HIGHT

Rounds	Difference	$\log_2 p_r$
0	80008AC28A01A0BB	-0
1	0000C2080128BB80	-6
2	000008E528E98000	-10
3	0000E5A8E9800000	-1
4	0000A82C80000000	-8
5	00002C8000000000	-2
6	0000800000000000	-3
7	0080000000000000	-0
8	80000000000000C3	-3
9	000000000072C380	-4
10	0000000C72E98000	-9
11	00A70CF2E9800000	-4
12	A700F22A80000002	-6
13	007B2A80009002A7	-5
Probability	2^{-61}	

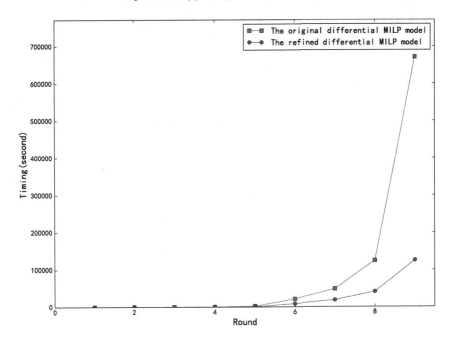

Fig. 8. Comparison of the solve time in the original and refined differential MILP model for the first 9-round HIGHT.

From Fig. 8 we know that the original differential MILP model can be solved within 671258 s for the 9-round HIGHT. Nevertheless, the refined differential MILP model just needed 125565 s. Comparatively speaking, the solve time of the refined MILP model is 5 times faster than the original MILP model.

5.2   The Linear Cryptanalysis for HIGHT

By solving the refined linear MILP model, the correlation of linear approxima- tion for reduced-round HIGHT obtained by our refined MILP model is listed in Table 7, the cor_r denotes the correlation of linear approximation for the r-round HIGHT.

For linear attack, from Table 7 we know that linear approximation for the 10-round HIGHT with correlation 2^{-25}, the correlation of the best linear approx- imation previously published covering 10-round is 2^{-26}. Moreover, we obtain the new linear approximation for 11-round HIGHT with correlation 2^{-31}, the max- imum linear bias is $\varepsilon^2 = 2^{-64}$, then the linear attack on the 11-round HIGHT require 2^{64} known plaintext, but the all-round HIGHT require plaintext is cer- tainly greater than 2^{64}. Therefore, it can be concluded that all-round HIGHT are sufficiently resistant to linear attack. Finally, the linear trail for 11-round HIGHT is listed in Table 8.

In order to clarify that the refined MILP model can solve more efficiently, we establish the MILP model for the first 8-round HIGHT, and solve the MILP

Table 7. The linear cryptanalysis results for refined MILP model application to HIGHT

Rounds	#variable	#constraint	$\log_2 cor_r$	Timing(s)
1	257	642	0	1
2	449	1282	-1	1
3	641	1922	-2	30
4	833	2562	-4	100
5	1025	3202	-6	207
6	1217	3842	-9	970
7	1409	4482	-12	8216
8	1601	5122	-16	165348
9	1793	5762	-22	464156
10	1985	6402	-25	986423
11	2177	7042	-31	1865719
12	2369	7682	-	>30 days

Table 8. Linear trail for 11-round HIGHT

Rounds	Mask	$\log_2 cor_r$
0	3863C24B000001C2	-0
1	01C200000001D638	-5
2	0000000001F65201	-4
3	0000000134530000	-4
4	0000013400000000	-2
5	0001000000000000	-1
6	0100000000000000	-0
7	C200000000000001	-1
8	0C000000000001C2	-3
9	160000000001F60C	-2
10	6000000001A44016	-6
11	B000000166601A60	-3
Correlation	2^{-31}	

model in the same experimental environment. In less than 10 days, the original linear MILP model of the first 8-round HIGHT was solved, and the optimal linear correlation is the same as Table 7. But the time expendure of round 1 to 8 is 1 s, 157 s, 882 s, 5029 s, 18653 s, 79523 s and 4264131 s, respectively. Figure 9 shows a comparison of the solve time between the original and the refined linear MILP model for the first 8-round HIGHT.

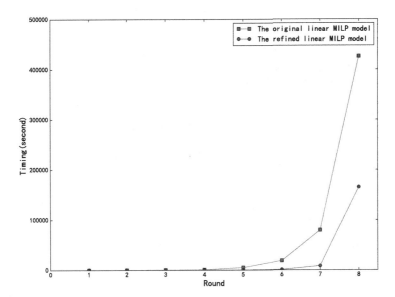

Fig. 9. Comparison of the solve time in the original and refined linear MILP model for the first 8-round.

From Fig. 9 we know that the original linear MILP model can be solved within 426413 s for the 8-round HIGHT. Nevertheless, the refined differential MILP model just needed 165348 s. By contrast, the solve time of the refined MILP model is 2.5 times faster than the original MILP model.

6   Conclusion

In this paper, we analyze the differential propagation for the 2-XOR operations in detail, and improve Sun *et al.*'s method for describing XOR operation with refined constraints. In the refined MILP model, the number of variables and constraints are reduced, which leads to quickly solve for the refined MILP model.

As an application, we implement our refined MILP model to the lightweight block cipher HIGHT. Compared with the existing attack results, the refined MILP model searches the optimal differential characteristic and linear approximation for HIGHT, the differential and linear trails increased to 13-round and 11-round. These results indicate that the refined MILP model is more efficient in practical cryptanalysis.

Acknowledgements. The authors would like to thank the anonymous reviewers for their helpful comments and suggestions. The work of this paper was supported by the National Natural Science Foundation of China (61772519, 61502532, 61379150, 61309016, 61502529), the Open Foundation of the Key State Key Laboratory of Mathematical Engineering and Advanced Computing (2016A02), and the State Key Laboratory of Information Security. The work of Jun Yin and Lijun Lyu is supported by the Youth Innovation Promotion Association of Chinese Academy of Sciences.

References

1. Wu, W., Zhang, L.: LBlock: a lightweight block cipher. In: Lopez, J., Tsudik, G. (eds.) ACNS 2011. LNCS, vol. 6715, pp. 327–344. Springer, Heidelberg (2011). https://doi.org/10.1007/978-3-642-21554-4_19
2. Bogdanov, A., Knudsen, L.R., Leander, G., Paar, C., Poschmann, A., Robshaw, M.J.B., Seurin, Y., Vikkelsoe, C.: PRESENT: an ultra-lightweight block cipher. In: Paillier, P., Verbauwhede, I. (eds.) CHES 2007. LNCS, vol. 4727, pp. 450–466. Springer, Heidelberg (2007). https://doi.org/10.1007/978-3-540-74735-2_31
3. Beierle, C., et al.: The SKINNY family of block ciphers and its low-latency variant MANTIS. In: Robshaw, M., Katz, J. (eds.) CRYPTO 2016. LNCS, vol. 9815, pp. 123–153. Springer, Heidelberg (2016). https://doi.org/10.1007/978-3-662-53008-5_5
4. Zhang, W., Bao, Z., Lin, D., Rijmen, V., Yang, B., Verbauwhede, I.: Rectangle: a bit-slice lightweight block cipher suitable for multiple platforms. Sci. Chin. Inf. Sci. 58(12), 1–15 (2015). https://doi.org/10.1007/s11432-015-5459-7
5. Hong, D., et al.: HIGHT: a new block cipher suitable for low-resource device. In: Goubin, L., Matsui, M. (eds.) CHES 2006. LNCS, vol. 4249, pp. 46–59. Springer, Heidelberg (2006). https://doi.org/10.1007/11894063_4
6. Wheeler, D.J., Needham, R.M.: TEA, a tiny encryption algorithm. In: Preneel, B. (ed.) FSE 1994. LNCS, vol. 1008, pp. 363–366. Springer, Heidelberg (1995). https://doi.org/10.1007/3-540-60590-8_29
7. Beaulieu, R., Shors, D., Smith, J., Treatman-Clark, S., Weeks, B., Wingers, L.: The simon and speck families of lightweight block ciphers. Cryptology ePrint Archive, Report 2013/404 (2013). http://eprint.iacr.org/2013/404
8. Dinu, D., Perrin, L., Udovenko, A., Velichkov, V., Großschädl, J., Biryukov, A.: Design strategies for ARX with provable bounds: SPARX and LAX. In: Cheon, J.H., Takagi, T. (eds.) ASIACRYPT 2016. LNCS, vol. 10031, pp. 484–513. Springer, Heidelberg (2016). https://doi.org/10.1007/978-3-662-53887-6_18
9. Biham, E., Shamir, A.: Differential cryptanalysis of DES-like cryptosystems. In: Menezes, A.J., Vanstone, S.A. (eds.) CRYPTO 1990. LNCS, vol. 537, pp. 2–21. Springer, Heidelberg (1991). https://doi.org/10.1007/3-540-38424-3_1
10. Matsui, M.: Linear cryptanalysis method for DES cipher. In: Helleseth, T. (ed.) EUROCRYPT 1993. LNCS, vol. 765, pp. 386–397. Springer, Heidelberg (1994). https://doi.org/10.1007/3-540-48285-7_33
11. Matsui, M.: On correlation between the order of S-boxes and the strength of DES. In: De Santis, A. (ed.) EUROCRYPT 1994. LNCS, vol. 950, pp. 366–375. Springer, Heidelberg (1995). https://doi.org/10.1007/BFb0053451
12. Chen, J., Miyaji, A., Su, C., Teh, J.S.: Accurate estimation of the full differential distribution for general feistel structures. In: Lin, D., Wang, X.F., Yung, M. (eds.) Inscrypt 2015. LNCS, vol. 9589, pp. 108–124. Springer, Cham (2016). https://doi.org/10.1007/978-3-319-38898-4_7
13. Chen, J., Miyaji, A., Su, C., Teh, J.: Improved differential characteristic searching methods. In: IEEE 2nd International Conference on Cyber Security and Cloud Computing, CSCloud 2015, New York, NY, USA, 3–5 November 2015, pp. 500–508 (2015). https://doi.org/10.1109/CSCloud.2015.42
14. Biryukov, A., Velichkov, V.: Automatic search for differential trails in ARX ciphers. In: Benaloh, J. (ed.) CT-RSA 2014. LNCS, vol. 8366, pp. 227–250. Springer, Cham (2014). https://doi.org/10.1007/978-3-319-04852-9_12

15. Mouha, N., Preneel, B.: Towards finding optimal differential characteristics for arx: Application to salsa20. Cryptology ePrint Archive, Report 2013/328 (2013). http://eprint.iacr.org/2013/328

16. Mouha, N., Wang, Q., Gu, D., Preneel, B.: Differential and linear cryptanalysis using mixed-integer linear programming. In: Wu, C.-K., Yung, M., Lin, D. (eds.) Inscrypt 2011. LNCS, vol. 7537, pp. 57–76. Springer, Heidelberg (2012). https://doi.org/10.1007/978-3-642-34704-7_5

17. Sun, S., Hu, L., Wang, P., Qiao, K., Ma, X., Song, L.: Automatic security evaluation and (related-key) differential characteristic search: application to SIMON, PRESENT, LBlock, DES(L) and other bit-oriented block ciphers. In: Sarkar, P., Iwata, T. (eds.) ASIACRYPT 2014. LNCS, vol. 8873, pp. 158–178. Springer, Heidelberg (2014). https://doi.org/10.1007/978-3-662-45611-8_9

18. Fu, K., Wang, M., Guo, Y., Sun, S., Hu, L.: MILP-based automatic search algorithms for differential and linear trails for speck. In: Peyrin, T. (ed.) FSE 2016. LNCS, vol. 9783, pp. 268–288. Springer, Heidelberg (2016). https://doi.org/10.1007/978-3-662-52993-5_14

19. Sun, S., Gerault, D., Lafourcade, P., Yang, Q., Todo, Y., Qiao, K., Hu, L.: Analysis of aes, skinny, and others with constraint programming. IACR Trans. Symmetric Cryptol. **2017**(1), 281–306 (2017). https://doi.org/10.13154/tosc.v2017.i1.281-306

20. International Organization for Standardization. ISO/IEC 18033-3: 2010. Information technology Security techniques Encryption algorithms Part 3: Block ciphers (2010)

21. Knudsen, L.R.: Truncated and higher order differentials. In: Preneel, B. (ed.) FSE 1994. LNCS, vol. 1008, pp. 196–211. Springer, Heidelberg (1995). https://doi.org/10.1007/3-540-60590-8_16

22. Wagner, D.: The boomerang attack. In: Knudsen, L. (ed.) FSE 1999. LNCS, vol. 1636, pp. 156–170. Springer, Heidelberg (1999). https://doi.org/10.1007/3-540-48519-8_12

23. Biryukov, A., Wagner, D.: Slide attacks. In: Knudsen, L. (ed.) FSE 1999. LNCS, vol. 1636, pp. 245–259. Springer, Heidelberg (1999). https://doi.org/10.1007/3-540-48519-8_18

24. Biham, E.: New types of cryptanalytic attacks using related keys. J. Cryptology **7**(4), 229–246 (1994). https://doi.org/10.1007/BF00203965

25. Lu, J.: Cryptanalysis of reduced versions of the hight block cipher from CHES 2006. In: Nam, K.-H., Rhee, G. (eds.) ICISC 2007. LNCS, vol. 4817, pp. 11–26. Springer, Heidelberg (2007). https://doi.org/10.1007/978-3-540-76788-6_2

26. Özen, O., Varıcı, K., Tezcan, C., Kocair, Ç.: Lightweight block ciphers revisited: cryptanalysis of reduced round PRESENT and HIGHT. In: Boyd, C., González Nieto, J. (eds.) ACISP 2009. LNCS, vol. 5594, pp. 90–107. Springer, Heidelberg (2009). https://doi.org/10.1007/978-3-642-02620-1_7

27. Chen, J., Wang, M., Preneel, B.: Impossible differential cryptanalysis of the lightweight block ciphers TEA, XTEA and HIGHT. In: Mitrokotsa, A., Vaudenay, S. (eds.) AFRICACRYPT 2012. LNCS, vol. 7374, pp. 117–137. Springer, Heidelberg (2012). https://doi.org/10.1007/978-3-642-31410-0_8

28. Cui, T., Jia, K., Fu, K., Chen, S., Wang, M.: New automatic search tool for impossible differentials and zero-correlation linear approximations. Cryptology ePrint Archive, Report 2016/689 (2016). http://eprint.iacr.org/2016/689

29. Koo, B., Hong, D., Kwon, D.: Related-key attack on the full HIGHT. In: Rhee, K.-H., Nyang, D.H. (eds.) ICISC 2010. LNCS, vol. 6829, pp. 49–67. Springer, Heidelberg (2011). https://doi.org/10.1007/978-3-642-24209-0_4

30. Igarashi, Y., Sueyoshi, R., Kaneko, T., Fuchida, T.: Meet-in-the-middle attack with splice-and-cut technique on the 19-round variant of block cipher HIGHT. In: Kim, K.J. (ed.) Information Science and Applications. LNEE, vol. 339, pp. 423–429. Springer, Heidelberg (2015). https://doi.org/10.1007/978-3-662-46578-3_50

31. Gurobi Optimazation, Gurobi optimizer reference manual. http://www.gurobi.com

32. CPLEX, Ibm software group: User-Manual CPLEX 12, https://www-01.ibm.com/software/commerce/optimization/cplex-optimizer/

33. Computational Algebra Group, School of Mathematics and Statistics, University of Sydney: Magma Computational Algebra System, http://magma.maths.usyd.edu.au

34. Sun, S., Hu, L., Wang, M., Wang, P., Qiao, K., Ma, X., Shi, D., Song, L.: Automatic enumeration of (related-key) differential and linear characteristics with predefined properties and its applications. IACR Cryptology ePrint Archive 2014, 747 (2014). http://eprint.iacr.org/2014/747

35. Wallén, J.: Linear approximations of addition modulo 2^n. In: Johansson, T. (ed.) FSE 2003. LNCS, vol. 2887, pp. 261–273. Springer, Heidelberg (2003). https://doi.org/10.1007/978-3-540-39887-5_20

Meet in the Middle Attack on Type-1 Feistel Construction

Yuanhao Deng$^{(\boxtimes)}$, Chenhui Jin, and Rongjia Li

Zhengzhou Information Science and Technology Institute, Zhengzhou 450000, China
515565349@qq.com

Abstract. We provide a key recovery attack on type-1 Feistel construction based on the meet-in-the-middle technique. This construction is described by Zheng, Matsumoto, and Imai in CRYPTO 1989. Type-1 Feistel structure is a well-known construction used to construct ciphers and hash functions, such as CAST-256 and Lesamnta. For Type-1 Feistel construction with n-bit blocks and d sub-blocks, we launch a $3d - 1$ rounds distinguisher by using a special truncated differential. We present an attack on $5d - 3$ rounds with the data complexity $2^{\frac{3}{d}n}$ chosen plaintexts, the memory complexity $2^{\frac{d-1}{d}n}$ blocks, each block is n bits, and the time complexity $2^{\frac{d-1}{d}n}$ encryptions, which is the best known generic key recovery attack on Type-1 Feistel construction. The attack is valid if the key length $k \geq n$.

Keywords: Type-1 Feistel construction · Meet in the middle attack
Key recovery attack · Generic attack

1 Introduction

Feistel construction is proposed by Feistel and Coppersmith [1] from IBM when designing Lucifer in 1973. On the one hand, the procedure of encryption and decryption are similar for this construction, on the other hand, it has few limitations to round function. So this construction is widely used in cipher designs. Many famous ciphers are based on Feistel construction, such as DES, Twofish and Camellia.

Based on Feistel construction, some extend constructions called generic Feistel scheme (GFS) are proposed, such as Type-1, Type-2 and Type-3 Feistel constructions [10], contracting and expanding Feistel constructions [14]. These constructions provide many new methods for designers and a plenty of ciphers are given out, for instance, CAST-256 (type-1), RC-6 (type-2), MARS (type-3). Type-1 and Type-2 Feistel are also respectively used in the construction of the hash functions Lesamnta and SHAvite-3_{512}. Since the widely use of the Feistel construction and GFS, the security of these constructions raised many researchers' interests. In SAC 2012, Sasaki et al. [3] proposed a kind of meet-in-the-middle distinguisher of Feistel construction. In SAC 2015, based on Fouque et al. [12] and Matsui et al. [13]'s work, Lin et al. [4] describe an algorithm for

© Springer International Publishing AG, part of Springer Nature 2018
X. Chen et al. (Eds.): Inscrypt 2017, LNCS 10726, pp. 427–444, 2018.
https://doi.org/10.1007/978-3-319-75160-3_25

searching the improved meet-in-the-middle distinguisher on Feistel construction and GFS by using recursion and greedy algorithm. In CRYPTO 2015, Dinur et al. [5] proposed a new meet-in-the-middle distinguisher for Feistel construction, which improved the former best attack results for Feistel construction. In CRYPTO 2016, Derbez et al. [6] describe an automatic meet-in-the-middle attack tool of Feistel construction.

The best attacking result of Feistel and Feistel-SP constructions, contracting and expanding Feistel constructions are proposed by Guo et al. In AisaCrypt 2014, Guo et al. [7] constructed 5-round meet-in-the-middle distinguisher and recover 6-round key for Feistel construction. As for Feistel-SP construction, they constructed 7-round meet-in-the-middle distinguisher and recover 14-round key. They extend their work for the circumstance that the key length over the block length in 2016 [8]. In FSE 2017 [9], they studied contracting and expanding Feistel constructions. By using the meet-in-the-middle technique and the similar skills from their former work, they give out the best known generic key-recovery attack for contracting and expanding Feistel constructions.

Compared with so many researches on Feistel construction, contracting and expanding Feistel constructions, there are few papers on generic attacks on Type-1, Type-2 and Type-3 Feistel constructions. In CANS 2013, Nachef et al. [16] present the best distinguish attack on Type-1, Type-2 and Type-3 Feistel constructions. However, this attack only distinguishes the scheme from a random permutation. It didn't recovery the key. In CSS 2014, Pudovkina et al. [17] provide upper and lower bounds for the maximum number of the rounds of impossible differential distinguisher on Type-1, Type-2 and Type-3 Feistel constructions. In FSE 2015, Blondeau et al. [15] analyze the relationship between impossible, integral and zero-correlation attacks on Type-2 Feistel construction. As for Type-1 Feistel construction, there is no generic key-recovery attack up till now and it is the aim of this paper.

Our Contribution. In this paper, we present a key-recovery attack on Type-1 Feistel. Our attack is derived from Guo et al. [7–9]'s work. We construct $3d - 1$ rounds distinguisher by using a special truncated differential. We present an attack on $5d - 3$ rounds Type-1 Feistel with the data complexity $2^{\frac{3}{d}n}$ chosen plaintexts, the memory complexity $2^{\frac{d-1}{d}n}$ blocks, each block is n bits, and the time complexity $2^{\frac{d-1}{d}n}$ encryptions, which is the best known generic attack on Type-1 Feistel construction. What's more, our attack is generic and we generalize the block number as d. In former work, the researchers always just considered the case that $d = 4$.

2 Preliminaries

2.1 Notation and Definition

Throughout the paper, we will use the following notation. We use d denote the number of sub-blocks for the Type-1 Feistel structure, n denote the block size, which is equal to the key length k. In the following, we define the Type-1 Feistel Construction:

Definition 1. *Type-1 Feistel Construction with d sub-blocks is the iteration of Fig. 1 for r times. Assume the input of the ith$(1 \leq i \leq r)$ round is $(X_{i,1}, X_{i,2}, \cdots, X_{i,d})$, $X_{i,j} \in \{0,1\}^{n/d}$, we call $X_{i,j}$ is the jth sub-block of the input of ith round. The round function is $F_i : \{0,1\}^{n/d} \rightarrow \{0,1\}^{n/d}$, and round sub-key is $K_i \in \{0,1\}^{n/d}$. The transformation of the ith round g_i is defined as follow:*

$$g_i(X_{i,1}, X_{i,2}, \cdots, X_{i,d}) = (F_i(X_{i,1} \oplus K_i) \oplus X_{i,2}, X_{i,3}, \cdots, X_{i,d}, X_{i,1})$$

Fig. 1. Definition of 1-round Type-1 Feistel

Moreover, we make some assumptions about the round function F_i: assume that the round function of each round is nonlinear and invertible.

In the following context, for the jth sub-block of the ith round, we use $X_{i,j}$ and $Y_{i,j}$ to denote the input and output, $\Delta X_{i,j}$ and $\Delta Y_{i,j}$ denote the input difference and output difference, the input and output of the round function F_i is F_i^I and F_i^O, the input difference and output difference is ΔF_i^I and ΔF_i^O.

Definition 2. *Let F be a $3d - 1$ rounds Type-1 Feistel Construction, defined: $F^\Delta(m, \delta) = Trunc(F(m) \oplus F[m \oplus (0, 0, \cdots, 0, \delta)])$, where $m \in \{0,1\}^n, \delta \in \{0,1\}^{\frac{n}{d}}$, $Trunc(x)$ is the truncation of the second sub-blocks of x.*

2.2 Properties

Property 1. Assume the sub-block numbers of the Feistel construction is d and the input of the round i is $(x_{i,1}, x_{i,2}, \cdots, x_{i,d})$, then

(1) $x_{i+1,1} = F_i(x_{i,1} \oplus K_i) \oplus x_{i,2}$, and we have $x_{i+1,j} = x_{i,(j+1) \bmod d}$ for each $2 \leq j \leq d$;

(2) We have $x_{i+t,j} = x_{i,(j+t) \bmod d}$ for $1 \leq t < d$ and $2 \leq j \leq d+1-t$.

Proof. (1) We can obtain this property from Definition 1.

(2) From (1) we can see that (2) holds when $t = 1$. Assume that (2) holds for $1 \leq t < m \leq d$, we will prove that (2) holds when $t = m$. Assume $2 \leq j \leq d+1-m$. we see that $m \geq 2$, $d+1-m \leq d$ and $2 \leq j \leq d$ since $1 \leq t < m$. So we obtain that $x_{i+m,j} = x_{i+m-1,(j+1) \bmod d}$ from (1). Because $j \leq d+1-m$, we know that $2 \leq j+1 \leq d+2-m = d+1-(m-1)$. then $d+1-(m-1) \leq d+1-(2-1) = d$ Due to $m \geq 2$, thus $(j+1) \bmod m = j+1 \leq d+1-(m-1)$. Moreover,

we find that $x_{i+m-1,(j+1) \bmod d} = x_{i,(j+m) \bmod d}$ by the assumption that (2) holds for $t = m - 1$, which means (2) holds for $t = m$. So we conclude that (2) holds by the means of mathematical induction.   □

Property 2. Assume that the input difference and the output difference of the round i is $(\Delta X_{i,1}, \Delta X_{i,2}, \cdots, \Delta X_{i,d})$ and $(\Delta Y_{i,1}, \Delta Y_{i,2}, \cdots, \Delta Y_{i,d})$ respectively. Thus, we have $\Delta F_i^O = \Delta F_{i-d+1}^I \oplus \Delta F_{i+1}^I$ for $i \geq d$.

Proof. By Property 1, we can deduce that $\Delta X_{i+1,1} = \Delta F_i^O \oplus \Delta X_{i,2}$, hence $\Delta F_i^O = \Delta X_{i,2} \oplus \Delta X_{i+1,1}$. Hence, $\Delta X_{i+t,j} = \Delta X_{i,(j+t) \bmod d}$ for $1 \leq t < d$ and $2 \leq j \leq d+1-t$ by (2) of Property 1. Since $i \geq d$, then $i-(d-1) \geq 1$. Therefore, we have: $\Delta X_{i,2} = \Delta X_{i-(d-1),[2-(d-1)] \bmod d} = \Delta X_{i-d+1,1}$. Moreover, $\Delta X_{i,2} \oplus \Delta X_{i+1,1} = \Delta X_{i-d+1,1} \oplus \Delta X_{i+1,1}$, Because $\Delta X_{i-d+1,1} = \Delta F_{i-d+1}^I, \Delta X_{i+1,1} = \Delta F_{i+1}^I$, then we have $\Delta F_i^O = \Delta F_{i-d+1}^I \oplus \Delta F_{i+1}^I$.   □

Property 3. For arbitrary nonlinear invertible function $F : \{0,1\}^m \rightarrow \{0,1\}^m$, given a set of input and output difference $\{(\Delta I_i, \Delta O_j)\}$. Denote $\Omega_F(\Delta I_i \rightarrow \Delta O_j) = \{X : F(X \oplus \Delta I_i) \oplus F(X) = \Delta O_j\}$. Then, on average the number of solutions of equations $F(X \oplus \Delta I_i) \oplus F(X) = \Delta O_j$ for each $(\Delta I_i, \Delta O_j)$ is $\frac{|\Omega_F(\Delta I_i \rightarrow \Delta O_j)|}{|\{(\Delta I_i, \Delta O_j)\}|}$, and $\frac{|\Omega_F(\Delta I_i \rightarrow \Delta O_j)|}{|\{(\Delta I_i, \Delta O_j)\}|} = 1$, if the set $\{(\Delta I_i, \Delta O_j)\}$ satisfy one of the following conditions: (1) The input difference ΔI_i fixed and the output difference ΔO_j takes each values of $\{0,1\}^m$ once. (2) The output difference ΔO_j fixed and the input difference ΔI_i takes each values of $\{0,1\}^m$ once. (3) Both the input difference ΔI_i and the output difference ΔO_j takes each values of $\{0,1\}^m$ once.

Proof. For each $(\Delta I_i, \Delta O_j)$, we have an equation. So, the number of equations $F(X \oplus \Delta I_i) \oplus F(X) = \Delta O_j$ is $|(\Delta I_i, \Delta O_j)|$. $\Omega_F(\Delta I_i \rightarrow \Delta O_j) = \{X : F(X \oplus \Delta I_i) \oplus F(X) = \Delta O_j\}$ denotes the solutions of $F(X \oplus \Delta I_i) \oplus F(X) = \Delta O_j$, so we can find that on average the number of solutions of equations $F(X \oplus \Delta I_i) \oplus F(X) = \Delta O_j$ for each $(\Delta I_i, \Delta O_j)$ is $\frac{|\Omega_F(\Delta I_i \rightarrow \Delta O_j)|}{|\{(\Delta I_i, \Delta O_j)\}|}$.

(1) If the input difference ΔI_i fixed and the output difference ΔO_j takes each values of $\{0,1\}^m$ once. Then $|\{(\Delta I_i, \Delta O_j)\}| = 2^m$, Hence

$$
\begin{aligned}
\frac{|\Omega_F(\Delta I_i \rightarrow \Delta O_j)|}{|\{(\Delta I_i, \Delta O_j)\}|} &= \frac{1}{2^m} \sum_{\Delta O_j \in \{0,1\}^m} |\Omega_F(\Delta I_i \rightarrow \Delta O_j)| \\
&= \frac{1}{2^m} \sum_{\Delta O_j \in \{0,1\}^m} |\{X : F(X \oplus \Delta I_i) \oplus F(X) = \Delta O_j\}| \\
&= \frac{1}{2^m} \left| \bigcup_{\Delta O_j \in \{0,1\}^m} \{X : F(X \oplus \Delta I_i) \oplus F(X) = \Delta O_j\} \right| \\
&= \frac{2^m}{2^m} = 1
\end{aligned}
$$

(2) If the output difference ΔO_j fixed and the input difference ΔI_i takes each values of $\{0,1\}^m$ once. Since the function F is invertible, then we can analysis the number of average solutions by detect the reverse-function F^{-1}. For each X satisfy the equation $F(X \oplus \Delta I_i) \oplus F(X) = \Delta O_j$, if and only if there exists a $Y \in \{0,1\}^m$ satisfy the equation $F^{-1}(Y \oplus \Delta O_j) \oplus F(Y) = \Delta I_i$. From (1) we can see that on average there is one solution for the equations $F^{-1}(Y \oplus \Delta O_j) \oplus F(Y) = \Delta I_i$, which means on average there is one solution for the equations $F(X \oplus \Delta I_i) \oplus F(X) = \Delta O_j$ too.

(3) If both the input difference ΔI_i and the output difference ΔO_j takes each values of $\{0,1\}^m$ once, then $|\{(\Delta I_i, \Delta O_j)\}| = 2^m \times 2^m$, hence

$$\frac{|\Omega_F(\Delta I_i \to \Delta O_j)|}{|\{(\Delta I_i, \Delta O_j)\}|} = \frac{1}{2^m} \times \frac{1}{2^m} \sum_{\Delta O_j \in \{0,1\}^m} \sum_{\Delta I_i \in \{0,1\}^m} |\Omega_F(\Delta I_i \to \Delta O_j)|$$

$$= \frac{1}{2^{2m}} \sum_{\Delta O_j \in \{0,1\}^m} \sum_{\Delta I_i \in \{0,1\}^m} |\{X : F(X \oplus \Delta I_i) \oplus F(X) = \Delta O_j\}|$$

$$= \frac{1}{2^{2m}} \left| \bigcup_{\Delta O_j \in \{0,1\}^m} \bigcup_{\Delta I_i \in \{0,1\}^m} \{X : F(X \oplus \Delta I_i) \oplus F(X) = \Delta O_j\} \right|$$

$$= \frac{2^{2m}}{2^{2m}} = 1$$

$\square$

Property 4. Let the input of the first round is $(X_{1,1}, X_{1,2}, \cdots, X_{1,d})$. The input of round function F_1 is F_1^I. Then we have the sub-key of the first round $K_1 = F_1^I \oplus X_{1,1}$.

Proof. Since $F_1^I = X_{1,1} \oplus K_1$, we can obtain that $K_1 = F_1^I \oplus X_{1,1}$. $\square$

Property 5. Let the input of the $4d$th round is $(X_{4d,1}, X_{4d,2}, \cdots, X_{4d,d})$ and the output of $5d - 3$th round is $(Y_{5d-3,1}, Y_{5d-3,2}, \cdots, Y_{5d-3,d})$. The input of round function F_{4d} is F_{4d}^I. Then we have the sub-key of the $4d$th round $K_{4d} = F_{4d}^I \oplus Y_{5d-3,3}$.

Proof. Since $F_{4d}^I = X_{4d,1} \oplus K_{4d}$, we have $K_{4d} = F_{4d}^I \oplus X_{4d,1}$. Hence, the output of $5d - 3$th round equals to the input of $5d - 2$th round, we can obtain that $Y_{5d-3,3} = X_{5d-2,3}$. From Property 1 (2), let $i = 4d, j = 3$ and $t = d - 2$, then we have $X_{5d-2,3} = X_{(i+t),j \bmod d} = X_{i,(j+t) \bmod d} = X_{4d,(3+d-2) \bmod d} = X_{4d,1}$. So $K_{4d} = F_{4d}^I \oplus X_{4d,1} = F_d^I \oplus X_{5d-2,3} = F_d^I \oplus Y_{5d-3,3}$. $\square$

3   Meet in the Middle Attacks on Type-1 Feistel Construction

3.1 $3d - 1$ Rounds Distinguisher of Type-1 Feistel Construction

To make the following analysis more general, we study $3d - 1$ rounds of the Tyep-1 Feistel construction which begin with the $t + 1$th round and end to the $t + 3d - 1$th round ($t \geq 0$).

Proposition 1. *Assume that $A, B, C \in \{0,1\}^{\frac{n}{d}} \setminus \{0\}$ and $A \neq B$. For any plaintext pair (m, m') with the input difference of the tth round is $(0,0,\cdots,0,A)$ and the output difference of the $t + 3d - 1$th round is $(B,C,?,\cdots,0)$. Let b be an arbitrary integer and $b \geq d$, then the sequence $\left(F^{\Delta}(m,1), F^{\Delta}(m,2),\cdots, F^{\Delta}(m,b)\right)$ can assume at most $2^{\frac{(d-1)}{d}n}$ possible values, and the values of the sequence is irrelevant to the plaintext m and m'. It is determined by the values of A, B, C.*

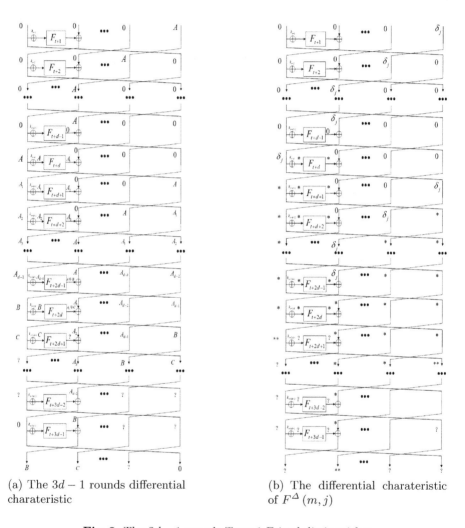

(a) The $3d - 1$ rounds differential charateristic

(b) The differential charateristic of $F^{\Delta}(m, j)$

Fig. 2. The $3d - 1$ rounds Type-1 Feistel distinguisher

Proof. Firstly, we show that, for any pair of plaintexts (m, m') which input difference of the tth round is $(0,0,\cdots 0, A)$ and the output difference of the $t + 3d - 1$th round is $(B, C, ?, \cdots, 0)$, the possible value of $\left(F^{I}_{t+d}, F^{I}_{t+d+1}, \cdots, F^{I}_{t+2d}\right)$

correspond to m is limited to $2^{\frac{d-1}{d}n}$ on average. As shown in Fig. 2, the input difference $(0, 0, \cdots 0, A)$ leads to output difference $(A, 0, \cdots 0, 0)$ after $d - 1$ round encryption. Namely, the input difference of the $t + d$th round is $(A, 0, \cdots, 0, 0)$, $\Delta F^I_{t+d} = A$. Note that $A_i = \Delta F^O_{i+t+d-1} (i = 1, 2, \cdots, d - 1)$, it follows that $\Delta F^O_i = A_{i-t-d+1} (i = t + d, \cdots, t + 2d - 2)$ and thus $\Delta F^I_i = \Delta X_{i,1} = \Delta F^O_{i-1} \oplus \Delta X_{i-1,2} = A_{i-t-d} \oplus 0 = A_{i-t-d}(i = t + d + 1, \cdots, t + 2d - 1)$. Similarly, because the output difference of the $t+3d-1$th round is $(B, C, ?, \cdots, 0)$, we can conclude that the input difference of the $t + 3d - 1$th round and the output difference of the $t + 3d - 2$th round is $(0, B, C, \cdots)$. Since the input difference of the $t + 3d - 1$th round is $(0, B, C, \cdots)$, from Property 1, we have $\Delta F^I_{t+2d} = \Delta X_{t+2d,1} = \cdots = \Delta X_{t+3d-1,2} = B$ and $\Delta F^I_{t+2d+1} = \Delta X_{t+2d+1,1} = \cdots = \Delta X_{t+3d-1,3} = C$. According to Property 2, we have $\Delta F^O_i = \Delta F^I_{i-d+1} \oplus \Delta F^I_{i+1}$, let $i = t + 2d - 1, t + 2d$, we can see that $\Delta F^O_{t+2d-1} = \Delta F^I_{t+d} \oplus \Delta F^I_{t+2d} = A \oplus B$ and $\Delta F^O_{t+2d} = \Delta F^I_{t+d+1} \oplus \Delta F^I_{t+2d+1} = A_1 \oplus C$.

Now we have derived the input and output difference of the $t + d, \cdots, t + 2d$th round function. It follows that $(\Delta F^I_{t+d}, \Delta F^O_{t+d}) = (A, A_1)$, $(\Delta F^I_{t+d+i}, \Delta F^O_{t+d+i}) = (A_i, A_{i+1}) (i = 1, 2, \cdots, d - 2)$, $(\Delta F^I_{t+2d-1}, \Delta F^O_{t+2d-1}) = (A_{d-1}, A \oplus B)$, $(\Delta F^I_{t+2d}, \Delta F^O_{t+2d}) = (B, A_1 \oplus C)$. Therefore, we can build an equation set which has $d + 1$ differential equations:

$$\Delta F_{t+d}(A) = A_1$$

$$\Delta F_{t+d+i}(A_i) = A_{i+1} (i = 1, 2, \cdots, d - 2)$$

$$\Delta F_{t+2d-1}(A_{d-1}) = A \oplus B$$

$$\Delta F_{t+2d}(B) = A_1 \oplus C$$

Since A, B, C are given non-zero values, upon the values of $A_i (i = 1, 2, \cdots, d - 1)$ are fixed, the input difference and output difference of the whole $d + 1$ differential equations are fixed. As $A_i (i = 1, 2, \cdots, d - 1)$ can take at most $2^{\frac{d-1}{d}n}$ different values, we can assume only $2^{\frac{d-1}{d}n}$ different equation set. According to Property 3, each differential equation has one solution on average, which means the value of $(F^I_{t+d}, F^I_{t+d+1}, \cdots, F^I_{t+2d})$ correspond to m can get only $2^{\frac{d-1}{d}n}$ different values on average. Since the solving process is irrelevant with the value of m and m', We should also point out that the possible value of $(F^I_{t+d}, F^I_{t+d+1}, \cdots, F^I_{t+2d})$ is only decided by the value of A, B, C.

Now we know that each value of $(A_1, A_2, \cdots, A_{d-1})$ correspond to a value of $(F^I_{t+d}, F^I_{t+d+1}, \cdots, F^I_{t+2d})$. We are going to proof that if we change the input difference from m' to $m \oplus (0, 0, \cdots, j)$, we are able to compute the truncation of the second sub-blocks of the output difference, as long as the value of $(A_1, A_2, \cdots, A_{d-1})$ is given. Moreover, the value of truncate difference is irrelevant to m. To distinguish the internal state while the plaintexts pair is $(m, m \oplus (0, 0, \cdots, j))$ from those for (m, m'), we denote ΔF^{*O}_i as the output difference of the round function F_i correspond to $(m, m \oplus (0, 0, \cdots, j))$.

As shown in Fig. 2(b), the input difference of $(m, m \oplus (0, 0, \cdots, j))$ is j, we can conclude that the input value of round function F_{t+d} correspond to $m \oplus$

$(0, 0, \cdots, j)$ is $F_{t+d}^I \oplus j$ and the output difference of the $t+d$th round is $\Delta F_{t+d}^{*O} = F_{t+d} \left(F_{t+d}^I \right) \oplus F_{t+d} \left(F_{t+d}^I \oplus j \right)$. Similarly, the output difference of the $t + d + $ 1th round is $\Delta F_{t+d+1}^{*O} = F_{t+d+1} \left(F_{t+d+1}^I \right) \oplus F_{t+d+1} \left(F_{t+d+1}^I \oplus \Delta F_{t+d}^{*O} \right)$. Hence, we are able to compute the output difference of the $t + d + 2, \cdots, t + 2d$th round: $\Delta F_i^{*O} = F_i \left(F_i^I \right) \oplus F_i \left(F_i^I \oplus \Delta F_{i-1}^{*O} \right) (i = t + d + 2, \cdots, t + 2d)$. Since $\Delta Y_{t+3d-1,2} = \Delta X_{t+2d+1,1} = \Delta F_{t+2d+1}^I$, according to Property 2, we can deduce that $\Delta F_{t+2d+1}^I = \Delta F_{t+d+1}^I \oplus \Delta F_{t+2d}^{*O} = \Delta F_{t+1}^I \oplus \Delta F_{t+d}^{*O} \oplus \Delta F_{t+2d}^{*O}$. Because $\Delta F_{t+1}^I = 0$, we can conclude that $\Delta Y_{t+3d-1,2} = \Delta F_{t+2d}^{*O} \oplus \Delta F_{t+d}^{*O}$, which means we can compute the truncation of the second sub-blocks of the output difference by computing the value of $\Delta F_{t+2d}^{*O}, \Delta F_{t+d}^{*O}$.

When the value of $\left(F_{t+d}^I, F_{t+d+1}^I, \cdots, F_{t+2d}^I \right)$ are fixed, we can compute the values of $\Delta F_{t+d}^{*O}, \Delta F_{t+d+1}^{*O}, \cdots, \Delta F_{t+2d}^{*O}$. Meanwhile, $j \to F^\Delta (mj)$ is a fixed map and the value of $F^\Delta (mj)$ can be computed in polynomial-level time. Besides, the solving process is suitable for any $j = 1, 2, \cdots, b$, so we can get the $\left(F^\Delta (m, 1), F^\Delta (m, 2), \cdots, F^\Delta (m, b) \right)$ sequence by repeat the solving process for b times. Since each $\left(F_{t+d}^I, F_{t+d+1}^I, \cdots, F_{t+2d}^I \right)$ correspond to a sequence and $\left(F_{t+d}^I, F_{t+d+1}^I, \cdots, F_{t+2d}^I \right)$ takes at most $2^{\frac{(d-1)}{d}n}$ values, therefore, the number of the sequence $\left(F^\Delta (m, 1), F^\Delta (m, 2), \cdots, F^\Delta (m, b) \right)$ is limited to $2^{\frac{(d-1)}{d}n}$. □

Notes. (1) Due to $F^\Delta (m, \delta_j) \in \{0, 1\}^{n/d}$, the sequence $\left(F^\Delta (m, 1), F^\Delta (m, 2) \cdots F^\Delta (m, b) \right)$ can assume $2^{\frac{b}{d}n}$ different values. However, according to the analysis of Proposition 1, we found that the value can be tightened to $2^{\frac{(d-1)}{d}n}$. In order to distinguish the possible value from the theoretical value, we need $b > d - 1$.

(2) To compute the sequence $\left(F^\Delta (m, 1), F^\Delta (m, 2), \cdots, F^\Delta (m, b) \right)$, we have to get the $2^{\frac{d-1}{d}n}$ different values of $\left(F_{t+d}^I, F_{t+d+1}^I, \cdots, F_{t+2d}^I \right)$. We find that we don't need solve all $d + 1$ differential equations. We give Algorithm 1 to compute the $2^{\frac{d-1}{d}n}$ different values of $\left(F_{t+d}^I, F_{t+d+1}^I, \cdots, F_{t+2d}^I \right)$. If we get the $2^{\frac{d-1}{d}n}$ different values of $\left(F_{t+d}^I, F_{t+d+1}^I, \cdots, F_{t+2d}^I \right)$, we can compute the $2^{\frac{(d-1)}{d}n}$ possible sequences of $\left(F^\Delta (m, 1), F^\Delta (m, 2) \cdots F^\Delta (m, b) \right)$. We use Algorithm 2 to compute these sequences.

From Proposition 1, we can find that if there is a pair of plaintext that follow the differential characteristic, then the sequence $\left(F^\Delta (m, 1), F^\Delta (m, 2) \cdots F^\Delta (m, b) \right)$ can assume $2^{\frac{(d-1)}{d}n}$ instead of $2^{\frac{b}{d}n}$ different values. In that way, we obtain a $3d - 1$ rounds distinguisher of type-1 Feistel construction.

3.2 $5d - 3$ Rounds Key Recovery Attack of Type-1 Feistel Construction

The aim of this section is to recover the 1st and 4dth round-key of the $5d - 3$ rounds Type-1 Feistel construction. As shown in Fig. 3, to make the $5d - 3$ rounds key recovery attack, we prepend d rounds at the top of the $3d - 1$ rounds

---

Algorithm 1. Find the values of $\left(F^I_{t+d}, F^I_{t+d+1}, \cdots, F^I_{t+2d}\right)$

Input: input parameters A, B, C
Output: output the values of $\left(F^I_{t+d}, F^I_{t+d+1}, \cdots, F^I_{t+2d}\right)$

1: For F_{t+2d-1}, make a table $T_1 = \{(x, D) : F_{t+2d-1}(x) \oplus F_{t+2d-1}(x \oplus D) = A \oplus B\}$;
2: For F_{t+2d}, make a table $T_2 = \{(x, D) : F_{t+2d}(x) \oplus F_{t+2d}(x \oplus B) = D \oplus C\}$;
3: **for** each x_{t+d} **do**
4:     Compute $A_1 = F_{t+d}(x_{t+d}) \oplus F_{t+d}(x_{t+d} \oplus A)$;
5:     Check whether there exist $(x, D) \in T_2$ such that $A_1 = D$,
6:     **if** yes **then**
7:         then let $x_{t+2d} = x$,
8:     **else**
9:         turn to Step 3 and check the next x_{t+d};
10:     **for** each x_{t+d+1} **do**
11:         Compute $A_2 = F_{t+d+1}(x_{t+d+1}) \oplus F_{t+d+1}(x_{t+d+1} \oplus A_1)$;
12:         **for** each x_{t+d+2} **do**
13:             Compute $A_3 = F_{t+d+2}(x_{t+d+2}) \oplus F_{t+d+2}(x_{t+d+2} \oplus A_2)$;
14:             ...
15:         ... ...
16:             **for** each x_{t+2d-2} **do**
17:                 Compute $A_{d-1} = F_{t+2d-2}(x_{t+2d-2}) \oplus F_{t+2d-2}(x_{t+2d-2} \oplus A_{d-2})$;
18:                 Check whether there exist $(x, D) \in T_1$ such that $A_{d-1} = D$
19:                 **if** yes **then**
20:                     then let $x_{t+2d-1} = x$, and store $(x_{t+d}, x_{t+d+1}, \cdots, x_{t+2d})$ in table T.
21:                 **else**
22:                     check the next x_{t+2d-2}.
23:         ... ...
24: **return** the obtained table T

---

distinguisher and append $d - 2$ rounds at the bottom of the distinguisher. If we can figure out the values of F^I_1 and F^I_{4d}, we are able to recover K_1 and K_{4d} from Properties 4 and 5 with knowing the corresponding plaintext and cipher. In order to get the values of F^I_1 and F^I_{4d}, we choose many pairs satisfy the differential $(*, *, 0, \cdots, 0, A) \rightarrow (*, 0, B, *, \cdots, *)$. But only the pairs satisfy the differential characteristic $(*, *, 0, \cdots, 0, A) \xrightarrow{d-round} (0, \cdots, 0, A) \xrightarrow{3d-1-round} (B, C, *, \cdots, , 0) \xrightarrow{d-2-round} (*, 0, B, *, \cdots, *)$ can be used to calculate the values of F^I_1 and F^I_{4d}. So we use Proposition 1 as a distinguisher to distinguish the pairs whether it satisfy the differential characteristic or not. Our attack consists of precomputation phase and online phase.

To clarify the attack, we propose following 3 propositions before the introduction of details of attack.

First of all, we proposed Proposition 2 to clarify the necessary and sufficient condition that the differential characteristic of the first d rounds is $(*, *, 0, \cdots, 0, A) \xrightarrow{d-round} (0, \cdots, 0, A)$.

Algorithm 2. Construct the sequences of $\left(F^{\Delta}(m,1), F^{\Delta}(m,2), \cdots, F^{\Delta}(m,b)\right)$

Input: input table T
Output: output the sequences of $\left(F^{\Delta}(m,1), F^{\Delta}(m,2), \cdots, F^{\Delta}(m,b)\right)$
1: **for** $\left(F^{I}_{t+d}, F^{I}_{t+d+1}, \cdots, F^{I}_{t+2d}\right)$ in table T, $j = 1, 2, \cdots, b$ **do**
2:      Let $\Delta F^{O}_{t+d-1} = j$;
3:      **for** each $i = t+d, t+d+1, \cdots, t+2d$ **do**
4:          compute $\Delta F^{O}_{i} = F_i(F^{I}_i) \oplus F_i(F^{I}_i \oplus \Delta F^{O}_{i-1})$;
5:      Let $F^{\Delta}(m,j) = \Delta F^{O}_{t+2d} \oplus \Delta F^{O}_{t+d}$;
6:      Store $F^{\Delta}(m,1), F^{\Delta}(m,2) \cdots F^{\Delta}(m,b)$ it in table T_{δ};
7: **return** the obtained table T_{δ}

Proposition 2. *Assume there is a pair of plaintexts of the d rounds type-1 Feistel (m, m'), where $m = (m_1, m_2, \cdots, m_d)$, $m' = (m_1', m_2', \cdots, m_d')$. Moreover, the input difference is $(\Delta m_1, \Delta m_2, \cdots, \Delta m_d)$ and $\Delta m_i \neq 0$ $(i = 1, 2)$, $\Delta m_j = 0$ $(j = 3, \cdots, d-1)$, $\Delta m_d = A \neq 0$. Then the output difference is $(0, 0, \cdots, 0, A)$, if and only if $F_1\left(F^{I}_1\right) \oplus F_1\left(F^{I'}_1\right) \oplus \Delta m_2 = 0$ and $F_d\left(F^{I}_d\right) \oplus F_d\left(F^{I'}_d\right) \oplus \Delta m_1 = 0$, F^{I}_i and $F^{I'}_i$ denotes the input of round function F_i correspond to m and m', respectively (Fig. 4).*

Proof. According to Property 1, we can see that $\Delta X_{d+1,j} = \Delta X_{j,1}$ $(j = 2, \cdots, d)$. Meanwhile, $\Delta X_{d+1,1} = \Delta F^{O}_d \oplus \Delta X_{d,2} = \Delta F^{O}_d \oplus \Delta X_{1,1} = F_d\left(F^{I}_d\right) \oplus F_d\left(F^{I'}_d\right) \oplus \Delta m_1$. Which means the output difference equals to $(0, 0, \cdots, 0, A)$, if and only if $F_d\left(F^{I}_d\right) \oplus F_d\left(F^{I'}_d\right) \oplus \Delta m_1 = 0$ and $\Delta X_{j,1} = 0$ $(j = 2, 3, \cdots, d)$. Because $\Delta X_{2,1} = \left[F_1\left(F^{I}_1\right) \oplus F_1\left(F^{I'}_1\right)\right] \oplus \Delta m_2$, $\Delta X_{2,1} = 0$ if and only if $F_1\left(F^{I}_1\right) \oplus F_1\left(F^{I'}_1\right) \oplus \Delta m_2 = 0$. Assume that $\Delta X_{2,1} = 0$, we can obtain that $\Delta F^{I}_2 = 0$, hence, we have $\Delta F^{O}_2 = 0$. Since the input difference is $(0, 0, \cdots, 0, A)$, from Property 1, we have $\Delta X_{2,2} = \Delta X_{1,3} = \Delta m_3 = 0$, moreover, we can see that $\Delta X_{3,1} = \Delta F^{O}_2 \oplus \Delta X_{2,2} = 0$. Similarly, we can prove that $\Delta X_{j,1} = 0$ $(j = 2, 3, \cdots, d)$ if and only if $\Delta X_{2,1} = 0$. To conclude, the output difference is $(0, 0, \cdots, 0, A)$, iff $F_1\left(F^{I}_1\right) \oplus F_1\left(F^{I'}_1\right) \oplus \Delta m_2 = 0$ and $F_d\left(F^{I}_d\right) \oplus F_d\left(F^{I'}_d\right) \oplus \Delta m_1 = 0$. $\qquad\square$

Next, we proposed Proposition 3 to clarify the necessary and sufficient condition that the differential characteristic of the last $d - 2$ rounds is $(B, C, *, \cdots, *, 0) \xrightarrow{d-2-round} (*, 0, B, *, \cdots, *)$.

Proposition 3. *Assume there is a pair of cipher (c, c') of the $5d - 3$ rounds type-1 Feistel, where $c = (c_1, c_2, \cdots, c_d)$ and the output difference is $(\Delta c_1, \Delta c_2, \cdots, \Delta c_d)$, Moreover, $\Delta c_2 = 0, \Delta c_3 = B$. Then $\Delta X_{4d,1} = B$, $\Delta X_{4d,2} = C$ and $\Delta X_{4d,d} = 0$ if and only if $F_{4d}\left(F^{I}_{4d}\right) \oplus F_{4d}\left(F^{I}_{4d} \oplus B\right) \oplus \Delta c_4 = C$, F^{I}_{4d} denotes the input of round function F_{4d} correspond to m (Fig. 5).*

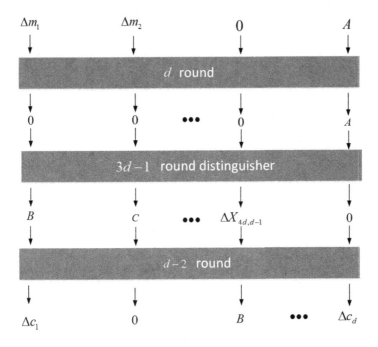

Fig. 3. $5d - 3$ rounds Type-1 Feistel attack model

Proof. According to Property 1, we can see that $\Delta X_{4d,1} = \Delta X_{5d-2,3} = \Delta c_3 = B$, $\Delta X_{4d,d} = \Delta X_{5d-2,2} = \Delta c_2 = 0$. Which means $\Delta X_{4d,1} = B$, $\Delta X_{4d,2} = C$ and $\Delta X_{4d,d} = 0$ if and only if $\Delta X_{4d,2} = C$. From Property 1, we have $\Delta X_{4d+1,1} = \Delta X_{5d-2,4} = \Delta c_4$. Since $\Delta X_{4d,2} = \left[F_{4d} \left(F_{4d}^I \right) \oplus F_{4d} \left(F_{4d}^{I\,'} \oplus B \right) \right] \oplus \Delta X_{4d+1,1} = F_1 \left(F_1^I \right) \oplus F_1 \left(F_1^{I\,'} \oplus B \right) \oplus \Delta c_4$, we can obtain that $\Delta X_{4d,2} = C$ if and only if $F_1 \left(F_1^I \right) \oplus F_1 \left(F_1^{I\,'} \oplus B \right) \oplus \Delta c_4 = C$. To conclude, the input difference satisfy that $\Delta X_{4d,1} = B$, $\Delta X_{4d,2} = C$ and $\Delta X_{4d,d} = 0$ if and only if $F_{4d} \left(F_{4d}^I \right) \oplus F_{4d} \left(F_{4d}^I \oplus B \right) \oplus \Delta c_4 = C$. □

In Proposition 4, we show for any m, if we get a candidate of $\left(F_1^I, F_d^I \right)$, how to construct a plaintext sequence $(M_1, M_2, \cdots, M_b)$ such that the differential characteristic in the input of $d+1$th round satisfy the conditions of Proposition 1, which can be used to justify the candidate of $\left(F_1^I, F_d^I \right)$ is right or not.

Proposition 4. *Assume $m = (m_1, m_2, \cdots, m_d)$ is a plaintext of the d rounds type-1 Feistel construction and the cipher is $c = (c_1, c_2, \cdots, c_d)$. Then we can construct plaintext sequence $(M_1, M_2, \cdots, M_b)$, $M_i = (m_{i,1}, m_{i,2}, \cdots, m_{i,d})$, which satisfy following equations: $m_{i,1} = m_1 \oplus F_d \left(F_d^I \right) \oplus F_d \left(F_d^I \oplus i \right)$, $m_{i,2} = m_2 \oplus F_1 \left(F_1^I \right) \oplus F_1 \left(F_1^I \oplus m_1 \oplus m_{i,1} \right)$, $m_{i,j} = m_j \, (j = 3, \cdots, d-1)$, $m_{i,d} = m_d \oplus i$. Such that the output difference of (m, M_i) after d round encryption is $(0, 0, \cdots, i)$ (Fig. 6).*

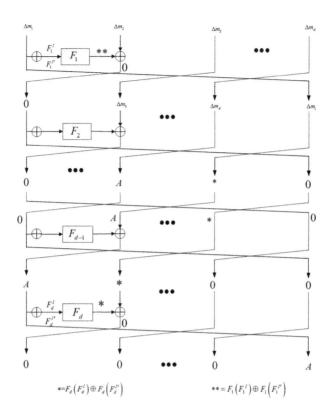

Fig. 4. d rounds prepended before the distinguisher (Fig. of Proposition 2)

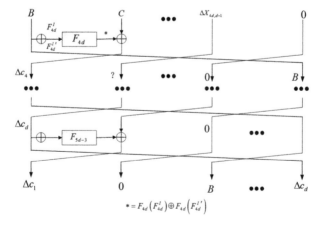

Fig. 5. $d - 2$ rounds appended to the distinguisher (Fig. of Proposition 3)

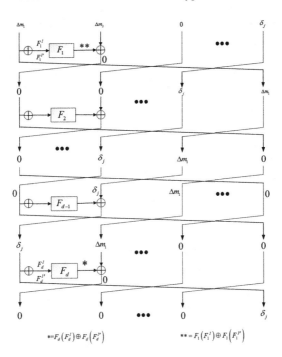

Fig. 6. Construct plaintext sequences $(M_1, M_2, \cdots, M_b)$ (Fig. of Proposition 4)

Proof. For each $1 \leq i \leq b$, since $\Delta m_j = m_j \oplus m_{i,j} = 0 \, (j = 3, \cdots, d-1)$, $\Delta m_d = i \neq 0$, from Proposition 2, we can see that the output difference of (m, M_i) after d round encryption is $(0, 0, \cdots, i)$ if and only if $F_1\left(F_1^I\right) \oplus F_1\left(F_1^{I'}\right) \oplus \Delta m_2 = 0$ and $F_d\left(F_d^I\right) \oplus F_d\left(F_d^{I'}\right) \oplus \Delta m_1 = 0$. Because $m_{i,1} = m_1 \oplus F_d\left(F_d^I\right) \oplus F_d\left(F_d^I \oplus i\right)$, $m_{i,2} = m_2 \oplus F_1\left(F_1^I\right) \oplus F_1\left(F_1^I \oplus m_1 \oplus m_{i,1}\right)$, hence, $\Delta m_1 = m_1 \oplus m_{i,1}$ and $\Delta m_2 = m_2 \oplus m_{i,2}$, we can obtain that the output difference is $(0, 0, \cdots, i)$. $\square$

We will present the specific attack process in the following part. Our attack composed of two phases: precomputation and online phase.

Precomputation phase. In this phase, we need to build 4 tables. Firstly, we build a table T_δ that includes $2^{\frac{(d-1)}{d}n}$ sequences of $\left(F^\Delta(m, 1), F^\Delta(m, 2) \cdots F^\Delta(m, b)\right)$ describe in Proposition 1. We divided the computation of the sequences into three steps. (1) Find the $2^{\frac{(d-1)}{d}n}$ different values of $\left(F_{t+d}^I, F_{t+d+1}^I, \cdots, F_{t+2d}^I\right)$. (2) Compute the $2^{\frac{(d-1)}{d}n}$ possible sequences of $\left(F^\Delta(m, 1), F^\Delta(m, 2), \cdots, F^\Delta(m, b)\right)$ and store it in T_δ. (3) Sort the sequences $\left(F^\Delta(m, 1), F^\Delta(m, 2), \cdots, F^\Delta(m, b)\right)$ in table T_δ so that lookup the table by binary search takes less time. The procedure to build table T_δ is described as follows:

1. Compute all $2^{\frac{(d-1)}{d}n}$ different values of $\left(F_{2d}^I, F_{2d+1}^I, \cdots, F_{3d}^I\right)$ and store it in T by using Algorithm 1 for parameter $t = d$.

2. For all $2^{\frac{d-1}{d}n}$ distinct $(F_{2d}^I, F_{2d+1}^I, \cdots, F_{3d}^I)$, compute all the $2^{\frac{d-1}{d}n}$ possible values of the sequences $(F^\Delta(m,1), F^\Delta(m,2), \cdots, F^\Delta(m,b))$ and store it in table T_δ by using Algorithm 2 for parameter $t = d$.

3. Sort the sequences in table T_δ according to the value of $(F^\Delta(m,1), F^\Delta(m,2), \cdots, F^\Delta(m,b))$ by using quick sort algorithm. Secondly, to complete the attack, we have to get access to the values of F_1^I, F_d^I and F_{4d}^I according to Propositions 2 and 3. To solve the following equations: $\Delta F_1(\Delta m_1) = \Delta m_2$, $\Delta F_d(A) = \Delta m_1$, $\Delta F_{4d}(B) = C \oplus \Delta c_4$ more efficient, we construct table T_1, T_d and T_{4d}. The procedure to build table T_1, T_d and T_{4d} is describe as follows:

1. For each $i = 0, 1, \cdots, 2^{n/d} - 1$, compute $\Delta F_d^O = F_d(i) \oplus F_d(i \oplus A)$ and $\Delta F_{4d}^O = F_{4d}(i) \oplus F_{4d}(i \oplus B)$. Store the values of $(i, \Delta F_d^O), (i, \Delta F_{4d}^O)$ in table T_d, T_{4d}, respectively.
2. Sorted the values in T_d, T_{4d} by the values of $\Delta F_d^O, \Delta F_{4d}^O$, respectively.
3. For $i = 0, 1, \cdots, 2^{n/d} - 1$, store the values of $(i, F_1(i))$ in a temple table tmp_1.
4. For all plaintext pairs $m = 0, 1, \cdots, 2^{n/d} - 1$, $m' = 0, 1, \cdots, 2^{n/d} - 1$, compute $\Delta F_1^O = tmp_1(m) \oplus tmp_1(m')$ and store $(m \oplus m',' \Delta F_1^O, m)$ in T_1.

Online Phase. This phase is divided into two parts: the pairs sieve phase and the keys recovery phase. We shall define a structure, and then sieve and store all plaintext pairs in table T that satisfy the plaintext difference and ciphertext difference of our truncated differential in the pairs sieve phase for a structure. Then it comes into the key recovery phase. We assume that each plaintext pair (m, m') satisfies our truncated differential, and construct $(F^\Delta(m,1), F^\Delta(m,2) \cdots, F^\Delta(m,b), F_1^I, F_d^I, F_{4d}^I)$. If the sequence $(F^\Delta(m,1), F^\Delta(m,2) \cdots, F^\Delta(m,b))$ is a precomputed one, then we get a candidate of (k_1, k_{4d}) from (F_1^I, F_{4d}^I). Otherwise, we conclude that (m, m') don't satisfy our truncated differential, and go ahead for next (m, m') in T. After all plaintext pairs in T are checked, we release the space occupied by T and do the same for the next structure.

The pairs sieve phase. At first, we define a structure $S(a) = S_0(a) \bigcup S_1(a)$ where $a = (a_3, a_4, \cdots, a_d)$ with $a_3, a_4, \cdots, a_d \in \{0, 1\}^{n/d}$ and

$$S_0(a) = \{(m_1, \cdots, m_d) : m_1, m_2 \in \{0, 1\}^{n/d}, m_i = a_i \ for \ i \geq 3\}$$

$$S_1(a) = \{(m_1, \cdots, m_d) \oplus (0, \cdots, 0, A) : (m_1, \cdots, m_d) \in S_0(a)\}$$

Thus, for each a, there needs $2^{\frac{2n}{d}+1}$ chosen plaintexts to construct a structure, and we can get $2^{4n/d}$ plaintext pairs $2^{4n/d}$ from $S(a)$ by $m \in S_0(a)$ and $m' \in S_1(a)$. For each $m \in S_0(a)$ and each $m' \in S_1(a)$, we have $m \oplus m' = (*, *, 0, \cdots, 0, A)$. So the plaintext pair (m, m') satisfies the condition on the plaintext difference of our truncated differential once the first and the second sub-blocks of $m \oplus m'$ are nonzero simultaneously. Next we may get all the plaintext pairs that satisfy the conditions on the plaintext difference and the ciphertext difference of our truncated differential for given a by the quick sort algorithm, and get about $2^{2n/d}$ pairs for a structure.

Theorem 1. *Let $S(a) = S_0(a) \bigcup S_1(a)$ is a structure. For each plaintext $m \in S_b(a)$ with $e \in \{0,1\}$, we define $h(m) = (c_2, c_3, e) = 2^{\frac{2n}{d}+1}c_2 + 2^{\frac{n}{d}+1}(c_3 \oplus e \times B) + e$ where $c = (c_1, c_2, \cdots, c_d)$ is the ciphertext of m. Let $\Omega_e(x) = \{m : m \in S_e(a) \text{ and } h(m) = (x, e)\}$ for each $e \in \{0,1\}$. Let $m, m' \in S(a)$, c and c' are the ciphertexts of m and m' respectively, $c' \oplus c = (\Delta_1, \Delta_2, \cdots, \Delta_d)$. Then $m_d \oplus m'_d = A$, $\Delta_2 = 0$ and $\Delta_3 = B$ iff there exists x and $e \in \{0,1\}$, such that $m \in \Omega_e(x)$ and $m' \in \Omega_{e \oplus 1}(x)$ respectively.*

Proof. Obviously, there exist x, x' and $e, e' \in \{0,1\}$, such that $m \in \Omega_e(x)$ and $m' \in \Omega_{e'}(x')$. Then we have $m_d \oplus m'_d = (e \oplus e')A$, and hence $m_d \oplus m'_d = A$ iff $e' = e \oplus 1$. Let $e' = e \oplus 1$, then $x \oplus x' = 2^{\frac{2n}{d}}(c_2 \oplus c'_2) + 2^{\frac{n}{d}}(c_3 \oplus c'_3 \oplus B)$, and hence $c_2 = c'_2$ and $c_3 \oplus c'_3 = B$ hold simultaneously iff $x = x'$, which concludes the proof. $\square$

By Theorem 1, we may construct all plaintext pairs produced by $S(a)$ that satisfy the conditions on the plaintext difference and the ciphertext difference of our truncated differential by the quick sort algorithm. We may rearrange the elements in $S(a)$ by the magnitude of $h(m)$ with $m \in S(a)$, and record the start points and the end points of any two successive runs that $h(m) = (x, 0)$ for each element m in the first run, and $h(m) = (x, 1)$ for each element m in the second run for some x. Then we pick out each plaintext pair (m, m') that m located in the first run, m' located in the second run and the first and the second sub-blocks of $m \oplus m'$ are nonzero simultaneously. At last, we save (m, m') in table T.

Hence, we should also compute how many plaintext structures we need to choose such that there are one pair satisfy the differential characteristic on average. For an arbitrary plaintexts pair whose input difference is $(*, *, 0, \cdots, 0, A)$, the output difference becomes $(*, 0, B, *, \cdots, *)$ with probability $2^{-2n/d}$. According to Proposition 2, the output difference is $(0, 0, \cdots, 0, A)$ after d round encryption, iff $F_1\left(F_1^I\right) \oplus F_1\left(F_1^{I'}\right) \oplus \Delta m_2 = 0$ and $F_d\left(F_d^I\right) \oplus F_d\left(F_d^{I'}\right) \oplus \Delta m_1 = 0$. These equations holds with probability $2^{-n/d} \times 2^{-n/d} = 2^{-2n/d}$. Similarly, from Proposition 3, if $\Delta c_2 = 0$, $\Delta c_3 = B$, then $\Delta X_{4d,1} = B$, $\Delta X_{4d,2} = C$ and $\Delta X_{4d,d} = 0$ if and only if $F_{4d}\left(F_{4d}^I\right) \oplus F_{4d}\left(F_{4d}^I \oplus B\right) \oplus \Delta c_4 = C$. This equation holds with probability $2^{-n/d}$. Which means for an arbitrary plaintexts pair whose input difference is $(*, *, 0, \cdots, 0, A)$, then this pair satisfy the whole differential characteristic $(*, *, 0, \cdots, 0, A) \xrightarrow{d-round} (0, \cdots, 0, A) \xrightarrow{3d-1-round} (B, C, *, \cdots, *, 0) \xrightarrow{d-2-round} (*, 0, B, *, \cdots, *)$ with probability $2^{-2n/d} \times 2^{-2n/d} \times 2^{-n/d} = 2^{-5n/d}$. So we should construct $2^{n/d} \left(= 2^{4n/d} \div 2^{-5n/d}\right)$ distinct plaintext structures by choosing $2^{n/d}$ different $a = (a_3, a_4, \cdots, a_d)$.

The keys recovery phase. As we choose $2^{\frac{n}{d}}$ plaintext structures, the previous phase result in $2^{n/d} \times 2^{4n/d} \times 2^{-2n/d} = 2^{3n/d}$ candidate pairs with a plaintexts difference $(*, *, 0, \cdots, 0, A)$ and a ciphertext difference $(*, 0, B, *, \cdots, *)$. In this phase, for each plaintext structure, we execute these operations:

1. For each candidate (m, m') in table T: (1) assume that the output difference is $(0, \cdots, 0, A)$ at round d and $(0, B, *, \cdots, *)$ at round $4d$. From Propositions 2 and 3, we have following three equations for each pair: $F_1\left(F_1^I\right) \oplus F_1\left(F_1^{I'}\right) \oplus \Delta m_2 = 0$, $F_d\left(F_d^I\right) \oplus F_d\left(F_d^{I'}\right) \oplus \Delta m_1 = 0$ and $F_{4d}\left(F_{4d}^I\right) \oplus F_{4d}\left(F_{4d}^I \oplus B\right) \oplus \Delta c_4 = C$.
   (2) Solve the equations by look up the tables T_1, T_d, T_{4d} by using binary search algorithm. According to Property 3, each equation has one solution and we can get a pair of inputs of the round function. So we can obtain about $2^{3n/d}$ candidates of $\left(F_1^I, F_d^I, F_{4d}^I\right)$.
   (3) Construct plaintext sequence $(M_1, M_2, \cdots, M_b)$ by knowing the values of F_1^I and F_d^I according to Proposition 4. Such that the output difference of (m, M_j) after d round encryption is $(0, 0, \cdots, j)$.
   (4) For each obtained $(M_1, M_2, \cdots, M_b)$, store it with corresponding $\left(F_1^I, F_d^I\right)$ as a sequence $(M_1, M_2, \cdots, M_b, F_1^I, F_d^I, F_{4d}^I)$.
2. For each sequence $(M_1, M_2, \cdots, M_b, F_1^I, F_d^I, F_{4d}^I)$, assume that cipher sequence $(C_1, C_2, \cdots, C_b)$ is encrypted by the plaintext sequence $(M_1, M_2, \cdots, M_b)$. $C_i = (c_{i,1}, c_{i,2}, \cdots, c_{i,d})$, denote that $\Delta C_i = (\Delta c_{i,1}, \Delta c_{i,2}, \cdots, \Delta c_{i,d}) = (c_1 \oplus c_{i,1}, c_2 \oplus c_{i,2}, \cdots, c_d \oplus c_{i,d})$. Then execute following operations:
   (1) For $i = 1, 2, \cdots, b$, compute $F^\Delta(m, i) = \Delta c_{i,4} \oplus F_{4d}\left(F_{4d}^I\right) \oplus F_{4d}\left[F_{4d}^I \oplus \Delta c_{i,3}\right]$. Thus, we can get $\left(F^\Delta(m, 1), F^\Delta(m, 2), \cdots, F^\Delta(m, b)\right)$.
   (2) Check whether we can find a match of the sequence $\left(F^\Delta(m, 1), F^\Delta(m, 2), \cdots, F^\Delta(m, b)\right)$ in table T_δ or not by using binary search method.
   (3) If we find a match, it means the assumption that the output difference of round d is $(0, 0, \cdots, 0, A)$ and the output difference of round $4d - 1$ is $(B, C, *, \cdots, 0)$ with high probability. So we can deduce the corresponding values of F_1^I, F_d^I while constructing sequence and F_{4d}^I while partial decryption the ciphertext are right. Thus, from Properties 4 and 5, by compute $K_1 = F_1^I \oplus m_1$, $K_{4d} = F_{4d}^I \oplus c_3$, we can get a correct guess of key (K_1, K_{4d}) with high probability. After that, return to step 2 and check the next sequence.
   (4) If we can't find a match, return to step 2 and check the next sequence.

3.3 Complexity Analysis

In the online phases, we choose $2^{\frac{n}{d}}$ different plaintext structure, each plaintext structure contains $2^{\frac{2}{d}n+1}$ plaintexts. So the data complexity of this attack is $2^{\frac{3}{d}n}$ chosen plaintext. The time complexity and memory complexity mainly concentrate on the precomputation phase when building precomputation table T_δ. To build the precomputation table T_δ, we have to compute and storage all $2^{\frac{d-1}{d}n}$ possible sequence of $\left(F^\Delta(m, 1), F^\Delta(m, 2) \cdots F^\Delta(m, b)\right)$. According to the Algorithm 1, for each $x_j \in (0, 1)^{\frac{n}{d}}$ $(j = i + d, \cdots, i + 2d - 2)$, we execute round function evaluation twice. So the time expenditure of Algorithm 1 is about $2(d - 1) \times 2^{\frac{(d-1)}{d}n}$ times round function evaluation. As for Algorithm 2, for each F_i^I $(i = t + d, t + d + 1, \cdots, t + 2d)$, we have to execute round function evaluation twice, meanwhile, this operation should repeat b times. So the time expenditure of Algorithm 2 is about $2b\,(d+1) \times 2^{\frac{(d-1)}{d}n}$ times round function evaluation. So the overall time expenditure of building the precomputation table T_δ is about $2bd \times 2^{\frac{(d-1)}{d}n} \left(\approx 2\,(d-1) \times 2^{\frac{(d-1)}{d}n} + 2b\,(d+1) \times 2^{\frac{(d-1)}{d}n}\right)$ times round function evaluations. Hence, we should store table T_δ, which includes $b \times 2^{\frac{d-1}{d}n}$ sub-blocks, each sub-block is n/d bits. The memory complexity of this phase is $b \times 2^{\frac{d-1}{d}n}$ sub-blocks, each sub-block is n/d bits.

Hence, we should to choose an optimal values of b. On the one hand , we need $b > d - 1$ according to the note of Proposition 1. On the other hand, the larger b is, the higher time and memory complexity this attack need. So we choose $b = d$ to make a balance. Then the time complexity becomes $2d^2 \times 2^{\frac{(d-1)}{d}n}$ times round function evaluations. Since the construction has $5d - 3$ rounds, an encryption takes about $5d$ times round function evaluations. We can convert the time complexity into $\frac{2}{5}d \times 2^{\frac{(d-1)}{d}n}$ times encryptions. Since $\frac{2}{5}d$ is a very small factor compared with $2^{\frac{d-1}{d}n}$, to make the formula briefer, we omit the factor $\frac{2}{5}d$ when talking about the time complexity. In this case, we get the time complexity with $2^{\frac{d-1}{d}n}$ encryptions. The memory complexity is $d \times 2^{\frac{d-1}{d}n}$ (since $b = d$) blocks, each block is n/d bits, which can convert into $2^{\frac{d-1}{d}n}$ blocks, each block is n bits. Finally, we can conclude that the data complexity is $2^{\frac{3}{d}n}$ chosen plaintext, the memory complexity is $2^{\frac{d-1}{d}n}$ blocks, each block is n bits. And the time complexity is $2^{\frac{d-1}{d}n}$ times encryptions.

4   Conclusion

In this paper, we present a key recovery attack on Type-1 Feistel with d sub blocks for $d \geq 4$. We construct a $3d - 1$ round distinguisher by using a special truncated differential. We present an attack on $5d - 3$ rounds when key length is no less than block length n. The data complexity is $2^{\frac{3}{d}n}$ chosen plaintexts, the memory complexity is $2^{\frac{d-1}{d}n}$ blocks, each block is n bits, and the time complexity is $2^{\frac{d-1}{d}n}$ encryptions, which is the best known generic attack on Type-1 Feistel construction.

Acknowledgements. The authors would like to thank editors and anonymous referees for their valuable suggestions. This work was supported by National Natural Science Foundation of China (Grant No.61772547, 61402523 and 61272488).

References

1. Feistel, H.: Cryptography and computer privacy. Sci. Am. **228**, 15–23 (1973)
2. Li, R.J., Jin, C.H.: Meet-in-the-middle attacks on 10-round AES-256. Des. Codes Crypt. **80**(3), 459–471 (2015)
3. Sasaki, Y., Wang, L.: Meet-in-the-middle technique for integral attacks against Feistel ciphers. In: Knudsen, L.R., Wu, H. (eds.) SAC 2012. LNCS, vol. 7707, pp. 234–251. Springer, Heidelberg (2013). https://doi.org/10.1007/978-3-642-35999-6_16

4. Lin, L., Wu, W., Zheng, Y.: Improved meet-in-the-middle distinguisher on Feistel schemes. In: Dunkelman, O., Keliher, L. (eds.) SAC 2015. LNCS, vol. 9566, pp. 122–142. Springer, Cham (2016). https://doi.org/10.1007/978-3-319-31301-6_7

5. Dinur, I., Dunkelman, O., Keller, N., Shamir, A.: New attacks on feistel structures with improved memory complexities. In: Gennaro, R., Robshaw, M. (eds.) CRYPTO 2015, Part I. LNCS, vol. 9215, pp. 433–454. Springer, Heidelberg (2015). https://doi.org/10.1007/978-3-662-47989-6_21

6. Derbez, P., Fouque, P.-A.: Automatic search of meet-in-the-middle and impossible differential attacks. In: Robshaw, M., Katz, J. (eds.) CRYPTO 2016, Part II. LNCS, vol. 9815, pp. 157–184. Springer, Heidelberg (2016). https://doi.org/10.1007/978-3-662-53008-5_6

7. Guo, J., Jean, J., Nikolić, I., Sasaki, Y.: Meet-in-the-middle attacks on generic Feistel constructions. In: Sarkar, P., Iwata, T. (eds.) ASIACRYPT 2014, Part I. LNCS, vol. 8873, pp. 458–477. Springer, Heidelberg (2014). https://doi.org/10.1007/978-3-662-45611-8_24

8. Guo, J., Jean, J., et al.: Extended meet-in-the-middle attacks on some Feistel constructions. Des. Codes Crypt. **80**(3), 587–618 (2016)

9. Guo, J., Jean, J., et al.: Meet-in-the-middle attacks on classes of contracting and expanding Feistel constructions. In: FSE 2017, IACR Transactions on Symmetric Cryptology, pp. 1–31 (2017)

10. Zheng, Y., Matsumoto, T., Imai, H.: On the construction of block ciphers provably secure and not relying on any unproved hypotheses. In: Brassard, G. (ed.) CRYPTO 1989. LNCS, vol. 435, pp. 461–480. Springer, New York (1990). https://doi.org/10.1007/0-387-34805-0_42

11. Nachef, V., Patarin, J., Volte, E.: Feistel Ciphers Security Proofs and Cryptanalysis. Springer, Heidelberg (2017)

12. Fouque, P.-A., Jean, J., Peyrin, T.: Structural evaluation of AES and chosen-key distinguisher of 9-round AES-128. In: Canetti, R., Garay, J.A. (eds.) CRYPTO 2013, Part I. LNCS, vol. 8042, pp. 183–203. Springer, Heidelberg (2013). https://doi.org/10.1007/978-3-642-40041-4_11

13. Matsui, M.: On correlation between the order of S-boxes and the strength of DES. In: De Santis, A. (ed.) EUROCRYPT 1994. LNCS, vol. 950, pp. 366–375. Springer, Heidelberg (1995). https://doi.org/10.1007/BFb0053451

14. Nyberg, K.: Generalized Feistel networks. In: Kim, K., Matsumoto, T. (eds.) ASIACRYPT 1996. LNCS, vol. 1163, pp. 91–104. Springer, Heidelberg (1996). https://doi.org/10.1007/BFb0034838

15. Blondeau, C., Minier, M.: Analysis of impossible, integral and zero-correlation attacks on type-II generalized Feistelnetworks using the matrix method. In: Leander, G. (ed.) FSE 2015. LNCS, vol. 9054, pp. 92–113. Springer, Heidelberg (2015). https://doi.org/10.1007/978-3-662-48116-5_5

16. Nachef, V., Volte, E., Patarin, J.: Differential attacks on generalized Feistel schemes. In: Abdalla, M., Nita-Rotaru, C., Dahab, R. (eds.) CANS 2013. LNCS, vol. 8257, pp. 1–19. Springer, Cham (2013). https://doi.org/10.1007/978-3-319-02937-5_1

17. Pudovkina, M., Toktarev, A.: Numerical semigroups and bounds on impossible differential attacks on generalized Feistel schemes. In: Kotulski, Z., Księżopolski, B., Mazur, K. (eds.) CSS 2014. CCIS, vol. 448, pp. 1–11. Springer, Heidelberg (2014). https://doi.org/10.1007/978-3-662-44893-9_1

Applications

Influence of Error on Hamming Weights for ASCA

Chujiao Ma[(⊠)], John Chandy, Laurent Michel, Fanghui Liu,
and Waldemar Cruz

Computer Science and Engineering Department, School of Engineering,
University of Connecticut, Storrs, CT 06269-4155, USA
chujiao.ma@uconn.edu

Abstract. Algebraic Side-Channel Attack (ASCA) models the cryptographic algorithm and side-channel leakage from the system as a set of equations and solves for the secret key. The attack has low data complexity and can succeed in unknown plaintext/ciphertext scenarios. However, it is susceptible to error and the complexity of the model may drastically increase the runtime as well as the memory consumption. In this paper, we explore the attack by examining the importance of various Hamming weights in terms of success of the attack, which also allows us to gain insights into possible areas of focus for countermeasures, as well as successfully launch ASCA on AES with a larger error tolerance.

Keywords: Algebraic side-channel attack · AES · Cryptography
Block cipher · Constraint programming

1 Introduction

Attacks such as side-channel analysis (SCA) observe leakages from the system such as power consumption and use them to break the algorithm. The power analysis attack assumes that different data and operations consume different amount of power. Since the input to the cryptographic algorithms consists of plaintext and a secret key, there is a correlation between the secret key and the leakage of any intermediate values. With enough data, the attacker is able to acquire the secret key using a divide-and-conquer strategy. The power analysis attack is first proposed by [7] and successfully performed on DES. Since then, it has also been proven to be successful on all common modes (ECB, CBC, CFB, OFB and CTR) of AES [6] as well as XTS-AES, an advanced mode of AES for data protection of sector-based devices that features two secret keys and an additional tweak for each data block [9]. Since the attack targets the implementation of the cryptographic algorithm, even the most mathematically secure algorithm may still be vulnerable to it.

However, the success of the attack depends on the quality and quantity of side-channel data. Due to measurement limitations and error from noise, the attack often requires hundreds and thousands of traces even with elaborate signal

© Springer International Publishing AG, part of Springer Nature 2018
X. Chen et al. (Eds.): Inscrypt 2017, LNCS 10726, pp. 447–460, 2018.
https://doi.org/10.1007/978-3-319-75160-3_26

processing methods [8,13,25,26]. One of the more recent methods is to combine SCA with algebraic analysis [20]. This attack models the cryptographic algorithm and the side-channel information as a system of equations. The equations are then put through a solver to solve for the secret key. Not only does ASCA requires less information than pure SCA, it can also succeed in the unknown plaintext/ciphertext scenario while SCA requires knowledge of the plaintext or ciphertext. ASCA can also exploit leakages from all rounds of the algorithm while SCA can only attack the first or last round. SCA employs a divide-and-conquer strategy and finds the key one byte at time while ASCA has low data complexity and recovers the whole key at once.

While ASCA utilizes less information than SCA, it is still susceptible to error due to noise from SCA information, which may give inaccurate solutions and increase the runtime. Depending on the size and complexity of the model or equations, it may also have high memory consumption. Pure ASCA is modeled as set of boolean equations and solved using a SAT solver. However, it is difficult to model the side-channel information with error as SAT instances, and non-linear operations greatly increase the complexity of the system. Instead of SAT, [14] models the equations as a non-linear pseudo-boolean optimization problem with error taken into account and is able to successfully perform the attack on Keeloq. The solver is able to find the key with 18.8% error and an average of 3.8 h in runtime. Another attack in [16] uses the constraint solver SCIPspx to choose the most likely intermediate state instead of the most likely key. Since the information has some redundancy, it can tolerate a lower level of accuracy in the recovery of individual state elements. This gives the correct key 100% of the time. The median for successful attack is 607 s and the maximum is 6 h. Another method, SASCA, is proposed in [27] where the AES Furious and template information are modeled as a factor graph that can be efficiently decoded by Belief Propagation Algorithm, which found the most possible key from low-density parity check codes. The performance of SASCA, ASCA and DPA based on real data are compared in [5]. SASCA performs better than ASCA but is generally more computationally intensive than divide-and-conquer (DPA) attacks. When there are a lot of traces available, the gain of SASCA over DPA is limited. If plaintext or ciphertext is unknown, then SASCA may be the best option. The amount and type of leakage samples exploitable are what made SASCA more or less powerful. The resolution time for ASCA depends on the quantity of information, whereas it is independent of this quantity in SASCA.

Our model utilizes constraint programming. By using bit-vectors with a constraint programming model, the complexity of the equation set for the cryptographic algorithm is greatly reduced, thus improving the runtime. Even if there are multiple solutions that satisfy the model, other solvers can only output one solution, which may or may not be correct while we use an Objective-C solver that outputs all solutions. Since the attack is no longer as disadvantageous due to memory and runtime as before, this allows us to explore the attack in more details. Aside from reducing runtime, another point of interest is the side-channel information needed for the attack. In this paper, we evaluate the tolerance of the model regarding the side-channel information, and how the information from different subrounds affects the performance of the attack. Not only will this give us a better understanding of the attack and let us know where error matters, it may also provide insights to what's most important to protect and help with future countermeasures.

The rest of the paper is organized as follows: Sect. 2 provides an overview of the Tolerant Algebraic Side-Channel Analysis (TASCA) attack. Section 3 provides a brief overview of the attack model. Section 4 explains the experimental setup for our implementation of the attack. Section 5 describes the experiments and results. Finally, Sect. 6 concludes the paper.

2   Tolerant Algebraic Side-Channel Analysis

The TASCA attack assumes the attacker has access to a Device Under Test (DUT) which emits a measurable side-channel leakage, such as power consumption, during encryption or decryption. The power consumption should be capturable with an oscilloscope with an error rate that the solver can handle. The TASCA methodology recovers the secret key from the power trace using the following steps:

- **Identify potential leaks**. The side-channel leakages used here are power consumption traces. Most smart card processors are implemented with CMOS circuits, where the switching of the gates causes a current flow that is observable. The Hamming weight model, which counts the number of switches, is used to model the relationship between the power trace and the intermediate variables which are correlated with the secret key. While most side-channel information is based on 8-bit micro-controllers, side-channel information can be acquired and used to attack most devices that use CMOS circuits. Previous works, [19,23,24], have shown that variations of power analysis attack can successfully break ECC, DES and RSA on FPGA. The attack has also been performed on AES implemented on an ASIC [18], STC89C52 Microprocessor [4] and a Cortex-M3 CPU [1].
- **Profile DUT and devise a decoding process.** The target cryptographic algorithm for this paper is AES-128. The side-channel leakages, or power traces, are acquired using a template attack. With a template attack, the attacker should have an identical experimental device to the DUT and is able to collect a large amount of side-channel information from it (usually thousands of power traces). With this information, the attacker uses a multivariate Gaussian model to devise a decoding process that maps the side-channel information to the Hamming weight of the intermediate values in the algorithm [16].
- **Acquire power traces from DUT and decode the Hamming weights.** In this phase, the attacker accesses the actual DUT to collect a few power traces from it with an oscilloscope. The template from the previous step is used as reference to extract Hamming weights from the power traces.

Since AES uses byte-wise operations, the Hamming weights of each leak are represented as an integer with a value between 0 and 8.

- **Model the DUT and side-channel information as set of algebraic equations.** The formal description of the DUT, or AES-128, the Hamming weights from the side-channel information as well as error variables are represented as a system of equations.
- **Solve for the secret key.** Given the model, as well as plaintext or ciphertext if desired, the solver finds the solution that satisfies the model, which should be the correct secret key.

The objectives of the attack are to acquire the key with a high success rate using the least amount of information and in a reasonable amount of time.

3   Attack Model

AES is one of the most widely used and commonly studied block ciphers. The variation studied in this paper is AES-128. It takes blocks of 128 bits as input (plaintext), arranges it into a state of 4 by 4 bytes, then combines it with a secret key. The output goes through 10 rounds of operations to produce the final ciphertext as illustrated in Fig. 1.

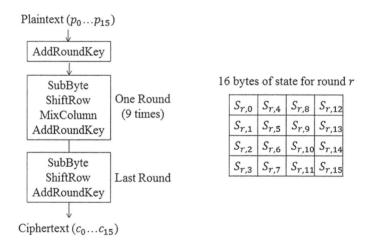

Fig. 1. AES-128.

Each round of AES consists of the following four subrounds, as illustrated in Fig. 2.

For more details on the structure of AES, please refer to [2]. The AES algorithm as well as side-channel information are written using bit-vectors with a constraint programming model [11]. With this, we can adopt a more direct and natural formulation that does not require linearizations.

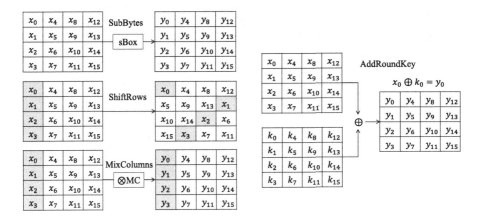

Fig. 2. The four subrounds of AES.

3.1 Variables

Each state variable is represented by 16 bit-vectors that are 8-bit wide. Many of the AES subrounds, such as SubByte and MixColumns, operate over 8-bit values. Therefore, 8-bit wide bit-vectors are used for this model. The side-channel information and the state of the AES algorithm are modeled as decision variables below:

- **State variables** $S_{sr,i,j}$ corresponds to each intermediate state. $S_{sr,i,j}$ denotes the value of bit j of state byte i at subround sr, where $sr \in [0, 40]$, $i \in [0, 15]$, $j \in [0, 7]$. S_0 represents the initial plaintext and S_{40} represents the ciphertext.
- **Key variables** $K_{r,i,j}$ corresponds to the 128-bit key. $K_{r,i,j}$ denotes the value of bit j of key byte i at round r, where $r \in [0, 10]$, $i \in [0, 15]$, $j \in [0, 7]$. K_0 refers to the cipher key and K_r ($r > 0$) refers to the round keys derived from the cipher key via key expansion.
- **Error variables** $E_{sr,i}$ relaxes the SCA constraints to account for noise in side-channel equations. The actual value of a state variable is allowed to deviate from the measured value by ± 1 for the initial experiment and ± 2 for the subsequent experiment to cover a more realistic range of errors. The variable $E_{sr,0}$ denotes the positive error and the variable $E_{sr,1}$ denotes the negative error.

3.2 Constraints

- **AddKey/AddRoundKey** XOR the state with the round key, which is derived from the secret key via key expansion. It is implemented as a bit-wise XOR operation between two input bit-vectors. Each byte of the state is represented by the constraint:

$$S_{sr,i} \oplus K_{sr,i} = S_{sr+1,i}, \forall i = \{0, ..., 15\}$$

– **SubByte** substitutes each byte of the state with another byte according to a look-up table or S-box. It is implemented as an element constraint over bit-vectors. The SubByte constraint models the permutation π with a single element constraint over bit-vectors:

$$SBox[A] = B$$

– **ShiftRows & MixColumns** shifts each of the four rows of the state 0, 1, 2, and 3 bytes to the left then multiplies each column with a matrix of constants (MC). The two operations are combined together and the 8-bit efficient MixColumns implementation is used in the CP encoding. Bit-vector constraints are used to capture XOR as well as $xtime$ resulting in the following for each column:

$$\beta_k = \texttt{xtime}(a_k \oplus a_{(k+1) \bmod 4}) \qquad \forall\, k \in 0..3$$

$$o_k = \left(\bigoplus_{i=0}^{3} a_i \right) \oplus \beta_k \oplus a_k \qquad \forall\, k \in 0..3$$

– **Key Expansion** is responsible for generating the round keys used during AddRoundKey from the cipher key. A set of bit-vectors variables representing round constants $rcon_i$ are created for each round. The SubByte and XOR constraints are used to generate the round keys. The following is the encoding for the Key Expansion:

$$K_{r,0} = SubByte(K_{r-1,13}) \oplus K_{r-1,0} \oplus RC_r$$
$$K_{r,1} = K_{r-1,1} \oplus SubByte(K_{r-1,14})$$
$$K_{r,2} = K_{r-1,2} \oplus SubByte(K_{r-1,15})$$
$$K_{r,3} = K_{r-1,3} \oplus SubByte(K_{r-1,12})$$
$$K_{r,i+4} = K_{r,i} \oplus K_{r-1,i+4}, \forall i = \{0, ..., 11\}$$

– **Side-Channel Constraints** are created from the Hamming weight vector. The *count* constraint counts the number of bits with the value of 1 in a bit-vector and represents the actual Hamming weights. The constraint $M_{sr,i}$ is represented as Hamming weights with errors included:

$$count(S_{sr,i}) + e^+_{sr,i} - e^-_{sr,i} = M_{sr,i}$$

– **Objective Function** is modeled as the total number of errors, same as the IP model:

$$Min : \sum e^+_{sr,i} + \sum e^-_{sr,i}$$

In addition to providing a set of Hamming weights (HW) to account for errors, TASCA goes a step further and picks the most likely HW to branch/search first as part of the goal function.

3.3   Search

Aside from the equations representing AES and the Hamming weights, a big part of the CP version of the TASCA model is the search, which optimizes the error from the Hamming weights. A Hamming weight is generated as the side-channel information for each byte of the state during AES encryption. A candidate value $v \in D(S_{r,i})$ has a *Hamming weight* $H(v) = \sum_{b \in 0..7} (v_{|b} = 1)$ capturing the number of bits at 1 in v. TASCA imposes that $D(S_{r,i})$ be restricted to values v for which

$$-1 \leq H(v) - M_{r,i} \leq 1$$

i.e., the discrepancy between the measurement $M_{r,i}$ and the value v does not exceed ± 1. The goal of the CP model is to minimize the total number of errors that may be observed for the Hamming weights. Branching on value assignments that yield the least amount of errors in the objective will be most effective to get high-quality solutions early on. With the objective equal to the sum of the errors, we have

$$\min \sum_{j=0}^{41} E(j, \sigma(S_j)) \text{ where } E(j, v) = \begin{cases} 0 & \text{if } H(v) = M_j \\ 1 & \text{if } H(v) = M_j \pm 1 \end{cases}$$

and σ is the current value assignment.

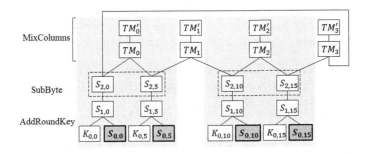

Fig. 3. Circuit for bytes $\{0, 1, 2, 3\}$.

A Circuit View: Considering a single round of the AES algorithm containing all 4 subrounds, AddRoundKey, SubBytes, and MixColumns/ShiftRows, Fig. 3 illustrates the subrounds for the first column of the state matrix. The structure is repeated four times for the entire state. The column $[S_{0,0} S_{0,5} S_{0,10} S_{0,15}]^T$ of the state and the column $[K_{0,0} K_{0,5} K_{0,10} K_{0,15}]^T$ of the sub-round key form the inputs at the bottom of the Figure. In round 1, $[S_{0,0} S_{0,5} S_{0,10} S_{0,15}]^T$ are known since they represent bytes of the plaintext. AddRoundKey and SubBytes apply bijective transformations once the state is known. Consequently, as soon as $S_{2,0}, S_{1,0}$ or $K_{0,0}$ is fixed, the others are fixed by propagation (This is also true for $S_{2,5}, S_{1,5}$ and $K_{0,5}$). The dashed box on the left that contains $S_{2,0}$ and $S_{2,5}$ highlights the inputs to an exclusive OR that yields the temporary value TM_0. The output of the circuit TM_0' is the first byte of the output, i.e., state $S_{4,0}$. The vertical light-gray column on the left is a description of the relations defining byte 0 of the output. The evaluation of TM_0 rests upon the availability of values for both $S_{2,0}$ and $S_{2,5}$. Observe that if, for instance, $S_{2,5}$ is fixed, one only needs to fix $S_{2,10}$ to get propagation up and down and fix both $K_{0,10}$ and TM_1'.

Variable Selection Heuristic: Branching on TM_0 will only trigger propagation "up" as the exclusive OR will not be able to push information "down". Similarly, branching on $S_{2,0}$ or even $K_{0,0}$ will have a limited propagation given the bijective nature of the relations in the bottom "legs". However, simultaneously assigning both variables in the dashed box will trigger propagation up *and* down.

This is a key insight into the variable selection strategy. The search should branch on *pairs of variables* that trigger propagation within an entire gray box. In Fig. 3 this implies four columns for a total of 16 pairs of variables to consider for the first branching decision (recall that this structure is replicated 4 times). The four columns are topped by TM_0', TM_1', TM_2' and TM_3'. Once a pair is selected, the search should create another pair by reusing one of the two variables from the first pair. For instance, if $\langle S_{2,0}, S_{2,5} \rangle$ is selected, it is tempting to consider the two pairs $\langle S_{2,5}, S_{2,10} \rangle$ and $\langle S_{2,15}, S_{2,0} \rangle$ as the domains of TM_1 and TM_3 are reduced by the first choice.

Finally, observe that not all pairs of values drawn from the domains of $S_{2,0}$ and $S_{2,5}$ are compatible. Some of these pairs may induce errors that exceed the ± 1 margin dictated by TASCA. It is therefore advisable to follow the first-fail and break ties among pairs of variables based on the *number of pairs* of values that yield assignments compatible with the error margins.

Value Selection Heuristic: The errors in the objective are driven by the sum of measurement errors on state variables. If the search considers a pair of values

$$\langle a, b \rangle \in D(S_{2,0}) \times D(S_{2,5})$$

it can assess the impact that the simultaneous assignments $S_{2,0} = a \land S_{2,5} = b$ would have on the errors at the state variables in the leftmost gray column. This assessment is an under-approximation of the true error induced by the assignments. Indeed TM_1' can expose errors caused by the choice of value b for $S_{2,5}$, but that falls outside the gray column and is therefore ignored. A sensible value selection heuristic considers pairs of values and assesses their quality with a scoring function C. Given a pair of values $\langle a, b \rangle$, the scoring is

$$C(\langle a, b \rangle) = C_{leg}(a, [S_{2,0}, S_{1,0}, K_{0,0}]) + C_{leg}(b, [S_{2,5}, S_{1,5}, K_{0,5}]) + C_{mc}(a \oplus b, [TM_0'])$$

The functions C_{leg} and C_{mc} model the errors attributable to a leg in the gray box, or the top of a gray box (the MixColumns operation) and $a \oplus b$ denotes the value inferred for TM_0 based on the connecting XOR constraint. C_{leg} and C_{mc} measure the differences between the value of the state variable and the expected Hamming weight. Given a pair of variables and the scoring function C, the value heuristic enumerates pairs that contribute the least to the objective function.

Optimality Pruning: Since the total contribution of errors due to Hamming weights accumulates as the search dives deeper, the total error can be used to further prune value-pairs whose contribution would bring the total beyond the total error for the incumbent solution.

4   Experimental Setup

We assume that the key expansion is done in advance and no leaks from the process are available. This corresponds to a more challenging scenario since it is shown in [10] that side-channel leakages from an 8-bit microcontroller implementation of AES during key expansion are sufficient to recover the complete key. Thus the attack will focus on the AES algorithm itself. Our analysis considers a simulated implementation of AES-128 in 8-bit microcontroller as device under test (DUT). For our attack, TASCA with CP model, the solver used in the experiments is Objective-CP [3].

The experiments runs on Mac with macOS Sierra 10.12.6, 3.1 GHz Intel Core i5 processor and 8 GB memory. For random instances, the plaintexts are chosen uniformly in $\{0, 1\}^{128}$ and the cipher keys are fixed. Each instance contains a known plaintext and 100 Hamming weights that correspond to the first round of AES.

4.1   Subround Comparison

Previous results from [17] have shown that if the leak count is reduced to less than a full round, the probability of success becomes vanishingly small. While [21] mentions that Hamming weights from MixColumn seem to be the most critical when solving the system, there have been no experiments done to explore the influence of side-channel information from different subrounds on the attack. One advantage of TASCA is the minimal data needed - only 1 round of Hamming weights from 1 trace in error-free situations. If the Hamming weights from certain subrounds are more important than others, then whether the attack will succeed when using less than 1 round of Hamming weights will also depend on which Hamming weights are left out. Since TASCA takes into account the amount of error, we can observe the effect of different subrounds on the success of the attack by introducing error on specific subround.

4.2   Error Tolerance

It has been demonstrated in [21] that side-channel information such as Hamming weight can be acquired with 80% accuracy from 1 power trace by using **error detection** (rejecting side-channel information that gives rise to incoherent input/output values for the S-boxes) and **likelihood rating** (only uses a subset of all the Hamming weights extracted with the templates, starting with the most likely ones). IASCA in [12] explores the error tolerance of the Hamming weight by examining the distribution from 2000 attack traces. The distribution of error is illustrated in Table 1.

Table 1. Error distribution from [12].

Error class	e_0	e_1	e_2	e_3	e_4
Occurrence	28%	44%	24%	4%	0%

The error variable e_0 to e_4 denotes the set of Hamming weights with 0 to 4 errors. The error class e_0 contains only the correct Hamming weight (HW), e_1 is a set of Hamming weights with one error (HW, HW+1), e_2 contains two errors (HW−1, HW, HW+1), e_3 contains 3 errors (HW−1, HW, HW+1, HW+2), and e_4 contains 5 errors (HW−2, HW−1, HW, HW+1, HW+2). For all error classes, it is assumed that the correct Hamming weight is within the set.

5 Results

The first experiment tests the effect of different subrounds on the success of the attack. The test is done on 10 randomly generated instances given the Hamming weights for the first round. There are a total of 16 bytes of Hamming weights for AddRoundKey, 16 bytes for SubByte, no Hamming weights for ShiftRow and 36 bytes for MixColumn. The experiment is conducted by comparing the runtime when introducing ±1 error on 16 bytes of AddKey, SubByte or MixColumn. Table 2 shows the average solving time, the solution pool (number of solutions that satisfies the model), and the success rate (how many instances solved for the key within 24 h).

Table 2. Influence of Hamming weights for subrounds on TASCA.

Subround with ±1 error	AddRoundKey	SubByte	MixColumn
Solving time	3,609.93 s	1,871.72 s	940.02 s
Solution pool	8,470	39,323	1,217
Success rate	80%	80%	60%

One unique aspect of our solver is that it outputs all possible solutions, so there are no wrong solutions. It either finds no solution (solution pool of 0), one solution or a set of possible solutions. [17] has shown that it only takes seconds to loop through set of solutions and finds the correct key, so the difference between the size of solution pool is not as important as the solving time and success rate. The experiment has shown that the attack is more difficult, or has a lower success rate, when all the errors are concentrated on information from MixColumn than on AddRoundKey and SubByte. However, out of the random instances that are solvable for the key within 24 h, MixColumn is the fastest followed by SubByte, then AddRoundKey.

Most error tolerant ASCA from previous works, such as [28], minimize error by using multiple traces and only model the Hamming weights with errors of ± 1, or up to error class e_2. However, this range cannot cover the complete range of Hamming weight errors in some cases. Expanding the error tolerance to e_3 or even e_4, may allow the attack to succeed with single trace or even with noisy measurements. Since TASCA-CP model delivers orders of magnitude improvement in runtime and memory usage for error class of up to e_2, [3], this paper extends the model to cover error class of e_3 and e_4 for a more practical attack. While we originally intend to run the code with 100% error rate, such a constraint negates the effect of the search function and the solver is unable to arrive at solutions within the 24 h time limit. Since it is possible to acquire Hamming weights with 80% accuracy with some data processing [21], we attempt error rate of 20% instead where 20% of the Hamming weights are assigned an error of -1, $+1$, -2 or $+2$. The attack is performed on 10 instances using only Hamming weights from the first round. The Hamming weight models with error class of e_3 and e_4 are explored in [12,15,22]. The results of ASCA with error tolerance beyond ± 1 (e_2) are shown in Table 3.

Table 3. Attack comparison.

	Error rate	Error class	Solving time	Success rate	Round	Traces
TASCA [15]	20%	e_3	1332.07 s	72.7%	R1	1
IASCA [12]	100%	e_3	84 s	100%	R1–R3	1
IASCA [12]	100%	e_4	100 s	50%	R1	1
ETASCA [22]	100%	e_4	33.29 s	-	R1	3
TASCA-CP	20%	e_4	8.6 h	50%	R1	1

All attacks are performed with known plaintext/ciphertext. The error rate refers to the percentage of Hamming weights with error. The success rate refers to the percentage of instances that outputs the correct key within a reasonable amount of time.

6   Conclusions

The results show that the side-channel information from some subrounds are indeed more influential than others for a successful attack. While accurate information from MixColumn is most important for a higher chance of success for TASCA, the instances with erroneous SubByte run faster than those with erroneous AddRoundKey. This information is useful for the attacker when given multiple traces with different errors, since choosing a trace or a round of Hamming weight with less error in the MixColumn subround may give a higher chance of successful attack. This experiment also provides a basis to help improve countermeasures for TASCA. Since current countermeasures, such as random masking, are very time consuming, it may be better to focus on masking side-channel information on specific subround instead. The fast runtime of TASCA in our case is strongly dependent on the search algorithm, which filters out less probable Hamming weights and errors by looking at the correlation of error between Hamming weights of different subrounds. Thus, when the error range is expand to ± 2 for Hamming weights, the search algorithm needs to be changed and the runtime drastically increased.

In addition to further improvement in search, we will also test error tolerant ASCA for unknown plaintext/ciphertext. No other ASCA with error tolerance has been successful with unknown plaintext/ciphertext, and yet, that is one of the big advantage of TASCA over SCA.

The most important points that define a successful attack are that it should run fast, use less data/information with lax requirements, and have a high success rate. A better error tolerance gives a better chance of attack succeeding even with lower quality data. However, the difference in results may be due to the different solver and model used in this attack rather than a property of the cryptographic algorithm itself. To prevent ASCA, either the algebraic complexity of the cryptographic algorithm needs to be increased, or countermeasures should be implemented to make it more difficult to collect accurate SCA.

In addition to what's presented in this paper, there are other areas of interest to explore. We can try different rounds instead of the first round, or see if the results still hold true with other profiling methods or leakage models. Aside from Hamming weights, vectors of probability is proposed as a better way to present side-channel information to minimize loss of data. We can attack AES-256 or other versions of AES (ECB/CBC etc.). Since SCA countermeasures are starting to become common among devices, it would be interesting to attempt the attack on a masked version (which requires knowledge of the exact point of Hamming weights in the power trace that correlates with the intermediate value and the type of masking).

References

1. Barenghi, A., Pelosi, G., Teglia, Y.: Improving first order differential power attacks through digital signal processing. In: Proceedings of the 3rd International Conference on Security of Information and Networks, SIN 2010, pp. 124–133. ACM, New York (2010). https://doi.org/10.1145/1854099.1854126
2. Daemen, J., Rijmen, V.: AES - The Advanced Encryption Standard. Springer, Heidelberg (2002)
3. Liu, F., Cruz, W., Ma, C., Johnson, G., Michel, L.: A tolerant algebraic side-channel attack on AES using CP. In: Beck, J.C. (ed.) CP 2017. LNCS, vol. 10416, pp. 189–205. Springer, Cham (2017). https://doi.org/10.1007/978-3-319-66158-2_13
4. Fei, H., Daheng, G.: Two kinds of correlation analysis method attack on implementations of advanced encryption standard software running inside STC89C52 microprocessor. In: 2016 2nd IEEE International Conference on Computer and Communications (ICCC), pp. 1265–1269, October 2016

5. Grosso, V., Standaert, F.-X.: ASCA, SASCA and DPA with enumeration: which one beats the other and when? In: Iwata, T., Cheon, J.H. (eds.) ASIACRYPT 2015. LNCS, vol. 9453, pp. 291–312. Springer, Heidelberg (2015). https://doi.org/10.1007/978-3-662-48800-3_12

6. Jayasinghe, D., Ragel, R., Ambrose, J.A., Ignjatovic, A., Parameswaran, S.: Advanced modes in AES: are they safe from power analysis based side channel attacks? In: 2014 IEEE 32nd International Conference on Computer Design (ICCD), pp. 173–180, October 2014

7. Kocher, P., Jaffe, J., Jun, B.: Differential power analysis. In: Wiener, M. (ed.) CRYPTO 1999. LNCS, vol. 1666, pp. 388–397. Springer, Heidelberg (1999). https://doi.org/10.1007/3-540-48405-1_25

8. Lu, Y., O'Neill, M.P., McCanny, J.V.: FPGA implementation and analysis of random delay insertion countermeasure against DPA. In: 2008 International Conference on Field-Programmable Technology, pp. 201–208, December 2008

9. Luo, C., Fei, Y., Ding, A.A.: Side-channel power analysis of XTS-AES. In: Design, Automation Test in Europe Conference Exhibition (DATE), pp. 1330–1335, March 2017

10. Mangard, S.: A simple power-analysis (SPA) attack on implementations of the AES key expansion. In: Lee, P.J., Lim, C.H. (eds.) ICISC 2002. LNCS, vol. 2587, pp. 343–358. Springer, Heidelberg (2003). https://doi.org/10.1007/3-540-36552-4_24

11. Michel, L.D., Van Hentenryck, P.: Constraint satisfaction over bit-vectors. In: Milano, M. (ed.) CP 2012. LNCS, pp. 527–543. Springer, Heidelberg (2012). https://doi.org/10.1007/978-3-642-33558-7_39

12. Mohamed, M.S.E., Bulygin, S., Zohner, M., Heuser, A., Walter, M., Buchmann, J.: Improved algebraic side-channel attack on AES. J. Cryptogr. Eng. **3**(3), 139–156 (2013). https://doi.org/10.1007/s13389-013-0059-1

13. Mpalane, K., Gasela, N., Esiefarienrhe, B.M., Tsague, H.D.: Vulnerability of advanced encryption standard algorithm to differential power analysis attacks implemented on ATmega-128 microcontroller. In: 2016 Third International Conference on Artificial Intelligence and Pattern Recognition (AIPR), pp. 1–5, September 2016

14. Oren, Y., Kirschbaum, M., Popp, T., Wool, A.: Algebraic side-channel analysis in the presence of errors. In: Mangard, S., Standaert, F.-X. (eds.) CHES 2010. LNCS, vol. 6225, pp. 428–442. Springer, Heidelberg (2010). https://doi.org/10.1007/978-3-642-15031-9_29

15. Oren, Y., Renauld, M., Standaert, F.-X., Wool, A.: Algebraic side-channel attacks beyond the hamming weight leakage model. In: Prouff, E., Schaumont, P. (eds.) CHES 2012. LNCS, vol. 7428, pp. 140–154. Springer, Heidelberg (2012). https://doi.org/10.1007/978-3-642-33027-8_9

16. Oren, Y., Weisse, O., Wool, A.: Practical template-algebraic side channel attacks with extremely low data complexity. In: Proceedings of the 2nd International Workshop on Hardware and Architectural Support for Security and Privacy, HASP 2013, pp. 7:1–7:8. ACM, New York (2013). https://doi.org/10.1145/2487726.2487733

17. Oren, Y., Wool, A.: Side-channel cryptographic attacks using pseudo-boolean optimization. Constraints **21**(4), 616–645 (2016). https://doi.org/10.1007/s10601-015-9237-3

18. Ors, S.B., Gurkaynak, F., Oswald, E., Preneel, B.: Power-analysis attack on an ASIC AES implementation. In: International Conference on Information Technology: Coding and Computing, Proceedings, ITCC 2004, vol. 2, pp. 546–552, April 2004

19. Örs, S.B., Oswald, E., Preneel, B.: Power-analysis attacks on an FPGA – first experimental results. In: Walter, C.D., Koç, Ç.K., Paar, C. (eds.) CHES 2003. LNCS, vol. 2779, pp. 35–50. Springer, Heidelberg (2003). https://doi.org/10.1007/978-3-540-45238-6_4

20. Renauld, M., Standaert, F.-X.: Algebraic side-channel attacks. In: Bao, F., Yung, M., Lin, D., Jing, J. (eds.) Inscrypt 2009. LNCS, vol. 6151, pp. 393–410. Springer, Heidelberg (2010). https://doi.org/10.1007/978-3-642-16342-5_29

21. Renauld, M., Standaert, F.-X., Veyrat-Charvillon, N.: Algebraic side-channel attacks on the AES: why time also matters in DPA. In: Clavier, C., Gaj, K. (eds.) CHES 2009. LNCS, vol. 5747, pp. 97–111. Springer, Heidelberg (2009). https://doi.org/10.1007/978-3-642-04138-9_8

22. Song, L., Hu, L., Sun, S., Zhang, Z., Shi, D., Hao, R.: Error-tolerant algebraic side-channel attacks using BEE. In: Hui, L.C.K., Qing, S.H., Shi, E., Yiu, S.M. (eds.) ICICS 2014. LNCS, vol. 8958, pp. 1–15. Springer, Cham (2015). https://doi.org/10.1007/978-3-319-21966-0_1

23. Standaert, F.-X., Mace, F., Peeters, E., Quisquater, J.-J.: Updates on the security of FPGAs against power analysis attacks. In: Bertels, K., Cardoso, J.M.P., Vassiliadis, S. (eds.) ARC 2006. LNCS, vol. 3985, pp. 335–346. Springer, Heidelberg (2006). https://doi.org/10.1007/11802839_42

24. Standaert, F.-X., van Oldeneel tot Oldenzeel, L., Samyde, D., Quisquater, J.-J.: Power analysis of FPGAs: how practical is the attack? In: Y. K. Cheung, P., Constantinides, G.A. (eds.) FPL 2003. LNCS, vol. 2778, pp. 701–710. Springer, Heidelberg (2003). https://doi.org/10.1007/978-3-540-45234-8_68

25. Standaert, F.-X., Örs, S.B., Preneel, B.: Power analysis of an FPGA. In: Joye, M., Quisquater, J.-J. (eds.) CHES 2004. LNCS, vol. 3156, pp. 30–44. Springer, Heidelberg (2004). https://doi.org/10.1007/978-3-540-28632-5_3

26. Standaert, O.X., Peeters, E., Rouvroy, G., Quisquater, J.J.: An overview of power analysis attacks against field programmable gate arrays. Proc. IEEE $94(2)$, 383–394 (2006)

27. Veyrat-Charvillon, N., Gérard, B., Standaert, F.-X.: Soft analytical side-channel attacks. In: Sarkar, P., Iwata, T. (eds.) ASIACRYPT 2014. LNCS, vol. 8873, pp. 282–296. Springer, Heidelberg (2014). https://doi.org/10.1007/978-3-662-45611-8_15

28. Zhao, X., Zhang, F., Guo, S., Wang, T., Shi, Z., Liu, H., Ji, K.: MDASCA: an enhanced algebraic side-channel attack for error tolerance and new leakage model exploitation. In: Schindler, W., Huss, S.A. (eds.) COSADE 2012. LNCS, vol. 7275, pp. 231–248. Springer, Heidelberg (2012). https://doi.org/10.1007/978-3-642-29912-4_17

State-of-the-Art: Security Competition in Talent Education

Xiu Zhang[1,2], Baoxu Liu[1], Xiaorui Gong[1], and Zhenyu Song[1(✉)]

[1] Institute of Information Engineering, Chinese Academy of Sciences,
Beijing, China
{zhangxiu,liubaoxu,gongxiaorui,songzhenyu}@iie.ac.cn
[2] School of Cyber Security, University of Chinese Academy of Sciences,
Beijing, China

Abstract. Security competitions have become increasingly popular events for recruitment, training, evaluation, and recreation in the field of computer security. And among these various exercises, Capture the flag (CTF) competitions have the widest audience. Participants in CTF of Jeopardy style focus on solving several specific challenges independently while participants in CTF of attack-defense mode concentrate on vulnerable service maintenance and vulnerability exploitation on an end-target box. However, according to a report published by TREND MICRO Corporation, there are six stages of a typical Targeted Attack: (1) Intelligence Gathering (2) Point of Entry (3) Command and Control Communication (4) Lateral Movement (5) Asset Discovery and (6) Data Exfiltration. Further, Lateral Movement is the key stage where threat actors move deeper into the network. Because of the lack of large-scale complex network environment, CTF cannot simulate a complete network penetration of the six stages, especially the Lateral Movement. It is indispensable to perform the Lateral Movement the skill of *Network Exploring* which is not included by security competitions at present. So we create *Explore-Exploit* which is an attack-defense mode competition that models the network penetration scenario, and promotes the participant's skill of *Network Exploring*. This paper is trying to convey a better methodology for teaching practical attack-defense techniques to participants through an alternative to CTF.

Keywords: Security competition · Talent education
Network penetration scenario · Attack-defense mode · *Explore-Exploit*

1 Introduction

The U.S. Commerce Department's National Institute of Standards and Technology (NIST) released the NICE [1] Cybersecurity Workforce Framework [2].

This paper was supported by the Key Laboratory of Network Evaluation Technology of China Academy of Sciences, and Beijing Key Laboratory of Network Security Protection Technology. This paper was funded by the Beijing Municipal Science and Technology Commission D161100001216001, Z161100002616032 project.

© Springer International Publishing AG, part of Springer Nature 2018
X. Chen et al. (Eds.): Inscrypt 2017, LNCS 10726, pp. 461–481, 2018.
https://doi.org/10.1007/978-3-319-75160-3_27

There are more than 50 work roles defined in the NCWE framework. Each work role is defined by extensive sets of related Knowledge, Skills, and Abilities (KSAs) and tasks. KSAs are the attributes required to perform a job and are generally demonstrated through relevant experience, education, or training. Furthermore, another Job Performance Model for Smart Grid Cybersecurity, Competency Box [3], was put forward. Competency Box is a three-dimensional model for understanding an individual's development and position along a learning trajectory from novice to master. Knowledge is defined as the understanding of strategy or procedure and is measured from shallow to deep. Skills are defined as the consistency of performance and are measured from inconsistent to consistent. Abilities are defined as the transfer across domains and are measured from narrow to broad.

From the perspective of talent cultivation, the textbook theoretical teaching is to handle the dimension of knowledge acquisition. The live security exercise is particularly well-suited to support hands-on experience, as they usually have both an attack and a defense component. Security competitions have become a popular way to foster security education by creating a competitive environment in which participants go beyond the effort usually required in traditional security courses. In other words, security competitions are trying to handle the remaining two dimensions, the proficiency of skills and the abilities to address the troubles in dynamic circumstance. In this paper, we pick out nine typical security competitions and conduct a comprehensive analysis of them. In view of the form, they are classified from seven angles. And in view of the content, eight indicators of the skill dimension are summarized.

Among multifarious security competitions, CTF is the most mature and the most popular one. As recorded in CTFtime [4], there are 129 CTF events which serve 9808 teams. Security competition of anti-penetration scenario is not yet widespread. But in America, there are some educational sponsored exercises for defensive purposes. At the high school level there is CyberPatriot [5], at the college level is the National Collegiate Cyber Defense Competition (NCCDC) [6] and for the U.S. Military Academies is the CyberDefenseExercise (CDX) [7]. The NCCDX is the nation's largest competition of this kind.

Participants in CTF of Jeopardy style focus on solving several specific challenges independently while Participants in CTF of attack-defense mode concentrate on vulnerable service maintenance and vulnerability exploitation on an end-target box. However, according to a report published by TREND MICRO Corporation [8], there are six stages of a typical Targeted Attack: (1) Intelligence Gathering (2) Point of Entry (3) Command and Control Communication (4) Lateral Movement (5) Asset Discovery and (6) Data Exfiltration. Further, Lateral Movement is the key stage where threat actors move deeper into the network. Because of the lack of large-scale complex network environment, CTF does not simulate a complete network penetration of the six stages, especially the Lateral Movement. Not to mention other competitions can do it.

Various security competitions cover eight dimensions of the skill: (1) Automated (2) Program Analysis and Reverse Engineering (3) Vulnerability Discovery and Exploiting (4) Web Security (5) Cryptography (6) Digital Forensic and Audit (7) System Management and Strategy (8) Social Engineering. Modern attacks are usually multi-stage, multi-host attacks, which combined the vulnerabilities existing of different machines. A determined attacker is not likely to stop at the machine he first compromises but can be expected to try to penetrate deeper into the network by jumping from one machine to another [9]. So it is indispensable the skill of *Network Exploring* which is not included by security competitions at present also because of the lack of large-scale complex network environment.

We create a new member to the family of security competition as an alternative to CTF. *Explore-Exploit* is short of *Network Exploring and Vulnerability Exploiting*. In view of the form, *Explore-Exploit* is an attack-defense mode competition that models the network penetration scenario. In view of the content, *Explore-Exploit* handles the issue of a new skill: *Network Exploring*. This paper is trying to convey a better methodology for teaching practical attack-defense techniques to participants through a new security competition.

In this paper, we make the following contributions:

– In this paper, we offer a state-of-the-art analysis of security competitions, including the seven kinds of classification, the eight dimensions of skill, and the attribute analysis and the skill analysis of nine representatives.
– We create a new type of scenarios for security competitions, *Explore-Exploit*, as an alternative to CTF. In view of the attribute, *Explore-Exploit* is a competition of attack-defense modeling the six stages of a network penetration scenario. In view of the skill, *Explore-Exploit* can handle the issue of a new skill: *Network Exploring*.

The rest of this article is structured as follows. Section 2 offers a state-of-the-art of security competitions, including attribute analysis and the skill analysis. Section 3 presents the overall design of *Explore-Exploit*. Experiences of running the first *Explore-Exploit* competition are shared in Sects. 4, 5 and 6. Section 4 describes the implementation. Section 5 gives a review of the actual progress. Section 6 offers the result analysis. Section 7 discussed the comprehensive evaluation of the first *Explore-Exploit* competition and the future state vision. Section 8 presents the conclusion.

2 State-of-the-Art: Security Competition

In this section, we offer a state-of-the-art analysis of security competitions, including the attribute analysis from seven angles and skill analysis from eight dimensions.

2.1   Introduction of Nine Representative Competitions

We pick out nine representative ones from a variety of security competitions. They are:

- (1) the Cyber Grand Challenge (CGC)
  The 2016 CGC [10] was a competition created by The Defense Advanced Research Projects Agency (DARPA) in order to develop automatic defensive systems capable of reasoning about flaws, formulating patches and deploying them on a network in real time. The organizers of the Cyber Grand Challenge have put much thought into designing a competition for automated binary analysis systems. For example, they addressed the environment model problem by creating a new OS specifically for the CGC: the DECREE OS. DECREE is an extremely simple operating system with just 7 system calls. Despite the simple environment model, the binaries provided by DARPA for the CGC have a wide range of complexity. They range from 4 KB to 10 MB in size and implement functionality ranging from simple echo servers to web servers, to image processing libraries [11].
- (2) DEFCON CTF Qualifier
  DEFCON is one of the world's largest hacker conventions that is held annually in Las Vegas, with the first DEF CON taking place in June 1993. DEFCON CTF Qualifier is an online Jeopardy style CTF where teams all over the world compete for several slots of the DEFCON Final. As usual, there were five categories of challenges: Pwnable, Reversing, Crypto, Web, and Misc.
- (3) DEFCON CTF Final
  DEFCON CTF Final is the world's top CTF event and is an attack-defense CTF. Different from CTF of Jeopardy style where the Flag is changeless along with the whole competition, in CTF of attack-defense mode, each service contains a Flag which is updated in every Tick (a period of time). In other words, participants should acquire the Flag in every Tick. This requires them to write an automated script to acquire the Flag and to submit for scoring.
- (4) DEFCON CTF Social Engineering CTF (SECTF)
  The SECTF [12] is an annual event and is held at the DEFCON Hacking Convention in Las Vegas. The competition was formed to demonstrate how serious social engineering threats could be to companies and how even novice individuals could use these skills to obtain damaging information.
- (5) iCTF 2017
  The iCTF is the world's largest and longest-running educational hacking competition that integrates both attack and defense aspects in a live setting. The iCTF 2017 [13] was run exclusively in the cloud.
- (6) Pwn2Own
  Pwn2Own [14] is held annually at the CanSecWest security conference, beginning in 2007. Contestants are challenged to exploit widely used software, such as Apple MacOS, Google Chrome, Microsoft Office Word. And the contestants are challenged to exploit Zero-day vulnerabilities.

- (7) GeekPwn
  GeekPwn [15] is the first worldwide security geek contest for smart life. Smart devices and IoT products that are available in public markets are all acceptable PWN targets (e.g., printer, camera, POS machine, drone, robot, smart watch).
- (8) Cyber-Defense Exercise (CDX)
  The CDX is sponsored by the Information Assurance Directorate of the US National Security Agency (NSA). The CDX 2016 challenges students to design, build, and defend the most sophisticated network they encounter during their time at the United States Military Academy. Furthermore, the CDX provides students an experience in leading a 26-man team.
- (9) Cyber-Physical System Based on Cyber Defense Competition (CPS-CDC)
  CPS-CDC [16] is focused on the protection and exploitation of critical infrastructures (e.g., power grid and water grid). The CPS-CDC emerged in 2016.

In particular, there is a kind of security competition that applies the idea of "Gamification" [17] to address specific issues about security. For instance, the contest outcomes of BiBiFi [18] shed light on factors that correlate with building secure software and breaking insecure software. This kind of competition does not focus on the training of talent cultivation, so it falls outside the scope of this paper.

There are also some noteworthy security competitions. A security Treasure Hunting [19] scenario was introduced in the iCTF 2008. Unlike CTF where each team sets up a vulnerable virtual server, in the iCTF 2008 the organizers created 39 identical, yet independent copies of a small network with 5 terminal nodes. Another interesting competition is NetKotH (Network King of the Hill) [20]. The NetKotH framework provides the source code for a basic scoring engine (ScoreBot) [20] and uses off the shelf pre-built challenges that can be downloaded freely from multiple sites such as Vulnhub [21]. These competitions raise some new ideas, but they have not evolved into a mature pattern. They have not been organized by a particular organization periodically, and they do not have a big impact on the security field. So they are also listed outside the scope of the analysis.

2.2 Attribute Analysis of Nine Representative Competitions

Based on these nine representations, security competitions are classified from seven aspects: Object-oriented, Access Type, Interactive Mode, Game Model, Scenario Type, Participant's Role, and Domain-oriented.

- (1) Object-oriented
  Security competitions are all human-oriented except CGC. CGC is the world's first all machine hacking tournament.
- (2) Access Type
  Usually, only CTF of Jeopardy style can be held online and all other competitions require contestants to be on-site. The iCTF 2017 is a distinction because a CTF-as-a-service system [13], ShellWePlayAGame [22], was developed by

Table 1. Seven kinds of classification of security competitions.

Foundation	Category
Object-oriented	Human
	Machine (Automation Attack-Defense)
Access Type	Online
	On-site
Interactive Mode	Interactive
	Challenge-based
Scenario Type	Capture the flag (CTF)
	Pwn
	Anti-penetration
	Penetration (*Explore-Exploit*)
Game Model	Jeopardy
	Attack-Defense
	Attack-only
	Defense-only
Participant's Role	Peer-to-peer
	Red-team and Blue-team
	Red-team
	Blue-team
Domain-oriented	Legacy Network Security
	Cyber-physical System
	Smart Life Equipment

the Shellphish Team. ShellWePlayAGame allows educators to instantiate the competition in the cloud so that the participating teams can seamlessly connect and play.

– (3) Interactive Mode
Precisely, security competitions can be divided into two modes: challenge-based mode (non-interactive mode) and interactive mode. Challenge-based competitions are a form of take-home test. They do not include any interaction with other teams. For example, CTF of Jeopardy style [23] is structured in a way that provides the participants with a number of challenges that address different skills at the different level of complexity. For another example, while in CDX 2016 [7] each Blue-team is actually competing against each other for points, they ostensibly form a single coalition network. Instead, interactive competitions focus precisely on the interaction between teams. For example, in CTF of attack-defense mode [23], every participant is given the same system. Their task is to identify flaws in their own copy of the server, patch their own services without breaking the service's functionality, and use the same knowledge to attack the other participants.

- (4) Scenario Type
  From the perspective of the scenario, security competitions are divided into three categories. The biggest feature of CTF scenario is that Flag is the basis for scoring. We take the CDX 2016 as a typical example of the anti-penetration scenario. CDX 2016 [7] divided its participants into four cells, labeled blue, white, gray, red. The exercise tasks the red cells with attacking the blue-cell networks. The blue-cell consists of the participating Academy teams who are tasked with defending the blue-cell networks. The scenario of Pwn is much simpler. The participants show the zero-day vulnerability exploitation in quite a short time period and are scored by a manual check of referees.
- (5) Game Model
  The categories of Game Model correspond to that of the Scenario Type. Among them, CTF scenario is divided into two categories: Jeopardy and attack-defense.
- (6) Participant's Role
  The categories of Participant's Role correspond to that of the Game Model. Since there is no interaction between participants in CTF of Jeopardy style and the participants do not obviously show any offensive and defensive intentions, we define the role of participants in CTF of Jeopardy style as Peer-to-peer.
- (7) Domain-oriented
  DEFCON CTF dates back to 1996. GeekPwn originates from 2014. CPS-CDC emerges in 2016. The three competitions reflect the evolutionary trend of cyber security.

Details of the foundation and category are summarized in Table 1. The attribute analysis of nine representative events is shown in Table 4 (at the end of the paper). DEFCON SECTF is an exception which is aimed at non-technical factors. The note NA is the abbreviation of Not Applicable.

2.3 Skill Analysis of Nine Representative Competitions

Eight dimensions are summarized: Automation, Program Analysis and Reverse Engineering, Vulnerability mining and Exploiting, Web Security, Cryptography, Digital Forensic and Audit, System Management and Strategy, and Social Engineering.

- (1) Automation
  As defined by the Areas of Excellence (AoE) of CGC final event [24], the hacking skill of this dimension means autonomous analysis, autonomous patching, autonomous vulnerability scanning, autonomous service resiliency, autonomous network defense.
- (2) Program Analysis and Reverse Engineering
  Program analysis is the process of automatically analyzing the behavior of computer programs regarding a property such as correctness, robustness, safety, and liveness. Reverse engineering is the process of taking a software program's binary code and recreating it so as to trace it back to the original source code.

- (3) Vulnerability Discovery and Exploiting
Vulnerability Discovery usually begins with the fuzzy testing. Fuzz testing (fuzzing) is a quality assurance technique used to discover coding errors and security loopholes in software, operating systems or networks. It involves inputting massive amounts of random data, called the fuzz, to the test subject in an attempt to make it crash. Vulnerability exploiting is an attack on a computer system, especially one that takes advantage of the particular vulnerability that the system offers to intruders. The objective of many exploits is to gain control over an asset.
- (4) Web Security
Website security audit identifies the security risks by looking for weaknesses in the website code, errors in the Web server settings and by detecting the results of viruses, trojans or worms. Web application security is the process of securing confidential data stored online from unauthorized access and modification. An attacker can take advantage of common vulnerabilities such as SQL injection, remote file inclusion (RFI), or cross-site scripting (XSS) in order to upload the Web shell. The common functionality of a Web shell includes shell command execution, code execution, database enumeration and file management.
- (5) Cryptography
The hacking skill of this dimension means the reduction of the encrypted information which is achieved by cracking the decryption algorithm, or by the side channel, or by the man-in-the-middle attack.
- (6) Digital Forensic and Audit
Memory forensics is the analysis of volatile data in a computer's memory dump and can provide unique insights into the runtime system activity, including open network connections and recently executed commands or processes. Network forensics is concerned with the monitoring and analysis of computer network traffic, both local and WAN/Internet, for the purposes of information gathering, evidence collection, or intrusion detection. The audit is the scrutiny of an organization's computer system to determine its level of vulnerability to threat actors.
- (7) System Management and Strategy
This dimension is for a particular role, called cyber-strategic leaders [25] who will lead, manage, and oversee cyber defense and cyber operations in this dynamic and ever-changing digital environment. These individuals need not have specific training in engineering or programming but must be equipped with the deep understanding of the cyber context in which they operate. As for security competitions, this dimension involves service maintenance, problem-finding, quick solution.
- (8) Social Engineering
Social Engineering [26] is defined as the use of deception to manipulate individuals into divulging confidential or personal information that may be used for fraudulent purposes.

One of the contributions of this paper is to offer a state-of-the-art analysis of security competitions as summarized in Tables 1, 4 and 5.

3   Explore-Exploit: Overview

Different from the current security competitions, the character traits of *Explore-Exploit* are reflected in four aspects: Network Scale, Scoring Mechanism, Winning Strategy, Weight Calculation.

3.1   Network Scale

Explore-Exploit is designed to simulate a complete network penetration of the six stages: (1) Intelligence Gathering (2) Point of Entry (3) Command and Control Communication (4) Lateral Movement (5) Asset Discovery and (6) Data Exfiltration. Especially, *Explore-Exploit* is designed to simulate the stage of Lateral Movement where threat actors move deeper into the network. The scenario of *Explore-Exploit* must involve a large-scale complex network environment. So the game network is composed of more than one subnetwork, a lot of GameBoxes, all kinds of weaknesses or vulnerabilities. We define **GameBox** as a Virtual Machine (VM) with vulnerabilities.

3.2   Scoring Mechanism

A consensus is that competition must be operational and measurable. Current competitions are either Capture the Flag or Plant the Flag (the act of writing the team tag in a specific location so the ScoreBot can see it) [27]. In a nutshell, Flag is the basis for scoring.

In *Explore-Exploit*, ControlRight is the sole basis for scoring. On the one hand, from the point of view of the threat actor, he could do anything he wanted provided that he had gained ControlRight of a GameBox. On the other hand, from the perspective of scoring, ControlRight is measurable. In addition, from the organizer's point of view, doing so saves time and effort. We define **ControlRight** as the administrative or root privilege that depends on the type of operating systems. More specifically, if the operating system of a computer belongs to Windows family, ControlRight is an administrative privilege, otherwise root privilege.

3.3   Winning Strategy

Explore-Exploit provides participants with two winning strategies corresponding to the two scenarios. As for security competitions at present, the challenges, the vulnerable services, or the vulnerabilities are not related. So a conservative strategy is that participants try to infiltrate GameBoxes as many as possible ignoring the connection between them. This is the scenario similar to the NetKotH.

In a real-world scenario, it is important to consider multi-stage, multi-host attacks. In particular, the combination of vulnerabilities of multi-host makes up an Attack Path [28]. And several attack paths make up an Attack Graph [28]. In *Explore-Exploit*, GameBoxes are interrelated. Despite the large number of GameBoxes, only one is the ultimate goal. An aggressive strategy is that the participants concentrate on GameBoxes that would contribute to approaching or infiltrating the target GameBox, without devoting many energies to those that are helpless for achieving their ultimate goal. This is a scenario similar to the Treasure Hunting.

However, *Explore-Exploit* can be a combination of the scenario of NetKotH and the scenario of Treasure Hunting. Among these massive GameBoxes, only one GameBox is the ultimate goal of the Target Attack. Participants must find it before compromised it. Thus *Explore-Exploit* can train the participant's skill of *Network Exploring*. If there was more than one team, they shall compete for the ControlRight of the ultimate goal.

3.4   Weight Calculation

As for security competitions at present, weight depends only on the difficulty. In *Explore-Exploit*, the weight of each GameBox is assessed on three dimensions: Difficulty of Vulnerability Exploiting, Network Position, and Value of Post-penetration.

- (1) Difficulty of Vulnerability Exploiting
  The assessment of the difficulty of vulnerability exploiting in *Explore-Exploit* is the same with that in security competitions at present.
- (2) Network Position
  As mentioned before, the game network is made up of a few subnetworks, each of which is characterized by a unique network connectivity. For each GameBox, the network location can determine its network properties: it is obvious that nodes located in one subnetwork can directly access each other, while nodes located in different subnetworks must do it through forwarding by upper layer devices, such as a firewall. So the firewall can determine the network accessibility of these nodes.
- (3) Value of Post-penetration
  The value of post-penetration of a GameBox is that an attacker can get an additional benefit on this GameBox after gain control of it. An example can be seen in Sect. 5.5. In the game network of *Explore-Exploit*, only a small number of GameBoxes were given the value of post-penetration deliberately by the organizers. And as for the participants, these GameBoxes play a decisive role in performing a successful multi-host attack to the ultimate goal.

Another contribution of this paper is to create a new type of scenarios for security competitions as summarized in Table 1. The attribute analysis of *Explore-Exploit* is summarized in Table 4. The skill analysis of *Explore-Exploit* is summarized Table 5.

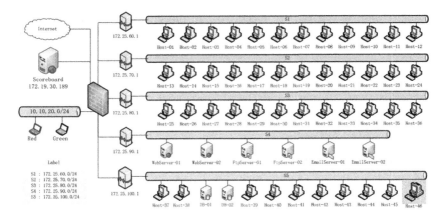

Fig. 1. The topology-by-design of the first *Explore-Exploit* competition. (Color figure online)

4   The First Explore-Exploit Competition: Design

Precisely, the first *Explore-Exploit* competition simulated the phase of Lateral Movement with an assumption that threat actors had gained a foothold of the corporation's internal network. But he did not know which GameBox is the ultimate goal and where it is exactly located. Depicted in the Fig. 1, the ultimate goal was colored in yellow. Participants of the first *Explore-Exploit* competition made up two teams: the Red and the Green.

4.1   Topology-by-Design

Depicted in Fig. 1, the game network consisted of 54 GameBoxes. Only one GameBox is the ultimate goal which can be identified by a peculiar account username *target*. These GameBoxes composed of five subnetworks and occupied one network segment (172.25.0.0/16). All participates were located on a different network segment (10.10.20.0 /24) which was the only segment that could access the Internet. A separate network segment (172.19.0.0/16) was allocated to the scoreboard server. All of these were connected to one firewall device which was used to achieve the access control of network layer based on the IP address.

More specifically, the game network was a prototype for a corporation's intranet. It was divided into five districts. There were three interconnected districts (S1, S2, S3) with similar structure, so two teams (Green, Red) were arranged to enter the game network separately. Another district (S4) was where six servers located. There were two Web servers, two E-mail servers, and two FTP servers. The last district (S5) was a physically isolated domain where the ultimate goal resided.

The firewall device would be responsible for packet forwarding in accordance with the rules of engagement. In the first three districts (S1, S2, S3), a third of

GameBoxes were permitted to access to the physically isolated domain (S5) so that discerning and infiltrating these GameBoxes provided a channel of approaching the ultimate goal. GameBoxes located at the physically isolated domain (S5) including the ultimate goal would visit these public servers periodically. So the six servers were also the perfect springboard to further penetrate the GameBox that was marked as the ultimate goal.

4.2   Vulnerable-by-Design

Different types of vulnerabilities were designed in the GameBoxes of the game network, so they allowed participants many more choices. A half of the Gameboxes were designed with customized vulnerable applications and services. These vulnerable applications and services were crafted by the experienced CTF players. So participants who have experience with CTF are more suitable for penetrating these GameBoxes. Another half of GameBoxes were designed by reproducing the context of vulnerabilities recorded in the Common Vulnerability Exposure (CVE). So participants who are adept at using penetration testing software (e.g., Metasploit Framework [29]) combined with public exploit database (e.g., The Exploit Database [30]) are more suitable for penetrating these GameBoxes.

Table 2 is a list of all vulnerabilities designed in GameBoxes of the game network. A vulnerability was listed only if we had verified that it did exist and could be exploited. Usually, more than one vulnerability of different types was designed in one GameBox. The version and quantity of operating systems involved are depicted in Fig. 3.

4.3   Interactive-by-Design

The first *Explore-Exploit* competition took the way that participant informed referees on his own initiative when he gained control of a GameBox and received visual feedback via the scoreboard. In this competition, referees were granted privileges to access all GameBoxes for the purpose of manual verification. The referees also managed an administrative interface of the scoreboard to type in score records.

5   The First Explore-Exploit Competition: Review

This section offers a detailed process of the first *Explore-Exploit* competition. The screen capture of the scoreboard at the end of this competition is Fig. 2.

5.1   Prior Training

Before the competition, each team was given an offering of four pages introducing *Explore-Exploit* as well as a few instructions. They were guided to add two

Table 2. Summary of all designed vulnerabilities in GameBoxes of the first *Explore-Exploit* competition.

Vulnerability Type (Based on the type of affected softwares)			Description or CVE-ID
Operating system	Windows Xp	SMB	Remote Code Execution
	Ubuntu	Overlayfs	Local Privilege Escalation
Platform	Java	JBoss	Java Deserialization Bug
	PHP	Joomla, phpStudy phpFileManager Discuz!, ComsenzEXP	Vulnerabilities (which can be exploited to upload Web shells)
Network service	Database	Redis, Mysql	Unauthorized Access or Authentication Bypass
	Remote access	SSH, Telnet, RDS	Weak Password
	File Share	FTP, samba	Defect Configuration
Client application	Document Viewer	Microsoft Office 2013 Adobe Acrobat Reader	CVE-2014-4114 CVE-2011-2462
	Browser and its Plug-in	Internet Explore Mozilla Firefox Adobe Flash Player	CVE-2014-6332 CVE-2015-5119 CVE-2011-2110
	Device manager	WebGate eDVR Manager	CVE-2015-2098

static routing rules to get an initial foothold. Because the particular vulnerability often affects the particular version of the operating system, the version of operating systems involved in the first *Explore-Exploit* competition was also disclosed to participants in the offering. This information as depicted in Fig. 3 could help participants make informed decisions and concentrate on more purposeful attempts. Participants were also trained to equip themselves with tools to compromise a GameBox, and to patch the vulnerabilities, and to deploy defensive strategies.

5.2 Vulnerability Exploiting: Uploading Web Shells

The Red team gained the first blood shortly after the start of the competition. They controlled Host-06 by uploading a Web shell. After cracking phpMyAdmin (a free software tool written in PHP, and intended to handle the administration of MySQL over the Web), a trojan named red.php was uploaded to the directory *C:/WWW/* and executed with an SQL query:

```
select '<?php @eval($\_POST[pass]);?>'
INFO OUTFILE 'C:/WWW/red.php'
```

A PHP Web shell would run with the same user and privileges, with which PHP is running. In this case, it was lucky that PHP was running with user

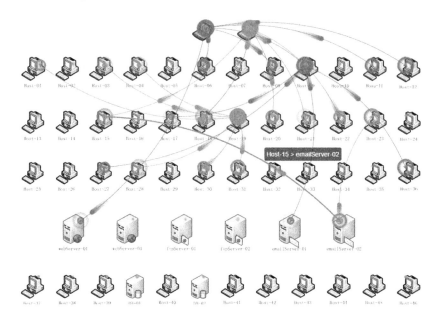

Fig. 2. The screen capture of the scoreboard at the end of the first *Explore-Exploit* competition. (Color figure online)

NT Authority System. In another case (Host-08), PHP was running with user *www-data*, so they had to escalate privilege locally by exploiting a vulnerability (CVE-2015-1328) in Ubuntu 12.04.

5.3   Network Exploring: Weak Password Cracking

When a member of the Red team sniffed an active OpenSSH service and fortunately established a session, he realized that *123456* might be the universal password. He took advantage of it to crack all active service he could sniff, including OpenSSH service, Telnet, and Remote Desktop Services (RDS). Eventually, he established another six OpenSSH session and one Telnet session.

By manipulating Iptables (a tool used to set up, maintain and inspect the tables of IP package filter rules in the Linux kernel), the Red team member ensured that the port of OpenSSH or Telnet was only accessible to their computers. It was forbidden by the Code of Conduct to replace the original password with a much stronger one.

5.4   Lateral Movement: Pivoting Techniques

After scanning with Nmap, the member of Green team found that the operating system of four GameBoxes was Windows XP SP2 English version. Then the notorious vulnerability numbered MS08-067 came to his mind. The exploiting

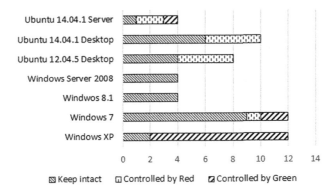

Fig. 3. Number of GameBoxes created, and classified according to the version of operating system, and controlled by the Red team and the Green team.

of this vulnerability is a classic case appearing in multifarious tutorials concerning penetration testing based on Metasploit Framework. A *meterpreter* session was established eventually, and another three GameBoxes located in the same subnetwork (S2) all got compromised soon.

Then they leveraged pivoting technique by executing *autoroute* script in *meterpreter* session established on one compromised GameBox (Host-19) for routing traffic from other inaccessible subnetworks. Pivoting technique aided them in finding out another six GameBoxes with the same vulnerability and infiltrating them. This breakthrough ended the backward state that Green Team stayed in. Corresponding to the scoreboard was the dense green curves among GameBoxes in the top three layers.

5.5 Asset Discovery: Value of Post-penetration

The value of post-penetration was elucidated by an interactive experiment that carried out near the end of this competition. On the scoreboard Fig. 2, there was a green curve, which started from EmailServer-02 and rebounded to Host-23 in the upper layer.

The authentication of this EmailServer-02 was bypassed due to defect configuration. After it was completely compromised by escalating privilege locally, Green team members observed that the FTP service was running. Thus, they created a Trojan for windows and uploaded it to the common directory. To cope with members of Green team, a staff was involved in synchronizing files using pre-install application FileZilla on Host-23 which was controlled once the Trojan executed.

In this experiment, EmailServer-02 was a springboard which was used to infect clients as a contaminated source. A springboard is distinguished with pivoting technique which was used to route traffic through a compromised GameBox.

6   The First Explore-Exploit Competition: Result

This section talks about the actual progress of the first *Explore-Exploit* competition, including the statistics, the achievements, and the weakness. Figure 3 shows the number of GameBoxes deployed and compromised based on the version of the operating system. At the end of the competition, the total number of GameBox that was compromised is 24. Among them, the Red team gained access to 11 GameBoxes while the Green team controlled 13 GameBoxes.

Table 3. Number and percentage of vulnerabilities initially designed, and that of vulnerabilities finally exploited.

Vulnerability type (Based on the type of the affected software)		Design number	Design percentage	Exploit number	Exploit percentage
Operating system	MS08-067	12	22%	10	80%
	CVE-2015-1328	22	41%	2	5%
Platform	Java deserialization bug	2	4%	0	0%
	PHP web shells	6	11%	4	67%
Network service	Authentication bypass	2	4%	1	50%
	Weak password	15	28%	8	53%
	Defect configuration	2	4%	1	50%
Client application		32	59%	0	0%

Table 3 counts the number and percentage of vulnerability that was designed at the beginning of the competition, and that of vulnerability which was exploited when the competition ended. Objectively speaking, the actual progress in this competition is listed as follows:

- Computers with a vulnerability in the server service of the operating system which allows remote code execution are most likely to be compromised. After scanning, such vulnerability can be easily identified. For the most part, the exploit has been collected in common exploit database. So even the *Script Kiddies* can infiltrate such vulnerable computers with the help of tools.
- The vulnerability in PHP components which can be exploited to upload Web shells is a convenient channel for further compromising a computer. A PHP Web shell would run with the same user and privileges, with which PHP is running. So vulnerability of this type would be exploited usually combined with a vulnerability of local privilege escalation.
- Network service penetration is also an effective channel to infiltrate a Game-Box. For defect configuration, an attacker should be familiar with the defected service. So he can constantly initiate malicious requests to discern potential improper configuration from the responses. This penetration strategy is

Table 4. Attribute analysis of nine typical security competitions.

	Object-oriented	Access Type	Interactive Mode	Game Model	Scenario Type	Participant's Role	Domain-oriented
CGC	Machine	On-site	Interactive	Attack-Defense	CTF	Red-team and Blue-team	Legacy Network Security
DEFCON CTF Qualifier	Human	Online	Challenge-based	Jeopardy	CTF	Peer-to-peer	Legacy Network Security
DEFCON CTF Final	Human	On-site	Interactive	Attack-Defense	CTF	Red-team and Blue-team	Legacy Network Security
DEFCON SECTF	Human	NA	NA	NA	NA	NA	NA
iCTF 2017	Human	Online	Interactive	Attack-Defense	CTF	Red-team and Blue-team	Legacy Network Security
Pwn2Own	Human	On-site	Challenge-based	Attack-only	Pwn	Red-team	Legacy Network Security
GeekPwn	Human	On-site	Challenge-based	Attack-only	Pwn	Red-team	Smart life equipment
CDX	Human	On-site	Challenge-based	Defense-only	Anti-penetration	Blue-team	Legacy Network Security
CPS-CDC	Human	On-site	Challenge-based	Defense-only	Anti-penetration	Blue-team	Cyber-physical System
Explore-Exploit	Human	On-site	Interactive	Attack-Defense	Penetration	Red-team and Blue-team	Legacy Network Security

Table 5. Skill analysis of nine typical security competitions.

	Automation	Program Analysis and Reverse Engineering	Vulnerability Discovery and Exploiting	Web Security	Cryptography	Digital Forensic and Audit	System Management and Strategy	Social Engineering	Network Exploring
CGC	★★★	★	★						
iCTF 2017	★	★★	★★	★	★	★	★		
DEFCON CTF Final		★★	★★	★	★	★			
DEFCON CTF Qualifier		★★	★★						
Pwn2Own		★★	★★★						
GeekPwn		★★	★★						
CDX		★				★★	★★★		
CPS-CDC						★★	★★	★	
DEFCON SECTF								★★★	
Explore-Exploit			★	★		★	★		★★★

equally applicable to unauthorized access and authentication bypass. Weak Password is a special category that would be easier to make use of once identified, but it takes the time to perform network scanning.

- None of the vulnerable client applications had been exploited owing to the lack of triggering behavior. All participants expressed the general truth that it was practically impossible to entirely bypass the graphical interface via command line without real user involved, especially for Windows operating system. For example, vulnerabilities in Web browsers needed to be triggered by clicking a malicious link.

All participants were so excited about *Explore-Exploit*. *Explore-Exploit* deepens the understanding of network topology and routing protocol. The complex network topology increases the enjoyment of exploring the unknown. It also provides them with an experimental environment to unrestrictedly attempt a variety of penetration testing tools. Moreover, there is no flag which avoids the trouble that cheating or plagiarism by sharing flags occurs in traditional CTF.

7   Discussion and Future Work

As for the first *Explore-Exploit* competition itself, two aspects need to be functionally enhanced. One motivation is that the next *Explore-Exploit* competition should more elaborately simulate a real-world intranet hacking scenario. Another motivation is to make this contest self-service thus the manual verification of referees will not be needed anymore. We have started to develop an Automatic Scoring System (ASS) since the beginning of 2016 and planned to put it into use in the next *Explore-Exploit* competition. An additional module, Virtual User Simulation (VUS), is integrated into the client of ASS which performs random operations to create an illusion as if someone is operating a computer.

In the long term, we are determined to develop a framework to implement *Explore-Exploit*-as-a-service similar with the framework CTF-as-a-service [22]. At present, the cost of holding the *Explore-Exploit* is too high and the workload is cumbersome. We are trying to establish an appropriate mechanism to hold *Explore-Exploit* periodically. Furthermore, if *Explore-Exploit* can be designed and implemented sufficiently complex and dynamic, *Explore-Exploit* is likely to be used to deal with the dimension of ability defined in the NCWF [2] better than current CTF.

8   Conclusion

We offer a state-of-the-art of security competitions, including the attribute analysis from seven angles and the skill analysis in eight dimensions. We create a new type of scenarios and contribute a new member to the family of security competitions. On account of a large-scale complex network environment, *Explore-Exploit* can model a network penetration scenario, especially the stage of Lateral Movement. And *Explore-Exploit* can handle the issue of training skill in *Network Exploring*. This paper shares the entire process as well as the preliminary outcome of first *Explore-Exploit* competition. For the competition itself, we are developing a framework to implement a more automated and sophisticated *Explore-Exploit* and trying to establish an appropriate mechanism to hold it periodically. We are also in the process of scaling the competition to collect larger data sets with the goal of making statistically significant correlations between various factors that influence the talent cultivation in the field of cyber security.

References

1. Paulsen, C., McDuffie, E., Newhouse, W., Toth, P.: Nice: creating a cybersecurity workforce and aware public. IEEE Secur. Privacy **10**, 76–79 (2012)
2. NIST: Nice Cybersecurity Workforce Framework, draft NIST Special Publication 800–181. http://csrc.nist.gov/publications/drafts/800-181/sp800_181_draft.pdf
3. O'Neil, L.R., Assante, M., Tobey, D.: Smart grid cybersecurity: Job performance model report. Technical report. Pacific Northwest National Laboratory (PNNL), Richland, WA (US) (2012)
4. CTF-Time. https://ctftime.org/
5. Cyberpatriot. https://www.uscyberpatriot.org/Pages/About/What-is-CyberPatriot.aspx
6. Nccdc. http://www.nationalccdc.org/
7. Petullo, W.M., Moses, K., Klimkowski, B., Hand, R., Olson, K.: The use of cyberdefense exercises in undergraduate computing education. In: ASE@ USENIX Security Symposium (2016)
8. How do threat actors move deeper into your network? http://about-threats.trendmicro.com/cloud-content/us/ent-primers/pdf/tlp_lateral_movement.pdf
9. Lufeng, Z., Hong, T., YiMing, C., JianBo, Z.: Network security evaluation through attack graph generation (2009)
10. Fraze, M.D.: Cyber Grand Challenge (CGC). https://www.darpa.mil/program/cyber-grand-challenge
11. Shoshitaishvili, Y., Invernizzi, L., Doupe, A., Vigna, G.: Do you feel lucky?: a large-scale analysis of risk-rewards trade-offs in cyber security. In: Proceedings of the 29th Annual ACM Symposium on Applied Computing, pp. 1649–1656. ACM (2014)
12. Hadnagy, M.F.C.: The def con 22 social-engineer capture the flagreport. https://www.social-engineer.org/wp-content/uploads/2014/10/SocialEngineerCaptureTheFlag_DEFCON22-2014.pdf
13. Doupé, A., Vigna, G.: Poster: Shell we play a game? CTF-as-a-service for security education
14. Wikipedia: Pwn2own. https://en.wikipedia.org/wiki/Pwn2Own
15. GeekPwn: Geekpwn. http://2017.geekpwn.org/1024/en/index.html
16. CPS-CDC: Cps-cdc. http://www.iserink.org/wp-content/uploads/2015/12/2016-CPS_CDC_invite.pdf
17. Deterding, S., Dixon, D., Khaled, R., Nacke, L.: From game design elements to gamefulness: defining gamification. In: Proceedings of the 15th International Academic MindTrek Conference: Envisioning Future Media Environments, pp. 9–15. ACM (2011)

18. Ruef, A., Hicks, M., Parker, J., Levin, D., Memon, A., Plane, J., Mardziel, P.: Build it break it: measuring and comparing development security. In: 8th Workshop on Cyber Security Experimentation and Test (CSET 2015) (2015)

19. Childers, N., Boe, B., Cavallaro, L., Cavedon, L., Cova, M., Egele, M., Vigna, G.: Organizing large scale hacking competitions. In: Kreibich, C., Jahnke, M. (eds.) DIMVA 2010. LNCS, vol. 6201, pp. 132–152. Springer, Heidelberg (2010). https://doi.org/10.1007/978-3-642-14215-4_8

20. Netkoth. http://archive.phreaknic.info/pn18z/content/netkoth.html

21. Vulnhub. https://www.vulnhub.com/

22. Shellweplayagame. https://shellweplayagame.org/

23. Vigna, G., Borgolte, K., Corbetta, J., Doupe, A., Fratantonio, Y., Invernizzi, L., Kirat, D., Shoshitaishvili, Y.: Ten years of ICTF: The good, the bad, and the ugly. In: 2014 USENIX Summit on Gaming, Games, and Gamification in Security Education (3GSE 2014) (2014)

24. CGC-rules. https://dtsn.darpa.mil/cybergrandchallenge/CyberGrandChallenge_Rules_v1.pdf

25. Connolly, C.: The cyber defense review. Technical report, vol. 1(1). Army Cyber Inst, West Point, NY, Spring 2016

26. Social engineering definition. https://en.oxforddictionaries.com/definition/social_engineering

27. Netkoth. https://netkoth.github.io/

28. Ou, X., Boyer, W.F., McQueen, M.A.: A scalable approach to attack graph generation. In: Proceedings of the 13th ACM Conference on Computer and Communications Security, pp. 336–345. ACM (2006)

29. The world's most used penetration testing software. https://www.metasploit.com/

30. The exploit database of offensive security. https://www.offensive-security.com/community-projects/the-exploit-database/

A Modified Fuzzy Fingerprint Vault Based on Pair-Polar Minutiae Structures

Xiangmin Li[1($\boxtimes$)], Ning Ding[1,2($\boxtimes$)], Haining Lu[3], Dawu Gu[1($\boxtimes$)],
Shanshan Wang[4], Beibei Xu[4], Yuan Yuan[4], and Siyun Yan[4]

[1] Department of Computer Science and Engineering, Shanghai Jiao Tong University,
Shanghai, China
{xiangminli,dwgu}@sjtu.edu.cn, dingning@cs.sjtu.edu.cn
[2] State Key Laboratory of Cryptology, P.O. Box 5159, Beijing 100878, China
[3] School of Cyber Security, Shanghai Jiao Tong University, Shanghai, China
hnlu@sjtu.edu.cn
[4] China Mobile (Hangzhou) Information Technology Co., Ltd., Hangzhou, China
{wangshanshan,xubeibei,yuanyuanhy,yansiyunhy}@cmhi.chinamobile.com

Abstract. Biometrics-based cryptosystems are helpful tools facilitating
our daily life. Examples are that people use their biometrics to identify themselves in clouds or protect passwords as well as other privacy
information in mobile phones connected to Internet-of-Things. Among
various biometrics, fingerprints are one used most widely. A promising
type of secure cryptosystems based on fingerprints is the so called fuzzy
fingerprint vaults which can bind a secret key to one fingerprint template and release the same key later with the help of a similar but not
necessarily fully identical fingerprint template. However, some aspects
such as the security and matching accuracy of existing fingerprint-based
fuzzy vaults are still not satisfactory due to troublesome issues such as
the alignment of fingerprints and minutiae quantization.

In this paper we propose a modified fuzzy fingerprint vault based on
pair-polar minutiae structures, aiming at achieving better performance.
Recall that Li et al. (2016) presented a fuzzy vault based on pair-polar
minutiae structures with two-level secure sketch in recovering the secret
key. Unlike their scheme, we compress the two-level structure into one
level in which minutia descriptors are exploited to match a vault and
thus lower the time complexity of decoding. Our experiment evaluation
shows that the proposed fuzzy fingerprint vault achieves high matching accuracy. Importantly, our vault is of strong security against existing
attacks such as the brute-force attack, false-accept attack and correlation
attack. Also this vault adopts the polynomial representation introduced
by Dodis and Reyzin (2006) which is thus able to reduce the storage
complexity.

Keywords: Biometric cryptosystems · Fuzzy vaults
Fingerprint alignment · Minutiae quantization
Pair-polar minutiae structures

X. Chen et al. (Eds.): Inscrypt 2017, LNCS 10726, pp. 482–499, 2018.
https://doi.org/10.1007/978-3-319-75160-3_28

1 Introduction

1.1 Background

Over decades, biometrics has been playing an important role in proving our identity. Nowadays biometrics-based identification is a new and promising research line for many security applications. Examples are that clients use biometrics to identify themselves in clouds or protect passwords as well as other privacy information in mobile phones connected to Internet-of-Things.

Among various biometrics, fingerprints are studied and applied most widely. Compared to traditional cryptosystems, fingerprint-based authentication systems offer more convenience. Fingerprints are born with us, which can equip authentication mechanisms with no memory burden stemming from passwords, tokens and passphrases etc.

Currently, fingerprint recognition techniques are mainly divided into two branches, in which the minutiae-based fingerprint recognition turns out to be more reliable than the texture-based one [11]. For the minutiae-based method, the matching score of two fingerprints is measured by the similarity of well-defined characteristics called ridge endings and bifurcations, which is proved to perform well in practice. Thus, we focus on minutiae-based fuzzy vaults in this paper.

Motivated by the potential of their usability, fingerprints are integrated into traditional cryptographic frameworks to build up so called biometric cryptosystems. These systems mainly aim at solving the key storage problem for cryptographic algorithms with the help of biometrics, of which fuzzy fingerprint vaults are the one best studied.

A typical application of fuzzy vaults is that in the enrollment phase the client runs the vault to encrypt a secret key, which is used for symmetric identification with the server, with his fingerprint; then in the identification phase, the client uses his fingerprint (cannot be exactly same as that used in enrollment) to decrypt the ciphertext to retrieve the secret key and then identify himself to the server using symmetric identification.

1.2 Previous Works

The groundwork of fuzzy vaults is laid by [6], which proposes the fundamental idea of building up a vault encoding some secret with a set A, and enables the later decoding with another set B if A and B have some overlapping. That is, a fuzzy vault can tolerate certain difference between the encoding set A and decoding set B, which is different from traditional cryptosystems that require the exactly same key in encryption and decryption.

The feature set in fuzzy vaults is designed to be compatible with the order invariance properties of fingerprints, and the fuzzy vault is applicable to fingerprint minutiae which have intra-variation during different captures. The alignment of fingerprint templates, minutiae quantization, chaff points generation are three critical steps in the workflow of minutiae-based fuzzy vaults. These steps have great impact on the performance and security of minutiae-based fuzzy vaults.

Generally, the genuine accept rate (GAR), false accept rate (FAR) and time complexity of attacks are measurements of interest for fuzzy vaults. GAR and FAR measure the matching accuracy of fuzzy vaults, where GAR is defined as the ratio of successful genuine trials to the overall genuine trials, and FAR is the ratio of successful imposter trials to the overall imposter trials. And the time complexity of the brute-force attack, false accept attack and correlation attack is used to measure the security strength of fuzzy vaults. For convenience, all fuzzy vaults mentioned later are referred to fuzzy fingerprint vaults.

Fuzzy vaults are first applied to fingerprints by [4], in which coordinates of fingerprint minutiae are encoded as finite field elements, and resulting elements, called the feature set, are embedded in a secret polynomial f derived from the secret key. Finally, chaff points are added to mix up with the genuine feature set for obscurity. However, the alignment issue of fingerprint templates is left out and the FAR of this fuzzy vault is 30% high.

Then researchers proposed two ways to solve the issue of the alignment of templates: the pre-alignment through reference points and the alignment-free method.

For the former, reference points are extracted from both the register and the query fingerprints (i.e. templates). The fuzzy vault in [25] uses the reliable minutia pair as the reference point to automate the alignment and its GAR is 83%. Another implementation in [15] makes use of both coordinates and orientations. Meanwhile, unlike the geometric hashing technique in [3,24] aligns templates with points of high curvature, which yields GAR larger than 90% and FAR less than 1%.

However, the complexity in estimating high curvatures points and matching is pretty high. [9] handles the alignment of templates using topological structures, yielding lower matching accuracy. [19,22] extract a unique directed reference point from a fingerprint and align all minutiae absolutely with respect to that point, and construct the helper data as a monic polynomial introduced in [5] to decrease the storage space. However, the directed reference point may not exist for some fingerprints and the workload of finding it is not low.

For those handling the alignment issue without using reference points, relative positions are utilized to construct alignment-free features. Generally, these features remain stable regardless of translations and rotations. Then [1] analyzes various features, such as the five-nearest neighbor, Voronoi neighbor and triangle structures, and shows that for the same/different finger, the ratio of matching pairs of the five-nearest neighbor, Voronoi neighbors and triangle structures are 70%/30%, 80%/60% and 75%/50%, respectively.

[10] proposes a fuzzy vault based on minutia descriptors and minutia local structures, which results in 92% GAR and 0% FAR. [26] extends the Delaunay triangle-based structures to Delaunay quadrangle-based structures, and achieves some improvement on matching. [8] generalizes the n-nearest neighbor to a global structure called pair-polar minutiae structures, on which fuzzy vaults based are shown to perform well in terms of matching accuracy. However, fuzzy vaults [8,10,26] achieve such promising matching accuracy at the expense of high space complexity.

As for the minutiae quantization, if interpreted geometrically, the distribution area of minutiae is generally divided into disjoint grids, such as rectangles [4, 15, 25], hexagons [22], sectors [8, 18]. Then, the position of each minutia is mapped to a pre-defined index of its containing grid. All the quantized data above are converted into some finite field elements, which generates a collection called the feature set.

Coming to the security aspect of fuzzy vaults, the brute-force attack, false accept (FA) attack [13] and correlation attack [7, 17] are three major attacks. The brute-force attack means trying all possible combinations of the finite field elements from minutiae quantization until the one recovering the secret polynomial is found. The FA attack attempts to crack fuzzy vaults with a large prepared fingerprint dataset based on the fact that there are few collisions of the quantized feature set between different fingerprints. The correlation attack takes advantage of the correlation among the helper data in more than two fuzzy vaults constructed from the same fingerprint.

As shown in [17], the constructions [4, 14, 15, 24, 25] are vulnerable to correlation attacks. Besides, the reference points assisting in the alignment of templates can leak information about fingerprints, which thus decreases the complexity for further attacks [10].

Then [13] uses a user-specific permutation for the feature set to eliminate the relation among different vaults from a same fingerprint, which is useful for resisting the correlation attack. [19, 22] incorporates the directed reference point to pre-align fingerprint templates, thus getting rid of the public helper data for aligning the templates. In [8] the correlation among features sets is decreased by minutia descriptors.

1.3   Our Contribution

In this paper, we propose a modified fuzzy vault scheme based on the pair-polar minutiae structures, which achieves strong security, high matching accuracy and low storage space.

As for the security, our vault can achieve higher brute-force security than existing vaults such as [8, 9, 15, 22], and better false-accept security than those in [9, 22] but lower than that of [15]. Besides, our vault can resist against the correlation attack as [4, 14, 15, 24, 25], securer than [9, 15, 22].

For the matching accuracy, we test our scheme on the FVC 2002 DB2-A fingerprint dataset and the test shows that our scheme can achieve GAR 93% (and FAR zero), higher than those of 91%, 89%, 91% in [9, 15, 22] respectively, while it is lower than those in [8, 10, 14], however, which require more space for storing helper data.

We also compare the time complexity of our vault to the open source implementation in [22] under the same experiment environment. The test turns out that ours is more efficient. (We do not provide comparison with other known works since the source codes of their vaults are not available.)

Our Techniques. In high-level, our vault makes use of the pair-polar minutiae structures [8] for alignment and adopts the polynomial representation of the vault [5]. The usage of the pair-polar minutiae structures can tackle the alignment issue of fingerprint templates, which eliminates reference points such as high curvatures points [14,15] and the directed reference point [22]. The usage of the polynomial representation can enhance the security (since the vault does not contain point pairs).

Unlike that of [8] which adopts a two-level structure of secure sketch, our vault adopts a one-level structure to recover the secret key. As indicated in [6], the secure sketch can be treated as a construction based on a generalized, error-tolerant form of Shamir secret sharing scheme. We use alignment-free minutia descriptors to pair the pair-polar minutiae structure to specific vaults during decoding. By adopting these modifications, we achieve lower overhead of unlocking the vault. Moreover, the potential correlation among different parts of the helper data is also removed by embedding them into different polynomials which are derived from the secret polynomial.

1.4 Organization

The rest of the paper is arranged as follows. Section 2 presents our modified fuzzy vault scheme. The experiment evaluation and analysis is given in Sect. 3. Section 4 concludes the paper.

2 Our Fuzzy Vault Scheme

Our scheme consists of five critical components, which are the pair-polar (P-P) minutiae construction, minutiae quantization, minutia descriptor estimation, encoding, and decoding. In this section, we first recall the P-P minutiae structures by [8], the scheme for quantizing minutiae to build up feature sets, and the minuitia descriptor by [23]. Then we present our encoding and decoding workflows in a theoretical way, and the instantiation and experiment will be shown in the next section.

2.1 Pair-Polar(P-P) Minutiae Structure Construction

Generally, a minutiae-based fingerprint template is described as a set of minutiae $M = \{m_i = (x_i, y_i, \theta_i)\}$, where (x_i, y_i) is the Cartesian coordinates of m_i and $\theta_i \in [0, 2\pi)$ is its orientation. To avoid the estimation of the reference point for aligning fingerprint templates, the pair-polar minutiae structure proposed by [8] is employed in our scheme, whose discriminative power has been tested. The P-P minutiae structure P_i for each $m_i \in M$ is constructed as follows. For any $m_j \in M(j \neq i)$, calculate its relative polar coordinate with respect to m_i as $p_{ij} = (r_{ij}, \varphi_{ij}, \theta_{ij})$ (shown in Fig. 1) where

$$r_{ij} = \sqrt{(x_j - x_i)^2 + (y_j - y_i)^2}$$
$$\varphi_{ij} = \mathrm{norm}(\arctan(y_j, x_j) - \arctan(y_i, x_i))^{\cdot} \tag{1}$$
$$\theta_{ij} = \mathrm{norm}(\theta_j - \theta_i)$$

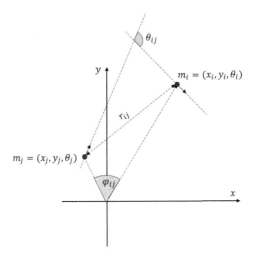

Fig. 1. An example of relative polar coordinates

in which $\arctan(y, x)$ is the counter-clockwise angle of vector (x, y) with respect to x-axis, and $\mathrm{norm}(\theta_1 - \theta_2) = \theta_1 - \theta_2$ if $\theta_1 > \theta_2$ and $\mathrm{norm}(\theta_1 - \theta_2) = \theta_1 - \theta_2 + 2\pi$ otherwise. The collection of such relative polar coordinates makes up the P-P minutiae structure $P_i = \{p_{ij}\}$ for m_i. And the P-P minutiae structure set of template M is denoted by $\boldsymbol{P}_M = \{P_i\}$.

2.2 Minutiae Quantization

In fuzzy vaults, the secret to protect is encoded as coefficients of a polynomial over the finite field, thus bringing about the requirement of finite field arithmetic operations in the subsequent encoding and decoding. However, fingerprint templates comprise minutiae whose information like coordinates is continuous, which cannot be bound directly with the secret polynomial. Hence, before being fed into encoding or decoding components, minutiae need discretization, i.e., converting into elements in the finite field. That discretization is termed *quantization* in fuzzy vaults.

Given a template M in our scheme, only the P-P minutiae structures of its first s minutiae of the highest quality are collected into the final P-P minutiae structure set $\boldsymbol{P}_M$. Each $P_i \in \boldsymbol{P}_M$ can be treated as a pre-aligned version template with m_i being the reference point. For a fingerprint image of width w and height h, the distribution area of minutiae in the pre-aligned template should be a circle with radius $r_m = \sqrt{w^2 + h^2}$. To make up the polar grid system for quantization, this circle is divied into l disjoint polar grids $\{\Omega_0, \ldots, \Omega_{l-1}\}$, where the quanta for quantization of its radius is $\hat{r}$ and the number of quanta for quantizing its angle is $\hat{\varphi}$, as demonstrated by Fig. 2 ($v = \lceil r_m/\hat{r} \rceil$). For each $P_i \in \boldsymbol{P}_M$, sort p_{ij} in descending order of the quality of its corresponding minutia m_j. Then each $p_{ij} = (r_{ij}, \varphi_{ij}, \theta_{ij}) \in P_i(j \neq i)$ is *quantized* into number as

$$a_{ij} = u + \lfloor \theta_{ij}/2\pi \cdot \hat{\theta} \rfloor \cdot (\lceil r_m/\hat{r} \rceil \cdot \hat{\varphi}). \tag{2}$$

where $u = \lceil r_{ij}/\hat{r} \rceil + \lceil r_m/\hat{r} \rceil \cdot \lfloor \varphi_{ij}/(2\pi) \cdot \hat{\varphi} \rfloor$ is the index of polar grid Ω_u containing point (r_{ij}, φ_{ij}) and $\hat{\theta}$ is a pre-defined number of quanta for θ_{ij}. To limit the size of the quantized set of P_i, only the first t unique a_{ij} are saved, denoted by A_i. Finally, the quantized version set of P-P minutiae structures of template M would be $\boldsymbol{A}_M = \{A_i = \{a_{ij} \mid j \neq i\}\}$, referred as the feature set later.

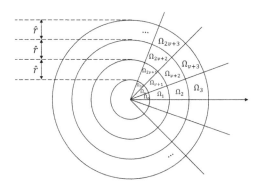

Fig. 2. The polar grid system for quantization

2.3   Minutia Descriptor Estimation

As a translation and rotation invariant feature, the minutia descriptor proposed in [23] is computed based on the relative orientation of sample points around the reference minutia as Fig. 3 shows. For a specific minutia of direction $\theta \in [0, 2\pi)$, 10, 16, 22 and 28 points equally distributed are sampled respectively in the 4 concentric circles of radius 27, 46, 63, 81. Points lying on each circle are ordered counter-clockwise where the first point are defined as the interection of the half-line along θ and the circle, and all points are collected into a vector $\boldsymbol{L} = (l_1, l_2, \ldots, l_{76})$ in order from inner most circle to the outer most one. Then a descriptor vector $\boldsymbol{O} = (o_1, o_2, \ldots, o_{76})$ can be estimated on each position l_i of orientation $\theta_i \in [0, \pi)$, where

$$o_i = \begin{cases} \Delta\theta_i + \pi & \text{if } -\pi < \Delta\theta_i \leq -\dfrac{\pi}{2} \\[2mm] \Delta\theta_i & \text{if } \dfrac{-\pi}{2} < \Delta\theta_i \leq \dfrac{\pi}{2} \\[2mm] \Delta\theta_i - \pi & \text{if } \dfrac{\pi}{2} < \Delta\theta_i \leq \dfrac{3\pi}{2} \\[2mm] \Delta\theta_i - 2\pi & \text{if } \dfrac{\pi}{2} < \Delta\theta_i < 2\pi \end{cases} \tag{3}$$

with $\Delta\theta_i = \theta_i - \theta$.

Fig. 3. An example of the descriptor (sampled at blue points) for a minutia (the red square) (Color figure online)

For matching minutiae in two templates, their descriptors should be more similar to each other than others. Hence, by binding the descriptor to the corresponding vault for each reference minutia in the P-P minutiae structures during encoding, these descriptors can help to coarsely filter out those query P-P minutiae structures unrelated to the specific vault. Similar to [8], each descriptor $\boldsymbol{O}_i$ is normalized into a value $d_i \in [0, 1]$ by formula

$$d_i = \frac{\boldsymbol{O}_i \cdot \boldsymbol{U}}{\|\boldsymbol{O}_i\| \cdot \|\boldsymbol{U}\|} = \sum_{j=1}^{76} o_j \cdot u_j$$

where $\boldsymbol{U} = (u_1, u_2, \ldots, u_{76})$ is a preset vector of length 76, and $\|\cdot\|$ is a function calculating the norm of a given vector. Obviously, this normalization process aims at reducing storage space and $\{d_i\}$ is used to assist in matching.

2.4   Encoding Stage

Unlike the two-level scheme [8] adapting to P-P minutiae structures, we compress the encoding of our vaults into one level. The encoding process first takes up a set of minutiae, constructs their P-P structures, and then embeds these structures into different polynomials derived from the secret polynomial, and finally saves the helper data consisting of the hash value of the secret polynomial and a collection of vault polynomials along with the corresponding minutia descriptor value of their reference minutiae.

An overview of the workflow of the encoding stage is depicted in Fig. 4. The detailed implementation is shown as follows.

1. **Template Extraction.** Given a fingerprint image, its template $M^T = \{m_i^T \mid m_i^T = (x_i^T, y_i^T, \theta_i^T)\}$ is extracted using VeriFinger 9.0 SDK [16].
2. **Minutia Descriptor Estimation.** For each minutia $m_i^T \in M^T$, estimate their descriptor value d_i^T according to the definition in Sect. 2.3 and extend m_i^T as $(x_i^T, y_i^T, \theta_i^T, d_i^T)$
3. **P-P Structure Estimation and Quantization.** Template M^T is used to estimate the feature set $\boldsymbol{A} = \{A_i \mid |A_i| = t\}_{i=1}^s$ where $A_i = \{a_{ij} \mid j \neq i\}$, according to the specification described in Sects. 2.1 and 2.2.

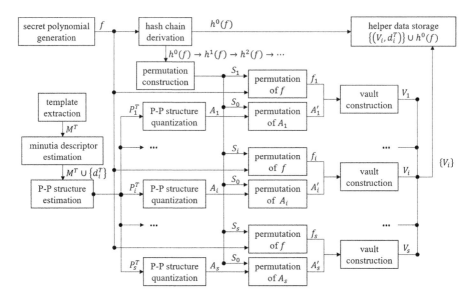

Fig. 4. The workflow of the encoding

4. **Secret Polynomial Generation.** Generate a random polynomial $f(x) = c_0 + c_1 x + \cdots + c_{k-1} x^{k-1} \in \mathbf{GF}[x]$, where the secret is parsed to $c_0 \mid c_1 \mid \cdots \mid c_{k-1}$ and "$\mid$" denotes concatenation.

5. **Hash Chain Derivation.** The hash value of the secret polynomial f is calculated as $h^0(f) = h(c_0 \mid c_1 \mid \cdots \mid c_{k-1})$. Then $h^0(f)$ is taken as the head of a hash chain to derive a hash sequence $h^1(f) = h(h^0(f)), h^2(f) = h(h^1(f)), \ldots$, where $h(\cdot)$ denotes a hash function and $h^i(x)$ is the output by applying $h(\cdot)$ of its given input x iteratively $(i+1)$ times. And $h(\cdot)$ is implemented as SHA-1 in our vault.

6. **Permutation Construction.** The hash value $h^0(f)$ is taken as the seed of deriving a random number sequence which helps in producing a random public permutation S_0 for all elements in the finite field $\mathbf{GF}[x]$ (refer to the thimble project open sourced in [22] for the instantiation details of generating the random number sequence and the random public permutation). Besides, $\{h^i(f) \mid i = 1, 2, \ldots, s\}$ is employed in the similar manner to construct a set of permutations $\{S_i \mid i = 1, 2, \ldots, s\}$ subject to the set $\{0, 1, \ldots, k-1\}$.

7. **Permutation of A_i.** As mentioned in [22], the permutation of the feature set can help to resist against the correlation attack [2]. For $A_i \in \mathbf{A}$, the permutation S_0 is applied to all its element $a_{ij} \in A_i$. Let $A_i' = \{a_{ij}'\}$ denote the re-ordered version of A_i.

8. **Permutation of f.** In our scheme, only one secret polynomial $f(x)$ is employed in essence, and several parts of the helper data are based on it. Only the re-ordered version of $f(x)$ will be used to construct the vault polynomial, which can eliminate the correlation among these several parts

of helper data. The re-ordering of $f(x)$ is generated as follows. For $S_i, i = 1, 2, \ldots, s$, re-order the secret polynomial as $f_i(x) = c_0^i + c_1^i x + \cdots + c_{k-1}^i x^{k-1}$, where $c_j^i = c_{S_i(j)}$.

The re-ordering operations above are to derive a set of secret polynomials $\{f_i(x) \mid i = 1, 2, \ldots, s\}$ which have the same coefficients as $f(x)$ but are of different order of these coefficients. Let $S_i(f)$ denote such re-ordering of f for all i.

9. **Vault Construction.** Here we adopt the construction in [5] to bind the feature set and the secret polynomial to the following monic polynomial which is of degree t.

$$V_i(X) = f_i(X) + \prod_{a'_{ij} \in A'_i} (X - a'_{ij}). \tag{4}$$

10. **Helper Data Storage.** Finally, the helper data is saved as $\{(V_i, d_i)\} \cup h^0(f)$ for later decoding.

2.5 Decoding Stage

The task of decoding is to recover the secret polynomial $f(x)$ hidden in the encoding process. The query fingerprint is fed into the decoding process, aiming at unlocking the helper data, whose overall workflow is shown in Fig. 5.

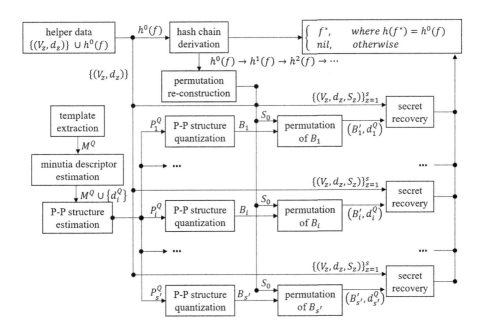

Fig. 5. The workflow of the decoding

We now give the detailed explanation of various steps in the decoding process as follows.

1. **Template Extraction.** Given another fingerprint image, its template $M^Q = \{m_i^Q \mid m_i^Q = (x_i^Q, y_i^Q, \theta_i^Q)\}$ is extracted using VeriFinger 9.0 SDK [16].
2. **Minutia Descriptor Estimation.** For each minutia $m_i^Q \in M^Q$, estimate their descriptor value d_i^Q according to the definition in Sect. 2.3 and extend m_i^Q as $(x_i^Q, y_i^Q, \theta_i^Q, d_i^Q)$
3. **P-P Structure Estimation and Quantization.** Unlike the corresponding step in encoding, the size limit for each quantized P-P structure is set as $t' < t$ to speed up subsequent procedures. Then the feature set $B = \{B_i = \{b_{ij} \mid j \neq i\} \mid |B_i| = t'\}$ is estimated from the query template M^Q as the specification in Sects. 2.1 and 2.2 shows.
4. **Hash Chain Derivation and Permutation Re-construction.** The hash chain $h^0(f) \to h^1(f) \to h^2(f) \to \cdots$ and the permutation set $S_0 \cup \{S_i\}$ are reconstructed in the same manner as those of the encoding stage.
5. **Permutation of B_i.** According to the permutation S_0, the feature set B_i is converted into its re-ordered version B_i' in the same way as the encoding stage.
6. **Secret Recovery.** Given the helper data $\{(V_z, d_z)\} \cup h^0(f)$, each B_i' satisfying $|d_i^Q - d_z| < \Delta d$ (Δd is a preset threshold) is employed in an attempt to unlock V_z for each pair (B_i', V_z) as follows.
   Do Lagrange Interpolation for all possible subsets of cardinality k for set $\{(b_{ij}', V_z(b_{ij}'))\}$ one by one, to recover some candidate re-ordered secret polynomials $f_z^*(x)$ of degree less than k. Let's define S_z^{-1} as the inverse of S_z, i.e., $S_z^{-1}(S_z(f))$ will reproduce the plain polynomial f from the re-ordered one $S_z(f)$. If any $f^* = S_z^{-1}(f_z^*)$ satisfying $h(f^*) = h^0(f)$ is found, the secret polynomial hidden in V_z is deemed to be f^*.
   The process above is repeated until the secret polynomial is recovered. In case of all pairs (B_i', V_z) tried out, report decoding failure and abort.

3   Experiment Evaluation and Analysis

3.1   Evaluation

The performance of the proposed fuzzy vault scheme is evaluated on FVC 2002 DB2-A [12] (which consists of 100 fingerprint groups and each group consists of 8 different images of a same fingerprint) according to the 1vs1 protocol. In the 1vs1 protocol only the first two impressions of each fingerprint group are involved.

According to its specification, the first impression of each group is used to encode a vault, then that vault is decoded with the second impression of the same group in the case of evaluating GAR and with the first impression of all other groups in the case of evaluating FAR. Here, for evaluating FAR, if an impression has been used for encoding, it will not be employed as a query for decoding any more. Hence, the number of encoding/decoding trials in evaluating GAR is 100 while that number is $100 \cdot 99/2 = 4950$ in evaluating FAR. The underlying parameters are summarized in Table 1.

Table 1. Parameters of the proposed fuzzy vault and their configuration value

Parameter	Meaning	Value
$\hat{r}$	The quanta for the polar radius	16
$\hat{\varphi}$	The number of quanta for the polar angle	20
$\hat{\theta}$	The number of quanta for minutia's orientation	30
s	The number of P-P minutiae structures during encoding	18
t	The size of the feature set during encoding	32
t'	The size of the feature set during decoding	15
k	$k - 1$ is the highest degree of the secret polynomial	7–10
Δd	Threshold for measuring similarity of minutia descriptors	0.1

We adopt the Galois Field $\mathbf{GF}(2^{16})$, which means the maximum number of bits to protect is 112, 128, 144, 160 for $k = 7, 8, 9, 10$ respectively. The vector U for normalization is generated randomly and fixed for all vaults.

The minutiae for each image in the database are extracted with Neurotechnology Ltd. VeriFinger SDK 9.0 [16], of which no failure of extractions is reported. With parameter configuration in Table 1, the result of evaluation under $k = 7, 8, 9, 10$ and its comparison with other fuzzy vaults are summarized in Table 2.

Table 2. GAR/FAR evaluation on our modified fuzzy vault

k	7	8	9	10
Nandakumar et al. [15]	91/0.13	91/0.01	-	86/0
Li et al. [9]	-	89/0.07	88/0.03	86/0.04
Tams et al. [22]	91/0.87	88/0.12	87/0.04	79/0
Our scheme	93/0	91/0	83/0	76/0

[1] all values of GAR and FAR are measured in percentage
[2] [9] evaluates its FAR on part of the database only
[3] [15] is vulnerable to the correlation attack as explained later

As indicated in Table 2, for FAR = 0, our modified scheme can yield GAR = 93%, higher than those listed. Especially, even if the FAR of our fuzzy vault is evaluated by employing all minutiae to build P-P structures of the same size to that of the encoding, only the FAR at $k = 7$ will be increased to 0.02% which means only one imposter succeeds in decoding. The FAR in [22] is estimated with a randomized decoder, where it is not the case for attackers. Besides, with the maximum number of decoding iterations increased, the FAR of their vault will be higher. Moreover, for [15], the estimation of the high curvature points and the ICP-based alignment of templates assisted by them involve pretty high time complexity. Both the topological structures [9] and the directed reference point [22] needed for alignment require the existence of core points in fingerprints, which are missing for some categories of fingerprints.

As for time complexity, we evaluate the decoding time for our scheme and that in [22] in a computer with an Intel(R) Core(TM) i5-3470 CPU @ 3.2 GHz proccessor, 4 cores and 8 GB RAM. Here, the genuine decoding time(GDT) is defined as the average decoding time in seconds for all successful genuine decoding trials, and the imposter decoding time(IDT) is defined as the average decoding time in seconds for all failed imposter decoding trials.

Note that the time for extracting templates is excluded from both GDT and IDT. Therefore, GDT in our scheme will be the total time from estimating minutia descriptors to finishing decoding, while in [22] it refers to the total time used for estimating the directed reference point and all other procedures after extracting templates. The result is summarized in Table 3.

Table 3. GDT/IDT evaluation on our modified fuzzy vault

Proposals	k			
	7	8	9	10
Tams et al. [22]	0.88/1.10	0.88/1.14	0.89/1.20	0.89/1.26
Our scheme	0.43/4.64	0.58/5.60	0.56/5.32	0.54/3.87

As shown in Table 3, both the GDT and IDT in our scheme are less than those in [22]. As k increases, it would require larger size $|A_i \cap B_j|$ of intersection between the query feature set $B_j \in \boldsymbol{B}$ to open the vault encoded by the reference feature set $A_i \in \boldsymbol{A}$, while the number of decoding trials (i.e., $\binom{t'}{k}$) for a single vault V_z by B_j decreases. Thus from $k = 7$ to 10, the major factor affecting GDT and IDT is the ratio of pairs (A_i, B_j) satisfying $|A_i \cap B_j| \geq k$ for $k = 7$ to 8, and then is the number of decoding trials for $k = 8$ to 10, as confirmed by the data above where both the GAT and IDT increase first and decrease then.

For [8,10,14] which are not listed in Table 2, they may possess better matching accuracy, but consume more storage space for helper data than our scheme. As indicated by their schemes, the helper data comprises reference points for alignment [14], or the genuine feature set along with another large set of chaff feature set [8,14]), or a set of non-normalized minutia descriptors [10], etc. In our scheme, the number of minutia descriptors equals to the number of vault polynomials, which is much smaller than the size of chaff feature set. Our helper data consists of all coefficients of vault polynomials and the corresponding normalized minutia descriptor values plus one hash value, thus taking up less space.

For example, according to the specified configuration, the helper data of [8] comprises 30 point sets $Vault_i$ where $|Vault_i| = 330$ and another 31 pairs, totally $19862(= (30 \times 330 + 31) \times 2)$ elements. Our vault is stored as $\{(V_i, d_i)\}_{i=1}^{18} \cup h^0(f)$ where $|V_i| = 33$ and there are $613(=(33 + 1) \times 18 + 1)$ elements in total, much smaller than 19862. Besides, the size of each element in our scheme is also smaller since ours works in the finite field $\mathbf{GF}(2^{16})$ while the scheme in [8] works in $\mathbf{GF}(2^{28})$ in implementation. Therefore, our vault is more efficient regarding the storage space.

3.2   Security Analysis

We now present time complexity of the brute-force attack, false-accept attack and correlation attack on our vault.

Brute-Force Attack. In our scheme, the time complexity of brute-force attack is decided by the number n of possible finite field elements mapped from minutiae, size t of the feature set during encoding and number k of coefficients of the secret polynomial. As implied by the parameter configuration specified in Table 1, the number of polar grids is $\lceil r_m/\hat{r} \rceil \cdot \hat{\varphi} = \lceil \sqrt{w^2 + h^2}/\hat{r} \rceil \cdot \hat{\varphi}$, and then $n = \lceil \frac{r_m}{\hat{r}} \rceil \cdot \hat{\varphi} \cdot \hat{\theta} = 24000$. Therefore, the brute-force security, using the formula given by [22], can be estimated as

$$\mathbf{bf}(n, t, k) = \binom{n}{k} \cdot \binom{t}{k}^{-1}. \tag{5}$$

Comparison of ours to several other fuzzy vaults is summarized in Table 4.

Table 4. Brute-force security analysis of the modified fuzzy vault

Proposals	(n, t)	k			
		7	8	9	10
Nandakumar et al. [15]	(224,24)	2^{28}	2^{32}	2^{36}	2^{40}
Li et al. [9]	(220,20)	2^{30}	2^{34}	2^{39}	2^{43}
Tams et al. [22]	(10398,44)	2^{56}	2^{65}	2^{73}	2^{81}
Li et al. [8]	(329,29)	2^{26}	2^{30}	2^{34}	2^{38}
Our scheme	(24000,32)	2^{68}	2^{78}	2^{88}	2^{98}

False-Accept Attack. The false-accept attack aims at cracking a vault by the possible overlapping between feature sets from different fingers. As we know, the secret polynomial will be recovered if the size of the intersection between the feature set A_i in encoding and B_j in decoding is higher than the degree $k - 1$ of the secret polynomial. Hence, given a large fingerprint dataset, it is possible for an attacker to exploit such potential collisions.

When FAR is known, the effort of performing a false-accept attack on the fuzzy vault is calculated as $\mathrm{FAR}^{-1} \cdot D$, where $D = \binom{t'}{k}$. In our scheme, FAR is nonzero as 0.02% at $k = 7$ only, which yields a false-accept security in 2^{28}. Besides, FAR at $k = 8, 9, 10$ is zero under the same evaluating protocol. Hence, our fuzzy vault is securer against the false-accept attack. Table 5 lists the security of ours and several other vaults.

Table 5. False-accept security analysis of our fuzzy vault and others

Proposals	(n, t)	k			
		7	8	9	10
Nandakumar et al. [15]	(224,24)	2^{30}	2^{34}	-	∞
Li et al. [9]	(220,20)	-	2^{22}	2^{23}	2^{23}
Tams et al. [22]	(10398,44)	2^{22}	2^{25}	2^{28}	2^{31}
Our proposal	(24000,32)	2^{28}	∞	∞	∞

[1] "–" means no data reported in the literature
[2] ∞ means resistance against the false-accept attack

Correlation Attack. For the fuzzy vaults in [4,14,15,24,25], attackers can take advantage of the similarities among genuine feature points from the same fingerprint to filter out chaff points shown in Fig. 6 [7,20].

However, in our scheme, minutiae are first mapped to some elements in a finite field, and these minutiae are discarded once the mapping is done. Then the final helper data is encoded as monic polynomials $\{V_i\}$, and each V_i can be treated as a point set $\{(a_j, V_i(a_j)) \mid a_j \in \mathbf{GF}\}$ where the set of abscissa values $\{a_j\}$ is the set of all elments in the finite field $\mathbf{GF}$. These monic polynomials erase the set difference of abscissa values in $\{(a_j, V_i(a_j))\}$ among different helper data. Besides, random public permutations based on the hash value of the secret

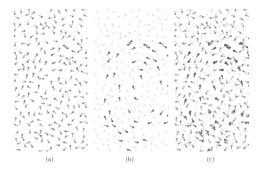

Fig. 6. cited from [7,20]. (a), (b) Two aligned vaults with chaff minutiae (gray and lightgray) and genuine minutiae (red and blue). (c) The genuine minutiae have a bias to be in agreement. (Color figure online)

polynomial are applied to the raw feature set before encoding, contributing to the removal of correlation between helper data from the same fingerprints as explained in [21].

If the same secret polynomial f is employed to make vault polynomials, two different vault polynomials V_i and V_j will yield the same ordinate value $V_i(a) = V_j(a) = f(a)$ in the case that the feature element a belongs to two different feature sets A_i and A_j, which is another potential correlation available to attackers. Therefore, rather than the original secret polynomial, different re-ordered versions derived from f are embedded in the vault polynomial V_i, contributing to the removal of correlation above further.

As seen, the alignment issue is handled better by pair-polar minutiae structures derived from minutiae, since these minutiae are more accurate than singularity points and other reference points. Therefore, our fuzzy vault achieves strong security against attacks while maintaining a high matching accuracy.

4   Conclusion

Fingerprints are a great medium for authentication nowadays. Among various fingerprint-based authentication systems, fuzzy fingerprint vaults are the most well constructed. However, issues such as the alignment of fingerprint templates and the quantization of minutiae are troublesome and still lack satisfactory solutions for minutiae-based fuzzy vaults, which impact greatly on their matching accuracy, storage space and security. This paper presents a modified fuzzy fingerprint vault, which integrates the alignment-free pair-polar minutiae structure [8] and proposes a one-level decoding strategy. We also exploit minutia descriptors for matching vaults and candidate P-P structures, and make use of the polynomial representation for vaults in [5] to lower the storage space. Our security analysis implies that our modified scheme can better resist against common attacks such as the brute-force attack, false-accept attack and correlation attack, while maintaining reasonable matching accuracy and lower storage space, when compared to existing fuzzy vaults.

Acknowledgments. We are grateful to the reviewers of Inscrypt 2017 for their useful comments. This work is supported by Research Fund of Ministry of Education of China and China Mobile (Grant No. MCM20150301).

References

1. Arakala, A., Jeffers, J., Horadam, K.J.: Fuzzy extractors for minutiae-based fingerprint authentication. In: Lee, S.-W., Li, S.Z. (eds.) ICB 2007. LNCS, vol. 4642, pp. 760–769. Springer, Heidelberg (2007). https://doi.org/10.1007/978-3-540-74549-5_80
2. Blanton, M., Aliasgari, M.: Analysis of reusability of secure sketches and fuzzy extractors. IEEE Trans. Inf. Forensics Secur. **8**(9), 1433–1445 (2013)

3. Chung, Y., Moon, D., Lee, S., Jung, S., Kim, T., Ahn, D.: Automatic alignment of fingerprint features for fuzzy fingerprint vault. In: Feng, D., Lin, D., Yung, M. (eds.) CISC 2005. LNCS, vol. 3822, pp. 358–369. Springer, Heidelberg (2005). https://doi.org/10.1007/11599548_31

4. Charles Clancy, T., Kiyavash, N., Lin, D.J.: Secure smartcardbased fingerprint authentication. In: ACM Sigmm Workshop on Biometrics Methods and Applications, pp. 45–52 (2003)

5. Dodis, Y., Reyzin, L.: Fuzzy extractors: how to generate strong keys from biometrics and other noisy data. SIAM J. Comput. **38**(1), 97–139 (2006)

6. Juels, A., Sudan, M.: A fuzzy vault scheme. In: Proceedings of IEEE International Symposium on Information Theory, p. 408 (2002)

7. Kholmatov, A., Yanikoglu, B.A.: Realization of correlation attack against the fuzzy vault scheme. In: Security, Forensics, Steganography, and Watermarking of Multimedia Contents (2008)

8. Li, C., Hu, J.: A security-enhanced alignment-free fuzzy vault-based fingerprint cryptosystem using pair-polar minutiae structures. IEEE Trans. Inf. Forensics Secur. **11**(3), 543–555 (2016)

9. Li, P., Yang, X., Cao, K., Shi, P., Tian, J.: Security-enhanced fuzzy fingerprint vault based on minutiae's local ridge information. In: Tistarelli, M., Nixon, M.S. (eds.) ICB 2009. LNCS, vol. 5558, pp. 930–939. Springer, Heidelberg (2009). https://doi.org/10.1007/978-3-642-01793-3_94

10. Li, P., Yang, X., Cao, K., Tao, X., Wang, R., Tian, J.: An alignment-free fingerprint cryptosystem based on fuzzy vault scheme. J. Netw. Comput. Appl. **33**(3), 207–220 (2010)

11. Liu, E., Zhao, H., Liang, J., Pang, L., Xie, M., Chen, H., Li, Y., Li, P., Tian, J.: A key binding system based on n-nearest minutiae structure of fingerprint. Pattern Recogn. Lett. **32**(5), 666–675 (2011)

12. Maio, D., Maltoni, D., Cappelli, R., Wayman, J.L.: Fvc 2002: second fingerprint verification competition. In: 16th International Conference on Pattern Recognition, p. 30811 (2002)

13. Merkle, J., Tams, B.: Security of the improved fuzzy vault scheme in the presence of record multiplicity (full version). Computer Science (2013)

14. Nagar, A., Nandakumar, K., Jain, A.K.: Securing fingerprint template: fuzzy vault with minutiae descriptors. In: International Conference on Pattern Recognition, pp. 1–4 (2009)

15. Nandakumar, K., Jain, A.K., Pankanti, S.: Fingerprint-based fuzzy vault: implementation and performance. IEEE Trans. Inf. Forensics Secur. **2**(4), 744–757 (2007)

16. Neurotechnology. Verifinger 9.0. (2016). http://www.neurotechnology.com/verifinger.html

17. Scheirer, W.J., Boult, T.E.: Cracking fuzzy vaults and biometric encryption. In: Biometrics Symposium, pp. 1–6 (2007)

18. Tams, B., Merkle, J., Rathgeb, C., Wagner, J.: Improved fuzzy vault scheme for alignment-free fingerprint features. In: International Conference of the Biometrics Special Interest Group, pp. 1–12 (2015)

19. Tams, B.: Absolute fingerprint pre-alignment in minutiae-based cryptosystems. In: International Conference of the Biometrics Special Interest Group, pp. 1–12 (2013)

20. Tams, B.: Attacks and countermeasures in fingerprint based biometric cryptosystems. Eprint Arxiv (2013)

21. Tams, B.: Unlinkable minutiae-based fuzzy vault for multiple fingerprints. Iet Biometrics **5**(3), 170–180 (2015)

22. Tams, B., Mihilescu, P., Munk, A.: Security considerations in minutiae-based fuzzy vaults. IEEE Trans. Inf. Forensics Secur. **10**(5), 985–998 (2015)
23. Tico, M., Kuosmanen, P.: Fingerprint matching using an orientation-based minutia descriptor. IEEE Trans. Pattern Anal. Mach. Intell. **25**(8), 1009–1014 (2003)
24. Uludag, U., Jain, A.: Securing fingerprint template: fuzzy vault with helper data. In: Proceedings of CVPR Workshop on Privacy Research in Vision, p. 163 (2006)
25. Yang, S., Verbauwhede, I.: Automatic secure fingerprint verification system based on fuzzy vault scheme. In: IEEE International Conference on Acoustics, pp. 609–612 (2005)
26. Yang, W., Jiankun, H., Wang, S.: A delaunay quadrangle-based fingerprint authentication system with template protection using topology code for local registration and security enhancement. IEEE Trans. Inf. Forensics Secur. **9**(7), 1179–1192 (2014)

NOR: Towards Non-intrusive, Real-Time and OS-agnostic Introspection for Virtual Machines in Cloud Environment

Chonghua Wang[1,2], Zhiyu Hao[1(✉)], and Xiaochun Yun[1,3]

[1] Institute of Information Engineering, Chinese Academy of Sciences, Beijing, China
`chonghuaw@live.com, haozhiyu@iie.ac.cn`
[2] School of Cyber Security, University of Chinese Academy of Sciences,
Beijing, China
[3] National Computer Network Emergency Response Technical Team/Coordination
Center of China, Beijing, China

Abstract. Cloud platforms of large enterprises are witnessing increasing adoption of the Virtual Machine Introspection (VMI) technology for building a wide range of VM monitoring applications including intrusion detection systems, virtual firewall, malware analysis, and live memory forensics. In our analysis and comparison of existing VMI systems, we found that most systems suffer one or more of the following problems: intrusiveness, time lag and OS-dependence, which are not well suited to clouds in practice. To address these problems, we present NOR, a non-intrusive, real-time and OS-agnostic introspection system for virtual machines in cloud environment. It employs event-driven monitoring and snapshot polling cooperatively to reconstruct the memory state of guest VMs. In our evaluation, we show NOR is capable of monitoring activities of guest VMs instantaneously with minor performance overhead. We also design some case studies to show that NOR is able to detect kernel rootkits and mitigate transient attacks for different Linux systems.

Keywords: Virtual machine introspection · Malware detection
Side-channel attacks · Cloud security

1 Introduction

In clouds, if a guest VM has been attacked and the malware is not detected in time, any intrusion occurred on the guest VM may be a compromise to the whole system (e.g., VM escape, data leakage, privilege escalation, VMM denial of service, etc.) [26]. Cloud platforms of large enterprises are witnessing increasing adoption of the virtual machine introspection (VMI) [16] technology for building a wide range of VM monitoring applications including intrusion detection systems [22], intrusion prevention [28], writable configuration and repair [14], malware analysis [10,32,43,44], live memory forensics [11,19] and so forth.

© Springer International Publishing AG, part of Springer Nature 2018
X. Chen et al. (Eds.): Inscrypt 2017, LNCS 10726, pp. 500–517, 2018.
https://doi.org/10.1007/978-3-319-75160-3_29

Using VMI, a VM monitoring application reconstructs the internal state of a monitored VM (e.g., running processes, opened files and active network connections, etc.) from the byte values in the VM's memory pages, which is more reliable than retrieving information within the guest OS. However, the semantic gap [20] exists because at the hypervisor layer, we have access only to the raw data of the hardware level state of a VM (e.g., CPU registers, physical memory or even instructions etc.). What we want is the semantic information about the guest OS state, such as the variables being accessed, variable types, and guest OS kernel events to obtain meaningful guest OS state information. Thus VMI systems must bridge the semantic gap problem.

State-of-the-Art. We conduct a systematic analysis and comparison of existing VMI systems, and we found that most systems suffer one or more of the following problems: *(1)* Intrusiveness, these systems [6, 15, 18, 28, 34] involve adding hooks or programs inside the guest to help for retrieving introspection information; *(2)* Time lag, these systems [22, 27] fail to guarantee the introspection information up-to-date by employing snapshot polling of guest states; *(3)* OS-dependence, these systems [22, 27] need to rebuild introspection tools according to detailed, up-to-date information about the internal workings of commodity OSes.

Unfortunately, these systems are not well suited to clouds in practice. Intrusiveness makes the introspection tools easy to be detected by advanced malware. Malware may camouflage itself when it has a sense of the existence of the introspection tools or even attack them. Time lag fails to stop stealthy attackers from launching transient attacks which cause damages during the interval and restore the modifications to normal before next snapshot polling. OS-dependence requires the administrator to maintain customized monitor VMs and introspection tools for different guest OSes. Considering that a cloud platform may provide thousands of VMs that differ in operating systems, kernel versions and patches, thus the kernel data structure layout and invariants are different accordingly. It is too trivial to maintain such work.

Our Approach. To address the above problems and complement existing work, we design NOR, a non-intrusive, real-time and OS-agnostic introspection system for virtual machines in cloud environment. It provides three important guarantees to monitor activities of guest VMs within the hypervisor. *(1) Non-intrusive*, no modifications to the guest OS and no additional software agents are needed in the monitored VM, thus the malware are unaware of the existence of the introspection tools; *(2) Real-time*, the introspection process is instantaneous to mitigate transient attacks that can detect the snapshot polling intervals; *(3) OS-agnostic*, the introspection still works when an agnostic guest OS is added in a cloud.

Our main challenge is how to capture the OS template information within the hypervisor instead of inserting an agent into the guest OS and make the introspection information up-to-date. To achieve the above guarantees and overcome the challenges, the basic idea of NOR is to launch snapshot polling and events-driven monitoring cooperatively. However, when event-driven monitoring is introduced into our system, how to reduce the monitoring overhead becomes our the third challenge..

To solve the first two challenges, NOR employs the following schemes and techniques. Specifically, events-driven monitoring traps system calls called inside the guest OS. The system calls monitored can deduce rich semantics of specific events initiated by kernel or malware. To capture critical events, NOR sets some strategies to track some particular system calls. Some system calls are used to explore OS templates (e.g., the layout and offset of interested kernel data structures,etc.) and some system calls are used to trigger the snapshot polling to reconstruct high level semantics instantaneously, as well as achieving completeness. Note that the OS templates are retrieved within the hypervisor in runtime and taken as the necessary input of snapshot polling. Snapshot polling is triggered based on the occurrences of designated events which could possibly deduce some suspicious activities conducted by malware. The designated events are the occurrence of some particular system calls that seem to be suspicious according to different monitoring requirements. For example, when the *sys_init_module* system call is captured by the event-driven monitoring means that the kernel loads a module into the system, which is treated as a trigger point for the snapshot polling. NOR could detect the kernel malware (e.g., the malware is inserted into the kernel in terms of loadable kernel module, etc.) non-intrusively and mitigate transient attacks by launching snapshot polling once the *sys_init_module* system call is captured. The third challenge will be overcome by our *selective system calls intercepting* technique described in Sect. 3.2.

In summary, we make the following contributions:

– We present NOR, a non-intrusive, real-time and OS-agnostic introspection system for virtual machines in cloud environments.
– We propose a novel approach to conduct event-driven monitoring and snapshot polling cooperatively to build high level semantics from memory states instantaneously, meanwhile achieving completeness for the semantics. We also introduce a novel approach called *selective system call interception* to intercept system calls selectively thus to reducing system overheads (Sect. 3).
– We have built a proof-of-concept prototype of NOR and designed some case studies to conduct extensive experiments and analysis (Sect. 4). Our evaluation results show that our system is capable of detecting kernel malware precisely and mitigating transient attacks for different Linux kernels with minor overheads (Sect. 5).

2   Background

The arise of cloud computing pushes our computing paradigm from multitasking to multi-OS, as well as pushing system monitoring from traditional in-VM monitoring to out-of-VM, hypervisor-based monitoring. Through extracting and reconstructing the guest OS states at the VMM layer, out-of-VM monitors become possible, empowering the monitoring system to control, isolate, interpose, inspect, secure, and manage a VM from the outside. Garfinke et al. [16] present a seminal paper to introduce the first hypervisor- based monitoring system, call[ed] this approach of inspecting a virtual machine from the outside for the purpose of analyzing the software running inside it VMI.

Systematic Study of VMI Techniques. There have been several efforts focusing on systemizing the knowledge of VMI and summarizing the general approaches that have been proposed for bridging the semantic gap. In particular, Bauman et al. [5] categorizes five major approaches to bridging the semantic gap: the manual approach [21,22,27,42], debugger-assisted approach [16], compiler assisted approach [30,35], binary analysis assisted approach [12,13], and guest assisted approach [28,32]. The first four approaches are classified based on the constraints that an implementer faces when building an out-of-VM monitor. At a high level, these constraints are based on whether there is access to guest OS data structures, debug information, source code, or binary code. In addition, there is an option to avoid the semantic gap altogether, at the cost of potentially sacrificing the security advantages from VMI. This approach, the guest-assisted approach, modifies a guest OS kernel or places a program inside to pass information to the out-of-VM monitors. Wang et al. [37] characterize the current VMI techniques into four categories: hardware-assisted [25], semantic gap avoiding [13,18,32,34], snapshot-triggered [21,22,27,42] and event-triggered [8,10,28,29], based on three orthogonal dimensions: *(1)* Whether hardware assistance is required. *(2)* How it solves the semantic gap problem. *(3)* How introspection is triggered. In addition, Denz and Taylor [9] summarized introspection from the perspective of cloud security. Jain et al. [20] summarized introspection from the perspective of trust, security concerns, and attacks. Suneja et al. [36] organizes the various existing VMI techniques into a taxonomy based upon their operational principles in term of performance, complexity and overhead.

Transient Attacks. Transient attacks often perform malware activities without incurring persistent changes to the victims system. Specifically, if attackers have a sense of monitor existence and know exactly the interval of snapshot polling, stealthy malware can be completed during the interval and restore the modification to normal before next snapshot polling. The kind of attack within the guest VM could determine the presence of a passive VMI system and its monitoring frequency. With snapshot polling, we note that whenever the hypervisor wants to perform a monitoring check on a guest VM, the VM has to be paused in order to obtain a consistent view of the hardware state. Wang et al. [38] present an approach to perform in-VM timing measurements (e.g., timing frequently recurring OS events that related to scheduling) to detect VM suspends.

Since the evidence of malicious modification with monitored system is visible for a short period, detecting such modification becomes difficult. Some approaches might raise the rate of snapshot polling to increase the probability of detection for lower false negatives. Because the polling rate of the VMI monitor is directly related to the performance overhead of the monitor, the polling rate must be chosen carefully so as to not introduce an unacceptable decrease in VM performance. HyperSentry [3] tries to meet both higher performance and security by confusing attackers to predict the snapshot polling time exactly impossible. Leveraging the knowledge of probability to randomizing the snapshot interval might be an assistant method to minimize the false negatives as possible. It seems attractive, but if malware leaves the traces as minimal as possible for a passing-by period, there is still a chance that it can avoid being detected by the snapshot.

Intrusiveness. Introspection systems may also involve adding hooks into the guest or simply running the monitoring system inside it. Since the program or the hook is running inside the OS, it has full access to all abstractions that normally are lost when moving the monitor outside the VM. However, this also potentially leaves it vulnerable to all of the same dangers as an in-host monitor, and thus all of the information it gives to the hypervisor- based monitor becomes unreliable if the system is compromised and no special care is taken to protect the inside component.

More than that, in cloud environment, the cloud platform may maintain thousands of virtual machines for users including enterprises or private person, inserting programs into the guest VM is not user friendly and it may be not allowed to modify guest kernel.

OS-agnostic. Since the OS template is specific to the operating system, we should maintain different introspection programs customized for different operating systems. Considering that a cloud platform may provide thousands of VMs that differ in operating systems, kernel versions and patches, thus the kernel data structure layout and invariants are different accordingly. Managing a diversity of signatures for various OSes is extremely complex, so that administrators tend to seek for a dynamic introspection tool to response to agnostic OSes.

3   Design

3.1   Threat Model and Assumptions

We assume the following adversary model when designing NOR: *(1)* The malware that has the highest privilege level inside the guest VM (e.g., the root privilege) and is capable of bypassing the monitoring tool inside the guest VM; *(2)* The kind of attack that is able to detect the interval of snapshot polling introspection technique by employing some advanced mechanism.

NOR guarantees that its privilege is higher than the malware and even though the malware is capable of employing transient attack as described in [38], NOR is capable of detecting the malicious activities instantaneously.

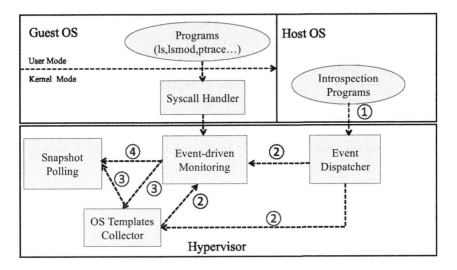

Fig. 1. The architecture of NOR.

Meanwhile, we assume a trusted VMM that provides strong VM isolation. The hypervisor is included in our Trusted Computing Base (TCB), the hypervisor level attack that may aim at the hypervisor vulnerabilities to devastate the fidelity or completeness of NOR is out of scope. Nevertheless, we can employ hypervisor integrity checking techniques such as [31,39] to ensure the intactness of the hypervisor before conducting introspection. We do not consider lower attack, such as hardware malware, the attacks before system loading and so forth.

3.2   Enabling Schemes and Techniques

NOR is designed to introspect virtual machines in cloud environments in a non-intrusive, real-time and OS-agnostic way. It is capable of obtaining a deep insight on system activities happened inside the guest VM. The architecture of NOR is composed of event dispatcher, event-driven monitoring, OS templates collector and snapshot polling as illustrated in Fig. 1. The basic idea of NOR is to launch events-driven monitoring and snapshot polling cooperatively. Specifically, there are four key steps involved during the execution of NOR:

Step ① : The introspection program initiates and sends commands to the event dispatcher according to different monitoring requirements.

Step ② : The event dispatcher determines the events that should be captured and OS templates information needed for the snapshot polling to meet the monitoring requirements, then notifies the event-driven monitoring and OS templates collector respectively. Meanwhile, the OS templates collector transfer the requirements to event-driven monitoring and wait for the response.

Step ③ : After the event-driven monitoring receives the requests, it immediately starts the monitoring process to trigger the determined event by event dispatcher and sends the OS template results to the OS templates collector. Sooner after that, OS templates collector sends the feedback to snapshot polling.

Step ④ : Once the event set by the event dispatcher happens, the snapshot polling will be triggered to start immediately and fetch introspection results to the introspection program.

The following sections will discuss the enabling schemes and techniques of NOR and how these schemes and techniques work cooperatively, as well as the challenges in detail.

Event Dispatcher. The event dispatcher is responsible for receiving the requests of the upon introspection programs, dispatching the events to event-driven monitoring. With event dispatcher, it is easy to extend NOR to adapt to different monitoring requirements.

Event-Driven Monitoring. Event-driven monitoring is responsible for capturing sensitive events, triggering snapshot polling, as well as analyzing OS template information for OS templates collector.

Event-driven monitoring employs system call interception to complete the above tasks. In most mainstream operating systems, system call is a very important interface to access kernel services. It is used by a user process to perform privileged operations. During normal execution, an application might issue millions of system calls in a short period, which can be regarded as a sufficient signature to distinguish between normal behavior and malicious attacks.

In order to trap system calls into NOR (part of the hypervisor), NOR configures the processor before a VM Entry so that execution of a system call entry or exit instruction causes a VM Exit and transfers control back to NOR. With this ability to intercept system call entries and exits, NOR monitors the arguments and return codes of system calls to gather insights about system activities inside the guest VM.

Selective System Call Interception: NOR avoids the overheads due to intercepting all guest applications in the guest VM by monitoring only when process context switches. NOR automatically monitors and un-monitors the children of the monitored processes by intercepting process-management related system calls such as *sys_fork*, *sys_clone* and *sys_kill*. NOR monitors all threads in a monitored process by default. Whenever a new guest process is scheduled on a virtual processor, NOR checks if the new process is monitored or not and turns system call interception on or off respectively.

To achieve the above aim, NOR should be able to identify processes and threads. NOR keeps track of the guest-OS assigned process identifiers (*PIDs*) and thread identifiers (*TIDs*) for the monitored processes and threads respectively. Processes are identified using *CR3* that contains the guest-physical address of the currently executing process's page directory base. Inspired by [23],

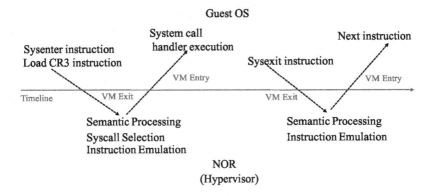

Fig. 2. Event-driven monitoring control flow of NOR.

NOR monitors *CR3* update to track each process context switch. During a system call interception, if NOR detects a different currently running *TID* from the one that last executed on the same virtual processor during the last system-call interception, it knows a thread switch has occurred. Figure 2 illustrates the control flow of the event-driven monitoring process. When the guest issue a system call instruction or load *CR3* instruction, it causes a VM exit and will be trapped into NOR. Then NOR starts to process the events and reconstruct upper level semantics, select system calls and emulation the instructions. After a VM entry, the guest continues to execute the system call handler. When the system call returns, the guest traps into NOR as well and NOR handles the following issues.

OS Templates Collector. OS templates are some knowledge like the layout and offset of interested kernel data structures. With these layout and offset, snapshot polling is capable of traversing kernel data structures form the memory bytes to reconstruct the upper semantics. Facilitated by OS templates collector, NOR do not need to inject program into guest VM to capture the OS templates for different kernels before conducting introspection. OS pattern exploring sends the request to event-driven monitoring, then event-driven monitoring retrieves the OS pattern related semantics through system call interception. NOR is capable of capturing these needed information on-the-fly meanwhile launching introspection. For example, in our prototype, the memory address of a loadable kernel module is retrieved by reconstructing the semantics from the *sys_init_moudle* related system calls intercepted and the initiated process (*init*) of Linux and pid_offset of a task_struct is retrieved by reconstructing the semantics from the *sys_fork* related system calls intercepted. With the OS templates captured by event-driven monitoring, NOR does not rely on knowledge of the guest OS itself to reconstruct semantics.

Snapshot Polling. The snapshot polling is triggered based on the occurrences of designated events which could possibly deduce some suspicious activities conducted by malware. The designated events are defined according to different introspect requirements, which can either be some system calls or *CR3* update. Triggered by these designated events, meanwhile acquiring the OS templates, the snapshot polling initiates, dumps the memory of the guest VM and traverses the kernel data structures from the memory bytes for reconstructing OS level semantics. The semantics reconstruction process is similar to the traditional snapshot polling introspection approach (e.g., LibVMI) does.

4   Implementation

The current NOR prototype is implemented on an x86 server that supports Intel VT technology, runs the KVM/QEMU hypervisor and is designed to support guest VMs running different kinds of Linux Kernels. The host machine runs a Linux along with a KVM kernel module. Each of the guest VMs is by itself a user-space process running the QEMU emulation program. In our implementation, the introspection modules including event dispatcher, event-driven monitoring, OS templates collector, snapshot polling are deployed within the KVM hypervisor.

4.1   Snapshot Polling Triggered by Designated Events Captured from Event-Driven Monitoring

All the experiments in this paper use 32-bit guest OSes and they run on an Intel processor. However, Inter VT only support trapping some specific events (e.g., MSR reads/writes, rdtsc, external interrupts, etc.), whereas it does not support some events (e.g., a system call) for security mechanisms. We should design some schemes to trap into the hypervisor on the event of a system call indirectly. Similar to previously known technique [10,29] for intercepting all x86-32 system call instructions, we force system interrupts(e.g., page faults, general protection faults, etc.) that are supported by the Intel VT to achieve the goal.

Intel x86-32 system uses the Sysenter/Sysexit and *INT 80* instructions for performing system call. For *INT 80*, NOR changing the interrupt descriptor table entry of interrupt to point to a non-present page in order to cause a page fault; For *Sysenter/Sysexit*, NOR sets the *SYSENTER_CS_MSR* to a null value, an address of a non-existent memory which will cause a general protection exception. The value of the *SYSENTER_CS_MSR* is copied into the *CS* register upon a call to Sysenter and an attempt to load the *CS* register with a null value results in a general protection exception. Both the page fault and general protection exception will be trapped into NOR. NOR sets a specific bit in the VM execution control register to capture the *CR3* update. As described in [7], if the *CR3-target count*is 0, *MOV to CR3* always causes a VM exit. Therefore NOR sets the value of *CR3_target_count* in the *VMCS* to 0 to track the *CR3* update. To retrieve semantics from the system calls, NOR resolves the arguments of the system calls which are stored in *eax, ebx, ecx, edx, esi, edi* respectively. In particular, NOR fetches the system call number from *eax*, and the return value from *eax* as well when a system call finishes. With the tracked *CR3* update and

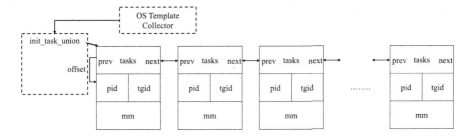

Fig. 3. Snapshot polling for semantics reconstruction (Processes list).

resolved semantics from the system calls, NOR is able to achieve *selective system call interception.*

We also leverage the LibVMI [27] library as the snapshot polling to facilitate our implementation. NOR uses the LibVMI to set up necessary memory mappings required to access the monitored VMs memory pages. Figure 3 shows the example of how NOR uses snapshot polling to list processes inside the guest VM. One union with *tasks, pid, tgid, mm* and other fields denotes a *task_struct* for a process in Linux. To reconstruct semantics of the complete processes list, NOR should locate the address of the initiate process running in Linux kernel and fetch the offset between the *tasks* field and *pid* field of a *task_struct*struct in advance. Then it traverses the linked list to resolve each process. NOR does not need to inject programs into guest OS to fetch the OS template information. Instead, it retrieves the OS template information from the OS Template Collector to avoid the intrusiveness and make our approach general to different kernels.

4.2 Flexible Monitoring

To adapt to different monitoring requirements, NOR captures different system calls as the sensitive events to trigger snapshot polling monitoring. Specifically, to monitor the kernel modules loaded into the guest kernel, NOR uses event-driven monitoring to trap *sys_init_module* related system calls and then trigger snapshot polling. To monitor the processes executed in the guest, NOR uses event-driven monitoring to trap *sys_fork* related system calls and then triggers snapshot polling. We list part of the related system calls for different monitoring requirements in the following (e.g., process/modules/files/directories monitoring, etc.) (Table 1).

To support more monitoring requirements, developers just need to write your own introspection programs and modify the event dispatcher to notify the event-driven monitoring of NOR. Then NOR traps different system calls to trigger the snapshot polling.

Table 1. Part of the related system calls for different monitoring requirements

Processes	*sys_fork, sys_execve, sys_clone, sys_kill*
Modules	*sys_init_module, sys_delete_module, sys_quiry_module*
Files Read/Write	*sys_read, sys_write, sys_getdents, sys_stat, sys_lstat, sys_fstat*
Files Open/Close	*sys_open, sys_close*
Directories	*sys_mkdir,sys_rmkir, sys_chdir, sys_linkat, sys_fchdir,* *sys_getdents, sys_getdents64, sys_unlinkat, sys_rename*
Connections	*sys_connect, sys_socket*

5   Evaluation

We have evaluated the effectiveness of our prototype to demonstrate the non-intrusive, real-time and OS-agnostic detection capabilities enabled by NOR. In particular, we show: (1) how the view comparison-based mechanism effectively detects on of the most stealthy kernel malware- rootkits (Sect. 5.1), (2) how transient attacks showed in [38] fail to evade NOR (Sect. 5.2) and (3) how NOR enables when new guest OS emerges (Sect. 5.3). Finally, we present performance overhead measurement results in Sect. 4. In our experiments, the host machine is an Intel Core i5 desktop running Ubuntu 14.04 and 8 GB of RAM. We use Linux kernels as the guest VM, which is configured with 1 GB of RAM.

5.1   Effectiveness Against Kernel Rootkits

We employ view comparison-based malware detection method to detect the very nature of rootikts - hiding attack processes, related files and the module itself. We have so far experimented with 10 real-world Linux rootkits samples (e.g., kbeast suterusu, knark, etc.) and the view comparison-based scheme is able to detect all the rootkits tested and pinpoint the corresponding hidden processes and/or file and/or the module itself.

Take Kbeast as an example, we run Kbeast in the guest OS with Ubuntu 12.04. KBeast is a kernel mode rootkit that loads as a kernel module. It hides the user-land component, as well as files, process and the module itself. It installs the rootkit in the "*/usr/h4x*" path and generate the "*_h4x_bd*" process. We run *ps* and *lsmod* with the command shell to find the hidden module and process, but it shows nothing in the inside view. By comparison, we run NOR to capture the external view. Once a kernel module is inserted into the kernel, NOR can be notified by intercepting the *sys_init_module* with event-driven monitoring and launches the snapshot polling to traverse the module lists from the memory bytes. As a result, NOR is capable of identifying the specific path of the hidden module and the process.

5.2   Effectiveness Against Transient Attacks

Wang et al. [38] present hypervisor introspection as a technique to determine the presence of and evade a snapshot polling monitoring system. To verify the effectiveness against transient attacks, we reimplement a backdoor shell similar to the one showed in [38] that repeatedly connects back to the attacker between monitoring checks to evade the snapshot polling monitoring.

Since the hypervisor introspection relies on fine-grained timing measurements to determine occurrences of VM suspends, it follows that making the suspending interval unpredictable could mitigate this kind of transient attacks. The suspending interval of NOR is determined by the occurrences of specific captured events with event-driven monitoring and it is hard to figure out. Our experiment shows that the backdoor implemented by [38] fails to evade NOR.

5.3   Effectiveness on Agnostic OSes

We have so far tested guest OS with Ubuntu (10.04, 11.04, 12.04, 13.04, 14.04, 15.04), CentOS 7.0, Debian 9 and Fedora 25 to verify the effectiveness on agnostic OSes. NOR is capable of retrieving the exact OS templates with the event-driven monitoring module to launch snapshot polling to adapt to different kinds of kernels, meanwhile detecting kernel rootkits and mitigating transient attacks.

5.4   Impact on Performance Overhead

To evaluate the impact on performance of NOR, the experiment results report as when running with NOR, as well as without NOR by disabling the NOR relevant code in our modified KVM module. To understand the direct computational overhead introduced by the NOR system on the test VMs, we used a combination of micro/synthetic benchmarks: PCMark05. PCMark05 is a commercial benchmark product which provides various CPU and disk drive tests. The adopted test cases include HTML rendering(page/s), file compression(MB/s), file decompression(MB/s), file encryption(MB/s), file decryption(MB/s), image compression (mpix/s) and file writing(MB/s). Note that, the main performance overhead come from the event-driven monitoring and snapshot polling of NOR. Since the snapshot polling won't be triggered unless there are some suspicious activities occur within the guest VM, the snapshot polling interval is unpredictable which is not like traditional LibVMI that depends on the intervals.

We only report the performance overhead of event-driven monitoring compared to the original system without introspection. Figure 4 shows the results of performance overhead running NOR with PCMark05 benchmarks. During each VM exit caused by a system call interception, the hypervisor code will be executed to figure out the exit reason, to handle the exit, to emulate the instruction that cause the VM exit and finally issues a VM entry. In addition to this, NOR performs some semantics computational work, checks whether the current process is monitored or not, and checks whether there has been a thread switch since the last system call interception for the same process. These work completed by

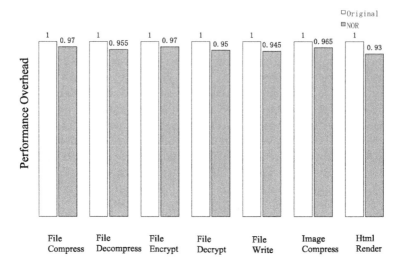

Fig. 4. Impact on performance overhead.

NOR are the origins of the performance degradation. With our *selective system call interception* to promote the performance, the average performance degradation of the benchmarks is 4.5%. The following section will discuss the scheme to enhance performance in further.

6 Discussion

We first discuss the limitations of our system and describe some attacks may target at NOR.

Limitations. Inspired by [2], NOR does not intercept all the system calls by with *Selective System Call Interception* to optimize overheads, but it fails to intercept specific system calls to adapt to users' needs. Many system calls are insignificant to be intercepted for reconstructing semantics to security issues. Before starting the event-driven monitoring, we should define a set of system calls to be trapped into the hypervisor according to some schemes. When those system calls that are not set to be trapped initiate, there should not have world switches between guest VM and hypervisor. With the mechanism, the overhead of NOR can be improved further.

The events set by NOR prototype to trigger snapshot polling is still limited, so that the semantics reconstructed to meet the presented guarantees is not abundant enough. NOR is capable of mitigating basic transient attacks that can detect the snapshot polling intervals, but we do not consider advanced side-channel attacks (e.g., cache side-channel attack [45], power analysis attack [24], etc.) to evade the snapshot polling. And NOR is our initial prototype that only support Linux kernels so far. But the basic idea is similar, it would take more engineering efforts to extend NOR with other OSes, such as closed-source Windows. We leave this to our future work.

Possible Attacks. As the same as most existing work, NOR can not resist Direct Kernel Structure Manipulation (DKSM) attacks [4]. DKSM is capable of compromising a guest such that the kernels use of any field of its data structures (or templates) could be potentially modified. Under DKSM, a malicious OS actively misleads VMI tools in order to violate a security policy. Jain et al. [20] define this as a strong semantic gap problem. A solution to strong semantic gap would not make any assumptions about the guest OS being benign to obtain any semantics (e.g., traversing the module list, etc.). If the strong semantic gap problem can be solved, it would be capable of monitoring the kernel functions and sensitive objects that could provide richer semantics to detect or prevent DKSM attacks. The most recent work [46] proposes the immersive execution environment (ImEE) with which the guest memory is accessed at native speed. Even if the guest kernel is compromised, it is capable of introspecting the live kernel states in the targets without modifying or suspending them. However, it only serves as the guest access engine for the VMI applications without involving kernel semantics, which could be integrated with our system to retrieve the higher semantics.

Note that virtualization is not hard to detected nowadays by attackers. In fact, a number of recent malware employs anti-VM or anti-sandbox techniques [1] to check whether they are running in a virtualization environment or a sandbox. If so, they would hide the malicious behavior to camouflage themselves. Enhancing the fidelity of hypervisor and isolation of VMs has a strong urgency and necessity. We assumed a trusted hypervisor to isolate the malware inside the guest. However, recent malware could leverage firmware and hardware [17] to attack the hypervisor and break the isolation of VMs. As a countermeasure, researchers are seeking to separate the hypervisor through same-privilege memory isolation to enforce cross-VM data and control-flow integrity [33].

7   Related Work

Virtual Machine Introspection. LibVMI [27] is a C library with Python bindings that makes it easy to monitor the low-level details of a running virtual machine by viewing its memory, trapping on hardware events, and accessing the vCPU registers. It is an open source monitoring library on both KVM and XEN, which is designed for general memory and disk introspection and provides a variety of utility functions that are useful to VMI developers.

Ether [10] is a transparent and external approach for malware analysis, which implements monitoring instruction execution, memory writes, system call execution, and limited scope of a chosen process by hardware virtualization extension. Recent work, Sky [2] gathers insights and information by intercepting system calls made by guest programs to specific processes and threads for capturing storage-specific information. With the specific I/O semantics, it is capable of enhancing virtualized storage performance.

Under a cloud environment, researchers and cloud administrators are seeking for the OS-agnostic nature. Work like ShadowContext [41], Hypershell [15] redirect the relative code and data that are kernel-dependent (mainly for system call) to guest VM for execution to avoid the semantic gap. It is attracting but makes the guest intrusive and can not guarantee the security of the process executed for help to retrieve introspection information inside the guest.

NOR is inspired by both event-driven monitoring approaches (e.g., Ether [10], Sky [2], Nitro [29], and etc.) and snapshot polling approaches (e.g., VMWatcher [22], LibVMI [27], and etc.) to conduct event-driven monitoring and snapshot polling cooperatively. The most related work to NOR is RTKDSM [19] and CloudMon [40]. The RTKDSM automates the process of kernel data structure extraction and enables VMI applications to monitor any real-time changes to selective guest OS data structures. It employs a periodic polling mode to check and compared the application- specified kernel data structures are once in every fixed-sized time interval. CloudMon [40] implements a self-adjusting algorithm for polling according to CPU utilization, memory utilization, and network usage, to mitigate transient attacks. As discussed in Sect. 2, this kind of randomizing or the snapshot interval might be an assistant method to minimize the false negatives as possible. It seems attractive, but if malware leaves the traces as minimal as possible for a passing-by period, there is still a chance that it can avoid being detected. NOR triggers the snapshot polling routine by event-driven monitoring instantaneously to make the interval of snapshot polling unpredictable.

In-VM Monitoring. SIM [32] forwards introspection tool back to guest VM within a protected address space. It leverages hardware memory protection and hardware virtualization features available in contemporary processors to create a hypervisor protected address space to enable introspection tools execute and access data in native speed. Gateway [35] is another in-VM monitoring system, which is designed for detecting kernel malware by monitoring kernel APIs invoked by device drivers. These in-VM monitoring systems achieve efficiency by executing and accessing data inside the guest VM meanwhile guaranteeing security of the monitoring tools by employing some isolation mechanisms, but bring intrusiveness to the guest and need to modify the guest kernel.

8   Conclusion

We develop NOR, a non-intrusive, real-time and OS-agnostic introspection system for virtual machines in clouds. It is capable of monitoring the activities of the guest OS instantaneously in an unaware way, as well as supporting different kernels. Such introspection technique is needed in cloud environment, especially for large enterprises. Due to the importance of introspection to address security issues, more efficient designs for VMI techniques are still highly needed.

Acknowledgement. We would like to thank the anonymous reviewers for their insightful comments that greatly helped to improve this paper. This work is is a part of the project supported by Beijing Natural Science Foundation (Y720011101). Any opinions, findings, and conclusions expressed in this material are those of the authors and do not necessarily reflect the views of these agencies.

References

1. Malicious documents leveraging new anti vm anti sandbox techniques. https://www.zscaler.com/blogs/research/malicious-documents-leveraging-new-anti-vm-anti-sandbox-techniques

2. Arulraj, L., Arpaci-Dusseau, A.C., Arpaci-Dusseau, R.H.: Improving virtualized storage performance with sky. In: Proceedings of ACM SIGPLAN/SIGOPS International Conference on Virtual Execution Environments (VEE), pp. 112–128 (2017)

3. Azab, A.M., Ning, P., Wang, Z., Jiang, X., Zhang, X., Skalsky, N.C.: Hypersentry: enabling stealthy in-context measurement of hypervisor integrity. In: Proceedings of ACM Conference on Computer and Communications Security (CCS), pp. 38–49 (2010)

4. Bahram, S., Jiang, X., Wang, Z., Grace, M., Li, J., Srinivasan, D., Rhee, J., Xu, D.: DKSM: subverting virtual machine introspection for fun and profit. In: Proceedings of IEEE Symposium on Reliable Distributed Systems (SRDS), pp. 82–91 (2010)

5. Bauman, E., Ayoade, G., Lin, Z.: A survey on hypervisor-based monitoring: approaches, applications, and evolutions. ACM Comput. Surv. **48**(1), 10:1–10:33 (2015)

6. Carbone, M., Conover, M., Montague, B., Lee, W.: Secure and robust monitoring of virtual machines through guest-assisted introspection. In: Balzarotti, D., Stolfo, S.J., Cova, M. (eds.) RAID 2012. LNCS, vol. 7462, pp. 22–41. Springer, Heidelberg (2012). https://doi.org/10.1007/978-3-642-33338-5_2

7. Intel Corporation. Intel 64 and ia-32 architectures software developer manuals

8. Deng, Z., Zhang, X., Xu, D.: Spider: stealthy binary program instrumentation and debugging via hardware virtualization. In: Proceedings of Annual Computer Security Applications Conference (ACSAC), pp. 289–298 (2013)

9. Denz, R., Taylor, S.: A survey on securing the virtual cloud. J. Cloud Comput. Adv. Syst. Appl. **2**(1), 17 (2013)

10. Dinaburg, A., Paul, P.R., Sharif, M., Lee, W.: Ether: malware analysis via hardware virtualization extensions. In: Proceedings of ACM Conference on Computer and Communications Security (CCS), pp. 51–62 (2008)

11. Dolan-Gavitt, B., Payneand, B., Lee, W.: Leveraging forensic tools for virtual machine introspection. In: Technical report GT-CS-11-05. Georgia Institute of Technology (2011)

12. Dolan-Gavitt, B., Leek, T., Zhivich, M., Giffin, J., Lee, W.: Virtuoso: narrowing the semantic gap in virtual machine introspection. In: Proceedings of IEEE Symposium on Security and Privacy (S&P), pp. 297–312 (2011)

13. Fu, Y., Lin, Z.: Space traveling across VM: Automatically bridging the semantic gap in virtual machine introspection via online kernel data redirection. In: Proceedings of IEEE Symposium on Security and Privacy (S&P), pp. 586–600 (2012)

14. Fu, Y., Lin, Z.: Exterior: using a dual-VM based external shell for guest-OS introspection, configuration, and recovery. In: Proceedings of ACM SIGPLAN/SIGOPS International Conference on Virtual Execution Environments (VEE), pp. 97–110 (2013)

15. Fu, Y., Zeng, J., Lin, Z.: Hypershell: a practical hypervisor layer guest OS shell for automated in-VM management. In: Proceedings of USENIX Annual Technical Conference (ATC), pp. 85–96 (2014)

16. Garfinkel, Z., Rosenblum, M.: A virtual machine introspection based architecture for intrusion detection. In: Proceedings of Network and Distributed System Security Symposium (NDSS), pp. 191–206 (2003)

17. Gorobets, M., Bazhaniuk, M., Matrosov, A., Furtak, A., Bulygin, Y.: Attacking hypervisors via firmware and hardware. In: Black Hat USA (2015)

18. Gu, Z., Deng, Z., Xu, Z., Jiang, X.: Process implanting: a new active introspection framework for virtualization. In: Proceedings of IEEE Symposium on Reliable Distributed Systems (SRDS), pp. 147–156 (2011)

19. Hizver, X., Chiueh, T.: Real-time deep virtual machine introspection and its applications. In: Proceedings of ACM SIGPLAN/SIGOPS International Conference on Virtual Execution Environments (VEE), pp. 3–14 (2014)

20. Jain, B., Baig, M.B., Zhang, D., Porter, D.E., Sion, R.: Sok: introspections on trust and the semantic gap. In: Proceedings of IEEE Symposium on Security and Privacy (S&P), pp. 605–620 (2014)

21. Jiang, X., Wang, X.: "Out-of-the-Box" monitoring of VM-based high-interaction honeypots. In: Kruegel, C., Lippmann, R., Clark, A. (eds.) RAID 2007. LNCS, vol. 4637, pp. 198–218. Springer, Heidelberg (2007). https://doi.org/10.1007/978-3-540-74320-0_11

22. Jiang, X., Wang, X., Xu, D.: Stealthy malware detection through VMM-based out-of-the-box semantic view reconstruction. In: Proceedings of ACM Conference on Computer and Communications Security (CCS), pp. 128–138 (2007)

23. Jones, S.T., Arpaci-Dusseau, A.C., Arpaci-Dusseau, R.H.: Antfarm: tracking processes in a virtual machine environment. In: Proceedings of USENIX Annual Technical Conference (ATC), pp. 1–14 (2006)

24. Kocher, P., Jaffe, J., Jun, B.: Differential power analysis. In: Wiener, M. (ed.) CRYPTO 1999. LNCS, vol. 1666, pp. 388–397. Springer, Heidelberg (1999). https://doi.org/10.1007/3-540-48405-1_25

25. Liu, Y., Xia, Y., Guan, H., Zang, B., Chen, H.: Concurrent and consistent virtual machine introspection with hardware transactional memory. In: Proceedings of IEEE International Symposium on High Performance Computer Architectur(HPCA), pp. 416–427 (2014)

26. Michael, P., Sherali, Z., Ray, H.: Virtualization: issues, security threats, and solutions. ACM Comput. Survey. 45(2), 17:1–17:39 (2013)

27. Payne, B.D.: Simplifying virtual machine introspection using LibVMI. In: Technical report SAND 2012-7818, Sandia National Laboratories (2012)

28. Payne, B.D., Carbone, M., Sharif, M., Lee, W.: Lares: an architecture for secure active monitoring using virtualization. In: Proceedings of IEEE Symposium on Security and Privacy (S&P), pp. 233–247 (2008)

29. Pfoh, J., Schneider, C., Eckert, C.: Nitro: hardware-based system call tracing for virtual machines. In: Iwata, T., Nishigaki, M. (eds.) IWSEC 2011. LNCS, vol. 7038, pp. 96–112. Springer, Heidelberg (2011). https://doi.org/10.1007/978-3-642-25141-2_7

30. Rhee, J., Riley, R., Xu, D., Jiang, X.: Kernel malware analysis with un-tampered and temporal views of dynamic kernel memory. In: Jha, S., Sommer, R., Kreibich, C. (eds.) RAID 2010. LNCS, vol. 6307, pp. 178–197. Springer, Heidelberg (2010). https://doi.org/10.1007/978-3-642-15512-3_10

31. Riley, R., Jiang, X., Xu, D.: Guest-transparent prevention of kernel rootkits with VMM-based memory shadowing. In: Lippmann, R., Kirda, E., Trachtenberg, A. (eds.) RAID 2008. LNCS, vol. 5230, pp. 1–20. Springer, Heidelberg (2008). https://doi.org/10.1007/978-3-540-87403-4_1

32. Sharif, M.I., Lee, M.I., Cui, W., Lanzi, A.: Secure in-VM monitoring using hardware virtualization. In: Proceedings of ACM Conference on Computer and Communications Security (CCS), pp. 477–487 (2009)

33. Shi, L., Wu, Y., Xia, Y., Dautenhahn, N., Chen, H., Zang, B., Guan, H., Li, J.L.: Deconstructing Xen (2017)

34. Srinivasan, D., Wang, Z., Jiang, X., Xu, D.: Process out-grafting: An efficient "out-of-VM" approach for fine-grained process execution monitoring. In: Proceedings of the 18th ACM Conference on Computer and Communications Security (CCS), pp. 363–374 (2011)

35. Srivastava, A., Giffin, J.: Efficient monitoring of untrusted kernel-mode execution. In: Proceedings of Network and Distributed System Security Symposium (NDSS) (2011)

36. Suneja, S., Isci, C., Lara, E., Bala, V.: Exploring Vm introspection: techniques and trade-offs. In: Proceedings of ACM SIGPLAN/SIGOPS International Conference on Virtual Execution Environments (VEE), pp. 133–146 (2015)

37. Wang, C., Yun, X., Hao, Z., Cui, L., Han, Y., Zou, Q.: Exploring efficient and robust virtual machine introspection techniques. In: Wang, G., Zomaya, A., Perez, G.M., Li, K. (eds.) ICA3PP 2015, Part III. LNCS, vol. 9530, pp. 429–448. Springer, Cham (2015). https://doi.org/10.1007/978-3-319-27137-8_32

38. Wang, G., Estrada, Z.J., Pham, C., Kalbarczyk, C., Iyer, R.K.: Hypervisor introspection: a technique for evading passive virtual machine monitoring. In: Proceedings of USENIX WOOT, pp. 12–19 (2015)

39. Wang, Z., Jiang, X.: Hypersafe: a lightweight approach to provide lifetime hypervisor control-flow integrity. In: Proceedings of IEEE Symposium on Security and Privacy (S&P), pp. 380–395 (2010)

40. Weng, C., Liu, Q., Li, K., Zou, D.: Cloudmon: monitoring virtual machines in clouds. IEEE Trans. Comput. **65**(12), 3787–3793 (2016)

41. Wu, R., Chen, P., Liu, P., Mao, B.: System call redirection: a practical approach to meeting real-world virtual machine introspection needs. In: Proceedings of Annual IEEE/IFIP International Conference on Dependable Systems and Networks, pp. 574–585 (2014)

42. Yan, K.L., Yin, H.: Droidscope: seamlessly reconstructing the OS and Dalvik semantic views for dynamic android malware analysis. In: Proceedings of USENIX Security, p. 29 (2012)

43. Yan, L., Jayachandra, M., Zhang, M., Yin, H.: V2E: combining hardware virtualization and softwareemulation for transparent and extensible malware analysis. In: Proceedings of ACM SIGPLAN/SIGOPS International Conference on Virtual Execution Environments (VEE), pp. 227–238 (2012)

44. Yin, H., Song, D., Egele, D., Kruegel, D., Kirda, E.: Panorama: capturing system-wide information flow for malware detection and analysis. In: Proceedings of ACM Conference on Computer and Communications Security (CCS), pp. 116–127 (2007)

45. Zhang, Q., Reiter, M.K.: Düppel: retrofitting commodity operating systems to mitigate cache side channels in the cloud. In: Proceedings of the 20th ACM SIGSAC Conference on Computer and Communications Security, pp. 827–838 (2013)

46. Zhao, S., Ding, X., Xu, W., Gu, D.: Seeing through the same lens: Introspecting guest address space at native speed. In: 26th USENIX Security Symposium (USENIX Security 2017), pp. 799–813 (2017)

A Method to Enlarge the Design Distance of BCH Codes and Some Classes of Infinite Optimal Cyclic Codes

Shanding Xu[1,2,3(✉)], Xiwang Cao[1], and Chunming Tang[3]

[1] Department of Mathematics, Nanjing University of Aeronautics and Astronautics,
Nanjing 210016, China
sdxzx11@163.com, xwcao@nuaa.edu.cn
[2] Department of Mathematics and Physics, Nanjing Institute of Technology,
Nanjing 211167, China
[3] Key Laboratory of Mathematics and Interdisciplinary Sciences,
Guangdong Higher Education Institutes, Guangzhou University,
Guangzhou 510006, China
ctang@gzhu.edu.cn

Abstract. Cyclic codes are a meaningful class of linearcodes due to their effective encoding and decoding algorithms. As a subclass of cyclic codes, Bose-Ray-Chaudhuri-Hocquenghem (BCH) codes have good error-correcting capability and are widely used in communication systems. As far as the design of cyclic codes is concerned, it is difficult to determine the minimum distance. It is well known that the minimum distance of a cyclic code of designed distance d is at least d. In this paper, by adjusting the generator polynomial slightly and using a concatenation technique, we present a method to enlarge the designed distance of cyclic codes and obtain two classes of $[pq, q-1, 2p]$ cyclic codes and $[pq, p-1, 2q]$ cyclic codes over $GF(2)$. As a consequence, a class of infinite optimal $[3p, 2, 2p]$ cyclic codes, where $p \equiv -1 \pmod 8$, with respect to the Plotkin bound over GF(2) is presented.

Keywords: Cyclic code · Cyclotomic sequence · Finite fields

1 Introduction

Let l be a power of a prime r and $GF(l)$ denote the finite field with l elements. And let $GF(l)^n$ be the n-dimensional linear space over $GF(l)$. A linear $[n, k, d]$ code C over $GF(l)$ is a k-dimensional subspace of $GF(l)^n$ with minimum distance d. A linear code C is called *cyclic* if $(c_0, c_1, \cdots, c_{n-1}) \in C$ implies $(c_{n-1}, c_0, \cdots, c_{n-2}) \in C$. Let $\gcd(l, n) = 1$. And we denote by R the ring $GF(l)[x]/(x^n - 1)$. For any vector $(c_0, c_1, \cdots, c_{n-1}) \in GF(l)^n$, we identify it with the polynomial

$$c_0 + c_1 x + \cdots + c_{n-1} x^{n-1} \in R.$$

© Springer International Publishing AG, part of Springer Nature 2018
X. Chen et al. (Eds.): Inscrypt 2017, LNCS 10726, pp. 518–528, 2018.
https://doi.org/10.1007/978-3-319-75160-3_30

Then any code C over GF(l) with length n corresponds to a subset of R. A linear code C over GF(l) is cyclic if and only if C is an ideal of R. If C is not trivial, there exists a unique monic polynomial $g(x) \in$ GF(l)[x] dividing $x^n - 1$ and $C = (g(x))$. Then we call $g(x)$ the *generator polynomial* of C and $h(x) = (x^n - 1)/g(x)$ the *parity $-$ check polynomial* of C. For more details of cyclic codes, please refer to [1].

The following bounds on the minimum distance of cyclic codes are proved in [1] and will be useful in this paper.

Lemma 1 *(BCH bound for cyclic codes). Let b be a nonnegative integer and let α be a n-th primitive root of unity in some extension field of GF(l). A BCH code over GF(l) with length n and designed distance d, $2 \le d \le n$, is a cyclic code defined by the roots*

$$\alpha^b, \ \alpha^{b+1}, \ \cdots, \ \alpha^{b+d-2},$$

of the the generator polynomial. Then the minimum distance of a BCH code of designed distance d is at least d.

Lemma 2 *(Plotkin Bound). For a linear $[n, k, d]$ code C over GF(l) we have*

$$d \le \frac{nl^{k-1}(l-1)}{l^k - 1}.$$

A linear $[n, k, d]$ code C over GF(l) is called optimal with respect to the Plotkin bound if the Plotkin bound in Lemma 2 is met with equality. In this case, we also say that C is optimal for short.

The total number of cyclic codes over GF(l) and their constructions are closely associated with the cyclotomic cosets modulo n. At present, there are many ways to construct cyclic codes. One way to construct cyclic codes over GF(l) with length n is to make use of the generator polynomial

$$\frac{x^n - 1}{\gcd(x^n - 1, S(x))} \tag{1}$$

where $s^{(n)} = (s_i)_{i=0}^{n-1}$ is a sequence of period n over GF(l) and

$$S(x) = \sum_{i=0}^{n-1} s_i x^i \in \text{GF}(l)[x]$$

is called the *generator polynomial* of the sequence $s^{(n)}$. Then the cyclic code C_s generated by (1) is called a *cyclic code defined by the sequence $s^{(n)}$*. Correspondingly, the sequence $s^{(n)}$ is called a *defining sequence of the cyclic code C_s*, see [5–7] for details.

In the past decades, a lot of important results in the field of cyclic codes from the cyclotomy or generalized cyclotomy have been presented in a series of papers (see, for example, [2–11]). A generalized cyclotomy with respect to pq, where p and q are two distinct primes, was introduced by Whiteman [12], whose motivation is to search for residue difference sets. The Whiteman's generalized cyclotomic sequence (WGCS) was introduced by Ding [13] and its coding properties were studied in [5,6,8,10]. More specifically, based on three types of generalized cyclotomy of order two, three classes of cyclic codes of length pq and dimension $(pq + 1)/2$ were presented and analyzed by Ding [5], some of which are best cyclic codes with respect to the minimum distance. Furthermore, Ding [6] has constructed a number of classes of cyclic codes over $\mathrm{GF}(l)$ with length pq from the extended WGCS with order two, and given lower bounds on the minimum distance of these cyclic codes. In addition, Sun et al. [8] and Kewat et al. [10] have produced several classes of cyclic codes using a class of WGCSs with order four and order six, respectively, and presented lower bounds on the nonzero minimum distance of these cyclic codes.

Linear complexity of a sequence is an important measure of its quality. It is defined to be the length of the shortest linear feedback shift register that can generate this sequence. By the Berlekamp-Massey algorithm, if linear complexity of a sequence of period n is greater than or equal to $n/2$, then the sequence is thought to be good from a view point of linear complexity. Pseudo-random sequences with high linear complexity have wide applications in cryptography and communication systems.

The contribution of this paper is twofold. Firstly, we construct a class of generalized cyclotimic sequences with order two (defined by (3)) of period $n = pq$ over $\mathrm{GF}(2)$, and investigate its linear complexity and minimal polynomial. It is showed that its linear complexity is higher than that of [6]. Secondly, as far as we know, there are only a few optimal cyclic codes from the generalized cyclotomy with respect to pq, see [5,6,8,10] for details. Here a natural question one would ask is: are there a lot of optimal cyclic codes from the generalized cyclotomy with respect to pq? Answering this question is our main concern. The hardest part for this question is how to enlarge the designed distance of cyclic codes. By using a concatenation technique, we obtain a double designed distance of that in [6], see Theorem 1 and Remark 3. To say it explicitly, if the roots set of a cyclic code contains two consecutive pieces with equal length, we can add a root to concatenate the two pieces such that the new cyclic code has a double designed distance. Consequently, we construct a class of infinite optimal $[3p, 2, 2p]$ cyclic codes with respect to the Plotkin bound over $\mathrm{GF}(2)$. The idea of constructing cyclic codes with special types of sequences employed in this paper comes from [5,6]. However, compared with the cyclic codes given by [5,6], our construction can generate a lot of optimal cyclic codes with new parameters. The ultimate goal of our work in the future is to construct optimal cyclic codes with more flexible parameters from the generalized cyclotomy with respect to pq. But it is a difficult job at present.

The rest of this paper is organized as follows. In Sect. 2, we introduce some preliminaries about sequences and generalized cyclotomy. In Sect. 3, we construct a class of cyclic codes using the general generalized cyclotomy of order two with respect to pq. Finally, Sect. 4 concludes this paper.

2   Preliminaries

In this section, we give basic notations and results of sequences and generalized cyclotomy that will be employed in subsequent sections.

2.1   Linear complexity and minimal polynomial

Let $s^{(n)} = s_0 s_1 \cdots s_{n-1}$ be a sequence of period n over a finite field $\mathrm{GF}(l)$. If there exist $L+1$ constants $c_0 = 1, c_1, \cdots, c_L \in \mathrm{GF}(l)$ such that

$$-c_0 s_i = c_1 s_{i-1} + c_2 s_{i-2} + \cdots + c_L s_{i-L}, \quad L \leq i < n,$$

then the sequence $s^{(n)}$ is called *linear feedback sequence* and the minimal positive integer L is defined as *linear complexity* of the sequence $s^{(n)}$. Such a polynomial $c(x) = c_0 + c_1 x + \cdots + c_L x^L$ is called a *feedback polynomial* of $s^{(n)}$. For a finite sequence, such a positive integer always exists. Any feedback polynomial of $s^{(n)}$ is also called a *characteristic polynomial* of $s^{(n)}$. The characteristic polynomial with the minimal length is called the *minimal polynomial* of $s^{(n)}$. The linear complexity of a periodic sequence is equal to the degree of its minimal polynomial. For the periodic sequences, there are many methods to compute the linear complexity and minimal polynomial. One of them is stated in the following lemma.

Lemma 3 [1]. *Let $s^{(n)}$ be a sequence of period n over $\mathrm{GF}(l)$. Define*

$$S(x) = s_0 + s_1 x + \cdots + s_{n-1} x^{n-1} \in \mathrm{GF}(l)[x].$$

Then the minimal polynomial $m(x)$ of $s^{(n)}$ is determined by

$$\frac{x^n - 1}{\gcd(x^n - 1, S(x))}$$

and the linear complexity L_s of $s^{(n)}$ is given by

$$L_s = n - \deg(\gcd(x^n - 1, S(x))).$$

2.2   The General Generalized Cyclotomy of Order Two Modulo pq and Generalized Cyclotomic Sequence

Now we introduce the definition and properties of general generalized cyclotomy of order two modulo pq. For more details, please refer to [6].

For a positive integer n, let $\mathbb{Z}_n$ be the residue class ring $\mathbb{Z}_n = \{0, 1, 2, ..., n-1\}$ and $\mathbb{Z}_n^*$ be the multiplicative group consisting of all elements relatively prime to n in $\mathbb{Z}_n$. An integer a is called a *primitive root* modulo n if the multiplicative order of a modulo n is equal to $\phi(n)$, where $\phi(n)$ is the Euler function and $\gcd(a, n) = 1$. Let H be a subset of $\mathbb{Z}_n$ and b be an element of $\mathbb{Z}_n$. Define

$$b + H = \{b + h : h \in H\}, \qquad b \cdot H = \{b \cdot h : h \in H\}.$$

From now on, it is always assumed that $n = pq$, where p and q are two distinct odd primes with $p \equiv -1 \pmod 8$ and $q \equiv 3 \pmod 8$, or $p \equiv 3 \pmod 8$ and $q \equiv -1 \pmod 8$, or $p \equiv -3 \pmod 8$ and $q \equiv 1 \pmod 8$, or $p \equiv 1 \pmod 8$ and $q \equiv -3 \pmod 8$. Let $N = \gcd(p-1, q-1)$ and $e = (p-1)(q-1)/N$. By the Chinese Remainder Theorem (CRT), there are common primitive roots modulo both p and q. Let g be a fixed common primitive root modulo both p and q and x be an integer satisfying

$$x \equiv g \pmod p, \; x \equiv 1 \pmod q.$$

Hence, we can get the general generalized cyclotomy of order two modulo pq as follows [6]

$$D_i^{(n)} = \{g^s x^{2t+i} : s = 0, 1, \cdots, e-1; t = 0, 1, \cdots, N/2 - 1\}, \; 0 \le i \le 1, \quad (2)$$

where the multiplication is performed modulo n.

The corresponding generalized cyclotomic numbers of order two modulo pq are defined as

$$(i, j)_2^{(n)} = |(D_i^{(n)} + 1) \cap D_j^{(n)}|, \text{where } 0 \le i, j \le 1.$$

Define

$$R = \{0\},$$
$$P = \{p, 2p, \cdots, (q-1)p\},$$
$$Q = \{q, 2q, \cdots, (p-1)q\},$$
$$C_0^{(n)} = P \cup D_0^{(n)},$$
$$C_1^{(n)} = Q \cup R \cup D_1^{(n)}.$$

Remark 1. From the above definitions, we have that $\{C_0^{(n)}, C_1^{(n)}\}$ is a partition of $\mathbb{Z}_n$. The objective of this partition is to concatenate the two pieces of roots of a generator polynomial such that the new cyclic code has a double designed distance.

The basic properties of the aforementioned generalized cyclotomy in [6] are given in the following lemmas and will be employed later.

Lemma 4

(1) $-1 \in D_0^{(n)}$;

(2) If $a \in D_j^{(n)}$ for some j, $0 \le j \le 1$, then $aD_i^{(n)} = D_{i+j \pmod 2}^{(n)}$.

Lemma 5. Let the symbols be the same as before. Then

$$(0,0)_2^{(n)} = \frac{(p-2)(q-2)+3}{4}, \quad (0,1)_2^{(n)} = (1,0)_2^{(n)} = (1,1)_2^{(n)} = \frac{(p-2)(q-2)-1}{4}.$$

Lemma 6. *Let the symbols be the same as before. Then*

$$|(D_0^{(n)} + \omega) \cap D_1^{(n)}| = \begin{cases} (0,1)_2^{(n)}, & \text{if } \omega \in D_0^{(n)} \\ (1,0)_2^{(n)}, & \text{if } \omega \in D_1^{(n)} \end{cases}$$

and

$$|(D_1^{(n)} + \omega) \cap D_1^{(n)}| = \begin{cases} (1,1)_2^{(n)}, & \text{if } \omega \in D_0^{(n)} \\ (0,0)_2^{(n)}, & \text{if } \omega \in D_1^{(n)}. \end{cases}$$

Lemma 7. *If $\omega \in P$, then*

$$|(D_i^{(n)} + \omega) \cap D_j^{(n)}| = \begin{cases} \frac{(p-1)(q-3)}{4}, & \text{if } i = j \\ \frac{(p-1)(q-1)}{4}, & \text{if } i \neq j. \end{cases}$$

If $\omega \in Q$, then

$$|(D_i^{(n)} + \omega) \cap D_j^{(n)}| = \begin{cases} \frac{(p-3)(q-1)}{4}, & \text{if } i = j \\ \frac{(p-1)(q-1)}{4}, & \text{if } i \neq j. \end{cases}$$

Definition 1. *The generalized cyclotomic sequence $s^{(n)} = (s_i)_{i=0}^{n-1}$ of period n is defined by*

$$s_i = \begin{cases} 0, & \text{if } i \pmod{n} \in C_0^{(n)} \\ 1, & \text{if } i \pmod{n} \in C_1^{(n)}. \end{cases} \tag{3}$$

Remark 2. In [14] and [6], Brandstatter and Ding independently defined a kind of generalized cyclotomic sequence with order two of period pq, where p and q are two distinct odd primes. However, our sequence is different from their sequence, which can be seen from Definition 1.

3 A Class of Optimal Cyclic Codes Derived from Generalized Cyclotomy

Henceforth, we always assume that $l = 2$. Let m be the order of 2 modulo n. Then the field $GF(2^m)$ has a n-th primitive root of unity ζ_n. Define

$$S(x) = \sum_{i \in C_1^{(n)}} x^i = 1 + \left(\sum_{i \in Q} + \sum_{i \in D_1^{(n)}} \right) x^i \in GF(2)[x]. \tag{4}$$

Our main objective in this section is to find the generator polynomial

$$g(x) = \frac{x^n - 1}{\gcd(x^n - 1, S(x))}$$

of the cyclic code C_s defined by the sequence $s^{(n)}$ in Definition 1, where $S(x)$ is the same as in Eq. (4). To compute the parameters of the cyclic code C_s defined by the sequence $s^{(n)}$, we need to compute $\gcd(x^n - 1, S(x))$. Hence, we need only to find such a's that $S(\zeta_n^a) = 0$, where $0 \leq a \leq n - 1$, since ζ_n is a n-th primitive root of unity. To this end, we need some auxiliary results. We have

$$\sum_{i \in P} \zeta_n^i = \sum_{i \in Q} \zeta_n^i = \sum_{i \in \mathbb{Z}_n^*} \zeta_n^i = 1. \tag{5}$$

Lemma 8 [6]. *Let the symbols be the same as before. Then*

$$\sum_{i \in D_1^{(n)}} \zeta_n^{ai} = \begin{cases} \frac{p-1}{2} & (\text{mod } 2), & if\ a \in P \\ \frac{q-1}{2} & (\text{mod } 2), & if\ a \in Q. \end{cases}$$

And let p be an odd prime with $p \equiv -1$ (mod 8). If $a \in P \cup Q$, then $\sum\limits_{i \in D_1^{(n)}} \zeta_n^{ai} = 1$.

Lemma 9. *Let the symbols be the same as above. Then*

$$S(\zeta_n^a) = \begin{cases} 1, & if\ a \in R \\ \frac{p+1}{2} & (\text{mod } 2), & if\ a \in P \\ \frac{q-1}{2} & (\text{mod } 2), & if\ a \in Q \\ S(\zeta_n), & if\ a \in D_0^{(n)} \\ S(\zeta_n) + 1, & if\ a \in D_1^{(n)}. \end{cases}$$

Proof. It follows immediately from Eq. (4) and Lemma 8, so we omit its proof.

Lemma 10. *Let the symbols be the same as before. Then we have that $S(\zeta_n)[S(\zeta_n) + 1] = 1$.*

Proof. By the definition of $S(x)$ and Eq. (5), we have $S(\zeta_n) = \sum\limits_{i \in D_1^{(n)}} \zeta_n^i$. Then we get

$$S(\zeta_n)[S(\zeta_n) + 1] = \sum_{i \in D_1^{(n)}} \sum_{j \in D_1^{(n)}} \zeta_n^{i+j} + \sum_{i \in D_1^{(n)}} \zeta_n^i. \tag{6}$$

By Lemma 4, $-1 \in D_0^{(n)}$ and $-D_1^{(n)} = D_1^{(n)}$. It follows from Eq. (5), Lemmas 5, 6 and 7 that

$$\sum_{i \in D_1^{(n)}} \sum_{j \in D_1^{(n)}} \zeta_n^{i+j} = \sum_{i \in D_1^{(n)}} \sum_{j \in D_1^{(n)}} \zeta_n^{i-j}$$

$$= \frac{(p-1)(q-3)}{4} \sum_{i \in P} \zeta_n^i + \frac{(p-3)(q-1)}{4} \sum_{i \in Q} \zeta_n^i + (1,1)_2^{(n)} \sum_{i \in D_0^{(n)}} \zeta_n^i$$

$$+ (0,0)_2^{(n)} \sum_{i \in D_1^{(n)}} \zeta_n^i$$

$$= 1 + \sum_{i \in D_1^{(n)}} \zeta_n^i. \tag{7}$$

Combining Eqs. (6) and (7) proves this lemma.

Let ζ_n be the same as before. Among the n-th roots of unity ζ_n^i, where $0 \leq i \leq n-1$, q elements ζ_n^i, $i \in P \cup R$, are q-th roots of unity, p elements ζ_n^i, $i \in Q \cup R$, are p-th roots of unity. Therefore

$$x^p - 1 = \prod_{i \in Q \cup R} (x - \zeta_n^i), \quad x^q - 1 = \prod_{i \in P \cup R} (x - \zeta_n^i).$$

Theorem 1. *Let $l = 2$. Then we have the following conclusions:*

(1) If $p \equiv -1 \pmod 8$ and $q \equiv 3 \pmod 8$, or $p \equiv 3 \pmod 8$ and $q \equiv -1 \pmod 8$, we have

$$m(x) = \frac{x^n - 1}{x^q - 1}(x - 1), \quad L_s = n + 1 - q.$$

In this case, the cyclic code C_s over $GF(2)$ defined by the sequence $s^{(n)}$ in Definition 1 has generator polynomial $m(x)$ above and parameters $[n, q - 1, 2p]$.

(2) If $p \equiv -3 \pmod 8$ and $q \equiv 1 \pmod 8$, or $p \equiv 1 \pmod 8$ and $q \equiv -3 \pmod 8$, we have

$$m(x) = \frac{x^n - 1}{x^p - 1}(x - 1), \quad L_s = n + 1 - p.$$

In this case, the cyclic code C_s over $GF(2)$ defined by the sequence $s^{(n)}$ in Definition 1 has generator polynomial $m(x)$ above and parameters $[n, p - 1, 2q]$.

Proof. We only give the proof of item (1) since the other is completely parallel. By Lemmas 9 and 10,

$$S(\zeta_n^a) \begin{cases} = 0, & a \in P \\ = 1, & a \in Q \cup R \\ \neq 0, & a \in \mathbb{Z}_n^*. \end{cases}$$

Hence, $\gcd(x^n - 1, S(x)) = \frac{x^q - 1}{x - 1}$ and $m(x) = \frac{x^n - 1}{x^q - 1}(x - 1)$. And the linear complexity of the sequence $s^{(n)}$ is equal to $\deg(m(x))$. The dimension of the code C_s follows from the definition of the code.

Now we shall prove only the minimum distance of C_s. Let ζ_n be a n-th primitive root of unity over the splitting field of $x^n - 1$ and the set $A = \{-(p - 1), -(p - 2), \cdots, -1, 0, 1, \cdots, p - 2, p - 1\}$. Then $m(\zeta_n^i) = 0$ for all $i \in A$. For any $1 \leq i \leq p - 1$ and $0 \leq j \leq p - 1$, $\zeta_n^{-i} \neq \zeta_n^j$. So it then follows from Lemma 1 that $d \geq 2p$.

On the other hand

$$m(x) = x^{n-2} + x^{n-3} + x^{n-5} + x^{n-6} + \cdots + x^4 + x^3 + x + 1$$

is a codeword of Hamming distance $2p$ in the code. It then follows that $d = 2p$. This completes the proof.

Corollary 1. *If p is an odd prime with $p \equiv -1 \pmod 8$, then the cyclic code C_s over GF(2) defined by the sequence $s^{(n)}$ in Definition 1 is optimal with respect to the Plotkin bound.*

Proof. It follows directly from Lemma 2, so we omit its proof.

Here we employ the following five examples to illuminate the construction of cyclic codes in Sect. 3.

Example 1. Let $(l, p, q) = (2, 7, 3)$. Then the linear complexity of the cyclotomic sequence is 19 and the minimal polynomial is $x^{19} + x^{18} + x^{16} + x^{15} + x^{13} + x^{12} + x^{10} + x^9 + x^7 + x^6 + x^4 + x^3 + x + 1$. The corresponding C_s is a $[21, 2, 14]$ cyclic code over GF(2). This cyclic code is an optimal linear code according to the Database [15].

Example 2. Let $(l, p, q) = (2, 3, 7)$. Then the linear complexity of the cyclotomic sequence is 15 and the minimal polynomial is $x^{15} + x^{14} + x^8 + x^7 + x + 1$. The corresponding C_s is a $[21, 6, 6]$ cyclic code over GF(2). The best binary linear code known with length 21 and dimension 6 has minimum distance 8.

Example 3. Let $(l, p) = (2, 23, 3)$. Then the linear complexity of the cyclotomic sequence is 67 and the minimal polynomial is $x^{67} + x^{66} + x^{64} + x^{63} + x^{61} + x^{60} + x^{58} + x^{57} + x^{55} + x^{54} + x^{52} + x^{51} + x^{49} + x^{48} + x^{46} + x^{45} + x^{43} + x^{42} + x^{40} + x^{39} + x^{37} + x^{36} + x^{34} + x^{33} + x^{31} + x^{30} + x^{28} + x^{27} + x^{25} + x^{24} + x^{22} + x^{21} + x^{19} + x^{18} + x^{16} + x^{15} + x^{13} + x^{12} + x^{10} + x^9 + x^7 + x^6 + x^4 + x^3 + x + 1$. The corresponding C_s is a $[69, 2, 46]$ cyclic code over GF(2). This cyclic code is an optimal linear code according to the Database [15].

Example 4. Let $(l, p, q) = (2, 5, 17)$. Then the linear complexity of the cyclotomic sequence is 81 and the minimal polynomial is $x^{81} + x^{80} + x^{76} + x^{75} + x^{71} + x^{70} + x^{66} + x^{65} + x^{61} + x^{60} + x^{56} + x^{55} + x^{51} + x^{50} + x^{46} + x^{45} + x^{41} + x^{40} + x^{36} + x^{35} + x^{31} + x^{30} + x^{26} + x^{25} + x^{21} + x^{20} + x^{16} + x^{15} + x^{11} + x^{10} + x^6 + x^5 + x + 1$. The corresponding C_s is a $[85, 4, 34]$ cyclic code over GF(2).

Example 5. Let $(l, p, q) = (2, 17, 5)$. Then the linear complexity of the cyclotomic sequence is 69 and the minimal polynomial is $x^{69} + x^{68} + x^{52} + x^{51} + x^{35} + x^{34} + x^{18} + x^{17} + x + 1$. The corresponding C_s is a $[85, 16, 10]$ cyclic code over GF(2).

Remark 3. Theorem 1 shows that the linear complexity of generalized cyclotomic sequences defined by Eq. (3) is very high. Moreover, our construction can produce new binary sequences with high linear complexity that cannot be produced by the earlier construction in [5], [6]and [14]. Furthermore, we obtain a double designed distance of cyclic codes than that of [6] (see Corollary 3.16) by using a concatenation technique and adjusting the generator polynomial. And more importantly, these sequences leads to a class of infinite optimal $[3p, 2, 2p]$ cyclic codes with respect to the Plotkin bound over GF(2), which can't be done by the former cyclotomic constructions of cyclic codes.

4   Conclusion

In this paper, we enlarge the designed distance of cyclic codes by using a concatenation technique and derive the minimum distance of cyclic codes. As a result, a class of infinite optimal cyclic codes with respect to the Plotkin bound are obtained. Our construction is straightforward, and one can generalize our framework to other cyclotomic constructions of cyclic codes. These work will be reported in our subsequent paper. Furthermore, it may be interesting to obtain optimal cyclic codes with more flexible parameters from the generalized cyclotomy and a concatenation technique. The reader are invited to attack these problems.

Acknowledgment. This work was partially supported by the National Natural Science Foundation of China (Grant No. 11771007, 11601177 and 61572027). The first author was also supported by the Funding of Jiangsu Innovation Program for Graduate Education (Grant No. KYZZ15_0090), the Funding for Outstanding Doctoral Dissertation in NUAA (Grant No. BCXJ16-08), the Open Project Program of Key Laboratory of Mathematics and Interdisciplinary Sciences of Guangdong Higher Education Institutes, Guangzhou University (Grant No. GDSXJCKX2016-07) and the Funding of Nanjing Institute of Technology (Grant No. CKJB201606).

References

1. Lidl, L., Niederreiter, H.: Finite Fields. Cambridge University Press, Cambridge (1997)
2. Ding, C., Helleseth, T.: New generalized cyclotomy and its applications. Finite Fields Appl. **4**, 140–166 (1998)
3. Ding, C., Pless, V.: Cyclotomy and duadic codes of prime lengths. IEEE Trans. Inf. Theory **45**, 453–466 (1999)
4. Ding, C., Helleseth, T.: Generalized cyclotomic codes of length $p_1^{e_1} \cdots p_t^{e_t}$. IEEE Trans. Inf. Theory **45**, 467–474 (1999)
5. Ding, C.: Cyclotomic constructions of cyclic codes with length being the product of two primes. IEEE Trans. Inf. Theory **58**, 2231–2236 (2012)
6. Ding, C.: Cyclic codes from the two-prime sequences. IEEE Trans. Inf. Theory **58**, 3881–3891 (2012)
7. Ding, C.: Cyclic codes from cyclotomic sequences of order four. Finite Fields Appl. **23**, 8–34 (2013)
8. Sun, Y., Yan, T., Li, H.: Cyclic code from the first class whiteman's generalized cyclotomic sequence with order 4. arXiv:1303.6378 (2013)
9. Ding, C., Du, X., Zhou, Z.: The bose and minimum distance of a class of BCH codes. IEEE Trans. Inf. Theory **61**, 2351–2356 (2015)
10. Kewat, P., Kumari, P.: Cyclic codes from the first class two-prime Whiteman's generalized cyclotomic sequence with order 6. arXiv:1509.07714 (2015)
11. Wang, Q.: Some cyclic codes with prime length from cyclotomy of order 4. Cryptogr. Commun. **9**(1), 85–92 (2016). https://doi.org/10.1007/s12095-016-0188-3
12. Whiteman, A.L.: A family of difference sets. Illinois J. Math. **6**, 107–121 (1962)

13. Ding, C.: Linear complexity of generalized cyclotomic binary sequences of order 2. Finite Fields Appl. **3**, 159–174 (1997)
14. Brandstatter, N., Winterhof, A.: Some notes on the two-prime generator of order 2. IEEE Trans. Inf. Theory **51**, 3654–3657 (2005)
15. Grassl, M.: Bounds on the minimum distance of linear codes. http://www.codetables.de

Author Index

Printed in the United States
By Bookmasters